Organic Coatings

SPE MONOGRAPHS

Injection Molding
Irvin I. Rubin

Nylon Plastics
Melvin I. Kohan

Introduction to Polymer Science and Technology: An SPE Textbook
Edited by Herman S. Kaufman and Joseph J. Falcetta

Principles of Polymer Processing
Zehev Tadmor and Costas G. Gogos

Coloring of Plastics
Edited by Thomas G. Webber

The Technology of Plasticizers
J. Kern Sears and Joseph R. Darby

Fundamental Principles of Polymeric Materials
Stephen L. Rosen

Plastics Polymer Science and Technology
Edited by M. D. Baijal

Plastics vs. Corrosives
Raymond B. Seymour

Aromatic High-Strength Fibers
H. H. Yang

Giant Molecules: An SPE Textbook
Raymond B. Seymour and Charles E. Carraher

Analysis and Performance of Fiber Composites, 2nd ed.
Bhagwan D. Agarwal and Lawrence J. Broutman

Impact Modifiers for PVS: The History and Practice
John T. Lutz, Jr. and David L. Dunkelbergber

The Development of Plastics Processing Machinery and Methods
Joseph Fred Chabot, Jr.

Coating and Drying Defects. Trouble-shooting Operating Problems
Edgar B. Gutoff and Edward D. Cohen

Handbook of Plastics Testing Technology, 2nd ed.
Vishu Shah

Organic Coatings

Science and Technology

Second Edition

ZENO W. WICKS, JR.

FRANK N. JONES

AND

S. PETER PAPPAS

WILEY–INTERSCIENCE

New York · Chichester · Weinheim · Brisbane · Singapore · Toronto

Library of Congress Cataloging-in-Publication Data:

Wicks, Zeno W.
    Organic coatings : science and technology / Zeno W. Wicks, Jr.,
  Frank N. Jones, and S. Peter Pappas. – 2nd ed.
        p.   cm. – (SPE monographs)
        "Wiley-Interscience publication."
        Includes index.
        ISBN 0-471-24507-0 (cloth : alk. paper)
        1. Plastic coatings. Wicks, Zeno W. Jr., I. Jones, Frank N., 1936– . II. Pappas,
  S. Peter (Socrates Peter), 1936– . III. Title. IV. Series.
  TP1175.S6W56   1998
  667'.9–dc 21   98-25870

Printed in the United States of America.
10 9 8 7 6 5 4 3 2 1

Contents

Foreword

The Society of Plastics Engineers is pleased to sponsor *Organic Coatings: Science and Technology*, Second Edition, by Zeno W. Wicks, Jr., Frank N. Jones, and S. Peter Pappas. This volume will serve as an excellent reference book for scientists, chemists, and engineers dealing with coatings.

SPE, through its Technical Volumes Committee, has long sponsored books on various aspects of plastics. Its involvement has ranged from identification of needed volumes and recruitment of authors to peer review and approval and publication of new books.

Technical competence pervades all SPE activities, not only in the publication of books, but also in other areas, such as sponsorship of technical conferences and educational programs. In addition, the Society publishes periodicals including *Plastics Engineering*, *Polymer Engineering and Science*, *The Journal of Injection Molding Technology*, *Journal of Vinyl & Additive Technology*, and *Polymer Composites*, as well as conference proceedings and other publications, all of which are subject to rigorous technical review procedures.

The resource of some 36,000 practicing plastics engineers, scientists, and technologists has made SPE the largest organization of its type worldwide. Further information is available from the Society at 14 Fairfield Drive, Brookfield, CT 06804.

MICHAEL R. CAPPELLETTI
Executive Director
Society of Plastics Engineers

Technical Volumes Committee:
Robert C. Portnoy, *Chairperson*
Exxon Chemical Company

Richard D. Cowell, *Reviewer*
Cowell & Associates

Preface

In the time since the publication of the first edition of *Organic Coatings: Science and Technology*, there have been many major developments. The text has been completely updated and rewritten as one volume instead of two. The change to one volume reduced duplication, and room for new material was made by condensing topics of mainly historical interest. In rewriting, we have changed the order of presentation, addressing key properties of coatings in the first seven chapters. In the first edition, we started with the raw materials used in coatings instead. There is no "correct" order in which to present the subject, but we believe the new chapter order can best help readers understand the scientific foundation.

Our purpose is still the same. It is to provide a text and reference book that summarizes coatings technology and relates it to current scientific understanding. We have taken care to define the jargon of coatings to help newcomers to the field understand its specialized language.

Entire books could be written about the subject of each chapter, and many have been. To be as comprehensive as possible in the limited space available, we have summarized each topic and selected references for readers seeking more detailed information. Each chapter includes references we believe to be reliable, and we have included an appendix that lists sources we have found to be broadly useful. We do not claim to provide a complete literature review on each topic, and many valuable sources are not cited. Readers are cautioned that the quality of the literature in the coatings field is uneven. Many published papers and monographs are excellent, but some are not; unfortunately, some authors did not fully understand the complexity of the field.

Coatings technology evolved empirically, by trial and error. The last few decades have seen a marked increase in scientific understanding of the applicable principles, but the complexities of the field are such that the formulator's art is still essential in developing and using coatings. The need to reduce air pollution while simultaneously maintaining and, preferably, improving coating performance requires radically new formulations on a short time scale. Our conviction is that increased understanding of the underlying science can help formulators work more effectively and that an appreciation of a formulator's craft is essential for scientists working in the field.

Like the first edition, this volume can be used as a textbook for a course on coatings science. As such, it is written for students who have had college-level chemistry courses at least through organic chemistry, but no coursework in polymer science is assumed. We are told that the first edition found wider use as a reference book and as a self-teaching text than as a classroom text. We tried to increase the reference value of this edition by increasing the number of literature citations. Almost half of the references cited are more recent than those in the first edition. While this book is written specifically about coatings, many of the principles involved apply to the related fields of printing inks, adhesives, and parts of the plastics industry.

Almost all of the chapters in this edition were reviewed by people with extensive working experience with the particular topic. These reviewers were Patrick Alutto, Colin

Anderson, Jamil Baghdachi, John Bentley, Gordon Bierwagen, Darlene Brezinski, Ashok Chattopadhyay, H. K. Chu, Werner Funke, John Gardon, John Gerlock, Loren Hill, Kenneth Hoy, Charles Hoyle, Josef Jilek, Walter Kühhirt, Jon Lawniczak, Eric Lynch, John Massingill, Thomas Miranda, T. A. Misev, Greg Muselman, Percy Pierce, George Pilcher, Paul Reynolds, Bradley Richards, Peter Robinson, Rose Ryntz, Clifford Schoff, Russell Schwartz, Robert Springate, Dean Webster, Keith Weiss, Douglas Wicks, Larry Williams, and Jim Woo. Many important suggestions and corrections came from the reviewers, and we are very grateful for their contributions. We are also grateful to Kelly Burnard, Jodie Pihalja, Laura Trapp, and Sandra Tanner for their help in preparing the manuscript. FNJ thanks Eastern Michigan University for a research fellowship that made available time to work on this volume.

ZENO W. WICKS, JR.
FRANK N. JONES
S. PETER PAPPAS

Symbols

A	Arrhenius preexponential term
C	Concentration—weight per unit volume of solution
c	Concentration—moles per liter
CPVC	Critical pigment–volume concentration (content)
CRH	Critical relative humidity
°C	Degrees Celsius
E	Modulus
E'	Storage modulus (elastic modulus)
E''	Loss modulus
E	Relative evaporation rate
E_a	Thermal coefficient of reaction rate (Arrhenius activation energy)
F	Functionality of a monomer
$\bar{F}$	Average functionality of a monomer mixture
f	Functionality of a polymer (resin)
$\bar{f}_n$	Number average functionality of a polymer (resin)
G	Free energy
G	Small's molar association constant
g	Gram
g	Gravitational constant
H	Enthalpy
i	Angle of incidence
K	Kelvin temperature
K	Absorption coefficient
K_E	Einstein (shape) constant
k	Rate constant
kg	Kilogram
KU	Krebs unit
L	Liter
M	Molecular weight
$\bar{M}_c$	Average molecular weight between cross-links
$\bar{M}_n$	Number average molecular weight
$\bar{M}_w$	Weight average molecular weight
MFFT	Minimum film formation temperature
mL	Milliliter
N	Newton
N	Number of moles
n	Refractive index
NVV	Nonvolatile volume—volume percent solids
NVW	Nonvolatile weight—weight percent solids
OA	Oil absorption
P	Vapor pressure

P	Degree of polymerization
$\bar{P}_n$	Number average degree of polymerization
$\bar{P}_w$	Weight average degree of polymerization
p	Extent of reaction
p_g	Extent of reaction at gelation onset
Pa	Pascal
Pa·s	Pascal second = 10 poise
mPa·s	MilliPascal second = 1 centipoise
PDI	Polydispersity Index = $\bar{M}_w/\bar{M}_n$
PVC	Pigment volume concentration (content)
R	Gas constant
r	Angle of reflection or angle of refraction
RH	Relative humidity
S	Entropy
S	Scattering coefficient
s	Second
T	Temperature (units are K if not otherwise specified)
T_b	Brittle–ductile transition temperature
T_g	Glass transition temperature
T_m	Melting point
t	Time
$\tan \delta$,	tan delta, loss tangent, E''/E'
V	Molar volume
V_i	Volume fraction of internal phase
VOC	Volatile organic compound
w	Weight fraction
X	Film thickness
x	Mole fraction
x	Optical path length
XLD	Cross-link density
γ	Surface tension
$\dot{\gamma}$	Shear rate
δ	Solubility parameter
δ	Phase shift in viscoelastic deformation
ε	Molar absorbance
η	Absolute shear viscosity
η_e	External phase viscosity
η_r	Relative viscosity = η/η_s
η_s	Viscosity of solvent
η^*	Extensional viscosity
$[\eta]$	Intrinsic viscosity
$[\eta]_w$	Weight-intrinsic viscosity
$[\eta]_\theta$	Intrinsic viscosity under theta conditions
θ	Contact angle
λ	Wave length
v	Kinematic viscosity
v_e	Mole of elastically effective network chains per cm^3
ρ	Density

σ	Shear stress
σ_0	Yield value
ϕ	Packing factor
χ	Activity coefficient

Organic Coatings

CHAPTER 1

What Are Coatings?

Look around you; coatings are everywhere. If you are indoors, there are coatings on the walls, refrigerator, cabinets, and furniture; less obviously, coatings are on the wires of electrical motors, printed circuits inside television sets, audiotapes and videotapes, and compact discs. If you are outdoors, coatings are on your house and car, as well as inside your car, under the hood, and on components of the automotive stereo and computer systems. Whether you drink beer or soft drinks, there are coatings on the inside and outside of the cans. The functional and decorative requirements of coatings span a broad spectrum. A diverse science and technology support the development, production, and use of coatings.

Students should realize that while coatings science is an old field, it is not a mature one; it offers exciting challenges and career opportunities. People entering the field will have opportunities to improve scientific understanding and to contribute to the major thrusts of coatings development: reducing emissions that cause air pollution, reducing energy requirements, and protecting metals against corrosion.

1.1. DEFINITIONS AND SCOPE

Coatings may be described by their appearance (e.g., clear, pigmented, metallic, or glossy) and by their function (e.g., corrosion protective, abrasion protective, skid resistant, decorative, or photosensitive). Coatings may be distinguished as organic or inorganic, although there is overlap. For example, many coatings consist of inorganic pigments dispersed in an organic matrix (the binder).

A confusing situation results from multiple meanings of the term coating. It is used to describe the material (usually a liquid) that is applied to a substrate, the resultant "dry" film, and the process of application. Usually, the intended meaning of the word coating can be inferred from the context.

We limit our discussion to coatings with organic binders that are applied purposefully to a substrate. Many kinds of coatings are not included. Porcelain enamels on kitchen ranges are coatings, but they do not have organic chemical binders. Electroplated copper, nickel, and zinc coatings are excluded for the same reason. We further restrict our discussion of organic coatings to those materials that can be historically traced back to paints. What is the difference between a coating and a paint? Not much—the terms are often used interchangeably. However, it has become common practice to use *coatings* as the broader term and to restrict *paints* to the familiar architectural and household coatings and sometimes to maintenance coatings for bridges and tanks. We follow this practice. Some prefer to call sophisticated materials that are used to coat automobiles and computer components coatings, perhaps sensing that paint sounds too lowbrow. Still another common term that is essentially a synonym for coating and paint is *finish*.

In limiting the scope of this book to organic coatings that can be related to historic paints, we exclude many materials that could be called coatings. Printing inks, polymers applied during production of paper and fabrics, coatings on photographic films, decals and other laminates, and cosmetics are but a few examples. However, many of the basic principles that are covered in this text are applicable to such materials. Restrictions of scope are necessary if the book is to be kept within one volume, but our restrictions are not entirely arbitrary. The way in which we are defining coatings is based on common usage of the term in worldwide business. It is close to the definition of organic coatings for statistical analyses of industrial output used by Bureau of Census of the U.S. Department of Commerce [1]. The Census Bureau defines three broad categories: (1) architectural coatings, (2) product coatings used by original equipment manufacturers (OEM coatings), and (3) special purpose coatings.

The worldwide production of coatings in 1996 was about 230 million metric tons, with 34% in European countries, 28% in North America, 7% in Japan, and 31% in the rest of the world [2]. The worldwide coatings market is estimated at about US$60 billion [3]. Shipments of paints and coatings by U.S. manufacturers in 1996 totaled 1.2 billion gallons (4.6×10^9 L) with a value of $14.9 billion, as broken down in Table 1.1 [2].

Architectural coatings include paints and varnishes (transparent paints) used to decorate and protect buildings, outside and inside. They also include other paints and varnishes sold for use in the home and by small businesses for application to such things as cabinets and household furniture (not those sold to furniture factories) and are often called *trade sales paints*. They are sold directly to painting contractors and do-it-yourself users through paint stores and other retail outlets. In 1996 in the United States, architectural coatings accounted for about 52% of the total volume of coatings; however, the unit value of these coatings was lower than for the other categories, so they made up about 42% of the total value. This market is the least cyclical of the three categories. While the annual amount of new construction drops during recessions, the resulting decrease in paint requirements tends to be offset by increased repainting of older housing, furniture, and so forth during at least mild recessions. Latex-based coatings make up 77% of architectural coatings [3].

Product coatings, also commonly called *industrial coatings* or *industrial finishes*, are applied in factories on products such as automobiles, appliances, magnet wire, aircraft,

Table 1.1. United States Coatings Shipments, 1996

Coatings	Liters $\times 10^{-9}$	Dollars $\times 10^{-9}$
Architectural	2.42	6.2
Product—OEM	1.41	5.4
Special Purpose	0.79	3.3
	—	—
	4.62	14.9

furniture, metal cans, chewing gum wrappers—the list is almost endless. This market is often called the OEM market, that is, the *original equipment manufacturer* market. In 1996 in the United States, product coatings were about 31% of the volume and 36% of the value of all coatings. The volume of product coatings depends directly on the level of manufacturing activity. This category of the business is cyclical, varying with OEM cycles. In most cases, product coatings are custom designed for a particular customer's manufacturing conditions and performance requirements. The number of products in this category is much larger than in the others; research and development requirements are also higher.

Special purpose coatings designates industrial coatings, which are applied outside a factory, and a few miscellaneous coatings, such as coatings packed in aerosol containers. It includes coatings for cars and trucks that are applied outside the OEM factory (usually in body repair shops), coatings for ships (they are too big to fit into a factory), and striping on highways and parking lots. It also includes *maintenance paints* for steel bridges, storage tanks, chemical factories, and so forth. In 1996 in the United States, special purpose coatings made about 17% of the total volume and 22% of the total value of all coatings.

Coatings are used for one or more of three reasons: (1) for decoration, (2) for protection, and/or (3) for some functional purpose. The low gloss paint on the ceiling of a room fills a decorative need, but it also has a function: It reflects and diffuses light to help provide even illumination. The coating on the outside of an automobile adds beauty to a car and also helps protect it from rusting. The coating on the inside of a beverage can has little or no decorative value, but it protects the beverage from the can. (Contact with metal affects flavor.) In some cases, the interior coating protects the can from the beverage. (Some soft drinks are so acidic that they can dissolve the metal.) Other coatings reduce the growth of algae and barnacles on ship bottoms, protect optical fibers for telecommunications against abrasion, serve as the recording medium on audiotapes and videotapes, and so on. While the public most commonly thinks of house paint when talking about coatings, all kinds of coatings are important throughout the economy, and they make essential contributions to most high-tech fields. For example, computer technology depends on microlithographic coatings to construct microprocessors.

Traditionally, coatings have changed relatively slowly in an evolutionary response to new performance requirements, new raw materials, and competitive pressures. An important reason for the relatively slow rate of change is the difficulty of predicting product performance on the basis of laboratory tests. It is less risky to make relatively small changes in composition and check actual field performance before making further changes. Since about 1965, however, the pace of technical change has increased. A major driving force for change has been the need to reduce VOC (volatile organic compound) emissions because of their detrimental effect on air pollution. Coatings have been second only to the gasoline-automobile complex as a source of VOC pollutants responsible for excess ozone in the air of many cities on many days of the year. This situation has resulted in increasingly stringent regulatory controls on such emissions. The drive to reduce VOC emissions has also been fueled by the rising cost of organic solvents. Other important factors have also accelerated the rate of change in coatings. Increasing concern about toxic hazards has led to the need to change many raw materials that were traditionally used in coatings. Furthermore, manufacturers often want their coatings modified so that they can be used at faster production rates, baked at lower temperatures, or changed in color. Product performance requirements have tended to increase; most notable is the need for increased effectiveness of corrosion protection by coatings.

1.2. COMPOSITION OF COATINGS

Organic coatings are complex mixtures of chemical substances that can be grouped into four broad categories: (1) binders, (2) volatile components, (3) pigments, and (4) additives.

Binders are the materials that form the continuous film that adheres to the *substrate* (the surface being coated), binds together the other substances in the coating to form a film, and that presents an adequately hard outer surface. The binders of coatings within the scope of this book are organic polymers. In some cases, these polymers are prepared and incorporated into the coating before application; in other cases, final polymerization takes place after the coating has been applied. The binder governs, to a large extent, the properties of the coating film.

Volatile components are included in a majority of all coatings. They play a major role in the process of applying coatings; they are liquids that make the coating fluid enough for application, and they evaporate during and after application. Until about 1945, almost all of the volatile components were low molecular weight organic solvents that dissolved the binder components. However, the term *solvent* has become potentially misleading because since 1945, many coatings have been developed for which the binder components are not fully soluble in the volatile components. Because of the need to reduce VOC emissions, a major continuing drive in the coatings field is to the reduce use of solvents by making the coatings more highly concentrated (*higher solids coatings*) or by using water as a major part of the volatile components (*water-borne coatings*). *Vehicle* is a commonly encountered term. It usually means the combination of the binder and the volatile components of a coating. Today, most coatings, including water-borne coatings, contain at least some volatile organic solvents. Exceptions are powder coatings and radiation curable coatings.

Pigments are finely divided insoluble solids that are dispersed in the vehicle and remain suspended in the binder after film formation. Generally, the primary purpose of pigments is to provide color and opacity to the coating film. However, they also have substantial effects on application characteristics and on film properties. While most coatings contain pigments, there are important types of coatings that contain little or no pigment, commonly called *clear coats*, or just *clears*. Clear coats for automobiles and transparent *varnishes* are examples.

Additives are materials that are included in small quantities to modify some property of the coating. Examples are catalysts for polymerization reactions, stabilizers, and flow modifiers.

Most coatings are complex mixtures. Many contain several substances from each of the four categories, and each substance is usually a chemical mixture. The number of possible combinations is limitless. The number of different applications is also limitless.

The person who selects the components from which to make a coating is a *formulator*, and the overall composition he or she designs is called a *formulation*. Formulation of paints started millennia ago as an empirical art or craft. Successive generations of formulators built on the experience of their predecessors and formulated coatings with increasingly better performance characteristics. Starting 50 or 60 years ago, formulators have been trying to understand the underlying scientific principles that control the performance of coatings. Most coating systems are so complex that our understanding of them today is still limited. Real progress has been made, but the formulator's art is still a critical element in developing high performance coatings. Demands on suppliers of coatings to develop new and better coatings are increasing at an accelerating pace so, time is now too limited to permit traditional trial and error formulation. Understanding the basic scientific principles can help a formulator design better coatings more quickly. In the chapters ahead, we present, to as great an extent as present knowledge permits, the current understanding of the scientific principles involved in coating science.

We also identify areas in which our basic understanding remains inadequate and discuss approaches to more efficient and effective formulation despite inadequate understanding. In some cases, in which no hypotheses have been published to explain certain phenomena, we offer speculations. Such speculations are based on our understanding of related phenomena and on our cumulative experience acquired over several decades in the field. We recognize the risk that speculation tends to increase in scientific stature with passing time and may even be cited as evidence or adopted as an experimentally supported hypothesis. It is our intent, rather, that such speculations promote the advancement of coatings science and technology by stimulating discussion that leads to experimentation designed to disprove or support the speculative proposal. We believe that the latter purpose outweighs the former risk, and we endeavor to identify the speculative proposals as such.

Cost is an essential consideration in formulation. Novice formulators are inclined to think that the best coating is the one that will last the longest time without any change in properties, but such a coating may be very expensive and unable to compete with a less expensive coating whose performance is adequate for the particular application. Furthermore, it is seldom possible to maximize all of the performance characteristics of a coating in one formulation. Some of the desirable properties are antagonistic with others; formulators must balance many performance variables while keeping costs as low as possible.

REFERENCES

1. U.S. Department of Commerce, Bureau of Census, *Current Industrial Reports—Paint, Varnish, and Lacquers*, Issued monthly.
2. M. S. Reisch, *Chem. Eng. News*, **75**, 36 (1997).
3. E. W. Bourguignon, *Paint Coat. Ind.*, October, 64 (1997).

CHAPTER 2

Polymerization and Film Formation

This chapter is designed to give an introduction to basic concepts of polymer chemistry and film formation.

2.1. POLYMERS

A polymer is a substance composed of *macromolecules*. Some authors reserve the term *polymer* to describe a substance and use the term *macromolecule* for the molecules making up the substance. This usage distinguishes between the material and the molecules, but is not common in the coatings field. We use the term polymer for both meanings; depending on the context, the term refers to either the molecules or the substance. The structure of polymers is a multiple repetition of units (mers) derived from molecules of low relative molecular weight (monomers). (The more rigorous designation of *molecular weight* is *molar mass*, but we use molecular weight because it is much more commonly used in the coatings field.) There is disagreement about how high the molecular weight has to be for a material to qualify as a polymer. Some people refer to materials with molecular weights as low as 1000 as polymers; others insist that only materials with molecular weights over 10,000 (or even 50,000) qualify. The term *oligomer*, meaning "few mers," is often used for materials having molecular weights of a few hundred to a few thousand. This additional term does not help the definition problem much because there is no clear-cut boundary between an oligomer and a polymer, but the term can be useful because it provides a name with which most can agree for materials containing 2 to about 20 mers.

Polymers occur widely in nature; familiar examples of biopolymers are proteins, starch, cellulose, and silk. In the coatings field, we are concerned mainly with synthetic polymers, although some chemically modified biopolymers are also used.

Synthetic polymers and oligomers are prepared by polymerization, in which small molecules are joined by covalent bonds. A polymer made from a single monomer is called a *homopolymer*. If it is made from a combination of monomers, it is often, but not always, called a *copolymer*. An example of a homopolymer is provided by the polymerization of vinyl chloride:

$$CH_2=CHCl \longrightarrow X-(CH_2-CHCl)_n-Y$$

Vinyl chloride monomer   Poly(vinyl chloride)

In this example, the $-(CH_2CHCl)-$ repeating unit is the mer, and n represents the number of mers joined together in the molecule. X and Y represent terminal groups on the ends of the chain of mers.

Polymers can be made in a variety of structures; three important classes are shown in Figure 2.1. When the mers are linked in chains, the polymers are called *linear polymers*, a term that is potentially misleading because the large molecules seldom form a straight line—they twist and coil. If there are forks in the chains, as in Figure 2.1(*b*), the polymer are called *branched polymers*. A class of importance in coatings results from the bonding of chains with each other at several sites to form *cross-linked*, or *network*, polymers, as in Figure 2.1(*c*). These polymers are branched polymers where the branches covalently are bound to other molecules, so the mass of polymer consists mainly of a single, interconnected molecule. Reactions that join polymer or oligomer molecules are called cross-linking reactions. Polymers and oligomers that can undergo such reactions are frequently called *thermosetting polymers*. Some confusion can result because the term thermosetting is applied not only to polymers that cross-link when heated, but also to those that can cross-link at ambient temperature. A polymer that does not undergo cross-linking reactions is called a *thermoplastic polymer*, because it becomes plastic (softens) when heated.

Copolymers of all three classes illustrated in Figure 2.1 are common. In linear copolymers, the different monomers may be distributed more or less at random throughout the chain (*random copolymers*), they may tend to alternate (*alternating copolymers*), or they may be grouped together (*block copolymers*). A polymer chain of one type of monomer having polymer branches of another type of monomer is called a *graft copolymer*. Analogous forms exist with cross-linked copolymers.

Another term commonly, but loosely, used in the coatings field is *resin*. This term overlaps the meanings of polymer and oligomer. Originally, the term meant hard, brittle materials derived from tree exudates, such as rosin, dammar, and elemi. A variety of these naturally occurring resins were used since prehistoric times to make coatings. In the 19th and early 20th centuries, resins were used with drying oils to make *varnishes*. (See Section 14.3.2.) The first synthetic polymers used in coatings were phenol-formaldehyde polymers

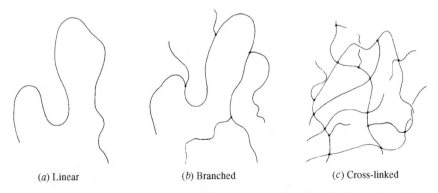

(*a*) Linear   (*b*) Branched   (*c*) Cross-linked

Figure 2.1. Schematic representations of three common classes of polymer structure.

(see Section 11.6), which replaced naturally occurring resins in many applications. Hence, it was natural to call them *phenol-formaldehyde resins* or *phenolic* resins. As more synthetic products were developed to replace naturally occurring resins, these products were also called resins.

When words do not have precise meanings, it is important to understand the context in which they are placed. Commonly, it is assumed, without much thought, that information that has been learned about high molecular weight polymers is also applicable to low molecular weight polymers or oligomers, because all are often called polymers. Many characteristics, however, depend on molecular weight. While much of the information available from studies of high molecular weight polymers can be useful in the coatings field, it must be used with caution, because the resins used in making coatings are commonly low molecular weight polymers or oligomers, even though they are frequently called polymers. In the next set of subsections, we describe some of the key characteristics of synthetic polymers and oligomers.

2.1.1. Molecular Weight

For most pure organic compounds, the concept of molecular weight is straightforward —each compound has **a** molecular weight. For synthetic polymers, however, the situation is more complex. All methods of synthesis lead to mixtures of molecules with different numbers of mers and, therefore, with different molecular weights. Even relatively simple thermoplastic homopolymers, such as polystyrene or poly(vinyl chloride), contain molecules with hundreds of different, although similar, structures. With copolymers, the number of different molecules present is much larger. There is some kind of a distribution of molecular weights in a synthetic polymer; molecular weights can be defined only by some sort of statistical calculation. In the simplest cases, the distribution of the number of molecules of each molecular weight resembles a skewed Gaussian distribution, but in other cases, the distribution may be quite complex. While many types of average molecular weight can be calculated, the two most widely used are number and weight average molecular weights.

Number average molecular weight $\bar{M}_n$ is the molecular weight average based on summing the products of the numbers of molecules and their molecular weights and dividing by the sum of the number of molecules in the sample. Mathematically, it is expressed by the following equation, where M_1, M_2, and M_i are the molecular weights of the first, second, and ith species, respectively, and the N values are the numbers of molecules of each species present:

$$\bar{M}_n = \frac{\sum N_1 M_1 + N_2 M_2 + \cdots}{\sum N_1 + N_2 + \cdots} = \frac{\sum_i N_i M_i}{\sum_i N_i}$$

$$\bar{P}_n = \frac{\sum_i N_i P_i}{\sum_i N_i}$$

A similar equation is often used to represent the *number average degree of polymerization* $\bar{P}_n$, where P is the number of mers in a molecule and P_i is the number of mers in the ith polymer. For homopolymers, $\bar{M}_n = \bar{P}_n$ times the molecular weight of each mer; for copolymers, a weighted average molecular weight of the mers is used. The differing weights of end groups can be neglected in calculating the $\bar{M}_n$ or $\bar{M}_w$ of high polymers, but not of oligomers, for which the effect can be appreciable.

Weight average molecular weight $\bar{M}_w$ is defined by the following equation, in which w_1, w_2, and w_i are the weights of molecules of species 1, 2, and i: Since $w_1 = N_1 M_1$, $\bar{M}_w$ can also be calculated from the numbers of molecules of the different species, as shown in the equation. *Weight average degree of polymerization* $\bar{P}_w$ is defined by analogous equations.

$$\bar{M}_w = \frac{w_1 M_1 + w_2 M_2 + \cdots}{w_1 + w_2 + \cdots} = \frac{\sum_i w_i M_i}{\sum_i w_i} = \frac{\sum_i N_i M_i^2}{\sum_i N_i M_i}$$

Figure 2.2 shows an idealized plot of weight fraction of molecules of each molecular weight as a function of degree of polymerization for oligomers made from the same monomer by three different processes. In relatively simple distributions of molecular weights, such as those shown in Figure 2.2, the value of $\bar{P}_n$ is at, or near, the peak of the weight fraction distribution curve. Because $\bar{M}_w$ and $\bar{P}_w$ give extra weight to higher molecular weight molecules, they are always larger than $\bar{M}_n$ and $\bar{P}_n$.

The breadth of the molecular weight distribution can have an important effect on the properties of a polymer and is often critical to achieving satisfactory performance of a coating. The ratio $\bar{M}_w/\bar{M}_n$ is widely used as an index of the breadth of distribution. In the case of high molecular weight polymers, $\bar{M}_w/\bar{M}_n = \bar{P}_w/\bar{P}_n$, but in the case of oligomers, differences in end groups can be significant and affect the equality of the ratios. These ratios are called *polydispersity* (PD), or sometimes, *polydispersity index* (PDI). We use the symbols $\bar{M}_w/\bar{M}_n$ and $\bar{P}_w/\bar{P}_n$. The ratios provide a convenient way to compare the

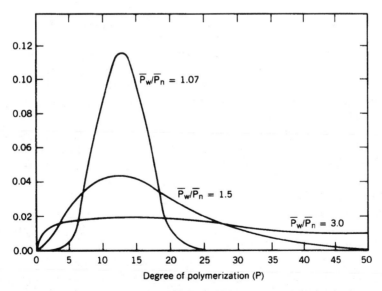

Figure 2.2. Degree of polymerization distribution plots calculated for three types of chain-growth polymers. $\bar{P}_w/\bar{P}_n = 1.07$ is for an ideal anionic polymerization, $\bar{P}_w/\bar{P}_n$ is 1.5 for an ideal free radical polymerization with termination by combination, and $\bar{P}_w/\bar{P}_n$ is 3.0 for a typical free radical polymerization. $\bar{P}_n$ is 12 for all plots, and $\bar{P}_w$ is 12.84, 18, and 36, respectively (from Ref. [1], with permission).

molecular weight distributions of different polymers. However, one must be cautious in the use of a single value to describe a possibly complex distribution. As shown in Figures 2.2 and 2.3, synthetic polymers commonly have broad distributions of molecular weights. As $\bar{M}_w/\bar{M}_n$ increases, the fractions of polymer at the extremes above and below the number average molecular weight increase. Even the oligomer with a number average of 12 mers and with $\bar{M}_w/\bar{M}_n = 1.07$ has substantial numbers of molecules containing 7 to 18 mers, and a polymer with a more typical $\bar{M}_w/\bar{M}_n$ of 3 has molecules spanning several orders of magnitude of molecular weight.

$\bar{M}_n$ is the molecular weight of most importance for relating stoichiometric ratios of reactants and for comparing certain physical properties. $\bar{M}_w$ often proves more useful than $\bar{M}_n$ when considering the relationship between molecular weight and many physical properties of polymers, including some of the properties that are crucial to coating performance.

Molecular weight measurement is difficult and beyond the scope of this book to discuss in detail. (See Ref. [2] for a discussion of the various ways of determining $\bar{M}_w$ and $\bar{M}_n$.) In practice, most scientists in the coatings field use *gel permeation chromatography* (GPC), more properly called *size exclusion chromatography* (SEC), to measure molecular weights. In this convenient method, a dilute solution of an oligomer or a polymer is pumped at high pressure through a series of columns containing porous gels. The molecules are "sorted" by sizes, with the largest ones coming out first and the smaller ones, which are slowed by entering and leaving more gel pores, coming out later. The concentration of polymer in the solvent is analyzed as it leaves the column and is plotted as a function of time. A computer program compares the plot to plots of standard polymers of known molecular weights and calculates $\bar{M}_n$, $\bar{M}_w$, and several other quantities that characterize the polymer. The results appear precise, but they are not accurate; errors of $\pm 10\%$ can be expected, and much larger errors are possible. Errors can result because the molecular weight is not measured directly—instead the size of the polymer molecules in solution is measured—and from differences in detector response to different compositions. Despite its imprecision, GPC is useful, especially for comparing polymers of similar structure.

The $\bar{M}_n$ of oligomers can be accurately measured by colligative methods, such as freezing point depression and vapor pressure osmometry. However, the accuracy decreases as molecular weight increases, and colligative methods are of little use above $\bar{M}_n = 50,000$. Mass spectroscopic methods are available that can precisely measure the molecular weights of individual molecules in oligomers, and even in fairly high polymers.

Some polymers and oligomers have molecular weight distributions approaching the idealized kind of distributions shown in Figure 2.2, as illustrated by the GPC trace of a polyester oligomer in Figure 2.3(a). However, many polymers used in coatings have complex distribution patterns as exemplified by the alkyds in Figure 2.3(b). The $\bar{M}_w$ and $\bar{M}_n$ can be calculated for the entire trace or for portions of complex traces. But, such polydispersity numbers must be used with caution for complex traces.

The molecular weight of resins is an important factor affecting the viscosity of coatings made with solutions of the resins: generally, the higher the molecular weight, the higher the viscosity. The molecular weight of oligomers used in higher solids coatings is especially critical. It is often desirable to make oligomers with as narrow a range of molecular weight as possible, since this minimizes the amount of very low and very high molecular weight resin. The low molecular weight fraction is generally undesirable from the standpoint of film properties, whereas the high molecular weight fraction increases resin solution viscosity disproportionately. However, alkyd resins having broad, complex

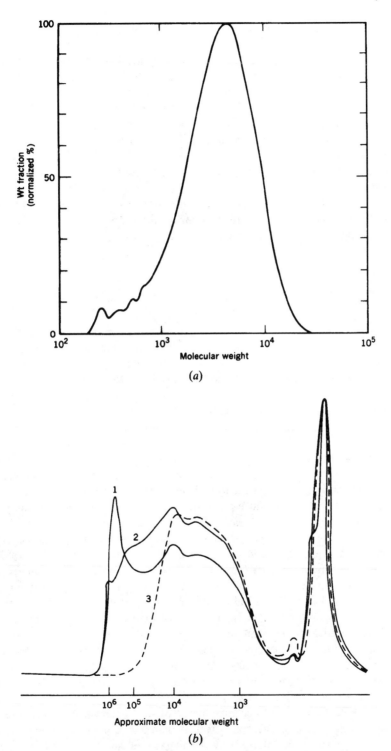

Figure 2.3. (*a*) Molecular weight distribution of a typical polyester resin (from Ref. [3] with permission). (*b*) Molecular weight distributions of three alkyd resins, as measured by GPC with a UV detector (from Ref. [4], with permission).

molecular weight distributions may perform better than alkyds with similar compositions having narrow distributions [4].

Molecular weight is often a critical factor controlling the strength of films that are not cross-linked. In general, the higher the molecular weight, the higher the tensile strength of such films, at least up to a point. The acrylic copolymer in automotive acrylic lacquers must have an $\bar{M}_w$ greater than about 75,000 for acceptable film properties, but less than 100,000 for acceptable application properties. For other lacquers the required molecular weight depends on polymer composition and application methods. Film property considerations argue in favor of using high molecular weight polymers in formulating coatings, but viscosity considerations favor low molecular weights. As is often the case in coatings, compromises are needed.

An important advantage of water-borne coatings is that molecular weight of the polymer generally does not directly affect viscosity, since the polymers are dispersions of polymer particles rather than solutions.

2.1.2. Morphology

Morphology is the study of the physical forms of materials. Like molecular weight, morphology is more complex with polymers than with small-molecule substances. Pure small molecules generally solidify to crystals if the temperature is sufficiently low. In contrast, few synthetic polymers crystallize completely, and many do not crystallize at all. Noncrystalline materials that appear to be solids are called *amorphous solids*. There are at least two reasons that synthetic polymers are at least partly amorphous. In general, synthetic polymers are not pure compounds, so it is difficult to achieve the completely regular structure characteristic of a crystalline material. In addition, the molecules are so large that the probability of complete crystallization is low. Part of a molecule can associate with part of a different molecule or with another part of the same molecule, reducing the odds of pure crystal formation. However, small crystalline domains are common in synthetic polymers; polymers with fairly regular structures, usually homopolymers, are most likely to crystallize partially. In these crystalline domains, fairly long segments of molecules associate with each other in a regular way. The remaining parts of the same molecules are unable to fit together regularly and remain amorphous. While polymers used in fibers and films (e.g., polyethylene and nylon) are often partly crystalline, polymers used in coating applications are, with few exceptions, completely amorphous.

Amorphous materials behave quite differently from crystalline materials. An important difference is shown in Figure 2.4(*a*) and (*b*), which schematically compare the changes in specific volume of crystalline and amorphous materials with temperature. In the case of a pure crystalline material, as temperature increases, initially there is a slow increase in specific volume, owing to increasing vibrations of the atoms and molecules. Then, at a specific temperature, the substance melts. The melting point T_m is the lowest temperature at which the vibrational forces pushing molecules apart exceed the attractive forces holding them together in crystals. With almost all substances (water is a notable exception), the molten compound occupies more volume at the same temperature than the crystals; because the molecules are freer to move in a molten compound, they "bounce" their neighbors out of the way, leading to an abrupt increase in specific volume at T_m. Above T_m, the specific volume of a liquid slowly increases with further increase in temperature.

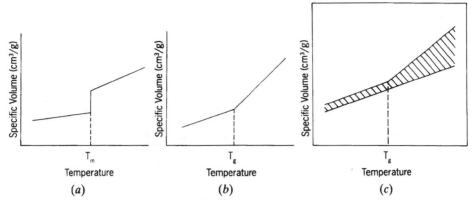

Figure 2.4. Specific volume as a function of temperature (*a*) for a crystalline material and (*b*) for an amorphous material; (*c*) shows free volume within an amorphous material as a function of temperature. Units of specific volume are volume per mass (usually cm³/g). (Adapted from Ref. [1], with permission.)

Amorphous materials behave differently, as shown in Figure 2.4(*b*). Starting from a low temperature, there is a slow increase in specific volume as temperature increases, but there is no temperature at which an abrupt change in volume occurs, there is no melting point. Rather, there is a temperature at which there is a change in the rate of increase of specific volume with temperature. Above that temperature, the thermal expansion coefficient is larger than below it. This change of slope is not a phase change; it is a second-order transition, that is, there is a discontinuity in a plot of the *derivative of volume change* as a function of temperature. The temperature at which it occurs is called the *glass transition temperature*, T_g. T_g is properly defined as the temperature at which there is an increase in the thermal expansion coefficient. By comparison, T_m is a first-order transition, that is, there is a discontinuity in change of volume as a function of temperature, corresponding to the solid-liquid phase change. Unfortunately, T_g is often improperly defined as the temperature below which a material is brittle and above which it is flexible. While there are many cases when this is true, there are cases when this definition is misleading. (See Section 4.2.) Perhaps, part of the reason for the misunderstanding is the connotation of the word *glass*, which we associate with a brittle material. Like all amorphous materials, glasses undergo a second-order transition. In fact, the phenomenon was first observed in the study of glasses—hence, the name *glass transition temperature*. The idea has proliferated that T_g is a phenomenon associated only with polymers. This assumption is not true; many small molecules can be supercooled without crystallization to form amorphous glasses that have a T_g. For example, the T_g of *m*-xylene is 125 K [5]. The T_g is always lower than T_m. Partially crystalline polymers show both a T_m and a T_g.

What is physically happening at T_g? As an amorphous material is heated, atoms in the molecules vibrate with increasing energy, colliding with neighbors and shoving molecules apart for brief periods of time. At T_g, a few of the short-lived "holes" between the molecules become large enough such that an adjacent molecule or a segment of a polymer molecule can fit between two molecules. Thus, T_g can be considered the lowest temperature at which segments of polymer molecules can move with some facility relative to neighboring segments. The increase in coefficient of thermal expansion above T_g results from the greater degree of freedom available to the molecule segments. The larger volume between molecules gives more degrees of freedom, so the same increase in temperature gives a greater increase in volume. As temperature rises, specific volume increases, but there is no more material—just the same material occupying more space. What is in this "extra" volume? Nothing. It is called *free volume*, represented by the hatched area in Figure 2.4(*c*). The molecular motions involved can be detected by spectroscopic techniques, such as solid state NMR, and change as polymers are heated through T_g [6,7].

While it is difficult to overemphasize the importance of the concepts of T_g and free volume in coatings science, our understanding of these parameters and our ability to measure them are limited. T_g values of a material that are determined by different methods may not agree with each other; one must be careful in comparing T_g values of different materials to be sure they are based on consistent test methods. T_g is often measured by dilatometry (measurement of specific volume as a function of temperature), by differential scanning calorimetry (DSC), or by dynamic mechanical analysis (DMA). The measured T_g depends on the measurement method and the conditions under which the measurement was made. Heating rate is an important variable. The faster the rate of heating during the determination, the higher the apparent T_g. When free volume is small, the rate of movement of molecules or segments is slow. If the rate of heating is slow, there is more time for movement and, hence, expansion, and the measured T_g is lower. (See Ref. [8] for discussion of various methods of measuring T_g.)

Some scientists believe that T_g is not a real thermodynamic parameter. They point out that if the determination of specific volume were done at a slow enough heating rate, no transition would be observed and that rather than two straight lines, as shown in Figure 2.4(*c*), there would be a smooth curve. Despite these difficulties, T_g is a very useful concept and is well understood in qualitative terms. Relationships between polymer structure and T_g are understood well enough that it is often possible to make reasonable predictions of T_g from knowledge of composition and $\bar{M}_n$. Important factors affecting the T_g of thermoplastic polymers include:

1. *Number average molecular weight.* T_g increases with increasing $\bar{M}_n$, finally approaching a constant value at $\bar{M}_n$ in the range of 25,000 to 75,000, depending on the polymer structure. It is logical that T_g is related to $\bar{M}_n$ because decreasing $\bar{M}_n$ results in an increasing fraction of chain ends to chain middles, since chain ends have more freedom of movement than chain middles. The relationship of T_g to $\bar{M}_n$ is shown in the following equation, where $T_{g\infty}$ is the T_g at infinite molecular weight and A is a constant:

$$T_g = T_{g\infty} - A/\bar{M}_n$$

2. *Polymer backbone flexibility.* T_g is affected by the ease of rotation about bonds in the polymer backbone. For example, the siloxane bond, Si–O–Si, rotates easily; the T_g of poly(dimethylsiloxane) is 146 K ($-127°$C) [9]. Aliphatic polyethers, such as poly(ethylene oxide), $-(CH_2-CH_2-O)_n-$, also have low T_g, generally in the range of 158 to 233 K, because there is considerable ease of rotation around the ether bond. The T_g of polyethylene varies because, although we commonly think of polyethylene as being just chains of methylene groups, the backbone is actually substituted to varying degrees with alkyl side chains, such as ethyl groups. Also, most grades of polyethylene are partially crystalline; only the amorphous areas show a T_g. However, all would agree that the T_g of a long linear aliphatic chain is low, perhaps less than 200 K. Substitution on the aliphatic chain by methyl groups—for example, polypropylene—increases the T_g, since ease of

rotation is decreased. Similarly, substitution of carboxymethyl groups on the chain–for example, poly(methyl acrylate) (PMA)—gives a higher T_g of 281 K. An even higher T_g of 378 K results when two groups are substituted on alternate carbons of the chain—for example, a methyl and a carboxymethyl group, poly (methyl methacrylate) (PMMA). The presence of rigid aromatic rings in the polymer backbone substantially increases T_g.

3. *Side chains.* Some polymers have side chains pendant from the backbone. If the side chains are several atoms long and flexible, they reduce T_g. An example is poly(n-butyl acrylate); its T_g is 219 K. However, if the side chain is short, bulky, and inflexible, it has less effect and, in some cases, raises T_g; for example, the T_g of poly(t-butyl methacrylate) is 347 K. Side chains containing rigid aromatic rings lead to high T_g—for example, 373 K for polystyrene.

Care must be taken when comparing T_g values to be sure that the determinations have been done under consistent conditions and that the molecular weights are high enough to eliminate molecular weight effects. Table 2.1 gives the T_g (determined by dilatometry) of high molecular weight homopolymers of a group of acrylic and methacrylic esters, as well as other monomers commonly used as comonomers in polymers for coatings.

Copolymers have T_g values intermediate between those of the homopolymers. It is common to use the *Fox equation* to estimate the T_g of copolymers, where w_1, w_2, w_3, etc. are the weight fractions of the various monomers in the copolymer and T_{g1}, T_{g2}, T_{g3}, etc. are the T_g (Kelvin) of their high molecular weight homopolymers:

$$\frac{1}{T_{g(copolymer)}} = \frac{w_1}{T_{g1}} + \frac{w_2}{T_{g2}} + \frac{w_3}{T_{g3}} + \cdots$$

Table 2.1. Glass Transition Temperatures (°C) for Homopolymers of Various Monomers[a]

	Acrylic and Methacrylic Acids and Esters	
Monomer	Methacrylate	Acrylate
Free acid	185	106
Methyl	105	8
Ethyl	65	−22
n-Butyl	20	−54
Isobutyl	64	−43
t-Butyl		74
2-Ethylhexyl	−10	−85
2-Hydroxyethyl	55	
2-Hydroxypropyl	73	

	Other Monomers	
Styrene	100	
Vinyl acetate	29	
Vinyl chloride	81	
Vinylidene chloride	−18	

[a]Data selected from Refs. [9,10].

Somewhat better approximations can be calculated using a different mixing equation, also devised by Fox, in which v_1, v_2, v_3, etc. are the volume fractions from the various monomers in the copolymer: This equation is not widely used because some of the homopolymer densities needed to calculate v_1, v_2, and v_3, etc. are not readily available.

$$T_{g(copolymer)} = v_1 T_{g1} + v_2 T_{g2} + v_3 T_{g3} + \cdots$$

Gupta [11] reports an extensive study of the estimation of T_g of acrylic copolymers. He recommends use of the van Krevelen equation for estimation of T_g, where M is the molecular weight of the repeat unit and Y_g is a molar glass transition factor: Gupta's values for the T_g of n-butyl methacrylate (10°C) and of 2-ethylhexyl acrylate (−63°C) vary considerably from the values given in Table 2.1, illustrating that different values are often found in the literature.

$$T_g = Y_g/M$$

The T_g of cross-linked polymers is controlled by four factors and their interactions: T_g of the segments of polymer between the cross-links, the cross-link density, the presence of dangling ends, and the presence of cyclic segments [12]. While generalized equations showing the relationships of these factors with T_g have been developed, the complex relationships are not fully understood. The T_g of the polymer segments between cross-links is governed by the chemical structures of the resin and the cross-linking agent, by the ratio of these components, and by the extent of the cross-linking reaction. The factors discussed in connection with thermoplastic polymers apply in terms of their effects on the T_g of the segments of the cross-linked polymer chains. Since cross-links restrict segmental mobility, T_g increases as cross-link density increases. On the other hand, T_g decreases with an increasing proportion of dangling ends–that is, chain segments with unreacted cross-link sites.

Solutions of polymer in solvent and of solvent in polymer have T_g values intermediate between the T_g of the polymer and that of the solvent. The T_g of solutions increases with increasing polymer concentration. When the weight fraction of solvent w_s is less than about 0.2, a simple mixing equation gives reasonable correlation between experimental and predicted results [13]:

$$T_{g(solution)} = T_{g(polymer)} - kw_s$$

Over a wider range of concentrations, this simple equation gives poor correlations. Solutions of oligomeric n-butyl methacrylate in m-xylene [5], Eq. 2.1 give a good fit between observed and predicted data over the whole range from pure solvent to solvent-free oligomer. Here, w_s and w_o are weight fractions and T_{gs} and T_{go} are the T_g of the solvent and the oligomer, respectively. While Eq. 2.1 has accurately described a limited number of oligomer and polymer solutions, its generality is not fully established.

$$\frac{1}{T_{g(solution)}} = \frac{w_s}{T_{gs}} + \frac{w_o}{T_{g0}} + Kw_s w_o \qquad (2.1)$$

An excellent general review of glass transition is available in Ref. [14]. Reference [15] provides a review of free volume considerations in coatings.

2.2. POLYMERIZATION

There are two major classes of polymerization reactions: *chain-growth* and *step-growth*. The common denominator of chain-growth polymerization is that reactions are chain reactions. Frequently, chain-growth polymerization is called *addition polymerization*, but this terminology is inadequate; while all chain-growth polymerizations involve addition reactions, not all addition polymerizations involve chain-growth reactions—some are step-growth reactions.

2.2.1. Chain-Growth Polymerization

Free radical initiated polymerization is the most commonly used chain-growth polymerization for making polymers for coatings. Its mechanisms and kinetics have been extensively studied. Reference [16] gives an extensive coverage of the topic, especially of the kinetics of the reactions. Reference [17] reviews recent work; with particular emphasis on side reactions. The free radical chain-growth polymerizations that are of most interest to coating applications are solution polymerization (see Chapter 12) and emulsion polymerization (see Chapter 8). Another process related to chain-growth polymerization that is important in coatings is the autoxidation involved in cross-linking drying oils and drying oil derivatives. (See Chapter 14.)

Three types of chemical reactions—*initiation, propagation*, and *termination*—are always involved in chain-growth polymerization, and a fourth, *chain transfer*, often plays a significant role. Initiation occurs when an *initiator* (I) reacts to form an initiating free radical (I ·), which, in turn, adds rapidly to a monomer molecule to form a second free radical, as shown by Eqs. 2.2 and 2.3:

$$I \longrightarrow I\cdot \qquad (2.2)$$

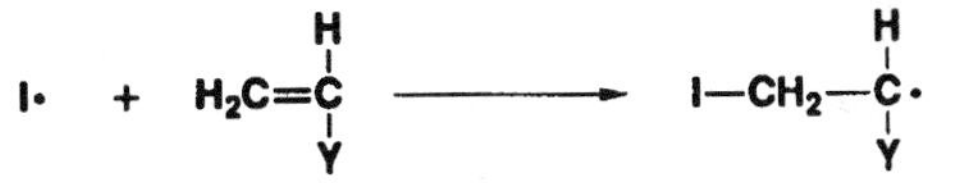

$$I\cdot \ + \ H_2C{=}\underset{\underset{Y}{|}}{\overset{\overset{H}{|}}{C}} \longrightarrow I{-}CH_2{-}\underset{\underset{Y}{|}}{\overset{\overset{H}{|}}{C}}\cdot \qquad (2.3)$$

The polymer chain grows by the propagation reaction, in which the monomer free radical adds to a second monomer molecule to extend the chain while forming a new free radical, as shown by Eq. 2.4.

$$I{-}CH_2{-}\underset{\underset{Y}{|}}{\overset{\overset{H}{|}}{C}}\cdot \ + \ H_2C{=}\underset{\underset{Y}{|}}{\overset{\overset{H}{|}}{C}} \longrightarrow I{-}CH_2{-}\underset{\underset{Y}{|}}{\overset{\overset{H}{|}}{C}}{-}CH_2{-}\underset{\underset{Y}{|}}{\overset{\overset{H}{|}}{C}}\cdot \qquad (2.4)$$

Propagation reactions are very fast—so fast that a chain with hundreds of mers can grow in a fraction of a second. At any moment, the concentrations of monomer and polymer greatly exceed the concentration of growing polymer molecules (circa 10^{-6} mL^{-1}).

The final stage is termination of the growing chain. Two common types of termination reactions are *combination*, shown in Eq. 2.5, and *disproportionation*, shown in Eq. 2.6. In most free radical initiated polymerizations, the rate of propagation, shown in Eq. 2.3, is faster than the rate of initiation, which is limited by the rate of Eq. 2.2.

$$(P)\text{---}CH_2\text{---}\overset{\overset{\text{H}}{|}}{\underset{\underset{\text{Y}}{|}}{C}}\cdot \ + \ \overset{\overset{\text{H}}{|}}{\underset{\underset{\text{Y}}{|}}{C}}\text{---}CH_2\text{---}(P) \longrightarrow$$

$$(P)\text{---}CH_2\text{---}\overset{\overset{\text{H}}{|}}{\underset{\underset{\text{Y}}{|}}{C}}\text{---}\overset{\overset{\text{H}}{|}}{\underset{\underset{\text{Y}}{|}}{C}}\text{---}CH_2\text{---}(P) \tag{2.5}$$

$$(P)\text{---}CH_2\text{---}\overset{\overset{\text{H}}{|}}{\underset{\underset{\text{Y}}{|}}{C}}\cdot \ + \ \overset{\overset{\text{H}}{|}}{\underset{\underset{\text{Y}}{|}}{C}}\text{---}CH_2\text{---}(P) \longrightarrow (P)\text{---}\overset{\overset{\text{H}}{|}}{\underset{\underset{\text{Y}}{|}}{C}}{=}\overset{\text{H}}{C}$$

$$+ \ H\text{---}\overset{\overset{\text{H}}{|}}{\underset{\underset{\text{Y}}{|}}{C}}\text{---}CH_2\text{---}(P) \tag{2.6}$$

Side reactions also occur; among the most important are chain transfer reactions, in which the free radical on the end of the propagating polymer chain abstracts a hydrogen atom from some substance X–H present in the polymerization reaction mixture, as shown in Eq. 2.7:

$$X\text{--}H \ + \ \cdot\overset{\overset{\text{H}}{|}}{\underset{\underset{\text{Y}}{|}}{C}}\text{---}CH_2\text{---}(P) \longrightarrow X\cdot \ + \ H\text{-}\overset{\overset{\text{H}}{|}}{\underset{\underset{\text{Y}}{|}}{C}}\text{---}CH_2\text{---}(P) \tag{2.7}$$

The net effect of chain transfer is to terminate the growing chain while generating a free radical, which may start a second chain growing. X–H may be a solvent, *chain transfer agent*, monomer, or molecule of polymer. When chain transfer is to a solvent or chain transfer agent, molecular weight is reduced. When chain transfer is to a polymer molecule, growth of one chain stops, but a branch grows on the polymer molecule; the result is a higher $\bar{M}_w/\bar{M}_n$.

Note that the structures of the propagating polymer chains show substitution on alternate carbon atoms. This structure results from the favored addition of free radicals to the CH_2 end of the monomer molecules: *head-to-tail addition*. With almost all monomers head-to-tail addition predominates, but a small fraction of *head-to-head addition* occurs. The result is a polymer with most of the substitution on alternating carbons in the chain, but with a few chain segments with substitution on adjacent carbons. The effect of a small fraction of head-to-head structure is generally negligible, but it sometimes has significant consequences for exterior durability and thermal stability.

Initiators, sometimes incorrectly called catalysts, are used in low concentration (usually in the range of 1–4 wt% (weight percent), but sometimes higher when low molecular weight is desired. A variety of free radical sources has been used. Two classes of initiators are used most often: azo compounds, such as azobisisobutyronitrile (AIBN) and peroxides such as benzoyl peroxide (BPO) or *t*-amyl peracetate. AIBN is fairly stable at 0°C, but it decomposes relatively rapidly when heated to 70–100°C to generate free radicals. A substantial fraction of the resulting radicals initiate polymerization. The half-life of AIBN is about five hours at 70°C and about seven minutes at 100°C.

BPO decomposes at similar temperatures—its half-life is about 20 minutes at 100°C. The benzoyl free radical generated can initiate polymerization; also, it can dissociate

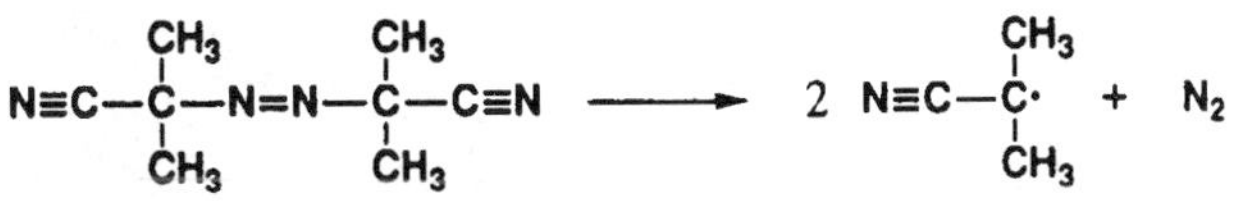

AIBN

(rapidly at higher temperatures, such as 130°C) to yield a highly reactive phenyl free radical and CO_2:

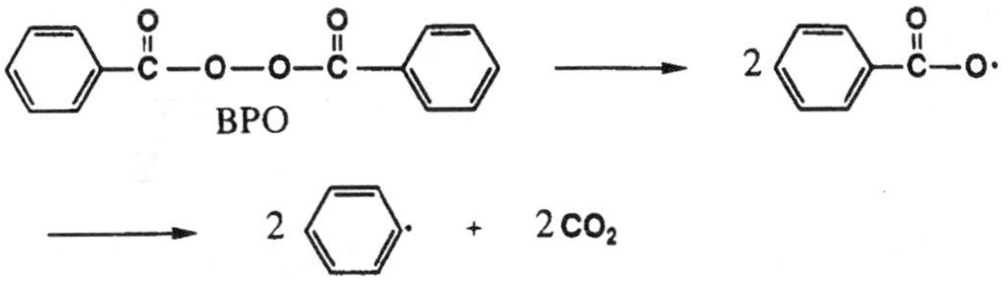

BPO

A range of monomers is capable of propagating a radical initiated chain reaction. Most are alkenes having an electron-withdrawing group; methyl acrylate (MA) and methyl methacrylate (MMA) are examples:

MA   MMA

Copolymers containing a preponderance of acrylic and methacrylic ester monomers are called *acrylic polymers*, or often, just *acrylics*. They are extensively used in coatings. Control of molecular weight and molecular weight distribution is critical in preparing polymers for coatings. There are four major factors that affect molecular weight with the same monomer, initiator, and solvent:

1. *Initiator concentration.* The higher the initiator concentration, the lower the molecular weight. When the initiator concentration is higher, more initiating free radicals are generated to react with the same total amount of monomer. More chains are initiated and terminated; the $\bar{M}_n$ and $\bar{M}_w$ of the polymer produced are lower.

2. *Temperature.* At higher temperatures, more initiator is converted into initiating free radicals in a given time, increasing the concentration of growing chains and the probability of a termination. As with increasing initiator concentration, the result is lower $\bar{M}_n$ and $\bar{M}_w$.

3. *Monomer concentration.* The higher the monomer concentration, the higher the $\bar{M}_n$ and $\bar{M}_w$. The highest molecular weight is obtained in a solvent-free reaction mixture. With the same concentration of growing free radical ends, a higher monomer concentration increases the probability of chain growth relative to termination.

4. *Solvent concentration.* The higher the solvent concentration, the lower the $\bar{M}_n$ and $\bar{M}_w$. There are for two reasons for this result. First, higher solvent concentration leads to lower monomer concentration, reducing molecular weight as described previously. Furthermore, most solvents undergo, to varying extents, chain-transfer reactions, which also reduce molecular weight.

To the extent that any of these factors change during a polymerization process, $\bar{M}_n$ and $\bar{M}_w$ of the polymer molecules also change. The result is a broader molecular weight distribution. Changes in monomers also give changes in molecular weight distribution. Consider the difference between MA and MMA. Since the free radicals at the ends of growing chains of poly(methyl methacrylate) (PMMA) are sterically hindered, termination by combination is impeded, and termination by disproportionation predominates. On the other hand, with MA, a major fraction of the termination reactions occur by combination. Theoretical calculations show that for high molecular weight polymers, the lowest $\bar{M}_w/\bar{M}_n$ attainable with termination by combination is 1.5, while the minimum with termination by disproportionation is 2.0. In actual polymerization processes, $\bar{M}_w/\bar{M}_n$ is usually higher, although with very high initiator concentrations, polydispersities tend to be lower. No basic studies have been reported to account for the low polydispersities with high initiator concentrations.

Chain transfer to polymer must also be considered. This reaction occurs to a degree in the polymerization of MMA, but is more important in the polymerization of MA. The hydrogen on the carbon to which the carboxymethyl group is attached is more susceptible to abstraction by free radicals than any other hydrogen in PMA or PMMA. The new free radical on the PMA chain can now add to a monomer molecule, initiating growth of a branch on the original polymer molecule. The result is a polymer containing branched molecules and having a larger $\bar{M}_w/\bar{M}_n$ than predicted for ideal linear polymerization. In extreme cases, chain transfer to polymer results in very broad molecular weight distributions and, ultimately, to formation of gel particles through cross-linking.

Branching can also result from the abstraction of hydrogen atoms from a polymer chain by initiating free radicals. The highly reactive phenyl free radicals from BPO, especially at temperatures over 130°C, are effective hydrogen abstracters, leading to substantial branching. BPO is a good choice when branching is desired, but in most cases, it is desirable to minimize branching. In these cases, azo initiators, such as AIBN, or aliphatic peroxy initiators are preferred over BPO. Initiator residues remain attached to the polymer chain ends. For high molecular weight polymers, they have a negligible effect on most properties; for oligomers, they may have an appreciable effect, particularly on exterior durability (see Section 12.2.1).

Molecular weight and molecular weight distribution also depend on solvent structure. For example, substituting xylene for toluene and keeping other variables constant leads to a decrease in molecular weight. Since each xylene molecule has six abstractable hydrogen atoms, while toluene has only three, the probability of chain transfer is increased; average molecular weight is decreased. If one wants to prepare a low molecular weight resin, one can add a compound that undergoes facile hydrogen abstraction as a chain-transfer agent. If the hydrogen atoms are sufficiently readily abstracted, the addition of even relatively low concentrations can lead to a substantial reduction in molecular weight. Mercaptans (RSH) are widely used as chain-transfer agents.

Other variables affecting molecular weight and molecular weight distribution are the decomposition rate of the initiator and the reactivity of the resulting free radicals. If one wishes to change the initiator and not to change the molecular weight, one would have to adjust the temperature to equalize the rates of decomposition of the initiators, but this change would probably alter the relative rates of propagation, termination, and chain transfer, perhaps leading to significant differences in the polymer.

To achieve a low $\bar{M}_w/\bar{M}_n$, concentrations of reactants must be kept as constant as possible through the polymerization. It is undesirable to simply charge all of the monomers, solvents, and initiators into a reactor and heat the mass to start the reaction. This procedure is sometimes used in small scale laboratory reactions, but never in production. At best, it yields a high $\bar{M}_w/\bar{M}_n$; at worst, the reaction may run violently out of control, because free radical polymerizations are highly exothermic. Instead, one charges some of the solvent into the reactor, heats to reaction temperature, and then adds monomer, solvent, and initiator solutions to the reactor at rates such that the monomer and initiator concentrations are kept as constant as possible. Adding monomer at a rate that maintains a constant temperature leads to a fairly constant monomer concentration. The appropriate rate of addition of the initiator solution can be calculated from the rate of its decomposition at the temperature being used. Keeping solvent concentration constant is more complex since as the polymerization proceeds, polymer is accumulating; in a sense, the polymer becomes a part of the "solvent" for the polymerization. Solvent is added at a decreasing rate so that the other concentrations stay as constant as possible. Perfect control is not possible, but careful attention to details makes an important difference in the $\bar{M}_w/\bar{M}_n$ of the polymer produced.

Copolymerization of mixtures of unsaturated monomers further complicates the situation. The rates of reaction involved in the various addition reactions depend on the structures of the monomers. If the rate constants for all of the possible reactions were the same, the monomers would react randomly and the average composition of molecules of substantial length would all be the same. However, the rate constants are not equal. If polymerization is carried out by putting all of the reactants in a flask and heating, the first molecules formed would contain more than proportional amounts of the most reactive monomer, and the last molecules formed would have an excess of the least reactive monomer. This situation is usually undesirable. Such effects have been extensively studied, and equations have been developed to predict the results with different monomer combinations. (See Ref. [16] for detailed discussion of copolymerization.) In actual practice, the problem is somewhat less complex. Reactions are not run in bulk, but rather, as mentioned above, monomers, solvent, and initiator solution are added gradually to the reaction mixture. If the additions are carefully controlled so that the rate of addition equals the rate of polymerization, copolymers having reasonably uniform composition corresponding to the feed ratio are obtained with most monomers. This procedure, called *monomer-starved conditions*, results in polymerization under conditions in which the concentration of monomers is low and fairly constant. Further process refinements are possible by adding individual reactants or mixtures of reactants in two or three streams at different rates. Computer modeling of the processes can help achieve the desired results.

Chain-growth polymerization can also be carried out by group-transfer and cationic polymerizations. Group-transfer polymerization is briefly discussed in Section 12.2.1. Cationic polymerization is the mechanism for homopolymerization of epoxy resins, as discussed in Sections 11.3.5 and 28.3.

2.2.2. Step-Growth Polymerization

A second class of polymerization that is important in the coatings field is *step-growth polymerization*. As the name indicates, the polymer is built up a step at a time. The term *condensation polymerization* has been used for this process because early examples involved condensation reactions–reactions in which a small molecule byproduct, such as water, is eliminated. While both terms are still used, *step-growth polymerization* is more appropriate because some of the reactions used are not condensation reactions. Step-growth polymerization reactions are used in two ways in coatings. One use is for preparation of resins for use as vehicles, and the other use is for cross-linking after the coating has been applied to a substrate. In this introductory section, polyester formation is used to illustrate the principles involved; polyesters are discussed more broadly in Chapter 13.

Of the many reactions that form esters, three are commonly used in making step-growth polymers and oligomers for coatings: direct esterification of an acid with an alcohol,

$$
\text{R}-\overset{\overset{\text{O}}{\|}}{\text{C}}-\text{OH} \;+\; \text{HO}-\text{R}' \;\underset{\longleftarrow}{\overset{\text{Cat.}}{\longrightarrow}}\; \text{R}-\overset{\overset{\text{O}}{\|}}{\text{C}}-\text{O}-\text{R}' \;+\; \text{H}_2\text{O}
$$

transesterification of an ester with an alcohol,

$$
\text{R}-\overset{\overset{\text{O}}{\|}}{\text{C}}-\text{O}-\text{R}'' \;+\; \text{HO}-\text{R}' \;\underset{\longleftarrow}{\overset{\text{Cat.}}{\longrightarrow}}\; \text{R}-\overset{\overset{\text{O}}{\|}}{\text{C}}-\text{O}-\text{R}' \;+\; \text{HO}-\text{R}''
$$

and reaction of an anhydride with an alcohol.

$$
\text{R}-\overset{\overset{\text{O}}{\|}}{\text{C}}-\text{O}-\overset{\overset{\text{O}}{\|}}{\text{C}}-\text{R} \;+\; \text{HO}-\text{R}' \;\longrightarrow\; \text{R}-\overset{\overset{\text{O}}{\|}}{\text{C}}-\text{O}-\text{R}' \;+\; \text{HO}-\overset{\overset{\text{O}}{\|}}{\text{C}}-\text{R}
$$

When one of the reactants is monofunctional, polymer cannot form. If, however, both reactants have two or more functional groups, a polymer can be made. When all monomers are difunctional, linear polymers form. Linear step-growth polymers are commonly used in fibers, films, and plastics, but there are few examples of their use in coatings. Most polyester resins used in coatings have relatively low molecular weights and are branched resins made using at least one monomer with three or more functional groups. After application of the coating, the terminal groups on the branch ends are reacted with a cross-linker to form the cured coating.

When a difunctional acid (AA) reacts with a difunctional alcohol (BB) in a direct esterification reaction, the molecular weight builds up gradually. Under ideal conditions, polymer chains averaging hundreds of mers per molecule can be made, but this can occur only if (a) the reactants AA and BB contain no monofunctional impurities, (b) exactly equimolar amounts of AA and BB are used, (c) the reaction is driven virtually to completion, and (d) side reactions are negligible. If one reactant is present in excess, terminal groups of the excess monomer predominate. The molecular weight of the completely reacted system is progressively lower as the difference from equal equivalents is increased. For example, if 7 moles of dibasic acid are completely reacted with 8 moles of a dihydroxy compound (a diol), the average molecule will have terminal hydroxyl groups as shown in the following equation (here, for convenience, AA and BB represent both the reactants and the mers in the polymer):

$$
7\text{AA} + 8\text{BB} \longrightarrow \text{BB}\left(\text{AA}-\text{BB}\right)_6\text{AA}-\text{BB} + 14\text{H}_2\text{O}
$$

The symbol F is used for the functionality of monomers, which is the number of reactive groups per molecule, for example:

Neopentyl Glycol   Glycerol   Pentaerythritol

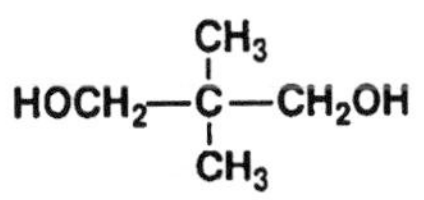

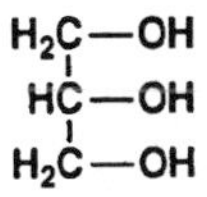

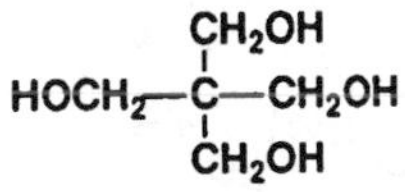

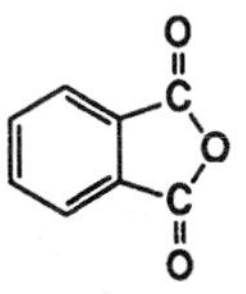

Phthalic Anhydride 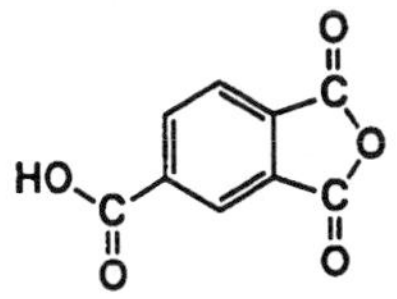   Trimellitic Anhydride

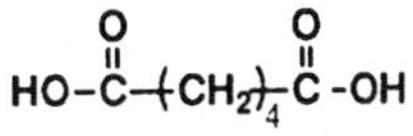

Adipic Acid

Note that the anhydride groups in phthalic anhydride and trimellitic anhydride count as two functional groups since they can form two ester groups during polymerization.

The average functionality, represented by $\bar{F}$, of a mixture of monomers containing equal equivalents of hydroxyl and carboxyl groups is calculated as follows:

$$\bar{F} = \frac{\text{total equivalents}}{\text{total moles}}$$

Most coating polyester resins are hydroxy-functional and are made using monomer mixtures with excess hydroxyl groups. Some of the hydroxyl groups have no carboxyl groups to react with, so the equation must be modified to reflect only the total number of equivalents that can react. In a resin with excess hydroxyl groups prepared from dicarboxylic acids, the total equivalents that can react correspond to twice the number of equivalents of carboxylic acid groups:

$$\bar{F} = \frac{\text{total equivalents that can react}}{\text{total moles}} = \frac{2(\text{equiv of COOH})}{\text{total moles}}$$

A simple formulation for a polyester oligomer is given in Table 2.2.

A further type of functionality important in designing resins is the functionality of the resin. To distinguish this functionality from that of the monomers and the monomer

Table 2.2. Polyester formulation

Component	mols	equivs
Adipic acid	0.9 mol	1.8 equiv
Phthalic anhydride	0.9 mol	1.8 equiv
Neopentyl glycol	1 mol	2 equiv
Glycerol	1 mol	3 equiv
Total	3.8 mol	8.6 equiv

$$\bar{F} = 2(3.6)/3.8 = 1.89$$

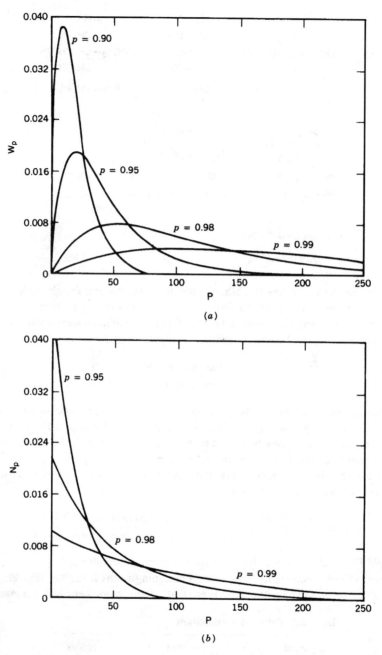

Figure 2.5. (*a*) Weight fraction distribution w_P of molecules in a linear step-growth polymer for several extents of reaction p. (*b*) Number, or mole fraction, distribution n_P. (From Ref. [22], with permission.)

mixture, the symbol f is used. Since almost all coating resins are made using some triol or tetrol, one usually uses a number average functionality $\bar{f}_n$:

$$\bar{f}_n = \frac{\text{number of functional groups in a sample}}{\text{number of molecules in the sample}}$$

The value of $\bar{f}_n$ can be calculated from $\bar{M}_n$ and the number of functional groups per sample weight obtained by analysis.

Esterification of a carboxylic acid with an alcohol is acid catalyzed. In the absence of a strong acid catalyst, the rate r is approximately third order in reactants, as shown in Eq. 2.8, with one carboxyl group reacting with the alcohol and a second catalyzing the reaction. Since water is generally removed rapidly from the reaction mixture, it is reasonable to use Eq. 2.8, which disregards the reverse reaction.

$$r = k\,[\text{RCOOH}]^2\,[\text{R}'\text{OH}] \tag{2.8}$$

Because of the second order dependence on acid concentration, the rate decreases precipitously as reaction proceeds. For example, polyesterification of equal moles of diethylene glycol with adipic acid at 160°C without a catalyst is 60% complete in 1 hour but requires 27 hours to reach 94.5% conversion and would require years to reach 99.8% conversion [18]. The reaction is catalyzed by strong acids. In many cases, conventional strong acids cause side reactions and discoloration, so the most appropriate catalysts are organotin compounds, such as monobutyltin dioxide, or titanate esters. It has been shown that both the organotin compound and the carboxylic acid act as catalysts [19].

The kinetics of ideal step-growth polyesterification for difunctional reactants can be analyzed in terms of p, the fractional extent of reaction; n_p, the number fraction of differing degrees of polymerization; P_n, the degree of polymerization; and w_p, the weight fraction of molecules [20, 21]. As p increases, the degree of polymerization builds up slowly at first— at $p = 0.5$ (corresponding to 50% conversion), $\bar{P}_n$ is only 2. The $\bar{P}_n$ is only 10 at $p = 0.9$, and $p = 0.998$ is required to reach a $\bar{P}_n$ of 500. Thus, with difunctional monomers, high molecular weight can only be attained when the mole ratio of COOH/OH is 1 and when esterification is driven beyond $p = 0.99$. This is difficult because of the decreasing reaction rate at high values of p. Note that, as shown in Figure 2.5(a), the number of unreacted monomer molecules remains higher than that of any other single species in the reaction mixture, no matter how high p becomes. As shown in Figure 2.5(b), $\bar{P}_n$, the peak of the P distribution curve, only reaches substantial values at high p values. In the case of high molecular weight linear polymers, under ideal conditions, the $\bar{M}_w/\bar{M}_n$ obtained in step-growth polymerizations is 2.

2.3. FILM FORMATION

Most coatings are liquids with a viscosity appropriate for the application method to be used, generally in the range of 0.05 to 1 Pa·s at high shear rates. After application, the liquid is converted to a "dry," that is, solid film. In powder coatings, the powder is liquified after application and then converted to a solid film.

If the applied coating were crystalline, there would be no difficulty in defining a solid film. The film would be solid if the temperature were below its freezing point; however, binders of coatings are almost always amorphous, with no melting point and no sharp demarcation between a liquid and a solid. A useful definition of a solid film is that it does not flow significantly under the pressures to which it is subjected during use. Thus, one can define whether a coating is a solid under a set of conditions by stating the minimum viscosity required such that flow does not exceed a specified extent under specified pressure and time. For example, it is reported that a film is *dry-to-touch* if its viscosity is greater than about 10^3 Pa·s [22]. However, if the definition of a solid is that the film resists *blocking*—that is, sticking together when two coated surfaces are put against each other for 2 seconds under a pressure of 1.4 kg cm^{-2} (20 psi)—the viscosity has to be greater than about 10^7 Pa·s.

For thermoplastic binders, we can use this information to predict polymer structures that could meet such tests. Using a simplified form (Eq. 2.9, in the equation T is in Kelvin) of the WLF equation (see Section 3.4), using "universal constants" and assuming that the viscosity at T_g is 10^{12} Pa·s, one can estimate the T_g of a binder required so that a film does not flow under some set of circumstances:

$$\ln \eta = 27.6 - \frac{40.2(T - T_g)}{51.6 + (T - T_g)} \tag{2.9}$$

Using Eq. (2.9), we can estimate the appropriate $(T - T_g)$ value required for a film to be dry-to-touch; that is, to have a viscosity of 10^3 Pa·s. The calculated $(T - T_g)$ value is 54°C, which corresponds to a T_g of −29°C for a film to be dry-to-touch at temperature T of 25°C. The T_g calculated for block resistance (at 1.4 kg cm^{-2} for 2 s at 25°C, i.e. for a viscosity of 10^7 Pa·s) is 4°C. Because there is considerable variation in the WLF "universal constants," these T_g values are not exact, but they can serve as a formulation guide. Since we have a reasonable idea of the relationships between structure and T_g (see Section 2.1.2), we can approximate the requirements to make a binder with the viscosity necessary to pass a particular test. If the coating has to pass a test at a higher temperature than 25°C, the T_g of the binder must be higher, since the free volume dependence is on $(T - T_g)$. If the pressure to which the film is to be subjected is higher or the time under pressure is to be longer, the T_g must be higher.

2.3.1. Film Formation by Solvent Evaporation from Solutions of Thermoplastic Binders

Films can be formed in a variety of ways. One of the simplest ways is to dissolve a polymer in solvent(s) at a concentration needed for application requirements, apply the coating, and allow the solvent to evaporate. Let us illustrate with a copolymer of vinyl chloride, vinyl acetate, and a hydroxyl-functional vinyl monomer with an $\bar{M}_n$ of 23,000 that is reported to give coatings with good mechanical properties without cross-linking [23]. The T_g of the copolymer is reported to be 79°C. A solution in methyl ethyl ketone (MEK) with a viscosity of 0.1 Pa·s required for spray application would have about 19 NVW (nonvolatile weight, that is, wt% solids) and an NVV [nonvolatile volume, that is, volume percent (vol%) solids] of about 12. MEK has a high vapor pressure at room temperature and evaporates rapidly from a thin layer. In fact, a sizable fraction of the MEK evaporates from the atomized spray droplets between the time they leave the spray gun and reach the substrate. As solvent evaporates from a film, viscosity increases, and the film will be dry-to-touch soon after application. Also, in a short time, the coating will not block under the conditions mentioned previously. Nevertheless, if the film is formed at 25°C, the "dry" film contains several percent of retained solvent. Why?

In the first stage of solvent evaporation from a film, the rate of evaporation is essentially independent of the presence of the polymer. Evaporation rate depends on the vapor pressure at the temperature, the ratio of surface area to volume, and the rate of air flow over the surface. However, as solvent evaporates, viscosity increases, T_g increases, free volume decreases, and the rate of loss of solvent becomes dependent on how rapidly solvent molecules can diffuse to the surface of a film so that they can evaporate. The solvent molecules must jump from free volume hole to free volume hole to reach the surface. As solvent loss continues, T_g increases, free volume decreases further, and solvent loss slows. If the film is formed at 25°C from a solution of a polymer that, when solvent free, has a T_g greater than 25°C (in this example, it is 79°C), the film retains considerable solvent even though it is a hard "dry" film. Solvent slowly leaves such a film but, it has been shown experimentally that 2–3% of solvent remains after several years at ambient temperature. To assure essentially complete removal of solvent in a reasonable period of time requires baking at a temperature significantly above the T_g of the solvent free polymer. Solvent loss from films is discussed in more detail in Section 17.3.4.

2.3.2. Film Formation from Solutions of Thermosetting Resins

A drawback of solution thermoplastic polymer based coatings is that the molecular weights required for film properties require high solvent levels (on the order of 80–90 vol%) to achieve the viscosity for application. Less solvent is needed for a coating based on solutions of lower molecular weight thermosetting resins. After application, the solvent evaporates, and chemical reactions cause polymerization and cross-linking to impart good film properties. Many combinations of chemical reactions are used in thermosetting coatings, as discussed in Chapters 8–16. A critical aspect of the design of a coating is the selection of components that give required mechanical properties; this is discussed in Chapter 4. In this section, we cover only the general principles of cross-linking reactions.

A problem with all thermosetting systems is the relationship between coating stability during storage and the time and temperature required to cure a film after application. Generally, it is desirable to be able to store a coating for many months, or even years, without a significant increase in viscosity resulting from reaction during the storage period. On the other hand, after application, one would like to have the cross-linking reaction proceed rapidly at the lowest possible temperature.

What controls the rate of a reaction? We can consider this question broadly using a general example of a reaction between two groups, represented by the symbols A and B that react to form a cross-link A–B:

$$A \ + \ B \ \longrightarrow \ A{-}B$$

In the simplest cases, one can express the rate of reaction r of A and B by Eq. 2.10, where k is the rate constant for the reaction between A and B at a specified temperature, and [A] and [B] represent the concentration of the functional groups in terms of equivalents per liter. The rate constant is the reaction rate when $[A] \times [B]$ equals 1 $eq^2\ L^{-1}$.

$$r = k[A][B] \tag{2.10}$$

[A] and [B] are reduced by dilution with solvent and increase as solvent evaporates; therefore, with all other factors being equal, cross-linking in the applied film after solvent evaporation is initially faster than during storage. As formulations are shifted to higher solids to reduce VOC emissions, there are higher concentrations of functional groups, and there is greater difficulty in formulating storage stable coatings. The problem results not only from the presence of less solvent, but also from the lower molecular weights and lower equivalent weights needed to achieve a cross-linked film of acceptable cross-link density. Both factors increase the concentration of functional groups in a stored coating. Concentration in a film after application and evaporation of solvent also increases, but since the molecular weight is lower, more reactions must occur to achieve the desired cross-linked film properties.

To minimize the temperature required for curing while maintaining adequate storage stability, it is desirable to select cross-linking reactions for which the rate depends strongly on temperature. This dependence is reflected in the rate equation by the dependence of k on temperature. It is commonly taught in introductory organic chemistry classes that rate constants double with each 10°C rise in temperature. That generalization is true for only a limited number of reactions within a narrow temperature range near room temperature. A better estimate of the temperature dependence of k is given by the empirical Arrhenius equation, Eq. 2.11, where A is the *preexponential term*, E_a is the thermal coefficient of reactivity (commonly labeled *activation energy*), R is the gas constant, and T is temperature (in Kelvin):

$$\ln k = \ln A - (E_a/RT) \tag{2.11}$$

Reaction rate data that fit the equation give straight lines when $\ln k$ is plotted against $1/T$, as illustrated in Figure 2.6. As seen in plot a of competing reactions (1) and (2), where $A(1) = A(2)$ and $E_a(1) > E_a(2)$, the temperature dependence of rate constants increases with increasing values of E_a. This effect can be counteracted by selecting a reaction with a higher A value, as shown in plot b, where $A(3) > A(1)$ and E_a for the two reactions is equal. If A and E_a both are sufficiently greater for one reaction than for another, the rate constant at storage temperature could be smaller, while the rate constant at a higher temperature could be larger, as shown schematically in plot c.

Term A is controlled predominantly by entropic factors, or more specifically, by changes in randomness or order, as the reaction proceeds to the activated complex in the transition state. Three important factors are: (1) Unimolecular reactions tend to exhibit larger A values than those with a higher molecular order, (2) ring opening reactions tend to have high A values, and (3) reactions in which reactants become less polar exhibit larger A values. The importance of these factors—in particular, factor (3)—depends on the reaction medium; accordingly, solvent selection can have a significant effect on storage stability.

While unimolecular reactions are desirable for high A values, cross-linking reactions are necessarily bimolecular. A way around this problem is to use a *blocked* reactant BX that thermally releases a reactant B by a unimolecular reaction—most desirably, with ring opening and decreasing polarity—followed by cross-linking between A and B:

$$BX \rightleftharpoons B + X$$

$$A + B \longrightarrow A{-}B$$

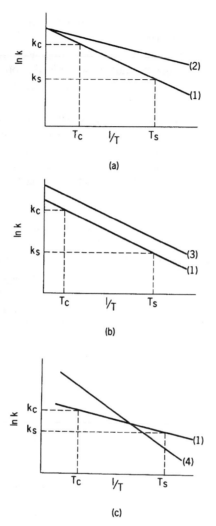

Figure 2.6. Arrhenius plots for competing reactions: (a) $A(1) = A(2)$, $E_a(1) > E_a(2)$; (b) $A(3) > A(1)$, $E_a(1) = E_a(3)$; (c) $A(4) > A(1)$, $E_a(4) > E_a(1)$. From Ref. [24], with permission.)

Another approach is to use a blocked catalyst CX, where C catalyzes the cross-linking of A and B:

$$CX \rightleftharpoons C + X$$

$$A + B \xrightarrow{C} A\text{–}B$$

An important consideration is that the cross-linking reaction, which follows unblocking, should be faster than the reverse reaction, which regenerates the blocked reactant or catalyst. While one often encounters "threshold" or "unblocking" temperatures for reactions in the literature, such minimum reaction temperatures do not exist in the kinetics of reactions. Reactions proceed at some rate at any temperature. Threshold or unblocking temperatures are actually the temperatures at which some observable extent of reaction has occurred within a specified time interval.

These concepts are helpful for understanding the differences we see between storage stabilities, but another reason for understanding these kinetic considerations is that they can be used to predict whether any chemical reaction will ever be found to meet some combination of stability and cure schedule that might be desired for a coating. Pappas and Hill have carried out calculations to permit such predictions [25]. They made a set of reasonable assumptions about reactive group concentrations during storage and in the applied film, permissible degrees of reaction during storage, and required extents of reaction during curing. Using these assumptions, they calculated the ratio of rate constants needed to permit any time of storage with any time of curing. In turn, this allows calculation of E_a and A values as a function of any combination of storage and curing temperatures. Kinetic parameters calculated for unimolecular blocked reactant systems that proceed 5% over a 6 month period at 30°C (storage temperature) and 90% in 10 minutes at various curing temperatures are listed in Table 2.3 [24].

Rate constants and kinetic parameters for unimolecular reactions are independent of concentration, but not for bimolecular reactions. Kinetic parameters calculated for bimolecular (i.e., second order) reactions using concentrations corresponding to high solids coatings were of similar magnitude to those for unimolecular reactions. Although the values in Table 2.3 represent order of magnitude calculation, they provide useful guidelines for avoiding development projects aimed at impossible goals and provide insights for design of thermosetting coatings. Kinetic parameters are known for many chemical reactions. As a point of reference, a reasonable upper limit of A values for unimolecular reactions is 10^{16} s^{-1}, which corresponds to an upper value for the frequency of simple vibrations. For bimolecular reactions, A values tend to be less than 10^{11} L mol^{-1} s^{-1}, an upper limit for the rate constant of diffusion, which must precede reaction. However, as shown in Table 2.3, a coating stable at 30°C would require an A value of 10^{24} s^{-1} if it were to cure within 10 minutes at 100°C. No such reaction is known or even conceivable. Users would like to have package stable coatings that cure in a short time at 80°C, a convenient temperature for low pressure steam heat, but it is pointless to pursue this goal via kinetic control. That is not to say that it is impossible to make such a coating, it means that the problem must be solved by an approach other than kinetic control.

Storage life can be extended by refrigeration, but users are seldom willing to incur the expense. More reactive combinations can be used in *two package coatings*, in which one package contains a resin with one of the reactive groups and the second package contains the component with the other reactive group. Alternatively, the second package could

Table 2.3. Kinetic Parameters as a Function of Cure Temperature

T(°C)	A(s^{-1})	E_a(kJ mol^{-1})[a]
175	10^{10}	109
150	10^{12}	121
125	10^{17}	146
100	10^{24}	188

[a] 1 kJ = 0.239 kcal
Source: See Ref. [24].

contain a catalyst for the reaction. The packages are mixed shortly before use. Two package coatings are often called *2K coatings* and single package coatings are sometimes called *1K coatings*. The *K* stands for the German word for component. Two package coatings are used on a large scale commercially, but are not generally desirable; they take extra time, material is usually wasted, they are generally more expensive, and there is a chance of error in mixing. Even *2K* coatings have the analogous problem of *pot life*—that is, the length of time after the two packages are mixed that the viscosity stays low enough for application. Pappas and have made similar calculations of *A* and E_a values for the shorter times involved [25].

There are several approaches to increasing package stability while permitting cure at ambient or moderately elevated temperature. One or more of these approaches are included in the discussions of many of the cross-linking reactions that are covered in later chapters. The following list gives some of the approaches to solving this "impossible" problem:

1. Use of a radiation-activated cross-linking reaction instead of a thermally activated reaction.

2. Use of a cross-linking reaction requiring an atmospheric component as a catalyst or reactant; reactions involving oxygen or water vapor in the air are examples. The same principle is involved in passing a coated article through a chamber containing a catalyst vapor.

3. Use of a volatile inhibitor that works when the coating is stored in a closed container, but volatilizes after application as a thin film, permitting the reaction to proceed. Examples include use of a volatile antioxidant in a coating that dries by oxidation and use of oxygen as an inhibitor in a composition that cures anaerobically.

4. Use of a cross-linking reaction that is a reversible condensation reaction involving loss of a volatile reaction product; some of the monofunctional volatile reactant can be used as a solvent in the coating. The equilibrium of the reaction favors the noncross-linked side during storage, but shifts to the cross-linked side after application, when the solvent evaporates. A similar approach can be used with blocked reactants and catalysts in which the blocking group is volatile.

5. Use of an encapsulated reactant or catalyst, with the capsules, being ruptured during application. Encapsulation has been useful in adhesives, but has not been very useful in coatings because residual capsule shells interfere with appearance and/or performance.

6. Use of a reactant that undergoes a phase change. While there are no threshold temperatures for kinetic reactions, phase changes can occur over narrow temperature ranges. A crystalline blocked reactant or catalyst, insoluble in the vehicle, could give an indefinitely stable coating; heating over the melting point permits the unblocking reaction to occur, releasing a soluble reactant or catalyst. Over a somewhat wider temperature range, the same idea can be used if the blocked reactant is amorphous, with a T_g about 50°C above the storage temperature and, perhaps, 30°C below the cure temperature.

Another consideration in selecting components for thermosetting coatings is the potential effect of the availability of free volume on reaction rates and reaction completion. For reactions to occur, the reacting groups must diffuse into a reaction volume to form an activated complex that can then form a stable bond. If the diffusion rate is greater than the reaction rate, the reaction will be kinetically controlled. If the diffusion rate is slow compared to the kinetic reaction rate, the rate of the reaction will be mobility controlled. The major factor controlling the diffusion rate is the availability of free volume. If the reaction is occurring at a temperature well in excess of T_g, the free volume is large and the rate of reaction is controlled by concentrations and kinetic parameters. If, however, the temperature is well below T_g, the free volume is so limited that the polymer chain motions needed to bring unreacted groups close together are very slow, and reaction virtually ceases. At intermediate temperatures, the reaction can proceed, but the reaction rate is controlled by the rate of diffusion—that is, by the mobility of the reactants.

Since cross-linking starts with low molecular weight components, T_g increases as the reaction proceeds. If the reaction temperature is well above the T_g of the fully reacted polymer, there will be no mobility effect on the reaction rate. However, if, as is often the case in ambient cure coatings, the initial T_g is below the ambient temperature and the T_g of the fully reacted polymer is above the ambient temperature, the reaction rate will become mobility controlled as cross-linking proceeds. As further reaction occurs, cross-linking may essentially stop before reaching completion. As T_g increases toward the temperature at which the reaction is occurring, reactions become slower. When T_g equals T, reactions become very slow and vitrification (glass formation) is said to occur. Unless the experiment is continued for a relatively long time, the reactions have been said to cease [26]. However, with extended reaction times, it can be seen that reactions continue slowly. One paper [27] reports that reaction rate constants drop by about 3 orders of magnitude when T_g equals T, but that the reaction continues at a slow rate until T_g increases to $T + 50°C$ [27]. It is interesting to compare this value to the 51.6 value for the universal B constant in the WLF equation. (See Eq. 2.9). $(T_g - B)$ is the temperature at which viscosity goes to infinity and free volume theoretically approaches zero. The effect of variables on mobility control of reaction rates has been studied by Dusek and Havlicek [28]. They studied the effects of temperature, polymer-solvent interaction, and solvent volatility on reaction rates and the extent of reaction of the diglycidyl ether of bisphenol A and 1,3-propanediamine. They also give a review of the theories involved.

It seems reasonable for a formulator to assume that cross-linking reactions begin to slow as T_g increases to about 10°C below the curing temperature and get progressively slower until T_g is about 50°C above the curing temperature, where reaction essentially ceases. The slow rates of reaction mean that properties of many ambient cure coatings can change substantially for several weeks, or even months. Caution is required, since T_g values are dependent on the method of determination and the rate of heating used. For this purpose, the most appropriate data to use are values obtained at very slow rates of heating and at low rates of application of stress. It has been pointed out that if reactions are very fast, equilibrium T_g may not be reached [28]. Another factor that may affect the development of mobility control is the size of the diffusing reactants. Small molecules may diffuse more readily to reaction sites than functional groups on a polymer chain. Water plasticizes coatings such as polyurethanes and epoxy-amines, lowering their T_g.

If the initial reaction temperature is well below the T_g of the solvent-free coating, it is possible that little or no reaction can occur after solvent evaporation and that a "dry" film forms merely due to solvent evaporation without much cross-linking. The result is a weak, brittle film. One must be careful when defining what is meant by a dry film, especially when dealing with ambient temperature cure coatings. One consideration is whether the film is dry-to-handle. This stage could be reached with little cross-linking if the T_g of the solvent free binder is high enough. Another consideration is whether some required extent of cross-linking has occurred. This must be tested by some method other than hardness, most easily by determining resistance to dissolution or by the extent of solvent swelling. (See Section 4.2.)

Mobility control is less likely to be encountered in baking coatings because in most cases, the final T_g of the film is below the baking temperature. Furthermore, the T_g is usually well above ambient temperatures so even if there are some unreacted groups, reactions do not continue after cooling to ambient temperature. In powder coatings, mobility control of reaction can be a limitation, since the initial T_g of the reactants has to be over 50°C so that the powder will not sinter during storage. (See Section 27.3.) To achieve a high extent of reaction, the baking temperature must be above the T_g of the fully reacted coating. Gilham and coworkers have studied factors influencing reaction rates in high T_g epoxy-amine systems; Ref. [29] summarizes Gilham's work and, in particular, points out its applicability to powder coatings.

2.3.3. Film Formation by Coalescence of Polymer Particles

In contrast to the processes of film formation from solutions of thermoplastic or thermosetting polymers, dispersions of insoluble polymer particles form films by coalescence (fusion) of the particles. After application and loss of volatile components, the particles form a continuous film. The largest volume of such coatings have latexes as a binder. A latex is a dispersion of high molecular weight polymer particles in water. (See Chapter 8.)

The mechanism of film formation from latexes has been extensively studied, but still is not fully understood; Refs. [30-32] review various theories. The film formation process may be divided, somewhat arbitrarily, into three overlapping steps: *evaporation* of water and water-soluble solvents that leads to a close packed layer of latex particles; *deformation* of the particles from their spherical shape that leads to a more or less continuous, but weak, film; and *coalescence*, a relatively slow process in which the polymer molecules *interdiffuse* across the particle boundaries and entangle, strengthening the film.

For a given latex, the lowest temperature at which coalescence occurs sufficiently to form a continuous film is called its *minimum film formation temperature* (MFFT); some authors call it the *minimum filming temperature* (MFT). MFFT is measured by placing samples on a heated metal bar with a temperature gradient. A major factor controlling MFFT is the T_g of the polymer in the particles. The T_g of poly(methyl methacrylate) (PMMA) is about 105°C, and one cannot form a useful film from a PMMA latex at room temperature; instead, one gets a layer of material that powders easily. Many latexes are designed to have layers of material with different T_g's within each particle (see Sections 8.1.3 and 8.2), making it difficult to directly relate MFFT to T_g.

Theoretical and experimental studies of coalescence suggest that it is only necessary for the molecules to interdiffuse a distance comparable to the radius of gyration of one molecule to develop maximum film strength [32]. This distance is considerably less than the diameter of a typical latex particle. The rate of interdiffusion is directly related to T_g and is, therefore, controlled by free volume availability [32]. The major factor affecting free volume is the difference between the temperature of film formation and the T_g of the particles. Presumably, the T_g of the material near the surface of the original particle is most important. As a general rule, coalescence will not occur unless the temperature is at least slightly higher than T_g. Latexes for paints are generally copolymers of acrylic and vinyl esters that have a T_g well below room temperature so that they can coalesce readily within the usual range of application temperatures.

Plasticizers can be added to the formulation to dissolve in the polymer and lower its T_g and MFFT. Since nonvolatile plasticizers permanently reduce T_g, most latex paints contain volatile plasticizers; they are called *coalescing solvents*, but they accelerate deformation as well as coalescence. A coalescing solvent must be soluble in the polymer and have a low, but appreciable, evaporation rate. It acts as a plasticizer to lower the MFFT, but after the film has formed, it diffuses to the surface of the film and evaporates. Since free volume in the film is relatively small, the rate of loss of the last of the coalescing solvent is very slow. Though the films feel dry in a short time, they will still block for days, or even weeks, after application. Effects of coalescing solvents on film formation have been quantitatively studied by fluorescence decay measurements to follow the extent of polymer diffusion in films as they coalesce [33]. The efficiency of coalescing solvents varies over a wide range; an example of a relatively efficient solvent is the acetate ester of propylene glycol monobutyl ether (P*n*BA) [34]. The rate of evaporation from films also varies; for example, dipropylene glycol dimethyl ether leaves a film more rapidly than P*n*BA, but somewhat more is required for film formation [34]. A widely used coalescing solvent is Texanol, the isobutyric ester of 2,2,4-trimethylpentane-1,3-diol.

The MFFT of latex particles can be affected by water, which can act as a plasticizer [31]. Reduction of T_g is largest with hydrophilic polymers. It has been shown that MFFT can be reduced by as much as 5°C by forming the film in a humid atmosphere [33]. Surfactant stabilizers can increase water absorption and also act as plasticizers for the polymer [31]. In general, higher surfactant content further reduces MFFT. Structure of the surfactant also affects MFFT; for example, nonyl phenol ethoxylates with fewer than 9 ethoxylate units reduced MFFT further than those with 20 or 40 units. Film properties of latex paints can be affected by the weather at the time of application. If paint is applied outdoors on a windy, low-humidity day, coalescence may be adversely affected, since the evaporation rate of water is accelerated, increasing the MFFT of the latex in the paint.

As will be discussed in Section 8.1.1, latexes are stabilized by charge repulsion and entropic repulsion. For film formation, the stabilization must be overcome. As water evaporates, the particles come closer together. It has been proposed that as the particles approach each other, the spaces between them act like the equivalent of a capillary, and capillary forces overcome the stabilizing repulsion [35]. It is estimated that such forces may generate as much as 3.5 MPa (5000 psi) where latex particles touch. While capillary forces may be a factor, Croll points out that the time span during which the forces are high is very short [36]. His data on drying rates support a proposal of Kendall and Padget that a major driving force for coalescence is surface energy reduction [37]. The surface area of a coalesced film is only a small fraction of the surface area of the particles, so the driving force resulting from reduction in surface area must be large. Other workers strongly adhere to the position that capillary pressure is the dominant driving stress for film formation [38]. Eckersley and Rudin show that both forces are involved and, consistent with the differences in surface area, small particle size latexes form films at somewhat lower temperatures than larger particle size ones [39]. However, with other latexes, larger size promotes coalescence. Still other work indicates that there is no effect of particle size on MFFT. It may well be that particle size distribution can affect MFFT. Possibly, these different conclusions are caused not just by particle size, but also by differences in the compositions of the latexes used. As previously discussed, T_g of the latex particles is an important factor controlling the rates of particle deformation and coalescence. Lower T_g particles have a lower modulus—that is, they are softer (see Section 4.1) and, hence, are more easily deformed. With the large number of variables involved in coalescence, one would expect interactions among the different variables.

While films form rapidly from latexes when the temperature is above MFFT, coalescence is a relatively slow process; in many cases, the film probably never equilibrates to become completely uniform. The rate of equilibration is affected by $(T - T_g)$. From a standpoint of rapid coalescence, it is desirable to have a latex with a T_g well below the temperature at which the film is to be formed. There are review papers that discuss factors affecting development of cohesive strength of films from latex particles [32,40]. The extent of coalescence has been studied by small angle neutron scattering, direct energy transfer of particles labeled with fluorescent dyes, and scanning probe microscopy [32,41,42].

Usually, architectural paints are formulated so that film formation occurs at temperatures as low as 2°C. Film formation at such a low temperature requires a low T_g latex. However, as discussed in the introductory paragraphs of Section 2.3, $(T - T_g)$ also affects whether the film obtained will be solid. It was estimated that for a film to withstand the relatively mild blocking test described, $(T - T_g)$ would have to be on the order of 21°C. If the film is to be exposed to the blocking test when the temperature is 50°C (not an unreasonable expectation during direct exposure to summer sun), the T_g should be about 29°C or higher. Thus, the paint formulator is faced with a difficult challenge to design a system that can form a film when applied at 2°C and yet resist blocking at 50°C.

Coalescing solvents help solve this problem, but environmental regulations are limiting permissible emissions of volatile organic compounds (VOC). A second approach is to design latex particles so that there is a gradient of T_g from a relatively high T_g in the center of the particles to relatively a low T_g at the outer periphery [43]. (See Sections 8.1.3 and 8.2.) The low T_g of the outer shell permits film formation at low temperature but, over time, the T_g of the coalesced film approaches the average T_g of the total polymer. This higher average T_g reduces the probability of blocking. One can use small amounts of coalescing solvents with such a latex. A third approach is to use blends of high and low T_g latexes, which can reduce MFFT without the presence of coalescing solvents [44]. It has been proposed that the films contain particles of high T_g polymer dispersed in a matrix of lower T_g polymer, reinforcing the matrix, increasing its modulus, and, hence, decreasing blocking. These studies were done without pigment; it would be interesting to compare the results with films made with a low T_g polymer with an equal volume content of pigment, which can also reinforce a film. A fourth approach involves use of core-shell latexes [45]. (See Section 8.1.3.)

Another way to reduce the MFFT of a latex and, hence, reduce (or eliminate) the need for coalescing solvent is to use low T_g thermosetting latexes. (See Section 8.1.4.) The low T_g permits film formation at ambient temperature without need for coalescing solvent. The cross-linking after application provides the increase in modulus required for such properties as blocking resistance. An important problem with many thermosetting latexes is reaction during storage, which leads to partially cross-linked polymer particles that may have poor coalescing properties. See Refs. [46] and [47] and Section 8.1.4 for examples. Cross-linkable baking coatings are relatively easy to formulate, but the greatest need is for coatings that cure at ambient temperature. The problem can be overcome by using two packages. This can be satisfactory for industrial applications, although pot life problems are still encountered, but it is undesirable for trade sales paints. Latexes substituted with allylic substitution have been developed that are package stable and that cross-link only after application and reaction with oxygen in the air. Also, latexes with trialkoxysilyl substitution are reported to be package stable, but cross-link after water evaporates from a film. (See Section 16.5.4.)

Most studies of the mechanism of film formation have been done with latexes, not with fully formulated paints. It is to be expected that other components of paints, such as pigments, pigment dispersing agents, and water soluble polymers used as thickening agents, among others, affect MFFT, rate of film formation, and blocking resistance. See Chapter 31 for further discussion of latex paints.

Other types of coatings that involve coalescence of particles are discussed in later chapters, including aqueous polyurethane dispersions, organosols, water-reducible resins, and powders.

GENERAL REFERENCES

G. W. Odian, *Principles of Polymerization*, 3rd. ed., Wiley-Interscience, New York, 1991.
J. E. Mark, Ed., *Physical Properties of Polymers Handbook*, American Institute of Physics, Woodbury, New York, 1996.
H.-G. Elias, *An Introduction to Polymer Science*, VCH, Weinheim, New York, 1997.

REFERENCES

1. L. W. Hill and Z. W. Wicks, Jr., *Prog. Org. Coat.*, **10**, 55 (1982).
2. H. G. Elias, "Structure and Properties," in *Macromolecules*, Plenum Press, New York, 1984, pp. 301–371.
3. C. J. Sullivan, D. C. Dehm, E. E. Reich, and M. E. Dillon, *J. Coat. Technol.*, **62** (791), 37 (1990).
4. J. Kumanotani, H. Hironori, and H. Masuda, *Org. Coat. Sci. Technol.*, **6**, 35 (1984).
5. Z. W. Wicks, Jr., G. F. Jacobs, I. C. Lin, E. H. Urruti, and L. G. Fitzgerald, *J. Coat. Technol.*, **57** (725), 51 (1986).
6. L. C. Dickinson, P. Morganelli, C. W. Chu, Z. Petrovic, W. J. Macknight, and J. C. W. Chien, *Macromolecules*, **21**, 338 (1988).
7. L. J. Mathias and R. F. Colletti, *Polym. Prepr.*, **30** (1), 304 (1989).
8. R. J. Roe, *Encyclopedia of Polymer Science and Technology*, 2nd. ed., Vol. 7, Wiley, New York, 1987, pp. 531–544.
9. W. A. Lee and R. A. Rutherford, "The Glass Transition Temperatures of Polymers," in J. Brandrup and E. H. Immergut, Eds., *Polymer Handbook*. 2nd. ed., Wiley, New York, 1975, pp. III–139–192.
10. P. M. Lesko and P. R. Sperry, "Acrylic and Styrene-Acrylic Polymers," in P. A. Lowell and M. S. El-Aasser, *Emulsion Polymerization and Emulsion Polymers*, Eds., Wiley, New York, 1997, pp. 622–623.
11. M. K. Gupta, *J. Coat. Technol.*, **67** (846), 53 (1995).
12. H. Stutz, K.-H. Illers, and J. Mertes, *J. Polym Sci.; B, Polym. Phys.*, **28**, 1483 (1990).
13. J. D. Ferry, *Viscoelastic Properties of Polymers*, 3rd. ed., Wiley, New York, 1980, p. 487.
14. A. Eisenberg, "The Glassy State and the Glass Transition" in J. E. Mark, A. Eisenberg, W. W. Graessley, L. Mandelkern, and J. L. Koenig, Eds., *Physical Properties of Polymers*, American Chemical Society, Washington, 1984, pp. 55–95.
15. Z. W. Wicks, Jr., *J. Coat. Technol.*, **58** (743), 23 (1986).
16. G. W. Odian, *Principles of Polymerization*, 3rd. ed., Wiley–Interscience, New York, 1991, pp. 198–334 and 452–531.
17. R. P.-T. Chung and D. H. Solomon, *Prog. Org. Coat.*, **21**, 227 (1992).
18. P. J. Flory, *J. Am. Chem. Soc.*, **61**, 3334 (1939).
19. W. L. Chang and T. Karalis, *J. Polym. Sci.; A, Polym. Chem.*, **31**, 493 (1993).

20. G. W. Odian, *Principles of Polymerization*, 3rd. ed., Wiley–Interscience, New York, 1991, pp. 40–125.
21. F. W. Billmeyer, Jr., *Textbook of Polymer Science*, 3rd. ed., Wiley–Interscience, New York, 1984, pp. 25–48.
22. H. Burrell, *Off. Digest*, **34** (445), 131 (1962).
23. W. P. Mayer and L. G. Kaufman, *XVII FATIPEC Congress Book I*, 110 (1984).
24. S. P. Pappas and H.-B. Feng, *Intl. Conf. Org. Coat. Sci. Technol.*, Athens, 1984, pp. 216–228.
25. S. P. Pappas and L. W. Hill, *J. Coat. Technol.*, **53** (675), 43 (1981).
26. M. T. Aronhime and J. K. Gilham, *J. Coat. Technol.*, **56** (718), 35 (1984).
27. H. E. Blair, *Polym. Prepr.*, **26** (1), 10 (1985).
28. K. Dusek and I. Havlicek, *Prog. Org. Coat.*, **22**, 145 (1993).
29. S. L. Simon and J. K. Gilham, *J. Coat. Technol.*, **65** (823), 57 (1993).
30. K. L. Hoy, *J. Coat. Technol.*, **68** (853), 33 (1996).
31. G. A. Vandezande and A. Rudin, *J. Coat. Technol.*, **68** (860), 63 (1996).
32. M. A. Winnik, "The Formation and Properties of Latex Films," in P. A. Lovell and M. S. El-Aasser, Eds., *Emulsion Polymerization and Emulsion Polymers*, John Wiley and Sons, New York, 1997, pp. 467–518.
33. M. A. Winnik, Y. Wang, and F. Haley, *J. Coat. Technol.*, **64** (811), 51 (1992).
34. C. Geel, *J. Oil Colour Chem. Assoc.*, **76**, 76 (1993).
35. D. P. Sheetz, *J. Appl. Polym. Sci.*, **9**, 3759 (1965).
36. S. G. Croll, *J. Coat. Technol.*, **58** (734), 41 (1987).
37. K. Kendall and J. C. Padget, *Intl. J. Adhesion Adhesives*, **2** (3), 149 (1982).
38. F. Lin and D. J. Meier, *Prog. Org. Coat.*, **29**, 139 (1996).
39. S. T. Eckersley and A. Rudin, *J. Coat. Technol.*, **62** (780), 89 (1990).
40. E. S. Daniels and A. Klein, *Prog. Org. Coat.*, **19**, 359 (1991).
41. R. M. Rynders, C. R. Hegedus, and A. G. Gilicinski, *J. Coat. Technol.*, **67** (845), 59 (1995).
42. H.-J. Butt and R. Kuropka, *J. Coat. Technol.*, **67** (848), 101 (1995).
43. K. L. Hoy, *J. Coat. Technol.*, **51** (651), 27 (1979); D. R. Bassett and K. L. Hoy, in D. R. Bassett and A. E. Hamielec, Eds., *ACS Symp. Series*, No. 165, 1981, p. 371.
44. M. A. Winnik and J. Feng, *J. Coat. Technol.*, **68** (852), 39 (1996).
45. D. Juhue and J. Lang, *Macromolecules*, **28**, 1306 (1995).
46. P. A. Geurink, L. van Dalen, L. G. J. van der Ven, and R. R. Lamping, *Prog. Org. Coat.*, **27**, 73 (1996).
47. J. M. Geurts, J. J. G. S. van Es, and A. L. German, *Prog. Org. Coat.*, **29**, 107 (1996).

CHAPTER 3

Flow

Rheology is the science of flow and deformation. This chapter deals only with flow; deformation aspects of rheology are discussed in Chapter 4. Flow properties of coatings are critical for proper application and appearance of films. For example, in brush application of a paint, the flow properties govern settling of pigment during storage, how much paint is picked up on the brush, film thickness applied, leveling of the applied film, and control of sagging of the film. Depending on how stress is applied to a fluid, there are several types of flow. Of major importance in coatings is flow under a shear stress. We consider shear flow first and then, more briefly, other types of flow.

3.1. SHEAR FLOW

To understand and define *shear flow*, consider the model shown in Figure 3.1. The lower plate is stationary, and the upper parallel plate is movable. The plates are separated by a layer of liquid of thickness x. Force F is applied to the top movable plate of area A, so the plate slides sidewise with velocity v. The model assumes that there is no slip at the interfaces and that there is no fluid inertia. When the plate moves, the liquid near the top moves with a velocity approaching that of the movable plate and the velocity of the liquid near the bottom approaches zero. The velocity gradient dv/dx for any section of the liquid is constant and, therefore, equals v/x. This ratio is defined as *shear rate* $\dot\gamma$. The units of shear rate are reciprocal seconds, s^{-1}.

$$\dot\gamma = \frac{dv}{dx} = \frac{v}{x}; \qquad \text{Units: } \frac{\text{cm s}^{-1}}{\text{cm}} = s^{-1}$$

Force F acting on the top plate of area A results in *shear stress* σ. (In older literature, τ is used as the symbol.) The units of shear stress are pascals (Pa).

$$\sigma = F/A; \qquad \text{Units: } \text{m/kg/s}^2/\text{m}^2 = \text{N/m}^2 = \text{Pa}$$

A liquid exerts a resistance to flow called *viscosity*, η, defined as the ratio of shear stress to shear rate. This type of viscosity is correctly called *simple shear viscosity*, but since it is

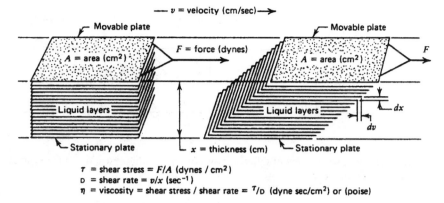

τ = shear stress = F/A (dynes / cm^2)
D = shear rate = v/x (sec^{-1})
η = viscosity = shear stress / shear rate = τ/D (dyne sec/cm^2) or (poise)

Figure 3.1. Schematic model of shear flow on an ideal liquid. (In current usage, the symbol for shear stress is σ and its units are Pa, the symbol for shear rate is $\dot{\gamma}$, and the units of η are Pa·s.) (From Ref. [1], with permission.)

the most widely encountered type of viscosity, it is usually just called viscosity. The units are pascal seconds (Pa·s). The older, and still commonly used, unit is poise (P). One Pa·s equals 10 P, and 1 mPa·s equals 1 cP.

$$\eta = \sigma/\dot{\gamma}; \qquad \text{Units: Pa/s}^{-1} = \text{Pa·s}$$

When a liquid flows through a hole or a capillary, part of the energy is diverted into kinetic energy; then the resistance to shear flow is called *kinematic viscosity v* with units of m^2 s^{-1}, formerly stokes, where 1 m^2 s^{-1} = 10^4 stokes. When the acceleration results from gravity, kinematic viscosity equals simple shear viscosity divided by density ρ of the liquid:

$$v = \frac{\eta}{\rho}; \qquad \text{Units: m}^2 \text{ s}^{-1}$$

3.2. TYPES OF SHEAR FLOW

When the ratio of shear stress to shear rate is constant—that is, viscosity is independent of shear rate (or shear stress)—liquids are called *Newtonian* liquids. A plot of shear rate as a function of shear stress is linear as shown in Figure 3.2(*a*). When plotted as shown in Figure 3.2, the slope equals the inverse of viscosity. Sometimes such plots appear in the literature with axes opposite to those shown; then the slope is the viscosity. Newtonian flow is exhibited by liquids composed of miscible small molecules. Many solutions of oligomeric resins also approximate Newtonian flow.

Some liquids are non-Newtonian; that is, the ratio of shear stress to shear rate is not constant. One class of non-Newtonian liquids exhibits decreasing viscosity as shear rate (or shear stress) increases; these liquids are called *shear thinning liquids*. Dispersed systems in which the dispersed phase is not rigid, such as emulsions, or in which there are particle-particle interactions, can exhibit shear thinning. In many cases, the curve becomes

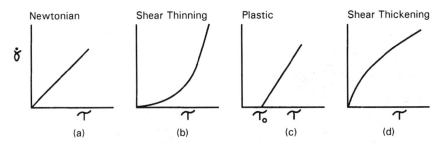

Figure 3.2. Schematic plots of flow of different types of liquids. (In present usage the symbol for shear rate is σ instead of τ.)

linear at higher shear rates, as shown in Figure 3.2(*b*). Shear thinning liquids showing this behavior are termed *pseudoplastic*. In some cases, no detectable flow occurs unless a minimum shear stress is exceeded. Such materials are sometimes called *Bingham bodies*; they are also said to exhibit *plastic flow*. The minimum shear stress required is called the *yield value*, or *yield stress*, and is designated by the symbol σ_0. A schematic flow diagram is shown in Figure 3.2(*c*). It is fairly common to extrapolate the linear part of the curve of a pseudoplastic liquid to the intercept with the shear stress axis and call the intercept a yield value.

Another class of liquids exhibits increasing viscosity as shear rate (or shear stress) increases. A schematic plot of such behavior is given in Figure 3.2(*d*). Such liquids are called *shear thickening* or *dilatant liquids*. Dilatant liquids increase in volume when shear is exerted on them. Dilatant fluids have dispersed phases that become less ordered and, hence, occupy more volume when exposed to increasing shear. Some pigment and resin dispersions exhibit shear thickening.

The Casson equation, Eq. 3.1, linearizes viscosity/shear rate data of shear thinning or thickening fluids; the slope of the line is the yield stress, and extrapolation gives the viscosity at infinite shear rate η_∞. In many cases, the value of *n* is 0.5, and commonly, the Casson equation is shown with just the half-power relationship. It is common to plot log viscosity against shear rate; in such plots the degree of curvature is related to the value of σ_0. A plot is given in Figure 3.3 in which the values of η and η_∞ are held constant to show the effect of changes in σ_0 on flow response. For a Newtonian fluid, σ_0 equals zero and the plot is a straight line parallel to the shear rate axis.

$$\eta^n = \eta_\infty^n + \frac{\sigma_0^n}{\dot{\gamma}} \tag{3.1}$$

Some fluids show time or shear history dependence of viscosity; the effect is illustrated in Figure 3.4(*a*). The curves in the figure result from shear stress readings taken at successively higher shear rates to some upper limit (right-hand curve), followed immediately by shear stress readings taken at successively lower shear rates (left-hand curve). At any shear rate on the initial curve, the stress would decrease with time to an equilibrium value between the two curves; that is, the viscosity would decrease. On the other hand, if such a system had been exposed to a high rate of shear and then the shear rate decreased, the shear stress would increase to an equilibrium value as the measurement was continued; that is, the viscosity would increase with time. This behavior is called *thixotropic* flow.

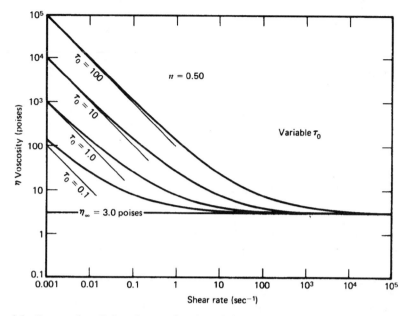

Figure 3.3. Casson plot of viscosity as a function of shear rate showing the dependence on σ_0 (old symbol is τ_0) with constant η and η_∞. (From Ref. [1], with permission.)

Thixotropic fluids are shear thinning fluids; their viscosity is also dependent on time and prior shear history. All thixotropic fluids are shear thinning fluids; but not all shear thinning systems are thixotropic. Unfortunately, the term thixotropy is often improperly used as a synonym for shear thinning.

Thixotropy usually results from reversible formation of a structure within a fluid; an example is association of dispersed particles held together by weak forces. It is commonly said that thixotropic structure is broken down by applying shear for a sufficient time and that the structure re-forms over time when shear is stopped. It is difficult to quantify a degree of thixotropy. Some thixotropic fluids undergo viscosity reduction to equilibrium values in short time periods and recover their viscosity rapidly when shearing is stopped;

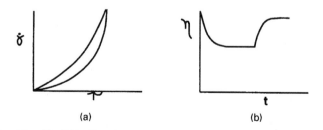

Figure 3.4. Schematic plots of systems exhibiting thixotropic flow. (*a*) The curve to the right is based on readings taken as shear rate was being increased and the curve to the left is based on readings taken as shear rate was then being decreased. (*b*) The viscosity drops as shear continues, then increases as the shear rate is decreased. (In current usage the symbol for shear rate is σ instead of τ.)

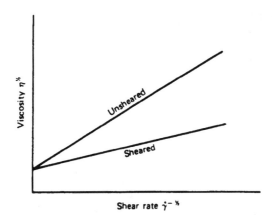

Figure 3.5. Schematic Casson plots of a sheared and unsheared thixotropic coating. The degree of divergence gives an estimate of the degree of thixotropy. (From Ref. [4], with permission.)

others change more slowly with time. In early work, areas within hysteresis loops, as shown in Figure 3.4(*a*), were compared. However, the areas of such loops are dependent on the time intervals between successive measurements. Another way to represent the effect is to plot the viscosity at a series of shear rates as a function of time, as illustrated in Figure 3.4(*b*).

Rheologists frequently discuss the properties of such systems in terms of viscoelasticity; that is, their flow is a combination of viscous flow and elastic deformation. (See Section 4.1 for discussion of viscoelasticity.) Such interpretations are valid and useful, but they have been seldom applied in the coatings industry. Time dependency can best be measured in terms of viscoelasticity; see Refs. [2] and [3] for two examples of such an analysis. Increasing use of viscoelastic flow analysis can be expected in the future.

Another way to show the effect of shear on a thixotropic fluid is by a different type of Casson plot, as shown in Figure 3.5. The square root of the viscosity is plotted against the square root of the reciprocal of the shear rate; the steeper the slope, the greater the degree of shear thinning. If the sample had been sheared until all of the thixotropic structure was broken down, and if the measurements could be made before any structure buildup occurred, the plot would be linear and parallel to the x axis. Although comparisons of the differences between the slopes of such lines give a qualitative expression of the extent of thixotropy, the slopes of the curves are dependent on prior shear history, the rate of acceleration of shear, and the length of time that the sample was exposed to the highest shear rate.

3.3. DETERMINATION OF SHEAR VISCOSITY

A variety of instruments is available to determine viscosity. They vary in cost, time required for measurements, operator skill required, susceptibility to abuse, precision, accuracy, and ability to measure shear rate variability or time dependency effects. Data obtained with different instruments and by different operators with the same instruments can vary substantially, especially for shear thinning liquids at low rates of shear [5]. Some of the variation can result from lack of attention to details, especially temperature control and possible solvent loss, and major errors can result from comparing samples that have had different shear histories.

Viscometers can be divided into three broad classes: (1) those that permit quite accurate viscosity determinations, (2) those that permit determination of reasonable approximations of viscosity, and (3) those that provide flow data casually related to viscosity. Space limitations restrict our discussion to major examples of measurement methods, but not the details of the instruments or the mathematical relationships involved in their use. The general references at the end of this chapter are good sources of further information.

Since viscosity depends strongly on temperature, it is critical that the sample has reached a constant, known temperature before measurement. When high viscosity fluids are sheared at high shear rates, heat is evolved and the temperature of the sample increases unless the heat exchange efficiency of the viscometer is adequate. If viscosities are determined as both shear rate and temperature are increasing, it is not possible to tell whether or not a fluid is shear thinning.

3.3.1. Capillary Viscometers

In *capillary viscometers*, the time required for a known amount of liquid to flow through a capillary tube is measured. Figure 3.6 shows an example of a capillary viscometer. While viscosities can be calculated based on the diameter of the capillary, usually each instrument is standardized with liquids of known viscosity; then, calculation is simply based on instrument constants and time. Flow is usually driven by gravity, so kinematic viscosity is

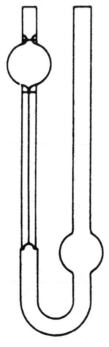

Figure 3.6. Ostwald capillary viscometer. (From Ref. [6], with permission.)

measured. Simple shear viscosity can be calculated by correcting for density. Capillary viscometers with a range of diameters permit the determination of viscosities from 10^{-7} to 10^{-1} m^2 s^{-1}. For liquids with a density of 1, these values correspond to a range of 1 mPa·s to 1000 Pa·s. For higher viscosity liquids, instruments have been designed that force the liquid through a capillary with pressure rather than just the force of gravity.

Capillary viscometers are applicable only to Newtonian liquids. The liquids must be sufficiently transparent in order to permit observation of the time at which the fluid meniscus passes the two volume marking lines. In such situations, capillary viscometers are the instruments of choice for research work, since the accuracy attained is high. They are appropriate for routine work; however, determinations are relatively time consuming especially when temperature dependence of viscosity data are desired. Temperature equilibration is slow because of the relatively large sample sizes and the low rate of heat transfer by glass. Capillary viscometers are particularly appropriate for use in determining the viscosity of volatile liquids or solutions in volatile solvents, since they are essentially closed systems.

3.3.2. Rheometers

For non-Newtonian liquids, including pigmented liquids, the highest accuracy over a wide range of shear rates is obtained with rotational rheometers exemplified by *cone and plate viscometers*. A schematic diagram is shown in Figure 3.7. The sample is placed on the plate that is then raised to a level with small clearance from the cone. The cone can be rotated at any desired number of revolutions per minute (rpm), and the torque is measured.

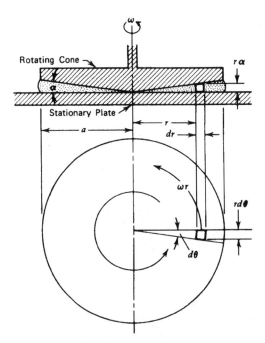

Figure 3.7. Schematic representation of cone and plate viscometer geometry. (From Ref. [1], with permission.)

The angle of the cone is designed to make the shear rate constant throughout the sample in the gap. Shear rate is proportional to the rpm and shear stress is related to the torque. Temperature is controlled by passing temperature-controlled water through the plate; temperature control problems are minimized by the small sample size.

A variety of rheometers is available, varying in the range of shear rates that can be used and the time required to increase or decrease shear rate. (See general references and Refs. [4] and [6] for further discussion.) The least expensive instruments are sufficiently rugged, simple to use, and fast enough for quality control applications. The most versatile ones are sensitive scientific instruments requiring skill in use and are most appropriate for research applications. When rheometers are used with solutions containing volatile solvents, the cone and plate unit should be shrouded in an atmosphere saturated with solvent vapor. Two types are available: controlled strain and controlled stress instruments. The latter type offers advantages for coatings, as it is generally superior for measurements at very low shear rates. In some instruments, the edge of the liquid sample is exposed to the atmosphere, and solvent can evaporate.

The viscosity of highly viscous materials can be determined at high rates of shear by the use of *mixing rheometers* that are small, heavy-duty mixers. The test sample is confined to a relatively small space and subjected to intense mixing by dual rotors in the form of sigma-shaped blades. A dynamometer measures the work input through a reaction torque that is converted to a strip chart readout. The speed is set by a tachometer. The latest instruments are computerized. These instruments were originally designed for studying the molding of plastics, but are also used in studying the effect of pigments on viscosity. Heat buildup can be substantial with high viscosity fluids.

3.3.3. Rotating Disk Viscometers

Rotating disk viscometers, such as a Brookfield Viscometer, have a motor that rotates a disk in a liquid over a range of rpm, and the resulting torque is measured. A schematic diagram is shown in Figure 3.8. The instruments must be calibrated with standards. Measurements should be made in a container of the same dimensions as that in which the standardization is carried out, since the distance of the disk below the surface of the liquid,

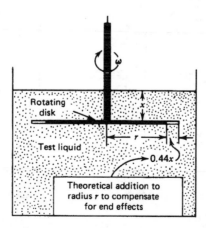

Figure 3.8. Schematic drawing of a disk viscometer. (From Ref. [1], with permission.)

above the bottom of the container, and from the side walls can affect the response. When properly used, the instruments provide relatively accurate viscosity measurements for Newtonian liquids. For non-Newtonian liquids, the viscosity reading represents an average response corresponding to the viscosities resulting from a span of shear stresses.

Rotating disk viscometers can detect whether a liquid is shear thinning or shear thickening by measurements carried out at different rpm settings. They can detect thixotropy by a change in response over time at the same rpm setting. They have limited value for comparisons of liquids with different degrees of shear thinning or thixotropy. They cannot provide viscosity data as a function of shear rate. Thus, they are appropriate for quality control work, but caution should be applied in their use for research and development purposes.

3.3.4. Bubble Viscometers

A bubble viscometer is based on the rate of rise of an air bubble in a tube of liquid; the higher the viscosity, the slower the bubble rises. A glass tube is filled with a liquid to a graduation mark and stoppered so that a definite amount of air is enclosed at the top as shown in Figure 3.9. The tube is placed in a thermostatic bath and there kept long enough for the temperature to equilibrate. Equilibration is slow, but it is essential if meaningful measurements are to be made. The tube is then inverted, and the time required for the air bubble to travel between two calibration marks on the tube is measured. Provided the length of the bubble is greater than its diameter, the rate of rise is independent of the bubble size. Density of the liquid affects the rate of rise of the bubble, so kinematic viscosity is measured. Sets of standard tubes are designated as A, B, C, and so on; after Z, tubes are designated as Z_1, Z_2, and so on. Kinematic viscosities range from about 10^{-5} to 0.1 m^2 s^{-1}. Sets of bubble tubes are widely used as control instruments in measuring the viscosities of resin solutions. (See Section 15.6.2.) They are only appropriate for Newtonian, transparent fluids. They are low in cost and simple to use.

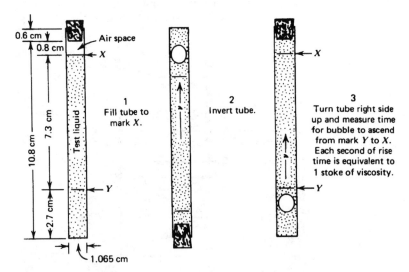

Figure 3.9. Determination of viscosity with a bubble tube. (From Ref. [1], with permission.)

3.3.5. Efflux Cups

The most widely used control device for measuring flow of industrial coatings, especially for spray application, is the *efflux cup*. A variety of efflux cups is used. (Refs. [4] and [6] provide a comparison of about two dozen that are used commercially.) A schematic diagram of one of the most common efflux cups, the Ford No. 4 cup, is shown in Figure 3.10. One holds a thumb over the hole in the bottom of the cup, fills the cup with coating, removes the thumb, and presses a stopwatch simultaneously. The stopwatch is stopped when the stream of coating flowing through the hole breaks. The result is expressed in seconds. The data should not be converted into kinematic viscosity numbers, since a significant amount of the force is converted into kinetic energy, especially with low viscosity coatings. The method is not appropriate for non-Newtonian liquids, although efflux cups are frequently used for coatings that exhibit a small degree of shear thinning. Despite their limitations, efflux cups are useful quality control devices. They are low in cost, rugged in construction, and easily cleaned. Results are simply and quickly obtained, but reproducibility is poor, reportedly in the range of only 18–20% [6].

The proper way to use an efflux cup to control viscosity for spraying, for example, is to adjust the viscosity of the coating by solvent addition until the coating sprays properly, and then measure the time it takes to flow through the efflux cup. This time can then be used as the standard for spraying that particular coating through that spray gun at that distance from the object being sprayed. Reversing this order—that is, adding solvent until the viscosity is such that the coating takes some fixed number of seconds to flow through the cup—is not appropriate. Proper efflux flow times for spraying vary with different coatings for a given application system and with different application systems for a given coating.

3.3.6. Paddle Viscometers

The most widely used "viscometer" in the United States in architectural paint formulation is the *Stormer Viscometer*. The instrument paddle is immersed in the paint and rotated at 200 rpm. The force required to maintain this rotation rate is measured by adding weights to a platform at the end of a cord over a pulley connected by a gear train to the paddle. A

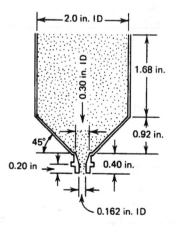

Figure 3.10. Schematic diagram of a Ford No. 4 efflux cup. (From Ref. [1], with permission.)

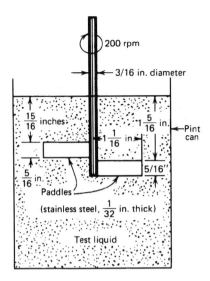

Figure 3.11. Schematic diagram of a paddle viscometer. (From Ref. [1], with permission.)

schematic diagram is shown in Figure 3.11. The weight loaded is converted into *Krebs Units* (KU) by an arbitrary conversion scale. Originally, a KU value of 100 was supposed to correspond to good brushing consistency. In current practice, paints are usually formulated with somewhat lower KU. The instrument is of little utility with Newtonian fluids, and the readings have no real meaning for non-Newtonian fluids, typical of most architectural paints. Even when used only for quality control, the paddle viscometer is not very satisfactory.

The proper way to establish the flow properties for a trade sales paint is to apply it with a brush or roller and adjust it until the best compromise of such properties as ease of brushing, leveling, sagging, settling, and so forth has been achieved. Having done this work, then a quality control test can be set up. Appropriate quality control instruments are cone and plate viscometers (some of which are low cost) or, lacking that, rotating disk viscometers. Stormer Viscometers are widely used, but that does not make them satisfactory. The director of research of a large manufacturer of trade sales paints in the United States said some years ago that the Stormer Viscometer was responsible for setting back the formulation of one-coat hiding latex paints by 20 years.

3.4. SHEAR VISCOSITY OF RESIN SOLUTIONS

The viscosity of liquids depends on free volume availability. To put it somewhat simplistically, there are free volume holes rapidly opening and closing in a liquid; molecules move randomly through these free volume holes. When a stress is applied, movements in the direction that relieves the stress are favored, and the liquid flows. Therefore, factors that control viscosity of resin solutions are those that control the availability of free volume. Many coatings are based on solutions of polymers or oligomers. The variables affecting the flow behavior of these concentrated solutions are not fully understood. The variables that govern flow of very dilute polymer solutions have been extensively studied and are better understood. Factors affecting the flow of dilute solutions are discussed in Section 3.4.2, and factors affecting the flow of more concentrated solutions are discussed in Section 3.4.3.

3.4.1. Temperature Dependence of Viscosity

Temperature dependence of viscosity for a range of low molecular weight resins and their solutions has been shown to fit a Williams Landel Ferry (WLF) equation [7–10]. (See Section 2.3.) In Eq. 3.2, T_r, the reference temperature, is the lowest temperature for which experimental data are available. Except for very dilute solutions, data fit Eq. 3.2 when the reference temperature is T_g and viscosity at T_g is assumed to be 10^{12} Pa·s [7].

$$\ln \eta = \ln \eta_r - \frac{c_1(T - T_r)}{c_2 + (T - T_r)} = 27.6 - \frac{A(T - T_g)}{B + (T - T_g)} \tag{3.2}$$

For higher molecular weight polymers at temperatures above $T_g + 100°$, the temperature dependence of viscosity fits an Arrhenius equation, shown in Eq. 3.3, where E_v is the activation energy for viscous flow. (Note that constant A is not the same constant as in Eq. 3.2.)

$$\ln \eta = K + B/T = \ln A + \frac{E_v}{RT} \tag{3.3}$$

For the data available for low molecular weight resins and their solutions, Arrhenius plots of $\ln \eta$ as a function of $1/T$ have been found to be curved and not linear [7,9]. On the other hand, the data do fit a WLF equation. From a practical viewpoint, the differences in the models are small if the temperature range is small. However, over a wide range of temperatures, the differences are relatively large. Figure 3.12 shows plots of the temperature dependence of viscosity of commercial standard liquid bisphenol A epoxy resin (see Section 11.1.1), calculated from both Arrhenius and WLF equations together with experimental data points; the data fit a WLF equation.

A major factor controlling viscosity of resin solutions is $(T - T_g)$, but it is not the only factor. When the differences between T_g's are small, differences in WLF constants A and B may overshadow the small difference in $(T - T_g)$. Constant A depends on the difference in thermal expansion coefficients above and below T_g, but no studies have been reported on the structural factors that control these coefficients. Constant B is the value of $(T_g - T)$ at which viscosity goes to infinity. The so-called universal value of this constant is $51.6°$, but the "constant" varies considerably with composition. No studies have been reported on the relationship between structure and the value of constant B.

Generally, in designing resins, it is reasonable to predict that a lower T_g will lead to a lower viscosity of the resins and their solutions. (See Section 2.1.2 for discussion of the factors controlling the T_g of polymers.) Linear poly(dimethylsiloxanes) have low T_g's and low viscosities. Linear polyethylene glycols have almost as low T_g's and viscosities. Poly(methyl methacrylate) resin solutions have higher T_g values and viscosities than comparable poly(methyl acrylate) resin solutions. BPA epoxy resins have higher T_g values and viscosities than corresponding hydrogenated derivatives. Exceptions to this generalization about the effect of T_g have been reported for some high solids acrylic resins. (See

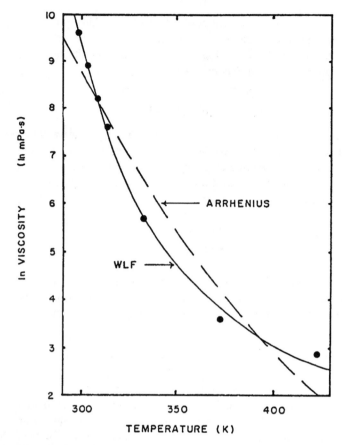

Figure 3.12. Viscosity dependence of standard liquid BPA epoxy resin on temperature. (From Ref. [7], with permission.)

Section 12.2.1.) Acrylic resins made with a comonomer that has a bulky group, such as 3,3,5-trimethylcyclohexyl methacrylate [11] or isobornyl methacrylate [12], are reported to have low viscosities at high solids even though they have high T_g values; no explanation of this effect has been advanced.

3.4.2. Dilute Solution Viscosity

If the concentration of a solution is low enough that individual polymer molecules and their associated solvent molecules are isolated from each other, the *relative viscosity* (η_r) of the solution can be expressed by the empirical Huggins equation, Eq. 3.4, where $[\eta]$ is *intrinsic viscosity* (sometimes called the *limiting viscosity number*) and C is the concentration of the polymer solution, expressed as weight of polymer per unit volume of polymer solution. Relative viscosity is the ratio (it is unitless) of solution viscosity to viscosity of the solvent. In recent literature, the units of C most commonly used are g polymer per mL solution, but in older literature, g per dL is common, so one must be careful to check units. When C is expressed in g mL^{-1}, the units of intrinsic viscosity are mL g^{-1}. Intrinsic viscosity is obtained by extrapolating a plot of ln η_r/C as a function of C

to zero concentration. It is related to the hydrodynamic volume of the sphere swept out by an isolated polymer molecule and its associated solvent as it moves through a dilute solution.

$$\ln \eta_r = [\eta]C + [\eta]^2 C^2 \tag{3.4}$$

Intrinsic viscosity depends on temperature. As temperature increases, the coil size of the polymer molecule usually increases, more solvent is entrapped, and intrinsic viscosity increases. Intrinsic viscosity is also affected by solvent–polymer interactions. The greater the extent of solvent association with a polymer molecule, the more the coil expands, and therefore, the higher the intrinsic viscosity. Intrinsic viscosity depends on molecular weight: the higher the molecular weight, the larger the intrinsic viscosity. Another factor is the rigidity of the chain. Everything else being equal, polymers with flexible, randomly kinked chains have lower intrinsic viscosities than those with rigid, rod-like structures. The relationship between intrinsic viscosity and molecular weight M is expressed by the Kuhn–Mark–Houwink–Sakurada equation, Eq. 3.5, in which K and a are constants:

$$[\eta] = KM^a \tag{3.5}$$

If the solvent is too poor or the temperature is too low, the polymer molecules precipitate, rather than staying in solution. The combination of minimum temperature and poorest solvent that just maintains solubility is called a *theta condition* (θ). Under theta conditions, intrinsic viscosity $[\eta]_\theta$ is at a minimum. If the chains are flexible, as for example with acrylic polymers, $[\eta]_\theta$ is proportional to the half-power of molecular weight, as shown in Eq. 3.6. Note that Eqs. 3.5 and 3.6 are based on polymers that have narrow molecular weight distributions.

$$[\eta]_\theta = K_\theta M^{1/2} \tag{3.6}$$

In better solvents the isolated polymer coils expand, intrinsic viscosity increases, and the exponent a in Eq. 3.5 increases to as high as 0.78 for flexible polymers, and even higher for rigid polymers.

3.4.3. Concentrated Solution Viscosity

There have been relatively few fundamental studies of the factors controlling viscosity of more concentrated solutions of polymers and resins, such as are used in the coatings field. Several empirical equations have been proposed to express the relationships. One such relationship for the concentration dependence of relative viscosity is Eq. 3.7, in which w_r is weight fraction resin and the k's are constants:

$$\ln \eta_r = \frac{w_r}{k_1 - k_2 w_r + k_3 w_r^2} \tag{3.7}$$

Nonlinear regression analysis of the limited number of sets of data available in the literature in 1985 fit Eq. 3.7 over a wide range of concentrations [7]. Even with this many terms, there is some systematic deviation from the model at very low concentrations. Constant k_1 is the reciprocal of *weight intrinsic viscosity*, $[\eta]_w$, which, although formally unitless, is based on the number of grams of solution containing a gram of resin. Weight intrinsic viscosity can be converted to the more familiar volume intrinsic viscosity $[\eta]$ by dividing by the density of the solution at the concentration $w_r = k_1$. No physical significance of the other two constants, k_2 and k_3, has been elucidated; they are presumably related to further solvent–resin interactions and to free volume availability.

Over narrower ranges of concentration, a simpler form of Eq. 3.7, Eq. 3.8, gives reasonable fits with experimental data. The even simpler Eq. 3.9 has been extensively used to calculate approximate relative viscosities over a narrow range of concentrations with viscosities from around 0.01 to 10 Pa·s.

$$\ln \eta_r = \frac{w_r}{k_1 - k_2 w_r} \tag{3.8}$$

$$\ln \eta_r = \frac{w_r}{k_1} = [\eta]_w w_r \tag{3.9}$$

As explained in Section 3.4.2, relative viscosity of dilute solutions of polymers increases as the solvent gets "better." However, in concentrated solutions, relative viscosity is higher in poor solvents than it is in good solvents. In the few cases reported in the literature, log of relative viscosity increases with the square root of molecular weight of resins dissolved in good solvents at these higher concentrations [7,13]. It is said that intrinsic viscosities of oligomers exhibit theta condition response; that is, relative viscosity of oligomer solutions is proportional to the square root of molecular weight [14], as shown in Eq. 3.10. This relationship appears to be true for solutions in good solvents of resins having relatively narrow molecular weight distributions and viscosities between about 0.01 and 10 Pa·s; but further research is needed.

$$\ln \eta_r = K w_r M^{1/2} \quad \text{or} \quad \ln \eta = \ln \eta_s + K w_r M^{1/2} \tag{3.10}$$

As shown in Eqs. 3.7–3.10, the viscosity of the solvent is a factor affecting the viscosity of resin solutions. At first glance, it might appear that a small difference in the viscosity of the solvent would have a trivial effect on the much higher viscosity of the resin solution. However, there are examples in which the difference in solvent viscosity is as little as 0.2 mPa·s, whereas the corresponding difference in viscosity of 50 wt% resin solutions in the same solvents is as great as 2 Pa·s.

Also important are the effects of resin–solvent interactions. In good solvents, there are stronger interactions between solvent molecules and resin molecules than in poor solvents. In very dilute solutions, this means that the chains become more extended and sweep out larger hydrodynamic volumes in good solvents than in poor solvents. However, in more concentrated solutions, resin molecules are constrained by the hydrodynamic volumes swept out by neighboring resin molecules. If interaction between solvent and resin is strong, there are weak resin–resin interactions, and the molecules can flow easily through the hydrodynamic volumes swept out by other molecules (provided free volume is adequate). On the other hand, in the case of a poor solvent–resin combination, there are stronger resin–resin interactions when resin molecules try to flow through the hydro-dynamic volume swept out by other resin molecules; more or less transient clusters of resin molecules form, and viscosity is higher. In solutions in good solvents, flow is generally Newtonian. In many cases, flow of more concentrated resin "solutions" in poor solvents behave somewhat like dispersed systems; they are non-Newtonian because shear can break up or distort resin clusters.

Although the difference in the viscosity of resin solutions in good and poor solvents is reasonably well understood, there is little definitive work in the literature on comparisons between solutions in various good solvents in which some of the solvents are "more good" than others. Erickson studied relative viscosities of solutions of several low molecular weight resins in a range of solvents [15]. He concluded that relative viscosities decrease as one changes from a very good solvent to a good solvent, pass through a minimum, and then increase rapidly in very poor solvents. As can be seen in Eqs. 3.7–3.9, which relate relative viscosity to concentration, the hydrodynamic volume of the isolated resin molecule and its associated solvent molecules is a factor in determining the viscosity not just of very dilute solutions, but also of more concentrated ones. As we change from a very good to a good solvent, intrinsic viscosity and, hence, relative viscosity should decrease; this prediction fits in with Erickson's hypothesis. The range of error in Erickson's work is not small enough to establish his conclusions beyond doubt. He may well be right, but there is need for further research.

Solvent effects on hydrogen bonding between resin molecules can be substantial. Figure 3.13 shows the viscosities of solutions of an acrylated epoxidized oil in three solvents chosen because of their similar viscosities, but very different potential effects on hydrogen

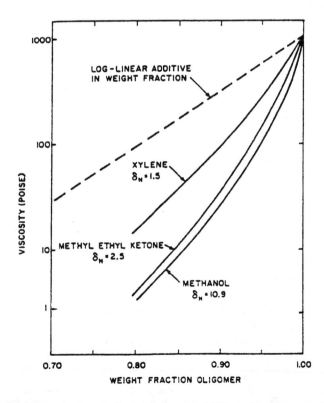

Figure 3.13. Viscosity reduction of a hydroxy-functional UV curable oligomer with xylene, MEK, and methyl alcohol compared to predicted viscosity if the viscosity reduction were a log-linear additive relationship by weight. (From Ref. [16], with permission.)

bonding. The resin molecules have multiple hydroxyl groups. Note that the viscosities of the solutions in xylene are highest. Xylene is a poor hydrogen-bond acceptor and hence promotes intermolecular hydrogen bonding between the resin molecules. Methyl ethyl ketone (MEK) is a good hydrogen-bond acceptor and reduces the viscosity more effectively than xylene by reducing intermolecular hydrogen bonding. Although methyl alcohol is a much stronger hydrogen-bonding solvent than MEK, it is only marginally better at reducing viscosity. Methyl alcohol is both a hydrogen-bond donor and acceptor. Possibly, methyl alcohol can bridge resin molecules by functioning as a hydrogen-bond donor with one resin molecule and a hydrogen-bond acceptor with the other; such bridging would counteract the effectiveness of viscosity reduction.

Intermolecular hydrogen bonding between carboxylic acid-functional resin molecules is particularly strong. Correspondingly, solvent effects on the viscosity of acid-substituted resins are particularly large [17]. It has been known for many years that simple carboxylic acids, such as acetic acid, exist as dimers in poor hydrogen-bond acceptor solvents like benzene, whereas, the dimers are dissociated in good hydrogen-bond acceptor solvents like acetone. The effect in resin solutions was demonstrated in a study of a primarily monocarboxylic acid-substituted acrylic oligomer [18]. Relative viscosity in xylene was higher than in methyl isobutyl ketone (MIBK). Molecular weight, as determined by vapor pressure depression, was lower in acetone than in benzene. In xylene, hydrogen bonding between carboxylic acid molecules was promoted; in MIBK, the predominant hydrogen bonding is between ketone and carboxylic acid groups, rather than between carboxylic acids.

Another solvent effect on viscosity is the solvent T_g. The T_g of resin solutions depends on the concentration and the T_g of both the resin and the solvent. This effect has been recognized in the addition of plasticizers to polymers, but has not been widely studied in resin solutions with concentrations and viscosities in the range of interest for coatings. In one study, it was found that the data fit Eq. 3.11, where T_{gs} is the T_g of the solvent and T_{gr} is the T_g of the solvent-free resin [7]. In this study, Eq. 3.11 fit the data over the whole range of concentrations from pure solvent to pure resin. Equation 3.11 needs to be tested with other systems.

$$\frac{1}{T_g} = \frac{w_s}{T_{gs}} + \frac{w_r}{T_{gr}} + kw_r w_s \tag{3.11}$$

Since it is common to use mixed solvents in coating formulations, further studies of the effects of mixed solvents on resin solution viscosity would be useful. It has been reported that the viscosity of mixtures of hydrogen-bond acceptor solvents like ketones and esters are nearly ideal in their effects on viscosity; that is, the viscosity of a mixture can be estimated quite well by calculating the weighted average viscosity from those of the components [19]. However, in the case of mixing alcohols with other solvents, the viscosities of mixtures varied substantially from ideal behavior. The deviation can be attributed to reduction of intermolecular hydrogen bonding of alcohols by the other solvents, and was particularly pronounced when water was one of the solvents in a mixture. While these results with solvent mixtures add to our understanding of the effects of intermolecular interactions on viscosity, they are not directly applicable to the problem of mixed solvent effects on viscosity of resin solutions. Since most resins have multiple hydrogen-bond donor and acceptor sites, the interactions with solvent are greater and more complex than in solvent

Table 3.1. Effects of Molecular Weight and Functional Group Content on Viscosity

Characteristics	SAA-I	SAA-II
$\bar{M}_n$	1600	1150
$\bar{M}_w/\bar{M}_n$	1.5	1.5
OH content (wt%)	5.7	7.7
Viscosities mPa·s		
80 wt% in MEK	10,000	6500
70 wt% in MEK	300	230
60 wt% in MEK	80	65
50 wt% in MEK	34	30
50 wt% in toluene	760	3840

blends. Little work has been published on this important question. In one paper, the authors suggest that the "best" solvent in the mixture dominates in determining the effect on the intrinsic viscosity of the resin solution [20]. The rationale is that the "best" solvent interacts most strongly with resin molecules hence controls the degree to which the resin molecules are extended. It is hoped that further research will be reported.

The relationships are further complicated because solvent/resin interactions can be further affected by resin molecular weight and the number of polar groups per molecule. For example, consider the data in Table 3.1 on the viscosity of solutions of a pair of styrene/allyl alcohol (SAA) copolymers in methyl ethyl ketone (MEK) and toluene [16]. SAA-I has a higher molecular weight, but a lower functional group content than SAA-II. In MEK, which accepts hydrogen bonds effectively, the effect on viscosity of the OH content is diminished so that the higher molecular weight of SAA-I results in a somewhat higher viscosity compared with SAA-2. In toluene, which does not hydrogen bond effectively, the difference in OH content dominates over the difference in molecular weight so that the SAA-II solutions have the higher viscosity. Comparison of the 50 wt% solutions in MEK and toluene shows that the hydrogen-bonding solvent is more effective for viscosity reduction of both SAAs. See Table 17.6 in Section 17.4 for examples of effects of solvents on viscosity of solutions of a high solids acrylic resin.

In reducing viscosity to spray viscosity by adding solvent to a coating, it is desirable to have an equation that combines the effects of temperature and concentration on viscosity. For the limited ranges of temperature and concentration involved, Eqs. 3.3 and 3.9 can be combined to give such a relationship in Eq. 3.12 [21]. (In Eq. 3.12, log of solvent viscosity is combined into constant K.)

$$\ln \eta = K + \frac{B}{T} + \frac{w_r}{k_1} \tag{3.12}$$

3.5. VISCOSITY OF LIQUIDS WITH DISPERSED PHASES

Since many coatings contain dispersed pigment and/or resin particles, it is important to consider the effect of dispersed phases on the viscosity of liquids. When a small amount of a dispersed phase is present, there is only a small effect on viscosity (unless the dispersed phase is flocculated); however, as the volume of dispersed phase increases, there is a sharply increasing effect. More energy is used to rotate particles, and the presence of the particles increasingly interferes with the ability of other particles to move. When the system becomes closely packed with particles, viscosity approaches infinity.

Mooney developed equations that model the effect of a dispersed phase on viscosity [22]. Equation 3.13 is a useful form of the Mooney equation for understanding the effects of variables on viscosity where η_e is the viscosity of the continuous or external phase, K_E is a shape constant, V_i is the volume fraction of internal phase, and ϕ is the packing factor.

$$\ln \eta = \ln \eta_e + \frac{K_E V_i}{1 - V_i/\phi} \tag{3.13}$$

The packing factor is the maximum volume fraction of internal phase that can be fit into the system when the particles are randomly close-packed and external phase just fills all the interstices between the particles. When V_i equals ϕ, the viscosity of the system approaches infinity. Figure 3.14 shows an example of a plot of the relationship between log viscosity and V_i for a dispersed-phase system. Two major assumptions are involved in the

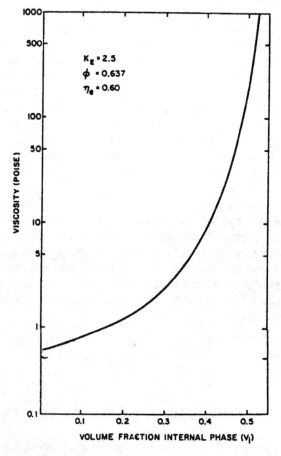

Figure 3.14. The effect of increasing volume fraction of noninteracting spherical particles on the viscosity of a dispersion. (From Ref. [16], with permission.)

Mooney equation: (1) the particles are rigid and (2) there are no particle-particle interactions.

The shape constant K_E for spheres is 2.5. Many of the particles in coatings are spheres or are reasonably close to being spheres, but there are exceptions. In the case of uniform diameter spheres—that is, monodisperse systems—the value of ϕ is 0.637. This value is the packing factor that has been calculated for a random mixture of cubical and hexagonal close-packed spheres and has been confirmed experimentally. To the surprise of many people first considering the question, the packing factor of monodisperse spheres is independent of particle size. The packing factor for basketballs is 0.637; the packing factor for marbles is 0.637; the packing factor for monodisperse latex particles is 0.637. However, marbles fit into the interstices between close-packed basketballs, and latex particles fit into the interstices between close-packed marbles. In other words, the packing factor depends strongly on particle size distribution—the broader the particle size distribution, the higher the packing factor.

The viscosity of dispersions of nonrigid particles does not follow the Mooney equation. When a shear stress is applied to such a dispersion (e.g., an emulsion), the particles can distort. When the particles are distorted, the shape constant changes to a lower value and the packing factor increases [23]; both changes lead to a decrease in viscosity. Commonly, such systems are thixotropic. This is logical since, depending on the difference between the viscosities of the internal and external phases, there would be time dependency of the distortion of the particles and, hence, a decrease in viscosity as a function of time at a given shear rate. There are modifications of the Mooney equation that consider viscosities of the two phases, but not time dependency [24]. Time dependency can be studied using viscoelastic deformation analysis [2,3].

Some fluids with readily distorted internal phases are: emulsions, water-reducible acrylic, polyester, and urethane dispersions; some latexes in which the outer layer of the latex particles and layers adsorbed on them are highly swollen by water, and some pigment dispersions with comparatively thick adsorbed layers of polymer swollen with solvent. While there may be other factors involved, many so-called thixotropic agents act by creating a swollen dispersed phase that can be distorted. For example, very small particle size SiO_2 adsorbs a layer of polymer swollen by solvent that is thick compared to the pigment and is distortable in a shear field. The degree of distortion, up to some point, increases as the shear stress increases and/or as the time of shearing increases. When shearing is stopped or decreased, the polymer layers recover their equilibrium shape and viscosity increases. In other cases, lightly cross-linked polymer gel particles are used; the particles swell with solvent, giving a distortable dispersed phase. The shear thinning behavior depends on the particle size, concentration, and internal viscosity of the dispersed phase. Smaller particles lead to higher shear rates for shear thinning. Shear thinning decreases with decreasing concentration and increasing internal viscosity of particles.

The viscosity of dispersions is also affected by particle-particle interactions. If clusters of particles form when stirring of a dispersion is stopped, the viscosity of the dispersion increases; if these clusters separate again when shear is exerted, the viscosity drops. Examples of such shear thinning systems are flocculated pigment dispersions and flocculated latexes. Another is the "gelation"—really flocculation—induced by water in coatings containing treated clay dispersions [25]. When clusters of particles form, continuous phase is trapped in the clusters; as a result, at low shear rates V_i is high. At high shear rates, when the clusters break up, the value of V_i is reduced to just that of

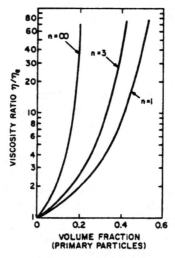

Figure 3.15. Effect of cluster formation on viscosity. (From Ref. [16], with permission.)

individual particles of the dispersion without trapped continuous phase. As V_i increases, viscosity increases and vice versa.

One can also consider the dispersion from just the point of view of the V_i of the primary particles, adjusting the K_E and ϕ in Eq. 3.13 to account for aggregation rather than adjusting the V_i. An example is shown in Figure 3.15 [26]. The vertical axis is the ratio of the viscosity to the viscosity of the external phase (η/η_e). The viscosity increases rapidly as the number of particles n in aggregates increases.

Polymer solutions containing dispersed phases are complex physical systems whose flow properties are still the subject of continuing research. In this discussion, we have used the Mooney equation; alternative treatments, such as those of Krieger and Dougherty [27] and Russel and Sperry [28], are also useful. Reference [29] includes a detailed analysis of effects of particle interactions on dispersion rheology.

3.6. OTHER MODES OF FLOW

Although flow from application of shear stress is the most common type of flow encountered in making and using coatings, other modes of flow are sometimes involved.

3.6.1. Turbulent Flow

Turbulent flow occurs at very high rates of shear or in irregularly shaped containers and pipes. At low shear rates, flow occurs in a laminar fashion, as illustrated in Figure 3.1. However, as shear rate increases, a critical point is reached where flow suddenly becomes chaotic. Laminar flow is disrupted; swirling eddies and vortices occur, and flow changes to turbulent flow. Even with Newtonian fluids, viscosity increases more than proportionally with shear rates above this critical value.

3.6.2. Normal Force Flow

When Newtonian fluids are stirred with a rotary stirrer, the liquid level becomes low in the center near the shaft of the stirrer and high on the walls of the vessel as a result of centrifugal force. This is shown schematically in Figure 3.16(*a*). However, some liquids climb the shaft of the stirrer, as shown in Figure 3.16(*b*), rather than the sides of the vessel. Such flow is normal (perpendicular) to the plane of force. This *normal force flow* behavior is typical of systems that are starting to gel. In the early stages of cooking resins, the flow pattern is as shown in Figure 3.16(*a*), but if cross-linking starts and gelation begins, the flow pattern can change abruptly to that shown in Figure 3.16(*b*). If this occurs, it is time to stop heating and dump the reaction mixture out as quickly as safely possible before the gel becomes intractable.

Normal force flow effects have been observed in the application of some coatings whose flow is viscoelastic. Under stress, the flow of these coatings has both elastic and viscous components. It seems logical to assume that a significant extent of elasticity could affect atomization in spraying, film splitting in roller coating, and leveling; however, there have been few studies of correlation between normal force flows and coating performance [6]. At least part of the reason for the lack of studies is that specialized, expensive instrumentation, such as an oscillatory motion rheometer, is required for measurements.

3.6.3. Extensional Flow

Another mode of flow encountered in some methods of coating application is *extensional flow*. Extensional flow occurs when fluid deformation is the result of stretching. Various types of stretching are possible. In spin coating, extension occurs in two dimensions. The extensional flow of greatest importance in most other coating processes is uniaxial—that is, in one direction. In uniaxial flow, the viscosity is properly called *dynamic uniaxial extensional viscosity* (DUEV); we simply use the term *extensional viscosity*, but it should be remembered that there are several types of extensional viscosity.

The difference between extensional flow and shear flow was first observed in fiber drawing. When the fiber material passes through the spinneret, the mode is shear flow. However, as the fiber is pulled after leaving the spinneret, there is no further shearing action; rather, the fiber is extended. The flow is extensional flow, and the resistance to flow

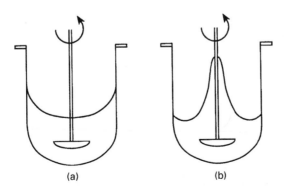

(a)   (b)

Figure 3.16. (*a*) Conventional compared to (*b*) normal direction flow of liquids on stirring.

Figure 3.17. Fiber development in roll coating a high extensional viscosity paint. (From Ref. [29], with permission.)

is extensional viscosity. The symbol used for extensional viscosity is η^*. In the case of Newtonian fluids, $\eta^*/\eta = 3$.

Extensional flow in coating application is encountered when applying coatings by direct roll coating. (See Section 22.4.) The material to be coated is passed through the nip between two rollers, one of which is covered with a layer of coating. In the nip, the coating is under pressure; as the coating comes out of the nip, the roller is moving up away from the film, and flow is extensional. As the film stretches, it splits; small imbalances of the pressures lead to variations in the timing of film splitting. If the extensional viscosity is

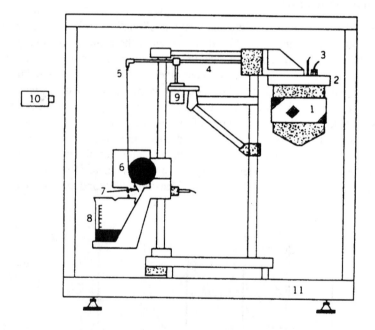

Figure 3.18. Extensional Viscometer: 1, reservoir; 2, mount; 3, compressed air line (for forcing sample through system); 4, thin-walled delivery tube; 5, nozzle (spinneret); 6, take-up drum; 7, scraper; 8, beaker; 9, transducer (to measure the deflection of the tube as a result of the forces generated during spinning); 10, video camera (permits determination of filament width); 11, environmental chamber (for temperature control). (From Ref. [4], with permission.)

relatively low, the film splits quickly, leaving a ridged film. However, with higher extensional viscosity, fibers grow; longer fibers tend to split in two places, resulting in formation of droplets, which are thrown out into the air. This is called misting or spattering. Figure 3.17 shows an extreme case of fiber development in roll coating of paint with an impractically high extensional viscosity [30]. References [31] and [32] discuss the relationship of variables and extensional viscosity effects in roll coating.

Extensional flow can also be encountered in spray application. If, for example, a solution of a thermoplastic acrylic resin with $\bar{M}_w$ above about 100,000 is sprayed, instead of droplets coming out of a spray gun orifice, fibers emerge. The "strength" of the solution is high enough that the stream of coating stays as a fiber rather than forming droplets. As the fiber extends, the mode of flow is extensional. The behavior is called cobwebbing. While cobwebbing is undesirable when painting a car, it can give a desirable effect for applying decoration to a Christmas tree. Reference [33] discusses other possible extensional viscosity phenomena in spray application.

A way of measuring extensional viscosity is by forming a fiber, wrapping it around a drum, and measuring the rate of extension and the force required for further extension. A schematic diagram of such a device that can be used with liquid paint is shown in Figure 3.18.

GENERAL REFERENCES

T. C. Patton, *Paint Flow and Pigment Dispersion*, 2nd. ed., Wiley-Interscience, New York, 1979.
C. K. Schoff, *Rheology*, Federation of Societies for Coatings Technology, Blue Bell, PA, 1991.
A. A. Collyer, *Techniques in Rheological Measurement*, Chapman and Hall, London, 1993.
P. A. Reynolds, "The Rheology of Coatings," in A. Marrion, Ed., *The Chemistry and Physics of Coatings*, Royal Society of Chemistry, London, 1994.

REFERENCES

1. T. C. Patton, *Paint Flow and Pigment Dispersion*, 2nd. ed., Wiley-Interscience, New York, 1979.
2. L. J. Boggs, M. Rivers, and S. G. Bike, *J. Coat Technol.*, **68** (855), 63 (1996).
3. R. D. Hester and D. R. Squire, Jr., *J. Coat. Technol.*, **69** (864), 109 (1997).
4. C. K. Schoff, "Rheological Measurements," in *Encyclopedia of Polymer Science and Engineering*, 2nd. ed., Vol. 14, Wiley, New York, 1988, pp. 454–540.
5. F. Anwari, B. J. Carlozzo, C. J. Knauss, R. B. Krafcik, P. Rozick, P. M. Slifko, and J. C. Weaver, *J. Coat. Technol.*, **61** (774), 41 (1989).
6. C. K. Schoff, *Rheology*, Federation of Societies for Coatings Technology, Blue Bell, PA, 1991.
7. Z. W. Wicks, Jr., G. F. Jacobs, I. C. Lin, E. H. Urruti, and L. G. Fitzgerald, *J. Coat. Technol.*, **57** (725), 51 (1985).
8. A. Toussaint and I. Szigetvari, *J. Coat. Technol.*, **59** (750), 49 (1987).
9. F. N. Jones, *J. Coat. Technol.*, **68** (852), 25 (1996).
10. S. Haseebuddin, K. V. S. N. Raju, and M. Yaseen, *Prog. Org. Coat.*, **30**, 25 (1997).
11. K. J. H. Kruithof and H. J. W. van den Haak, *J. Coat. Technol.*, **62** (790), 47 (1990).
12. A. J. Wright, *Eur. Coat. J.*, **32**, 696 (1996).
13. P. R. Sperry and A. Mercurio, *ACS Coat. Plast. Chem. Prepr.*, **43**, 427 (1978).
14. W. A. Lee and R. A. Rutherford, "The Glass Transition Temperatures of Polymers," in J. Branderup and E. H. Immergut, Eds., *Polymer Handbook*, 2nd. Wiley, New York, 1975, p. III–141.

15. J. R. Erickson, *J. Coat. Technol.*, **48** (620), 58 (1976).
16. L. W. Hill and Z. W. Wicks, Jr., *Prog. Org. Coat.*, **10**, 55 (1982).
17. M. A. Sherwin, J. V. Koleske, and R. A. Taller, *J. Coat. Technol.*, **53** (683), 35 (1981).
18. Z. W. Wicks, Jr. and L. G. Fitzgerald, *J. Coat. Technol.*, **57** (730), 45 (1985).
19. A. L. Rocklin and G. D. Edwards, *J. Coat. Technol.*, **48** (620), 68 (1976).
20. J. R. Erickson and A. W. Garner, *ACS Org. Coat. Plast. Chem. Prepr.*, **37** (1), 447 (1977).
21. M. J. Eiseman, *J. Coat. Technol.*, **67** (840), 47 (1995).
22. M. Mooney, *J. Colloid Sci.*, **6**, 162 (1951).
23. D. A. R. Jones, B. Leary, and D. V. Boger, *J. Colloid Interface Sci.*, **150** (1), 84 (1992).
24. L. E. Nielsen, *Polymer Rheology*, Marcel Dekker, New York, 1977, pp. 56–61.
25. S. J. Kemnetz, A. L. Still, C. A. Cody, and R. Schwindt, *J. Coat. Technol.*, **61** (776), 47 (1989).
26. T. B. Lewis and L. E. Nielsen, *Trans Soc. Rheology.*, **12**, 421 (1968).
27. I. M. Krieger and T. J. Dougherty, *Trans. Soc. Rheology*, **III**, 137 (1959); G. M. Choi and I. M. Krieger, *J. Colloid Interface Sci.*, **113**, 94, 101 (1986).
28. W. B. Russel and P. R. Sperry, *Prog. Org. Coat..*, **23**, 305 (1994).
29. J. W. Goodwin and R. W. Hughes, in J. E. Glass, Ed., *Technology for Waterborne Coatings*, ACS Symposium Series 663, American Chemical Society, 1997, pp. 94–125.
30. J. E. Glass, *J. Coat. Technol.*, **50** (641), 56 (1978).
31. D. A. Soules, R. H. Fernando, and J. E. Glass, *J. Rheology*, **32**, 181 (1988).
32. R. H. Fernando and J. E. Glass, *J. Rheology*, **32**, 199 (1988).
33. D. A. Soules, G. P. Dinga, and J. E. Glass, in J. E. Glass, Ed., *Polymers as Rheology Modifiers*, American Chemical Society, Washington, 1991, pp. 322–332.

CHAPTER 4

Mechanical Properties

The critical properties of most coating films relate to their ability to withstand use without damage. The range of potential mechanical damage is large. The coating on the outside of an automobile should withstand being hit by a piece of flying gravel without film rupture. The coating on the outside of a beer can must be able to withstand abrasion when cans rub against each other during shipment in a railroad car. The coating on wood furniture should not crack when the wood expands and contracts as a result of changing temperatures during winter shipment or due to swelling and shrinkage resulting from changes in moisture content of the wood. The coating on aluminum siding must be flexible enough to withstand fabrication of the siding and must resist scratching during installation on a house. In addition, many coatings must also withstand the effects of weather (see Chapter 5), retain adhesion (see Chapter 6), and protect metals from corrosion (see Chapter 7).

The introductory part of this chapter applies to all aspects of durability; it discusses some of the broad problems of developing, evaluating, and testing coatings. Development of coatings with adequate durability is made complex by the wide range of conditions to which coatings are exposed. It is safe to generalize that the only way to know how a coating will perform in actual use is to apply the coating to the final product, use the product over its lifetime, and see whether the coating performs satisfactorily. But, in many cases, the lifetime of the product can be very long. The coating on the outside of an automobile should maintain its integrity and appearance for well over five years. The coating on furniture should perform satisfactorily for 20 or more years. No laboratory tests are available that permit satisfactory product performance predictions for many applications; however, the formulator must have some way of judging the merits of a new formulation.

The most powerful tool is a data bank of field use performance of previous formulations. Formulators have made judgments about the effects of formulation changes on durability based on their years of experience; experienced formulators try to pass on their accumulated experience to novice formulators. Historically, new formulations were relatively small modifications of formulations with known field performance. If the change was significant, initial field use might be limited. In the automobile field, for example, it used to be common that after a promising new formulation had been developed, to coat just a few cars with it. Then, the next year, if no problems were encountered, the new formulation might be adopted for one color on one model of automobile. The following year, the use might be extended to three or four colors on two or three models. Finally, if all of this history was satisfactory, the new formulation might be widely adopted. This gradual approach to formulating and testing worked quite successfully. However, in recent years, there has been pressure to accelerate the process. This has resulted from increasing performance requirements, pressures to reduce costs, and the need to meet regulation requirements. Reduction of VOC emissions has been a major driving force, but other factors, such as increasing recognition of possible toxic hazards—especially from long term exposure to relatively low levels of some chemicals—have required changes in relatively short time spans.

Increasingly, databases created from actual field use are being accumulated. For example, for many years, teams of representatives of automobile manufacturers and suppliers have surveyed cars in parking lots in various parts of the country. The serial number, which can be seen through the windshield, permits identification of the coatings on that car. In other fields, performance of pipelines, exterior siding, and offshore oil rigs is monitored. Abrasion resistance of exterior beer can coatings can be related to shipment variables. Computers make possible analysis of masses of data correlating actual performance with coating composition and application variables. The computer makes available to all formulators in a company all records, rather than just the memory of one formulator. Data on coatings that fail are as important as data on coatings that are satisfactory. More use of this approach in the future is critical to future progress in formulating superior coatings. A word of caution: Some technical people think in terms of maximizing only performance, but economic factors are also important. It is foolish to apply an expensive coating that will last 20 years on a product that will last only five years. A 5-year life coating can be fully satisfactory and usually costs less than a 20-year life coating.

Accumulating a database takes time. Meanwhile, formulators need tools to work with. The formulator has three kinds of needs when considering the mechanical properties of films. To select the most appropriate components of the coating, the formulator needs to understand the relationships between composition and properties; research instruments can help develop this understanding. Laboratory tests are needed to follow the effects of changes in formulation. Quality control tests are needed to compare production batches to a standard. The accuracy and the appropriate uses of laboratory and quality control tests are often quite limited. Unfortunately, many people working with coatings do not appreciate the limitations of the tests they use. It is common for formulators and users of coatings to assume that an empirical quality control test or an accelerated laboratory test that has been correlated with use performance for one type of coating can predict performance for coatings of a different type. That is almost never the case.

Dickie has proposed a methodology for systematically considering the factors involved in service life prediction [1]. He suggests that predictive models can provide the framework for assessing the importance and relevance of available information and may give insight as to what information may be missing from the evaluation of a given material or application. A monograph [2] provides a discussion of the problems of predicting service lives and proposes reliability theory methodology for database collection and analysis.

4.1. BASIC MECHANICAL PROPERTIES

Understanding relationships between composition and basic mechanical properties of films can provide a basis for more intelligent formulation. Most coating formulators were educated as chemists, not as engineers, and few have had any education on mechanical properties. Terms like *loss modulus*, *storage modulus*, and *tan delta*, may have little meaning to chemists, but the plastics, rubber, and fiber industries have used such concepts for many years in developing products with superior performance. Because of the diversity of coatings, the relatively small volume of most types of coatings, and this lack of understanding of the physical behavior of films, the coatings industry has lagged behind in applying these concepts. In 1977, Hill published a review paper discussing stress analysis as a tool for understanding coatings performance [3]. He did an excellent job of presenting an introduction to stress analysis in terms that a coatings formulator could understand. Much of the information presented had to be based on examples from plastics, rubber, and fiber work; there were few such papers that dealt with coatings at the time. Twenty years later, application of stress analysis to coatings has mushroomed. References [4–6] are more recent review papers. Further rapid progress in broadening the understanding can be anticipated.

Basic to understanding the study of mechanical properties of coatings is recognition that coating films are *viscoelastic materials*. The mode of deformation of any solid can be elastic and/or viscous. In ideal *elastic deformation*, or *Hookean deformation*, a material elongates under a tensile stress in direct proportion to the stress applied in conformance with Hooke's law, as exemplified by a steel spring. When the stress is released, the material returns to its original dimensions essentially instantaneously. An ideal viscous material, a Newtonian fluid, also elongates when a stress is applied in direct proportion to the stress, but, in contrast to an ideal elastic material, it does not return to, or even toward, its original dimensions when the stress is released; the deformation is permanent. Almost all coating films are viscoelastic—that is, they exhibit intermediate behavior. Thermoplastic films frequently do not completely recover their original shape after deformation; the viscous flow part of the deformation is permanent. In cross-linked films, if there is no yield point, the recovery of the original dimensions may be complete, even though there was viscous flow. The stress on the cross-links supplies the force to reverse the viscous flow. If there is a yield point, there is partial, but incomplete recovery, of the original dimensions.

Figure 4.1 shows a schematic plot of the results of a stress-strain test, in which a coating film is elongated (*strain*) at a constant rate and the resulting *stress* is recorded. (Methods for measurement of mechanical properties are discussed in Section 4.4.). By convention, the stress (force per unit of cross-sectional area) is based on the original dimensions. Strain is expressed in terms of percent elongation. The slope A of the initial, essentially straight line, portion of the graph is the *modulus*—that is, the ratio of stress to strain. One must be careful to know how the term modulus is used in a specific case. In the initial part of this plot, modulus is independent of strain. However, as strain increases, the ratio is no longer constant, and the modulus depends on the strain. The end of the curve signifies that the sample has broken. This point is defined in two ways: (1) *elongation-at-break* is a measure of how much strain E is withstood before breaking; and (2) the tensile strength, or *tensile-at-break*, is a measure of the stress B when the sample breaks. The area under the curve represents the *work-to-break* (energy vol^{-1}). Quite commonly, as shown in Figure 4.1, at an intermediate strain, the stress required for further elongation decreases. The maximum stress C at that point is called the *yield point*. Yield point can also be designated in two ways: *elongation-at-yield* (D) and *yield strength* (C).

An ideal elastic material deforms virtually instantaneously when a stress is applied and recovers its original shape virtually instantaneously when the stress is released. Elastic deformation is, over a wide range, almost independent of temperature. In contrast, viscous flow is time dependent; the flow continues as long as a stress is applied. The rate of

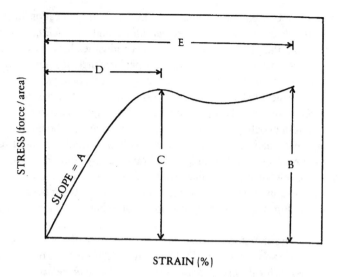

Figure 4.1. Stress-strain plot. (From Ref. [4], with permission.)

deformation depends on the viscosity of the material, as well as on the temperature. As a result, viscoelastic deformation is dependent on the temperature and the rate at which a stress is applied. If the rate of application of stress is rapid, the response can be primarily elastic; if the rate of application of stress is low, the viscous component of the response increases and the elastic response is correspondingly lower. Similarly, if the temperature is low, the response can be primarily elastic, whereas at a higher temperature, the viscous response is correspondingly greater.

These differences are illustrated in Figure 4.2(a) and (b). Plot (a) shows, schematically, the results of elongating a film at two different rates at the same temperature. Curve A is the same curve shown in Figure 4.1; curve B results from a more rapid application of stress. In curve A, there is time for the sample to undergo some viscous flow along with the elastic deformation. In curve B, the stress was applied at such a rapid rate that there was

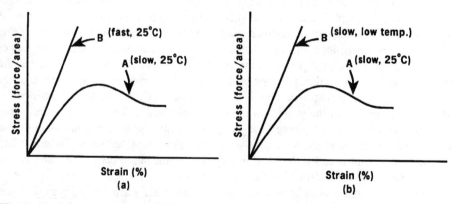

Figure 4.2. Schematic diagrams of the effects of (a) rate of application of stress and (b) temperature on stress-strain responses.

little time for viscous flow, and the elastic response dominated. Note that, as commonly occurs in real samples, elongation-at-break is less and tensile-at-break is greater when the rate of application of stress is higher. Plot (*b*) shows, schematically, the results of pulling the films at the same rate—equal to the slower rate in plot (*a*), but at two different temperatures. Curve A is at the same temperature as in plot (*a*); curve B is at a lower temperature. At the lower temperature, the viscosity was higher so that even at the slower rate of extension, there was essentially no viscous flow, and elastic deformation dominated. At the higher temperature, the viscosity was low enough to permit substantial viscous flow during stretching. As commonly occurs in real samples, elongation-at-break is less and tensile-at-break is higher when the temperature is lower.

Plots (*a*) and (*b*) are identical. The rates of application of stress and temperatures were chosen so that the change in the viscous response would be the same. In viscoelastic materials, the effects of higher rates of application of stress and lower temperatures are in the same directions. It is possible to do time-temperature superpositioning of curves mathematically. If one's instrument cannot operate at as high a rate of application of stress as one would want in order to evaluate stress-strain behavior, one can operate at a lower temperature and then calculate data points at higher stress rates, as discussed in Ref. [4].

Stress-strain determinations are also run in creep or relaxation modes, rather than a tensile mode. In creep experiments, a constant stress is applied and the resulting strain is determined. An ideal elastic material undergoes an instantaneous strain, with stress application at time t_1, that stays constant until the strain is removed at time t_2, as shown in Figure 4.3 (*a*). A Newtonian fluid subjected to a creep test shows linearly increasing strain with time, as shown in Figure 4.3(*b*). For a viscoelastic sample, strain is observed to increase over time in a nonlinear manner. Typical responses for cross-linked and thermoplastic coatings (viscoelastic) are shown in Figure 4.3(*c*) and (*d*), respectively. In a relaxation test, one applies an instantaneous strain, elongating the sample, then follows the change in stress with time. The stress stays constant with time for an ideal elastic material, and a Newtonian liquid exerts no stress. For viscoelastic samples, the stress is initially high and drops to lower values (relaxes) with time.

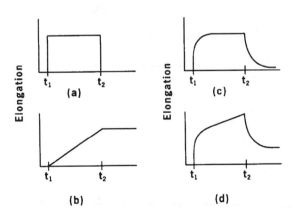

Figure 4.3. Schematic plots of creep test responses. A constant stress is applied at t_1 and removed at t_2 for (*a*) an ideal elastic solid, (*b*) a Newtonian fluid, (*c*) a viscoelastic sample with cross-links that result in complete recovery, and (*d*) a viscoelastic sample that shows incomplete recovery. (Adapted from Ref. [4], with permission.)

Stress-strain analysis can also be done dynamically by using instruments that apply an oscillating strain at a specific frequency. The stress and strain vary according to sine waves, due to the alternation from the oscillations. The stress and the phase angle difference between applied strain and resultant measured stress are determined. For an ideal elastic material, the maximums and minimums occur at the same angles since there is an instantaneous stress response to an applied strain; the phase shift is 0°. For a Newtonian fluid, there would be a phase shift of 90°. Viscoelastic materials, on the other hand, show an intermediate response, as illustrated in Figure 4.4. If the elastic component is high, the phase shift δ is small; if the elastic component is low compared to the viscous component, the phase shift is large. The phase shift, along with the maximum applied strain ε_0 and the maximum measured stress σ_0 are used to calculate the dynamic properties.

Storage modulus E', sometimes called *elastic modulus*, a measure of elastic response, equals $(\sigma_0 \cos \delta)/\varepsilon_0$. Its magnitude and physical significance are similar to modulus obtained from the intial straight line slope of a stress-strain curve, such as shown in Figure 4.1. The term storage modulus reflects the fact that E' measures the recoverable portion of the energy imparted by the applied strain. The *loss modulus E''* is a measure of the viscous response; and $E'' = (\sigma_0 \sin \delta)/\varepsilon_0$. The term loss modulus reflects the fact that viscous flow leads to dissipation (as heat) of part of the energy imparted by the applied strain. The square of the total modulus equals the sum of the squares of the storage and loss moduli. The ratio E''/E' is called the *loss tangent*, since all of the terms cancel except the ratio $\sin \delta / \cos \delta$, corresponding to the tangent of an angle, $\tan \delta$, commonly called *tan delta*.

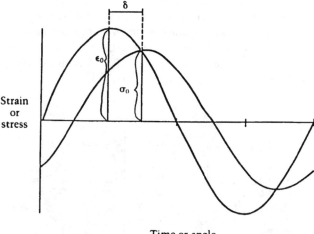

$$\text{Storage modulus} = E' = \frac{\sigma_0 \cos \delta}{\varepsilon_0}$$

$$\text{Loss modulus} = E'' = \frac{\sigma_0 \sin \delta}{\varepsilon_0}$$

$$\text{Loss tangent} = \frac{E''}{E'} = \tan \delta$$

Figure 4.4. Dynamic mechanical analysis plot as a sinusoidal strain is applied and a sinusoidal stress is determined. (From Ref. [4], with permission.)

Dynamic mechanical analysis has the advantage over stress-strain studies, such as shown in Figure 4.1, that the elastic and viscous components of a modulus can be separated. The frequency of oscillation is a variable related to the rate of application of strain. The higher the frequency, the greater the elastic response—that is, the smaller the phase angle; the lower the frequency, the greater the viscous response—that is, the larger the phase angle. The higher the frequency, the less time there is for viscous flow; hence, elastic response dominates. Similarly, lowering temperature reduces viscous flow, decreasing the phase angle, and raising temperature gives greater viscous flow, increasing the phase angle. Generally, it is possible to run experiments over a range of frequencies in dynamic tests that is wider than the range of rates of application of stress possible in linear stress-strain experiments. Furthermore, dynamic testing can be done over a wider range of temperatures and rates of heating. In dynamic tests, it is not possible to determine tensile-at-break (tensile strength), elongation-at-break, or work-to-break, since the sample must remain unbroken in order to run the test.

Stress-strain analysis discussed above is based on elongation of samples by application of tensile stress. Oscillating (dynamic) and linear (static) stress-strain analysis can also be carried out by application of shear forces. In shear tests, the stress is applied sideways—analogous to shear viscosity tests. The ratio of shear stress to shear strain is called *shear modulus* and is represented by the symbol G. For many coatings, the tensile modulus E' equals three times the shear modulus G''. Recall from Section 3.6.3 that the analogous extensional and shear viscosity have the same relationship.

4.2. FORMABILITY AND FLEXIBILITY

In many cases, a coated metal object is subjected to mechanical forces either to make a product, as in forming bottle caps or metal siding, or in use, as when a piece of gravel strikes the surface of a car with sufficient force to deform the steel substrate. To avoid film cracking during such distensions, the elongation-at-break must be greater than the extension of the film under the conditions of fabrication or distortion.

To illustrate some of the variables involved, let us consider the simpler case of a plastic, poly(methyl methacrylate) (PMMA), for which data on stress-strain relationships are available. Tensile stress-strain curves of PMMA ($T_g = 378$ K) as a function of temperature are shown in Figure 4.5. At low temperatures, there is no yield point, modulus is high, and elongation-at-break is low. Breakage at low temperature is called *brittle failure*. The terminology confuses some people; at low temperature, PMMA approaches being an ideal elastic material, but its failure is classified as brittle because the elongation-at-break is low. At higher temperatures, but still below T_g, greater elongations without breaking are possible, and there is a yield point.

Wu studied the modulus, elongation-at-yield, and elongation-at-break of PMMA and other plastics as a function of temperature [7]. Figure 4.6 shows plots of these three properties versus temperature. The abrupt drop in modulus at T_g is typical of amorphous high molecular weight thermoplastic homopolymers. Similar data are in the literature for many such polymers. Fewer examples of the plots of elongation-at-break appear. Note that elongation-at-break starts to increase rapidly at a temperature below the T_g of PMMA. Wu defines the temperature at the intercept of the elongation-at-break and elongation-at-yield plots as the *brittle-ductile transition temperature*, T_b. (T_b can also be defined as the intercept of plots of tensile strength and yield strength versus temperature.) Below T_b, the

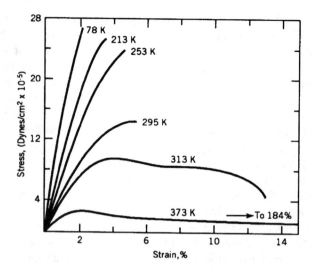

Figure 4.5. Stress-strain plots of PMMA as a function of temperature. (From Ref. [3], with permission.)

polymer is brittle; between T_b and T_g, it is hard and ductile; above T_g, the polymer becomes increasingly soft. Plastics can be deep-drawn above T_b; it is not necessary to be above T_g, which is contrary to the common, but erroneous, definition of T_g as the temperature below which an amorphous material is brittle.

Copolymers of MMA with methacrylates of longer chain alcohols, such as *n*-butyl methacrylate, have lower T_g and lower T_b values. There is substantial variation in the difference between T_g and T_b of various thermoplastic polymers, as illustrated by the data

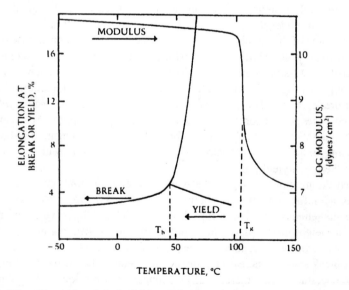

Figure 4.6. Elongation-at-break and -yield as functions of temperature, superimposed on a storage modulus, temperature plot. (From Ref. [4], with permission.)

Table 4.1. Transition Temperatures of Homopolymers

Polymers	T_g (°C)	T_b (°C)	$(T_g - T_b)$
Polystyrene	100	90	10
Poly(methyl methacrylate)	105	45	60
Poly(vinyl chloride)	80	10	70
Bisphenol A polycarbonate	150	−200	350

in Table 4.1. Since these polymers are viscoelastic, the deformation is dependent on rate of application of stress, as well as on temperature. Both T_g and T_b are dependent on the rate of application of stress: the higher the rate, the higher the T_g and T_b values.

Brittle-ductile transition temperature has been studied less with thermosetting systems. The limited data available indicate that polymers with a low degree of cross-linking show a differential between T_b and T_g but the differential decreases as cross-link density (XLD) increases. Probably, most thermoset coatings show little, if any, differential between T_g and T_b, but little data has been published. The T_g of cross-linked polymers depends on several factors: structure of the segments between cross-links, XLD, amount of dangling chain ends, and the extent, if any, of cyclization of the backbone [8].

It has been known for many years that there is a proportional relationship between XLD and modulus of low XLD elastomers above T_g, but it had been uncertain whether the relationship could be extended to relatively higher XLD coatings. Hill demonstrated that this relationship also holds for melamine-formaldehyde (MF) cross-linked acrylic and polyester coating films [6,9]. The relationship is surprisingly simple, as shown in Eq. 4.1, in which v_e is XLD expressed as the number of moles of *elastically effective network chains* per cubic centimeter of film. An elastically effective network chain is one that is connected at both ends to the network at different junction points—short cyclical chains and dangling ends are not elastically effective. The terms E and G are tensile and shear storage moduli, respectively. Since E'' and G'' are low at temperatures well above T_g, $E \approx E'$ and $G \approx G'$. It has been suggested that the simple form of Eq. 4.1 is favored when network chains are short and when E' is obtained with very small deformations, as used in DMA [10].

$$E = 3v_e RT \qquad \text{(or } G = v_e RT\text{)} \ (T \gg T_g) \qquad (4.1)$$

Thus, at least for the classes of films studied, it is possible to calculate XLD (v_e) from modulus against temperature plots [9]. Note that v_e divided by the film density provides the moles of network chains per gram. The inverse, grams per mole of network chains, corresponds to the average molecular weight of network chains, frequently called average molecular weight between cross-links $\bar{M}_c$. Commonly, $\bar{M}_c$ is erroneously defined and used as the molecular weight per branch point; one must be careful when reading any paper dealing with molecular weight of network chains and XLD. Cross-link density can also be calculated, with an assumed interaction parameter, from the extent of swelling of a film by solvent [6]. While cross-linked films do not dissolve in solvent, solvent does dissolve in a cross-linked film. As cross-links get closer together—that is, as XLD increases—the extent of swelling decreases.

Equation (4.1) can also be used to predict the storage modulus above T_g from the XLD. In a system with stoichiometric amounts of two reactants whose functional groups react completely, one can estimate the XLD from the equivalent weights and the average functionality. If the reactant mixture contains molecules of several different functionalities, calculation becomes more difficult. A more general approach is provided by the Scanlan equation, Eq. 4.2 [6]:

$$v_e = \frac{3}{2}C_3 + \frac{4}{2}C_4 + \frac{5}{2}C_5 + \cdots \tag{4.2}$$

The C values are the concentrations of reactants with functionality of 3 to 5 (or more), expressed in units of moles per cubic centimeter of final cured film. The volume of the final film depends on density of the cured film and loss of volatile byproducts of the reaction. Equation (4.2) does not include a term for difunctional reactants because these reactants do not create junction points in a network; they only extend chains. Although the Scanlan equation is convenient for stoichiometric reactions, it does not apply to other cases. Recent modifications permit consideration of incomplete conversion [10]. For nonstoichiometric mixtures and/or incomplete conversion, Miller–Macosko equations are useful general equations. Bauer selected the Miller–Macosko equations most useful for coatings, gave examples of their applications, and provided a computer program [11].

Properties are affected by the extent to which cross-linking has been carried to completion. Incomplete reaction leads to lower XLD and, hence, lower storage modulus above T_g. The extent of reaction can be followed by determining storage modulus as a function of time [12]. As cross-linking continues, storage modulus increases until a terminal value is reached.

Thus, one can, at least in theory, design a cross-linked network to have a desired storage modulus above T_g by selecting an appropriate ratio of reactants of appropriate functionality. By proper selection of the structures between cross-links and cross-link density, one can design the T_g of the cross-linked network.

Determination of dynamic mechanical properties has proved to be a valuable tool for studying cross-linking of hydroxy-functional resins with MF resins [6,9]. As the stoichiometric ratio of methoxymethyl groups from the MF resin to hydroxyl groups was raised from values less than 1, the storage modulus above T_g of the fully cured films increased up to the point that the ratio became one. As discussed in Section 9.3.2, these results show that all of the functional groups on the MF resin can react with hydroxyl groups and the reaction is not limited by steric hindrance, as previously thought. When the amount of MF resin was increased so that excess methoxymethyl groups were present, storage modulus above T_g increased at higher temperatures during dynamic mechanical testing. This behavior is explained on the basis that the excess methoxymethyl groups can undergo self-condensation reactions during the determination. The self-condensation reaction is relatively slow and was incomplete during the baking cycle used in preparing the film; hence, the reaction continued at the higher temperatures used in dynamic mechanical analysis, leading to the higher storage modulus. Self-condensation during baking of coatings also occurs when excess MF resin is used; then, the extent of self-condensation increases as baking time and temperature increase.

As done in Figure 4.7, it is common to assign the peak of the tan delta curve as the T_g. Some authors prefer to assign the peak of the loss modulus plot as the T_g. As can be seen, there is a substantial difference between the two. The peak of the loss modulus plot is

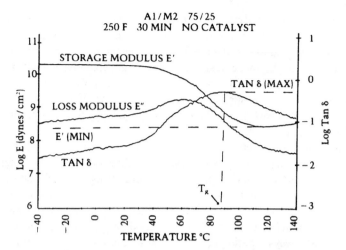

Figure 4.7. Dynamic properties of a highly cross-linked acrylic coating film (solid lines). Dashed lines indicate values E'_-(min), tan δ(max), and T_g that are measures of the extent cure. The values of E'(min) and T_g tend to increase, whereas tan δ(max) tends to decrease as extent of cure increases. (From Ref. [4], with permission.)

usually nearer to the T_g determined using DSC. As noted before, T_g is dependent on the rate of heating in DSC (see Section 2.1.2) and on the rate of application of stress—the frequency of oscillation—as well as the rate of heating in a dynamic test. This dependence on experimental technique might lead one to wonder if T_g means anything. It does, but one must always be careful to compare T_g values determined in the same way. In using T_g with regard to fabrication or deformation of coatings, the most relevant T_g is the one determined at a rate close to the rates to be encountered in deformation of the coating.

An additional factor that can affect the mechanical properties of polymeric materials is the breadth of the T_g transition region [13]. In some materials, the slope of the transition region is steep, as shown in Figure 4.6. In others, this slope is shallower, as shown in Figure 4.7. The same effect can be seen in tan delta plots, which exhibit various breadths. The breadth of the T_g transition region can also be estimated by differential scanning calorimetry (DSC), but dynamic mechanical analysis gives a clearer picture.

Factors controlling the breadth of the T_g transition have only been partly elucidated, but broad tan δ peaks are frequently associated with heterogeneous polymeric materials. Blends of different thermoplastic resins often display two distinct T_g's, presumably because of phase separation. Other blends of thermoplastics have a single, often broad, T_g, presumably when phase separation is indistinct or when the phases are very small. For thermosetting polymers, the T_g transition region is generally broader than for thermoplastics, and the breadth may vary considerably. Breadth of the distribution of chain lengths between cross-links is a factor, and blends of thermosetting resins such as acrylics and polyesters often display a single, broad T_g transition [9]. As a rule—to which there are probably exceptions—materials with broad and/or multiple T_g's have better impact resistance than comparable polymers with a sharp, single T_g.

Also related to impact resistance is the presence in some homogeneous polymers of one or more, usually small, tan delta peaks at temperatures below T_g. These peaks are called *low temperature loss peaks*, or β- and γ-transitions. They are thought to result from the onset of some specific small-scale motions of parts of the polymer molecules as temperature is increased. These transitions have been detected in a number of polymers and, in some cases, related to specific motions of molecule segments [14]. In plastics, it is well established that tough, impact resistant materials generally have low temperature loss peaks; polymers made from bisphenol A (epoxies and polycarbonates) are common examples. It is reasonable to speculate that coatings with low temperature loss peaks may have good impact resistance (if adhesion is good), but the relationship is not well documented in the literature.

Mechanical properties of coatings are generally more complex than those of most plastics. One reason is that coatings are used as thin films on a substrate. Interaction with the substrate affects the mechanical properties of thin films. The substrate can limit the extent of deformation that occurs. The substrate can act as an energy sink to dissipate the energy so that there is less effect on the coating film. Adhesion can have a profound affect on ability to withstand fabrication. If adhesion is good, fabrication and impact resistance of the films are almost always superior, perhaps because less stress is concentrated in the film. The extent of permanence of the deformation can have important effects. Thermoplastic films are more likely to be permanently deformed than cross-linked films. When a cross-linked film on a metal substrate is deformed by fabrication, it is held in the deformed state by the metal substrate. As a result, there is a stress within the film acting to pull the film off the substrate. This recoverable extension has an associated restoring force that can overcome the adhesion. Coatings have been known to pop off postformed metal bottle caps when a jar is sitting on a supermarket shelf. A permanent deformation may be better for postformability because there is no associated restoring force. Stress within films can also arise during the last stages of solvent loss and/or cross-linking of films [15]. Both solvent loss and cross-linking result in shrinkage. If this shrinkage occurs when the temperature is near the T_g of the film, the resulting internal stresses may persist indefinitely. (See Section 6.2 for discussion of the effects of internal stresses on adhesion.)

Film thickness is also a factor in the ability of a coating film to withstand fabrication without cracking. Thin films can be used for deeper draws than thick films. In making coated exterior siding, the hardness of the film can be increased without encountering cracking by limiting the film thickness. Of course, thinner films of pigmented coatings give poorer hiding; a common compromise in this case is at 20–25 μm film thickness. Two-piece fish cans are coated as flat sheets with a relatively highly cross-linked phenolic coating that is quite brittle in order to minimize swelling with fish oil. Such cans can only be successfully formed without cracking if the film thickness is about 5 μm or less.

Relatively little basic work has been published on the effects of pigmentation of films on mechanical properties. In many cases, as the pigment volume concentration of films is increased up to the critical pigment volume concentration, the tensile strength of the films increases. (See Section 21.1.) It is also possible that imperfections resulting from some types of pigmentation may lead to defects from which crack propagation is facile, as discussed in connection with fracture mechanical adhesion failure in Section 6.2. In some cases, the storage modulus above T_g is increased by the presence of pigment. For pigmented coatings, one cannot expect a direct proportional relationship between storage modulus above T_g and XLD [6]; however, if the pigment content is constant in a series of films varying in the polymeric portion, the relative values of E' should still indicate relative XLDs for the pigmented films.

Another important variable can be the timing of fabrication or flexing after curing a coating. It is common for coatings to become less flexible as time goes on. Particularly in air dry coatings, some solvent may be retained in films. Since most coatings have T_g values near or a little above room temperature, solvent loss may be very slow. (See Section 17.3.4.) Solvents generally act as plasticizers, so as solvent is lost, T_g and storage modulus increase, and coatings tend to become less flexible. In cross-linkable coatings, if the cross-linking reaction was not complete, the reaction may continue slowly, increasing XLD and, hence, storage modulus, and decreasing flexibility. Continued cross-linking is particularly likely to occur in air dry coatings, since reaction rates are likely to become mobility rate controlled, and, hence, the last part of the reaction is slow, as discussed in Section 2.1.3. Reactions during the use life of the coating, especially during exterior exposure (see Section 5.1), can result in embrittlement. Dynamic mechanical analyses of a variety of clear coats after Florida exposure and QUV testing, show the changes in mechanical properties resulting from exposure [16].

Hardening of baked cross-linked coatings over time is also rather commonly observed. Although in some cases, further volatile loss or continued cross-linking may be responsible, another possible factor is *densification*. If a polymer is heated above its T_g and then cooled rapidly (quenched), the density is commonly found to be lower than if the sample had been cooled slowly. During rapid cooling, more and/or larger free volume holes are frozen into the matrix than with slower cooling, which provides greater chances for molecular motion. On storage, the molecules in quenched films slowly move even though the temperature is somewhat below T_g and free volume decreases, causing densification; since this process results in changes in properties with aging with no chemical change, it is also called *physical aging*. With the decrease in free volume and the increase in density, cracking during fabrication is more likely. This phenomenon has been widely observed in plastics, but is only beginning to be considered in coatings. It may occur when coatings are baked on metal at high temperatures and then cooled rapidly after coming out of the oven. It has been suggested that densification may be a common cause of embrittlement during aging of baked coating films. Greidanus has studied physical aging at 30°C of polyester/MF films that had been baked at 180°C and then quenched to 30°C [17]. There was a small, but reproducible, increase in modulus with time at 30°C. The aging rate (i.e., the rate of increase of modulus) decreased with time. If the sample was heated again at 180°C and again quenched to 30°C, the modulus returned to its lower value and underwent physical aging again. Further work is needed, but it is evident that physical aging can be an important phenomenon.

When coatings are baked industrially, there are further complications involved. Oven temperatures can vary, not only the air temperature in the oven as a whole, but also within the oven. The rate of heating of a coating can depend on the substrate thickness. For example, the temperature of the coating on the sheet metal roof of a car increases faster than the temperature of the coating on joints, where the metal thickness is greater. To achieve the desired properties, some minimum time at a temperature is required, but overbaking can lead to excessive cross-linking. There is a *cure window* for any baked coating, within this set of time and temperature satisfactory properties are obtained. As discussed in Section 9.3, high solids acrylic melamine coatings have narrower cure windows than conventional solids coatings had. The effects of some of the variables in thermal history have been modeled [18].

4.3. ABRASION AND MAR RESISTANCE

Abrasion is the wearing away of a surface, whereas marring is a disturbance of a surface that alters its appearance. Both phenomena are included in the field of *tribology*, the science of surfaces in sliding contact. Terminology is not standardized, and terms such as *scratching*, *buffing*, *gouging*, and *wearing* are used with meanings that sometimes overlap *abrasion* and *marring*.

4.3.1. Abrasion Resistance

One might suppose that hard materials are less likely to fail by abrasion than soft materials. In some cases, this assumption is true, but in many other cases, softer materials are more abrasion resistant; for example, rubber tires resist abrasion far better than steel tires.

Evans studied the mechanical properties of a series of floor coatings with known actual wear life [19]. He determined tensile-at-break, elongation-at-break, and work-to-break. His data are given in Table 4.2; the coatings are listed in order of increasing wear life. One might suppose that higher tensile strength would give higher abrasion resistance; the data show the reverse. (It should not be assumed from these limited data that abrasion resistance is always inversely related to tensile strength.) Elongation-at-break values gave the proper rank order, but Evans concluded that work-to-break values best represented the relative wear lives. Intuitively, it seems reasonable that abrasion resistance would be related to work-to-break. Work-to-break values vary with the rate of application of stress and should be determined at a rate comparable to that to be expected in use.

In studies on another series of coatings, Evans and Fogel determined that work-to-break did not always correlate with abrasion resistance determined by loss of gloss in a ball mill abrasion tester when the stress-strain tests were carried out at ambient temperatures [20]. They reasoned that the strain rate of their instrument was too low relative to the stress application in the test. Using a time-temperature superposition relationship, they calculated that the tests at an accessible strain rate should be carried out at $-10°C$ in order to compensate for the instrumentally inaccessible high rate of stress application at ambient temperature. The resulting work-to-break values did correlate with abrasion resistance for urethane films with a T_g equal to or greater than $-10°C$.

In studies of erosive wear of clear coats for automobiles, it has been shown that wear resistance increases as energy-to-break of films increases [21]. Erosion rate is also affected by the substrate; for example, clear coats applied directly to steel showed significantly less durability than when applied over primer and base coats. Wear tends to increase as the angle of application of stress decreases.

Urethane coatings generally exhibit superior abrasion resistance combined with solvent resistance. This combination of properties may result from the presence of intersegment hydrogen bonds in addition to the covalent bonds. At low levels of stress, hydrogen bonds

Table 4.2. Mechanical Properties of Floor Coatings

Floor Coating	Tensile Strength (psi)	Elongation-at-break (%)	Work-to-break (in-lb/in^3)	Taber[a] (rev/mil)
Hard Epoxy	9000	8	380	48×10^3
Medium Epoxy	4700	19	600	33×10^3
Soft Epoxy	1100	95	800	23×10^3
Urethane Elastomer	280	480	2000	36×10^3

act like cross-links, reducing swelling on exposure to solvent. At higher levels of stress, the hydrogen bonds can dissociate, permitting the molecules to extend without rupturing covalent bonds. When the stress is released, the molecules relax and new hydrogen bonds form. Urethanes are used as wear layers for flooring, as well as top coats in aerospace applications, where this combination of properties is desirable.

Factors in addition to work-to-break are involved in abrasion resistance. The coefficient of friction of the coating can be an important variable. For example, abrasion of the coating on the exterior of beer cans during shipment can be minimized by incorporation of a small amount of incompatible wax or fluorosurfactant in the coating. When the two coated surfaces rub against each other, the reduced surface tension, resulting from the additive, reduces the coefficient of friction so that transmission of shear force from one surface to the other is minimized and abrasion is reduced.

Another variable is surface contact area. Incorporation of a small amount of a small particle size SiO_2 pigment in a thin silicone coating applied to plastic lens eyeglasses reduces abrasion, adding to the effect of the low surface tension of the silicone surface. The pigment particles reduce contact area, permitting the glasses to slide more easily over a surface. Another example of the same principle is the incorporation of a small amount of coarse SiO_2 inert pigment in wall paints to reduce *burnishing*. If a wall paint without such a pigment is frequently rubbed, as around a light switch, it abrades to a smoother glossier surface—that is, it burnishes. The coarse inert pigment reduces burnishing by reducing contact area.

An approach that has been used for many years in resin-bonded pigment print colors on textiles is to incorporate rubber latex in the print paste. The latex particles are not soluble in the resin and end up as individual particles in the resin, along with the pigment particles. The abrasion resistance is markedly improved by the latex addition. Similar work is now being done to improve the abrasion resistance of continuous coatings. Presumably, the relatively soft rubber particles act to dissipate stresses on the film, minimizing the chance of a stress concentration leading to film rupture. Glass microspheres have been shown to increase abrasion resistance of epoxy coatings by damping the energy released by impacts on the surface [22]. Lee reviewed abrasion resistance as one type of wear in a broad approach to fracture and surface energetics of polymer wear [23].

4.3.2. Mar Resistance

Mar resistance is related to abrasion resistance, but there is an important difference. Abrasion may go deeply into the coating, while marring is usually a near-surface phenomenon; mars less than $0.5 \mu m$ deep can degrade appearance. Marring is a major problem with automobile coatings, particularly those with a clear coat as the final top coat. (See Section 29.1.2.) In going through automatic car washes, the surfaces of some clear coats are visibly marred and may lose gloss due to marring [24,25]. Mar resistance is also a critical requirement in coatings for floors and for transparent plastics—for example on polycarbonate glazing or eyeglasses.

The physics of marring is complex. Different authors use different terms to describe the phenomena involved. Various models have been proposed to describe what happens to a viscoelastic material when a hard object is drawn over its surface. One such model classifies the response of the material as elastic, plastic, or fracture. Since the elastic response recovers essentially instantaneously, only plastic deformation and fracture lead to marring. Qualitatively, mars caused by plastic deformation have shoulders, while those caused by fracture do not. While it is simplistic, this model has the advantage that the three responses can be quantitatively measured by scanning probe microscopy [26]. Most coatings exhibit a mixture of responses. Different coatings vary widely in their responses, and the same coating may respond quite differently as the force or rate of marring stress changes. A further complication is that mars in some coatings can slowly heal by reflow (creep).

Efforts are underway to relate mar resistance to the chemical structure of coatings, but relatively few systematic studies have been published. In general, MF cross-linked acrylic clear coats are more resistant to marring than isocyanate cross-linked coatings (urethanes), but MF cross-linked coatings have poorer environmental etch resistance. [MF cross-linked polycarbamates are an exception, combining etch and mar resistance. (See Section 9.3.4.)] Since urethanes generally have superior abrasion resistance, it is surprising that they have inferior mar resistance; this might be explained by differences between surface and bulk properties. A study of marring of clear coats by a scanning probe microscope indicated that an acrylic polyurethane had a thin layer of deformable plastic material on its surface, while an acrylic-MF clear coat had a layer of elastic material [26]. New instruments, *nanomechanical analyzers*, that can characterize near-surface mechanical properties are now available and may help with studies of surface properties.

Two strategies are available for the design of coatings with exceptional mar resistance: They can be made hard enough that the marring object does not penetrate far into the surface, or they can be made elastic enough to bounce back after the marring stress is removed. If the hardness strategy is chosen, the coating must have a minimum hardness. If the T_b (see Section 4.2) is above the testing temperature, groove formation is minimized; however, such coatings may fail by fracture. Film flexibility is an important factor influencing fracture resistance. Use of 4-hydroxybutyl acrylate instead of 2-hydroxyethyl acrylate in an acrylic resin cross-linked with MF resin gave improved results, as did use of polyol modified hexamethylenediisocyanate isocyanurate instead of isophoronediisocyanate isocyanurate in cross-linking urethane coatings [27]. Further improvement was obtained using silicone-modified acrylic resins. Courter proposes that maximum mar resistance will be obtained with coatings having as high a yield stress as possible without being brittle [28]. In this way, high yield stress minimizes plastic flow and avoidance of brittleness thereby minimizes fracture. Courter's paper provides a good review of attempts to relate bulk mechanical properties of coatings to their mar resistance, but these studies have not led to a broadly applicable theory of marring. This is understandable, since the mechanical properties near a coatings surface may be quite different from the mechanical properties of the bulk material.

A further problem related to mar resistance is *metal marking*. When a metal edge is rubbed across a coating, a black line is sometimes left on the coating; some of the metal has rubbed off on the surface of the coating. A common test for metal marking resistance is to draw a coin across the surface of a coating to see if a dark streak is left. Metal marking usually occurs with relatively hard coatings. The problem can be reduced or eliminated by reducing the surface tension of the coating so the coefficient of friction is low; then, the metal slips over the surface. Additives can be incorporated to increase slip. Modified polysiloxanes have been reported to be particularly effective [29]. Care must be exercised in selecting the particular grade of silicone additive and the amount of the additive used so as to minimize marring, scratching, and metal marking without causing other defects, such as crawling.

4.4. MEASUREMENT OF MECHANICAL PROPERTIES

Most instruments require free films for measurement of mechanical properties. Two major disadvantages to using free films are that (1) the interaction of the film with the substrate can have major effects on some film properties and (2) free films are sometimes difficult to prepare and handle. Test results are generally less variable with thick films than with thin films; however, thick films may not provide data applicable to thin films. Preparation of thin unsupported films can be difficult. In some cases, it is possible to make a film by drawing down a coating on a release paper with a wire wound bar. Release papers are coated with low surface tension materials to minimize adhesion; but if the surface tension is lower than the surface tension of the coating being applied, there is the possibility of crawling; that is, the coating tries to minimize surface free energy by drawing up into a ball. (See Section 23.4 for discussion of crawling.) One tries to find a release paper with a low enough surface tension so that adhesion is poor, but high enough so that crawling does not occur. A generally more effective method is to apply the coatings to tin-plated steel panels. After curing, one end of the panel is placed in a shallow pool of mercury. Mercury creeps under the coating, forming an amalgam with the tin, and the film comes free of the panel. Mercury vapor is toxic and care must be taken to minimize the hazard. The safety regulations of some laboratories do not permit such use of mercury. After the film is freed from the substrate, a specimen is cut from it. Cutting free films may result in nicks or cracks along the edge of the film. When subjected to stress, such films commonly tear easily, starting at the imperfection, leading to meaningless results. Handling films with a T_g above room temperature is especially difficult; they tend to be brittle and easily broken.

One must be careful about changes that may occur during storage before testing, such as loss of residual solvent, chemical changes, or physical aging. Test results will correspondingly change with time. Storage conditions can be critical. Most films absorb some water from the atmosphere. If the T_g is near room temperature and especially if the film has groups such as urethanes that hydrogen bond strongly with water, the T_g and film properties can be strongly affected by the humidity conditions in storage, since water acts as a plasticizer. Comparisons should be done with samples that have been stored at the same temperature and humidity. In actual use, films encounter a variety of humidity conditions and, hence, show a variation in properties.

Several types of instruments are available for determining mechanical properties. The Instron Tester is used for tensile (nondynamic) experiments. The sample is mounted between two jaws of the tester; care must be taken to ensure that the film is in line with the direction of pull. The instrument can be run with a range of rates of jaw separation, but even the highest rates are slow compared to the rates of stress application found in many real situations. This problem can be partially overcome by running the tests at low temperatures. This method has the advantage that stress can be increased until the film fails, making possible determination of tensile strength, tensile modulus, elongation-at-break, and work-to-break. However, one cannot separate the viscous and elastic components of the mechanical properties.

A thermal mechanical analyzer (TMA) is a penetrometer that measures indentation versus time and temperature. An advantage that it has over most tensile instruments is that a TMA includes a furnace and temperature programmer so that heating, cooling, and isothermal operations are possible. TMAs can be used with films on a substrate. An example of a use of TMAs is measuring *softening point*, which is related to the extent of cure of cross-linking films. Figure 4.8 shows a plot of probe penetration as a function of

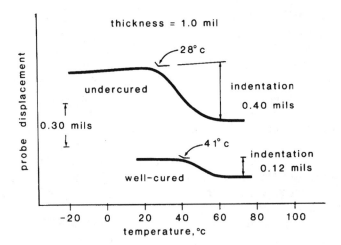

Figure 4.8. TMA plot of probe displacement against temperature for an undercured and a well-cured acrylic coil coating. (From Ref. [5], with permission.)

temperature for an undercured and a well-cured 25 μm acrylic coil coating. The softening points for the two samples are marked on the graph. The softening point is related to, but not identical to T_g; it is frequently used as an index of flexibility [5].

Various dynamic mechanical analyzers (DMA) are also available [30]. The most versatile are those in which the sample is subjected to an oscillating strain by attachment under tension to a fixed clamp on one end and a vibrating clamp on the other. Oscillating stresses are imparted to the sample. A range of frequencies can be used, and properties can be determined over a wide range of temperatures. The most sophisticated instruments are set up in line with a computer that analyzes the data and provides storage and loss modulus and tan delta figures and plots as functions of temperature. A schematic diagram is shown in Figure 4.9.

Another type of instrument used for dynamic mechanical analysis is a torsional pendulum. In its simplest form, one end of a film is fastened in jaws, and the lower end is attached to a disk to which weights can be added. The lower weight is twisted, setting up a pendulum motion whose decay can be analyzed to give the dynamic properties. Torsional pendulums have been most widely used not with film, but with a fiber braid that is

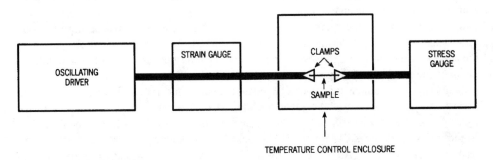

Figure 4.9. Schematic diagram of a dynamic mechanical analyzer.

saturated with liquid cross-linking polymers. The instrument has the advantage that it can be used to follow changes in dynamic properties starting with liquid coatings as reactions occur on the braid. It has the disadvantages that the sample is not a film and that there are large surface areas of fiber-polymer interface that may affect properties. A torsional pendulum apparatus has been modified so that shrinkage can be measured simultaneously with mechanical properties [31]. Since the volume change is largely proportional to the extent of cure, it can be used as a measure of the progress of cure, while simultaneously measuring changes in mechanical properties.

The range of tensile modulus that can be measured with a torsional pendulum without changing the weight on the pendulum is limited. Furthermore, since the resonance frequency is used as a measure of viscoelastic behavior, modulus results and the measuring frequency are not independent. These difficulties can be overcome by using a DMA instrument with glass or carbon fiber braid support [32].

4.5. TESTS ON COATINGS

A variety of test methods has been established to characterize properties of coatings. In general, these methods do not permit calculation of the basic mechanical properties, but rather, test some combination of properties of the coating. There are two categories of such tests: One type can be appropriate for prediction of actual use performance, and the second type is suitable for quality control. There are very real needs for both types of tests. It is too common that paint tests, which may be appropriate for quality control, are used to predict performance, even though they are not capable of providing results that permit performance predictions. There are three broad classes of coating tests: field exposure tests, laboratory simulation tests, and empirical tests.

4.5.1. Field Exposure Tests

As stated in the introduction to this chapter, the only reliable way of knowing how a coating will perform when used is to use it and see how it performs. But, there is a need for tests to predict performance before large scale use is made of the coating. The next best approach to actual use is to apply and use the coating in field applications on a smaller scale and under especially stringent conditions that may accelerate possible failure. The more limited the test and the greater the degree of acceleration, the less reliable are the predictions, but carefully designed and analyzed tests of this type can be very useful. There are many examples; we cite a few to illustrate the principles.

Highway marking paints can be tested by painting stripes across the lanes of traffic instead of parallel to the traffic flow. In this way, the exposure of the paint to wear is greater, and many paints can be tested and compared on a short length of highway, where they will receive the same amount of traffic wear. Controls with known use performance should be tested alongside new coatings. Tests should be done at different times of the year because the effects of hot sun, snow plows, salt application, and so forth must be considered. Tests should be set up on different types of highway materials, such as concrete and asphalt. Floor paints can be similarly evaluated in busy hallways.

Test automobiles painted with new coatings are driven on *torture tracks* with stretches of gravel, through water, under different climate conditions, and so on. Sample packs of canned goods are made; the linings are examined for failure and the contents evaluated for flavor after various lengths of storage.

4.5.2. Laboratory Simulation Tests

Many tests have been developed to simulate use conditions in the laboratory. The value of these tests depends on how well use conditions are simulated and how thorough a validation procedure has been used. An important key to the use of any test for performance prediction is the simultaneous evaluation of standards with known performances that cover the range from poor to excellent performance. It is not enough to use only the extremes of standards; while such information may be a first step for checking the possible applicability of a test, performance prediction requires more than two standard data points. Enough replica tests must be run to decide the number of repeat tests that must be run to give results within desired confidence limits. Chemists commonly think of standard deviations, but these are only 67% confidence limits; the odds are 33% that the proper value is outside of the standard deviation range. See Ref. [2] for an extended discussion of problems in predicting service life.

An example of a well-validated test is the laboratory shaker test devised to simulate the abrasion of can coatings when six-packs of beer are shipped by railcar [33]. Six-packs are loaded onto a shaker designed to simulate the pressures, rate of shaking, range of motion, etc. actually encountered in rail shipments. The test was validated with cans with known field performance. The time to reach different degrees of abrasion failure were established. In unpublished work, in laboratories of several coatings suppliers and can manufacturers, the results were compared with known performance and with the results of standard abrasion tests. It was found that none of the standard laboratory abrasion tests gave satisfactory predictions, but this test gave results that could be used for performance prediction.

The automobile industry uses *gravelometers* to evaluate resistance of coatings to chipping when struck by flying gravel. Pieces of standard gravel or shot are propelled at the coated surface by compressed air under standard conditions. The tests have been standardized by comparison to a range of actual results and have been found to give reasonably good predictions of actual performance. A more sophisticated instrument, a precision paint collider, which permits variations in angle and velocity of impact and temperature has been described [34]. It was used in studies of impact failure of various coatings on automotive bumpers made with various plastic polymers. In some cases, failure was cohesive failure in the upper layers of the plastic and in other cases, delamination of the coating from the plastic.

Several laboratory devices are available that approximately reproduce stamping or other forming operations to test the ability of coated metal to withstand fabrication. Individual companies design such tests to match the conditions of their factory's forming operations as closely as possible.

Commonly, such simulation tests check only one or a few of the performance requirements, so they must be used along with other tests to predict overall performance. For example, the shaker test for beer cans obviously can give no information on the important ability of the coating to withstand the pasteurization procedure, but separate simulation tests are available for testing pasteurization performance.

In most cases, simulation tests are designed for use in performance prediction rather than quality control. Generally, the sample preparation and testing time are too long for checking whether production batches are equal to the standard.

4.5.3. Empirical Tests

A range of empirical tests is used to test coatings. In some cases, the results can be used as part of the data to predict performance, particularly when comparing formulations that are very similar to standards with known performance. In most cases, empirical tests are more appropriate for quality control. Fairly commonly, they are subject to a considerable range of error, and many replicates should be run. They are frequently required in product specifications; but specifications are sometimes used not just for quality control, but also as requirements for new coatings.

We can only mention a few of the most widely used tests. ASTM annually publishes books describing tests; most of the tests of importance to the coatings field are in Volumes 06.01, 06.02, and 06.03: *Paint—Tests for Formulated Products and Applied Coatings*. While the books are published annually, most of the methods are not changed, although they are supposed to be reviewed every four years. Each method has a number, such as ASTM D-2832-92 (Reapproved 1994). The "92" means that the test was approved in 1992. In this particular case, the test was reevaluated and reapproved in 1994. If one finds a reference to a test as D-1876-71 and then refers to a 1997 ASTM book and finds it listed as D-1876-88, it means that the test was last reviewed, and possibly revised, in 1988. In general, one should use the new test method. Sometimes, a method will be designated as D-459a. The "a" means that there was a minor rewriting that did not change the basic method. Sometimes, people feel that because a test method is given an ASTM designation, it can be used not just for quality control, but also to predict performance. This assumption is frequently not the case. However, following ASTM procedures does provide the best chance for obtaining comparable test results. Some ASTM methods include precision statements, usually based on repeatability and reproducibility studies involving different laboratories. The precision statements should not be ignored; many people believe the tests are more precise than they proved to be in ASTM round robin tests [35].

An excellent reference book is the *Paint and Coating Testing Manual* (*Gardner-Sward Handbook*) [36]. It provides descriptions of a wide range of test methods and summaries of each major class of properties, as well as background information and comparisons of the utility of various tests. Hill provides an informative, brief discussion of the more important tests in his monograph on mechanical properties [4].

4.5.3.1. Hardness

Measuring hardness of a viscoelastic material is not as straightforward as it might appear. Units of hardness, force per unit area (N/M^2), are the same as those of modulus. When interpreting hardness and modulus data, it is important to know how the force was applied (tensile, shear, bending, or compressive?), what area (before, during, or after deformation?) was measured, the rate of application of stress, and the temperature. As reviewed by Guevin, three types of empirical hardness tests are used for coatings: indentation, scratch, and pendulum tests [37].

One indentation test is run with a Tukon Indentation Tester. An indenter with a diamond-shaped tip is pressed into the film with a fixed weight for a fixed time. The indenter is raised, and the indentation left in the film is measured with a calibrated microscope. The results are expressed in Knoop Hardness Numbers (KHN), which are related to the weight divided by the area of the indentation. Results are affected by film thickness; thin films on hard substrates give higher values than thicker films of the same coating. Meaningful results are obtained only with high T_g films. Intermediate T_g materials may have partial recovery of the indentation in the time needed to move the sample under the microscope and make the measurement. Low T_g films give considerable response variation, and the indenter may leave no indentation at all on a rubbery material. A careless tester may conclude that a rubbery material is very hard, even though it is obviously soft. The Tukon method is most appropriate for baked coatings, since they are more likely to have a T_g above the testing temperature.

A widely used scratch test is the pencil hardness test. Pencils with hardnesses varying from 6B (softest) to 9H (hardest) are available. The "lead"—actually graphite and clay—in the pencil is not sharpened as for writing, but is squared off by rubbing perpendicularly on abrasive paper. For the test, the pencil is held at a 45° angle to the panel and pushed forward with a pressure just below that which will break the lead. Hardness is reported as the grade of pencil that does not cause any marring of the surface. Experienced testers can reproduce results to ±1 hardness grade. Probably, the test reflects some combination of modulus, tensile strength, and adhesion.

A pendulum test, particularly used in the United States, is the Sward rocker. The rocker is a circular device made up of two rings joined with a glass level gauge and is weighted off center. The circumference of the rings rests on the panel. The rocker is rolled to a marked angle and released. The number of rocks (swings) required to dampen the motion down to a smaller fixed angle is determined. The rocker is calibrated to give 50 rocks (a hardness reading of 100) on polished plate glass. Hard coatings give higher readings (but less than 100) than soft coatings. Dampening is caused by rolling friction as well as by mechanical loss. The results are dependent on film thickness and surface smoothness. The Sward rocker is probably most useful for following the increase of hardness of a coated panel during drying of an ambient cure coating. It has only limited validity for comparing hardness of different coatings.

In Europe, Koenig and Persoz pendulums are used. In these tests, a pendulum makes contact with a coated panel through two steel balls. As the pendulum swings back and forth through a small angle, movement of the balls requires some deformation of the coating near the surface. Test results are reported as the time (in seconds) for the swing to be dampened from a higher to a lower angle (from the perpendicular)—from 6 to 3° in the case of the Koenig pendulum and from 12 to 4° in the Persoz test. Usually, harder coatings give longer times. However, soft, rubbery coatings may also give longer times. Based on the reasonable assumption that the main contribution to dampening the pendulum is absorption of mechanical energy by the film, these apparently conflicting results can be explained in terms of the loss modulus. As shown in Figure 4.7, loss modulus values are low in both the regions below and well above T_g and are highest in the transition region. Low loss modulus could account for longer dampening times for both soft, rubbery films, with T_g values below ambient temperature and hard films, with T_g values well above ambient temperature. This hypothesis predicts that dampening times for coatings in the transition region at ambient temperature may be very sensitive to temperature, since loss modulus goes through a maximum in this region.

4.5.3.2. Formability and Flexibility

One flexibility test is a *mandrel bend test*, in which a coated panel is bent around a rod or cone (the mandrel). The panel is bent with the coated side away from the mandrel. Any crack in the coating on the bend is reported as a failure. In the cylindrical mandrel test, a series of different mandrels is used, and the smallest diameter that permits a bend without failure is reported as the mandrel flexibility. In a *conical mandrel test*, one end of the panel is clamped, and a lever is used to bend the panel over a cone. The distance from the small end of the cone to the end of a crack is reported. This distance, which is proportional to the radius of curvature of the mandrel at that point, can be used to estimate the elongation-at-break [36]. Thicker films crack more easily than thinner films because the elongation at the same distance along the mandrel is greater. The bent edge should be inspected with a magnifying glass to see hairline cracks, and the panels should also be inspected the next day because sometimes cracks appear later. If adhesion is poor, the film comes loose from the panel. The severity of the test can be increased by testing at low temperature (by putting the panel and tester in a freezer before testing). The severity of the test is also affected by the rate of bending.

Another formability test, widely used for testing coil coatings, is the *T-bend test*. The coated panel is bent back on itself with the coating side out. If there is no crack at the edge, the result is reported as 0T. The zero means there is no additional layer of metal inside the bend. If the coating cracks, the panel is bent back on itself again. Repeated bends back over the original bend are made until the coating does not crack. The radius of curvature gets greater as the number of bends increases. The results are reported as 0T, 1T, 2T, 3T, and so forth, counting the number of thicknesses of metal inside the bend. The severity of the test is affected by temperature and rate of bending, and the panels should be re-examined after some time has elapsed.

4.5.3.3. Impact Resistance

Impact tests evaluate the ability of a coating to withstand extension without cracking when the deformation is applied rapidly. A weight is dropped down a guide tube onto a hemispherical indenter that rests on the coated panel. An opening opposite the indenter in the base support on which the panel rests permits deformation of the panel. If the coated side is up—that is, it is directly hit by the indenter—the test is called a *direct impact test*. If the back of the panel is up, the test is called a *reverse impact test*. The weight is dropped from greater and greater heights until the coating on the panel cracks. In the United States, the results are reported in in-lbs—that is, the number of inches the weight falls times its weight. In the most common apparatus, the maximum is 160 in-lb. Generally, reverse impact tests, in which the coating is extended, are more severe than direct impact tests in which the coating is compressed. The thickness, mechanical properties, and surface of the substrate substantially affect the results. If the substrate is thick enough, it is not distorted by the impact, and almost any coating will pass. While one laboratory may run the test on one metal substrate, another laboratory may use another substrate, and comparison of the results would be meaningless. Different lots of test panels of the same type can have subtle differences in their surfaces that can affect test results.

4.5.3.4. Abrasion Resistance

In the *Taber Abraser* test, abrasive wheels roll on a panel, creating a circular wear path. The test is continued until the coating is worn through. Results are reported as the number of rotations required to wear through one mil (25 μm) of coating. The results often do not correlate with field use tests. For example, in the Taber abrasion results for the four floor coatings described in Table 4.2, the "hard epoxy" coating had the poorest abrasion resistance in actual field use, but it exhibited the highest rating in the Taber test. Another example, in which correlation was poor is the beer can abrasion problem mentioned previously. On the other hand, some authors report that Taber tests correlate with visual observation of abrasion of clear coats on cars by automatic car washes [25], although others disagree. Generally, softer coatings tend to give poorer Taber Abraser results, probably because the abrasion disks rotate at a constant speed and, therefore, impart more energy to softer coatings. However, very soft coatings may clog the abrasive on the wheel and give spurious results.

Another abrasion test is the *falling sand test*, in which sand falls out of a hopper through a tube onto a coated panel held at a 45° angle to the stream of sand. The result is reported in liters of sand required to wear through a unit thickness of coating. A more sophisticated version of the falling sand test has been reported [21]. A gas blast erosion apparatus has been shown to give reproducible results. When the stream of particles is perpendicular to the surface, the erosion scar produced is circular; the radius of the scar provides an accurate measure of abrasion. The velocity of the particles and the angle of impact can be varied.

A further method of determining abrasion uses a ball rotating in a slurry of small abrasive particles to measure abrasive wear of a small area of coating in the upper 30 μm of a film with good reproducibilty [21].

4.5.3.5. Solvent Resistance

Solvent resistance is not a mechanical property, but it is included here because it is one of the properties that must be balanced with mechanical properties for many applications. It is also appropriate to mention here because resistance to swelling of cross-linked films is related to XLD, which affects many mechanical properties.

The most common test of solvent resistance is the methyl ethyl ketone (MEK) double rub test. This test can be done by rubbing a tissue soaked with MEK on the panel, but it is more conveniently done with a felt tip marker pen filled with solvent. The test can be mechanized so that there is one stroke back and forth (one double rub) on the film each second, and a timer can be used as a counter. A soft thermoplastic coating will rub off with very few rubs. In the case of thermosetting coatings, the number of rubs that the coating withstands increases as the degree of reaction increases. The test is sensitive to the development of low cross-link density, but insensitive to changes as the cross-link density gets higher. Usually, the test is stopped after 200 double rubs, and therefore, a series of highly cross-linked coatings may all be reported to give 200+ double rub resistance, even though there may be differences in the extent of cross-linking. At higher cross-link densities, DMA or solvent swelling can be used to determine the extent of cross-linking, as discussed in Section 4.2.

Another solvent resistance test is to expose the coating to the solvent for a certain period of time—say 15 minutes—and then carry out a pencil hardness test on the exposed area. For example, a test for aircraft top coats specifies that the coating shall not lose more than two pencil hardness units after exposure to hydraulic fluid for 15 minutes.

GENERAL REFERENCES

L. W. Hill, *Mechanical Properties of Coatings*, Federation of Societies for Coatings Technology, Blue Bell, PA, 1987.

D. J. Skrovanek and C. K. Schoff, *Prog. Org. Coat.*, **16**, 135 (1988).

J. V. Koleske, Ed., *Paint and Coating Testing Manual* (*Gardner-Sward Handbook*), 14th ed., ASTM, Philadelphia, 1995. (Particularly, L. W. Hill, "Dynamic Mechanical and Tensile Properties," Chapter 46, pp. 534–546.)

J. E. Mark, Ed., *Physical Properties of Polymers Handbook*, American Institute of Physics, Woodbury, New York, 1996. (Particularly, W. Brostow, J. Kubat, and M. M. Kubat, "Mechanical Properties," pp. 313-334.)

REFERENCES

1. R. A. Dickie, *J. Coat. Technol.*, **64** (809), 61 (1992).
2. J. W. Martin, S. C. Saunders, F. L. Floyd, and J. P. Wineburg, *Methodologies for Predicting Service Lives of Coating Systems*, Federation of Societies for Coatings Technology, Blue Bell, PA, 1996.
3. L. W. Hill, *Prog. Org. Coat.*, **5**, 277 (1977).
4. L. W. Hill, *Mechanical Properties of Coatings*, Federation of Societies for Coatings Technology, Blue Bell, PA, 1987.
5. D. J. Skrovanek and C. K. Schoff, *Prog. Org. Coat.*, **16**, 135 (1988).
6. L. W. Hill, *J. Coat. Technol.*, **64** (808), 29 (1992).
7. S. J. Wu, *J. Appl. Polym. Sci.*, **20**, 327 (1976).
8. H. Stutz, K.-H. Illers, and J. Mertes, *J. Polym. Sci., B: Polym. Phys.*, **28**, 1483 (1990).
9. L. W. Hill, *J. Coat. Technol.*, **59** (751), 63 (1987).
10. L. W. Hill, *Prog. Org. Coat.*, **31**, 235 (1997).
11. D. R. Bauer, *J. Coat. Technol.*, **60** (758), 53 (1988).
12. D. J. Skrovanek, *Prog. Org. Coat.*, **18**, 89 (1990).
13. M. B. Roller, *J. Coat. Technol.*, **54** (691), 33 (1982).
14. J. R. Fried, "Sub-T_g Transitions," in J. E. Mark, Ed., *Physical Properties of Polymers Handbook*, American Institute of Physics, Woodbury, New York, 1996, pp. 161–175.
15. D. Y. Perera and P. Schutyser, *FATIPEC Congress Book*, Vol. I, 1994, p. 25.
16. L. W. Hill, H. M. Korzeniowski, M. Ojunga-Andrew, and R. C. Wilson, *Prog. Org. Coat.*, **24**, 147 (1994).
17. P. J. Greidanus, *FATIPEC Congress Book*, Vol. I, 1988, p. 485.
18. R. A. Dickie, D. R. Bauer, S. M. Ward, and D. A. Wagner, *Prog. Org. Coat.*, **31**, 209 (1997).
19. R. M. Evans, in R. R. Myers and J. S. Long, Eds., *Treatise on Coatings*, Vol. 2, Part I, Marcel Dekker, New York, 1969, pp. 13–190.
20. R. M. Evans and J. Fogel, *J. Coat. Technol.*, **47** (639), 50 (1977).
21. K. L. Rutherford, R. I. Trezona, A. C. Ramamurthy, and I. M. Hutchings, *Wear*, **203–204**, 325 (1997).
22. D. Kotnarowska, *Prog. Org. Coat.*, **31** 325 (1997).
23. L. H. Lee, *Polym. Mat. Sci. Eng.*, **50**, 65 (1984).
24. T. Hamada, H. Kanai, T. Koike, and M. Fuda, *Prog. Org. Coat.* **30**, 271 (1997).
25. P. Betz and A. Bartelt, *Prog. Org. Coat.*, **22**, 27 (1993).
26. F. N. Jones, W. Shen, S. M. Smith, Z. Huang, and R. A. Ryntz, *Prog. Org. Coat.* **34**, 119, (1998).
27. B. V. Gregorovich and I. Hazan, *Prog. Org. Coat.*, **24**, 131 (1994).
28. J. L. Courter, *J. Coat. Technol.*, **69** (866), 57 (1997).
29. F. Fink, W. Heilen, R. Berger, and J. Adams, *J. Coat. Technol.*, **62** (791), 47 (1990).
30. L. W. Hill, "Dynamic Mechanical and Tensile Properties," in J. V. Koleske, Ed., *Paint and Coating Testing Manual*, 14th ed. ASTM, Philadelphia, 1995, p. 534.
31. J. Lange, H. Andersson, J. A. E. Manson, and A. Hult, *Surface Coat. Intl.*, **79** 486 (1996).
32. Th. Frey, K.-H. Grosse-Brinkhaus, and U. Rockrath, *Prog. Org. Coat.*, **27**, 59 (1996).

33. G. A. Vandermeersche, *Closed Loop*, **3**, April (1981).

34. R. A. Ryntz, A. C. Ramamurthy, and J. W. Holubka, *J. Coat. Technol.*, **67** (842), 23 (1995).

35. R. D. Athey, Jr., *Amer. Paint Coat. J.*, December 7, 38 (1992).

36. J. V. Koleske, *Paint and Coating Testing Manual*, 14th ed., ASTM, Philadelphia, 1995.

37. P. R. Guevin, Jr., *J. Coat. Technol.*, **67** (840), 61 (1995).

CHAPTER 5

Exterior Durability

Exterior durability of coatings refers to their resistance to change during outdooor exposure; such changes include changes of modulus, loss of strength, embrittlement, discoloration, loss of adhesion, chalking, loss of gloss, and environmental etching. Thus, both aesthetic and functional properties are involved. The terms *outdoor durability* and *weatherability* are also used. Corrosion protection by coatings is discussed in Chapter 7.

The most common chemical processes leading to degradation of coatings are photo-initiated oxidation and hydrolysis resulting from exposure to sunlight, air, and water. These processes are interrelated, including enhanced photoxidative degradation in high humidity and enhanced hydrolytic degradation during photoexposure. Furthermore, both processes are accelerated by higher temperatures. Hydrolytic degradation may be enhanced by exposure to acid, as from acid rain. Other atmospheric degradants include ozone and oxides of nitrogen and sulfur. Changes in temperature and humidity may result in cracking, which arises from the expansion and contraction of coatings or substrates. Rates at which these processes occur vary, depending on exposure site(s), time of year, coating composition, and substrate.

5.1. PHOTOINITIATED OXIDATIVE DEGRADATION

Coatings formulated for exterior durability should exclude or minimize resin components that absorb UV radiation at wavelengths longer than 290 nm or that are readily oxidized. Photoinitiated oxidation of polymers by a chain reaction is outlined in Scheme 5.1. Absorption of UV by a polymer (P) or other coating component produces highly energetic photoexcited states (P*) that undergo bond cleavage to yield free radicals (P·). Free radicals undergo a chain reaction with O_2 (autoxidation), leading to polymer degradation. Hydroperoxides (POOH) and peroxides (POOP) are unstable products of photoinitiated oxidation; they dissociate with sunlight and moderate heat to yield alkoxy (PO·) and hydroxy (HO·) radicals, so the degradation reactions are autocatalytic. These radicals are highly reactive toward hydrogen abstraction and yield polymer radicals (P·), which enter into the propagation stage of polymer degradation. Tertiary alkoxy radicals dissociate into ketones and a lower molecular weight polymer radical (P'·), resulting in scission of the polymer. As shown in Scheme 5.1, chain propagation leading to oxidative degradation proceeds by hydrogen abstraction from the polymer autocatalytically.

Functional groups in a coating that promote hydrogen abstraction should be minimized. A general ordering of $\underline{CH}$ groups having decreasing vulnerability to free radical abstraction is:

$$\text{Amines:} -\underline{CH}_2-NR_2 >$$

$$\text{Diallyl:} -CH=CH-\underline{CH}_2-CH=CH \longrightarrow >$$

$$\text{Ethers, alcohols:} -\underline{CH}_2-O-R(H) >$$

$$\text{Benzyl, allyl, tertiary alkyl:} Ph-\underline{CH}_2-, -CH=CH-\underline{CH}_2-, R_3\underline{CH}$$

$$\text{Urethanes:} -\underline{CH}_2-NH\text{-}CO\text{-}OR >$$

$$\text{Esters:} -\underline{CH}_2-CO\text{-}O\text{-}\underline{CH}_2- >$$

$$\text{Secondary, primary alkyl:} -\underline{CH}_2- > \underline{CH}_3,$$

$$\text{Methylsiloxanes:} -Si(\underline{CH}_3)_2-O-$$

None of these functional groups absorbs sunlight directly. In the absence of absorbing aromatic groups, absorption of sunlight occurs primarily by inadvertently present peroxide and ketone groups, both of which absorb UV above 290 nm. Essentially, all resins used in coatings contain some hydroperoxides, as do most organic substances. Peroxides, ketones, and aldehydes are also formed in photoinitiated oxidation, see Scheme 5.1. Photolysis of

Scheme 5.1

Initiation

$$\text{Polymer(P)} \xrightarrow{\text{Sunlight}} P^* \tag{5.1}$$

$$P^* \longrightarrow \text{Free Radicals (P·)} \tag{5.2}$$

Propagation

$$P· + O_2 \longrightarrow POO· \tag{5.3}$$

$$POO· + \text{Polymer(P--H)} \longrightarrow POOH + P· \tag{5.4}$$

$$PO·(·OH) + \text{Polymer(P--H)} \longrightarrow POH(H_2O) + P·$$

Chain Termination

$$2POO· \longrightarrow POOP + O_2$$

$$2P· \longrightarrow P\text{--}P \text{ or Disproportionation products}$$

$$POO· + P· \longrightarrow POOP \text{ or Disproportionation products}$$

$$2POO· \longrightarrow \text{Ketones (Aldehydes)} + \text{Alcohols}$$

Polymer Scission

$$PO· \longrightarrow \text{Ketones} + P'·$$

Autocatalysis

$$POOH(P) \xrightarrow{\text{Sunlight}} PO· + ·OH(P)$$

Scheme 5.2

Oxidation of Aldehydes and Ketones (Peracid Formation)

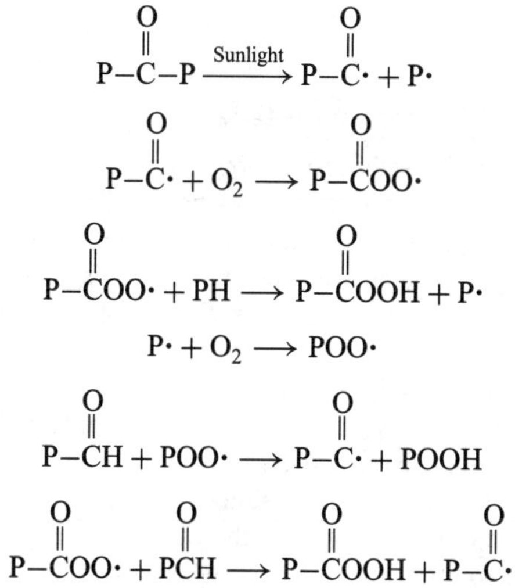

$$
\overset{O}{\underset{\|}{P-C-P}} \xrightarrow{\text{Sunlight}} \overset{O}{\underset{\|}{P-C\cdot}} + P\cdot
$$

$$
\overset{O}{\underset{\|}{P-C\cdot}} + O_2 \longrightarrow \overset{O}{\underset{\|}{P-COO\cdot}}
$$

$$
\overset{O}{\underset{\|}{P-COO\cdot}} + PH \longrightarrow \overset{O}{\underset{\|}{P-COOH}} + P\cdot
$$

$$
P\cdot + O_2 \longrightarrow POO\cdot
$$

$$
\overset{O}{\underset{\|}{P-CH}} + POO\cdot \longrightarrow \overset{O}{\underset{\|}{P-C\cdot}} + POOH
$$

$$
\overset{O}{\underset{\|}{P-COO\cdot}} + \overset{O}{\underset{\|}{PCH}} \longrightarrow \overset{O}{\underset{\|}{P-COOH}} + \overset{O}{\underset{\|}{P-C\cdot}}
$$

peroxides and ketones is shown in Scheme 5.1 and Scheme 5.2, respectively. Photoxidation of aldehydes and ketones, as shown in Scheme 5.2, yields peracids, which are strong organic oxidants. Peracids may play a significant role in oxidative degradation.

Methyl-substituted silicones and silicone-modified resins exhibit high photoxidation stability, generally in proportion to the silicone content. The excellent exterior durability of fluorinated resins may be attributed, at least in part, to the absence (or reduced level) of C–H groups.

Aromatic groups with directly attached heteroatoms, as found in aromatic urethanes (Ar–NH–CO–OR) and bisphenol A (BPA) epoxies (Ar–O–R), absorb UV above 290 nm and undergo direct photocleavage to yield free radicals that can participate in oxidative degradation. Coatings made using aromatic isocyanates yellow badly after only short exposures to UV radiation. Coatings based on BPA epoxies generally chalk rapidly on exposure outdoors. Since ketones absorb UV, they should be avoided. When acrylic resins are polymerized in ketone solvents such as methyl amyl ketone, ketone groups can be incorporated in the resin by chain transfer [1]; it is preferable to use polymerization solvents such as esters or toluene to avoid introduction of ketone groups.

Highly chlorinated resins such as vinyl chloride copolymers, vinylidene chloride copolymers, and chlorinated rubber degrade by autocatalytic dehydrochlorination on exposure to either heat or UV. They must be formulated with stabilizers, as discussed in Section 5.5.

5.2. PHOTOSTABILIZATION

Defenses against photoinitiated oxidative degradation include: UV absorbers to reduce UV absorption by the polymer (Scheme 5.1, Eq. 5.1), excited state quenchers to compete with bond cleavage of P* (Scheme 5.1, Eq. 5.2), and antioxidants to reduce hydroperoxide decomposition and oxidative degradation (Scheme 5.1, Eqs. 5.3 and 5.4). Review articles and books on photostabilization and thermal stabilization of coatings, including degradative pathways, are available [2].

5.2.1. UV Absorbers and Excited State Quenchers

Important characteristics of UV stabilizers—both absorbers (A) and quenchers (Q)—are photostability and chemical and physical permanence. Photostability requires that the photoexcited stabilizer (A* or Q*) can return to the ground state by converting the UV energy into thermal energy, as shown in Scheme 5.3. This process generally occurs by reversible intramolecular hydrogen transfer or E-Z (cis-trans) isomerization of double bonds.

An important feature of excited state quenchers is their effective quenching volume, within which quenching of photoexcited polymers occurs efficiently. Effective quenching volume depends on the mechanism of energy transfer. These considerations have been discussed and related to the prospects for stabilizing aromatic polymers [3]. Stabilization by UV absorbers requires strong absorption in the wavelength region in which the polymer and/or trace impurities also absorb. Stabilizers, which function as UV absorbers in some coatings, may perform as excited state quenchers in other coatings. The most effective stabilizers probably function by both roles.

One cannot eliminate UV absorption by the resin by adding a UV absorber; it merely reduces absorption by the binder to slow the rate of photodegradation reactions. Since absorption increases as the path length increases, UV absorbers are most effective in protecting the lower parts of a film or substrate—for example, a base coat, wood, or plastic under a clear top coat containing an absorber—and least effective in protecting the layer at the air interface. It follows that the thickness of a clear topcoat can be an important variable affecting the protection of base coats or plastics under them, since thicker films transmit less radiation. A critical consideration in the design and selection of UV absorbers is their absorption spectra. In general terms, one would like to have very high absorption of UV radiation, from 290 through 380 nm. To avoid color effects by an absorber, ideally, there would be no absorption above 380 nm.

Scheme 5.3

UV Absorber (A):

$$A \xrightarrow{\text{Sunlight}} A*$$

$$A* \longrightarrow A + \text{Heat}$$

Overall: Radiation $\longrightarrow$ Heat

Excited State Quencher (Q):

$$P \xrightarrow{\text{Sunlight}} P*$$

$$P* + Q \longrightarrow P + Q*$$

$$Q* \longrightarrow Q + \text{Heat}$$

Overall : Radiation $\longrightarrow$ Heat

Substituted 2-hydroxybenzophenones, 2-(2-hydroxyphenyl)-2*H*-benztriazoles, 2-(2-hydroxyphenyl)-4,6-phenyl-1,3,5-triazines, benzylidenemalonates, and oxalanilides are classes of UV stabilizers. Specific members of these classes may act as UV absorbers, as excited state quenchers, or as both, but they are frequently called "UV absorbers" (UVA) in the coatings literature.

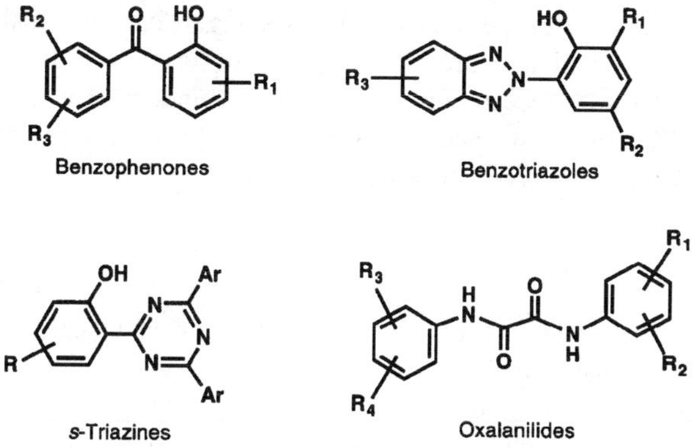

Benzophenones   Benzotriazoles

s-Triazines   Oxalanilides

These UV stabilizers convert UV energy into heat by intramolecular hydrogen transfer or cis-trans isomerization. For example, UV absorption or excited state quenching by a 2-hydroxyphenyl-substituted stabilizer yields a photoexcited state that converts the excess electronic energy into chemical energy by undergoing intramolecular hydrogen transfer to yield an unstable intermediate. The unstable intermediate spontaneously undergoes reverse hydrogen transfer to regenerate the UV stabilizer with conversion of the chemical energy into heat. The process is illustrated with 2-hydroxybenzophenone.

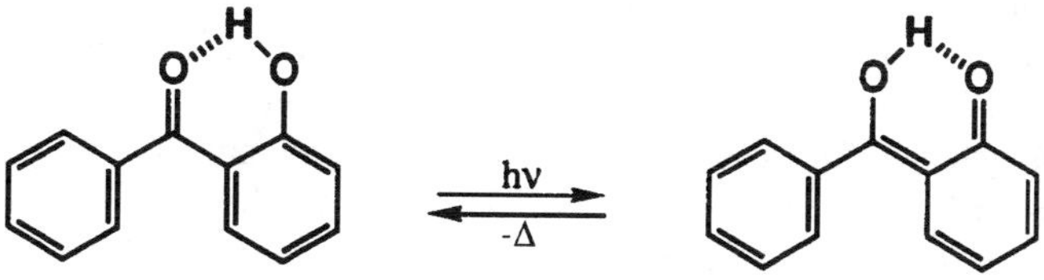

The UV stabilizer must be soluble in the coating film; several grades of the various stabilizers are available with different substituents on the aromatic rings that provide for solubility in different polymer systems. Commonly, the stabilizer is added to the top coat of a multicoat system. However, especially in baking systems, migration may result in the stabilizer being distributed through the whole coating, reducing the concentration in the top coat. This effect has been demonstrated by analyzing sections through a film of clear coat-base coat automotive finishes [4]. In one combination of coatings, in which UV stabilizer was added only to the clear coat, the content through the whole film, both clear coat and base coat, was essentially uniform. In a second case, with a different type of base coat, a major fraction of the stabilizer stayed in the clear coat. In coatings for plastics, it has been shown that stabilizers can migrate into the plastic [5].

A critical requirement of a UV stabilizer is permanence. Loss of stabilizer can result by two mechanisms. There can be physical loss by vaporization, leaching, or migration and/or there can be chemical loss by deterioration of absorbance by photochemical reactions of the stabilizer. If a UV stabilizer has even a small vapor pressure, it slowly volatilizes from the surface over the long term period for which durability is desired. Analysis of one UV stabilizer as a function of depth into the film initially and after one year exterior exposure [on a Florida black box (see Section 5.6.1)] showed significant losses, particularly near the surface of the film [4].

Hydroxyphenyltriazines have very low vapor pressures. Longer term physical permanence may be achieved by using oligomeric photostabilizers. Polymer-bound stabilizers can also be used assuring physical permanence. UV stabilizers can chemically degrade and lose effectiveness. The degradation presumably proceeds via free radical intermediates, and therefore, degradation proceeds most rapidly in binders that are most readily photodegraded by a free radical mechanism. Reference [6] provides an extensive discussion of the complex factors involved in permanence.

5.2.2. Antioxidants

Antioxidants may be classified into two groups: preventive and chain-breaking antioxidants. Preventive antioxidants include peroxide decomposers, which reduce hydroperoxides to alcohols and become oxidized into harmless products. Examples of peroxide decomposers are sulfides and phosphites that are initially oxidized to sulfoxides and phosphates, respectively, as shown in Eqs. 5.5 and 5.6 for dilauryl thiodipropionate (LTDP) and triphenylphosphite. Further reactions may occur.

$$POOH + S(CH_2CH_2CO-OC_{12}H_{25})_2 \rightarrow POH + O{=}S(CH_2CH_2CO-OC_{12}H_{25})_2 \qquad (5.5)$$
$$LTDP$$

$$POOH + (PhO)_3P \rightarrow POH + (PhO)_3P{=}O \qquad (5.6)$$

Metal complexing agents are a second type of preventive antioxidant. They tie up transition metal ions that are present as contaminants and that catalyze the undesired conversion of hydroperoxides into peroxy and alkoxy radicals by redox reactions, as exemplified in Scheme 5.4. The resulting peroxy and alkoxy radicals are undesirable because they promote oxidative degradation, as shown in Scheme 5.1. For oxidative cure of drying oils and alkyds (see Section 14.2.2), it is desired to accelerate autooxidation; transition metal ions are deliberately added as catalysts (driers). Thus, the reactions in Scheme 5.4 are involved in both oxidative cross-linking and oxidative degradation. It follows that drier concentrations should be minimized to reduce subsequent degradation during exterior exposure.

The tetrafunctional bidentate imine derived from o-hydroxybenzaldehyde (salicylaldehyde) and tetraaminomethylmethane is reported to complex a large number of transition metal ions effectively, including Co, Cu, Fe, Mn, and Ni (represented by M^{n+} in complex **1**) [7].

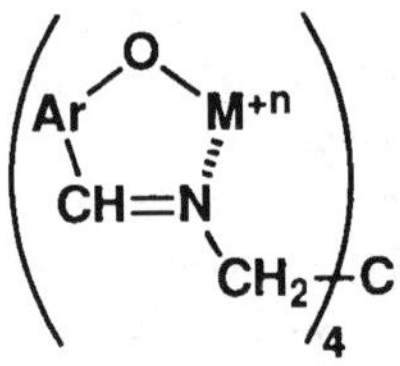

(Ar = 1,2-disubstituted benzene)

1

Scheme 5.4

$$POOH + Co^{2+} \longrightarrow POO\cdot + H^+ + Co^{3+}$$

$$POOH + Co^{3+} \longrightarrow PO\cdot + HO^- + Co^{2+}$$

$$\text{Overall}: 2POOH \longrightarrow POO\cdot + PO\cdot + H_2O$$

Chain-breaking antioxidants function by interfering directly with chain propagation steps of autoxidation, shown in Scheme 5.1. An example is a hindered phenol, 2,2'-methylenebis(4-methyl-6-tert-butylphenol), which reacts with peroxy radicals in competition with hydrogen abstraction from a polymer (PH) to yield a resonance-stabilized, less reactive phenoxy radical, as shown in Eq. 5.7.

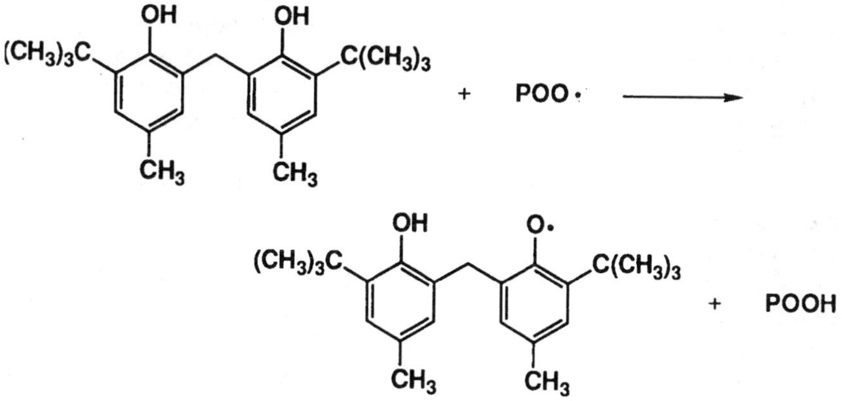

$$(5.7)$$

Note that in Eq. 5.7, hydroperoxides are generated; this is a basis for synergistic stabilization of polymers by a combination of peroxide decomposers and chain-breaking phenolic antioxidants. Synergistic stabilization means that the combination of stabilizers is more effective than the additive effect of each stabilizer by itself.

5.2.3. Hindered Amine Light Stabilizers

Hindered amine light stabilizers (HALS) are amines with two methyl groups on each of two alpha carbons; most are derivatives of 2,2,6,6-tetramethylpiperidine, as shown in general formula **2**. They are reported to function both as chain-breaking antioxidants [8] and transition metal complexing agents [9]. The former function appears to be most important. Note that the 2,2,6,6-methyl groups prevent oxidation of the ring carbons attached to the nitrogen.

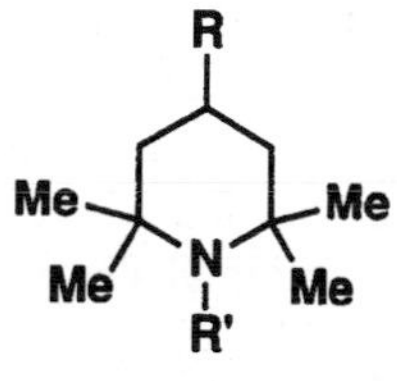

(R'= H, alkyl, alkanoyl, alkoxy)

2

Scheme 5.5

$$R_2N-R' + O_2 \xrightarrow{\text{UV}} R_2NO\cdot + \text{other products}$$

$$2R_2NO\cdot + 2P\cdot \longrightarrow R_2NOH + R_2NOP + P(\text{minus H}) \qquad (5.8)$$

$$R_2NOH(P) + POO\cdot \longrightarrow R_2NO\cdot + POOH(P) \qquad (5.9)$$

HALS derivatives undergo photoxidative conversion into nitroxyl radicals ($R_2NO\cdot$) that react with carbon-centered radicals by disproportionation and combination to yield corresponding hydroxylamines and ethers, respectively, as shown in Eq. 5.8 in Scheme 5.5. The hydroxylamines and ethers, in turn, react with peroxy radicals to regenerate nitroxyl radicals, shown in Eq. 5.9 in Scheme 5.5. In this manner, HALS derivatives interfere with propagation steps involving both carbon-centered and peroxy radicals in autoxidation. (See Scheme 5.1.) In contrast to nitroxyl radicals, hindered phenols do not react with carbon-centered radicals. The chain breaking antioxidant activity of hindered phenols is limited to reaction with peroxy radicals, and oxygen is required as a costabilizer.

To be effective, HALS derivatives must undergo rapid photoxidation to form nitroxyl free radicals on exterior exposure of the coating. In a photoexposed film, only a small fraction (about 1%) is present as the nitroxyl radical, the major storage components being the corresponding hydroxylamine (R_2NOH) and ethers (R_2NOP). Continued stabilization requires the presence of the nitroxyl radical, and its disappearance is followed shortly by rapid polymer degradation. Probably, the ultimate demise of HALS occurs, at least in part, by oxidation of the nitroxyl radical, accompanied by opening of the piperidine ring. Transition metal ions and peracids are potential oxidants for this process.

A variety of HALS compounds is available. The "R" in general formula **2** is often a diester group that joins two piperidine rings; this increases the molecular weight, decreasing volatility. The first commercial HALS compounds, still used to a degree, had $R' = H$. Later versions with $R' = $ alkyl exhibit better long-term stability. Both of these types are basic and interfere with acid catalyzed cross-linking reactions, such as those involving melamine-formaldehyde resins. Alkanoyl HALS compounds with $R' = C(=O)R''$ are not basic, but their initial reaction to form nitroxyl free radicals is slower (see Section 5.6.3) [2(c)]. More recently, hydroxylamine ethers ($R' = OR''$) have gained wide acceptance. The octyl ether provides a HALS with low basicity that converts rapidly to nitroxyl free radicals [10]. HALS compounds, especially with $R' = H$, can accelerate degradation of polycarbonate plastics, possibly by base catalyzed hydrolysis.

Combinations of UV absorbers and HALS compounds can act synergistically [2(c),11]. The UV absorber reduces the rate of generation of radicals, whereas the HALS compound reduces the rate of oxidative degradation by the radicals. A further factor is that UV absorbers are inefficient at protecting the outer surface of a film; in contrast, HALS compounds can effectively scavenge free radicals at the surface. Analysis of films after exterior exposure shows significant amounts of HALS derivatives remain after two years black box Florida exterior exposure (see Section 5.6.1) [4].

The N-2,2,6,6-pentamethylhydroxypiperidinyl diester of 4-methoxyphenylmethylene-malonic acid combines the UV absorber function of a benzylidene malonic ester and the antioxidant properties of HALS derivatives [12]. Furthermore, on exposure of films, this HALS derivative grafts onto the polymer chains, eliminating volatility and migration losses.

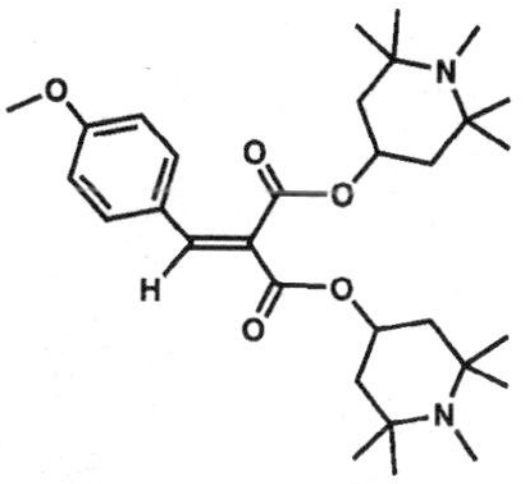

5.2.4. Pigmentation Effects

Many pigments absorb UV radiation. The strongest UV absorber known is fine particle size carbon black. Many carbon blacks have structures with multiple aromatic rings and, in some cases, phenol groups on the pigment surface. Such black pigments are both UV absorbers and antioxidants. Enhanced exterior durability is obtained with carbon black pigmented coatings. When Henry Ford said, "The customer can have any color car he wants, as long as it is black," he was not just being crotchety. Black was by far the most durable paint available at the time.

Other pigments absorb UV radiation to varying degrees. For example, 50 μm coatings pigmented with fine particle size, transparent iron oxide pigments absorb virtually all radiation below about 420 nm [13]. This strong absorption is particularly useful in stains for use over wood, since the pigmented transparent coating protects the wood from photodegradation.

Rutile TiO_2 absorbs UV strongly. Absorption is a function not only of wavelength and concentration, but also of particle size of the pigment [14]. Optimum particle size for absorption of UV by rutile TiO_2 increases from 0.05 μm for 300 nm radiation to 0.12 μm for 400 nm radiation. This size is smaller than the optimum particle size of 0.19 μm for hiding. (See Section 18.2.3.) Rutile TiO_2 white pigment with an average particle size of 0.23 μm still absorbs UV strongly. Anatase TiO_2 also absorbs UV strongly, although not as strongly as rutile TiO_2 in the near UV. Thus, TiO_2, especially rutile TiO_2, functions as a UV absorber in coatings. However, TiO_2 can accelerate photodegradation of films on exterior exposure, causing *chalking* of coatings—that is, degradation of the organic binder and exposure of unbound pigment particles on the film surface that rub off easily, like chalk off a blackboard. Degradation of the binder is enhanced by interaction of photo-excited TiO_2 with oxygen and water to yield oxidants, as shown in Scheme 5.6 [15]. Photoexcitation of TiO_2 results in promoting a low energy valence band electron into the higher energy conduction band, creating a separated electron (e)/hole(p) pair, signified by

Scheme 5.6

$$TiO_2 + Light \longrightarrow TiO_2^*(e/p)$$

$$TiO_2^*(e/p) + O_2 \longrightarrow TiO_2(p) + O_2\cdot^-$$

$$TiO_2(p) + H_2O \longrightarrow TiO_2 + H^+ + HO\cdot$$

$$H^+ + O_2\cdot^- \longrightarrow HOO\cdot$$

$$2HOO\cdot \longrightarrow H_2O_2 + O_2$$

$$TiO_2^*(e/p) + H_2O_2 \longrightarrow TiO_2 + 2\ HO\cdot$$

TiO_2^* (e/p) in Scheme 5.6. Electron capture by O_2 (reduction) and hole capture by H_2O (oxidation) result in regeneration of ground-state TiO_2 and lead to hydroperoxy and hydroxy radicals that can participate in oxidative degradation, as shown in Scheme 5.1. Anatase TiO_2 is more active in promoting oxidative degradation than rutile TiO_2.

The photoactivity of TiO_2 pigments is reduced by coating the pigment particles with a thin layer of silica and/or alumina (see Section 19.1.1) to form a barrier layer against the electron transfer redox reactions. Treated rutile pigments are available that accelerate chalking to very minor extents, and these grades are used in exterior coatings. A laboratory test has been developed to compare the photoreactivity of various grades of TiO_2 [16]. Various stabilizing additives, including HALS, have also been reported [17].

Chalking of exterior house paints eventually leads to complete erosion of the film. Chalking also leads to reduction in gloss, since the film becomes rougher. However, loss of gloss does not necessarily correlate with ease of chalking [18]. It was shown that initial gloss loss in some TiO_2 pigmented coatings resulted from film shrinkage, which, in some cases, was greater with more resistant grades of TiO_2. In paints containing both TiO_2 and color pigments, chalking results in color changes as a result of the gloss loss; the higher surface reflectance of the low gloss films gives an appearance of weaker colors.

Formulating coatings for exterior use can be complicated by pigment-binder interactions. In some cases, a pigment shows excellent color retention after exterior exposure when formulated with one class of resins, but poor durability when formulated with another class. For example, thioindigo maroon had excellent color retention when used in nitrocellulose lacquers, but poor color stability when used in acrylic lacquers. While one can use experience in other systems as an initial basis for selecting pigments for use with a new class of resins, field tests are required to assure that each combination of pigments and resins is suitable.

5.3. DEGRADATION OF CHLORINATED RESINS

Highly chlorinated resins undergo dehydrochlorination on exposure to either heat or UV. The reactions are autocatalytic; the ultimate products are conjugated polyenes. As the number of conjugated double bonds in series increases, the polymer progressively discolors, finally becoming black. The resulting highly unsaturated polymer undergoes autoxidation, resulting in cross-linking and embrittlement.

The mechanism of degradation of poly(vinyl chloride) has been extensively studied. Dehydrochlorination is promoted when chlorine is situated on a tertiary carbon or on a carbon allylic to a double bond. It has been proposed that at least one major weak point results from addition of a vinyl chloride monomer in a head-to-head fashion, as in Eq. 5.10 in Scheme 5.7, to the growing polymer chain, followed by chain transfer of Cl to monomer as in Eq. 5.11 in Scheme 5.7 [19]. The resulting allylic chloride is highly susceptible to dehydrochlorination, which generates a new allylic chloride with two conjugated double bonds. Progressive dehydrochlorination is favored because the growing number of conjugated double bonds increases the lability of allylic chlorides down the chain.

A variety of stabilizing agents is used. Since dehydrochlorination is catalyzed by hydrogen chloride, HCl traps, such as epoxy compounds and basic pigments, are useful. Diels-Alder dienophiles can act as stabilizers; the Diels-Alder addition breaks up the chain of conjugated double bonds. Dibutyltin diesters are effective stabilizers. It is proposed that the activated chlorine atoms are interchanged with ester groups of the tin compounds to

Scheme 5.7

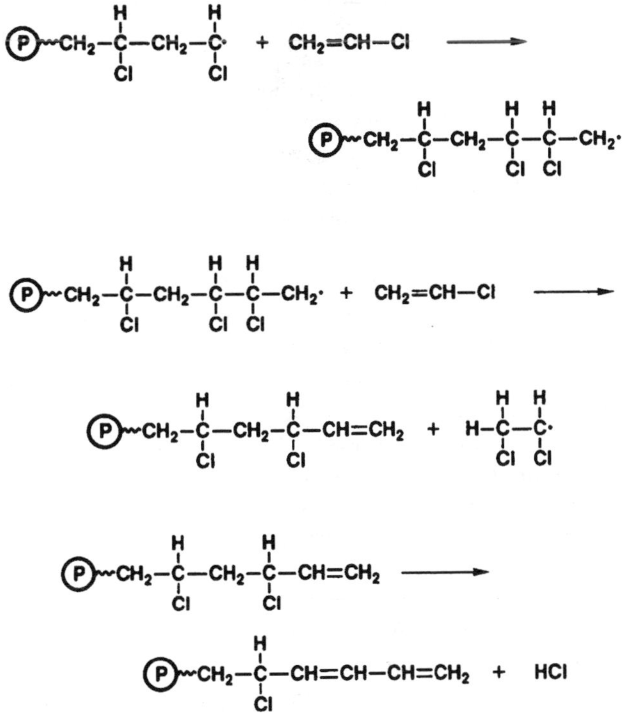

$$P\text{-}CH_2\text{-}\underset{\underset{Cl}{|}}{\overset{\overset{H}{|}}{C}}\text{-}CH_2\text{-}\underset{\underset{Cl}{|}}{\overset{\overset{H}{|}}{C}}\cdot \quad + \quad CH_2=CH\text{-}Cl \quad \longrightarrow$$

(5.10)

$$P\text{-}CH_2\text{-}\underset{\underset{Cl}{|}}{\overset{\overset{H}{|}}{C}}\text{-}CH_2\text{-}\underset{\underset{Cl}{|}}{\overset{\overset{H}{|}}{C}}\text{-}\underset{\underset{Cl}{|}}{\overset{\overset{H}{|}}{C}}\text{-}CH_2\cdot$$

$$P\text{-}CH_2\text{-}\underset{\underset{Cl}{|}}{\overset{\overset{H}{|}}{C}}\text{-}CH_2\text{-}\underset{\underset{Cl}{|}}{\overset{\overset{H}{|}}{C}}\text{-}\underset{\underset{Cl}{|}}{\overset{\overset{H}{|}}{C}}\text{-}CH_2\cdot \quad + \quad CH_2=CH\text{-}Cl \quad \longrightarrow$$

(5.11)

$$P\text{-}CH_2\text{-}\underset{\underset{Cl}{|}}{\overset{\overset{H}{|}}{C}}\text{-}CH_2\text{-}\underset{\underset{Cl}{|}}{\overset{\overset{H}{|}}{C}}\text{-}CH=CH_2 \quad + \quad H\text{-}\underset{\underset{Cl}{|}}{\overset{\overset{H}{|}}{C}}\text{-}\underset{\underset{Cl}{|}}{\overset{\overset{H}{|}}{C}}\cdot$$

$$P\text{-}CH_2\text{-}\underset{\underset{Cl}{|}}{\overset{\overset{H}{|}}{C}}\text{-}CH_2\text{-}\underset{\underset{Cl}{|}}{\overset{\overset{H}{|}}{C}}\text{-}CH=CH_2 \quad \longrightarrow$$

$$P\text{-}CH_2\text{-}\underset{\underset{Cl}{|}}{\overset{\overset{H}{|}}{C}}\text{-}CH=CH\text{-}CH=CH_2 \quad + \quad HCl$$

form the more stable ester-substituted polymer molecules. Dibutyltin maleate is a particularly effective stabilizer, since it acts both as an ester interchange compound and a dienophile. Barium, cadmium, and strontium soaps act as stabilizers. Choice of stabilizer combinations can be system specific, especially depending on whether stabilization is needed against heat, UV, or both. In the case of UV stabilization, UV absorbers can further enhance stability.

5.4. HYDROLYTIC DEGRADATION

A general ordering of functional groups subject to hydrolysis is: esters > ureas > urethanes ≫ ethers, although activated ethers, such as in melamine-formaldehyde (MF) (see Chapter 9) cross-linked hydroxy-functional resins, are more reactive than ureas and urethanes. The tendency of each type of group to hydrolyze can be reduced by steric hindrance—for example, by placement of alkyl groups in the vicinity of the susceptible groups, such as esters. (See Section 13.1.1.) The alkyl groups may also reduce hydrolysis by decreasing solubility of water, hence reducing rates of hydrolysis. Studies have shown that the lower the water solubility of the diacid or diol used to make a polyester, the greater the resistance to hydrolysis [20]. Rates of hydrolysis are also influenced by neighboring groups. For example, phthalate half esters, in which the groups are ortho, are more readily hydrolyzed under acidic conditions (often encountered during exterior exposure), than isophthalate half esters, in which the groups are meta. (See Section 13.1.2.) Hydrolysis of polyesters results in backbone degradation. On the other hand, the backbones of (meth)acrylic resins are completely resistant to hydrolysis, since the linkages are carbon-carbon bonds. Acrylate—and particularly methacrylate—ester side groups are very resistant to hydrolysis, owing to the steric effects of the acrylic backbone.

Hydrolysis of MF cross-linked hydroxy-functional resins is enhanced by residual acid catalysts (generally sulfonic acids) used to catalyze cross-linking. (See Section 9.3.1.) Curing temperatures can be reduced by increasing the concentration of sulfonic acid in the coating, but the sulfonic acid remaining in the cured film enhances susceptibility to hydrolysis. An apparently ideal solution would be use of transient, or fugitive, acid catalysts that either leave the film or become neutralized after cure.

Base coat-clear coat finishes for automobiles are subject to *environmental etching*. Small, unsightly spots appear in the clear coat surface, sometimes within days, in a warm, moist climate with acidic rain or dew, like Jacksonville, Florida. The spots are uneven, shallow depressions in the clear coat surface. Presumably, they result from hydrolytic erosion of resin in the area of a droplet of water containing a significant acid concentration. Several factors are involved in differences in resistance to environmental etching [21,22]. Since urethane linkages are more resistant to acid hydrolysis than the activated ether cross-links obtained with MF cross-linked hydroxy-functional resins, generally, urethane-polyol clear coats are less susceptible to environmental etching than many MF-polyol clear coats. When MF resins are used to cross-link carbamate-functional resins (see Section 9.3.4), good resistance to environmental etching can be combined with mar resistance. Temperature and T_g are also important, as is surface tension of the clear coat. A variety of approaches have been undertaken to minimize the problem; see Section 29.1.2 for further discussion.

Acrylic-urethane coatings are reported to be more effectively stabilized by HALS than are acrylic-melamine coatings [23]. This may reflect. at least in part, the tendency of acrylic-melamines to degrade by hydrolysis; HALS derivatives do not stabilize against hydrolytic degradation and may promote hydrolysis by basic catalysis. Both urethane and melamine coatings undergo oxidative degradation. Hydrogens on carbons adjacent to urethane nitrogen groups are expected to be activated toward abstraction by free radicals, as are hydrogens on carbons adjacent to nitrogen and ether oxygen groups in melamines. However, evidence has been presented that hydroperoxide levels are significantly lower in melamine than in urethane cross-linked coatings, indicative of lower free radical formation rates and, therefore, lower susceptibility to stabilization of melamines by HALS. The results have been attributed to the ability of melamines to decompose hydroperoxides [24].

Hydrolytic degradation of acrylic-urethane [23], as well as acrylic-melamine [25], coatings are reported to be accelerated by UV exposure. This may result from the increase of hydrophilic groups, such as hydroperoxides, alcohols, ketones, and carboxylic acids, resulting from photoxidation, which increases solubility of water in the coating. Photoxidation may also occur at specific sites to generate groups that are more susceptible to hydrolysis. Photodegradation of acrylic-melamine coatings is also reported to be accelerated in high humidity [25]. This result has been attributed to conversion of formaldehyde (from hydrolysis) into performic acid, a strong oxidant.

Silicone coatings, which are highly resistant to photodegradation, are subject to hydrolysis at cross-linked sites, where silicon is attached to three oxygens [26,27]. Apparently, the electronegative oxygens facilitate nucleophilic attack at Si by water. (See Section 16.5.3.) The reaction is reversible, so the cross-links can hydrolyze and reform. If a silicone-modified acrylic or polyester coating is exposed to water over long periods or is used in a climate with very high humidity, the coating can get softer. It is common to include some MF resin as a supplemental cross-linker in the formulation. Apparently, the MF cross-links with the acrylic or polyester are more hydrolytically stable than the bonds between the silicone resin and the polyester or acrylic.

5.5. OTHER MODES OF FAILURE ON EXTERIOR EXPOSURE

While exposure to UV and the hydrolytic effects of rainfall and humidity are major causes of exterior failure of coatings, many other phenomena can occur.

Automotive finishes can undergo microcracking when there is a rapid change of temperature—for example, when a car is driven out of a heated garage on a very cold winter day. The stress built up by the differential of coefficients of expansion of coating and steel as the temperature drops rapidly can lead to shrinkage in excess of the elongation-to-break of the coating.

When paint is applied to wood, it must be able to withstand the elongation that results from the uneven expansion of wood grain when it absorbs moisture; otherwise, *grain cracking*—that is, cracking parallel to the grain—occurs. This failure mode can occur with interior coatings; however, it is more likely to happen with exterior coatings, particularly with alkyds or drying oils that embrittle with exterior exposure. As a result of their greater exterior durability and extensibility, acrylic latex paint films seldom fail this way.

Another problem of exterior, oil-based house paints on wood siding is blistering. The blistering results from accumulation of water in the wood beneath the paint layer. The vapor pressure of the water increases with heating by the sun, and blisters form to relieve the pressure. Since latex paints have higher moisture vapor permeability than oil-based paints, the water vapor can pass through a latex paint film, relieving the pressure before blisters grow. However, the high moisture vapor permeability of latex paint films can lead to failures of other types. For example, if calcium carbonate fillers are used in an exterior latex paint, *frosting* can occur. Water and carbon dioxide permeate into the film, dissolving calcium carbonate by forming soluble calcium bicarbonate, a solution of which can diffuse out of the film. At the surface of the film, the equilibrium changes direction, and the calcium bicarbonate is converted back to a deposit of calcium carbonate.

Dirt retention can be a difficult problem with latex house paints. Latex paints must be designed to coalesce at relatively low application temperatures. At warmer temperatures, soot and dirt particles that land on the paint surface may stick tenaciously and not be washed off by rain. This problem is minimized with low gloss paints because the high pigment content increases the viscosity of the paint film surface. The problem is particularly severe in areas where soft coal is burned. As would be expected, dirt pickup is less for paints formulated with higher T_g polymers [28]. It was also shown that the more hydrophobic styrene/acrylic latex paints picked up less dirt than acrylic latex paints having the same T_g. Cross-linking of functional latexes also reduced dirt pickup.

Growth of mildew can lead to blotchy, dark deposits on the surface of coatings. Prevention of mildew growth is discussed in Section 31.1.

Often, one needs to be a detective to determine the origin of a paint failure. For example, there were suddenly numerous reports of ugly dirt spots on houses in Bismarck, North Dakota. It turned out that the flight pattern of airplanes taking off from the Bismarck airport had been changed. Oily droplets from plane emissions were landing on the paint, softening its surface to make it a good adhesive for dirt. The wide range of things that can happen to coatings outdoors makes the problem of predicting performance from the results of any simplified test conditions difficult.

5.6. TESTING FOR EXTERIOR DURABILITY

Unfortunately, no test is available that reliably predicts the exterior durability of all coatings. The best way to determine whether a coating will be durable for some length of time in a particular environment is to apply the coating to the product and use the coated product in that environment to determine its lifetime. Even this is difficult, due to the wide variety of environments and variability in coatings and application conditions. The limitations of accelerated tests, the need for data based on actual field experience, and methods of building a database are described in the introduction to Chapter 4 and more extensively in Ref. [29]. Bauer [30] recommends use of reliability theory using statistical distribution functions of material, process, and exposure parameters for predicting exterior durability of automotive coatings. Valuable data can be obtained by analysis of customer complaints; they are especially useful for defining performance standards. By analyzing field experience and customer complaint data over many years, coatings and automobile manufacturers have accumulated an increasingly useful data bank. The information can be used to test hypotheses relating structure to performance. Established structure-performance relationships are useful starting points for designing new coatings. The data can also be used to evaluate laboratory tests. Similar databases are accumulated by some manufacturers and users of maintenance paints, marine coatings, and coil coatings for exterior siding.

5.6.1. Accelerated Outdoor Testing

Accelerated test methods are desirable to permit prediction of performance in shorter times than possible by actual field use. Reference [31] provides an analysis of various test methods. The most reliable accelerated tests are outdoor fence exposures of coated panels, especially if they are carried out in several locations with quite different environments. Reference [32] provides a review of different means of exposure and testing of the exposed panels. Even at one location, there are sometimes substantial variations in conditions that lead to variability of test results [29,30,33]. The most commonly used exposure for architectural paints and industrial coatings is in southern Florida with panels facing south at a 45° angle; automotive coatings are commonly tested at 5° from horizontal. It is important to maintain records of the weather throughout the exposure period. Modern practice is to use photocells to record the cumulative amount of radiant energy striking the test specimen. Since different colors absorb different amounts of infrared radiation, the maximum temperatures reached by coatings during exposure depends on their color; this difference can affect exposure test results.

Southern Florida has a subtropical climate with high humidity, temperature, and sunshine level. Arizona has more hours of sunshine per year and a higher average daily high temperature, but lower humidity. Martin [34] presents a review of the differences between exposure conditions in Florida and Arizona. Cities in the northeastern states and in northern Florida have higher levels of acid in the atmosphere; the Denver area has a higher UV intensity because of the high altitude and the change of temperature from day to night is greater. Bauer estimates that for automotive coatings, the lower humidity in Arizona approximately offsets the higher UV intensity and temperature, and that one year of exposure in Florida has a comparable effect to two years in Michigan, where the UV intensity and temperature are lower, but the humidity is higher [30].

Test specimens are examined periodically to compare their appearance before and after exposure. Usually, at least part of the coating surface is cleaned for the comparison. Ease of cleaning, change in gloss, change in color, degree of chalking, and any gross film failures are reported. Changes in color are particularly difficult to assess, since change in gloss or chalking changes the color even if there has been no change in the color of the components of the coating. The effects of changes in gloss and chalking on the color of the components can be minimized by cleaning the panels and then coating part of each panel with a thin layer of clear gloss lacquer. The lacquer layer minimizes the effect of differences in surface reflection on the color. Test fence exposures can eliminate some formulations as inadequate after a few months of exposure; however, two to five years may be required to permit one to conclude that resistance to exposure at that location under those conditions is adequate.

Film degradation can be accelerated by using *black box* exposure. Panels are mounted at 5° to the horizontal on black boxes, rather than with the backs of the panels open to the air. This method is widely used for testing automotive top coats. A substantial increase in the temperature of the coating when the weather is sunny accelerates degradation. The temperature increase and, therefore, the extent of acceleration can vary substantially from coating to coating, especially with color. A variable that can affect the extent of change due to black box exposure is the relationship between the temperatures and the T_g of the film.

Loss of gloss was formerly the primary mode of failure of automotive top coats, but since the adoption of clear coat-base coat finishes, a major mode of failure has become clear coat cracking in combination with delamination between top coats and base coat, primer, or plastic substrate [30]. Application of sufficient film thickness and proper choice of UV stabilizers and HALS can be critical to avoiding delamination.

Results can be obtained in substantially shorter times by using Fresnel reflectors to concentrate sunlight on test panels. High intensity is achieved by reflecting sunlight from moving mirrors that follow the sun to maintain a position perpendicular to the sun's direct beam radiation. Fresnel reflector devices with trade names such as *EMMAQUA* (Equatorial Mount with Mirrors for Acceleration plus water) *FRECKLE*, *EMMA*, and *Sun-10* are operated in the Arizona desert [34]. They enhance the intensity of sunlight on the panel surfaces by a factor of 8 over direct exposure; it is said to accelerate degradation rates 4 to 16 times the rate for non-accelerated exposure [32]. Panels are cooled by blowing air both over and under the samples. Since rainfall and humidity in the desert are low, the test facilities permit periodic spraying of water on the panel surfaces. In any highly accelerated test, it must be remembered that the rate of volatilization of UV stabilizers depends more on temperature than on radiant energy intensity; stabilizer losses may be smaller during accelerated testing than during long term exposure.

Fresnel reflector weathering devices record actual exposure energies per unit area, expressed in MJ m^{-2}. For comparison, the results can be approximately translated into Arizona or Florida exposure times based on relative solar intensity—that is, energy per unit time. Comparisons based on intensity of exposure to a 10 nm band of UV around 313 nm give results that correlate most closely with field exposures. In addition, spraying with water at night has been found to give results that more closely correspond to field exposure than spraying only during the daylight hours, as was done in earlier procedures.

5.6.2. Analysis of Chemical and Mechanical Changes

Analysis of chemical changes occurring in coatings that have been subjected to exterior exposure is a powerful tool. Chemical changes begin long before the physical changes (gloss loss, delamination, etc.) that characterize failure become evident. Analysis can permit early detection of impending failure after relatively brief exposure, shortening evaluation time for new coatings. However, to have predictive value, chemical changes must be correlated with actual field experience for each type of coating, and only a few correlations have been reported. Studies of chemical changes can help determine the mechanism of failure, providing a basis for formulating more resistant coatings. Early mechanical changes, measured by scanning electron microscopy (SEM) or by dynamic mechanical analysis, can also be correlated with chemical changes and with field experience [35]. References [10] and [31] provide reviews of various approaches.

Electron spin resonance spectrometry (ESR) can monitor changes in free radical concentrations within a coating. The rate of disappearance of stable nitroxyl radicals has been correlated with loss of gloss in long-term Florida exposure [11]. In this technique, nitroxyl radical precursors are incorporated into TiO_2 pigmented acrylic-melamine coatings subjected to ambient or accelerated short-term UV exposure. Similarly, the rate of buildup of phenoxy radicals from bisphenol A epoxy resins incorporated into acrylic-melamine coatings exposed to UV radiation in an ESR spectrometer cavity has been correlated with cracking of coating films exposed in a QUV weathering device [36]. The latter technique was used to evaluate UV stabilizers.

Use of ESR spectrometry to monitor the rate of disappearance of nitroxyl radicals in acrylic-melamine coatings allows calculation of photoinitiation rates (PR) of free radical formation, which were found to correlate with rates of gloss loss (GLR): GLR $\propto$ (PR)$^{1/2}$. This proportional relationship of GLR with (PR)$^{1/2}$ is consistent with a free radical process and results from termination by second-order radical-radical reactions [11]. Photoinitiation rates, determined by this method, have also been used to evaluate experimental conditions for the synthesis of acrylic polyols by free radical polymerization, including the effect of initiator, temperature, and solvent on the projected exterior durability of the resulting acrylic-melamine coating [37].

The nitroxyl early detection method has also been used to investigate the synergistic stabilizing effect of a benzotriazole UV stabilizer and a HALS derivative in acrylic-urethane coatings. The UV stabilizer reduces the photoinitiation rate (PR) of free radical formation, whereas the HALS derivative reduces the propagation rate by lowering the concentration of free radicals.

A somewhat more direct application of ESR to studying photostability of coating films is the determination of free radical concentration after UV irradation of films at a temperature of 140 K, well below T_g [38]. Under these conditions, free radicals are stable. The method is appropriate for evaluation of stablizers such as HALS compounds by comparing the radical concentrations with and without stabilizer. Useful comparisons can be made in three hours [11].

ESR instruments are expensive; an alternative approach is to measure the buildup of hydroperoxides by exposing coatings, cryogenically grinding the films to a fine powder, and titrating them with periodate. A linear relationship between photoxidation rates (PR) and hydroperoxide concentrations in acrylic-urethane and acrylic-MF coatings was reported [37,39]. The relationship GLR $\propto$ (PR)$^{1/2}$, established by ESR as previously shown, is thought to apply to data obtained by titration, but the data needed to fully evaluate the method have not yet been published.

Fourier transform infrared spectroscopy (FTIR) is another powerful tool for following chemical changes on a surface [33]. Photoacoustic FTIR has also been used, it has the advantages that the sample does not have to be removed from the substrate and that it can be used to analyze not only the surface, but also at different depths within a film [40].

A variation on accelerating degradation of a binder in TiO_2 pigmented films is to substitute the surface-treated rutile TiO_2 that would be used in a commercial coating with photoactive, untreated anatase TiO_2. Similarly, one can study delamination failures of automotive coatings by exposing panels with no UV stabilizer in the clear coat over various substrates [29]. The results could permit one to decide whether a major effort should be put into improving the resistance of the substrate to photoxidation or aid in predicting the required minimum film thickness of a clear coat containing a UV stabilizer to protect the substrate.

5.6.3. Accelerated Weathering Devices

Many laboratory devices for accelerating degradation are available. See Ref. [41] for descriptions of various devices and their advantages and disadvantages. The various devices expose panels to UV sources with different wavelength distributions. Also, the panels are subjected to cycles of water spray (or high humidity). Although these tests are widely used, results frequently do not correlate with actual exposure results. A general problem with accelerated weathering methods is the difficulty in accelerating the effects of radiation, heat, and moisture uniformly, not to mention the effects of other atmospheric degradants. The predictive value of accelerated weathering with artificial light sources is particularly questionable when a light source includes wavelengths less than 290 nm. Variability of performance of the test instruments can also be a problem, especially when comparing results from laboratory to laboratory [42].

An evaluation of accelerated weathering devices for a polyester-urethane coating, using photoacoustic-FTIR spectroscopy, concluded that none of the conventional devices were suitable, including an Atlas Weather-O-Meter housing a carbon arc with Corex D filters, an Atlas xenon arc with borosilicate inner and outer filters, and a Q-Panel QUV weathering device with FS-40 fluorescent bulbs [43]. All artificial light sources resulted in the loss of isophthalate groups in the coating, which was not observed during accelerated weathering in Florida (5° south) and Arizona (EMMAQUA). These results correlate with the general wisdom that polyesters, presumably with phthalate or isophthalate groups, perform worse than acrylics in accelerated weathering devices relative to their performance in Florida or Arizona exposure. The unnatural weathering that resulted in loss of isophthalate groups was attributed to excessive amounts of short wavelength light (lower than 290 nm) from the artificial light sources. A comparison of spectra of sunlight (Miami average optimum) and artificial weathering devices that illustrates this disparity is provided in Figure 5.1. As illustrated in Figure 5.1, the FS-40 fluorescent bulbs, utilized in Atlas UVCON and Q-Panel QUV devices exhibit strong unnatural emission below 300 nm. Filtered xenon Weather-O-Meters that closely match the short UV content of sunlight have recently been made available.

Many examples of reversals of results comparing coatings with known exterior durability with laboratory tests have been found. For example, poor results in early evaluations of HALS by QUV testing might have led to their abandonment. Fortunately, remarkable outdoor durability results were obtained, which prompted continued development. Another example is that some automotive base coat-clear coats containing alkanoyl HALS (see Section 5.2.3) performed very well in QUV tests, but had poor resistance to

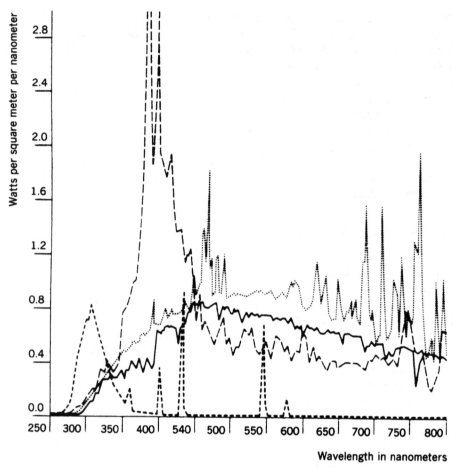

Figure 5.1. Wavelength distribution of radiation from various sources. (Courtesy of Atlas Electric Devices, Chicago, IL.)

delamination in the field. It is suspected that this type of HALS is quickly activated by short wavelength UV, but that sunlight turns it on too slowly. In this case, reliance on the QUV test to predict performance had severe economic consequences. Not only are there examples of poorer coatings showing better laboratory test results and vice versa, but also, there are disparities and reversals when comparing different laboratory test methods. Thus, one may optimize a coating formulation for performance in a particular weathering device only to find that it performs poorly in a different weathering device or outdoors. More serious, as might have occurred in the original work on HALS, is the possibility of rejecting a superior material on the basis of accelerated laboratory testing. The argument that laboratory tests are at least useful for eliminating coatings that exhibit clearly inadequate performance is specious, since accumulated knowledge of failure mechanisms is adequate to eliminate such formulations without any testing.

Although accelerated weathering devices have been improved by closer correspondence of UV emissions with sunlight, it appears unlikely that a completely satisfactory device will be developed because the variations in exterior environments to which coatings are exposed are so great. As with any other highly accelerated test, stabilizer loss by volatilization may well be insignificant during the test period, but very important over the long time periods of actual use. This situation serves to emphasize the need to accumulate data banks of real field use results and to correlate these results with structure to develop a basic understanding of the mechanisms of failure.

GENERAL REFERENCE

A. Valet, *Light Stabilizers for Paints*, translated by M. S. Welling, Vincentz, Hannover, Germany, 1997.

REFERENCES

1. J. L. Gerlock, D. F. Mielewski, D. R. Bauer, and K. R.Carduner, *Macromolecules*, **21**, 1604 (1988).
2. (*a*) W. L. Hawkins, "Stabilization," in *Encyclopedia of Polymer Science and Technology*, 2nd. ed., Vol. 15, Wiley, New York, 1989, pp. 539–583; (*b*) M. Dexter, "Antioxidants," in *Kirk-Othmer Encyclopedia of Chemical Technology*, 4th. ed., Vol. 3, Wiley, New York, 1992, pp. 424–447; (*c*) A. Valet, *Light Stabilizers for Paints*, translated by M. S. Welling, Vincentz, Hannover, Germany, 1997.
3. S. P. Pappas, L. R. Gatechair, E. L. Breskman, and R. M. Fischer, in S. P. Pappas and F. H. Winslow, Eds., *Photodegradation and Photostabilization of Coatings*, ACS Symp. Series, No. 151, Am. Chem. Soc., Washington, DC, 1981, pp. 109–116.
4. H. Bohnke, L. Avar, and E. Hess, *J. Coat. Technol.*, **63** (799), 53 (1991).
5. G. Haacke, F. F. Andrawes, and B. H. Campbell, *J. Coat. Technol.*, **68** (855), 57 (1996).
6. A. Valet, *Light Stabilizers for Paints*, translated by M. S. Welling, Vincentz, Hannover, 1997, pp. 64–70, 116–124.
7. J. D. Shelton, "Stabilization Against Thermal Oxidation," in, W. L. Hawkins, Ed., *Polymer Stabilization*, New York, Wiley-Interscience, 1971, pp. 80–84.
8. D. J. Carlsson, J. P. T. Jensen, and D. M. Wiles, *Makromol. Chem., Suppl.*, **8**, 79 (1984).
9. S. P. Fairgrieve and J. R. MacCallum, *Polym. Degrad. Stab.*, **8**, 107 (1984).
10. D. R. Bauer, *J. Coat. Technol.*, **66** (835), 57 (1994).
11. J. L. Gerlock, D. R. Bauer, L. M. Briggs, and R. A. Dickie, *J. Coat. Technol.*, **57** (722), 37 (1985).
12. Anonymous, *Technical Bulletin, Sanduvor PR-31*, Clairiant Corp., Charlotte, NC, 1996.
13. R. F. Sharrock, *J. Coat. Technol.*, **62** (789), 125 (1990).

14. P. Stamatakis, B. R. Palmer, C. F. Boren, G. C. Salzman, and T. B. Allen, *J. Coat. Technol.*, **62** (789), 95 (1990).
15. H. G. Voelz, G. Kaempf, H. G. Fitzky, and A. Klaeren, "The Chemical Nature of Chalking in the Presence of Titanium Dioxide Pigments," in S. P. Pappas and F. H. Winslow, Eds., *Photodegradation and Photostabilization of Coatings*, ACS Symp. Series, No. 151, Am. Chem. Soc., Washington, DC, 1981, pp. 163–182.
16. G. Irick, Jr., G. C. Newland, and R. H. S. Wang, "Effect of Metal Salts on the Photoactivity of Titanium Dioxide: Stabilization and Sensitization Processes," in S. P. Pappas and F. H. Winslow, Eds., *Photodegradation and Photostabilization of Coatings*, ACS Symp. Series, No. 151, Am. Chem. Soc., Washington, DC, 1981, pp. 147–162.
17. J. H. Braun, *J. Coat. Technol.*, **62** (785), 37 (1990).
18. J. H. Braun and D. P. Cobranchi, *J. Coat. Technol.*, **67** (851), 55 (1995).
19. W. H. Starnes, Jr., *Pure Appl. Chem.*, **57**, 1001 (1985). See also G. Georgiev, L. Christiv, and T. Gancheva, *J. Macromol. Sci., Chem.*, **A27**, 987 (1990).
20. T. E. Jones and J. M. McCarthy, *J. Coat. Technol.*, **67**, (844), 57 (1995).
21. P. Betz and A. Bartelt, *Prog. Org. Coat.*, **22**, 27 (1993).
22. B. V. Gregorovich and I. Hazan, *Prog. Org. Coat.*, **24**, 131 (1994).
23. D. R. Bauer, M. J. Dean, and J. L. Gerlock, *Ind. Eng. Chem. Res.*, **27**, 65 (1988).
24. D. F. Mielewski, D. R. Bauer, and J. L. Gerlock, *Polym. Degrad. Stab.*, **33**, 93 (1991).
25. D. R. Bauer, *Prog. Org. Coat.*, **14**, 193 (1986).
26. L. H. Brown, "Silicones in Protective Coatings," in R. R. Myers and J. S. Long, Eds., *Treatise on Coatings*, Vol. I, Part III, Marcel Dekker, New York, 1972, pp. 536–563.
27. Y.-C. Hsaio, L. W. Hill, and S. P. Pappas, *J. Appl. Polym. Sci.*, **19**, 2817 (1975). See also S. P. Pappas and R. L. Just, *J. Polym. Sci., Polym. Chem. Ed.*, **18**, 527 (1980).
28. A. Smith and O. Wagner, *J. Coat. Technol.*, **68** (862), 37 (1996).
29. J. W. Martin, S. C. Saunders, F. L. Floyd, and J. P. Wineburg, *Methodologies of Predicting Service Lives of Coatings Systems*, Federation of Societies for Coatings Technology, Blue Bell, PA, 1996.
30. D. R. Bauer, *J. Coat. Technol.*, **69** (864), 85 (1997).
31. B. W. Johnson and R. McIntyre, *Prog. Org. Coat.*, **27**, 95 (1996).
32. L. S. Hicks and M. J. Crewdson, "Natural Weathering," in J. V. Koleske, Ed., *Paint and Coatings Testing Manual*, 14th ed., ASTM, Philadelphia, 1995, p. 619.
33. R. M. Fischer, W. P. Murray, and W. D. Ketola, *Prog. Org. Coat.*, **19**, 151 (1991).
34. J. L. Martin, *Proc. Adv. Coat. Technol. Conf. Eng. Soc.*, Society of Automotive Engineeers, Detroit, 1991, p. 219.
35. L. W. Hill, H. M. Korzeniowski, M. Ojunga-Andrew, and R. C. Wilson, *Prog. Org. Coat.*, **24**, 147 (1994).
36. S. Okamoto, K. Hikita, and H. Ohya-Nishiguchi, in *Proc. FATIPEC Congress*, 1986, pp. 239–255.
37. J. L. Gerlock, D. R. Bauer, L. M. Briggs, and J. K. Hudgens, *Prog. Org. Coat.*, **15**, 197 (1987).
38. A. Sommer, E. Zirngiebl, L. Kahl, and M. Schonfelder, *Prog. Org. Coat.*, **19**, 79 (1991).
39. D. R. Bauer, *Prog. Org. Coat.*, **23**, 105 (1993).
40. D. R. Bauer, D. F. Mielewski, and J. L. Gerlock, *Polym. Degrad. Stab.*, **38**, 57 (1992).
41. V. D. Sherbondy, "Accelerated Weathering," in J. V. Koleske, Ed., *Paint and Coating Testing Manual*, 14th ed., ASTM, Philadelphia, 1995, p. 643.
42. R. M. Fischer, W. D. Ketola, and W. P. Murray, *Prog. Org. Coat.*, **19**, 165 (1991).
43. D. R. Bauer, M. C. P. Peck, and R. O. Carter, III, *J. Coat. Technol.*, **59** (755), 103 (1987).

CHAPTER 6

Adhesion

Adhesion is an essential characteristic of most coatings. Unfortunately, there is inadequate basic scientific understanding of the variables affecting adhesion. As a result, some of the comments in this discussion are based only on reasonable deductions that fit in with accumulated experience.

One difficulty in dealing with the subject is defining what adhesion means. In most cases, a coatings formulator thinks of adhesion in terms of the question: How hard is it to remove the coating? But, a physical chemist would think in terms of the work required to separate two interfaces that are adhering. These perspectives can be different considerations; the latter is only one aspect of the former. Removal of a coating requires breaking or cutting through the coating and pushing the coating out of the way, as well as separating the coating from the substrate. Consider as an extreme example, adhesion of a plastic covered wire used in electrical connections. The covering must have good "adhesion" to stay on the wire to protect against short circuits and shocks. However, to attach the wire to a fixture, one needs to make a cut through the covering and easily slip the coating off the metal. It is desirable to have minimal interactive forces between the plastic and the copper, but considerable toughness within the coating so that accidental removal is unlikely.

6.1. SURFACE MECHANICAL EFFECTS ON ADHESION

Resistance to separation of coating and substrate can be affected by mechanical interlocking. Consider the schematic representations in Figure 6.1. With a very smooth interface between coating and substrate, as shown in sketch A, the only forces holding the substrate and coating together are the interfacial attractive forces per unit of geometric

A   B   C

Figure 6.1. Schematic representations of the geometries of surface interactions between a coating and a substrate. **A**: smooth interface between coating and substrate. **B**: rough surface on a microscopic scale. **C**: rough surface with incomplete penetration of coating.

area. With a rough surface on a microscopic scale, as represented in sketch B, two other factors are important. In some places, there are undercuts in the substrate; to pull the coating off the substrate one would either have to break the substrate or break the coating to separate them. The situation is analogous to using a dovetail joint to hold two pieces of wood together. Another factor is that the actual contact area between the coating and the rough substrate is substantially larger than the geometric area.

Better adhesion can generally be obtained if the surface of a substrate is roughened before coating; however, as can be seen in sketch C, surface roughness can be a disadvantage. If the coating does not completely penetrate into the microscopic pores and crevices in the surface, dovetail effects are not realized, and the actual interfacial contact area can be smaller than the geometric area. Furthermore, when water permeates through the film to the substrate, there will be areas of contact of water with uncovered substrate, which can be a major problem, especially if the coating is to protect steel against corrosion.

The importance of surface roughness is widely recognized, but its effect on coatings has not been subjected to many scientific studies. The scale of the roughness of surfaces can vary from macroscopic to microscopic to submicroscopic. It is important to consider the situation on a microscopic and submicroscopic scale. What factors control the rate of penetration of a liquid into such pores and crevices? The situation is analogous to penetration of a liquid into a capillary. Equation 6.1 shows the variables affecting distance of penetration L (cm) into a capillary of radius r (cm) in time t (s), where γ is surface tension (mN m^{-1}), θ is contact angle, and η is viscosity (Pa $\cdot$ s):

$$L = 2.24 \left[\left(\frac{\gamma}{\eta} \right) (r \cos \theta) t \right]^{1/2} \tag{6.1}$$

Rate of penetration Lt^{-1} is greatest if surface tension of the coating is high. However, there is an upper limit to this surface tension effect because the rate is strongly affected by contact angle. The rate is fastest when cosine of the contact angle is 1—that is, when the contact angle is zero. The cosine can only be 1 if the surface tension of the liquid is less than that of the solid substrate. The radius of the capillary is a variable of the substrate, not of the coating.

The variable over which the formulator has greatest control is viscosity. On the scale of microscopic and submicroscopic crevices and pores, pigment and polymer particles in the coating are large compared to at least some surface irregularities, and therefore, the critical viscosity is that of the continuous (external) phase of the coating, not the bulk viscosity of the coating. The lower the viscosity of the external phase, the more rapid the penetration. Since in most cases, viscosity of the vehicle increases after application; it is important to keep the viscosity low for a long enough time for penetration to approach completion. Since viscosity of resin solutions increases with molecular weight, one would expect that lower molecular weight resins would provide superior adhesion after cross-linking, everything else being equal. This hypothesis has been confirmed in the case of epoxy resin coatings on steel [1]. Another possible advantage of low molecular weight resins is that their molecules may be able to penetrate into smaller crevices than high molecular weight molecules. Coatings with low viscosity external phases, slow evaporating solvents, and relatively slow cross-linking rates have been found, in general, to give better adhesion. In general, baked coatings give better adhesion than do air dry coatings. (The term *air dry* is widely used, but is potentially confusing. It generally means just that the film is formed at ambient temperature, but does not necessarily mean that oxygen is required for cross-linking.) When the coated article goes into an oven, temperature increases, viscosity of the external phase decreases, and penetration into surface irregularities increases. This is only one of several possible explanations for the advantages of baked coatings when adhesion is critical.

6.2. EFFECTS OF INTERNAL STRESS AND FRACTURE MECHANICS

Internal stresses in coatings amount to forces that counteract adhesion; less external force is required to disrupt the adhesive bond. The common statement that internal stresses result from shrinkage is misleading; rather, internal stresses result from the inability of coatings to shrink as they form films on rigid substrates. When solvent evaporates from a thermoplastic coating (lacquer), in the early stages, the polymers can accommodate the resulting voids by relaxation; thus, shrinkage occurs. However, as film formation proceeds, T_g rises and free volume is reduced; it becomes more difficult for the polymer to accommodate the voids from solvent evaporation, so it becomes fixed in unstable conformations, and internal energy (stress) increases. This phenomenon is particularly likely to occur in coatings in which the T_g approaches the film-forming temperature [2]. In some cases, the stresses build up sufficiently that spontaneous delamination occurs [3]. Stresses can result not only from volume contractions, but also from volume expansions, such as swelling of films by exposure to high humidity [3] or water immersion [4].

In thermosetting coatings, cross-linking reactions lead to formation of covalent bonds that are shorter than the distance between two molecules before they react. When such reactions occur at temperatures near the T_g of the film, stresses result from the inability of the coating to undergo shrinkage. As the rate of cross-linking increases, stresses also tend to increase, since less time is available for polymer relaxation to occur. An extreme example is UV curing of acrylated resins by free radical polymerization that occurs in a fraction of a second at ambient temperatures. (See Section 28.2.) Shrinkage, measured by thermomechanical analysis (TMA), has been shown to lag significantly behind polymerization [5]. The high rates of polymerization, together with the relatively large shrinkage that accompanies polymerization of double bonds, contribute to the generally observed poor adhesion of UV cure acrylated resins to smooth metal surfaces. Heating after UV curing relaxes the cross-link network and often improves adhesion.

Nonuniform curing, particularly in the later stages, as well as film defects or imperfections in the film, can lead to localized stresses that exert adverse effects on adhesion [6]. Localization of stresses at imperfections falls within the discipline of *fracture mechanics*. The effects of fracture on abrasion resistance are considered in Section 4.3. The phenomena have been most extensively studied in adhesive bonding of substrates [7], but also must affect adhesion of coatings. If there is a local imperfection in a film, any stress applied to that part of the film could be concentrated at the imperfection. Localized imperfections result in greater stress (force per unit area) and greater probability of forming a crack. Once a crack starts, the stress concentrates at the point of the crack, leading to crack propagation. If the crack propagates to the coating-substrate interface, stress concentration can cause the film to delaminate.

While the principle has been recognized, identification of causes of imperfections is more difficult. Pigment particles with sharp crystal corners and air bubbles are examples of potential sites for concentration of stresses. On the other hand, as noted in Section 4.3, it has been proposed that incorporation of particles of rubber may lead to dissipation of stresses. Presumably, such stress dissipation would reduce the probability of fracture mechanical adhesive failure.

6.3. RELATIONSHIPS BETWEEN WETTING AND ADHESION

Wetting is a major, and perhaps limiting, factor in adhesion. If a coating does not spread spontaneously over a substrate surface so that there is intermolecular contact between the substrate surface and the coating, there cannot be interactions, and hence, no contribution to adhesion. The relationships between wetting and adhesion were extensively studied by Zisman [8]. A liquid spreads spontaneously on a substrate if the surface tension of the liquid is lower than the surface free energy of the solid. (Surface free energies of solids have the same dimensions as surface tension; see Section 23.1.) If the surface tension of a liquid is too high, a drop of the liquid stays as a drop on the surface of the solid—it has a contact angle of 180°. If a liquid has a sufficiently low surface tension, it spreads spontaneously on the substrate—it has a contact angle of 0°. At intermediate surface tensions, there are intermediate contact angles. A schematic drawing of a drop of liquid with an intermediate surface tension is shown in Figure 6.2.

The relationship between contact angle θ and the surface free energy of the substrate γ_S, the surface tension of the liquid γ_L, and the interfacial tension between the solid and the liquid γ_{SL} for a planar surface is given in Eq. 6.2:

$$\cos \theta = \frac{(\gamma_S - \gamma_{SL})}{\gamma_L} \tag{6.2}$$

Maximum adhesion requires a contact angle of 0°. Experimental determination of contact angles of complex systems like coatings, especially on rough substrates with a heterogeneous composition, is difficult [9]. In general, it is sufficient to conceptualize the relationships by saying that adhesion requires that the liquid have a lower surface tension than the surface free energy of the substrate to be coated. From a practical standpoint, it is useful to do cruder, but easier, experiments. One can apply a drop of coating on a substrate, put the sample in an atmosphere saturated with the solvents in the coating, and watch the spreading. If the droplet of coating stays as a small ball, spreading is poor and adhesion problems should be expected. If the drop spreads out to a thin, wide circle, the coating meets at least one criterion for good adhesion. It is also useful to carry out a second

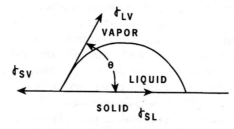

Figure 6.2. Schematic representation of contact angle.

type of experiment; in which a coating is mechanically spread on a substrate, under a solvent saturated atmosphere and then observed. Generally, a liquid that spreads spontaneously in the first experiment will remain spread out in the second experiment. Sometimes, a liquid that spreads on the surface will draw up into droplets, or at least pull away in sections from the substrate when standing, without solvent evaporation.

Consider, for example, the application of *n*-octyl alcohol to a clean steel surface. The surface tension of *n*-octyl alcohol is lower than the surface free energy of steel, and it spontaneously spreads on steel. However, if one spreads out a film of *n*-octyl alcohol on steel, the film draws up to form droplets on the surface of the steel. The low surface tension of the *n*-octyl alcohol results from the linear hydrocarbon chain; however, after spreading on the polar surface of the steel, the hydroxyl groups of the *n*-octyl alcohol molecules interact with the surface so that a monolayer of oriented *n*-octyl alcohol molecules forms on the surface. This makes a new surface, aliphatic hydrocarbon, which has a low surface tension, lower than the surface tension of *n*-octyl alcohol. The *n*-octyl alcohol above a monolayer *dewets*. The behavior of *n*-octyl alcohol illustrates a principle important in formulating coatings: One must be careful about using additives with single polar groups and long hydrocarbon chains in coatings to be used directly on metals. An example is the poor adhesion to steel that can result from use of dodecylbenzenesulfonic acid as a catalyst. (See Section 9.3.1.1.) Adhesion of latex films can be affected by a layer of surfactant forming at the interface between the coating and the substrate [10].

6.4. ADHESION TO METAL SURFACES

The metal and the surface characteristics of the metal can have major effects on adhesion; Ref. [11] provides a review of metal surface characteristics, cleaning, and treatments. The surface tension of a clean metal surface (usually metal oxide) is higher than that of any potential coating. However, metal surfaces are frequently contaminated with oil, and such surfaces can have very low surface tensions. Whenever possible, it is desirable to clean the surface of the metal before applying a coating. Sometimes, the metal is wiped with rags wet with solvent. A more effective method is *vapor degreasing*, in which the object is hung from a conveyor that carries it into a tank above a boiling chlorinated solvent. The cold steel surface acts as a condenser, condensing solvent on the surface of the steel, where it dissolves oils. The solution drips off, removing the oil. The solvent is purified by distillation for reuse. Surfactant solutions are also used for cleaning oil from metals [11]. Care must be exercised in selecting surfactants and in rinsing the surface after cleaning. It is possible for some surfactants to adsorb on the surface, creating a hydrocarbon layer on the surface of the metal.

Steel can be cleaned by abrasive particle blasting, such as sandblasting. (See Ref. [11] for discussion.) The surface of the steel, including rust, is removed, leaving a rough surface. Sandblasting is widely used for steel structures like bridges and tanks, but it leaves the steel surface too rough for products such as automobiles and appliances. Sandblasting is effective, but there are health hazards to workers from inhalation of silica dust. A variety of alternative blasting techniques are being used or evaluated [12]. These include use of other dry, abrasive materials, such as steel grit or water-soluble abrasives—including sodium bicarbonate and salt—to replace sand; cryogenic cleaning by blasting with dry ice pellets; vacuum blasting using aluminum oxide; and, for cleaning softer metals, such as aluminum, plastic pellet blasting. Ultra-high pressure hydroblasting at pressures above

175 mPa (25,000) psi is very effective in removing oil and surface contaminants, such as salt.

Formulations are tested on laboratory panels, but the surfaces of these panels are not the same as those of the product to which the coating will be applied. Furthermore, commercial test panels vary—for example, the side of the panel next to the wrapping of a package of panels was shown to have a different surface analysis than inner panels from the same package [13]. Washing the panels with warm water and rinsing with acetone before coating generally improved adhesion. Laboratory tests should also be carried out on actual ultimate substrate, or at least on sample pieces of metal to be used in production, simulating, as closely as possible, factory cleaning and treating procedures.

6.4.1. Surface Preparation

When good adhesion, corrosion protection, and a relatively smooth surface are required, it is common to chemically treat the surface of the metal. The treatments for steel are called *conversion coatings*, or *chemical pretreatments*. A variety of phosphate based conversion coatings is used. One such coating involves spray or immersion application of a phosphoric acid based "iron phosphate" solution. This method provides a mild etch of the steel surface and precipitation of a ferrous/ferric phosphate monolayer. Adhesion of coatings is markedly increased; however, corrosion protection is only slightly enhanced. In zinc phosphate conversion coating, the steel object is immersed in a bath of zinc acid phosphate solution. A coprecipitate of zinc and ferric phosphates is formed on the steel surface, as shown in Scheme 6.1, enhancing both adhesion and corrosion protection.

The coprecipitate forms a mesh of crystals that adheres tightly to the surface, increasing the surface area on a microscopic scale. Depending on zinc concentration in the treatment bath, different crystals can be deposited. At relatively high zinc concentrations, the crystals are predominantly hydrated zinc phosphate, $Zn_3(PO_4)_2 \cdot 4H_2O$, called hopeite. Under zinc-starved conditions, the crystals have been identified as phosphophyllite, $Zn_2Fe(PO_4)_2 \cdot 4H_2O$ [14]. The performance of the conversion coating is affected by the uniformity and degree of surface treatment. Zinc phosphate coatings are generally applied in the range of 1.5–4.5 g m^{-2}. A variety of other phosphate coatings is also used. (See Ref. [11] and Sections 7.3.1.1, 26.1, 26.2, and 31.1.1 for further discussion of conversion coating.)

The reactions shown in Scheme 6.1 are straightforward. Achieving the high rates of reaction required to permit minimum dwell times for treatment is more difficult. Proprietary formulations reduce times to the order of minutes or seconds. Treatment quality is dependent on time, temperature, and pH; these and other variables must be

Scheme 6.1

$$HPO_4^{2-} \rightleftharpoons H^+ + PO_4^{3-}$$

$$2H^+ + Fe \longrightarrow Fe^{2+} + H_2$$

$$Fe^{2+} + [Ox] \longrightarrow Fe^{3+}$$

$$Fe^{3+} + PO_4^{3-} \longrightarrow FePO_4 \downarrow$$

$$3Zn^{2+} + 2PO_4^{3-} \longrightarrow Zn_3(PO_4)_2 \downarrow$$

closely controlled to assure that the desired type and dimensions of crystals are formed. The treated surface must be thoroughly rinsed to remove any soluble salts, since these salts could lead to blister formation when water vapor permeates through a coating film applied over the soluble salts; rinsing also removes loosely adhering crystals. It has been common for the last rinse to contain a low concentration of chromic acid to protect against corrosion. Due to the toxic risks of hexavalent chromium, replacements for the chromate rinse are being sought. Over zinc phosphate, a rinse of 0.5% methyltrimethoxy silane with sufficient H_2ZrF_6 to give a pH of about 4 is reported to give better performance than a chromic acid rinse [15]. However, over iron phosphate, the silane-H_2ZrF_6 rinse gave inferior performance compared to a chromic acid rinse.

The mechanism of action of the phosphate crystal layer is not fully understood. The coating penetrates into the crystal mesh, giving a mechanical attachment to the crystals attached to the surface. The interfacial area for interaction is greater than for a relatively smooth steel surface. It is also possible that hydrogen-bond interactions between these crystals and the resin molecules are stronger (that is, less readily displaced by water) than those between the steel surface and the resin molecules.

The surface of aluminum is a thin, dense, coherent layer of aluminum oxide. For many applications, no treatment other than cleaning is required. However, for applications in which there might be exposure to salt, surface treatment is necessary. Most treatments for aluminum have been chromate treatments. An example of one such treatment is an acid bath containing chromate, fluorides, and a ferricyanide salt as an accelerator. The resultant coating is said to have the following composition: $6Cr(OH)_3 \cdot H_2CrO_4 \cdot 4Al_2O_3 \cdot 8H_2O$. See Ref. [11] for further discussion. In the past several years, many chrome- and cyanide-free proprietary aluminum conversion coatings have been developed with equivalent performance to chrome coatings.

To provide greater protection against corrosion, steel coated with zinc is widely used in construction and automobiles. Several types of zinc-coated steel are used; the best known is galvanized steel. There can be large variations in adhesion depending on the condition of the zinc layer of the galvanized steel. If zinc-coated steel has been exposed to rain or high humidity before coating, there may have been some degree of surface oxidation, leading to formation of a combination of ZnO, $Zn(OH)_2$, and $ZnCO_3$; all of these are basic and somewhat soluble in water. Therefore, it is important to use saponification resistant resins in primers for galvanized steel. Resins such as alkyds, which tend to saponify, are likely to give poor adhesion in service. Zinc surfaces on automobile bodies are treated with zinc, manganese, nickel phosphate conversion coatings before cationic electrodeposition of primer [16]. The effectiveness of phosphate coatings is affected by the presence of small amounts of other metals in the zinc coating [17]. Reference [17] provides a review of the coating of galvanized steel.

Stainless steel, because of its smoothness and lack of oxide and hydroxide groups on the surface, is difficult to adhere to. In some cases, roughening of the surface provides a basis for anchoring; in others, pretreatment with an electrochemically produced flash of chromium/chromium oxide is required [18].

Plasma treatment is a new approach being investigated for cleaning and treating the surface of cold-rolled steel [19]. Initially, the steel is cleaned by plasma discharge, and then, trimethylsilane is introduced into the plasma chamber leading to polymerization of a thin layer of polymer bonded to the surface. Laboratory tests indicated superior adhesion of electrodeposition coatings (see Chapter 26) to the surface as compared to a conventionally pretreated galvanized steel substrate. If the process can be successfully scaled up for production, substantial benefits can be foreseen. Cold-rolled steel is less expensive and more readily recycled than electrogalvanized steel, and the waste disposal problems associated with electrogalvanizing and phosphating would be eliminated.

Another approach being investigated is use of bis(trialkoxysilyl)alkanes to react with the surface of steel; see Section 6.4.3.

6.4.2. Coating-Substrate Interactions

The surface of clean steel is not iron; rather, hydrated iron oxides are present as a monolayer on the iron. The layer is not a layer of rust particles, but a monolayer of hydrated oxide [20]. Adhesion to this surface is promoted by developing hydrogen bonds between groups on the resin molecules and the oxide and hydroxide groups on the surface of the steel. (In the case of phosphate conversion coated steel, there is the further possible hydrogen bond interaction with the phosphate groups.) Some authors prefer to interpret the interactions in terms of association between soft acids and soft bases [21].

It follows that adhesion is promoted by using resins having multiple hydrogen-bond donating and accepting groups. Adhesion is promoted by such groups as carboxylic acid (strongly hydrogen donating), amine (strongly hydrogen accepting), hydroxyl, urethane, amide (the latter three being both hydrogen donating and accepting), and phosphate. One might assume that a large number of such substituents on a molecule would be desirable. However, it is known from adsorption studies that if there are large numbers of polar groups, at equilibrium the adsorbed layer can be very thin. The principle can be illustrated by considering a polymer molecule with an aliphatic backbone chain with polar groups on every other carbon atom. At equilibrium, adsorption of adjacent polar groups is favored sterically, resulting in a thin adsorbed layer with the polar groups on the steel surface and only hydrocarbon groups exposed to the rest of the coating. The interactions between the rest of the coating and the hydrocarbon groups would be expected to be weak, resulting in a weak boundary layer and poor cohesion. If a smaller number of hydrogen-bond donating groups are scattered along a resin chain, adsorption of resin molecules may occur with loops and tails sticking up from the surface so that some of the polar groups are adsorbed on the surface and some are on the loops and tails, where they can interact with the rest of the coating. On those parts of the resin molecule in the loops and tails, there can be groups to hydrogen bond with molecules in the coating, as well as functional groups, which can react with a cross-linker in the coating.

BPA epoxy resins (see Sections 11.1.1 and 15.8) and their derivatives commonly provide excellent adhesion to steel. These resins have hydroxyl groups and ether groups along the chain, which can provide for interactions with both the steel surface and other molecules in the coating. It may also be important that the backbone consists of alternating flexible 1,3-glyceryl ether and rigid bisphenol A groups. It seems logical that such a combination could provide the flexibility necessary to permit multiple adsorption of hydroxyl groups on the surface of the steel, along with the rigidity to prevent adsorption of all of the hydroxyl groups. The remaining hydroxyl groups can participate in cross-linking reactions or hydrogen bond with the rest of the coating. References [1] and [22] discuss the effects of variations in epoxy resin composition on adhesion.

There is need for further research on interactions and orientation at the interface of coatings and substrates. There have been many studies of adsorption of polymer molecules on metal surfaces, and, in general terms, the results are consistent with the picture given above. However, many of these studies involve adsorption from dilute solutions. Observations are made over relatively long time intervals, permitting equilibrium conditions to develop. In our example of a polymer molecule with polar groups on alternate carbon atoms, initially, many molecules could be adsorbed, forming a thick adsorbed layer, but the equilibrium condition would favor adsorption of all, or substantially all, of the polar groups of individual molecules, leading to the thin adsorbed layers described previously. With polydisperse molecular weight adsorbents, low molecular weight species are adsorbed first, but, at equilibrium, they are displaced by higher molecular weight molecules with larger numbers of polar groups. What really happens when a coating is applied? The resin is in a relatively concentrated solution, and the solvent evaporates in a relatively short time. There may not be time for equilibrium to be established. Depending on the coating, those groups that happened to be near the surface when the film was applied might remain there and could lead to poor adhesion, even if the same resins could provide good adhesion given the opportunity for appropriate orientation and equilibration to occur. Such a scenario is compatible with the improved adhesion using slow evaporating solvents, which can also permit more complete penetration into surface crevices. Perhaps another reason that baking coatings commonly leads to improved adhesion since there is greater opportunity for orientation of molecules at the steel/coating interface at the higher temperature.

Surface analysis can be useful in understanding factors affecting adhesion. The surface of steels have been studied by Auger analysis; organic compounds have been detected on the surface of some cold-rolled steels. These organic compounds apparently become imbedded in the surface of the steel during coil annealing. If this happens, it becomes difficult to obtain high quality phosphate conversion treatments on the steel [23]. Such steels more commonly lead to adhesion failures and to inferior corrosion protection by coatings.

X-ray photoelectron spectroscopy (XPS) can also be used to study the surface of steel from which a coating has been removed and the underside of the coating that was in contact with the steel. This technique can be particularly powerful for showing where failure occurred—that is, whether failure was between the steel and the coating or between the main body of the coating and a monolayer (or a very thin layer) of material on the surface of the steel. Other valuable analytical procedures for thin surface layers is attenuated total reflectance (ATR) and FTIR. Such techniques have been most useful for diagnosing problems; further use of these and other analytical techniques for understanding the mechanism of adhesion can be anticipated.

While strong interaction between coating and steel is critical for achieving good adhesion, it is also important to develop interactions that cannot be easily displaced by water. A reason that the presence of multiple groups on resin molecules such as hydroxyl groups on epoxy resins may be desirable is that some may remain bonded to the steel while others are reversibly displaced by water. This phenomenon has been termed *cooperative adhesion* [24]. It has been found empirically that amine groups on the cross-linked resin molecules promote corrosion protection. Explanations for the effect are controversial; one hypothesis is that the amine groups interact strongly with the steel surface and are not as easily displaced by water from the surface as are hydroxyl groups. Phosphate groups are another substituent group that has been found to impart improved adhesion in the presence of water (i.e., wet adhesion). For example, the use of epoxy phosphates (see Section 11.6) in epoxy coatings has been shown to improve both adhesion and wet adhesion [22].

6.4.3. Covalent Bonding to Glass and Metal Substrates

Stronger interactions with the substrate surface should be possible by forming covalent bonds instead of the more readily displaced hydrogen bonds. One such approach is use of reactive silanes, see Section 16.5.4; they are very effective in enhancing adhesion of coatings to glass [25]. A variety of reactive silanes are available; they all have a trialkoxysilyl group attached to a short hydrocarbon chain, the other end of which has a functional group such as amine, mercaptan, epoxy, vinyl, and so forth. The alkoxysilyl group can react with hydroxyl groups on the surface of glass and with other alkoxysilyl groups after hydrolysis, so the surface of the glass becomes covalently bonded to a series of hydrocarbon tails substituted with reactive groups that can cross-link with the coating being applied.

For example, in formulating an epoxy-amine coating for glass, one could add 3-aminopropyltrimethoxysilane to the amine package of the two package coating. After application, the trimethoxysilyl group can react with silanol groups on the surface of the glass to generate siloxane bonds, as shown in first step of Scheme 6.2. The trimethoxysilyl groups can also react with water to produce silanol groups that can, in turn, react with remaining silyl methoxy groups to generate polysiloxane groups at the glass surface, see the second step in Scheme 6.2. The terminal amine groups can react with epoxy groups in the resin so that the coating is multiply bonded to the surface of the glass, as shown in the third step in Scheme 6.2.

When water vapor penetrates through the coating to the glass interface, hydrolysis of some of the interfacial Si—O bonds occurs. However, with multiple interfacial bonds, some of the bonds remain intact and prevent the coating from delaminating. Furthermore, hydrolysis is reversible, so the hydrolyzed bonds can re-form. Before the advent of reactive silanes, it was difficult to formulate coatings that maintained adhesion to glass after exposure to a humid atmosphere.

Scheme 6.2

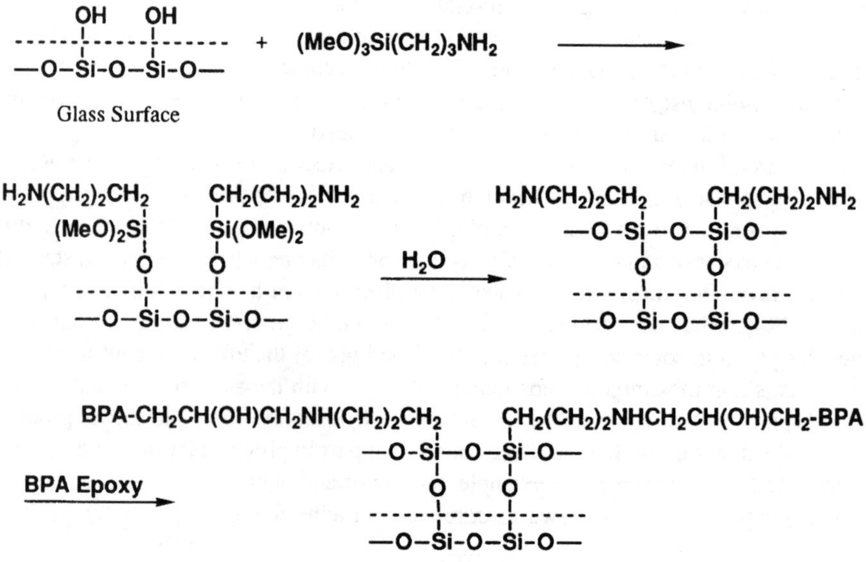

Reactive silanes have also been added to coatings with the objective of improving adhesion to steel surfaces [25]. There has been some evidence of improvement. The trialkoxysilyl group can react with hydroxyl groups attached to iron. However, reactive silanes have not been widely adopted as additives to improve the adhesion of coatings to steel. Perhaps, the lack of widespread use results from the lower stability of the bonds formed with iron as compared with glass and other metals such as aluminum [26]. In the case of amino silanes, the amine group may be preferentially adsorbed on iron. An extensive review of the journal and patent literature is available [27].

Recently, investigation of the use of bis(trilkoxysilyl)alkanes for the treatment of metals has been reported [28]. For example, clean steel is rinsed with water, and then the wet steel is dipped in an aqueous solution of bis(trimethoxysilyl)ethane (BTSE). The BTSE hydrolyzes in the water and is then adsorbed on the surface of the steel reacting with hydroxyl groups and cross-linking with other silylethane molecules. The multiple covalent attachments form a coating that is stable when immersed in water. The surface can then be treated with a reactive silane that reacts with silanol groups and also provides a reactive group to react with a coating binder. Many variables affect the treatment; they are currently being investigated in the laboratory, and commercial use is projected. Laboratory tests indicate excellent adhesion in the presence of water and good corrosion protection.

Another approach to achieving chemical bonding to steel is the use of resins containing groups that can form coordination complexes with ferric compounds. For example, one can make resins with acetoacetic ester substituents. (See Section 16.6.2.) Such esters are highly enolized and can coordinate with metal ions, including ferric salts. Preliminary reports indicate improvement in adhesion and corrosion protection [29]. Because of the potential hydrolysis of acetoacetic esters, evaluation over relatively long time intervals will be required to assess their commercial utility.

6.5. ADHESION TO PLASTICS AND TO COATINGS

In contrast to clean steel and other metals, it is possible that there will be a problem wetting the surface of plastic substrates with a coating. Wetting and adhesion can be affected by the presence of mold release agents on a molded plastic part. Mold release agents should be avoided if at all possible. If essential, release agents should be selected that are relatively easily removed from the molded part and care should be exercised to remove all traces. Even after cleaning, the surface free energies of some plastics are lower than the surface tensions of many coatings. The contact angle between coating and substrate should be 0° to permit spreading. Determination of contact angle is experimentally difficult, especially due to surface roughness and inhomogeneity [9].

Polyolefins in general have low surface free energies, and therefore, it can be difficult to achieve good adhesion of coatings to them. Attainment of satisfactory adhesion to polyolefins generally requires treatment of the surface to increase its surface free energy. This can be done by oxidation of the surface to generate polar groups such as hydroxyl, carboxylic acid, and ketone groups. The presence of these groups not only increases surface free energy so that wetting is possible with a wider range of coating materials, but also provides hydrogen-bond acceptor and donor groups for interaction with complementary groups on coating resin molecules. A variety of processes can be used to treat the surface [30,31]. The surfaces of films, flat sheets, and cylindrical objects can be oxidized by flame treatment with gas burners using air-gas ratios such that the flames are oxidizing. Oxidation can also be accomplished by subjecting the surface to a corona discharge atmosphere; the ions and free radicals generated in the air by the electron emission serve to oxidize the surface of the plastic. Various chemical oxidizing treatments are effective; the most widely used treatment has been an aqueous potassium dichromate-sulfuric acid solution, however, hazardous waste disposal is a serious problem.

Adhesion to untreated polyolefins can be assisted by applying a thin *tie coat* of a low solids solution of a chlorinated polyolefin or chlorinated rubber. Ryntz has reviewed the various approaches and provided the results of various types of surface analysis [31]. See Section 32.3 and Ref. [32] for further discussion of surface treatments.

Both theoretical and experimental studies show that the molecules at the surface of a polymer are more mobile than those in the bulk material [33]. One theoretical study indicated that the layer of mobile molecule segments is about 2 nm in thickness [34]. Because polymer surfaces are dynamic, they adjust to the environment; polar groups will, with time, move away from a polymer-air interface and toward a polymer-water interface. Thus, the polar groups placed on the surface by oxidation treatments can be expected to move to the interior within hours or days after treatment [33].

While adhesion between coatings and plastic substrates can be enhanced by hydrogen-bond interactions, still further enhancement can be obtained if the temperature is above the T_g of the plastic substrate. At temperatures above T_g, there is adequate free volume to permit resin molecules from the coating to move into the surface of the plastic and vice versa. Presumably, penetration can be deeper than the thin layer of mobile molecule segments at the surface. When the solvent evaporates, the intermingled molecules increase adhesion. This interaction can be enhanced by having the structure of the coating resin be sufficiently similar to the structure of the plastic so that the resin molecules somewhat soluble in the plastic substrate. In some cases, promotion of adhesion by heating the plastic substrate above its T_g is not feasible because the plastic substrate may undergo heat distortion.

Solvents in the coating that are soluble in the plastic can enhance adhesion. The solvent swells the plastic, lowering its T_g, and facilitating penetration of coating resin molecules into the surface of the plastic. The solvents should evaporate slowly to permit time for penetration to occur. Fast evaporating solvents, like acetone, can cause *crazing* of the surface of high T_g thermoplastics, like polystyrene and poly(methyl methacrylate). Crazing is the development of large numbers of minute surface cracks; see Section 32.3 for further discussion of crazing and coatings for plastics.

Adhesion to other coatings, commonly called *intercoat adhesion*, is another example of adhesion to plastics. The same principles apply. The surface tension of the coating being applied must be lower than the surface free energy of the substrate coating to permit wetting. The presence of polar groups in both coatings permits hydrogen bonding; in the case of thermosetting coatings, covalent bonding enhances intercoat adhesion. It has been found empirically that the presence of relatively small amounts of amine groups on resins commonly gives coatings with superior intercoat adhesion. Such comonomers as 2-(*N*,*N*-dimethylamino)ethyl methacrylate and 2-aziridinylethyl methacrylate have been used to make acrylic resins with enhanced intercoat adhesion.

Curing temperatures above T_g increase the probability of satisfactory adhesion. Use of compatible resins in the substrate coating and top coat also increases the probability of satisfactory adhesion. Using solvents in the coating that can swell the substrate coating is a commonly used technique for enhancing intercoat adhesion. Coatings with lower cross-link density are more swollen by solvents and, in general, are easier to adhere to than are coatings with high cross-link density. Sometimes, one can undercure the primer thus having a lower cross-link density when the top coat is applied. Cure of the primer is completed when the top coat is cured.

Adhesion to high gloss coatings is difficult to achieve because of their surface smoothness. Gloss coatings that have undergone excessive cross-linking on aging are particularly difficult surfaces to which to apply an adherent coating. Sanding to increase surface roughness may be necessary to achieve intercoat adhesion. One reason for formulating primers with low gloss is because they have rougher surfaces and, hence, are easier to adhere to. When possible, increasing the pigment loading of a primer above critical pigment volume concentration (CPVC) (see Chapter 23) facilitates adhesion of a top coat. Above CPVC, the dry film contains pores. When a top coat is applied, vehicle from the top coat can penetrate into the pores in the primer, providing a mechanical anchor to promote intercoat adhesion. Care must be exercised not to have PVC too much higher than CPVC or so much vehicle will be drained away from the top coat that the PVC of the top coat will increase, leading to a loss of gloss.

An essential requirement of many industrial coatings is *recoat adhesion*—that is, the ability of a coating to adhere to itself well enough that flawed or damaged objects can be repainted without extensive preparation. This requirement can be difficult to satisfy, especially with highly cross-linked gloss enamels. Additives to overcome film defects during application (see Section 25.3) may interfere with recoat adhesion.

6.6. TESTING FOR ADHESION

In view of the complexity of adhesion phenomena, it is not surprising that there is difficulty in devising suitable tests for adhesion. As is so often the case in coatings, the only really conclusive way of telling whether adhesion of a coating is satisfactory is to use the product and see whether the coating adheres over its useful life.

A common method formulators use to evaluate adhesion is to see how easily a penknife can scrape a coating from a substrate. By comparing the resistance of a new coating-substrate combination to combinations with known field performance, the formulator has some basis for performance prediction. While a penknife in the hand of an experienced person can be a valuable tool, it has major disadvantages as a test method. The experience is not easily transferred from one person to another; even the technique for the test is not easily transferred. Also, there is no good way of assigning a numerical value to the results. Thus, it does not provide a basis for following small changes in adhesion as a result of changes in composition to aid in developing hypotheses to relate composition and adhesion.

Relatively satisfactory test methods for evaluating adhesives have been developed, but few of these methods are applicable to coatings. Many investigators have worked on a variety of different methods in attempts to devise meaningful tests for evaluating the adhesion of coatings [35,36]. None of these tests are very satisfactory. For research purposes, the most useful technique is a direct pull test. A rod is fastened perpendicular to the upper surface of the coated sample with an adhesive. The panel is fastened to a support with a perpendicular rod on its back so that the two perpendicular rods are lined up exactly opposite each other. The assembly is put into the jaws of an Instron Tester (see Section 4.4), and the tensile force required to pull the coating off the substrate is recorded. Since the procedure is subject to considerable experimental error, multiple determinations must be made. Experienced operators can achieve precisions of ±15%. The adhesive must bond the rod to the coating surface more strongly than the coating is bonded to the substrate. It is also essential that the adhesive not penetrate into the coating to perturb the coating-substrate interface. Cyanoacrylate adhesives are generally satisfactory. The rods must be aligned exactly with each other and perpendicular to the coating. If the rod is at even a slight angle to the surface, stress is concentrated on only part of the substrate-coating interface, and less force is required to break the bond. Sometimes, the weakest component is the substrate—this may be nice for advertising purposes, but it does not provide a measure of the adhesive strength.

Another potential complication is cohesive failure of the coating; again, no information on adhesion is obtained. In reading the literature, one sometimes finds data from tests in which cohesive, adhesive, and mixed cohesive-adhesive failures have occurred. The authors may then discuss the improvement in adhesion from some change that resulted in a greater force to get adhesive failure, as compared to another sample that failed cohesively. Clearly, such comparisons are invalid. When there is cohesive failure, all that is known is that the adhesive strength is above the measured value (within experimental error).

One must use caution in interpreting the results even when the sample appears to have failed adhesively at the substrate-coating interface. Sometimes, when no coating can be seen on the substrate surface after the test, there is a monolayer (or thin layer) of material from the coating left on the substrate surface. In this event, failure was not at the substrate surface, but between the material adsorbed on the surface and the rest of the coating. Surface analysis is useful in determining the locale of failure and the identity of the adsorbed material. Fairly often, there is a combination of adhesive and cohesive failure. A possible explanation of such failures is that there was a fractural failure starting at some imperfection within the film, and the initial crack propagated down to the interface. Tensile values from samples that fail in this way cannot be compared to the tensile values of samples that failed adhesively.

The direct pull test does not evaluate the potentially important differences between the difficulty of breaking through a coating film and of shoving it out of the way, as mentioned in the beginning of the chapter. In spite of all the difficulties, direct pull tests are the most useful available. Instruments have been devised for direct pull tests under field conditions. The method is quite widely used for quality control in high performance maintenance and marine coatings. Serious disadvantages for use on actual products are that the test is destructive and the tested area must be repainted. See Ref. [35] for a more detailed discussion of the effect of variables on the test results.

Adhesion can be affected by the angle of application of stress. An instrument called STATRAM II has been devised to combine a normal load and lateral traction to measure friction induced damage [37]. Optical measurements are combined with measurements of total energy consumed during the scraping process. The test has been used to study delamination of coatings when plastic automobile bumpers rub together or scrape against solid objects. In many cases, cohesive failure of the plastic occurred near the surface of thermoplastic olefin (TPO), rather than adhesive failure between the coating and the substrate. Composition of coatings, especially solvents, can affect the structure of the upper layer of the plastic. See Section 30.3.2 for further discussion of adhesion of coatings to plastics.

Probably, the most widely used specification test is the *cross-hatch adhesion* test. Using a device with 6 or 11 sharp blades, a scratch mark pattern is made across the sample, followed by a second set cut perpendicular to the first. A strip of pressure sensitive adhesive tape is pressed over the pattern of squares and pulled off. Adhesion is assessed qualitatively on a 5 to 0 scale by comparing to a set of photographs, ranging from trace removal along the incisions to removal of most of the area. The test is subject to many sources of error; one being the rate at which the cuts are made. If the cuts are made slowly, they are likely to be even. However, if the cuts are made rapidly, it is possible that there will be cracks proceeding out from the sides of the cuts, due to more brittle behavior at higher rates of application of stress. Other important variables are the adhesive tape; the pressure with which it is applied; the angle and rate at which the tape is pulled off the surface; bending, if any, of the substrate during the test; and the surface of the coating to which the tape is applied. Some additives that appear to improve adhesion may actually only improve the test results by decreasing adhesion of the tape to the coating. The test may be useful for distinguishing between samples having poor adhesion and those having fairly good adhesion, but is not very useful in distinguishing among higher levels of adhesion. See Ref. [35] for a more detailed discussion of the variables affecting the test.

See Sections 7.3.1.1 and 7.5, which deal with corrosion, for further discussion of factors affecting adhesion and testing for adhesion in the presence of water.

GENERAL REFERENCES

K. L. Mittal, *Adhesion Aspects of Polymeric Coatings*, Plenum, New York, 1983.

S. R. Hartshorn, *Structural Adhesives: Chemistry and Technology*, Plenum, New York, 1986.

G. L. Nelson, "Adhesion," in J. V. Koleske, Ed., *Paint and Coatings Testing Manual*, 14th. ed., ASTM, Philadelphia, 1995, pp. 513–524.

J. A. Baghdachi, *Adhesion Aspects of Polymeric Coatings*, Federation of Societies for Coatings Technology, Blue Bell, PA, 1996.

REFERENCES

1. P. S. Sheih and J. L. Massingill, *J. Coat. Technol.*, **62** (781), 25 (1990).
2. D. Y. Perrera and D. Van den Eynden, *J. Coat. Technol.*, **59** (748), 55 (1987).
3. D. Y. Perera, *Prog. Org. Coat.*, **28**, 21 (1996).
4. O. Negele and W. Funke, *Prog. Org. Coat.*, **28**, 285 (1995).
5. J. G. Kloosterboer, *Adv. Polym. Sci.*, **84**, 1 (1988).
6. V. E. Basin, *Prog. Org. Coat.*, **12**, 213 (1984).
7. A. V. Pocius, "Fundamentals of Structural Adhesive Bonding," in S. R. Hartshorn, Ed., *Structural Adhesives: Chemistry and Technology*, Plenum, New York, 1986, pp. 23–68; G. B. Portelli, "Testing, Analysis, and Design," in S. R. Hartshorn, Ed., *Structural Adhesives: Chemistry and Technology*, Plenum, New York, 1986, pp. 407–449.
8. W. A. Zisman, *J. Coat. Technol.*, **44** (564), 42 (1972).
9. M. Yekta-Fard and A. B. Ponter, *J. Adhesion Sci. Technology.*, **6**, 253 (1992).
10. J. Y. Charmeau, E. Kientz, and Y. Holl, *Prog. Org. Coat.*, **27**, 87 (1996).
11. B. M. Perfetti, *Metal Surface Characteristics Affecting Organic Coatings*, Federation of Societies for Coatings Technology, Blue Bell, PA, 1994.
12. J. Rex, *J. Protective Coat. Linings*, October, 50 (1990).
13. B. S. Skerry, W. J. Culhane, D. T. Smith, and A. Alavi, *J. Coat. Technol.*, **62** (788), 55 (1991).
14. M. J. Dyett, *J. Oil Colour Chem. Assoc.*, **72**, 132 (1989).
15. N. Tang, W. I. Van Ooij, and G. Gorecki, *Prog. Org. Coat.*, **30**, 255 (1997).

16. C. K. Schoff, *J. Coat. Technol.*, **62**, (789), 115 (1990).
17. S. Maeda, *Prog. Org. Coat.*, **28**, 227 (1996).
18. L. Lori, A. Tamba, F. Deflorian, L. Fedrizzi, and P. L. Bonora, *Prog. Org. Coat.*, **27**, 17 (1996).
19. T. J. Lin, J. A. Antonelli, D. J. Yang, H. K. Yasuda, and F. T. Wang, *Prog. Org. Chem.*, **31**, 351 (1997).
20. G. Reinhard, *Prog. Org. Coat.*, **15**, 125 (1987).
21. H. J. Jacobasch, *Prog. Org. Coat.*, **17**, 115 (1989).
22. J. L. Massingill, P. S. Sheih, R. C. Whiteside, D. E. Benton, and D. K. Morisse-Arnold, *J. Coat. Technol.*, **62** (781), 31 (1990).
23. S. Maeda, *J. Coat. Technol.*, **55** (707), 43 (1983).
24. W. Funke, *J. Coat. Technol.*, **55** (705), 31 (1983).
25. E. P. Pluddemann, *Prog. Org. Coat.*, **11**, 297 (1983).
26. G. L. Witucki, *J. Coat. Technol.*, **65** (822), 57 (1993).
27. M. N. Sathyanaarayana and M. Yaseen, *Prog. Org. Coat.*, **26**, 275 (1995).
28. W. J. van Ooij and T. Child, *Chemtech*, Februrary, 26 (1998).
29. F. D. Rector, W. W. Blount, and D. R. Leonard, *J. Coat. Technol.*, **61** (771), 31 (1989).
30. R. A. Ryntz, *Polym. Mat. Sci. Eng.*, **67**, 119 (1992).
31. J. M. Lane and D. J. Hourston, *Prog. Org. Coat.*, **21**, 269 (1993).
32. R. A. Ryntz, *Painting of Plastics*, Federation of Societies for Coatings Technology, Blue Bell, PA, 1994.
33. F. Garbassi, M. Morra, and E. Occhiello, *Polymer Surfaces from Physics to Technology*, Revised and Updated Ed., John Wiley & Sons, New York, 1998, pp. 49–68.
34. K. F. Mansfield and D. N. Theodorou, *Macromolecules*, **24**, 6283 (1991).
35. G. L. Nelson, "Adhesion," in J. V. Koleske, Ed., *Paint and Coating Testing Manual*, 14th. ed., ASTM, Philadelphia, 1995, pp. 513–524.
36. T. R. Bullett and J. L. Prosser, *Prog. Org. Coat.*, **1**, 45 (1972).
37. A. C. Ramamurthy, J. A. Charest, M. D. Lilly, D. J. Mihora, and J. W. Freese, *Wear*, **203–204**, 350 (1997); D. J. Mihora and A. C. Ramamurthy, *Wear*, **203–204**, 362 (1997).

CHAPTER 7

Corrosion Protection by Coatings

Corrosion is a process by which materials, especially metals, are worn away by electro-chemical and chemical actions. In this chapter, we discuss the principles of corrosion and the protective role of organic coatings; specific types of coatings for corrosion control are covered in Chapters 26, 29, and 32.

7.1. CORROSION OF UNCOATED STEEL

The major economic losses are from the electrochemical corrosion of steel and most of the discussion in this chapter deals with protection of steel by use of organic coatings. Before we consider the role of coatings, it is important to understand the corrosion of uncoated steel.

In the corrosion of metals, electrochemical and chemical reactions are involved. An electrochemical (galvanic) element is formed when two pieces of different metals are connected with a conductive wire and partly immersed in an electrolyte, generally water containing some dissolved salts. An electrochemical reaction begins spontaneously with the oxidation of metal atoms and dissolution of ions at the anodic parts of the corrosion elements. Metals may be arranged in an electromotive series such that the noblest (least easily oxidized) metal forms the cathode. However, because polarization processes may change the character of a metal, for example by formation of thin layers of corrosion products or by the adsorption of gases such as hydrogen, the practical electromotive series differs somewhat from the theoretical one. The theoretical series for the metals involved in the discussion in this chapter, in decreasing order of ease of oxidation, is: magnesium, aluminum, zinc, iron, tin, and copper.

There are many kinds of steel; all are alloys of iron and carbon with other metals. Various kinds of steel corrode at different rates, depending on their composition and on the presence of mechanical stresses. The composition varies from location to location on the surface; as a result, some areas are anodic relative to other areas that are cathodic. Stresses and morphological structure of the metal surface can also be factors for setting up anode-cathode pairs. Cold-rolled steel has more internal stresses than hot-rolled steel and is generally more susceptible to corrosion; but, it is widely used because it is stronger. Internal stresses can also be created during fabrication or by the impact of a piece of gravel on an auto body.

Steel adsorbs on its surface a thin layer of water, which dissolves traces of soluble salt, thus providing the electrolyte for electrochemical reactions. In the absence of oxygen, ferrous ions are the primary anodic corrosion products of steel. At cathodes, hydrogen is formed, which stops further dissolution of iron by cathodic polarization, except when the pH of the electrolyte is low.

$$\text{Anode: Fe} \rightarrow \text{Fe}^{2+} + 2e^-$$

$$\text{Cathode: } 2H^+ + 2e^- \rightarrow H_2$$

However, in the presence of oxygen, depolarization of the cathode takes place, hydroxyl anions are formed at the cathodes, and dissolution of iron continues.

$$O_2 + H_2O + 4e^- \rightarrow 4OH^-$$

$$2Fe + O_2 + 2H_2O \rightarrow 2Fe^{2+} + 4OH^-$$

The rate of corrosion of steel depends on the concentration of oxygen dissolved in the water at the steel surface, as shown in Figure 7.1. At low concentrations, the rate increases with increasing dissolved oxygen concentration. At high concentrations, the rate declines because of *passivation*. (See Section 7.2.1.) The equilibrium concentration of oxygen in water exposed to the atmosphere at 25°C is about 6 mL L^{-1}.

Corrosion can occur at a significant rate only if there is a complete electrical circuit. The rate of corrosion depends on the conductivity of the water at the steel surface. Dissolved salts increase conductivity, which is one reason that the presence of salts increases the rate of corrosion of steel. Effects of salts on corrosion rates are complex; the reader is referred to Ref. [1], or other general texts on corrosion, for detailed discussions. The relationship between NaCl concentration and corrosion rate is shown in Figure 7.2. The dashed vertical line in the figure indicates the salt concentration in seawater. At higher salt contents, the

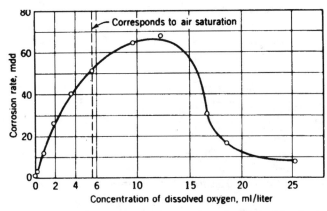

Figure 7.1. Effect of oxygen concentration on corrosion of mild steel in slowly moving distilled water, 48 hour test, 25°C. (From Ref. [1], with permission.)

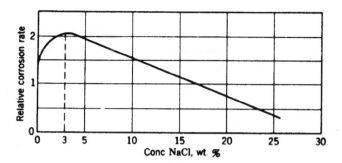

Figure 7.2. Effect of sodium chloride on corrosion of iron in aerated solutions at room temperature (composite of data from several investigators). (From Ref. [1], with permission.)

rate of corrosion decreases since solubility of oxygen decreases as NaCl concentration increases.

The rate of corrosion also depends on pH, as shown in Figure 7.3. Since iron dissolves in strong acid even without electrochemical action, it is not surprising that corrosion is most rapid at a low pH. Corrosion rate is nearly independent of pH between about 4 and 10. In this pH region, the initial corrosion causes a layer of ferrous hydroxide to precipitate near the anode. Subsequently, the rate is controlled by the rate of oxygen diffusion through the layer. Underneath, the surface of the iron is in contact with an alkaline solution with a pH of about 9.5. When the environmental pH is above 10, the increasing alkalinity raises the pH at the iron surface. The corrosion rate then decreases because of passivation. (See Section 7.2.1.)

Corrosion rate also depends on temperature, as shown in Figure 7.4. The reactions proceed more rapidly at higher temperatures, as indicated by the increase in corrosion rate

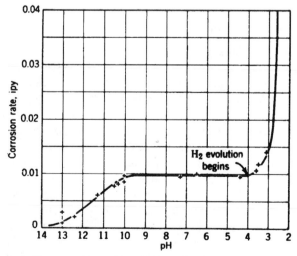

Figure 7.3. Effect of pH on corrosion of iron in aerated soft water at room temperature. (From Ref. [1], with permission.)

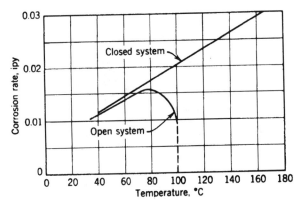

Figure 7.4. Effect of temperature on corrosion of iron in water containing dissolved oxygen. (From Ref. [1], with permission.)

in a closed system. However, the solubility of oxygen in water decreases as temperature increases, so in an open system, where the oxygen can escape, the rate of corrosion goes through a maximum at some intermediate temperature. The temperature at which corrosion rate maximizes is system dependent.

7.2. CORROSION PROTECTION OF METALS

Three strategies are employed to control electrochemical corrosion without the use of organic coatings. First, one can suppress the anodic reaction, as discussed in Section 7.2.1. Second, one can suppress the cathodic reaction, as discussed in Section 7.2.2. Third, one can cover steel with a barrier coat to prevent water and oxygen from contacting the surface, as discussed in Section 7.2.3.

7.2.1. Inhibition and Passivation

There are inhibitors that suppress corrosion. An important class acts by retarding the anodic reaction, they are called *passivators*. A passivator suppresses corrosion above some critical concentration, but may accelerate corrosion at lower concentrations by cathodic depolarization. This phenomenon is illustrated by the effect of oxygen concentration on corrosion rate as shown in Figure 7.1. Increasing oxygen concentration up to about 12 mL L^{-1} increases corrosion rate because it acts to depolarize the cathode. At higher concentrations, more oxygen reaches the surface than is reduced by the cathodic reaction; beyond that concentration, oxygen is a passivating agent. The mechanism of passivation has not been fully elucidated. According to one theory, if the oxygen concentration near the anode is high enough, ferrous ions are oxidized to ferric ions soon after they are formed at the anodic surfaces. Since ferric hydroxide is less soluble in water than ferrous hydroxide, a barrier of hydrated ferric oxide forms over the anodic areas. The iron is said to be passivated.

The critical oxygen concentration for passivation depends on conditions. It increases with dissolved salt concentration and with temperature, and it decreases with increases in pH and velocity of water flow over the surface. At about pH 10, the critical oxygen concentration reaches the value for air-saturated water (6 mL L^{-1}) and is still lower at higher pHs. As a result, iron is passivated against corrosion by the oxygen in air at sufficiently high pH values. It is impractical to control corrosion by oxygen passivation below about pH 10, since the concentrations needed are in excess of those dissolved in water in equilibrium with air. However, a variety of oxidizing agents can act as passivators. Chromate, nitrite, molybate, plumbate, and tungstate salts are examples. As with oxygen, a critical concentration of these oxidizing agents is needed to achieve passivation. The reactions with chromate salts have been most extensively studied. Partially hydrated mixed ferric and chromic oxides are deposited on the surface, where they presumably act as a barrier to halt the anodic reaction.

Certain nonoxidizing salts, such as alkali metal salts of boric, carbonic, phosphoric, and benzoic acids, also act as passivating agents. Their passivating action may result from their basicity. By increasing pH, they may reduce the critical oxygen concentration for passivation below the level reached in equilibrium with air. Alternatively, it has been suggested that the anions of these salts may combine with ferrous or ferric ions to precipitate complex salts to form a barrier coating at the anode. Possibly, both mechanisms operate to some extent.

Many organic compounds are corrosion inhibitors for steel. Most are polar substances that tend to adsorb on high energy surfaces [2]. Amines are particularly widely used. Clean steel wrapped in paper impregnated with a volatile amine or the amine salt of a weak acid is protected against corrosion. Amines are also used in boiler water to minimize corrosion. The reason for their effectiveness is not clear. They may act as inhibitors because they are bases and neutralize acids. It may be that amines are strongly adsorbed on the surface of the steel by hydrogen bonding or salt formation with acidic sites on the surface of the steel. This adsorbed layer then may act as a barrier to prevent oxygen and water from reaching the surface of the steel.

Aluminum is higher in the electromotive series than iron and is more easily oxidized. Yet, aluminum generally corrodes more slowly than steel. A freshly exposed surface of aluminum oxidizes quickly to form a dense, coherent layer of aluminum oxide. In other words, aluminum is passivated by oxygen at concentrations in equilibrium with air. On the other hand, aluminum corrodes more rapidly than iron under either highly acidic or highly basic conditions. Also, salt affects the corrosion of aluminum even more than it affects the corrosion of iron; aluminum corrodes rapidly in the presence of sea water.

7.2.2. Cathodic Protection

If steel is connected to the positive pole of a battery or a direct current source while the negative pole is connected to a carbon electrode, and both electrodes are immersed in salt water, the steel does not corrode. The impressed electrical potential makes the whole steel surface cathodic relative to the carbon anode. The result is electrolysis of water, rather than corrosion of steel. This is an example of cathodic protection.

A related method is to connect the steel electrically to a piece of metal higher in the electromotive series than iron—for example, magnesium, aluminum, or zinc. When a block of one of these metals is connected to steel and immersed in an electrolyte, the metal is the only anode, in the circuit, and all corrosion takes place at that anode. The metal is called a sacrificial anode. This method is often used to protect pipelines and steel hulls of ships. The sacrificial anode is gradually used up and must be replaced periodically. Zinc and magnesium are generally the preferred sacrificial metals. Aluminum is often ineffective because a barrier layer of aluminum oxide forms on its surface. (See Section 7.2.1.) However, aluminum is appropriate for marine applications, since it corrodes readily in salt water.

Another method of cathodic protection is coating steel with zinc to make galvanized steel. The steel is protected in two ways: Zinc functions as a sacrificial anode, and also acts as a barrier preventing water and oxygen from reaching the steel surface. Since zinc is easily oxidized, it is passivated by oxygen at concentrations below 6 mL L^{-1}. If the surface of a galvanized sheet is damaged and bare steel and zinc are exposed, the zinc corrodes, but not the steel. After exposure to the atmosphere, the surface of the zinc becomes coated with a mixture of zinc hydroxide and zinc carbonate. Both are somewhat soluble in water and strongly basic.

7.2.3. Barrier Protection

For steel to corrode, oxygen and water must be in direct molecular contact with the surface of the steel. Barriers that can prevent oxygen and water from reaching the surface prevent corrosion. The zinc layer on galvanized steel acts as a barrier. It may even be considered that a layer of irreversibly adsorbed small molecules can act as a barrier.

It is often incorrectly assumed that the tin coating on steel in tin cans acts electrochemically similarly to zinc in galvanized steel. Tin is lower in the electromotive series than iron, so the iron is the anode and the tin the cathode. Before a can is opened, the tin coating is intact and acts as a barrier so that no water or oxygen reaches the steel. After a can has been opened, the cut bare edges expose both steel and tin to water and oxygen, and the steel corrodes relatively rapidly.

7.3. CORROSION PROTECTION BY INTACT COATINGS

Coatings can be effective barriers to protect steel when it is anticipated that the coating can be applied to cover essentially all of the substrate surface and when the film remains intact in service. However, when it is anticipated that there will not be complete coverage of the substrate or that the film will be ruptured in service, alternative strategies using coatings that can suppress electrochemical reactions involved in corrosion may be preferable; they are discussed in Section 7.4. It is seldom effective to try to use both strategies at the same time—one must choose one or the other.

7.3.1. Critical Factors

Until about 1950, coatings were generally believed to protect steel by acting as a barrier to keep water and oxygen away from the steel surface. Then it was found by Mayne [3] that the permeability of paint films was so high that the concentration of water and oxygen coming through the films would be higher than the rate of consumption of water and oxygen in the corrosion of uncoated steel. Mayne concluded that barrier action could not explain the effectiveness of coatings and proposed that the electrical conductivity of coating films is the variable that controls the degree of corrosion protection. Presumably, coatings with high conductivity would give poor protection, as compared to coatings of lower conductivity. It was confirmed experimentally that coatings having very high conductivity afforded poor corrosion protection. However, in comparisons of films with relatively low conductivity, little correlation between conductivity and protection has been found. It may be that high conductivity films fail because they also have high water permeability; some investigators believe that conductivity of all coatings is at least a factor in corrosion protection [4,5].

Current understanding of protection of steel against corrosion by intact films is based, to a significant degree on the work of Funke [6–9]. He found that an important factor not given sufficient emphasis in earlier work was adhesion of a coating to steel in the presence of water. Funke proposed that water permeating through an intact film could displace areas of film from steel. In such cases, the film shows poor *wet adhesion*. Water and oxygen dissolved in the water would then be in direct contact with the steel surface; hence, corrosion would start. As corrosion proceeds, ferrous and hydroxide ions are generated, leading to formation of an osmotic cell under the coating film. Osmotic pressure can provide a force to remove more coating from the substrate. Osmotic pressure can be expected to range between 2500 and 3000 kPa, whereas the resistance of organic coatings to deformational forces is lower, ranging from 6 to 40 kPa [6]. Thus, blisters form and expand, exposing more unprotected steel surface. It has also been proposed that blisters can grow by a nonosmotic mechanism [10]. The suggestion has been made that water absorbed by a coating induces in-plane compressive stress within the coating and elastically extends the interfacial bonds between the coating and the steel substrate [11]. At a point of weak adhesion between the coating and the substrate, the stress can lead to disbondment.

In either osmotic or nonosmotic mechanisms, the key to maintaining corrosion protection by a coating is sufficient adhesion to resist displacement forces. Both mechanisms predict that if the coating covers the entire surface of the steel on a microscopic, as well as macroscopic scale, and if perfect wet adhesion could be achieved at all areas of the interface, the coating would indefinitely protect steel against corrosion. It is difficult to achieve both of these requirements in applying coatings, so a high level of wet adhesion becomes an important, but not the only factor affecting corrosion protection by coatings. Funke found that in addition to wet adhesion, low water and oxygen permeability help increase corrosion protection [8]. In any case, if wet adhesion is poor, corrosion protection is also poor. However, if the adhesion is fairly good, a low rate of water and oxygen permeation may delay loss of adhesion long enough so that there is adequate corrosion protection for many practical conditions.

Primers made with saponification resistant vehicles commonly give better corrosion protection than primers made with vehicles that saponify readily [12,13]. When water and oxygen permeate through a film and water displaces some of the adsorbed groups of the coating resin from the surface of the steel, corrosion starts. Hydroxide ions are generated at the cathodic areas. Hydroxide ions catalyze hydrolysis (saponification) of such groups as esters. If the backbone of the vehicle resin is connected by ester groups, hydrolysis results in polymer network degradation, leading to poorer wet adhesion and, ultimately catastrophic failure.

7.3.2. Adhesion for Corrosion Protection

Chapter 6 deals broadly with adhesion, but wet adhesion, which is especially critical to corrosion protection, is reviewed here. Good dry adhesion must be taken as a given for achieving corrosion protection. If there is no coating left on the substrate, it cannot protect the steel. It has not been so obvious, however, that good wet adhesion is required. Good wet adhesion means that the adsorbed layer of the coating will not desorb when water permeates through the film.

The first step to obtain good wet adhesion is to clean the steel surface, especially to remove any oils and salts. Application of phosphate conversion coatings (see Section 6.4.1) gives further advantages. Various types of steel and coated steel may require different cleaning and treatment methods [14]. New experimental surface treatments with silyl derivatives show promise for replacing phosphate conversion treatments. (See Sections 6.4.1 and 6.4.3.) After cleaning and treating, the surface should not be touched and should be coated as soon as possible. Finger prints leave oil and salt on the surface. After exposure to high humidity, fine blisters can form, disclosing the identity of the miscreant by the finger prints. A rusty handprint was once observed on a ship after only one ocean and lake passage [2]. When coating surfaces near the ocean, it is critical to avoid having any salt on the metal surface when the coating is applied.

It is also critical to achieve as nearly complete penetration into the micropores and irregularities in the surface of the steel as possible. If any steel is left uncoated, when water and oxygen reach the surface, corrosion will start, generating an osmotic cell that can lead to blistering. As discussed in Section 6.1, an important factor for achieving penetration is that the viscosity of the external phase be as low as possible and remain low long enough to permit complete penetration. It is desirable to use slow evaporating solvents, slow cross-linking coatings, and, when possible, baking primers. Macromolecules may be large compared to the size of small crevices, so lower molecular weight components may give better protection.

Wet adhesion requires that the coating not only be adsorbed strongly on the surface of the steel, but also that it not be desorbed by water that permeates through the coating. Empirically, it is found that wet adhesion is enhanced by having several adsorbing groups scattered along the resin chain, with parts of the resin backbone being flexible enough to permit relatively easy orientation and other parts rigid enough to assure that there are loops and tails sticking up from the surface for interaction with the rest of the coating. Another reason another baking primers commonly provide superior corrosion protection is that at the higher temperature, there may be greater opportunity for orientation of resin molecules at the steel interface. Amine groups are particularly effective polar substituents for promoting wet adhesion. Perhaps, water is less likely to displace amines than other groups from the surface. Phosphate groups also promote wet adhesion. For example, epoxy phosphates have been used to enhance the adhesion of epoxy coatings on steel [15]. Carboxylic acid-substituted polymers, such as polyacrylic acid, also promote wet adhesion by forming salt groups [16]. However, if the coating is ruptured, the ready availability of water and the base generated by corrosion make polyacrylic acid ineffective. There is need for further research to understand the relationships between resin structure and wet adhesion.

Saponification resistance is another important factor. Corrosion generates hydroxide ions at the cathode, raising pH levels as high as 14. Ester groups in the backbone of a binder can be saponified, degrading the polymer near the interface and reducing wet adhesion. Epoxy-phenolic primers are an example of high bake primers that are completely resistant to hydrolysis. In some epoxy-amine primers, there are no hydrolyzable groups. Amine-terminated polyamides, which are widely used in air dry primers to react with epoxy resins, have amide groups in the backbone that can hydrolyze. However, amides are more resistant to base-catalyzed hydrolysis than are esters. Alkyd resins are used when only moderate corrosion protection is required and low cost is important. Epoxy ester primers show greater resistance to saponification than do alkyd primers.

Water-soluble components that may stay in primer films should be avoided because they can lead to blister formation. For example, zinc oxide is generally an undesirable pigment to use in primers. Its surface interacts with water and carbon dioxide to form zinc hydroxide and zinc carbonate, which are somewhat soluble in water and can lead to osmotic blistering. Passivating pigments, discussed in Section 7.4.2, cannot function unless they are somewhat soluble in water; their presence in coating films can, therefore, lead to blistering. Funke showed that hydrophilic solvents, which become immiscble in the drying film as other solvents evaporate, can be retained as a separate phase and lead to blister formation [6].

7.3.3. Factors Affecting Oxygen and Water Permeability

Many factors affect permeability of coating films to water and oxygen [17]. Water and oxygen can permeate, to at least some extent, through any amorphous polymer film, even though the film has no imperfections such as cracks or pores. Small molecules travel through the film by jumping from free volume hole to free volume hole. Free volume increases as temperature increases above T_g. Therefore, normally, one wants to design coatings with a T_g above the temperature at which corrosion protection is desired. Since cross-linking reactions become slow as the increasing T_g of the cross-linking polymer approaches the temperature at which the reaction occurs and become very slow at $T < T_g$, air dry films cannot have T_g values much above ambient temperatures. Water can act as a plasticizer for coatings such as epoxy-amines and polyurethanes; the swelling caused by the water increases internal stress that can lead to delamination [18]. The internal stresses increase when a film is cycled through wet and dry stages. If the T_g is higher than the temperature of the water, water absorption is decreased, so internal stresses do not build up. The higher T_g values that can be reached with baked coatings may be another factor in their generally superior corrosion protection. In general, higher cross-link density leads to lower permeability. Both T_g and cross-link density affect other coating properties; so that some compromise between T_g and cross-link density must be accepted.

Permeability is also affected by the solubility of oxygen and water in a film. The variation in oxygen solubility is probably small, but variation in water solubility can be large. Salt groups on a polymer lead to high solubility of water in films. This makes it difficult to formulate high performance air dry, water-reducible coatings that are solubilized in water by amine salts of carboxylic acids. Although to a lesser degree than salts, resins made with polyethylene oxide backbones are likely to give high water permeabilities, as are silicone resins. On the other hand, water has low solubility in halogenated polymers; hence vinyl chloride and vinylidene chloride copolymers and chlorinated rubber are commonly used in formulating top coats for corrosion resistance. Fluorinated polymers have low permeabilities and good wetting properties; hydroxy-functional poly(vinylidene fluoride) cross-linked with polyisocyanates is reported to give good corrosion protection even as a single coat [19].

Pigmentation can have significant effects on water and oxygen permeability. Oxygen and water molecules cannot pass through pigment particles; therefore, permeability decreases as PVC increases. However, if the PVC exceeds CPVC, there are voids in the film, and passage of water and oxygen through the film is facilitated. Some pigments have high-polarity surfaces that adsorb water, and in cases in which water can displace polymer adsorbed on such surfaces, water permeability can be expected to increase with increasing pigment content. Pigments should be used that are as free as possible of water-soluble impurities and use of hydrophylic pigment dispersants should be avoided, or at least minimized.

Pigments with platelet shaped particles can reduce permeability rates as much as fivefold when they are aligned parallel to the coating surface [7,20]. A factor favoring alignment is shrinkage during solvent evaporation. Since oxygen and water vapor cannot pass through the pigment particles, the presence of aligned platelets can reduce the rate of vapor permeation through a film. The alignment is critical to the action of the platelets; if they are not aligned, permeability may be increased, especially if the film thickness is small relative to the size of the platelets. (See Section 29.1.2 for discussion of factors affecting platelet alignment.) Mica, talc, micaceous iron oxide, glass flakes, and metal flakes are examples of such pigments. Aluminum flake is widely used; stainless steel and nickel platelets, while more expensive, have greater resistance to extremes of pH. When appearance permits, use of leafing aluminum pigment (see Section 19.2.5) in the top coat is particularly effective. Leafing aluminum is surface treated, so its surface free energy is very low. As a result, the platelets come to the surface during film formation, creating an almost continuous barrier. In formulating coatings with leafing aluminum, it is necessary to avoid resins and solvents that displace the surface treatment from the flakes.

A Monte Carlo simulation model of the effect of several variables on diffusion through pigmented coatings has been devised [21]. The model indicates, as would be expected, that finely dispersed, lamellar pigment particles at a concentration near, but below CPVC give the best barrier performance.

There are advantages to applying multiple layers of coatings. The primer can be designed so that it has excellent penetration into the substrate surface and has excellent wet adhesion without particular concern about other properties. The top coat(s) can provide for minimum permeability and other required properties. The primer film does not need to be thick, as long as the top coat is providing barrier properties; the lower limit is probably controlled by the need to assure coverage of the entire surface. Funke has reported good results with 0.2 μm primer thickness [9], or even a 10 nm layer of a wet adhesion promoting polymer [16]. Another advantage of applying multiple coats is the decrease in probability that any area of the substrate will escape having any coating applied.

Film thickness affects the time necessary for permeation through films. Thicker films are expected to delay somewhat arrival of water and oxygen at the interface, but are not expected to affect the equilibrium condition. The corrosion protection afforded by intact films would be expected to be essentially independent of film thickness. However, since film thickness affects the mechanical performance of films, there may be some optimum film thickness for the maintenance of an intact film. For example, erosion losses would take longer to expose bare metal as film thickness increases, but the probability of cracking on bending increases as film thickness increases. However, in air dry, heavy duty maintenance coatings, there is generally a film thickness, dependent on the coating, that provides a more than proportional increase in corrosion protection relative to thinner films. Commonly, this film thickness is as much as 400 μm or more. Funke suggests that below certain coating thicknesses, there may be microscopic defects extending down through the film to the substrate [22]. The film may look intact, but there may be microscopic defects that are large compared to the free volume holes through which permeation in fully intact films occurs. A potential source of such defects is cracks resulting from shrinkage of films as the last solvent is lost from a coating, with the T_g of the solvent-free system around ambient temperature. Funke suggests that if the film is thick enough, such defects may not reach the substrate, hence substantially reducing passage of water and oxygen. This hypothesis is consistent with the general observation that greater protection is achieved by applying more coats to reach the same film thickness. In line with this proposal, the use of barrier platelet pigments permits a reduction in the required film thickness without loss of protection. The platelets may minimize the probability of defects propagating through the film to the substrate. Such defects are less likely to occur in baked films, and this may be another factor in the generally superior corrosion protection afforded by baked films, even though thinner film thicknesses are used.

7.4. CORROSION PROTECTION BY NONINTACT FILMS

Even with coatings designed to minimize the probability of mechanical failure, in many end uses, there will be breaks in the films during their service life. There are situations in which it is not possible to have full coverage of all the steel surface. In such cases, it is generally desirable to design coatings to suppress electrochemical reactions, rather than primarily for their barrier properties.

7.4.1. Minimizing Growth of Imperfections

If there are gouges through the film down to bare metal, water and oxygen reach the metal and corrosion starts. If the wet adhesion of the primer to the metal is not adequate, water creeps under the coating, and the coating comes loose from the metal over a wider and wider area. Poor hydrolytic stability can be expected to exacerbate the situation. This mode of failure is called cathodic delamination. Control of cathodic delamination requires wet adhesion and saponification resistance. It has also been shown that blisters are likely to develop under a film near the location of a gouge [2,7].

When wet adhesion varies on a local scale, *filiform corrosion* can occur [23]. It is characterized by development of thin threads of corrosion wandering randomly under the film, but never crossing another track. Formation of these threads often starts from the edge of a scratch. At the head of the thread, oxygen permeates through the film, and cathodic delamination occurs. The head grows following the directions of poorest wet adhesion. Behind the head, oxygen is consumed by oxidation of ferrous ions and ferric hydroxide precipitates, passivating the area, explaining why threads never cross. Since the ion concentration decreases, osmotic pressure drops, and the thread collapses, but it leaves a discernable rust track. Filiform corrosion can be difficult to see through pigmented films.

7.4.2. Primers with Passivating Pigments

Passivating pigments promote formation of a barrier layer over anodic areas, passivating the surface. (See Section 7.2.1.) To be effective, such pigments must have some minimum solubility. However, if the solubility is too high, the pigment would leach out of the coating film too rapidly, limiting the time that it is available to inhibit corrosion. For the pigment to be effective, the binder must permit diffusion of water to dissolve the pigment. Therefore, the use of passivating pigments may lead to blistering after exposure to humid conditions. Such pigments are most useful in applications in which the need to protect the steel substrate after film rupture has occurred outweighs the desirability of minimizing the probability of blistering. They are also useful when it is not possible to remove all surface contamination (blistering will probably occur anyway) or when it is not possible to achieve complete coverage of the steel by the coating.

Red lead pigment, Pb_3O_4 containing 2–15% PbO, has been used as a passivating pigment since the mid-19th century. Red lead in oil primers are used for air dry application over rusty, oily steel. (See Section 34.1.3.) The mechanisms of action are not fully understood [2]. They presumably include oxidation of ferrous ions to ferric ions followed by coprecipitation of mixed iron-lead salts or oxides. The somewhat soluble PbO raises the pH and neutralizes any fatty acids formed over time by hydrolysis of the drying oil. Toxic hazards of red lead restrict its use to certain industrial and special purpose applications, and regulations can be expected to prohibit its use.

The utility of chromate pigments for passivation is well established. Various mechanisms have been proposed to explain their effectiveness [2]. All the proposed mechanisms require that the chromate ions be in aqueous solution. Like all passivators, chromate ions accelerate corrosion at low concentrations. The critical minimum concentration for passivation at 25°C is approximately 10^{-3} mols CrO_4^{2-} L^{-1}. The critical concentration increases with increasing temperature and increasing NaCl concentration. Sodium dichromate is an effective passivating agent, but would be a poor passivating pigment; its solubility in water (3.3 mols CrO_4^{2-} L^{-1}) is too high. It would be rapidly leached out of a film and would probably cause massive blistering. At the other extreme, lead chromate (chrome yellow) is so insoluble (5×10^{-7} mols CrO_4^{2-} L^{-1}) that it has no electrochemical action.

"Zinc chromates" have been widely used as passivating pigments. The terminology is poor, since zinc chromate itself is too insoluble and could promote corrosion, rather than passivate. Zinc yellow pigment is [$K_2CrO_4 \cdot 3ZnCrO_4 \cdot Zn(OH)_2 \cdot 2H_2O$], (alternative ways of giving the same composition are $Zn_4K_2Cr_4O_{20}H_6$ and ($4ZnO \cdot K_2O \cdot 4CrO_3 \cdot 3H_2O$). It has an appropriate solubility (1.1×10^{-2} mols CrO_4^{2-} L^{-1} at 25°C) and has been widely used in primers. Zinc tetroxychromate [$ZnCrO_4 \cdot 4Zn(OH)_2$ also written as $Zn_5CrO_{12}H_8$] has a solubility lower than desirable (2×10^{-4} mols CrO_4^{2-} L^{-1}), but is used in *wash primers*. Phosphoric acid is added to wash primers before application; it may be that this changes the solubility so that the chromate ion concentration is raised to an appropriate level. Strontium chromate ($SrCrO_4$) has an appropriate solubility in water (5×10^{-3} mols CrO_4^{2-} L^{-1}) and is sometimes used in primers, especially latex paint primers, in which the more soluble zinc yellow can cause problems of package stability.

It has been established that zinc chromates—and, presumably, other soluble chromates—are carcinogenic to humans. They must, therefore, be handled with appropriate caution. In some countries, their use has been prohibited and prohibition worldwide is probable in the future. Substantial efforts have been undertaken to develop less hazardous passivating pigments [24]. However, it is difficult to conclude from the available literature and supplier technical bulletins how these pigments compare with each other and with zinc yellow. In some cases, a formulation that has been optimized for one pigment is compared to a formulation containing another pigment that may not be the optimum formulation for that pigment. (A common example is the substitution of one pigment for another on an equal weight basis, rather than formulating to the same ratio of PVC to CPVC; the results could be very misleading, since primer performance is quite sensitive to the PVC-CPVC ratio.) Much of the published data is based on comparing corrosion resistance in salt fog chamber tests (or other laboratory tests) rather than on actual field experience. A problem with accelerated tests is that an important factor in the performance of a pigment is the rate at which it is leached from a film. As discussed in Section 7.5, there is no laboratory test available that provides reliable predictions of field performance.

Basic zinc and zinc-calcium molybdates are said to act as passivating agents in the presence of oxygen, apparently leading to precipitation of a ferric molybdic oxide barrier layer on the anodic areas. Barium metaborate is the salt of a strong base and a weak acid. It may act by increasing the pH, thus lowering the critical concentration of oxygen required for passivation. To reduce its solubility in water, the pigment grade is coated with silicon dioxide. Even then, some workers feel that the solubility is too high for use in long-term exposure conditions. Zinc phosphate, $Zn_3(PO_4)_2 \cdot 2H_2O$, has been used in corrosion protective primers and may act by forming barrier precipitates on the anodic areas. There is considerable difference of opinion as to its effectiveness. Calcium and barium phosphosilicates and borosilicates are being used increasingly; they may act by increasing pH.

These pigments are all inorganic pigments; a wider range of potential oxidizing agents would seem to be available if organic pigments were used. An example of a commercially available organic pigment is the zinc salt of 5-nitroisophthalic acid. It is said to be as effective as zinc yellow at lower pigment levels. (Since over a long time, any effective passivating pigment will be lost by leaching, it seems doubtful that an equal lifetime could be achieved at a substantially lower pigment content.) The zinc salt of 2-benzothia-zoylthiosuccinic acid has been recommended as a passivating agent.

7.4.3. Cathodic Protection by Zinc-Rich Primers

Zinc-rich primers are another approach to protecting steel with nonintact coatings. They are designed to provide the protection given by galvanized steel, but to be applied to a steel structure after fabrication [25]. The primers contain high levels of powdered zinc—over 80% by weight is usual. On a volume basis, the zinc content exceeds CPVC to assure good electrical contact between the zinc particles and with the steel. Furthermore, when PVC is greater than CPVC, the film is porous, permitting water to enter, completing the electrical circuit. The CPVCs of zinc powders vary, depending primarily on particle size and particle size distribution; values on the order of 67% have been reported [26]. The zinc serves as a sacrificial anode, and zinc hydroxide is generated in the pores.

Vehicles for zinc-rich primers must be saponification resistant; alkyds are not appropriate resins for this application. Among organic binders, epoxies are the most commonly used. However, the most widely used vehicles are tetraethyl orthosilicate and oligomers derived from it by controlled partial hydrolysis with a small amount of water. (See Section 16.5.5.) Ethyl or isopropyl alcohol is used as the principal solvent, since an alcohol helps maintain package stability. After application, the alcohol evaporates, and water from the air completes the hydrolysis of the oligomer to yield a film of polysilicic acid partially converted to zinc salts. Cross-linking is affected by relative humidity (RH); properties can be adversely affected if the RH is low at the time of application. Such a primer is referred to as an *inorganic zinc-rich primer*. The need to reduce VOC emissions has led to the development of water-borne zinc-rich primers. These use sodium, potassium and/or lithium silicate solutions in water as the vehicle. In many applications, they have been found to be as useful as solvent-borne zinc-rich primers.

Properly formulated and applied, zinc-rich primers are very effective in protecting steel against corrosion. Their useful lifetime is not completely limited by the amount of zinc present, as one might first assume. Initially, the amount of free zinc decreases from the electrochemical reaction; later, loss of zinc metal becomes slow, but the primer continues to protect the steel. Possibly, the partially hydrated zinc oxide formed in the initial stages of corrosion of the zinc fills the pores and, together with the remaining zinc, acts as a barrier coating [27]. It is also possible that the zinc hydroxide raises the pH to the level at which oxygen can passivate the steel.

Zinc is expensive, especially on a volume basis. Early attempts to replace even 10% of the zinc with low-cost inert pigment caused a serious decrease in performance, presumably due to decrease in metal to metal contact, even though the PVC was above CPVC. A relatively conductive inert pigment, iron phosphide (Fe_2P) has shown promise [28]. It has been reported that in ethyl silicate based coatings up to 25% of the zinc can be replaced with Fe_2P; however, with epoxy-polyamide coatings replacement of part of the Zn with Fe_2P leads to a reduction in protection [29]. The Steel Structures Painting Council is comparing field exposure performance of Fe_2P extended primers to that of conventional zinc-rich primers.

Zinc-rich primers are frequently top coated to minimize corrosion of the zinc, protect against physical damage, and improve appearance. Formulation and application of top coats must be done with care. If the vehicle of the top coat penetrates into the pores in the primer film, conductivity of the primer may be substantially reduced, rendering it ineffective. See Section 32.1.2 for further discussion.

7.5. EVALUATION AND TESTING

There is no laboratory test available that can be used to predict corrosion protection performance of a new coating system. This unfortunate situation is an enormous obstacle to research and development of new coatings, but it must be recognized and accommodated.

Use testing is the only reliable test of a coating system—that is, to apply it and then observe its condition over years of actual use. The major suppliers and end users of coatings for such applications as bridges, ships, chemical plants, and automobiles have collected data correlating performance of different systems over many years. These data provide a basis for selection of current coatings systems for particular applications. They also provide insight into how new coatings could be formulated to improve chances of success.

Simulated tests are the next most reliable for predicting performance. One common approach is to expose laboratory prepared panels on test fences in inland south Florida or on beaches in south Florida or North Carolina. The difficulties in developing tests to simulate corrosion in marine environments are discussed in Ref. [30]. Test conditions must simulate actual use conditions as closely as possible. For example, exposure at higher temperatures may accelerate corrosion reactions; however, oxygen and water permeability can be affected by $(T - T_g)$. If actual use will be at temperatures below T_g, but the tests are run above T_g, no correlation should be expected.

Variables in preparation of test panels are frequently underestimated. The steel used is a critical variable [31]. Also significant are how the steel is prepared for coating and how the coating is applied. Film thickness, evenness of application, flash off time, baking time and temperature, and many other variables affect performance. Results obtained with carefully prepared and standardized laboratory panels can be quite different than results with actual production products. In view of these problems, it is desirable, when possible, to paint test sections on ships, bridges, chemical storage tanks, and so forth, and to observe their condition over the years. The long times required for evaluation are undesirable, but the results can be expected to correlate reasonably with actual use.

Since wet adhesion is so critical to corrosion protection, techniques for studying wet adhesion can be very useful. *Electrochemical impedance spectroscopy* (EIS) is widely used to study coatings on steel. Many papers are available covering various applications of EIS; References [32] and [33] are reviews that provide extensive discussion of the theory and interpretation of data. Impedance is the apparent opposition to flow of an alternating electrical current and is the inverse of apparent capacitance. When a coating film begins to delaminate, there is an increase in apparent capacitance. The rate of increase of capacitance is proportional to the amount of area delaminated by wet adhesion loss. High performance systems show slow rates of increase of capacitance, so tests must be continued for long time periods. Onset of delamination can be determined by EIS studies [34]. As with many tests of coatings, results of EIS tests are subject to considerable variation; it has been recommended that a minimum of five replicate tests should be run [35]. The method is very sensitive for detecting defects, but no information is obtained as to whether the defect is characteristic of the coating system or a consequence of poor application. A series of other problems involved in EIS are discussed in Ref. [36].

There have been many attempts to develop laboratory tests to predict corrosion protection by coatings, and these efforts continue. However, available tests have limited reliability in predicting performance; nonetheless, they are widely used. The most widely used test method for corrosion resistance is the salt spray (fog) test (ASTM Method B-117-95). Coated steel panels are scribed (cut) through the coating in a standardized fashion exposing bare steel and hung in a chamber where they are exposed to a mist of 5% salt solution at 100% relative humidity at 35°C. Periodically, the nonscribed areas are examined for blistering, and the scribe is examined to see how far from the scribe mark the coating has been undercut or has lost adhesion. It has been repeatedly shown that there is little, if any, correlation between results from salt spray tests and actual performance of coatings in use [37–40].

Many factors are probably involved in the unreliability of the salt spray test. Outdoor exposure can have a significant effect on film properties, and environmental factors such as acid rain vary substantially from location to location. The application of the scribe mark can be an important variable; narrow cuts generally affect corrosion less than broader ones. Also, if the scribe mark is cut rapidly, there may be chattering of cracks out from the main cut, whereas slow cutting may lead to a smooth cut. A passivating pigment with high solubility might be very effective in a laboratory test, but may provide protection for only a limited time under field conditions, owing to loss of passivating pigment by leaching.

Since with intact films, it is common for the first failure to be blister formation, humidity resistance tests are also widely used (ASTM Method D-2247-94). The face of a panel is exposed to 100% relative humidity at 38°C, while the back of the panel is exposed to room temperature. Thus, water continuously condenses on the coating surface. This humidity test is a more severe test for blistering than the salt fog test because pure water on the film generates higher osmotic pressures with osmotic cells under the film than the salt solution used in the salt fog test. It is common to run the test at 60°C "because it is a more severe test." The pitfalls of this approach are obvious in view of the previous discussion of the importance of $(T - T_g)$. Humidity tests do not provide a prediction of the life of corrosion protection, but may provide useful comparisons of wet adhesion. Funke recommends testing for wet adhesion by scribing panels after various exposure times in a humidity chamber, immediately followed by applying pressure sensitive tape across the scribe mark, and then pulling the tape off the panel [22]. A peel adhesion test for wet adhesion has been described [41]. Wet adhesion can also be checked after storing panels in water [42].

It is often observed that alternating high and low humidity causes faster blistering than continuous exposure to high humidity. A possible explanation of this is that intermediate corrosion products form colloidal membranes, causing polarization and temporary inhibition of corrosion. The membranes are not stable enough to survive drying out and aging. Another factor may be the increase in internal stress that has been reported by cycling through wet and dry periods [18]. A large number of humidity cycling tests has been described, commonly involving repeated immersion in warm water and removal for several hours. In some industries, such tests have become accepted methods of screening coatings, although their predictive value is questionable. Simply correlating them with salt fog tests proves little.

A testing regimen called *Prohesion* (a trademark of BP Chemicals) has been reported to correlate better with actual performance than the standard salt spray test [43]. The procedure combines care in selection of substrates that will reflect real products, use of thin films, emphasis on adhesion checks, and a modified salt mist exposure procedure. Instead of 5% NaCl solution, a solution of 0.4% ammonium sulfate and 0.05% NaCl is used. Scribed panels are sprayed with the mixed salt solution cycling over 24 hours, six 3-hour periods alternating with six 1-hour drying periods using ambient air. During these cycles, water can penetrate through the film to a greater extent than in salt fog chamber testing in which the humidity is always at 100% and the 5% salt solution minimizes water penetration because of reverse osmotic pressure. In some laboratories, a QUV exposure cycle is being included in the cycling regime.

Another cycling test in which the automotive industry is gaining confidence is the Society of Automotive Engineers test SAE J-2334. A variety of other cycling tests are also being investigated as alternatives to salt spray tests [44]

Neither salt fog nor humidity tests have good reproducibility. It is common for differences between duplicate panels to be larger than differences between panels with different coatings. Precision can be improved by testing several replicate panels of each system. (Commonly, decisions are based on the results from testing two or three panels.) The problem is further complicated by the difficulty of rating the degree of severity of failure. A rating system and an approach to statistical analysis of data have been published [45]. The study was based on panels with an acrylic clear coating and a pigmented alkyd coating (neither coating would be expected to have good corrosion protection properties) exposed to 95% relative humidity at 60, 70, and 80°C. The times to failure were extrapolated down to ambient temperatures. In light of the effect of T_g on permeability, the validity of such extrapolations is doubtful.

A further problem of evaluating panels for corrosion protection is the difficulty of detecting small blisters and rust areas underneath a pigmented coating film without removing the film. Infrared thermography has been recommended as a nondestructive testing procedure [46].

A great deal of effort has been expended on electrical conductivity tests of paint films and electrochemical tests of coated panels. (See Ref. [47] for an extensive review.)

A variety of cathodic disbonding tests specifically for testing of pipeline coatings has been established by ASTM: G-8-96, G-42-96, and G-80-88 (reapproved 1992). In these tests, a hole is made through the coating and the pipe is made the cathode of a cell in water with dissolved salts at a basic pH. Disbonding (loss of adhesion) as a function of time is followed. While there is considerable variability inherent in such tests and their utility for predicting field performance is doubtful, useful guidance in following progress in modifying wet adhesion may be obtained. Such tests may be useful more broadly than just for pipeline coatings. For a discussion of research on cathodic delamination, including investigation of the migration of cations through or under coating films, see Ref. [48].

Appleman has reported the results of an extensive survey of accelerated test methods for anticorrosive coating performance [49]. The need for everyone in the industry to become aware of the current testing situation and to work cooperatively to develop more meaningful methods of testing is emphasized. The lack of laboratory test methods that reliably predict performance puts a premium on collection of databases permitting analysis of interactions between actual performance and application and formulation variables. It is especially critical to incorporate data on premature failures in the database. Availability of such a database can be a powerful tool for a formulator and may be especially useful in testing the validity of theories about factors controlling corrosion. In time, it may be possible to predict performance better from a knowledge of the underlying theories than from laboratory tests. Many workers feel this is already true in comparison with salt fog chamber tests.

GENERAL REFERENCES

R. A. Dickie and F. L. Floyd, Eds., *Polymeric Materials for Corrosion Control*, ACS Symp. Ser. No. 322, American Chemical Society, Washington, DC, 1986.

C. C. Munger, *Corrosion Prevention by Protective Coatings*, National Association of Corrosion Engineers, Houston, TX, 1997.

REFERENCES

1. H. H. Uhlig, *Corrosion and Corrosion Control*, 2nd. ed., Wiley, New York, 1971.
2. H. Leidheiser, Jr., *J. Coat. Technol.*, **53** (678), 29 (1981).
3. J. E. O. Mayne, *Official Digest*, **24** (325), 127 (1952).
4. J. E. O. Mayne in L.L. Shreir, Ed., *Corrosion*, Vol. 2, Butterworth, Boston, 1976, pp. 15:24–15:37.
5. H. Leidheiser, Jr., *Prog. Org. Coat.*, **7**, 79 (1979).
6. W. Funke, *J. Oil Colour Chem. Assoc.*, **62**, 63 (1979).
7. W. Funke, *J. Coat. Technol.*, **55** (705), 31 (1983).
8. W. Funke, *J. Oil Colour Chem. Assoc.*, **68**, 229 (1985).
9. W. Funke, *Farbe Lack*, **93**, 721 (1987).
10. J. W. Martin, E. Embree, and W. Tsao, *J. Coat. Technol.*, **62** (790), 25 (1990).
11. D. Y. Perera, *Prog. Org. Coat.*, **28**, 21, 1996.
12. J. W. Holubka, J. S. Hammond, J. E. DeVries, and R. A. Dickie, *J. Coat. Technol.*, **52** (670), 63 (1980).
13. J. W. Holubka and R. A. Dickie, *J. Coat. Technol.*, **56** (714), 43 (1984).
14. B. M. Perfetti, *Metal Surface Characteristics Affecting Organic Coatings*, Federation of Societies for Coatings Technology, Blue Bell, PA, 1994.
15. J. L. Massingill, P. S. Sheih, R. C. Whiteside, D. E. Benton, and D. K. Morisse-Arnold, *J. Coat. Technol.*, **62** (781), 31 (1990).
16. W. Funke, *Prog. Org. Coat.*, **28**, 3 (1996).

17. N. L. Thomas, *Prog. Org. Coat.*, **19**, 101 (1991).
18. O. Negele and W. Funke, *Prog. Org. Coat.*, **28**, 285 (1995).
19. A. Barbucci, E. Pedroni, J. L. Perillon, G. Cerisola, *Prog. Org. Coat.*, **29**, 7, (1996).
20. B. Bieganska, M. Zubielewicz, and E. Smieszek, *Prog. Org. Coat.*, **16**, 219 (1988).
21. D. P. Bentz and T. Nguyen, *J. Coat. Technol.*, **62** (783), 57 (1990).
22. W. Funke, private communication, 1997.
23. A. Bautista, *Prog. Org. Coat.*, **28**, 49 (1996).
24. A. Smith, *Inorganic Primer Pigments*, Federation of Societies for Coatings Technology, Blue Bell, PA, 1988.
25. J. E. O. Mayne and U. R. Evans, *Chem. Ind.*, **63**, 109 (1944).
26. M. Leclercq, *Materials Technology*, March, 57 (1990).
27. S. Feliu, R. Barajas, J. M. Bastidas, and M. Morcillo, *J. Coat. Technol.*, **61** (775), 63, 71 (1989).
28. N. C. Fawcett, C. E. Stearns, and B. G. Bufkin, *J. Coat. Technol.*, **56** (714), 49 (1984).
29. S. Feliu, Jr., M. Morcillo, J. M. Bastidas, and S. Feliu, *J. Coat. Technol.*, **63** (793), 31 (1991).
30. T. S. Lee and K. L. Money, *Mater. Perform.*, **23**, 28 (1984).
31. R. G. Groseclose, C. M. Frey, and F. L. Floyd, *J. Coat. Technol.*, **56** (714), 31 (1984).
32. J. R. Scully, *Electrochemical Impedance Spectroscopy for Evaluation of Organic Coating Deterioration and Underfilm Corrosion—A State of the Art Technical Review*, Report No. DTNSRDC/SME-86/006, D. W. Taylor Naval Ship Research and Development Center, Bethesda, MD, 1986.
33. U. Rammelt and G. Reinhard, *Prog Org. Coat.*, **21**, 205 (1991).
34. E. P. M. van Westing, G. M. Ferrari, F. M. Geenen, and J. H. W. de Wit, *Prog. Org. Coat.*, **23**, 89 (1993).
35. W. S. Tait, *J. Coat. Technol.*, **66** (834), 59 (1994).
36. H. J. Prause and W. Funke, *Farbe und Lack*, **101**, 96 (1995).
37. R. Athey, R. Duncan, E. Harmon, D. Izak, K. Nakabe, J. Ochoa, P. Shaw, T. Specht, P. Tostenson, and R. Warness, *J. Coat. Technol.*, **57** (726), 71 (1985).
38. J. Mazia, *Met. Finish.*, **75** (5), 77 (1977).
39. R. D. Wyvill, *Met. Finish.*, **80** (1), 21 (1982).
40. W. Funke, in H. Leidheiser, Jr., Ed., *Corrosion Control by Coatings*, Science Press, Princeton, NJ, 1979, pp. 35–45.
41. M. E. McKnight, J. F. Seiler, T. Nguyen, and W. F. Rossiter, *J. Prot. Coat. Linings*, **12**, 82 (1995).
42. M. Hemmelrath and W. Funke, *FATIPEC Congress Book*, Vol. IV, 137 (1988).
43. F. D. Timmins, *J. Oil Colour Chem. Assoc.*, **62**, 131 (1979).
44. B. R. Appleman, *J. Prot. Coat. Linings*, November, 71 (1989).
45. J. W. Martin and M. E. McKnight, *J. Coat. Technol.*, **57** (724), 31, 39, 49 (1985).
46. M. E. McKnight and J. W. Martin, *J. Coat. Technol.*, **61** (775), 57 (1989).
47. J. N. Murray, *Prog. Org. Coat.*, **30**, 225 (1997); **31**, 255 (1997); **31**, 375 (1997).
48. J. Parks and H. Leidheiser, Jr., *FATIPEC Congress Book*, Vol. II, 317 (1984).
49. B. R. Appleman, *J. Coat. Technol.*, **62** (787), 57 (1990).

CHAPTER **8**

Latexes

A *latex* is a dispersion of polymer particles in water. Most latexes are made by a free radical initiated chain-growth polymerization process in which the monomers are emulsified in water at the start; hence, the process is called *emulsion polymerization*. It is not surprising that latex paints are sometimes misnamed *emulsion paints*, but this terminology is best avoided, since it can cause confusion with systems that are true emulsions. Aqueous dispersions of polymers are also prepared by other methods, for example, polyurethane dispersions in water are prepared by step-growth polymerization. Usually, these materials are called aqueous dispersions rather than latexes; they are discussed in Section 10.8.

Molecular weights of polymers prepared by emulsion polymerization are generally high; $\bar{M}_w$ of 1,000,000 or higher is common. However, unlike solution polymers, the molecular weight of the polymer in latex particles does not affect the viscosity of the latex. Instead, latex viscosity is governed by the viscosity of the medium in which the polymer particles are dispersed (the continuous phase), by the volume fraction of particles, and by their packing factor (see discussion of the Mooney equation in Section 3.5). This lack of dependence of viscosity on molecular weight makes it possible to formulate latex coatings at higher solids than solutions of polymers having high molecular weight. Latexes are used as the principal vehicle in a large majority of architectural coatings in the United States. A small, but growing, part of the OEM product and special purpose coatings markets is latex based. Most latex paints form films by coalescence at ambient temperatures. (See Section 2.3.3.) A limitation of most latex-based coatings is the inability to achieve high gloss. On the other hand, the durability of the films is superior to films formed from drying oil and alkyd paints, and VOC emissions are generally lower.

8.1. EMULSION POLYMERIZATION

Emulsion polymerization is carried out in water using monomer(s), surfactant(s), and a water-soluble initiator. Many of the same monomers are used as in solution polymerization, and the reactions are broadly similar at the molecular level. However, the physical circumstances of polymerization are different, affecting the polymerization chemistry. The process is easily varied, and the properties of the latex are affected by changes in physical conditions of the polymerization.

There have been many studies of the mechanisms of emulsion polymerization; however, there are so many variables that no general theory has been developed that can predict the results obtained in all emulsion polymerizations. Monomer structure, solubility, and concentration; surfactant(s) structure and concentration; initiator concentration and rate of radical generation; presence and concentration of added electrolytes; and temperature are some of the critical variables. The rate of agitation and reactor design can affect the results as well. Discussion of the details of proposed mechanisms is beyond the scope of this text; the reader is referred to the general references provided at the end of the chapter for extensive discussion and proposed equations for predicting the number of particles and particle sizes of latexes.

Many early laboratory studies were carried out by a small-scale *batch process*, in which all of the ingredients were put into a sealed bottle that was then shaken in a temperature controlled water bath. This process is often called the *pop-bottle process*. Many papers describing the mechanism of emulsion polymerization are based on such studies. However, batch processes cannot be used commercially because the heat evolved by the exothermic polymerization process would be uncontrollable in a large vessel. Instead, commercial latexes are produced by a *semicontinuous batch process*, using apparatus such as shown schematically in Figure 8.1. Monomers and initiators are added in proportions and rates such that rapid polymerization occurs. In this way, the monomer concentration at any time is low; the polymerization is said to be done under *monomer-starved conditions*. This facilitates temperature control. Furthermore, the composition of copolymers formed under monomer-starved conditions approximately equals the composition of the monomer feed, regardless of the relative reactivity of the various monomers. This procedure also permits changing monomer composition during the course of a polymerization, as discussed in Section 8.1.3.

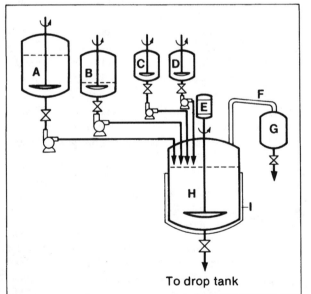

A — Main momer feed tank
B — Auxiliary monomer feed tank
C,D — Initiator feed tanks
E — Stirring motor
F — Condensor
G — Receiver
H — Reactor
I — Heating/cooling jacket

Not shown: Safety vents, pressure and temperature gauges, filters, coolers for feed tanks.

Figure 8.1. Schematic diagram of a semicontinuous batch process production unit. (Adapted from Ref. [1], with permission.)

The products obtained from batch and semicontinuous emulsion polymerization are often different. The differences are particularly large when copolymers are prepared by the two methods. (See Sections 8.1.2 and 8.3.) Care must be exercised in reading the literature since conclusions derived from studies on the batch process are often misleading when applied to the semicontinuous process, and vice versa. Since many variables affect the composition and properties of the latex produced, it is vital to plan laboratory procedures to simulate eventual production operations as closely as possible to minimize problems in scaling up to production equipment.

8.1.1. Raw Materials for Emulsion Polymerization

1. *Monomers*. Many monomers have been converted to latexes by emulsion polymerization. The main requirements are that the monomers can undergo free radical chain polymerization and that they do not react with water. Water-insoluble monomers and monomers with limited solubility in water are most useful, although water-miscible comonomers can be used in small amounts. The two major classes of latexes used in coatings are based on acrylic esters (see Section 8.2) and on vinyl esters (see Section 8.3). Vinylidene chloride/acrylic ester copolymer latexes give films with exceptionally low water permeability. (See Section 8.2.) The earliest latexes used in coatings were copolymers of styrene with butadiene; they are seldom used today in architectural coatings, but are used in coatings for paper. Various functional monomers are also used. Methacrylic acid (MAA) and acrylic acid (AA) improve colloidal stability, affect flow properties, improve adhesion, and give carboxylic acid groups to serve as cross-linking sites. Hydroxyethyl acrylate (HEA) and hydroxyethyl methacrylate (HEMA) provide hydroxyl groups for cross-linking. Monomers have been developed that promote wet adhesion; methacrylamidoethylethyleneurea is widely used for this purpose.

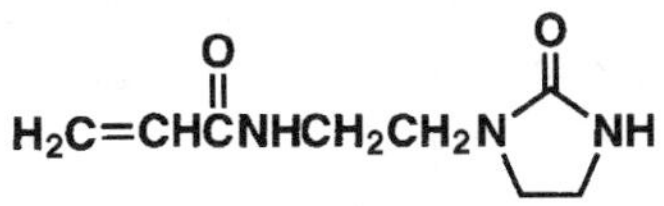

$$H_2C=CHCNHCH_2CH_2N\overset{O}{\underset{}{\parallel}}\quad NH$$

Methacrylamidoethylethyleneurea

2. *Initiators*. The principal initiators used in emulsion polymerization are soluble in water. The most common initiators are persulfate salts, especially ammonium persulfate (more correctly, ammonium peroxydisulfate). Persulfates cleave thermally in the water phase to sulfate ion radicals (see Eq. 8.1) that initiate polymerization, as shown with ethyl acrylate in Eq. 8.2. Note that the terminal group is a sulfate half ester ion.

$$^-O_3S-O-O-SO_3^- \longrightarrow 2\ ^-O_3S-O\cdot \qquad (8.1)$$

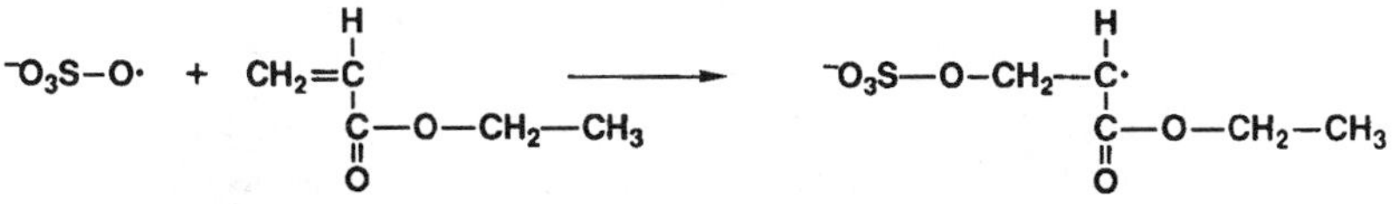

$$^-O_3S-O\cdot\ +\ CH_2=\overset{H}{\underset{\underset{O}{\overset{\parallel}{C}}-O-CH_2-CH_3}{C}} \longrightarrow\ ^-O_3S-O-CH_2-\overset{H}{\underset{\underset{O}{\overset{\parallel}{C}}-O-CH_2-CH_3}{C}}\cdot \qquad (8.2)$$

The sulfate ion radical can also abstract hydrogen from water, leading to the formation of an acid sulfate ion and a hydroxy free radical. In this case, initiation leaves a terminal hydroxyl group. While this reaction is disfavored by the strength of the O—H bond toward cleavage, it occurs to some extent under conditions in which monomer concentrations are low. Bisulfate ions are generated lowering the pH; it is often necessary to add a buffering agent.

$$^-O_3S{-}O\cdot \ + \ H_2O \ \longrightarrow \ HSO_4^- \ + \ HO\cdot$$

The half life of persulfate is such that a 0.01 M solution of $(NH_4)_2S_2O_8$ at pH 10 produces 8.4×10^{12} radicals per mL per s at 50°C and 2.5×10^{15} radicals per mL per s at 90°C. For rapid polymerization at lower temperatures, radical production can be accelerated by reducing agents. For example, mixtures of ferrous, thiosulfate $(S_2O_3^{2-})$, and persulfate salts react faster than persulfate alone as a result of a sequence of oxidation and reduction reactions. The processes using this type of initiation are called *redox* emulsion *polymerization*. The reactions involved are complex; the following are said to be representative:

$$^-O_3SOOSO_3{}^- + Fe^{2+} \longrightarrow {}^-O_3SO\cdot + SO_4{}^{2-} + Fe^{3+}$$

$$S_2O_8{}^{2-} + 2S_2O_3{}^{2-} \longrightarrow S_4O_6{}^{2-} + 2{}^-O_3SO\cdot$$

$$2Fe^{3+} + 2S_2O_3{}^{2-} \longrightarrow 2Fe^{2+} + S_4O_6{}^{2-}$$

Using redox systems, polymerization can be initiated at room temperature; the exothermic polymerization reaction warms the reaction mixture to the desired temperature (often 50–80°C). Cooling is required to prevent overheating. Sometimes, redox initiation is used in the first stage of polymerization, and the latter stages are continued using thermal initiation.

It is common to add a second, more lipophylic "chaser" initiator in the late stages of the semicontinuous process to facilitate conversion of monomer to high levels, preferably greater than 99%. An initiator such as *t*-butyl hydroperoxide, which is more soluble in the polymer particles than in water, is more effective than ammonium persulfate in the late stages, when most of the unreacted monomer is dissolved in the polymer particles. Even when an auxiliary initiator is used, latexes still contain some unreacted monomer.

3. *Surfactants*. Surfactants are an important component in emulsion polymerizations. A variety of both anionic and nonionic surfactants is used. An example of an anionic surfactant is sodium lauryl sulfate (often incorrectly called sodium lauryl sulfonate):

$$H_3CCH_2CH_2CH_2CH_2CH_2CH_2CH_2CH_2CH_2CH_2OSO_3{}^-Na^+$$

A typical nonionic surfactant is a nonylphenylethoxylate:

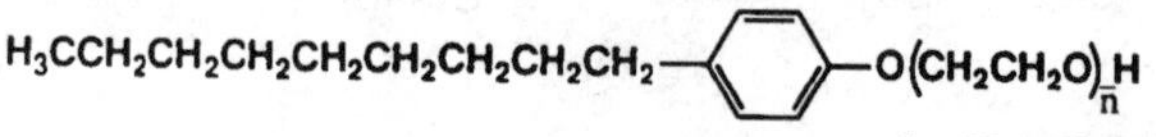

$$\bar{n} = 20 \text{ to } 40$$

Surfactants generally have limited solubility in water. Above that solubility limit, the *critical micelle concentration* (CMC), nonpolar ends of surfactant molecules associate with each other in clusters called *micelles*. Micelles are submicroscopic aggregates typically containing 30–100 surfactant molecules, with the hydrophobic parts of each molecule oriented toward the center and the hydrophilic parts oriented outward in contact with water. The CMCs of different surfactants vary over a wide range from about 10^{-7} to 10^{-3} gL^{-1}. See Ref. [2] for a review of surfactants. Surfactants play a critical role in stabilization of the dispersion of polymer particles and preventing coagulation of the latex on standing. If the particles approach each other closely, van der Waals forces tend to hold them together; the particles *flocculate*. As noted in Section 3.5, flocculation of a dispersed phase system leads to changes in flow properties; viscosity increases and flow becomes shear thinning.

There are two general mechanisms by which stabilization of dispersions is affected: *charge repulsion* and *entropic repulsion*. With charge repulsion, the surfaces of the particles have an excess of one electrostatic charge, most commonly negative. For example, a latex can be stabilized by an anionic surfactant adsorbed on the surface of the polymer particles. The surfactant molecules orient with the long hydrophobic hydrocarbon tails in the polymer and the hydrophilic salt groups on the periphery associated with the water. As a result, the surface of the particles is covered with anions; each anion has an associated cation. This surface layer of cations is called a *Stern layer*; it is somewhat rigid and behaves as if it were physically part of the particle. Its presence induces a second diffuse layer of anions to surround the particle, giving it a negatively charged so-called *double layer*. When two particles approach each other, their diffuse, negatively charged layers electrostatically repulse each other. The stability of such dispersions can be affected by the addition of salts. The stability is particularly sensitive to the addition of multivalent ions with the opposite charge of the stabilizing charge; hence deionized water is generally used.

The second mechanism for stabilization is by repulsion resulting from the outer layers of the particle. Three terms are widely used in the literature to characterize this type of stabilization: entropic, steric, and osmotic repulsion. If the outer surface of the particles is hydrophilic, water is absorbed and swells the surface. If this swollen layer is thick enough, the particles are not able to approach each other closely enough to flocculate; hence the term *steric repulsion*. If, for example, a nonionic surfactant—that is, a surfactant with a hydrophobic nonpolar segment and a relatively long chain of repeating ethylene ether groups as a hydrophilic segment—is adsorbed on the surface of a latex particle, the hydrophilic ether groups will be on the outer surface. The ether groups hydrogen bond with water, and more water is absorbed in the layer. The absorbed layer can exist in a very large number of conformations, and water molecules can move in and out of the layer. When two particles approach each other, the layers are compressed leading to a reduction in the number of conformations that molecules in the layers can assume. As a result, the system becomes less random, corresponding to a reduction in entropy. Resistance to the reduction in entropy leads to repulsion, hence the term *entropic repulsion*.

Some authors prefer to focus on the reduction in the amount of water in the layer if compression occurs. There is a tendency for the water to return to the equilibrium concentration in the layer; some consider this to be analogous to osmosis and call the result *osmotic repulsion*. Since the amount of water in the layer was at equilibrium, expulsion of water also requires a decrease in entropy, resulting in a further element of entropic repulsion. Some authors prefer to avoid the controversy over terms by simply calling the repulsion, *steric repulsion*. We use the term *entropic repulsion*, but it should be understood that factors other than entropy may well be involved. In industrial practice, stabilization of the dispersion of polymer particles in a latex is found to be most effective when charge repulsion and entropic repulsion are combined.

The surfactants control the particle size and particle size distribution of latexes. (See Section 8.1.2.) In general, anionic surfactants are used at levels of 0.5 to 2 wt% based on polymer; they cost less than nonionic surfactants, and they reduce particle size more efficiently on a weight basis. On the other hand, nonionic surfactants, used at levels of 2 to 6 wt%, are more effective in stabilizing the latex against coagulation during freeze-thaw cycling, make it less sensitive to coagulation by salts (especially salts of polyvalent cations), less sensitive to changes in pH, and sometimes less likely to stabilize foam. In general terms, anionic surfactants stabilize primarily by charge repulsion, and nonionic surfactants stabilize primarily by entropic repulsion. The two surfactant types impart different rheological characteristics to the latex. Anionic surfactants lead to formation of essentially rigid particles. Such latexes can have low viscosity at relatively high solids. On the other hand, with nonionic surfactants, the thicker, swollen layers on the surface of entropically stabilized particles lead to lower polymer solids or higher viscosity at a given solids. The surface layer of an entropically stabilized latex particle is not rigid and, hence, can be distorted when shear stress is applied, imparting shear thinning characteristics. Commonly, both anionic and nonionic surfactants are used in an emulsion polymerization recipe. More explicit generalizations about the use of surfactants in emulsion polymerization are treacherous because there are exceptions to almost any statement that can be made.

All surfactants impart some water sensitivity to films from latex paints. This sensitivity is reflected, for example, in a tendency of house paints to waterspot if rained on soon after application and in the somewhat limited corrosion resistance of most latex coatings on steel. Selection of surfactant combinations that minimize such shortcomings is an ongoing subject of research. So-called "soap-free" latexes have been developed in which emulsion polymerization is effected without use of conventional surfactants. Such approaches may reduce the water sensitivity associated with surfactants. Reference [3] provides an extensive review of emulsifier-free latexes. The most widely used approach is to incorporate a hydrophilic comonomer, such as (meth)acrylic acid or ammonium hydroxyethyl acrylate sulfate. Copolymerization of these monomers creates molecules that function as polymeric surfactants.

Another approach to the problem of water sensitivity is to use polymerizable surfactants. For example, it has been shown that a surfactant composed of allyl alcohol, butylene oxide, ethylene oxide, and a sulfonate salt terminal group can be used to make vinyl acetate/butyl acrylate latexes [4]. The surfactant is reported to be completely consumed during the polymerization. Presumably, a hydrogen is abstracted from the surfactant yielding a free radical that initiates growth of another polymer chain, covalently bonding the surfactant to a polymer molecule.

Choice of surfactant can also affect film formation temperature, since some nonionic surfactants plasticize the latex polymer, leading to a lower T_g and, hence, a lower temperature for coalescence. For example, nonylphenylethoxylate nonionic surfactants with less than nine ethoxyl units reduce film formation temperature, as compared to those with 20 to 40 ethoxylate units, but the higher ethoxylated surfactants are more effective latex stabilizers [5]. Nonionic surfactants have many easily abstractable hydrogens and can form graft copolymers.

Additional entropic stabilization can be achieved by including in the polymerization recipe a small amount of water-soluble polymer (sometimes called a *protective colloid*) that can form graft copolymers readily. A polymer commonly used for this purpose is poly(vinyl alcohol) (PVA), which has many easily abstractable hydrogens. During polymerization, initiator radical ions abstract a hydrogen from PVA to form a free radical on a PVA chain. A graft of the monomers being polymerized grows on the PVA chain; more than one such graft can grow on a single PVA molecule. As the graft chain becomes longer, it becomes hydrophobic and associates with other polymer molecules in particles, carrying the PVA part of the molecule to the particle surface. The PVA part associates with water to form an entropic stabilizing layer analogous to that formed by nonionic surfactants. Such layers can be thicker than the layer from adsorbed surfactant molecules and are effective in promoting stability. Water sensitivity of films may be reduced because less water-soluble surfactant is needed. Hydroxyethyl cellulose (HEC) can also be used. It has been shown that grafting occurs and that the resulting latex exhibits thixotropy and is more shear thinning than a latex prepared with no water-soluble polymer [6]. While monomer-HEC grafting improves latex stability, it favors large particle size, broad particle size distribution, and low film clarity [7].

4. *Water and other additives.* Variations in water quality can cause a variety of problems, especially when anionic surfactants are used. Therefore, deionized water is used in emulsion polymerization and is usually used in latex paint production. Water is somewhat soluble in some polymers and acts as a plasticizer, reducing film formation temperature [5]. Other ingredients sometimes used in emulsion polymerization processes include buffers, which protect sensitive monomers from hydrolysis and sensitive surfactants from deactivation, and thickeners, which control viscosity.

8.1.2. Emulsion Polymerization Variables

As noted earlier, many theories for the mechanism of emulsion polymerization have been proposed, but none has proven to be applicable to all systems because of the wide range of variables involved. Reference [8] provides a concise review of various theories and mathematical models that have been proposed. Since none of the models have proven to be widely applicable, a qualitative discussion of the important considerations follows.

The monomers are emulsified in water with surfactant molecules oriented at the surface of the emulsion droplets, stabilizing the emulsion. There is also surfactant dissolved in the water and excess surfactant present in micelles. In most systems of interest in coatings, at least one of the monomers is somewhat soluble in water, so there are monomer molecules in solution as well as in the emulsion droplets. For example, the solubility of methyl methacrylate (MMA) and ethyl acrylate (EA) is approximately 1% in water. The initiator is also dissolved in the water.

There are three potential sites for initiation and initial propagation of the polymerization: (1) in the monomer swollen micelles, (2) in the aqueous phase (so-called *homogeneous nucleation*), and (3) in the emulsion droplets. It is suggested that all three of these modes may occur in any given polymerization, but that the ratio of the three modes is system dependent [8]; monomer solubility in water, surfactant composition, and emulsion droplet size may be particularly important variables. The predicted number of particles and particle size is different in each of the three cases (or combinations of them). However, qualitatively, one can say that particle size decreases and number of particles increases as surfactant concentration increases.

When at least one of the monomers is somewhat soluble in water, it is commonly assumed that initiation occurs predominantly in the water phase. The initiating radical reacts with a monomer molecule, forming a new free radical that is soluble in water; these free radicals can react with further dissolved monomer molecules to form oligomers with terminal free radicals. As the molecular weight of the growing oligomer chain increases, its solubility in water decreases, and one of three things can happen: The growing chain may enter a micelle, where propagation continues, as monomer molecules also enter the micelle; surfactant from the solution may be adsorbed on the surface of the growing oligomer molecule; or the growing chain can enter a monomer droplet. In any case, a surfactant-stabilized polymer particle containing a single free radical is formed. Such particles rapidly imbibe monomer, and propagation continues within the particles. When all of the monomers are low in solubility, initiation in micelles may predominate. The monomer in the micelle polymerizes rapidly to form a surfactant-stabilized polymer particle containing a single free radical.

Termination (by combination or disproportionation, as described in Section 2.2.1) occurs only after a second free radical enters a particle. After termination, the surfactant-stabilized polymer particle is temporarily inactive because it contains no free radicals. More monomer molecules then migrate into the inactive polymer particle. A third free radical (IMMMMM·) enters the monomer-swollen particle and initiates polymerization. Chain growth continues until a fourth free radical enters the particle and causes termination. Dissolution of monomer from the monomer droplets replenishes the monomer in solution. The process is repeated over and over, leading to formation of additional polymer molecules in the particle. (This model does not apply to cases in which the T_g of the polymer is higher than the polymerization temperature, but this is not usually the case.)

As the polymer particles grow, the surface expansion requires further surfactant, the need for which is satisfied by adsorbing more surfactant molecules from solution, which, in turn, are replenished by dissolution of micelles in which polymerization is not occurring. The result is growth of the particles that acquired a polymer molecule early on and disappearance of micelles that did not. Polymer particle growth continues until unreacted monomer or initiator is used up; enough initiator is used to assure high conversion of monomer. After the early stages of the process, much of the polymerization occurs within a fixed number of polymer particles; the number of particles per unit volume depends on the concentration of micelles. If the concentration of micelles is high, the number of particles is high, and hence, the particle size at the end of the polymerization is small. At the same surfactant concentration, a surfactant with a low CMC gives a smaller particle size latex than a surfactant with a higher CMC.

In the homogeneous nucleation mechanism, surfactant is absorbed on the surface of the growing oligomer chains. As the number of particles and their size increase, the amount of surfactant becomes too small to stabilize them, and particles coalesce to form fewer and larger particles until the surfactant concentration is just high enough to stabilize the particles.

A third mechanism is initiation in the droplets of monomer emulsion. Since in most cases, the size of the droplets of monomer emulsion is relatively large and their surface area is small relative to the surface area of the more numerous micelles, the probability of an initiating radical or a growing oligomer radical entering an emulsion droplet is small. However, if the rate of agitation is very high and/or a high surfactant concentration is used, emulsion droplet size becomes smaller, and the probability of propagation in emulsion droplets increases. If a water-insoluble initiator is used, most of the polymerization occurs in the emulsion droplets, resulting in large, unstable particles. This method is the basis for a different polymerization method, called *suspension polymerization*.

The number of particles and, hence, their particle sizes are also affected by the rate of initiation (a function of the type of initiator, its concentration, and the temperature). The faster the rate of formation of initiating radicals, the larger the number and the smaller the size of the particles, everything else being equal.

Since much of the polymerization occurs when there is only one free radical in a particle, molecular weights obtained in emulsion polymerization are generally high. Also, since the growing free radical is in the presence of a limited amount of monomer and a high concentration of polymer, chain transfer to polymer is more likely than in solution polymerization. Chain transfer to polymer results in growth of branched chain molecules and broad molecular weight distribution.

Latexes are frequently made starting with a *seed latex*, that is, a small particle size latex [9]. Monomer and initiator are added slowly, and the polymerization occurs primarily in the seed particles, resulting in a constant number of particles as polymerization proceeds. The seed latex can be prepared as the first step in a polymerization so that it contains 10–20% solids. Alternatively, it is common practice to produce a large batch of seed latex and subdivide it for subsequent use in many batches. The seed latex is diluted to 3 to 10% solids. Increasing the number of seed particles, with other factors being equal, yields a latex of smaller average particle size. This practice can improve batch-to-batch reproducibility and is particularly useful in sequential polymerizations.

The $\bar{M}_w$ obtained by both batch and semicontinuous processes is generally very high; an $\bar{M}_w$ over 1,000,000 is common. In many cases, there is a broad distribution of molecular weights. High $\bar{M}_w$ results from several factors. First, during polymerization, the fraction of polymer particles containing more than one free radical at a given instant is low. As a result, the rate of termination reactions within the particle is reduced, and a single growing chain may consume most of the monomer present in the particle before it terminates. Second, the viscosity inside the particles is very high, reducing the mobility of the polymer chains. It is then easier for more mobile monomer molecules to diffuse and to add to the growing chain and increase the molecular weight than for two less mobile growing chains to react with each other and terminate growth. Third, hydrogen atoms on polymer chains can be abstracted, leading to chain transfer to polymer and formation of branched chain molecules. The environment within the polymer particle favors chain transfer to polymer because the concentration of free radicals is low (often, one per particle) while the concentration of polymer approaches 100%. Chain transfer to polymer increases $\bar{M}_w$ and $\bar{M}_w/\bar{M}_n$ because it is statistically probable that when growth of short chains is ended by chain transfer, the new chain will grow as a branch on a polymer molecule that already has a high molecular weight.

In a study comparing batch and semicontinuous homopolymerization and copolymerization of vinyl acetate and butyl acrylate, $\bar{M}_w/\bar{M}_n$ for batch homopolymers and copolymers was broad, ranging from 15 to 21, and the semicontinuous process gave even broader distributions, ranging from 9 to 175 [10]. The semicontinuous process commonly afforded bimodal distributions with considerable amounts of relatively low molecular weight polymer. Semicontinuous poly(vinyl acetate) homopolymer had the broadest distribution; GPC indicated substantial fractions of material with M below 660 and above 1,700,000.

Particle size distribution is an important variable in formulation of latex paints. Particle size distribution is commonly measured by hydrodynamic chromatography, ultracentrifuge, or transmission electron microscopy and may be expressed as number average diameter ($\bar{D}_n$) and volume average diameter ($\bar{D}_v$). The ratio $\bar{D}_v/\bar{D}_n$ is a convenient index of particle size distribution. A broad or bipolar distribution is often advantageous because the packing factor is affected by particle size distribution. The volume fraction of latex particles that can be present at a standard viscosity is highest with broad particle size distribution.

The particle size of a latex is determined by the number of particles relative to the amount of monomer. The number of particles is governed primarily by the amount and type of surfactant(s) and initiator(s). Anionic surfactants tend to produce more particles than nonionic surfactants when used on an equal weight basis and, hence, lead to smaller particles. The specific anionic surfactant used affects the particle size; for example, use of $NP(OCH_2CH_2)_{20}OSO_3^-$ NH_4^+ (NP stands for the nonpolar part of the molecule) in making an acrylic latex gave a particle diameter of 190 nm, while use of the less hydrophilic $NP(OCH_2CH_2)_9OSO_3^-$ NH_4^+ gave a diameter of 100 nm [11]. In general, particle size distribution is broader with anionic surfactants and more nearly approaches monodisperse particle size with nonionic surfactants. The effect of changing the ratio of nonionic surfactant nonylphenol ethoxylate with 40 mol of ethylene oxide to anionic surfactant sodium dodecylbenzenesulfate on particle size and particle size distribution with a constant amount of total surfactant in the polymerization of a vinyl acetate/dibutyl maleate copolymer has been studied [10]. The lowest ratio gave the smallest particle size and the broadest distribution of particle size.

Emulsion polymerization usually produces latexes with essentially spherical particles. During the 1980s, latexes were commercialized in which groups of a few particles are fused together into nonspherical lobed particles [12]. At equal concentration, the viscosity of these latexes is higher than the viscosity of comparable latexes with spherical particles, since the packing factor is smaller than for a spherical particle latex. Paints formulated with a lobed latex are shear thinning and give a higher high shear viscosity at the same low shear viscosity than do conventional latexes. The higher viscosity at shear rates experienced during application leads to thicker films with low levels of thickener and, hence, lower cost. Adhesion to chalky surfaces is said to be improved as compared to paints formulated with conventional latexes.

8.1.3. Sequential Polymerization

When latexes are prepared by semicontinuous polymerization under monomer-starved conditions, the composition of the monomer feed can be changed during the course of the polymerization. Depending on a number of variables, the resulting latex particles can have a variety of morphologies. In some cases, latexes having *core-shell* morphology can be made by shifting from the initial monomer feed composition to another part way through the process. The composition of the polymer in the center of each particle (the core) reflects the initial monomer feed composition, while the composition near the surface (the shell) reflects the second monomer feed composition. Many such core-shell latexes display two different T_g's after coalescence. In other cases, an *inverse core-shell* morphology can be obtained with the composition of the core polymer being that of the second monomer feed and the shell polymer derived from the initial monomer feed. In other cases, a homogeneous composition is obtained in spite of a change in monomer feed composition. In still other cases, it is possible to make a gradient morphology by continually changing the monomer feed composition during the polymerization. There is an extensive literature on sequential polymerization. Reference [13] provides a brief review of literature applicable to latexes of interest in coatings and discussion of the variables involved.

Several factors affect the morphology of particles obtained in sequential polymerization. Formation of outer layers resulting in the lowest interfacial tension with the water phase is favored; thus, when different monomer compositions are fed, the one containing the most polar monomers is likely to end up as the shell, regardless of its place in the sequence. Morphology can be affected by the amount of free monomer present during the polymerization, since free monomer acts as a plasticizer; hence, formation of core-shell morphology tends to be favored under highly monomer-starved conditions. Presence of di- or trifunctional cross-linking monomers in at least one of the monomer combinations increases the probability of phase separated morphology. In latexes designed for coatings, generally only limited amounts of polyfunctional monomers can be used because of the need for coalescence for good film formation. But, it is a common practice to use a low level of cross-linking monomer in the first composition in order to increase the likelihood that it will remain in the core. Examples of applications of sequential polymerization are given in later sections of this chapter.

8.1.4. Thermosetting Latexes

While the majority of latexes are thermoplastic polymer latexes, there are applications for thermosetting latexes as well. Many types of thermosetting latexes have been reported [14,15]. In general, a lower T_g latex can be used in coatings, which permits coalescence without addition of a coalescing solvent, and/or at lower film formation temperatures. After film formation, cross-linking increases modulus to give block resistance and other desirable final film properties. If a significant degree of cross-linking occurs before application, coalescence will be adversely affected. In many cases, two package coatings are required, but they are only useful for industrial applications; a few stable systems have been developed that can be used for architectural paints.

For two package coatings, a variety of cross-linkers for hydroxy-functional and carboxylic acid-functional polymers can be used. Hydroxy-functional polymers are readily prepared by using hydroxyethyl (meth)acrylate as a comonomer. The resultant latexes can be formulated with urea-formaldehyde (UF) or melamine-formaldehyde (MF) resins as cross-linkers. UF resins tend to evolve high levels of formaldehyde from hydrolysis; MF resins evolve lesser amounts of formaldehyde. The amino resins penetrate slowly into the latex particles, resulting in uneven cross-linking. To overcome this problem, the emulsion polymerization has been carried out by dissolving an MF resin in the monomer mixture before polymerization [16]. Premature cross-linking is minimized by controlling pH at above 5. With addition of catalyst, films cross-link; the pot life of the resin after adding catalyst is 1–2 days. The MF resin acts as a plasticizer before curing, reducing film formation temperature, as compared to a similar latex without MF resin.

Carboxylic acid-functional latexes are prepared using (meth)acrylic acid. These latexes can be cross-linked with ammonium complexes of zinc or zirconium salts; when a film dries, the ammonia evaporates, giving the salt cross-links. Carbodiimides (see Section 16.6.4) can be used as cross-linkers [17]. Polyfunctional aziridines (see Section 16.6.3), such as the propyleneimine addition product with pentaerythritol triacrylate, serve as cross-linkers for carboxylic acid-functional latexes [18]. Pot lives of 48–72 hours are reported. Carboxylic acid-functional latexes can also be cross-linked with epoxy silanes, such as β-(3,4-epoxycyclohexyl)ethyltriethoxysilane [19]. Addition of emulsions of this cross-linker increased solvent resistance and hardness, especially when baked 10 minutes at 116°C. Latexes with high COOH content and equivalent amounts of epoxy silane gave the largest improvements. Package stability is reported to be at least one year. Cross-linking can be catalyzed—for example, using 1-(2-trimethylsilyl)propyl-1H-imidazole.

m-Isopropenyl-α,α-dimethylbenzyl isocyanate (TMI) (see Section 10.3.2) reacts slowly with water and can be used to make thermosetting latexes [20]. Acetoacetate-functional (see Section 16.6.2) latexes can be cross-linked with polyamines, but pot life is short before coalescence is adversely affected [21].

There have been fewer reports of thermosetting latexes that cross-link at room temperature and are storage stable for the long times required for architectural paints. One approach is to use a cross-linking mechanism based on autoxidation. A latex with allylic substitution cross-links on exposure to air after application [22,23]. Hybrid alkyd/acrylic latexes have been prepared by dissolving an oxidizing alkyd (see Section 15.1) in the monomers used in emulsion polymerization, yielding a latex with alkyd grafted on the acrylic backbone [24,25]. Stable thermosetting latexes can be prepared using triisobutoxysilylpropyl methacrylate as a comonomer [19]. The isobutoxy derivative, in contrast to the ethoxy derivative, has sufficient hydrolytic stability to permit emulsion polymerization at 80–90°C without extensive reaction with water. The resultant latex has a package stability of over a year, yet cross-links after application in a week using organotin catalysts. This unusual combination of storage stability and reactivity may mean that the trialkoxysilyl groups hydrolyze during storage, but in the presence of large amounts of water, do not cross-link until the water evaporates after the coating is applied.

8.2. ACRYLIC LATEXES

Acrylic latexes are widely used for exterior paints because of their resistance to photodegradation. They are also more resistant to hydrolysis and saponification than are vinyl acetate latexes. These properties are critical for exterior paints; they also make acrylic latex paints useful for alkaline substrates such as masonry and galvanized metal and for applications in which there is exposure to high humidity. Acrylic and styrene-acrylic latexes are being used increasingly for industrial maintenance coatings. Acrylic latexes are finding increasing interest for kitchen cabinet finishes and for OEM automotive applications. For example, base coats based partially on latexes are being developed for OEM automotive base coat-clear coat coatings.

A critical decision in designing a latex is monomer selection. An important considera-tion is to select a monomer combination that produces a copolymer with the appropriate T_g. Monomer cost is also important. The paint business is highly price competitive, for lower priced paints adequate film properties must be attained using low cost monomers. The T_g must be low enough to permit coalescence of the latex at the lowest anticipated application temperature, yet high enough to assure adequate film hardness and toughness. T_g in the vicinity of 5 to 15°C is common for exterior house paints, which may be applied at temperatures as low as 2°C. The *minimum film forming temperature* (MFFT) (see Section 2.3.3) of latexes is related to T_g, but is influenced by other factors such as particle size, phase separation within the latex particles, and the plasticizing effects of water and surfactants. The MFFT tends to be somewhat lower than T_g. A T_g and MFFT in the desired range are usually attained by copolymerizing monomers whose homopolymers have much higher and much lower T_g values than the target values.

As a high T_g comonomer methyl methacrylate (MMA, homopolymer $T_g = 105$°C) imparts excellent exterior durability and hydrolytic stability at moderate cost. Styrene is often substituted partly or completely for MMA because it costs less and provides a similar

T_g effect (homopolymer $T_g = 100°C$). Such products are sometimes called acrylic/styrene latexes but usually just acrylic latexes. Styrene imparts excellent hydrolytic stability, but its effect on overall exterior durability is not clear cut. Styrene homopolymer degrades relatively rapidly outdoors. However, in a copolymer, some MMA can be replaced with styrene without measurably decreasing exterior durability. The amount of styrene that can be substituted apparently varies, depending on the other comonomers and probably on other variables, such as process conditions. For example, processes that would tend to produce blocks of homopolystyrene are suspect. It is safest to test each composition thoroughly outdoors before use. Other properties may also be affected. For example, it cannot be assumed that the mechanical properties of the copolymer will be equal when styrene is substituted for MMA, even though its T_g may be about the same. Polystyrene has a significantly higher brittle-ductile transition temperature than poly(MMA) (see Section 4.2), even though it has a slightly lower T_g.

Acrylic esters are generally used as the low T_g monomer(s), and the choice of which acrylic ester(s) to use is affected by cost considerations. If the cost of the common acrylic esters is intermediate between that of MMA and of styrene, it is more economical to use ethyl acrylate (EA; homopolymer $T_g = -24°C$) than butyl acrylate (BA; homopolymer $T_g = -54°C$) in styrene-free latexes in order to get as high a proportion of EA in the composition as possible. However, with styrene/acrylic latexes, BA may be more economical when styrene is less expensive than EA or BA. Presumably, the different acrylic esters impart different film properties at a given T_g, but published data is inadequate to evaluate the effects.

An example of a laboratory semicontinuous batch process for preparation of a 40 : 59 : 1 MMA/EA/MAA copolymer latex is provided in Table 8.1. Reference [26] provides many more examples. Note from Table 8.1 that the measured MFFT of the latex at 9°C, is somewhat lower than the T_g of 17°C estimated by the Fox equation for a 40 : 60 MMA/EA copolymer. Also, note that changing to a 50 : 50 MMA/EA ratio increases MFFT to 21°C, while the Fox equation predicts a T_g value of 28°C. These figures give a general picture for how T_g and MFFT relate to composition.

Commercial suppliers of latexes provide little information as to the composition of the latexes they offer. While a simple acrylic latex made as described in Table 8.1 could be used to make a house paint, it probably would not be competitive. Commercial production involves many process nuances that are held confidential by the latex producing companies. It is difficult to replicate commercial latexes, even though the monomer compositions can be precisely analyzed.

Acrylic acid (AA), or the somewhat less water-soluble methacrylic acid (MAA), is commonly used in preparation of acrylic latexes, usually on the order of 1 to 2 wt% of the monomer charge. The carboxyl functionality incorporated with these monomers enhances mechanical stability of latex paint, reducing the amount of surfactant needed. The effects of acid monomers on stability and viscosity are maximized when they are incorporated in the last part of the monomer feed and the polymerization medium is acidic [1,27]. The viscosity of a latex made in this way depends on pH. As ammonia is added, there is little change in viscosity until the pH reaches about 7. Above pH 7, viscosity increases steeply until pH reaches 9 or 10; at still higher pH levels, it decreases. The pH of acrylic latex paints is usually adjusted to about 9, where there is a substantial viscosity effect: not only is the viscosity high, the paint exhibits shear thinning, which is sometimes a desirable application characteristic (see Chapter 31).

Table 8.1. Laboratory Procedure for Preparation of a MMA/EA/MAA Copolymer Latex

Materials

1 L	Deionized water
96 g	Triton X-200 (sodium salt of an alkylaryl polyether sulfate ester) (Union Carbide Co.)
320 g	Methyl methacrylate (10 ppm MEHQ stabilizer) (MEHQ is monomethylether of hydroquinone.)
480 g	Ethyl acrylate (15 ppm MEHQ stabilizer)
8 g	Methacrylic acid (100 ppm MEHQ stabilizer)
1.6 g	Ammonium persulfate

Procedure

Prepare an emulsion of all reactants in 800 mL of water. Place 200 mL of the emulsion and 200 mL of water in a 3-L flask fitted with an inert gas inlet tube, a thermometer, a stirrer, an addition funnel, and a reflux condenser. Heat in a 92°C water bath while stirring until the internal temperature reaches 82°C. The mixture will begin to reflux, and its temperature will rise to about 90°C within a few minutes, indicating rapid polymerization. When refluxing subsides, add the remaining emulsion continuously over a period of 1.5 hours. Heat to maintain refluxing; the internal temperature will be 88–94°C. After monomer addition is complete, heat to 97°C to complete conversion of monomer. Cool and strain.

Properties

NVW 42.9 (calculated 43.1); pH 2.7; Viscosity (Brookfield), 11.5 mPa·s; MFFT 9°C; Film hardness (Tukon), 1.2 KHN. A similarly prepared 50:50 MMA/EA copolymer latex (no MAA), had similar properties, except that MFFT was 21°C and film hardness was 6.2 KHN.

Source: See Ref. [26].

Hoy describes a latex that shows this effect [1]; it is a MMA/EA/BA/AA (40:52:6:2 by weight) copolymer, in which the acrylic acid was added late in the process. The T_g of this copolymer was 15°C. Neutralization with ammonia to pH 9 caused the surface layer of the particles to expand, so the diameter of the particles was about 1.8 times that at pH 7. The corresponding increase in volume was almost sixfold. The expansion resulted from association of water with the highly polar salt groups in the polymer near the particle surface. The presence of the expanded layer at pH 9 increased viscosity at low shear rates. However, because this layer could be distorted by shear forces, the viscosity increase at high shear rates was less marked.

In the same paper, Hoy describes the effects of using sequential polymerization during semicontinuous batch polymerization [1]. In one experiment, an MMA/BA ratio of 40:60 was used, with 2 wt% of AA added late in the process. When the 40:60 ratio was maintained throughout monomer addition, the product, called a *uniform feed* latex, had a T_g of 20°C. A second *staged feed* latex of the same overall composition was made by a process designed to yield core-shell particles (see Section 8.1.3) [1]. In the first half of the monomer addition, the MMA/BA ratio was 70:30, and in the second half, it was 10:90. The resulting latex showed two T_g's, one at 60°C and the other at −10°C. Because the shell of this latex has a low T_g, its MFFT is substantially lower than that of the uniform feed latex. However, coalesced films are cloudy. Apparently the high T_g cores remain uncoalesced; the composition of the cores and the shells is sufficiently different that the polymers are not miscible with one another. Because the refractive index of the two polymers is different, light passing through the film is scattered, making the film cloudy. Since the T_g of the continuous phase is low, the blocking resistance of a film of the staged feed latex is inferior to a film of the uniform feed latex.

A third latex with the same monomer composition was made by a process called *linear power feed addition*, using the apparatus sketched in Figure 8.2 [1]. In this process there are two stirred tanks for monomer mixtures. The near tank was loaded with half of the monomer charge at an MMA/BA ratio of 70:30, and the far tank was loaded with the other half at a 10:90 ratio. During polymerization, monomer was pumped from the near tank into the reactor in the usual way. At the same time, monomer was being pumped at the same rate from the far tank to the near tank. Again, 2 wt% of AA was added in the late stage of the process. In this procedure, the monomer feed starts at 70:30 MMA/BA and ends at 10:90, but changes continuously through the process. Each latex particle presumably contains high T_g polymer at its center and low T_g polymer (with COOH groups) near its surface. It differs from the usual core-shell latex in that there is a linear gradient of composition and T_g from the center to the surface. Films cast from this latex have equal clarity and the same T_g, about 20°C, as those cast from the uniform feed latex. However, the glass transition is much broader for the linear power feed latex than for the uniform feed latex. The linear power feed latex has a lower MFFT than the uniform feed latex and also has a higher blocking resistance temperature.

Exterior house paints for large surfaces are usually formulated at low gloss in the United States; the formulations are heavily pigmented. Block resistance requirements are modest, and the pigment helps improve this property; T_g values about 5–15°C are adequate. Latex paint films are more permeable to water than films from oil or alkyd paints, an advantage because it reduces blistering of paints applied to wood surfaces, but a disadvantage for coatings applied over metal.

Applications such as gloss trim and door paints, as well as interior trim paints (for kitchen cabinets, windowsills, etc.) have different requirements; to be glossy, they must be formulated at lower pigment levels, yet block resistance is required. This situation presents a challenge to designers of latexes. The problems have been at least partly overcome, as described by Mercurio [28]. A latex T_g of about 55°C is reported to be required to achieve adequate block resistance. Coalescence of acrylic latexes with this T_g can be attained by

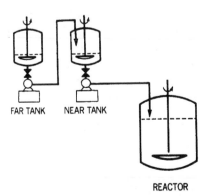

FAR TANK   NEAR TANK

REACTOR

Figure 8.2. Schematic representation of an apparatus for semicontinuous latex polymerization using linear power feed addition. (Adapted from Ref. [1], with permission.)

using a small particle size latex with substantial amounts of carefully selected coalescing solvent(s). The coalescing solvent increases VOC to near the current regulatory limit of 250 g of VOC/L of paint as applied, excluding water. Commercial products based on these principles are available; details of how they are made are proprietary.

While the VOC of latex paints is usually lower than that of solvent-borne paints, there is increasing pressure to reduce VOC further. Reducing the amount of coalescing solvent required is the most evident approach to this end. Sequencing the composition of the monomer feed during polymerization so that the last part of the monomer feed has a higher fraction of monomers giving a low T_g, as in the power feed approach discussed previously would permit reduction of coalescing solvent. (However, see Section 8.1.3 for discussion of factors that influence which polymers end up on the surface.) It is also suggested addition of acrylic acid in the later part of the comonomer feed may permit reduction of coalescing solvent. Water associates with a salt of the carboxylic acid, plasticizing the surface of the particles, promoting coalescence, and reducing the need for coalescing solvent. Another method is to use cross-linkable latexes with a lower T_g; cross-linking after film formation increases block resistance to offset the effect of the lower T_g. (See Section 8.1.4). It may also be possible to design a coalescing solvent that would cross-link after film formation. Another approach for eliminating the need for coalescing solvent is to use blends of high and low T_g latexes [29,30]. When a film of such a blend dries, the high T_g latex does not coalesce, but is dispersed in the continuous phase from the low T_g polymer. The hard particles act to reinforce the low T_g polymer film increasing the modulus of the film so that its block resistance is superior to that of a film from only the low T_g latex. It would be interesting to compare the results with those of a film from the low T_g latex with a volume content of pigment equal to that of the high T_g latex.

For latexes to be used over metal, it is desirable to have low moisture vapor permeability. Acrylic latexes give films with quite high permeability. One approach to reducing water permeability is to use vinylidene chloride as a comonomer. Reference [31] provides details of procedures for the copolymerization and characterizations of several vinylidene chloride/acrylate latexes. Since the reactivities of vinylidene chloride and acrylic ester monomers are quite different, monomer-starved polymerization conditions are used to achieve reasonably uniform compositions. While the moisture vapor permeability is greatly reduced, stabilizing the chlorinated copolymer against photodegradation may be difficult.

A different use for an acrylic emulsion polymer is for the preparation of microgel particles. For example, acrylic latexes containing some cross-links by using a small fraction of divinylbenzene as a comonomer were prepared by emulsion polymerization using a water-soluble azo initiator [4,4′-azobis(cyanovaleric acid)] [32]. The latex polymer was isolated as a powder by precipitation with an equal volume of isopropyl alcohol, washing with 4 : 1 methyl alcohol/water, and drying under vacuum at 50°C. The resultant powder readily dispersed in aliphatic hydrocarbon to yield a dispersion of an acrylic microgel. Among other applications, acrylic microgels are used as additives to control sagging of high solids coatings. (See Section 23.3.)

8.3. VINYL ESTER LATEXES

Vinyl acetate (VAc) is less expensive than (meth)acrylic ester monomers. However, VAc latexes are inferior to acrylic latexess in both photochemical stability and resistance to hydrolysis. When an acetate group on a PVAc chain hydrolyzes, the liberated acetic acid catalyzes hydrolysis of more acetate groups, and the hydroxyl group may exert a neighboring group (anchimeric) effect that promotes hydrolysis of adjacent acetate groups. Hence, PVAc latexes find their principal use in interior coatings that do not have to withstand high humidity exposure or frequent wetting. One such use, flat wall paints, is the largest volume type of paint sold in the U.S. market. The T_g of PVAc homopolymer is 29°C, too high for film formation under ambient conditions. Thus, one must either formulate PVAc with a plasticizer or copolymerize VAc with a monomer that reduces T_g. A widely used comonomer is n-butyl acrylate (BA), although other comonomers, such as 2-ethylhexyl acrylate and di-n-butyl maleate, have been used. Longer chain vinyl esters are being used as comonomers; the resultant latexes have superior hydrolytic stability and exterior durability.

Copolymerization of VAc with BA poses potential problems because the rate of reaction of a free radical on a terminal BA group with another molecule of BA is greater than its rate of reaction with VAc; furthermore, the rate of reaction of a free radical on a terminal VAc group is much greater with BA than with VAc. Thus a 50 : 50 mixture of BA and VAc will produce a copolymer very rich in BA at the outset of polymerization and very rich in VAc in the late stages. This problem is overcome by use of the semicontinuous batch process, wherein the monomer mixture is added at a rate equal to the polymerization rate. A steady state is reached in which the concentration of unreacted VAc is low, but the concentration of BA is even lower. Under these monomer-starved conditions, a relatively uniform copolymer forms with a composition similar to the ratio of monomers being fed. Such copolymers are made commercially on a large scale. An extensive study of the copolymerization of VAc with BA by batch and semicontinuous batch processes and characterization of the copolymers has been reported [10,33] and is discussed in part in Section 8.1. Since the reactions rates are relatively low, the rate of monomer addition must be quite slow. If the initial charge is high in the least reactive monomer, the addition can be made faster, while still producing reasonably uniform copolymerization. Mathematical simulations can be run to predict the effects of different addition modes [34].

Vinyl acetate is more soluble in water than acrylic esters; this factor also affects the polymerization. The initial oligomeric free radicals formed are soluble with more extended chains than with acrylic esters, increasing the probability of termination while still in aqueous solution, leading to low molecular weight fractions. The high water solubility of VAc increases the need for effective initiator systems to remove residual monomer. See Ref. [35] for a review of vinyl acetate latexes, especially regarding the effect of different initiator combinations. Chain transfer to polymer is more facile with polyvinyl acetate, leading to high molecular weight and broad molecular weight distribution.

An essential component of the recipe for polymerization of VAc is a buffer, such as sodium bicarbonate, to maintain the pH near 7. A buffer is essential because VAc monomer hydrolyzes at appreciable rates under either acidic or basic conditions. The hydrolysis is irreversible because the products are acetaldehyde and acetic acid. At pH less than 7, hydrolysis is autocatalytic because acetic acid catalyzes further reaction. Furthermore, acetaldehyde is oxidized by persulfate, which consumes initiator and generates still more acetic acid to catalyze hydrolysis.

$$CH_2=CH-O-\overset{O}{\overset{\|}{C}}-CH_3 \ + \ H_2O \ \longrightarrow \ CH_3-\overset{O}{\overset{\|}{C}}-OH \ + \ CH_3-\overset{O}{\overset{\|}{C}}-H$$

Copolymer latexes of VAc and (meth)acrylic monomers have been designed for dual-purpose use in both exterior and interior flat paints. The objective is to save cost in two ways—lower raw material cost for exterior paints and lower inventory and storage costs by basing all production on a single resin. This approach has been commercially successful, although the latexes undoubtedly represent a compromise between the best resins for exterior and interior use. The amount of vinyl acetate in an acrylic latex can be increased by sequential polymerization to form inverse core-shell latexes, lowering the cost without loss of properties [36]. For example, a seed latex of MMA/BA/MAA made with persulfate initiator was further reacted with vinyl acetate using azobisisobutyrylnitrile as initiator. Since the initial polymer has polar acrylate salt groups, the acrylate polymer becomes the shell with a polyvinyl acetate core.

A vinyl ester of a C_{10} branched acid, vinyl versatate, is widely used in Europe and, to a lesser extent in North America. Copolymer latexes of this monomer with VAc are used in both interior and exterior paints.

$$H_2C=CH-O-\overset{\overset{\displaystyle O}{\|}}{C}-\overset{\overset{\displaystyle R_1}{|}}{\underset{\underset{\displaystyle R_3}{|}}{C}}-R_2$$

Vinyl Versatate (R_1, R_2, R_3 are alkyl groups totaling to C_8H_{19})

Reference [37] discusses the use of a variety of vinyl esters, including vinyl pivalate and vinyl 2-ethylhexanoate, in latexes. Such monomers yield polymers that are more hydrophobic than vinyl acetate homopolymers and that have superior hydrolytic stability and scrub resistance. Reference [38] reports the advantages of using vinyl versatate in both vinyl acetate and acrylic copolymers.

GENERAL REFERENCES

D. C. Blackley, *Emulsion Polymerization, Theory and Practice*, Wiley, New York, 1975.
J. W. Vanderhoff, *J. Polym. Sci.: Polym. Symp.*, **72**, 161–198 (1985).
M. S. El-Aasser and P. A. Lovell, Eds., *Emulsion Polymerization and Emulsion Polymers*, Wiley, New York, 1997.

REFERENCES

1. K. L. Hoy, *J. Coat. Technol.*, **51** (651), 27 1979).
2. J. L. Lynn, Jr. and B. H. Bory, *Kirk-Othmer Encyc. of Chem. Tech.*, Vol. 23, 4th ed. Wiley, 1997, pp. 478–541.
3. R. Aslamazova, *Prog. Org. Coat.*, **25**, 109 (1995).
4. K. Holmberg, *Prog. Org, Coat.*, **20**, 325 (1992).
5. G. A. Vandezande and A. Rudin, *J. Coat Technol.*, **68** (860), 63 (1996).
6. B. D. Nguyen and A. Rudin, *J. Coat. Technol.*, **58** (736), 53 (1986).
7. D. R. Craig, *J. Coat. Technol.*, **61** (779), 49 (1989).
8. F. K. Hansen, "Is There Life Beyond Micelles?," in E. S. Daniels, E. D. Sudol, and M. S. El-Aasser, Eds., *Polymer Latexes*, American Chemical Society, Washington, DC, 1992, pp. 12–27.
9. J. L.Gardon, "A perspective on Resins for Aqueous Coatings," in J. E. Glass, Ed., *Technology for Waterborne Coatings*, ACS Symp. Series, 663, American Chemical Society, Washington, DC, 1997, p. 27.

10. M. S. El-Aasser, T. Makgawinata, and J. W. Vanderhoff, *J. Polym. Sci.: Polym. Chem. Ed.*, **21**, 2363 (1983).

11. K. Alahapperuma and J. E. Glass, *Prog. Org. Coat.*, **21**, 53 (1992).

12. C.-S. Chou, A. Kowalski, J. M. Rokowski, and E. J. Schaller, *J. Coat. Technol.*, **59** (755), 93 (1987); Anonymous, *Technical Bulletin, Rhoplex Multilobe 200*, Rohm and Haas Co. (1992).

13. J. C. Padget, *J. Coat. Technol.*, **66**, (839), 89 (1994).

14. B. G. Bufkin and J. R. Grawe, *J. Coat. Technol.*, **50** (641), 41; **50** (643), 67; **50** (644), 83; **50** (645), 70; **50** (647), 65 (1978); **51** (649), 34 (1979).

15. E. S. Daniels and A. Klein, *Prog. Org. Coat.*, **19**, 359 (1991).

16. Y. Huang and F. N. Jones, *Prog. Org. Coat.*, **28**, 133 (1996).

17. J. W. Taylor and D. W. Bassett, in J. E. Glass, Ed., *Technology for Waterborne Coatings*, American Chemical Society, Washington, DC, 1997, p. 137.

18. G. Pollano, *Polym. Mater. Sci. Eng.*, **77**, 73 (1996).

19. M. J. Chen, F. D. Osterholtz, A. Chaves, P. E. Ramdatt, and B. A. Waldman, *J. Coat. Technol.*, **69** (875), 49 (1997).

20. Y. Inaba, E. S. Daniels, and M. S. El-Aasser, *J. Coat. Technol.*, **66** (833), 63 (1994).

21. P. J. A. Geurink, L. van Daden, L. G. J. van der Ven, and R. R. Lamping, *Prog. Org. Coat.*, **27**, 73 (1996).

22. G. Monaghan, *Polym. Mater. Sci. Eng.*, **76**, 178 (1997).

23. M. J. Collins, J. W. Taylor, and R. A. Martin, *Polym. Mater. Sci. Eng.*, **76**, 172 (1997).

24. T. Nabuurs, R. A. Baijards, and A. L. German, *Prog. Org. Coat.* **27**, 163 (1996).

25. J. W. Gooch, S. T. Wang, F. J. Schork, and G. W. Poehlein, *Proc. Waterborne, High-Solids, Powder Coat. Symp.*, New Orleans, 1997, p. 366.

26. Anonymous, *Emulsion Polymerization of Acrylic Monomers, Tech, Bull.* CM-104 A/cf, Rohm & Haas Co., Philadelphia.

27. B. Emelie, C. Pichot, and J. Guillot, *Makromol. Chem.*, **189**, 1879 (1988).

28. A. Mercurio, K. Kronberger, and J. Friel, *J. Oil Colour Chem. Assoc.*, **65**, 227 (1982).

29. M. A. Winnik and J. Feng, *J. Coat. Technol.*, **68** (852), 39 (1996).

30. S. A. Eckersley and B. J. Helmer, *J. Coat. Technol.*, **69** (864), 97 (1997).

31. H. R. Friedli and C. M. Keillor, *J. Coat. Technol.*, **59** (748), 65 (1987).

32. M. Nair, *Prog. Org. Coat.*, **20**, 53 (1992).

33. S. C. Misra, C. Pichot, M. S. El-Aasser, and J. W. Vanderhoff, *J. Poly. Sci.; Polym. Chem. Ed.*, **21**, 2383 (1983).

34. G. Arzamendi and J. M. Asua, *J. Appl. Polym Sci.*, **38**, 2019 (1989).

35. G. A. Vandezande, O. W. Smith, and D. R. Bassett, "Vinyl Acetate Polymerization," in *Emulsion Polymerization and Emulsion Polymers*, P. A. Lovell and M. S. El-Aasser, Eds., Wiley, New York, 1997, pp. 563-587.

36. G. A. Vandezande and A. Rudin, *J. Coat. Technol.*, **66** (828), 99 (1994).

37. R. A. Prior, W. R. Hinson, O. W. Smith, and D. R. Bassett, *Prog. Org. Coat.*, **29**, 209 (1996).

38. F. Decocq, D. Heymans, M. Slinckx, S. Spanhove, and C. Nootens, *Proc. Waterborne, High-Solids, Powder Coat, Symp.*, New Orleans, 1997, p. 168.

CHAPTER 9

Amino Resins _____

Amino resins, also called aminoplast resins, are major cross-linking agents for baked thermosetting coatings; the amino resins most commonly used are derived from melamine-that is, 2,4,6-triamino-1,3,5-triazine. Benzoguanamine, urea, glycoluril, and copolymers of (meth)acrylamide are also used.

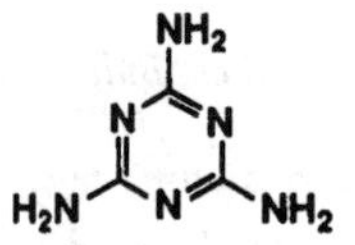

Melamine

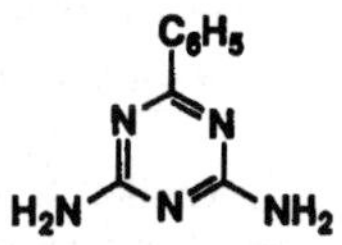

Benzoguanamine

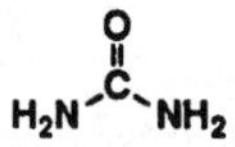

Urea

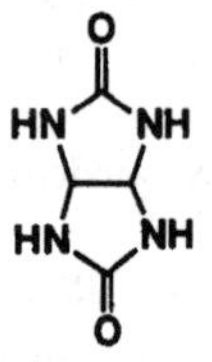

Glycoluril

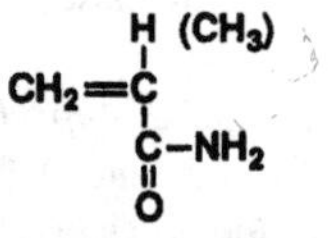

(Meth)Acrylamide

Amino resins are made by reacting one of these compounds with formaldehyde ($H_2C=O$) and, subsequently, with alcohols (ROH) to yield ethers with the general structure $>NCH_2OR$. Ethers of amino resins are activated toward nucleophilic substitution by the neighboring N and are much more reactive than aliphatic ethers. When the nucleophile is the alcohol of a polyol (POH), transetherification can occur, as shown in Eq. 9.1, resulting in formation of a cross-linked polymer. The reaction is catalyzed by acid. Carboxylic acids, urethanes, and phenols with an unsubstituted ortho position also react, as shown in Eqs. 9.2, 9.3, and 9.4.

$$>N-CH_2-OR + P-OH \rightleftharpoons >N-CH_2-OP + ROH \qquad (9.1)$$
$$\mathbf{1}$$

$$\mathbf{1} + P-COOH \rightleftharpoons PCOO-CH_2-N< + ROH \qquad (9.2)$$

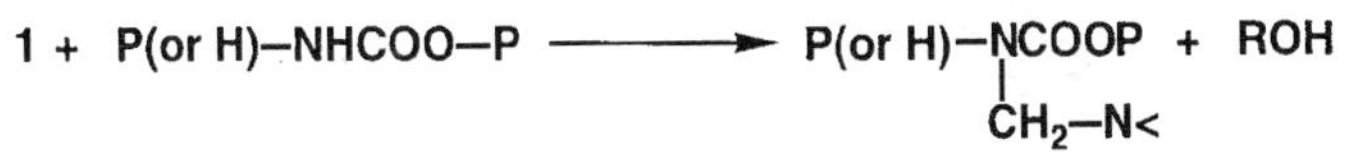

$$1 + \text{P(or H)--NHCOO--P} \longrightarrow \underset{\overset{|}{\text{CH}_2\text{--N}<}}{\text{P(or H)--NCOOP}} + \text{ROH} \qquad (9.3)$$

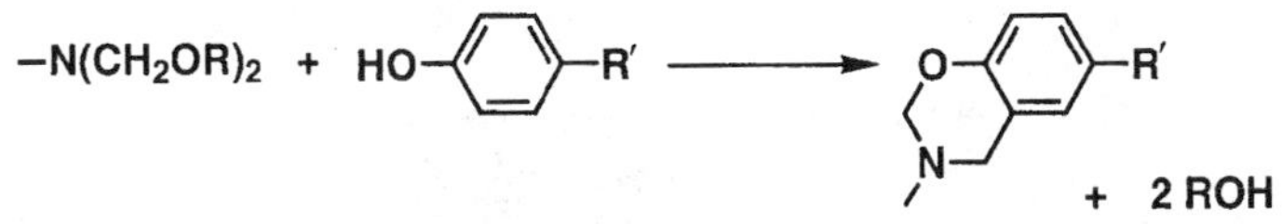

$$-\text{N(CH}_2\text{OR)}_2 + \text{HO--}\langle \text{C}_6\text{H}_4 \rangle\text{--R}' \longrightarrow \text{(benzoxazine)R}' + 2\text{ ROH} \qquad (9.4)$$

9.1. SYNTHESIS OF MELAMINE-FORMALDEHYDE RESINS

The first step in synthesis of melamine-formaldehyde (MF) resins is methylolation, the reaction of melamine with formaldehyde under basic conditions. With excess formaldehyde, the reaction can be driven to form predominantly hexamethylolmelamine (1), where R is H. A mixture of partially methylolated derivatives, including species such as symmetrical trimethylolmelamine (2), where R is H, is formed with less than the stoichiometric 6 mol of formaldehyde per mol of melamine.

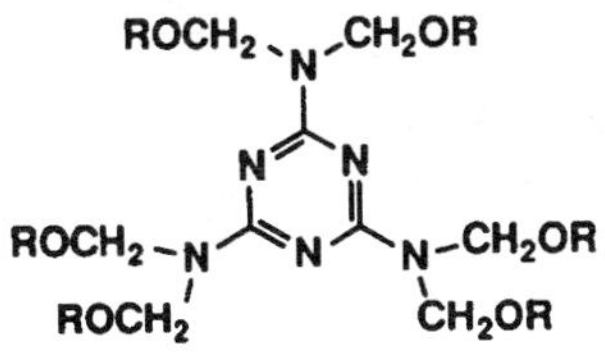

1 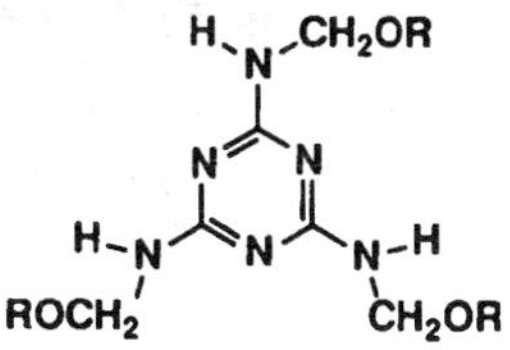   2

The second step is acid-catalyzed etherification of methylolmelamines with an alcohol, such as methyl or butyl alcohol. Complete etherification of methylolmelamines (1) and (2), where R is H, yields alkoxymethyl derivatives (1) and (2), where R is alkyl. Many commercial MF resins are only partially etherified. In addition to monomeric species, commercial MF resins contain oligomeric species in which triazine rings are linked by methylene ($>\text{NCH}_2\text{N}<$) and dimethylene ether bridges ($>\text{NCH}_2\text{OCH}_2\text{N}<$).

9.1.1. The Methylolation Reaction

A probable mechanism for base catalyzed reaction of melamine with formaldehyde is outlined in Scheme 9.1, where $-\text{NH}_2$ represents the melamine amino groups. The first step involves nucleophilic attack by the amino group on the electrophilic C of formaldehyde, which is facilitated by removal of a proton from N by the base (B^-). This is followed by proton transfer from the resulting B--H to the negative oxygen atom, which yields the methylolated product and regenerates the base catalyst; both steps are reversible.

Studies of the kinetics of the reaction indicate that the presence of one methylol group on N deactivates the group for a second reaction by a factor of 0.6. On the other hand, substitution on one amino group has little effect on the reactivity of the other amino groups. These kinetic factors favor formation of the symmetrical trimethylolmelamine (TMMM) (2), where R is H. However, the preference is not strong enough to overcome the thermodynamic tendency to produce mixtures of products. Thus, at equilibrium, the reaction of 6 mol of formaldehyde with 1 mol of melamine yields a mixture of products

Scheme 1

Main Reaction:

$$—NH_2 + H_2C=O + B^- \rightleftharpoons —NH—CH_2—O^- + B—H$$

$$—NH—CH_2—O^- + B—H \rightleftharpoons —NH—CH_2—OH + B^-$$

Side Reaction:

$$—NH—CH_2—O^- + H_2C=O \rightleftharpoons —NH—CH_2—O—CH_2—O^-$$

$$—NH—CH_2—O—CH_2—O^- + B—H \rightleftharpoons —NH—CH_2—O—CH_2—OH + B^-$$

including all levels of methylolation and free formaldehyde [1]. A side reaction is methylolation on oxygen, as shown in Scheme 9.1.

9.1.2. The Etherification Reaction

Following methylolation, the base catalyst is neutralized with excess acid and the appropriate alcohol is added. The acid-catalyzed reversible reaction leads to formation of alkoxymethyl groups. Nitric acid is commonly used as a catalyst because nitrate salt byproducts are relatively easily removed.

There is disagreement in the literature about the mechanism of the etherification reaction, as well as of the closely related transetherification reaction, shown in Eq. 9.1. The contending mechanisms for these substitution reactions are S_N1 and S_N2, both of which are outlined in Scheme 9.2, where R is H and R′OH is methyl or butyl alcohol. Note that the conjugate base (A^-) is omitted from the intermediate steps.

The distinguishing feature is whether the protonated methylol group ionizes to an intermediate, resonance-stabilized carbocation, as in Eq. 9.5, before reacting with the alcohol, characteristic of the S_N1 mechanism, or reacts directly with the alcohol, as in Eq. 9.6, characteristic of the S_N2 mechanism. The controversy about mechanism in MF resin synthesis also applies to the transetherification reaction involved in cross-linking MF resins with polyols, where R is methyl or butyl and R′OH is the polyol. Experimental evidence bearing on this controversy has been acquired primarily in studies of coatings and is discussed in Section 9.3.2. Based on the available evidence, we favor the S_N1 mechanism.

Etherification of methylol groups on a singly substituted N (that is, an N that also bears a H atom) is proposed to follow the mechanism outlined in Scheme 9.3, where R is H and R^1OH is methyl or butyl alcohol.

The distinguishing feature is formation of the uncharged imine intermediate, shown in Eq. 9.7, which is possible due to the presence of the N—H group. Imines form by elimination of water catalyzed by relatively weak acids, such as carboxylic acids. Complexation of the methylolmelamine by an acid (A—H) is sufficient to yield the imine, either in a concerted push-pull mechanism, as shown in Scheme 9.3, or a step-wise mechanism. Complexation of the reactive imine by acid is also sufficient to effect subsequent addition of alcohol, resulting in overall substitution of water by alcohol. This reaction is also reversible. Since catalysis in Scheme 9.3 involves complexation,

Scheme 9.2

S_N1 Mechanism

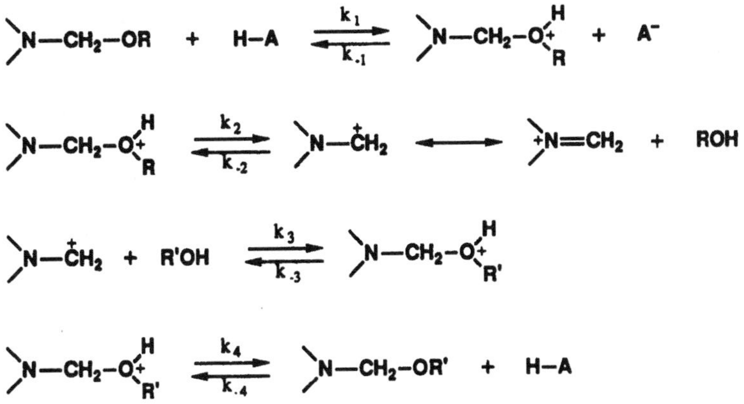

$$\begin{array}{c}\rangle N{-}CH_2{-}OR \;+\; H{-}A \; \underset{k_{-1}}{\overset{k_1}{\rightleftharpoons}} \; \rangle N{-}CH_2{-}O{\overset{H}{\underset{R}{+}}} \;+\; A^-\end{array}$$

$$\begin{array}{c}\rangle N{-}CH_2{-}O{\overset{H}{\underset{R}{+}}} \; \underset{k_{-2}}{\overset{k_2}{\rightleftharpoons}} \; \rangle N{-}\overset{+}{C}H_2 \; \longleftrightarrow \; \rangle \overset{+}{N}{=}CH_2 \;+\; ROH\end{array} \qquad (9.5)$$

$$\begin{array}{c}\rangle N{-}\overset{+}{C}H_2 \;+\; R'OH \; \underset{k_{-3}}{\overset{k_3}{\rightleftharpoons}} \; \rangle N{-}CH_2{-}O{\overset{H}{\underset{R'}{+}}}\end{array}$$

$$\begin{array}{c}\rangle N{-}CH_2{-}O{\overset{H}{\underset{R'}{+}}} \; \underset{k_{-4}}{\overset{k_4}{\rightleftharpoons}} \; \rangle N{-}CH_2{-}OR' \;+\; H{-}A\end{array}$$

S_N2 Mechanism

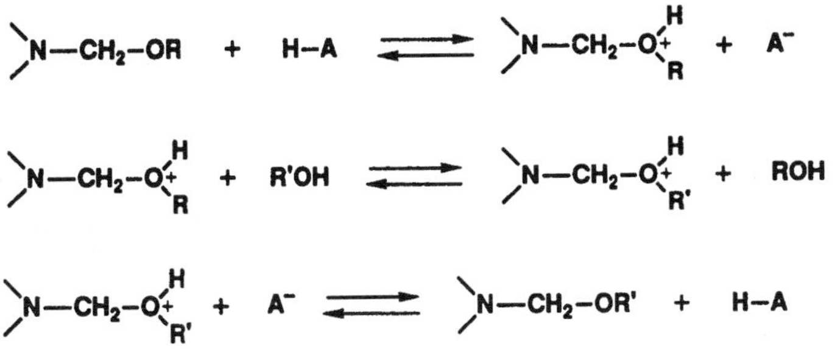

$$\begin{array}{c}\rangle N{-}CH_2{-}OR \;+\; H{-}A \; \rightleftharpoons \; \rangle N{-}CH_2{-}O{\overset{H}{\underset{R}{+}}} \;+\; A^-\end{array}$$

$$\begin{array}{c}\rangle N{-}CH_2{-}O{\overset{H}{\underset{R}{+}}} \;+\; R'OH \; \rightleftharpoons \; \rangle N{-}CH_2{-}O{\overset{H}{\underset{R'}{+}}} \;+\; ROH\end{array} \qquad (9.6)$$

$$\begin{array}{c}\rangle N{-}CH_2{-}O{\overset{H}{\underset{R'}{+}}} \;+\; A^- \; \rightleftharpoons \; \rangle N{-}CH_2{-}OR' \;+\; H{-}A\end{array}$$

rather than protonation by the acid, the rate is dependent on the nature of H−A, as well as its concentration. Indeed, both the more positive H end and the more negative A end of H−A participate by facilitating breaking of both the C−O and the N−H bonds, respectively. By convention, this is called *general acid catalysis*. The mechanism in Scheme 9.3 also applies to cross-linking of polyols by MF resins with N−H groups by transetherification, where R is methyl or butyl and R′OH represents the polyol.

In contrast, elimination of water from methylolmelamines lacking N−H groups, either by the S_N1 or S_N2 mechanism (shown in Scheme 9.2), requires protonation of the methylol group, necessitating a strong, ionizing acid, such as nitric, sulfuric, or a sulfonic acid. Such catalysis, requiring protonation, is called *specific acid catalysis*, signifying that the rate is dependent only on the concentration of protons (H^+) and is independent of the conjugate base (A^-).

9.1.3. Self-Condensation Reactions

Self-condensation refers to reactions that lead to the formation of bridges between triazine rings, resulting in dimers, trimers, and higher oligomers. The extent of such reactions

Scheme 9.3

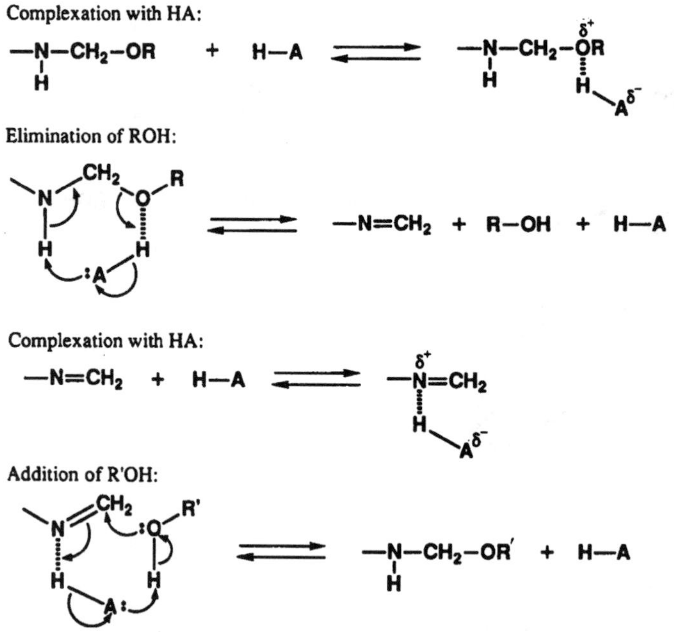

Complexation with HA:

$$—N—CH_2—OR \quad + \quad H—A \quad \rightleftarrows \quad —N—CH_2—\overset{\delta^+}{O}R$$

Elimination of ROH:

$$\rightleftarrows \quad —N{=}CH_2 \quad + \quad R—OH \quad + \quad H—A \tag{9.7}$$

Complexation with HA:

$$—N{=}CH_2 \quad + \quad H—A \quad \rightleftarrows \quad \overset{\delta^+}{—N}{=}CH_2$$

Addition of R'OH:

$$\rightleftarrows \quad —N—CH_2—OR' \quad + \quad H—A$$

depends on process factors including pH, ratio of reactants, reaction temperature, rate of removal of water, and probably others. Two types of linkages occur between triazines: methylene bridges ($>NCH_2N<$) and dimethylene ether bridges ($>NCH_2OCH_2N<$) [2,3]. Acid catalysis is said to favor dimethylene ether bridges and base catalysis to favor methylene bridges; furthermore, formation of methylene bridges may occur by base-catalyzed scission of dimethylene ether bridges [4].

Formation of methylene bridges has been demonstrated in the acid-catalyzed reaction of a model compound N,N-dimethoxymethyl-N',N',N,N-tetramethylmelamine (3) and water [5]. In the presence of p-toluenesulfonic acid (pTSA) and water at elevated temperatures, a cyclic trimer (4) with methylene groups connecting the three nitrogens is formed. Formation of the trimer is reversible [6]; when trimer (4) is heated with methyl alcohol and formaldehyde at 130°C, compound (3) is regenerated.

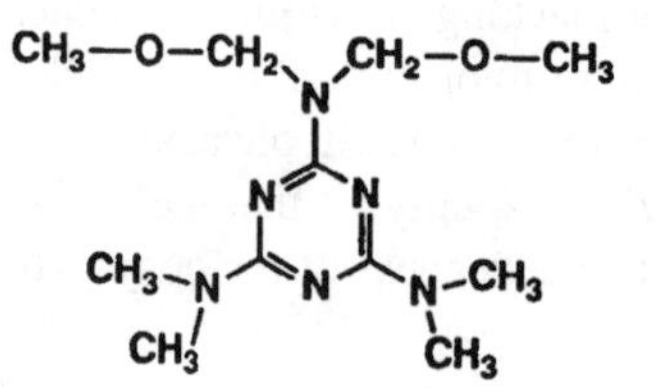

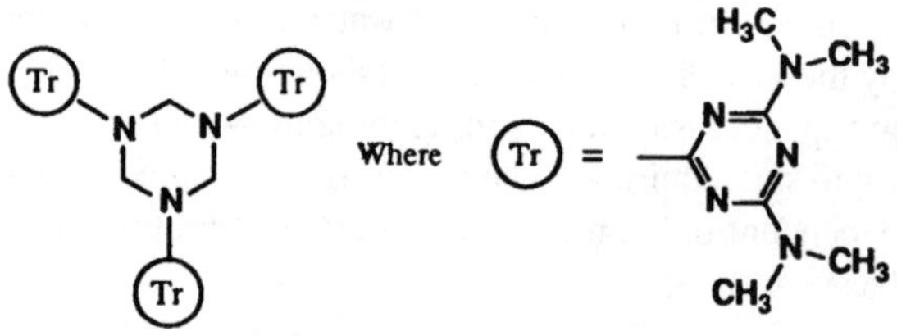

The monomethoxymethyl model compound (5) is more reactive than (3) [7]. It rapidly forms the cyclic trimer (4) in the absence of water at 100°C. In solution with an acid catalyst, it equilibrates rapidly at 25°C with the dimethylene ether bridge dimer (6).

$$\text{Tr}-\overset{\overset{\text{H}}{|}}{\text{N}}\text{CH}_2\text{OCH}_3 \underset{\substack{25\,^{\circ}\text{C} \\ -\text{CH}_3\text{OH}}}{\overset{\text{H}^+}{\rightleftharpoons}} \text{Tr}-\overset{\overset{\text{H}}{|}}{\text{N}}\text{CH}_2\overset{\overset{\text{CH}_2\text{OCH}_3}{|}}{\text{N}}-\text{Tr} \underset{100\,^{\circ}\text{C}}{\overset{\text{H}^+}{\rightleftharpoons}} \quad 4$$

5   **6**

9.2. TYPES OF MF RESINS

A variety of MF resins is made commercially, with differences in the ratio of functional groups, the alcohol used for etherification, and average degree of polymerization $\bar{P}$. To facilitate discussion, MF resins are classified into two broad classes, I and II. Class I resins are made with relatively high ratios of formaldehyde to melamine, and hence, most of the nitrogens have two alkoxymethyl substituents. Class II resins are made with smaller ratios of formaldehyde, and many of the nitrogens have only one substituent. A variety of each class of resins is produced commercially, including methyl and butyl ethers with different $\bar{P}$'s, as well as mixed ether resins, which are commonly mixtures of methyl with n-butyl, isobutyl, or isooctyl ethers.

From about 1940 through the 1950s, the predominant melamine resins were Class II types used for cross-linking alkyds. These resins have enough bridging so that $\bar{P}$ is three or more, and the alcohol is usually n-butyl or isobutyl alcohol. Such resins are economical, they are readily miscible in alkyd formulations, and they give a wide latitude; that is, coatings do not require exacting control of formulation to produce acceptable application characteristics and film properties.

Class I MF resins were commercialized in the 1950s. Methylated Class I resins are more compatible than butylated resins with certain coreactant resins, and in some cases, provide tougher films. Introduction of water-borne and high solids coatings during the 1970s accelerated the shift toward Class I resins. Methylated MF resins are more miscible with the water, solvent, resin blend used in water-borne coatings, and Class I resins with low $\bar{P}$ give lower viscosity to high solids coatings. These generalities are not universal, some Class II resins are used in low-solvent coatings.

A commercial methylated MF Class I resin has been reported to contain 62% monomers (i.e., one triazine ring), 23% dimers, together with 15% trimers and higher oligomers [8]. The equivalent weight of pure hexamethoxymethylmelamine (HMMM) is 65 g eq^{-1}, but the equivalent weight of this resin is 80 g eq^{-1}, resulting from the presence of dimers, trimers, and higher oligomers. (In estimating equivalent weight, it is necessary to make an assumption as to whether $>\text{NCH}_2\text{N}<$ or $>\text{NCH}_2\text{OCH}_2\text{N}<$ bridging groups react to form cross-links. Here, it is assumed that they do not.) Chromatographic analysis of a different, but similar, resin of this type is shown in Figure 9.1. The chromatograms show at least 30 different chemical species present in significant amounts [2,9]. While the predominant groups are $-\text{N}(\text{CH}_2\text{OCH}_3)_2$, various species containing incompletely etherified groups $[-\text{N}(\text{CH}_2\text{OCH}_3)(\text{CH}_2\text{OH})]$ and incompletely formylated groups $[-\text{NH}(\text{CH}_2\text{OCH}_3)]$ are present. These methylated MF resins are frequently called *HMMM resins*, but the terminology *high HMMM resin* is more appropriate.

Other Class I resins are available with different degrees of polymerization $\bar{P}$ and with different extents of formylation and etherification. In addition to methylated derivatives, butylated, as well as mixed methyl/n-butyl, methyl/isobutyl, methyl/ethyl, and

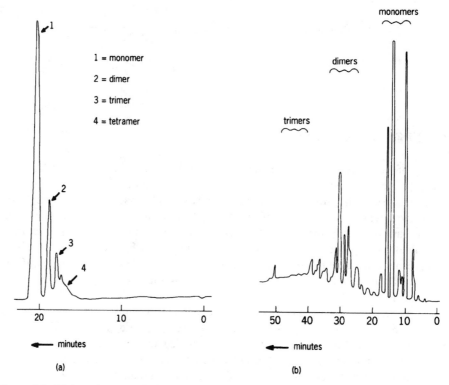

Figure 9.1. High-performance liquid chromatograms (HPLC) of a typical Class I, high HMMM resin. (*a*) An SEC chromatogram and (*b*) a gradient HPLC chromatogram. (From Ref. [9], with permission.)

methyl/isooctyl derivatives are also available. Class I resins generally have lower $\bar{P}$ and, therefore, lower viscosity than Class II resins at the same percent solids. The viscosity is also reduced by the presence of fewer polar groups. At a given $\bar{P}$, butyl ethers exhibit even lower viscosities, attributable to lower T_g values, than the corresponding methyl ethers.

A broader range of Class II resins is available because the ratio of formaldehyde/ melamine and of alcohol/formaldehyde can be varied through a wider range. It is harder to suppress bridge-forming reactions during synthesis of Class II resins, and very low $\bar{P}$ resins cannot be prepared. The predominant reactive group present in Class II resins is $-NHCH_2OR$. Hence, they are frequently called *high NH resins*. They also contain $-NH_2$, $>NCH_2OH$, as well as $-N(CH_2OR)_2$, groups. To minimize viscosity and maximize reactivity, resin producers strive to reduce $\bar{P}$ and maximize the amount of symmetrical TMMM (**3**), where R is Me. This is possible to a degree because methylolation of $-NH_2$ groups, yielding $-NHCH_2OH$, is kinetically favored over methylolation of $NHCH_2OH$, yielding $-N(CH_2OH)_2$. Owing to advances in HPLC analysis techniques, it is possible to follow the effects of small changes in process conditions on composition. Continuing refinement of processes has permitted commercial production of resins with as high as 50% TMMM. A variety of alcohols are used to make different resin grades.

9.3. REACTIONS OF MF RESINS IN COATINGS

Melamine-formaldehyde resins are used to cross-link coreactant resins having hydroxyl, carboxylic acid, urethane (carbamate), and/or amide groups. Acrylic (see Chapter 12), polyester (see Chapter 13), alkyd (see Chapter 15), epoxy (see Chapter 11), and polyurethane (see Chapter 10) resins are the most important classes of coreactant resins.

Amino resin cross-linked coatings give off small amounts of formaldehyde vapor during application and cross-linking, posing a potential hazard to workers who are exposed daily. Some resins are better than others in this respect. Formaldehyde exposure levels are strictly regulated in the United States, and current technology appears adequate to satisfy these regulations. While the regulations already appear conservative, it is possible that allowable exposure levels will be reduced further. Intensive research on ways to reduce formaldehyde emissions has been underway. Besides improving ventilation, approaches include changing resins, changing formulations, and introducing formaldehyde scavengers.

Automotive top coats made with hydroxy-functional acrylics and MF resins are more or less vulnerable to environmental etching—that is, formation of small blemishes on the coating surface on exposure to acid rain. (See Sections 5.5 and 29.1.2 for discussion of environmental etching). Use of Class I resins to cross-link carbamate-functional resins gives good resistance. (See Section 9.3.4.)

Polyols (hydroxy-functional acrylics, polyesters, etc.) are the resins most commonly cross-linked by MF resins. The hydroxyl groups of polyols react by either transetherification with the activated alkoxymethyl groups or by etherification of methylol groups of MF resins to form new ether cross-links. The reaction with phenols, as in Eq. 9.4, has the advantage that a C−C bond is formed so that the product is stable to hydrolysis; see Section 13.3 for an example of the use of this reaction. Strong acid catalysts, such as sulfonic acids, are used for Class I resins, and weak acid catalysts, such as carboxylic acids, for Class II resins. The reactions are reversible, but are driven towards cross-linking by volatilization of the monofunctional alcohol or water produced, as shown in Eq. 9.8, where R is an alkyl or H and POH is the polyol.

$$\ce{>N-CH2-OR} \quad + \quad \textcircled{P}\text{--OH} \quad \underset{\xrightarrow{\hspace{1cm}}}{\overset{H-A}{\xleftarrow{\hspace{1cm}}}}$$

$$\ce{>N-CH2-O}\text{--}\textcircled{P} \quad + \quad R-OH$$

$$(9.8)$$

Carboxylic acid-functional resins react with MF resins to form the corresponding ester derivatives (see Eq. 9.2); the reaction is slower than with hydroxyl groups. Carboxylic acid groups are present in substantial levels in water-borne hydroxy-functional resins to enhance dispersibility, as well as in some solvent-borne polyols, at low levels, to enhance the stability of pigment dispersions. When significant levels of carboxylic acid groups are present, it is desirable to cure under conditions that ensure they react, since residual carboxylic acid groups increase water sensitivity of cured films.

Rates of reaction with hydroxyl groups depend on the structure of the polyol and the MF resin, the type and amount of catalyst, and the temperature. The rate at which MF resins cross-link polyols increases with increasing volatility of the alcohol, according to the following order: methyl>ethyl>n-butyl. These rates are probably influenced by the rate of diffusion of the alcohol from the reactive site and by evaporation from the film.

Factors in addition to cure response must be considered when selecting the alcohol. For example, viscosity at a given percent solids content can be reduced by substituting Class I mixed ether resins for HMMM types. The alcohols in mixed ether resins are frequently 1 : 6 to 1 : 3 n-butyl or isobutyl alcohol to methyl alcohol. Only modest reductions of VOC are attained because mixed ether resins contain a higher weight fraction of volatile byproducts; butyl groups represent a higher weight fraction of resins than do methyl groups. Perhaps, a greater advantage of mixed ether resins is that they impart lower surface tension to coatings than high HMMM types. High surface tension is often associated with development of film defects such as crawling, cratering, leveling (see Chapter 23), and poor intercoat adhesion (see Section 6.5) [10].

It has been commonly said that polyols with primary hydroxyl groups react faster than those with secondary hydroxyl groups, since the less sterically hindered primary alcohols would be expected to react faster. One study indicates that the reactivity of primary and secondary alcohols with a high HMMM resin is the same; however, the reverse reaction is faster with secondary alcohols [11]. This finding is a valid alternative explanation for the apparent higher reactivity of primary alcohols, since the curing reactions occur under reversible conditions. This proposal is consistent with the S_N1 mechanism for transetherification (shown in Scheme 9.2). Unfortunately, an oversimplified rate expression was used in this study, raising a question about the conclusions; the rate constants were determined from a rate expression that was inappropriate for the equivalent amounts of alcohol and HMMM resin utilized.

The rates of development of solvent resistance and film hardness when a fractionated Class I resin with about 95% HMMM was used to cross-link polyester resins made with cyclohexanedimethanol (CHDM), neopentyl glycol (NPG), and hexylene glycol (HG) (1,6-hexanediol) have been reported [7,12]. CHDM polyesters are most reactive, NPG polyesters are a close second, and HG polyesters are least reactive. It was also found that acrylic polyols generally reacted more rapidly than polyester polyols.

During acid-catalyzed cross-linking with polyols, reactions between MF resin molecules also occur. These *self-condensation reactions* include formation of methylene bridges and dimethylene ether bridges. Self-condensation reactions result in cross-linking, as do reactions of MF resins with polyol (or another coreactant), which have been termed *cocondensation reactions* [13]. Both self- and cocondensation reactions contribute to the structure of the cross-linked polymer network and to its film properties.

With strong acid catalysis, the apparent rate at which Class I resins react with most polyols by cocondensation is faster than self-condensation. However, with Class II resins, the apparent rates of cocondensation and self-condensation are similar. The $-NHCH_2OR$ and $>NCH_2OH$ groups in Class II resins promote self-condensation reactions. The methylol groups can react analogously with a polyol, and the $-NHCH_2OR$ groups can eliminate alcohol to yield reactive melamine imine ($-N=CH_2$) groups, which can react to form dimethylene ether bridges with $-NCH_2OH$ groups and methylene bridges by reaction with $>NH$ groups.

Class I resins are used with an excess of alkoxymethyl groups over hydroxyl groups, so final development of film properties depends on the extent of self-condensation reactions as well as on cocondensation. In some cases, this high stoichiometric ratio is dictated by the lower cost of MF resins as compared to some polyols. It may also be that self-condensation cross-links give some advantage in film properties. The amounts of the MF resin and catalyst are determined empirically and are optimized for a specified baking time and temperature. Formulations are designed to achieve optimum properties when cocondensation is nearly complete and self-condensation is partly complete. If the conditions actually experienced in the use of the coating differ from those for which the formulation was designed, the extent of self-condensation and cocondensation, will differ from the optimum levels. Coating properties such as hardness, adhesion, exterior durability, and impact resistance may be affected.

In high solids coatings in which the hydroxy equivalent weight is higher and the average functionality of the polyol is lower in comparison with low solids, higher molecular weight polyols, the results are particularly sensitive to variations in cure temperature and time. These considerations have been addressed in terms of a *cure window*, corresponding to the range of cure times and temperatures that provides films of acceptable properties [14]. The cure window was shown to be smaller for high solids MF-polyol coatings, especially when Class II resins were used.

9.3.1. Catalysis of MF-Polyol Reactions

With strong acid catalysts, commonly aryl sulfonic acids, usually in the range of 0.5 to 1 wt% of MF resin, reactions of MF Class I resins with polyols give cured films in 10–30 minutes at 110–130°C. Cocondensation of MF Class I resins with polyols can be catalyzed by weak acids, such as carboxylic acids; however, elevated cure temperatures, generally greater than 140°C, are required [15]. In the case of Class II resins, carboxylic acids are more effective in catalyzing cocondensation in accordance with the mechanism provided in Scheme 9.3. Since many polyol resins contain some carboxylic acid groups, which are present to promote adhesion and facilitate pigment dispersion, no added catalyst may be needed for cross-linking with Class II resins. At the high temperatures (air temperatures as high as 375°C) used for short times in coil coating ovens, strong acid catalysts are needed even with Class II resins.

Cure time and/or temperature may be reduced by increasing the catalyst concentration. However, storage (or package) stability is reduced by this approach, since the reaction at ambient temperature is also catalyzed by acid. Another danger in increasing acid catalyst concentration to reduce cure time and/or temperature relates to the durability of the cured coating. In addition to catalyzing transetherification (and etherification) reactions of MF resins with polyols, acid residues catalyze hydrolysis of cross-links in cured coatings. The course of the hydrolysis reaction can be followed by reference to Schemes 9.2 and 9.3, where $>N-CH_2-OR$ represents a cross-linked polymer and $R'OH$ represents water. The hydrolysis reaction breaks cross-linked bonds and generates methylol groups that, at least in part, eliminate formaldehyde as shown in Eq. 9.9 [13].

$$\text{>N}-CH_2-OH \quad \rightleftharpoons \quad \text{>NH} \quad + \quad H_2C=O \tag{9.9}$$

Sunlight enhances acid hydrolysis, which reduces exterior durability of MF-polyol top coats [14]. From the standpoint of exterior durability, it is desirable to bake at relatively high temperatures and/or for relatively long times so that the level of acid catalyst can be minimized. This assertion applies both to long-term durability and to the resistance of relatively new cars to environmental etching. However, other important considerations, such as energy savings, production rates, and the increasing use of plastic components motivate use of low temperature, short bakes. The optimum compromise involves many factors.

Free acids such as *p*-toluene sulfonic acid (pTSA or TsOH) catalyze cross-linking under ambient conditions at a rate that may increase the viscosity of a liquid coating above the range for application in less than 6 months. One-package systems commonly use acid catalysts that are deactivated—variously called *acid precursors*, *latent acids*, and *blocked acids* [16]. Aryl sulfonic acids are strong acids (the pK_a of pTSA $= -6$); their tertiary amine salts are weak acids ($pK_a = 8$ to 10) and do not catalyze the reactions of Class I MF resins. As shown in Eq. 9.10, there is an equilibrium between pTSA amine salt and the protonated MF resin that is the active species for catalysis.

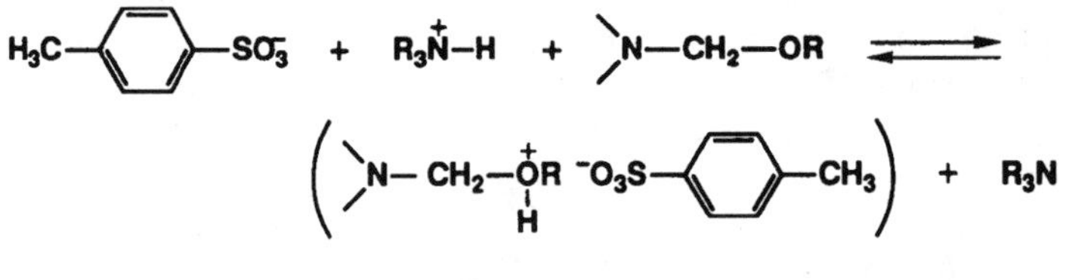

$$H_3C-\!\!\!\left\langle\!\!\!\bigcirc\!\!\!\right\rangle\!\!\!-SO_3^- \;+\; R_3\overset{+}{N}\!-\!H \;+\; {\scriptstyle\diagdown}N\!-\!CH_2\!-\!OR \;\rightleftharpoons$$

$$\left({\scriptstyle\diagdown}N\!-\!CH_2\!-\!\underset{H}{\overset{+}{O}}R \;\; ^-O_3S-\!\!\!\left\langle\!\!\!\bigcirc\!\!\!\right\rangle\!\!\!-CH_3\right) \;+\; R_3N \qquad (9.10)$$

While the equilibrium favors structures on the left, it may be shifted to the right by volatilization of the amine, increasing the concentration of protonated MF resin, the initial intermediate in the transetherification reaction; see Scheme 9.2. Generally, storage stability of a coating with blocked catalyst approaches the stability of an uncatalyzed coating. In some cases, the cure rate approaches that of a coating catalyzed with the free sulfonic acid, while in others, the curing rate is somewhat reduced [17]. The balance of storage stability and cure rate is reported to be particularly favorable for *N*-benzyl-*N*,*N*-dimethylanilinium sulfonates [18].

Certain esters of sulfonic acids liberate free acid when heated either by alcoholysis with polyols or by decomposition reactions. An example of a tosylate ester that has proven useful as a catalyst is the oxime ester shown in Eq. 9.11. When heated, the ester decomposes in a complex reaction to liberate TsOH. Unlike amine salts, tosylate esters do not increase the conductivity of coatings, an important consideration for coatings that are to be applied by electrostatic spray.

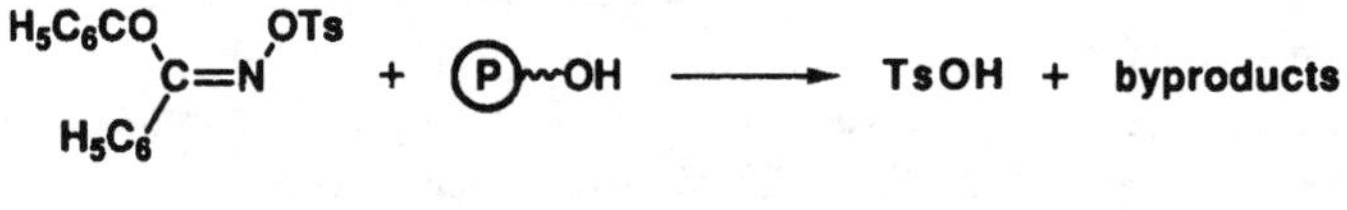

$$\begin{array}{c} H_5C_6CO\quad OTs \\ \diagdown\!\!C\!\!=\!\!N\diagup \\ H_5C_6\diagup \end{array} \;+\; \text{(P)}\!\!\sim\!\!OH \;\longrightarrow\; \mathbf{TsOH} \;+\; \mathbf{byproducts} \qquad (9.11)$$

The acid strength of the medium can be no stronger than that of protonated melamine resin, which exerts a leveling effect on acid strength. Just as in water, the strongest acid is the hydronium ion. Apparently, pTSA (TsOH) is completely ionized by MF resins, which explains why even stronger acids, such as hexafluorophosphoric acid, exhibit similar catalytic activity. On the other hand, pTSA is more effective for Class I MF resins than weaker acids, such as butylphosphoric and carboxylic acids. With pTSA, the rate of cocondensation of an HMMM resin and acrylic polyol is reported to be proportional to the square root of the acid concentration [19].

In water-borne MF-polyol coatings, the amine utilized to neutralize the solubilizing carboxylic acid groups in the polyol (see Section 12.3 for discussion) also neutralizes the sulfonic acid. It was shown that commonly utilized hydroxy-functional amines, such as *N*,*N*-dimethylethanolamine (DMEA), may also participate in transetherification of MF resins and/or transesterification of ester groups in the polyol during storage or initial stages of curing, which may retard or prevent their volatilization during cure, causing loss of cure response [20]. It was also shown that 2-amino-2-methylpropanol (AMP) gives faster cure since the primary amine can react with an acrylic resin to form amides, reducing the basicity. Formation of an amide, together with transesterification by the hydroxyl group of AMP, forms cross-links. Furthermore, AMP has been shown to react with formaldehyde to form an oxazolidine, also reducing basicity so that it does not inhibit cure like DMAE [21]. Dynamic mechanical analysis also shows the lesser extent of inhibition by AMP and supports auxiliary cross-linking by AMP [22].

While the acid strength of a strong acid catalyst makes little difference in cure rates, the choice of acid (or blocked acid) catalyst can lead to differences in film properties of cured coatings. The most widely used catalyst is pTSA. The grade of pTSA is important, since some grades contain significant quantities of sulfuric acid, which can lead to pronounced yellowing of films during baking. It has been found that water resistance of films can be improved by use of a more hydrophobic sulfonic acid, such as dinonylnaphthalene disulfonic acid (DNNDSA), which is reported to produce films that are less likely to blister when exposed to high humidity conditions [23]. DNNDSA is particularly effective in coatings applied directly to metal that, with pTSA, are prone to blistering on humidity exposure.

p-Dodecylbenzenesulfonic acid (DDBSA) provides similar catalytic activity and film properties to those of DNNDSA when used in coatings applied over a prime coat. However, when a DDBSA catalyzed coating is used directly on steel, adhesion of the coating is likely to be poor. It seems probable that the sulfonic acid group on DDBSA is strongly adsorbed on the steel surface, causing the surface to become covered with dodecyl groups. The low surface tension of the long hydrocarbon dodecyl groups on DDBSA may cause dewetting by the rest of the coating or formation of a weak boundary layer that reduces adhesion. The better adhesion with DNNDSA catalyzed coatings may be related to the presence of two sulfonic acid groups or weaker surfactant properties.

The activity of acid (or blocked acid) catalysts can be affected by pigmentation of the coating. Some grades of titanium dioxide pigments lead to loss of catalyst activity with storage time. The loss is related to the composition of surface treatments on the TiO_2. Silicon dioxide treated TiO_2 is preferable to aluminum oxide treated TiO_2 in this respect, probably owing to the basicity of alumina, which tends to neutralize the acid. When alumina treated TiO_2 is required for other reasons, a higher concentration of acid catalyst may be necessary to counteract the effect.

9.3.2. Kinetics and Mechanism of MF-Polyol Cocondensation

Many studies have been directed at elucidating the mechanism(s) of reactions between MF resins and polyols. Until recently, most of these studies were based on following the time necessary to reach some degree of film hardness. However, both co- and self-condensation reactions contribute to hardness [8]. Therefore, one must view with caution conclusions from such studies applied specifically to the cocondensation reaction between hydroxyl groups and MF resins. Based on the time and temperature required to achieve hard films, MF Class I and II resins have been classified as high and low cure temperature resins, respectively. This conclusion is usually explained by a statement that the $-NH-CH_2OR$ group reacts more rapidly with hydroxyl groups than does the $-N(CH_2OR)_2$ group. However, direct evidence for this statement is in short supply. It is well established that Class II resins self-condense more rapidly than Class I resins, but there is no published evidence of which we are aware that the rate constant for cocondensation with Class II resins is higher than with Class I resins. It may be higher, but evidence for this conclusion is lacking.

It is desirable to develop methods for measuring the extent of cocondensation directly. Infrared (IR) and Fourier transform infrared (FTIR) spectroscopy have been used to follow changes in functional group concentrations using Class I MF resins as a function of time and temperature [19]; particular care must be taken in interpreting results because of band overlap in the hydroxy region. The rate of evolution of volatile reaction products, including formaldehyde and methyl alcohol, has been followed by gas chromatography [13,15] and by thermogravimetric analysis [17]. Major contributions have been made by application of dynamic mechanical analysis to MF-polyol cross-linking coatings (see Section 5.1) [8,24]. Oscillating plate rheometry has also been used [25].

A misconception resulting from using hardness results to study kinetics is that the cocondensation reaction of Class I resins was thought to be much slower than it has now been found to be. Disappearance of polyol hydroxyl groups, as followed by FTIR, showed that the reaction is more rapid than development of film properties [15,25,26]. Possibly, the relatively slow development of optimal film properties reflects the longer time required for the self-condensation reactions, which may be necessary to achieve the optimal cross-link density.

As noted in Section 9.1.2, the mechanism of cocondensation of MF Class I resins with polyols remains controversial with experimental evidence being interpreted in support of (or against) the S_N1 or S_N2 mechanisms, provided in Scheme 9.2. Based on studies with the model analog (4) and monofunctional alcohols, it was proposed that both S_N1 and S_N2 mechanisms occur [5]. Subsequently, Meijer studied the reactivity of a high HMMM resin with chiral monofunctional alcohols [11]. He concluded that the S_N1 mechanism dominates and also reinterpreted the earlier results of Ref. [5] in favor of the S_N1 mechanism. Unfortunately, these conclusions remain in question, owing to the utilization of an oversimplified rate expression.

Using an HMMM-acrylic composition and following the cocondensation reaction by FTIR, Bauer provides support for the S_N1 mechanism pointing out the role of methyl alcohol in the kinetic order of the reaction and the rate dependence on polyol [19]. A key point is that the rate of cocondensation by the S_N1 mechanism is expected to be dependent on the concentration of polyol if the methyl alcohol, formed from HMMM, competes with the polyol for the intermediate carbocation; refer to Scheme 9.2. Water also competes with the polyol for this carbocation. This valid interpretation of the S_N1 mechanism is an important consideration, since dependence of the rate on the polyol had mistakenly been interpreted as evidence against the S_N1 mechanism. Competition for the intermediate carbocation by methyl alcohol is expected to gain relative importance as the reaction proceeds because the methyl alcohol concentration increases, at least temporarily, whereas the polyol concentration decreases with increasing conversion. Thus, kinetic studies based on hardness or other film properties, which tend to develop only at high conversions, are expected to exhibit strong dependence on the concentration and nature of the polyol. Such results had generally been interpreted in favor of the S_N2 mechanism, but are equally consistent with the S_N1 mechanism.

A widely believed, but mistaken, conclusion is that steric hindrance limits the number of active ether groups on high HMMM resins that can participate in cross-linking of films to three, or at most, four of the six groups [1]. This conclusion was based on the need to use an excess over the stoichiometric amount of MF Class I resin relative to polyol to achieve desired film properties at baking schedules such as 30 minutes at 120°C, using around 0.3

wt% (on the total weight of polymers) of pTSA. However, Hill and Kozlowski provide strong evidence in support of an essentially complete reaction of $-NCH_2OCH_3$ groups with certain polyols, showing that all $-NCH_2OCH_3$ groups of high HMMM resins can cocondense if there are enough hydroxyl groups present for them to react with [8]. Their conclusions were based on studies of changes in mechanical properties, as discussed in Section 5.1. These results further substantiate the hypothesis that the excess MF resin used in most formulations is needed not to complete the cocondensation cross-linking, but to permit some degree of self-condensation cross-linking that may be necessary to achieve the desired film properties. It has been shown that very high (95%) HMMM Class I resins give good film properties more rapidly than lower HMMM (62%) resins [7]. It is speculated that the relatively basic $-NHCH_2OCH_3$ groups present in most HMMM resins tie up a portion of the catalyst.

A further complication is the strong possibility that cross-linking with MF resins leads to a gradient of compositions and properties within films. While relatively few films have been studied in this way, most studies detected a gradient. For example, Haacke et al. used a microtome to separate layers in acrylic/MF clear coats and found that different layers had different compositions [27]. Furthermore, they found that T_g was as much as 15°C higher at the surface than deep in the film and that cross-link density was almost twice as high near the surface. They attributed the gradient to the differing effects of escaping byproduct alcohol on the cross-linking equilibrium reactions at different levels in the film. There is the additional possibility that catalyst concentrations may not be uniform within the film, especially when catalysts blocked with volatile amines are used. Other studies have suggested that there may be a very thin layer of material at the surface that is quite different in composition and properties than the bulk [28]. These scattered reports suggest that to fully understand the performance of amino resin cross-linked coatings (and perhaps of other types as well) it may be necessary to appreciate the effects of composition and property gradients within the films.

Considerable progress has been made toward understanding the complex combinations of reactions that occur during the curing of MF cross-linked coatings, but the process is not yet fully understood; further research is needed.

9.3.3. Package Stability Considerations

Package stability of coatings containing MF resins is affected by other factors in addition to catalyst discussed in Section 9.3.1. Class II resins generally give poorer package stability than Class I due to the preponderance of $>NH$ and $>NCH_2OH$ groups, which participate in slow weak acid catalyzed reactions during storage at room temperature. Stability is somewhat improved by addition of small quantities of tertiary amines. Primary or secondary amines, which react with formaldehyde, should not be used with Class II resins. Free formaldehyde is in equilibrium with melamine methylol groups in these resins. Reaction of the formaldehyde with primary or secondary amines displaces the equilibrium reaction in favor of formaldehyde formation (i.e., demethylolation), which reduces the methylol group functionality of the MF resins.

An important approach for increasing package stability is to use as much monofunctional alcohol as possible in formulations containing either class of MF resins. The presence of monofunctional alcohol in the formulation extends the storage stability since its reaction with the MF resin does not lead to cross-linking. On the other hand, cross-linking and viscosity buildup occur when the MF resin self-condenses or reacts with the polyol. Generally, it is desirable to utilize the same alcohol that is used to synthesize the MF resin. If a different alcohol is used, undesirable changes may occur. For example, if *n*-butyl alcohol is used in the solvent with a methoxymethylmelamine resin, the cure response gradually becomes slower as the proportion of butyl ether increases. The viscosity of a coating may decrease during storage because butoxymethylmelamine is less viscous (it has a lower T_g) than the corresponding methoxymethylmelamine. Viscosity decrease may also result from breaking dimethylene ether bridges by excess alcohol.

9.3.4. Reaction with Urethanes and with Malonate-Blocked Isocyanates

As shown in Eq. 9.3, MF resins react with urethane groups. Polyurethanes containing $-OC(=O)NH-$ structures (secondary carbamates) and no hydroxyl groups can be cross-linked with MF resins at temperatures only slightly higher than those used to cross-link polyols [29]. This result indicates that when hydroxy-terminated urethanes are used with MF resins, both groups are involved in cross-linking. Both groups react when polyurethane polyols are used as reactive diluents [30]. A variety of MF resins can cross-link secondary carbamates; Class I resins are more reactive than Class II resins [29]. Secondary carbamates are nearly as reactive as hydroxyl groups despite their greater steric hindrance. A possible explanation of this surprising result is that the carbamate reaction is irreversible [31]. Thus, the forward reaction might be substantially slower than the forward reaction of hydroxyl groups, but the overall rate could be close to equal if the reverse reaction does not occur with carbonates.

Polymers with pendant $-OC(=O)NH_2$ structures (primary carbamates) are less hindered than secondary carbamates and would be expected to cross-link more rapidly. Conclusive evidence for this expectation has not been published, but patent examples suggest that it is true; polymers with primary carbamates are cured with Class I MF resins under the same conditions as polymers with primary $-OH$ groups [32,33]. Primary carbamate-functional resins can be made by reacting isocyanate-terminated resins with hydroxypropylcarbamate [32] or from acrylic resins [33]. The isocyanate based resins can be cross-linked with Class I MF resins, but the cross-link density is too high for some applications. Functionality can be reduced by blocking half of the functional groups of Class I resins with butyl carbamate [32]. Automotive clear coats made from such combinations combine the usual high mar resistance of MF cross-linked coatings with high resistance to environmental etching.

Class I MF resins also react with malonic ester derivatives of isocyanates at rates similar to their reactions with $-OH$ groups [34]. NMR study of a model compound confirmed that environmental resistant carbon-carbon cross-links are irreversibly formed in the process. As a result, environmental etch resistance of a coating made from a malonic ester derivative of a triisocyanate and a Class I MF resin is excellent [34]. Malonic ester derivatives of isocyanates can also function as cross-linkers for hydroxy-functional resins, as discussed in Section 10.5.

9.4. OTHER AMINO RESINS

Other amino resins are used on a smaller scale. The structures of the other starting materials are provided in the beginning of this chapter. While the chemistry of these resins is similar to that of MF resins, there are differences, particularly as a result of differences in basicity of the resins [35].

9.4.1. Benzoguanamine-Formaldehyde Resins

Using benzoguanamine as a starting material, a range of resins analogous to MF resins can be prepared. As with melamine, benzoguanamine may be methoxymethylated and/or butoxymethylated to various extents. The average functionality is lower because there are only two $-NH_2$ groups per molecule. Etherified benzoguanamine-formaldehyde (BF) resins yield cross-linked films with greater resistance to alkali and to alkaline detergents, such as sodium tripolyphosphate, compared with MF resins. Exterior durability of BF-based coatings is poorer than MF-based coatings. Thus, BF resins are used for applications such as washing machines and dishwashers in which resistance to alkaline detergents is more important than exterior durability. The reduced exterior durability of BF-based coatings probably reflects lower photostability arising from the presence of the phenyl group on the triazine ring of benzoguanamine, although supporting evidence for this reasonable hypothesis does not appear to be available. Aside from the hydrophobicity of the phenyl groups, a convincing explanation for enhanced alkali resistance is not intuitively forthcoming.

9.4.2. Urea-Formaldehyde Resins

Urea also reacts with formaldehyde to form methylol derivatives. The first and second formaldehyde units add easily, but trimethylol and tetramethylol derivatives of urea have never been isolated. Methylolated ureas can be etherified by reaction with alcohols analogously to methylolated melamines. Self-condensation reactions also occur under both acidic and basic conditions. A variety of etherified urea-formaldehyde (UF) resins are made with different ratios of formaldehyde to urea and different alcohols for etherification.

In general, UF resins are the most economical amino resins; they are also the most reactive. With sufficient acid catalyst, coatings formulated with UF resins can cure at ambient or mildly elevated temperatures. However, such coatings have poor exterior durability, probably because the cross-links are relatively reactive and have poor resistance to hydrolysis. MF or BF resins are preferred even in coatings for indoor use, when good hydrolytic resistance is required, as is usually the case for coatings for steel. UF resins are used in coatings for temperature sensitive substrates such as wood furniture, paneling, and cabinetry. In such applications, low temperature baking is essential, and corrosion resistance (related to the hydrolytic stability of the cross-links) is unimportant.

9.4.3. Glycoluril-Formaldehyde Resins

Glycoluril reacts with formaldehyde to give tetramethylolglycoluril (TMGU) [35]. In the range of pH 4 to 8, aqueous solutions of TMGU show an equilibrium level of about 3.6 methylol groups per glycoluril, with 0.4 parts free formaldehyde. Above pH 8.5, demethylolation is more favored. For example, at pH 11, there are approximately equal parts of methylol groups and free formaldehyde. Below pH 3, self-condensation occurs to form glycoluril dimers with methylene ether bridges; see Eq. 9.12, where $>N-CH_2-OH$ represents methylolated glycoluril groups.

$$\text{\textbackslash}N\text{—}CH_2\text{—}OH \underset{\longleftarrow}{\overset{pH < 4}{\longrightarrow}} \text{\textbackslash}N\text{—}CH_2\text{—}O\text{—}CH_2\text{—}N\text{/} + H_2O \qquad (9.12)$$

This behavior is unlike that of MF or UF resins. While TMGU does not self-condense above pH 4, methylolated melamines and ureas self-condense at pH 7 and higher.

Tetramethylolglycoluril reacts with alcohols in the presence of strong acid catalyst to form tetraalkoxymethylglycoluril (GF) resins. Tetramethoxymethylglycoluril is a relatively high melting solid and is used as a cross-linker in powder coatings (see Section 27.1.3). In solution coatings, (dimethoxymethyldiethoxymethyl)glycoluril and tetrabutoxymethylglycoluril are used, since they are liquids and are more readily handled. The mixed methyl/ethyl ether is water soluble.

Relative to other amino resins, GF resins produce coatings that exhibit greater flexibility at a similar cross-link density. Hence, GF resins are used in applications such as coil coatings and can coatings, where flexibility is important. GF resins possess the further advantage that less formaldehyde is evolved during cure, as compared with MF resins. Also the glycoluril cross-linked polyols are more resistant to hydrolysis under acidic conditions than MF cross-linked polyols [35]. The higher cost of GF resins limits their use.

9.4.4. (Meth)acrylamide-Formaldehyde Resins

Acrylic copolymers of N-isobutoxymethylacrylamide can be prepared by two routes: (1) synthesis of N-isobutoxymethylacrylamide monomer (from a step-wise reaction of acrylamide with formaldehyde and isobutyl alcohol), followed by copolymerization with acrylate monomers and (2) copolymerization of acrylamide, followed by step-wise reaction with formaldehyde and isobutyl alcohol. Analogous reactions can be carried out with other alcohols and with methacrylamide.

Such alkoxymethyl(meth)acrylamide amino resins are used in coil coatings, where, with proper design, they are weather resistant and more flexible than coatings based on MF-acrylic polyol resins. One can speculate that their flexibility is enhanced by the absence of densely cross-linked clusters of self-condensed MF resins with their rigid triazine rings.

REFERENCES

1. J. O. Santer, *Prog. Org. Coat.*, **12**, 309 (1984).
2. T. T. Chang, *Prog. Org. Coat.*, **29**, 45 (1996).
3. R. P. Subrayan and F. N. Jones, *J. Appl. Polym. Sci.*, **62**, 1237 (1996).
4. R. Nastke, K. Dietrich, G. Reinisch, and G. Rafler, *J. Macromolecular Sci.*, **A23**, 579 (1986).
5. Z. W. Wicks, Jr. and D. Y. Y. Hsia, *J. Coat. Technol.*, **55** (702), 29 (1983).
6. U. Samaraweera and F. N. Jones, *J. Coat. Technol.*, **64** (804), 69 (1992).
7. F. N. Jones, G, Chu, and U. Samaraweera, *Prog. Org. Coat.*, **24**, 189 (1994).
8. L. W. Hill and K. Kozlowski, *J. Coat. Technol.*, **59** (751), 63 (1987).
9. J. H. van Dijk, A. S. van Brakel, W. Dankelman, and C. J. Groenenboom, *FATIPEC Congress Book*, Vol. II, 1980, p. 326.
10. N. Albrecht, *Proc. Water-Borne, Higher-Solids Coat. Symp.*, New Orleans, 1986, p. 200.
11. E. W. Meijer, *J. Polym. Sci., A: Polym. Chem.*, **24**, 2199 (1986).

12. G. Chu and F. N. Jones, *J. Coat. Technol.*, **65** (819), 43 (1993).
13. W. J. Blank, *J. Coat. Technol.*, **51** (656), 61 (1979).
14. D. R. Bauer and R. A. Dickie, *J. Coat. Technol.*, **54** (685), 57 (1982).
15. M. G. Lazzara, *J. Coat. Technol.*, **56** (710), 19 (1984).
16. S. P. Pappas, B. C. Pappas, X.-Y. Hong, R. Kirchmayr, and G. Berner, *Proc. Water-Borne, Higher-Solids Coat. Symp.*, New Orleans, 1988, p. 24.
17. W. J. Mijs, W. J. Muizebelt, and J. B. Reesink, *J. Coat. Technol.*, **55** (697), 45 (1983).
18. T. Morimoto and S. Nakano, *J. Coat. Technol.*, **66** (833), 75 (1994).
19. D. R. Bauer, *Prog. Org. Coat.*, **14**, 193 (1986).
20. Z. W. Wicks, Jr., and G. F. Chen, *J. Coat. Technol.*, **50** (638), 39 (1979).
21. P. E. Ferrell, J. J. Gummeson, L. W. Hill, and L. J. Truesdell-Snider, *J. Coat. Technol.*, **67** (851), 63 (1995).
22. L. W. Hill, P. E. Ferrell, and J. J. Gummeson, in T. A. Provder, M. A. Winnik, and M. W. Urban, Eds., *Film Formation in Water-borne Coatings*, American Chemical Society, Washington, DC, 1996.
23. L. J. Calbo, *J. Coat. Technol.*, **52** (660), 75 (1980).
24. L. W. Hill and K. Kozlowski, *Proc. Intl. Symp. Coat. Sci. Technol.*, Athens, 1986, p. 129.
25. T. Yamamoto, T. Nakamichi, and O. Ohe, *J. Coat. Technol.*, **60** (762), 51 (1988).
26. T. Nakamichi, *Prog. Org. Coat.*, **14**, 23 (1986).
27. G. Haacke, J. S. Brinen, and P. J. Larkin, *J. Coat. Technol.*, **67** (843), 29 (1995).
28. F. N. Jones, W. Shen, S.M. Smith, Z. Huang, and R. A. Ryntz, *Prog. Org. Coat.*, **34**, 119 (1998).
29. H. P. Higginbottom, G. R. Bowers, P. E. Ferrell, and L. W. Hill, *Proc. Waterborne, High-Solids and Powder Coat. Symp.*, New Orleans, 1998, p. 527.
30. L. W. Hill, *Polym. Mater. Sci. Eng.*, **77**, 387 (1997).
31. W. J. Blank, Z. A. He, E. T. Hessell, and R. A. Abramshe, *Polym. Mater. Sci. Eng.*, **77**, 391 (1997).
32. J. W. Rehfuss and W. H. Ohrbom, *U. S. Patent* 5373069 (1994).
33. J. W. Rehfuss and D. L. St. Aubin, *U. S. Patent* 5356669 (1994).
34. Z. A. He and W. J. Blank, *Proc. Waterborne, High-Solids and Powder Coat. Symp.*, New Orleans, 1998, p. 21.
35. G. G. Parekh, *J. Coat. Technol.*, **51** (658), 101 (1979).

CHAPTER 10

Binders Based on Isocyanates: Polyurethanes

Polyurethanes are polymers containing urethane ($-NH-CO-O-$) linkages; carbamate is a synonym for urethane. Urethanes are usually formed by reaction of an alcohol with an isocyanate, but they can also be made by other methods. Commonly, urethanes made by a nonisocyanate route are called carbamates. To add to the confusing terminology, the terms *urethane* and *polyurethane* are applied to almost any binder derived from isocyanates even though only part, if any, of the reaction products are urethanes. Isocyanates are used to make urethane-modified alkyds (uralkyds) as discussed in Section 15.7.

Urethane groups form intermolecular hydrogen bonds between polymer molecules; they may be acyclic and/or cyclic:

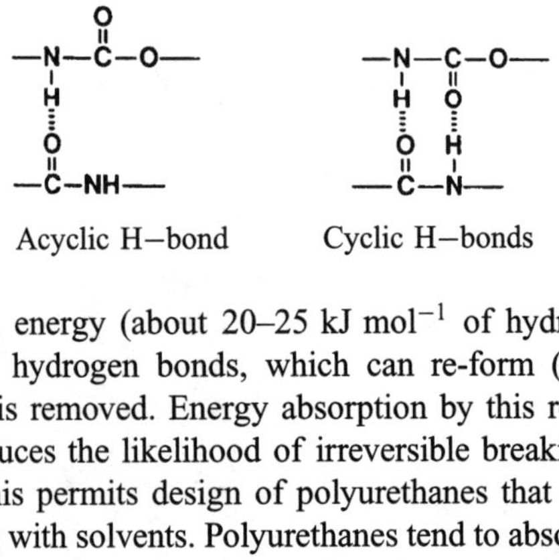

Acyclic H-bond   Cyclic H-bonds

Under mechanical stresses, energy (about 20–25 kJ mol^{-1} of hydrogen bonds) may be absorbed by separation of hydrogen bonds, which can re-form (probably in different positions) when the stress is removed. Energy absorption by this reversible bond-breaking/re-forming process reduces the likelihood of irreversible breaking of covalent bonds leading, to degradation. This permits design of polyurethanes that are abrasion resistant while still resisting swelling with solvents. Polyurethanes tend to absorb water by hydrogen bonding with water, which plasticizes the coatings.

The isocyanate group is highly reactive, so polyisocyanates can be used to make coatings that cure at ambient temperature or at moderately elevated temperatures. Coatings based on aliphatic diisocyanates exhibit exceptional exterior durability when stabilized with hindered amine light stabilizers (see Section 5.3.3). Resistance of urethane coatings to environmental etching is superior to that of many melamine-formaldehyde (MF) cross-linked coatings [1].

180

The principal limitations of isocyanates are cost and toxicity (particularly of relatively low molecular weight compounds). Any cross-linker that reacts with hydroxyl, amine, and/or carboxylic acid groups near room temperature is likely to be toxic, since the body contains proteins and other materials with such substituents. The important question is not toxicity as much as toxic hazard. Since higher molecular weight reduces vapor pressure and permeability through body membranes, toxic hazard decreases, in general, as molecular weight increases. With adequate ventilation and use of air masks and protective clothing, as commonly done in chemical factories and resin manufacturing plants, even relatively low molecular weight isocyanates can be handled safely. Most low molecular weight diisocyanates are sensitizers; that is, after exposure some people become allergic to further exposure to isocyanate. The exposure leading to sensitization varies over a wide time period for individuals and may occur only after years of repeated exposure for some. The most common symptoms are hives and asthma. In extreme cases sensitized people cannot be in the same room, or even the same building, in which isocyanates are handled. Isocyanate suppliers supply safety instructions for handling and using isocyanates. Coating applicators, especially amateur users, are less likely than professional chemists to have appropriate facilities or even to follow safety instructions. In using any highly reactive cross-linker, the competence of the user to handle toxic materials should be taken into consideration in establishing formulations. To minimize the hazard, most isocyanates used in coatings are oligomeric or polymeric derivatives.

10.1. REACTIONS OF ISOCYANATES

Isocyanates react with any active hydrogen compound. Alcohols and phenols react with isocyanates to form urethanes, as in the following reaction, in which R and R' can be aromatic or aliphatic groups. The reaction is reversible at elevated temperatures.

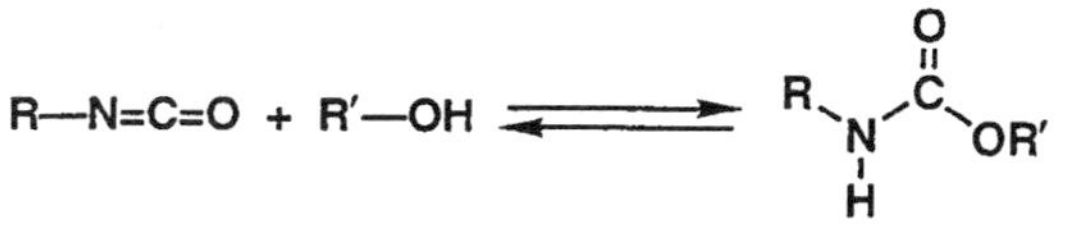

$$R{-}N{=}C{=}O \ + \ R'{-}OH \ \rightleftharpoons \ R{-}\underset{H}{\overset{\displaystyle O}{\underset{|}{N}}}\overset{\displaystyle \parallel}{\underset{}{C}}{-}OR'$$

In general, rates of urethane formation decrease in the following order: primary alcohols > secondary alcohols > 2-alkoxyethanols > 1-alkoxy-2-propanols > tertiary alcohols. As a rule, ease of reversion is the inverse of reactivity. Urethanes from tertiary alcohols are relatively unstable and may decompose with heat to give alkenes, carbon dioxide, and amines, rather than alcohols and isocyanates.

Urethanes react with isocyanates to form allophanates. This reaction is much slower than the reaction of isocyanate with alcohol.

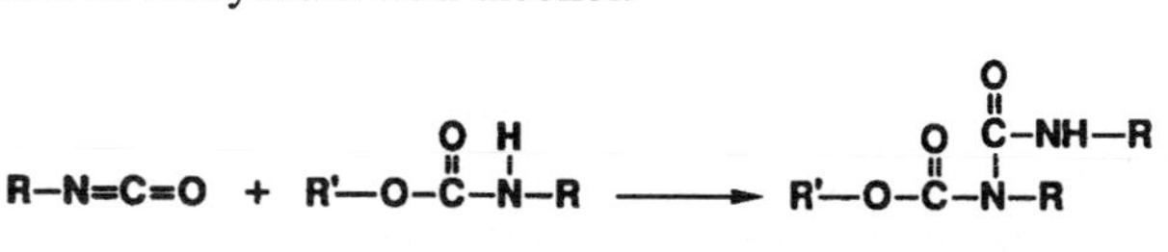

$$R{-}N{=}C{=}O \ + \ R'{-}O{-}\overset{O}{\overset{\parallel}{C}}{-}\overset{H}{\overset{|}{N}}{-}R \ \longrightarrow \ R'{-}O{-}\overset{O}{\overset{\parallel}{C}}{-}\overset{\overset{\displaystyle O}{\overset{\parallel}{C}{-}NH{-}R}}{\underset{}{N}}{-}R$$

Isocyanates react rapidly with primary and secondary amines to form ureas. The reaction is much faster than the reaction of isocyanates with alcohols. For many coating applications, reactions are too rapid; however, hindered amines have been developed that react sufficiently slowly to permit use in two package (2K) coatings (see Section 10.4) [2,3].

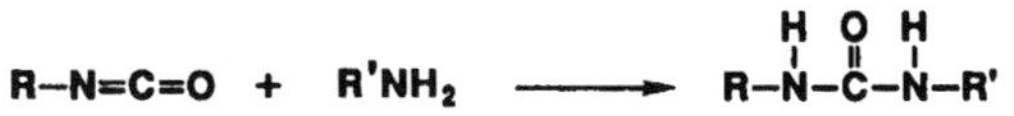

$$R-N=C=O \;+\; R'NH_2 \;\longrightarrow\; R-N-C-N-R'$$

Isocyanates react with ureas to form biurets. Biuret formation is slower than urethane formation, but faster than allophanate formation.

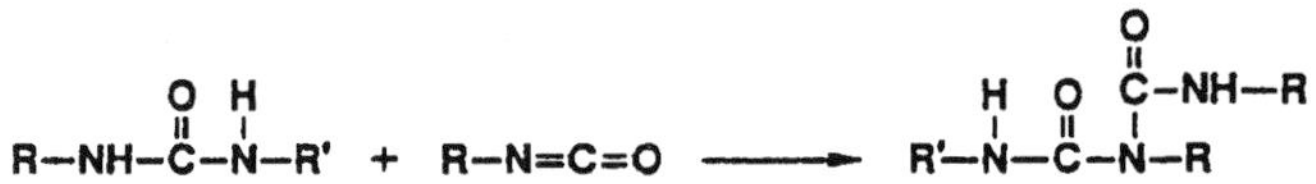

$$R-NH-C-N-R' \;+\; R-N=C=O \;\longrightarrow\; R'-N-C-N-R$$

Isocyanates react with water to form unstable carbamic acids, which dissociate into carbon dioxide and an amine. The amine is so much more reactive than water that it reacts with a second isocyanate (in preference to water) to form a urea. The reactivity of water with isocyanates is somewhat slower than that of secondary alcohols, but more rapid than that of ureas.

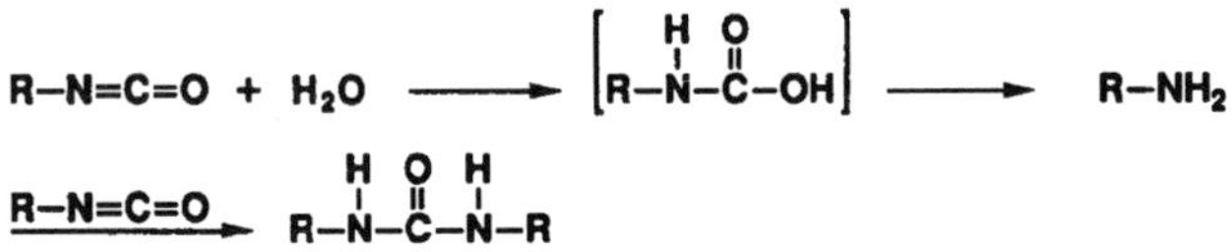

$$R-N=C=O \;+\; H_2O \;\longrightarrow\; \left[R-N-C-OH \right] \;\longrightarrow\; R-NH_2$$

$$R-N=C=O \;\xrightarrow{\quad} R-N-C-N-R$$

Oxazolidines hydrolyze with water to yield free amine and hydroxyl groups that react with isocyanate to form urea and urethane linkages. Difunctional oxazolidines yield higher functionality [4].

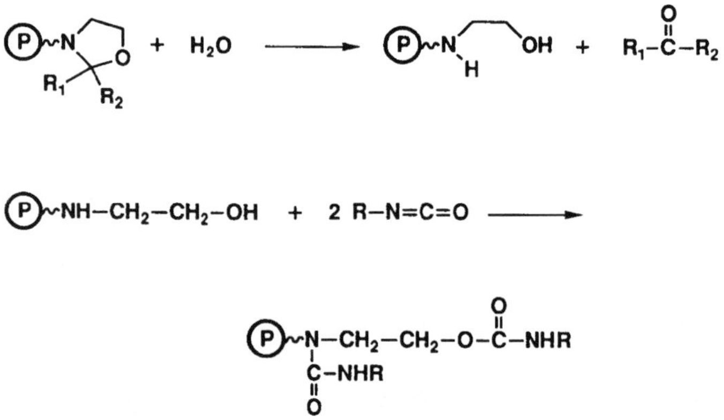

Imines act as blocked amines because they hydrolyze to yield free amines, which react with an isocyanate. Ketimines also react directly with isocyanates to yield a variety of products, depending on the particular reactants and conditions. In the absence of water, isobutyl isocyanate reacts with the ketimine derived from methylamine and acetone in 3 hours at 60°C to yield isobutylmethyl urea and a cyclic unsaturated substituted urea [3,5].

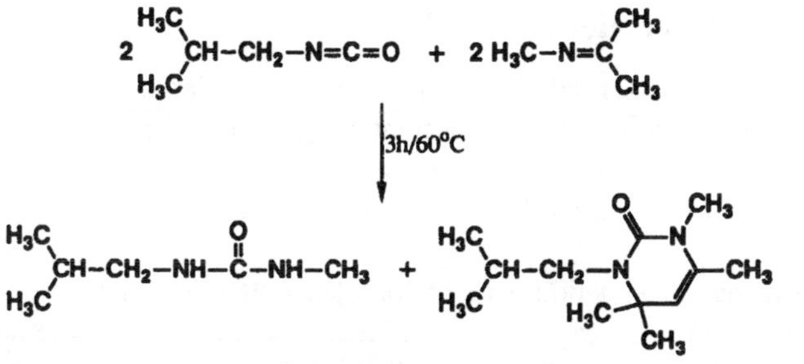

Aldimines react analogously with isocyanates to yield unsaturated substituted ureas [3]. Since aldimines are more stable to hydrolysis than ketimines, the fraction undergoing direct reaction with isocyanate in the presence of water is greater than that with ketimines.

Carboxylic acids react relatively slowly to form amides and CO_2. Hindered carboxylic acid groups, such as in 2,2-dimethylolpropionic acid react, very slowly.

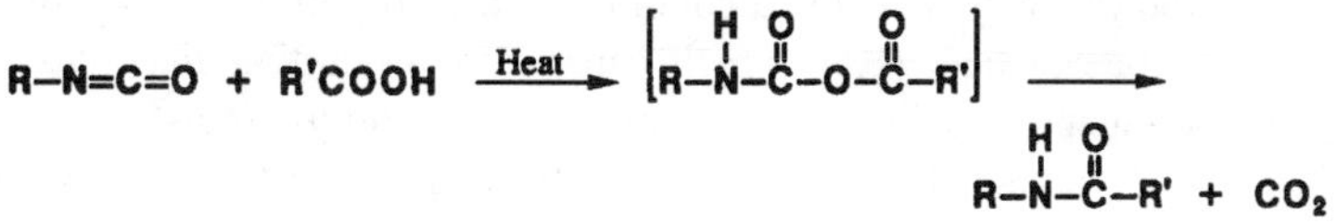

Isocyanates also react with each other to form dimers (uretdiones) and trimers (isocyanurates). Formation of uretdiones is catalyzed by phosphines. Formation of isocyanurates is catalyzed by quaternary ammonium compounds; trimerization of aromatic isocyanates is catalyzed by tertiary amines. Uretdiones decompose thermally to regenerate isocyanates and are used as blocked isocyanates (see Section 10.5). Isocyanurates are stable and are extensively used as multifunctional isocyanates (see Section 10.3).

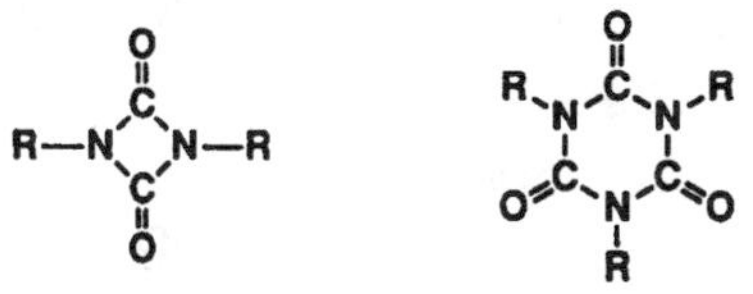

Uretdione   Isocyanurate

10.2. KINETICS OF ISOCYANATE REACTIONS WITH ALCOHOLS

The mechanism and catalysis of reaction of isocyanates with alcohols are not fully understood. One must use caution when evaluating rate data from the literature, especially in interpreting tables of rate constants compiled from different sources. Original literature should be consulted to make sure that comparisons are based on studies done in the same solvents and initial concentrations, since reaction rates depend on these variables. Also, rates of urethane formation are commonly determined by following disappearance of isocyanate. If other products, such as allophanates and isocyanurates, are also formed, isocyanate disappearance does not correspond to urethane formation.

10.2.1. Noncatalyzed Reactions

One might expect urethane formation to follow second-order kinetics, with the rate proportional to the concentration of each reactant, as shown in Eq. 10.1.

$$\text{rate} = k[\text{R}-\text{N}=\text{C}=\text{O}][\text{R}'-\text{OH}] \qquad (10.1)$$

Kinetic studies show that the situation is more complex; often the kinetic order changes as the reaction progresses. In some cases, the reaction follows third-order kinetics, at least in its early stages, first order in isocyanate and second order in alcohol concentration. Rate equation 10.2 would apply to such cases. The implication of Eq. 10.2 is that two molecules of alcohol are involved in the reaction with one molecule of isocyanate.

$$\text{rate} = k[\text{R}-\text{N}=\text{C}=\text{O}][\text{R}'-\text{OH}]^2 \qquad (10.2)$$

Such results can be reasonably interpreted in terms of the mechanism provided in Scheme 10.1. Reaction of isocyanate and alcohol (at a rate proportional to k_1) produces a zwitterionic reactive intermediate (**RI**), which can revert to starting materials (k_{-1}) or proceed to product, urethane (k_2). Product formation requires proton transfer from oxygen to nitrogen. The second molecule of alcohol facilitates this transfer by way of a six-membered cyclic activated complex (**A**).

Equation 10.3 is a rate expression corresponding to the sequence of reactions in Scheme 10.1.

$$\text{rate} = k_1[\text{RNCO}][\text{R}'\text{OH}]\frac{k_2[\text{R}'\text{OH}]}{k_2[\text{R}'\text{OH}]+k_{-1}} \qquad (10.3)$$

If the term $k_2[\text{R}'\text{OH}]$ is substantially greater than k_{-1}, Eq. 10.3 simplifies to Eq. 10.1. This situation is favored when the alcohol concentration is high, such as at the start of the reaction. If, on the other hand, the term k_{-1} is substantially larger than $k_2[\text{R}'\text{OH}]$, Eq. 10.3

Scheme 10.1

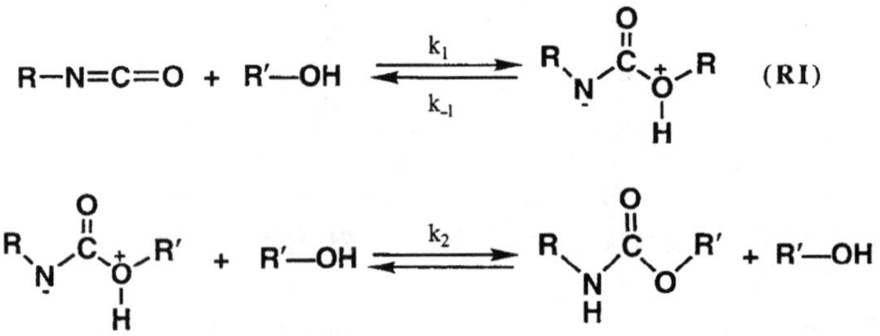

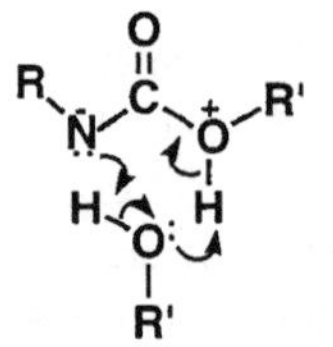

simplifies to Eq. 10.2 and second-order dependence on alcohol concentration is observed. The latter circumstance is more likely as alcohol is consumed during the reaction, reducing [R'OH]. As a result, the reaction rate diminishes rapidly as the reaction proceeds (since the rate is proportional to the square of a small, diminishing alcohol concentration). The mechanism in Scheme 10.1 provides an explanation for the strong solvent effects observed. Rates decrease as the hydrogen-bond acceptor potential of solvents increase in the order: aliphatic hydrocarbons, aromatic hydrocarbons, esters and ketones, ethers, and glycol diethers. Rates in aliphatic hydrocarbons can be two orders of magnitude faster than in glycol diethers. Hydrogen bonding of the zwitterionic reactive intermediate (**RI**) and/or the alcohol with solvent may reduce the rate by reducing the concentration of cyclic activated complex (**A**).

The mechanism in Scheme 10.1 is also consistent with the observed higher reactivity of aromatic isocyanates. When R is an aromatic group, the negative charge on nitrogen in (**RI**) is delocalized in the *pi*-electron system of the aromatic ring, resulting in a lower energy (by resonance stabilization) for (**RI**) and a faster rate of formation relative to aliphatic isocyanates.

Rate constants reported in the literature for reactions of alcohols with isocyanates are most commonly apparent initial second-order rate constants. Initial rate constants provide a useful way of comparing reactivities without becoming involved in the complexities of the kinetics, provided comparisons are made in the same solvent and the same initial concentrations of alcohols and isocyanates. Unfortunately, rate constants are sometimes compared which were determined in different solvents and/or at different initial concentrations. Such comparisons can be misleading.

The kinetics of the isocyanate-alcohol reaction are even more complex than discussed thus far, since the reaction is autocatalyzed by the urethane being formed. Another cyclic transition state (**B**), could be involved, in which a molecule of urethane facilitates proton transfer from oxygen to nitrogen in the zwitterionic reactive intermediate (**RI**).

(B)

The autocatalyzed reaction would follow third-order kinetics, but would be first order in alcohol. The decline in alcohol concentration would be offset by the increase in urethane concentration as the reaction proceeds. The apparent rate constant would change through the course of the reaction depending, on the rate constant for the reaction involving activated complex (**A**) in comparison with that for (**B**). Sato studied the reaction of various isocyanates with methyl alcohol [6]. In the absence of added catalyst, his results can be expressed by Eq. 10.4, where a and b are the initial concentrations of isocyanate and methyl alcohol, x is the concentration of product (urethane), and k_2 represents the rate constant of the autocatalytic reaction:

$$\frac{dx}{dt} = k_1(a-x)(b-x)^2 + k_2(a-x)(b-x) \tag{10.4}$$

It is assumed that there are no side reactions; therefore, $(a - x)$ in Eq. 10.4 equals $[R-N=C=O]$ in Eq. 10.2, and $(b - x)$ equals $[R'OH]$. In most cases studied by Sato, k_2 was larger than k_1, but in a few cases, it was comparable. When k_2 is substantially larger, the second term in Eq. 10.4 dominates in the later stages of the reaction (as the urethane concentration builds up). Sato's studies were done in di-n-butyl ether; it would be of interest to know what changes in the rate constants would result from changes in solvent. Sato reported that autocatalysis is more important for aliphatic than aromatic isocyanates, although the rate constants for autocatalysis were similar [6]. This results from the higher reactivity of aromatic isocyanates with alcohols (k_1), which reduces the relative importance of autocatalysis.

10.2.2. Catalysts

Reactions of isocyanates with alcohols are catalyzed by a variety of compounds, including bases (tertiary amines, alkoxides, carboxylates), metal salts and chelates, organometallic compounds, acids, and urethanes. The most widely used catalysts in coatings are tertiary amines, commonly diazabicyclo[2.2.2]octane (DABCO—a trademark of Air Products), and organotin (IV) compounds, most commonly, dibutyltin dilaurate (DBTDL). Combinations of DABCO and DBTDL often act synergistically; that is, the effect of the combination is greater than would be predicted by the sum of the individual effects of the two catalysts.

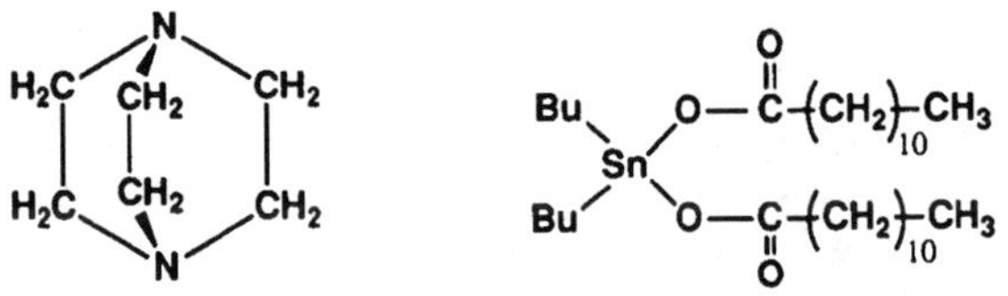

DABCO   DBTDL

The mechanisms by which these and other catalysts operate are controversial. A reasonable explanation for catalysis by amines is that they facilitate proton transfer from the alcohol to the isocyanate (and similarly to alcohols and urethanes). Proton removal from the alcohol may occur during reaction with the isocyanate, thereby avoiding formation of the positive charge on oxygen and lowering the energy of the reactive intermediate, which may proceed to product by a proton transfer from the protonated amine, as shown in Scheme 10.2. Proton removal at an earlier stage by amines than by alcohols and urethanes is reasonable, owing to the greater basicity of amines.

Scheme 10.2

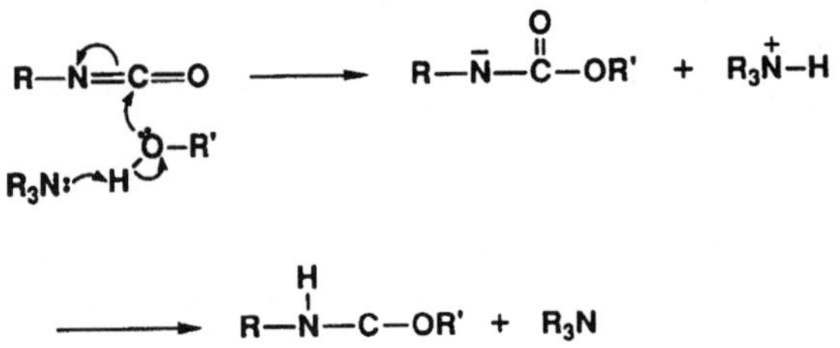

Sato included catalysts in the study described in Section 10.2.1 [6]. For triethylamine catalyzed reactions of isocyanates with methyl alcohol, Sato's data fit Eq. 10.5, where k_3 and (cat) represent the rate constant of the catalyzed reaction and the concentration of catalyst, respectively:

$$\frac{dx}{dt} = k_1(a-x)(b-x)^2 + k_2(a-x)(b-x) + k_3(\text{cat})(a-x)(b-x) \qquad (10.5)$$

If the catalyst is effective, k_3 is larger than k_1 or k_2, and when the catalyst concentration is sufficient, the rate is governed by the third term and is first order in alcohol. This is generally observed in amine catalyzed reactions. The mechanism in Scheme 10.2 is consistent with the third term of rate Eq. 10.5. Amine basicity is not the only factor since DABCO, a weaker base, is a more active catalyst than triethylamine. The nitrogen electron pairs are more readily accessible with DABCO than triethylamine, which could account for higher catalytic activity. This possibility, that both nitrogen basicity and electron accessibility are important factors, is supported by the even higher catalytic activity (relative to DABCO) of 1-azabicyclo[2.2.2]octane (quinuclidine), which has both the nitrogen accessibility of DABCO and the high basicity of triethylamine.

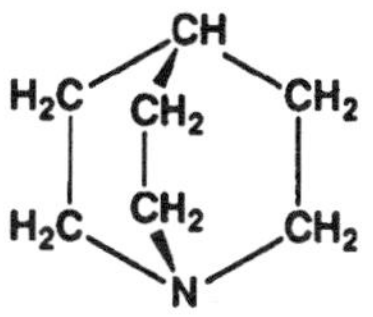

Quinuclidene 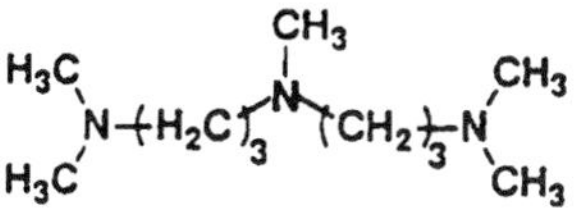   PMPTA

Comparisons are complicated because tertiary amines also catalyze allophanate formation and trimerization of aromatic isocyanates to form isocyanurates. For example, reaction of phenyl isocyanate with *n*-butyl alcohol (at 50°C in acetonitrile) in the presence of pentamethyldipropylenetriamine (PMPTA) yielded 30% urethane, while 70% of the isocyanate was converted into triphenylisocyanurate [7]. On the other hand, when DABCO was used as a catalyst, the urethane was the principal product, with a small amount of an allophanate also formed. Possibly, urethane formation is favored by sterically accessible amines (e.g., DABCO), and isocyanurate formation is less sensitive to this factor. The zinc complex of 2,4-pentanedione (Zn acac), tin octoate, and quaternary ammonium compounds, such as tetramethylammonium octoate, specifically catalyze allophanate formation [8]. Reference [8] provides rate constants for the formation of urethanes, allophanates, and isocyanurates for a variety of catalysts and cocatalysts.

Acids also catalyze the reaction, perhaps, by protonating the isocyanate group. Carboxylic acids act as catalysts [9], and stronger acids are even more effective [10]. Phenyl acid phosphate is less effective as a catalyst than DBTDL at temperatures below 100°C, but is more effective at 130°C. Blocking the acids by forming amine salts extends pot life without reducing the cure rate at 130°C.

Of the many metal derivative catalysts, DBTDL is the most widely used in coatings. It is soluble in a wide range of solvents, comparatively low in cost, colorless, and, in general, highly effective at levels on the order of 0.05 wt%. DBTDL promotes urethane formation without promoting allophanate formation [11] or trimerization [7]. Dimethyltin diacetate

(DMTDA) is usually a somewhat more effective catalyst than DBTDL and is particularly useful with sterically hindered isocyanates. Tin compounds are effective catalysts for reaction of alcohols with aromatic isocyanates and are even more effective with aliphatic isocyanates. While aromatic isocyanates are more reactive than aliphatic isocyanates in uncatalyzed reactions with alcohols, the reactivity of aliphatics can be roughly equal with DBTDL. On the other hand, amine catalysts are more effective with aromatic than aliphatic isocyanates.

Isocyanates also react with water; if this happens in a solvent-borne coating during application, CO_2 is generated, which may reduce gloss or result in bubbling. Catalyst selection affects the relative rate of reaction of isocyanates with hydroxyl groups and water. In the reaction of n-butyl alcohol or water with H_{12}MDI, dimethyltin dilaurate gave a significantly higher ratio of reaction rate of alcohol to water than dibutyltin dilaurate [12]. A zirconium acetoacetate complex (Zr acac) catalyst is also said to be more selective than DBTDL [13]. Triphenylbismuth (and other organobismuth catalysts) also catalyze reaction with alcohols much more than reaction with water [13,14].

Many mechanisms have been proposed for the catalytic activity of tin compounds, but none has been universally accepted. A plausible proposal is based on studies of reaction of phenyl isocyanate with excess methyl alcohol using dibutyltin diacetate (DBTDA) as catalyst [15]; see also Ref. [14]. The reaction rate is first order in isocyanate and half order in both alcohol and catalyst concentration. Restating the kinetic equation in the same form used thus far gives Eq. 10.6:

$$\frac{dx}{dt} = k_3 (\text{cat})^{1/2}(a - x)(b - x)^{1/2} \tag{10.6}$$

Based on kinetic results and the observation that the rate was suppressed by addition of acid, the mechanism in Scheme 10.3 was suggested. The mechanism involves sequential complexation of alcohol (with loss of H^+) and isocyanate to the tin. The proposed proton loss is consistent with the observed reduction in activity of tin catalysts in the presence of carboxylic acids. Presumably, addition of H^+ favors decomplexation of alcohol and reversion to starting materials. This proposal is also consistent with the observation that tin compounds are not effective catalysts for reactions of isocyanates with amines.

Coordination of the isocyanate with tin is consistent with the observed roughly equivalent reactivity of aliphatic and aromatic isocyanates with tin catalysts, since stabilization of the reactive intermediate by electron delocalization with the aromatic ring is eliminated as a major factor. Furthermore, tin activates both the alcohol and isocyanate, whereas amine catalysts probably activate only the alcohol by facilitating proton removal. The rate dependence on alcohol concentration has important implications on the pot life versus curing schedule with isocyanates. If the dependence changes from first to second order (characteristic of the uncatalysed reaction), the rate will diminish rapidly as the reaction nears completion. With less sensitive half-order dependence, as reported for tin catalysis, the rate at high concentrations (during storage) will be lower and will not slow down as much as the reaction proceeds to completion, thereby favoring both a longer pot life and shorter cure time (or lower cure temperature). The first-order dependence with amine catalysis results in an intermediate situation. Unfortunately, there are only limited reports in the literature of the effect of solvent on rates of catalyzed reactions.

Scheme 3

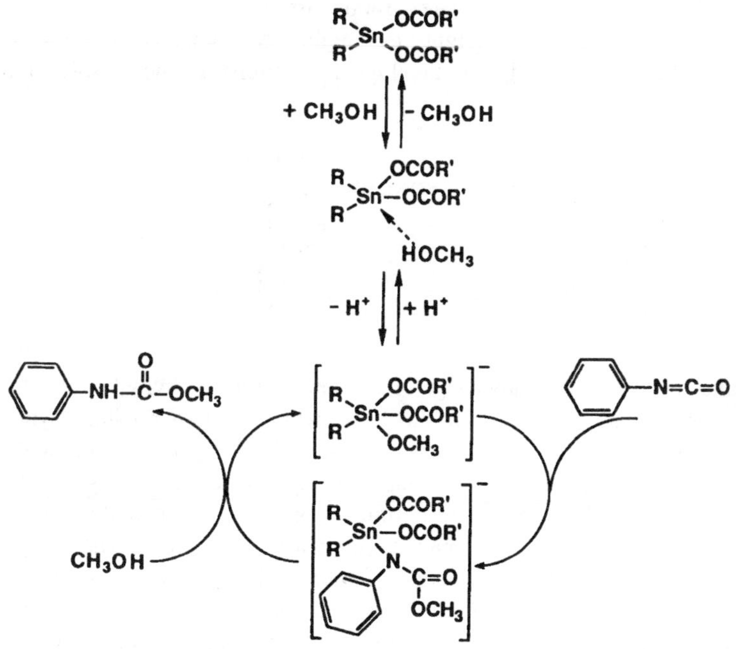

The reactions of amines and imines with isocyanates are catalyzed by carboxylic acids and water. Since organotin (IV) compounds and amines complex with acids, both these types of additives decrease reactivity [3].

10.3. ISOCYANATES USED IN COATINGS

Aromatic and aliphatic isocyanates are used in coatings; the former are less expensive and the latter provide films with better color retention and exterior durability.

10.3.1. Aromatic Isocyanates

The aromatic diisocyanate most widely used in coatings is MDI. MDI is available in several grades: bis(4-isocyanatophenyl)methane, a mixture of 55% of the 2,4' isomer and 45% of the 4,4' isomer; and several oligomeric (frequently called polymeric) MDI with longer chains of methylene phenyl groups. The cost is lower than toluene diisocyanate (TDI) and the volatility (particularly of the oligomeric grades) is low enough to reduce toxic hazard.

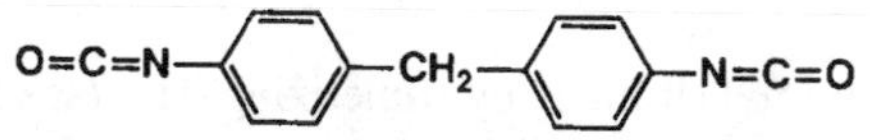

The most common grade of commercial TDI consists of a mixture of about 80% 2,4- and 20% 2,6-diisocyanato isomers. Nearly pure 2,4-TDI is also available at a premium price. Due to toxic hazards, TDI is not used as such in final coating formulations. For coatings in which unreacted isocyanate groups are needed, TDI is converted into derivatives of higher molecular weight and higher functionality. Higher molecular weight reduces the toxic hazard, and the higher functionality yields solvent resistant films more rapidly.

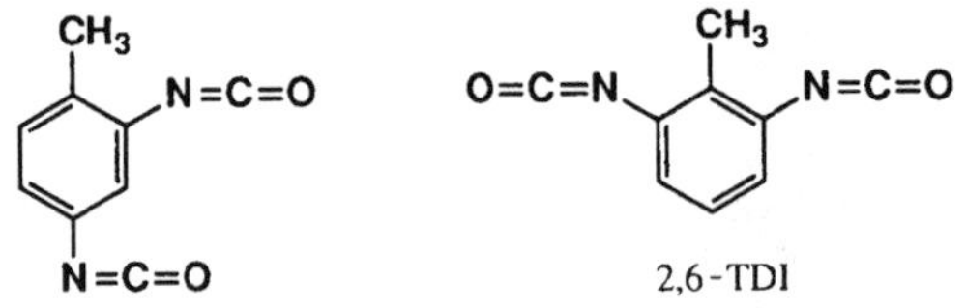

2,4-TDI   2,6-TDI

Toluene diisocyanate has the advantage of a differential in reactivity between the ortho- and the para-isocyanate groups with alcohols, which makes possible synthesis of isocyanurates and prepolymers with narrower molecular weight distribution than with diisocyanates in which the isocyanate groups are equally reactive. At 40°C, the *para*-isocyanate group of TDI is about seven times more reactive than the *ortho*- group. Furthermore, no matter which isocyanate group reacts first, the second group is less reactive than the first. Overall, after reaction of the *para*-isocyanate, the remaining *ortho*-group is 20 times less reactive than a *para*-isocyanate on a second TDI. The difference in reactivity decreases as the temperature increases; at temperatures above 100°C, the *ortho*- and *para*-isocyanate groups have similar reactivities. Thus, for maximum selectivity a prepolymer should be prepared at a low temperature; of course, this means the reaction is slow. Catalysts can be used, but the catalyst stays in the product and, therefore, sets a lower limit on the amount of catalyst in the final formulation.

Any polyhydroxy compound can be reacted with TDI to make prepolymers. Low molecular weight hydroxy-terminated polyesters or mixtures of diols and triols are commonly used. For safety, the levels of unreacted TDI in the prepolymer must be very low. Low levels of TDI can be assured by using a ratio of NCO : OH less than 2 : 1 and pushing the reaction to completion, but chain extension (caused by reaction of both N=C=O groups of some TDI molecules) increases the molecular weight of the product. An alternative process, used when low molecular weight prepolymers are needed, is to react the polyhydroxy compound, often trimethylolpropane, with a large excess of 2,4-TDI, then remove excess TDI using a vacuum wiped-film evaporator. Very low levels of free TDI and minimal chain extension are attainable, yielding low molecular weight products suitable for high solids coatings.

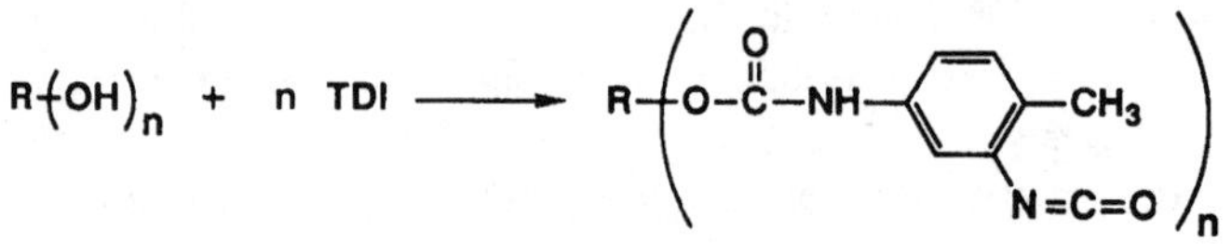

TDI Prepolymer

The isocyanurate derived from TDI made by trimerizing TDI (see Section 10.1) has a lower toxic hazard than monomeric TDI. The trimerization reaction occurs exclusively with the *para*-isocyanate group.

10.3.2. Aliphatic Isocyanates

The principal aliphatic isocyanates used are 1,6-hexamethylene diisocyanate (HDI), bis(4-isocyanatocyclohexyl)methane ($H_{12}MDI$), isophorone diisocyanate (IPDI), tetramethyl-*m*-xylidene diisocyanate (TMXDI), *m*-isopropenyl-α,α-dimethylbenzylisocyanate (TMI), and 2,2,5,-trimethylhexane diisocyanate (TMHDI). Diisocyanates are often converted to derivatives before use in coatings to increase functionality and reduce toxic hazard.

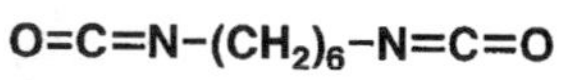

HDI

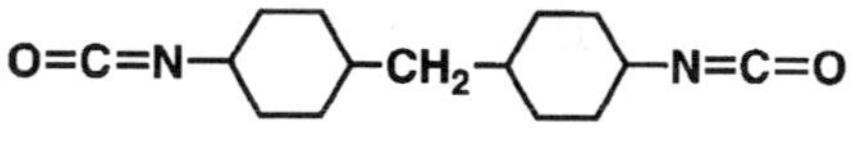

$H_{12}MDI$

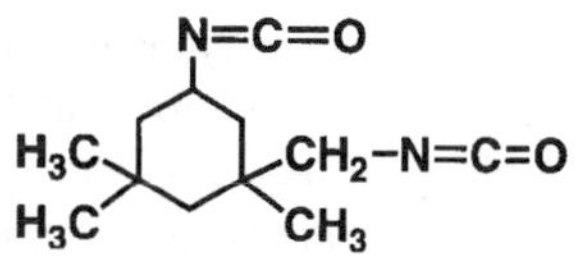

IPDI

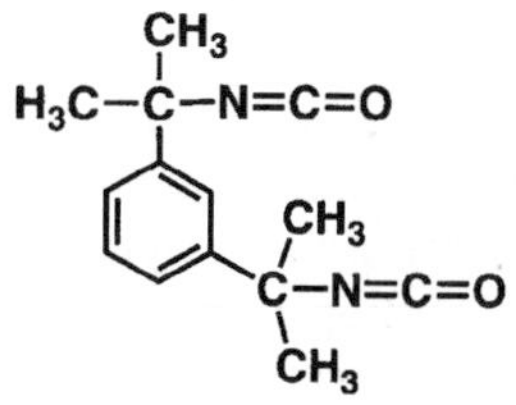

TMXDI

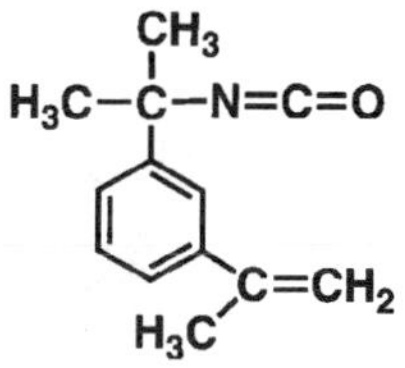

TMI

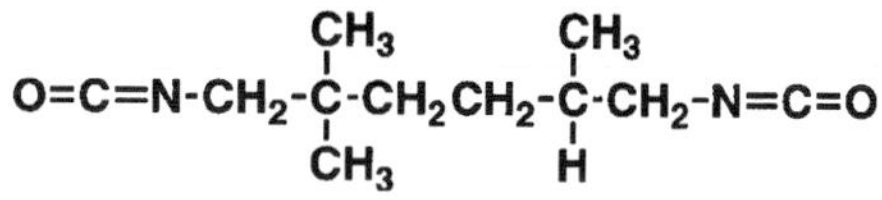

TMHDI

HDI is especially hazardous and is handled on a large scale only in chemical plants. The first less hazardous derivative was a biuret, which can be made by reacting HDI with a small amount of water and removing excess HDI. The structure of HDI biuret shown below is idealized; commercial products contain varying fractions of oligomeric biurets. (The presence of oligomeric biurets makes the average functionality higher than 3.) These polyfunctional isocyanates give coatings with good color retention and weather resistance. The viscosity of an early commercial product was about 11.5 Pa · s at 20°C. Grades with lower average molecular weights (and average functionalities nearer to 3) are available with viscosities as low as 1.4 Pa · s.

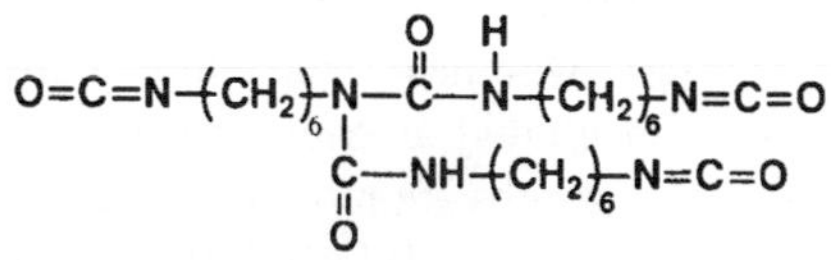

HDI-Biuret

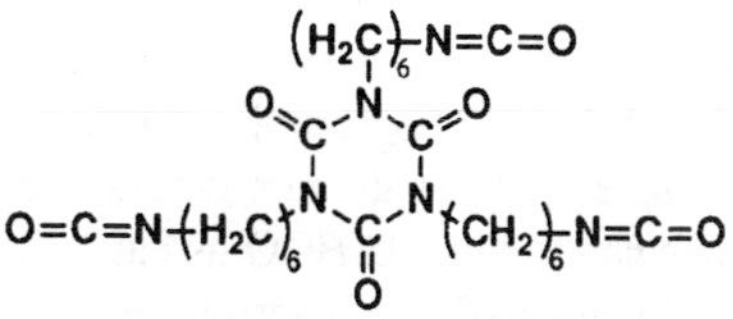

HDI-Isocyanurate

HDI isocyanurates are used on a larger scale. The isocyanurate gives coatings with greater heat resistance and even better long-term exterior durability than HDI biuret. Commercial products contain oligomeric material, and average functionality is over 3. Grades with lower oligomer content with viscosities as low as 1 Pa·s are available.

With ammonium fluoride as a catalyst, an isomeric trimer, an iminooxadiazenedione (called a nonsymmetrical trimer) of HDI is made as an approximately 50–50 mixture with HDI isocyanurate. It has the advantage that viscosity at the same oligomer content is lower than the corresponding HDI isocyanurate: 1 Pa·s versus 3 Pa·s [16].

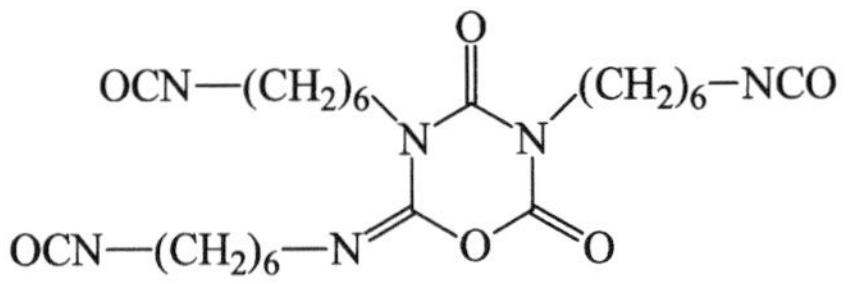

Allophanate derivatives of HDI and IPDI are another type of polyfunctional isocyanate. They are made by reacting an alcohol or diol with excess isocyanate, then removing unreacted diisocyanate with a wiped film evaporator giving an isocyanate-terminated allophonate [17]. The properties can be varied by using different alcohols to make the starting urethane to make compounds with different R substituents shown in the structure. For example, the cetyl alcohol urethane from HDI yields an allophanate diisocyanate that is soluble in aliphatic hydrocarbons. Derivatives with higher functionality are made by reacting a glycol with excess diisocyanate.

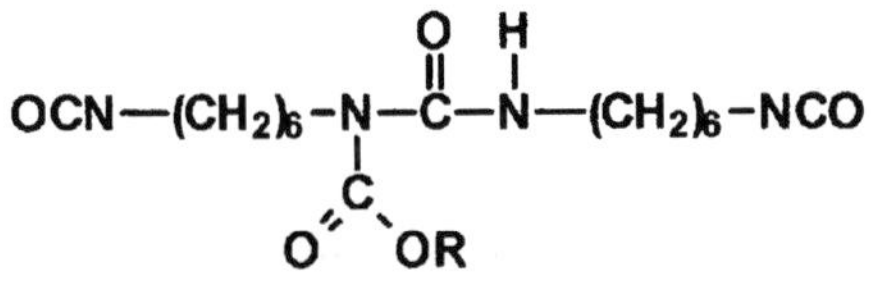

The uretdione dimer of HDI also has lower volatility with low viscosity and can be used for cross-linking in ambient cure coatings. Very low viscosity grades (< 100 mPa·s) are reported to permit formulation of very low VOC coatings [18].

TMHDI is used in place of HDI to make isocyanurate and other derivatives, which have the advantage of flexibility without the crystallization that can occur with HDI derivatives.

Bis(4-isocyanatocyclohexyl)methane (H_{12}MDI) is less volatile than HDI and is sometimes used as a free diisocyanate in coatings to be applied by roll coating, but not by spray coating. It is a mixture of stereoisomers; since both isocyanato groups are secondary, reactivity is lower than HDI or IPDI. Combined allophanate-isocyanurate derivatives are available with low free monomer content. Reference [19] covers the chemistry and uses of H_{12}MDI.

Commercial IPDI is a mixture of Z-(cis) and E-(trans) isomers in a 75:25 ratio. The isomers are difficult to separate. Isophorone diisocyanate has two different types of N=C=O groups. Studies performed under different conditions show that with DBTDL catalysis, the secondary N=C=O group of both Z and E isomers are more reactive than the primary N=C=O group [20,21,22]. The selectivity decreases with increasing temperature, and selectivity is greater with sec-butyl alcohol than with n-butyl alcohol [20]. On the other hand, with DABCO as catalyst, the primary N=C=O groups are more reactive [21] but this inversion of reactivity does not occur with other amine catalysts [22]. Isocyanurate derivatives of IPDI are widely used.

TMXDI and TMI have aromatic rings, but give color retention and exterior durability equivalent to aliphatic isocyanates. The exterior durability of TMXDI-derived urethanes probably results from the absence of isocyanate groups directly substituted on the aromatic ring, as well as from the absence of abstractable hydrogens on the carbons adjacent to nitrogen. Since the isocyanato group is on a tertiary carbon, the reactivity is lower than that of less sterically hindered aliphatic isocyanates. This difference can be offset by using higher catalyst levels and sterically accessible tin catalysts such as DMTDA instead of DBTDL. TMXDI is offered as a low molecular weight, essentially diisocyanate-free, prepolymer with trimethylolpropane. TMI is used as a comonomer with acrylic esters to make 2,000–4,000 molecular weight copolymers with 40–50 mole percent TMI; thus each molecule has several isocyanate groups [23].

10.4. TWO PACKAGE (2K) URETHANE COATINGS

The largest volume of urethane coatings is two package (2K) coatings that are mixed just before application. One package contains the polyol (or other coreactant), pigments, solvents, catalyst(s), and additives; the other contains the polyisocyanate and moisture-free solvents. Sometimes, the catalyst is in a separate third package so that cure rate can be adjusted for variations in ambient conditions. While the major reaction is formation of urethane cross-links, some urea cross-links result from reaction of atmospheric water with the isocyanate.

Any hydroxy-functional coreactant can be used; hydroxy-terminated polyester and hydroxy-substituted acrylic resins are most common. In general terms, polyesters permit higher solids and give films with greater solvent resistance and better adhesion to metals. Acrylics provide faster dry, lower cost (since their equivalent weight is usually higher than that of polyesters, resulting in the need for less of the more expensive isocyanate), and better exterior durability, resulting from superior hydrolytic and photochemical stability. Since alkyd resins have unreacted hydroxyl groups, their rate of dry can be accelerated by adding a polyisocyanate like the isocyanurate trimer of IPDI just before application. Nitrocellulose (shipped wet with plasticizer instead of ethyl or isopropyl alcohol) is used in formulating cross-linking furniture lacquers. Bisphenol A epoxy resins are also cross-linked with isocyanates through their hydroxyl groups. While hydroxy-terminated polyethers are widely used in urethane foams, they have limited usage, since the resulting coatings show high moisture vapor permeability, relatively poor exterior durability, and are soft as a result of the low T_g of the polyethers.

Urethane coatings for maintenance paint applications are generally cured at ambient temperatures, while those for automobile refinishing and aircraft applications are cured at ambient or modestly elevated temperatures. If the T_g of the partially reacted system is near the temperature at which the cross-linking must occur, the rate of urethane formation becomes mobility controlled and cure rate is slow. If the T_g of the fully reacted system is significantly above the cure temperature, the reaction virtually stops before the cross-linking reaction has gone to completion [24]. Water can continue to diffuse longer through such a matrix, and the last stages of reaction are predominately with water. Since in many ambient cure coatings, one wants a coating with a T_g somewhat above the curing temperature, selection of a combination of polyisocyanate and polyol that provides an appropriate final T_g is critical. A polyisocyanate with relatively flexible aliphatic chains (such as the isocyanurate of HDI) is used with a relatively high T_g acrylic or polyester. On the other hand, a polyisocyanate that yields relatively rigid cross-linked segments (such as the trimethylol-propane/TMXDI prepolymer) requires a lower T_g acrylic or polyester. Another approach to balancing T_g is to use mixtures of IPDI and HDI isocyanurates [25].

An important variable in 2K polyisocyanate/polyol coatings is the ratio of N=C=O/OH used. In ambient cure systems, it is often found that a ratio of the order of 1.1 : 1 gives better film performance than a 1 : 1 ratio. A probable reason is that part of the N=C=O reacts with water from solvent, pigment, or air to give urea cross-links. To the extent this happens, two hydroxyls are unreacted for each water molecule; use of excess N=C=O minimizes residual unreacted hydroxyl groups. Solvent resistance is also improved. Since polyisocyanates are typically lower in viscosity than the polyol, excess NCO gives lower VOC [3]. Some finishes, for example aircraft finishes, are formulated with N=C=O/OH ratios as high as 2 : 1. The resulting reduced hydroxyl concentration gives longer pot life. It is possible that the high mobility of water and the reactivity of amine groups resulting from the water/isocyanate reaction favor faster reaction rates at higher T_g values than possible with only the hydroxyl/isocyanate reaction.

There is always a compromise between pot life and curing time. (See Section 2.3.2.) Several formulating variables are available to increase pot life with little effect on cure time. Concentrations of reactive groups should be kept as low as possible; this becomes more difficult as one formulates to increasingly higher solids. As noted in the preceding paragraph, using mixed polyol/moisture curing systems helps. Since isocyanate/alcohol reactions proceed most slowly in media with high levels of hydrogen-bond accepting groups, solvents should be selected to the extent possible which are strong hydrogen-bond acceptors, and resins should be designed, if possible, with low levels of hydrogen-bond accepting groups. After application, as the solvent evaporates, the medium becomes less hydrogen-bond accepting and the reaction rate increases accordingly. This strategy serves the purpose of both extending pot life and promoting reactivity after application. Since solvents with greater hydrogen-bond accepting strength tend to have higher viscosities, higher solids and longer pot lives are obtained with solvents having intermediate hydrogen-bond accepting strength [9].

Organotin catalysts are generally preferable to amine catalysts (see Section 10.2.2) because reaction rates depend on alcohol concentration to the one-half power with organotin catalysts and to the first power with amine catalysts. Since the effectiveness of tin catalysts is reduced by carboxylic acids, one can add a volatile acid such as acetic or formic acid to the formulation. The acid inhibits reaction during the pot life stage, but evaporates during application so that the inhibiting effect disappears. 2,4-Pentadione increases pot life by chelating with tin compounds. An aluminum complex catalyst with 2,4-pentadione is said to provide longer pot life than DBTDL at equal cure time [13].

Availability of lower viscosity isocyanate cross-linkers (see Section 10.3.2) makes possible formulation of higher solids 2K coatings. However, one must consider not just the viscosity of the cross-linker, but also that of the combination of polyol and cross-linker. In some cases, a lower viscosity cross-linker also has a lower equivalent weight; this means that the weight ratio of cross-linker to polyol must be reduced to maintain the same stoichiometric ratio of N=C=O/OH [26]. Since the polyol often has a much higher viscosity, this reduced ratio can give a higher viscosity, in spite of the lower viscosity of the cross-linker. Higher solids can be obtained using a somewhat higher viscosity polyiso-cyanate with a higher equivalent weight.

If very fast cure at relatively low temperatures is needed, reactive coreactants and/or high catalyst levels are used, and the pot life is short. Such formulations can be applied using special spray equipment in which the two packages are fed to a spray gun by separate proportioning pumps and mixed inside the gun just before they are sprayed. Care is required to assure that the proper ratios are fed and thoroughly mixed; such spray equipment is expensive.

Most amines react too rapidly for use in 2K coatings, however, hindered amines have been developed that permit their use. Polyaspartic esters [2,3] are hindered amines that are used in high solids coatings.

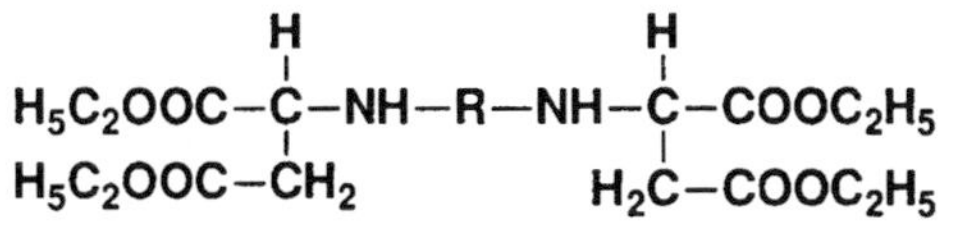

$$H_5C_2OOC-\underset{\underset{H_5C_2OOC-CH_2}{|}}{\overset{\overset{H}{|}}{C}}-NH-R-NH-\underset{\underset{H_2C-COOC_2H_5}{|}}{\overset{\overset{H}{|}}{C}}-COOC_2H_5$$

As shown in Table 10.1, the gel time of polyisocyanate/polyaspartic ester combinations depends on the structure of the polyaspartic ester. The pot life of these combinations is increased by addition of DBTDL [3]. On the other hand carboxylic acids and water accelerate reaction rates. By proper selection of isocyanate and aspartate and adjustment of catalysts, it is possible to formulate very high solids 2K coatings with a reasonable pot life and fast curing.

Imines react directly with isocyanates (see Section 10.1) or with water; the ratio of the two reactions depends on relative humidity, the time between application and cure, and the curing temperature [3]. The reaction with water releases amine to react with isocyanate and carbonyl compound, which volatilizes. In the direct reaction, carbonyl compound from the imine is not released, and hence, VOC emissions are lower. The ratio of the two reactions depends on the relative humidity and on the time between application and baking. The direct reaction is catalyzed more than the water reactions by carboxylic acids, and water and tertiary amines reduce the rate of reaction. The imines have very low viscosities; for example, the bis-(methyl isobutyl ketone) ketimine of ethylene diamine has a viscosity of

Table 10.1. Viscosities (100% solids, mPa · s at 23°C) and Gel Times[a] of Substituted Polyaspartic Acid Ethyl Esters

R	Viscosity	Gel Time
$(CH_2)_6$	150	<5 min
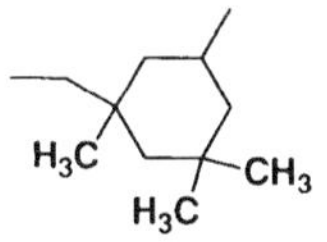	800	2–3 hr
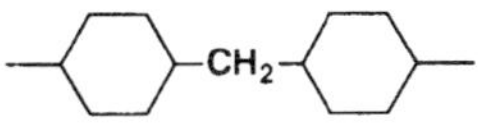	1000	2–3 hr
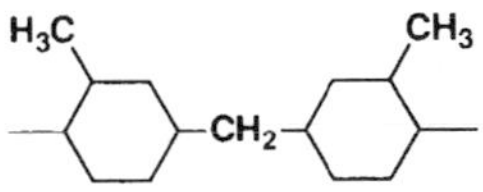	1500	>24 hr

[a]The values given are for when the esters are mixed with equal equivalents of a hexamethylene diisocyanate isocyanurate cross-linker at 65 percent solids in butyl acetate and hiflash naptha.

5 mPa · s and an equivalent weight of 112. Direct reaction is more favored with aldimines, especially above 60°C. Imines permit formulation of 2K coatings with very high solids, better pot life, and faster drying than with polyols [3]. They can be used alone or in combination with a hydroxy-functional polyester or acrylic [27].

10.5. BLOCKED ISOCYANATES

In early work on isocyanates in coatings, it was recognized that there would be advantages to develop systems supplied in one package with minimal toxic hazard. These objectives led to work on *blocked isocyanates*, taking advantage of reversible reactions of isocyanates. A di- or polyisocyanate is reacted with a monofunctional active hydrogen compound, and the adduct (a blocked isocyanate) is formulated into a coating with a coreactant resin. The coating is stable at ambient temperature, but when baked, the monofunctional reactant is released (and usually volatilized), and the coreactant crosslinks. Many blocked isocyanates are sufficiently stable in the presence of water at storage temperatures for use in water-borne systems. However, use of blocked isocyanates has two drawbacks: relatively high temperatures are required for cross-linking and, in some cases, release of the blocking agent can cause a pollution problem.

Blocked isocyanates are synthesized by reacting a blocking group (BH) with an isocyanate:

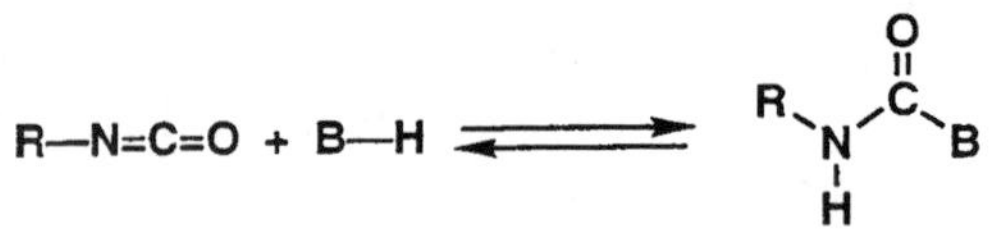

$$\text{R}{-}\text{N}{=}\text{C}{=}\text{O} + \text{B}{-}\text{H} \rightleftharpoons \underset{\text{H}}{\overset{\overset{\displaystyle O}{\overset{\|}{\text{C}}}}{\text{R}{-}\text{N}}}{-}\text{B}$$

A large number of blocking agents has been studied [28,29,30]. The ones most widely used commercially are phenols, oximes, alcohols, ε-caprolactam, 3,5-dimethylpyrazole, 1,2,4-triazole, and diethyl malonate. It is common to see tables of "unblocking temperatures." The implication is that there is some "magic temperature" above which the various blocked isocyanates abruptly unblock; there is not. Rates of unblocking reactions are governed by Arrhenius constraints (see Section 2.3.2). Published unblocking temperature data are based on some measurement of extent of reaction under a specific set of conditions. Examples of such criteria are the lowest temperature at which free isocyanate can be detected spectroscopically or the temperature required to achieve a specified degree of solvent resistance in cross-linking a coreactant resin in a specified time. Provder showed how instrumentation can be used to determine the kinetics of unblocking reactions and the rate of development of cure [31]. The values depend on rate of heating as well as the time at a temperature.

The reaction pathway for cross-linking is often written as the dissociation of the blocked isocyanate to give blocking agent and free isocyanate, which then reacts with a hydroxyl group (i.e., elimination followed by addition). Alternatively, the reaction could proceed by addition of the alcohol to the blocked isocyanate to yield a tetrahedral intermediate followed by elimination of the blocking agent (i.e., addition followed by elimination). Both pathways are shown in Scheme 10.4, in which B—H represents the blocking group. Evidence has been presented for both pathways with elimination-addition favored by increasing temperature. In some cases, k_2 is small compared to k_{-1}, but

Scheme 10.4

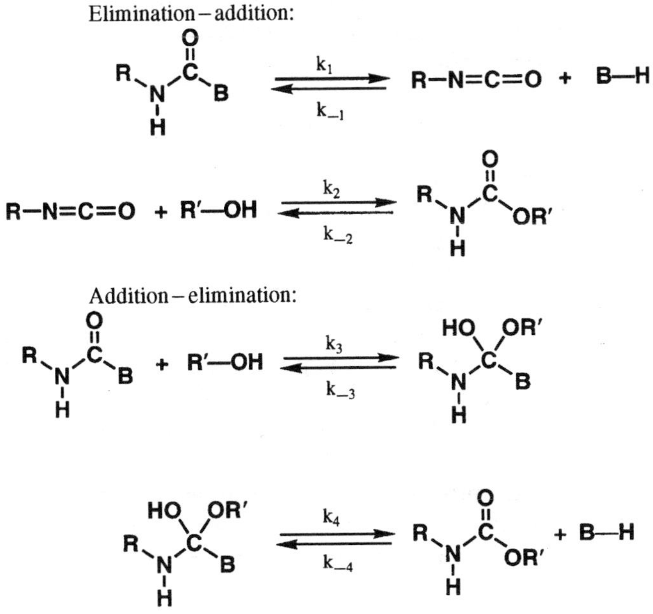

Elimination – addition:

$$R-\underset{\underset{H}{|}}{N}-\underset{\underset{O}{\parallel}}{C}-B \underset{k_{-1}}{\overset{k_1}{\rightleftharpoons}} R-N=C=O + B-H$$

$$R-N=C=O + R'-OH \underset{k_{-2}}{\overset{k_2}{\rightleftharpoons}} R-\underset{\underset{H}{|}}{N}-\underset{\underset{O}{\parallel}}{C}-OR'$$

Addition – elimination:

$$R-\underset{\underset{H}{|}}{N}-\underset{\underset{O}{\parallel}}{C}-B + R'-OH \underset{k_{-3}}{\overset{k_3}{\rightleftharpoons}} R-\underset{\underset{H}{|}}{N}-\underset{\underset{OH\ OR'}{\ }}{C}-B$$

$$R-\underset{\underset{H}{|}}{N}-\underset{\underset{HO\ OR'}{\ }}{C}-B \underset{k_{-4}}{\overset{k_4}{\rightleftharpoons}} R-\underset{\underset{H}{|}}{N}-\underset{\underset{O}{\parallel}}{C}-OR' + B-H$$

volatilization of the blocking agent displaces the equilibria toward the cross-linked product.

A variety of catalysts is used: organotin compounds, DABCO, and zinc octoate are among them. Various organotin compounds catalyze reactions more or less effectively [32]. In primers, where discoloration is not a problem, copper and cobalt naphthenates have been found effective. It is not usually known which reaction is being catalyzed. It could be decomposition of blocked isocyanate to give isocyanate. But it also could be catalysis of the reaction of the nucleophile with the isocyanate so k_2 increases relative to k_{-1}. It could also be catalysis of some stage of the addition-elimination pathway.

Phenols were the first blocking agents used commercially, despite their environmental problems. Blocked isocyanates made with phenol or alkyl-substituted phenols are less thermally stable than those blocked with alcohols. Blocked aromatic isocyanates are more reactive than blocked aliphatic isocyanates. A typical curing schedule with a phenol blocked aromatic polyisocyanate and a hydroxy-functional resin is 30 minutes at 160°C. Phenol blocked isocyanates are used for wire coatings, applied by running the wire through a coating bath and then through an exit die appropriately larger than the diameter of the wire. This process requires a coating with one-package stability. Since wire is small and can be heated to high temperatures without major energy cost, high cure temperature is not a major disadvantage. Phenols had been used for many years as solvents in wire coating, and ovens were already engineered to incinerate the phenol, eliminating the pollution problem and providing some fuel value. The excellent abrasion resistance of polyurethane wire coatings justifies the cost. Phenol blocked isocyanates react with amines at room temperature. They are used as flexibilizing reactive additives in amine cured epoxy coatings.

Oximes, such as methyl ethyl ketone oxime (MEKO), form less thermally stable blocked isocyanates than phenols. Their reactivity is such that reasonably stable one package coatings can be formulated with hydroxy-functional resins that can be cured in 30 minutes at 130°C or somewhat lower.

$$R\text{—}N{=}C{=}O \quad + \quad \underset{R''}{\overset{R'}{\diagdown}}C{=}N\text{—}OH \quad \rightleftharpoons \quad R\text{—}\overset{H}{\underset{|}{N}}\text{—}\overset{O}{\overset{\|}{C}}\text{-}O\text{—}N{=}\underset{R''}{\overset{R'}{C}}$$

Oximes are used as blocking agents in magnetic metal oxide coatings for application to oriented polyester tape. Such coatings must be cured in a short time at temperatures no higher than 80°C to avoid distortion of the plastic tape and yet must have sufficient pot life (at least a few hours) to permit application with a roll coater. An oxime blocked diisocyanate with an amine-terminated polyamide coreactant gives sufficient pot life, but still cures rapidly at 80°C. The cured coatings may contain some unreacted amine groups, which could discolor in time, but this is not a disadvantage for magnetic tapes. The excellent abrasion resistance and flexibility of polyurethane coatings are major advantages in this application.

The more rapid reaction of amines than alcohols with blocked isocyanates has been interpreted as evidence for the addition-elimination pathway. The intuitive reasoning is that the more nucleophilic amines are expected to be more reactive than alcohols in addition to blocked isocyanates, but are not expected to promote the elimination reaction. However, the higher reactivity of amines can also be rationalized in terms of the elimination-addition reaction when one takes into account that the elimination step is reversible and that amines are better competitors than alcohols for the intermediate isocyanate relative to the blocking group. In terms of Scheme 10.4, k_2 is much higher for amines than for alcohols.

Strong evidence for the elimination-addition pathway in reactions of amines and alcohols with oxime-blocked aromatic and aliphatic isocyanates has been obtained [32,33]. With amines, the rate of reaction was found to be independent of amine concentration, thereby ruling out the addition mechanism and indicating that essentially all intermediate isocyanate proceeds to product (a urea). (Bear in mind that the addition-elimination pathway requires dependence of the reaction rate on amine concentration.) In contrast, the rate of reaction with alcohols was substantially slower than with amines and was dependent on the concentration of alcohol. With a very large excess of alcohol rates with alcohols and amines were almost equivalent. These results indicate that with alcohols, a substantial fraction of intermediate isocyanate reverts to starting materials, resulting in an overall slower rate of conversion to urethane. The conclusions are in line with the order of reactivity with isocyanates: amines $\gg$ oximes $\gg$ alcohols. The rate expression for elimination-addition (refer to Scheme 10.4) is provided in Eq. 10.7, where BI, BH, and NuH represent the blocked isocyanate, blocking group, and nucleophile, respectively.

$$\text{rate} = k_1[\text{BI}]\ \frac{k_2[\text{NuH}]}{k_2[\text{NuH}] + k_{-1}[\text{BH}]} \tag{10.7}$$

When $k_2[\text{NuH}] \gg k_{-1}[\text{BH}]$, the ratio in Eq. 10.7 approaches 1, and the equation simplifies to Eq. 10.8. In this case, the rate is independent of the nucleophile concentration, as observed in the reaction of oxime blocked isocyanates with amines.

$$\text{rate} = k_1[\text{BI}] \tag{10.8}$$

When $k_{-1}[\text{BH}]$ is greater than or equivalent to $k_2[\text{NuH}]$, the ratio in Eq. 10.7 becomes less than 1 resulting in a slower reaction rate. When $k_{-1}[\text{BH}]$ is substantially larger than

k_2[NuH], Eq. 10.7 simplifies to Eq. 10.9. This is the situation with alcohols since oximes react faster with isocyanates than alcohols $(k_{-1} \gg k_1)$.

$$\text{rate} = k_1[\text{BI}]\frac{k_2[\text{NuH}]}{k_{-1}[\text{BH}]} \tag{10.9}$$

Steric crowding on either oxime or isocyanate enhances the dissociation rates. Isocyanates blocked with diisopropyl and diisobutyl ketoximes dissociate faster than the same isocyanates blocked with MEKO [34]. Oxime blocked TMXDI dissociates faster than other less sterically crowded aliphatic isocyanates and with comparable rates to aromatic isocyanates. Generally, blocked aromatic isocyanates are more reactive than the corresponding blocked aliphatic isocyanates. On the other hand, coatings derived from aliphatic isocyanates exhibit substantially better exterior durability. Thus, the high reactivity of blocked TMXDI (and TMI) derivatives, similar to that of aromatic isocyanates, and their exterior durability, characteristic of aliphatic isocyanates, are potentially important advantages. Blocked TMI cross-linkers can be synthesized by first copolymerizing the TMI with acrylate esters, then reacting the polyfunctional acrylate with an oxime such as MEKO. Alternatively, one can react TMI with the oxime and then copolymerize the blocked monomer with acrylate esters [35].

Oximes oxidize at elevated temperatures to yield colored products. As a result, in some cases, films made with oxime blocked isocyanates yellow during baking. The extent of yellowing is related to the rate at which the oxime can escape from a film; hence, MEKO gives less yellowing than diisopropyl ketoxime in spite of the lower baking temperatures possible with the latter. The yellowing can be minimized by including a reducing agent such as a hydrazide derivative, for example, an oligomeric hydrazine adipate, in the formulation [36]. Class II melamine-formaldehyde resins also inhibit yellowing. Isocyanates blocked with 3,5-dimethylpyrazole (DMP) yellow less than with MEKO and give somewhat faster curing [37,38]. The curing of MEKO blocked isocyanate clear coats over MF cross-linked base coats is reported to be inhibited by the acid catalyst in the base coat, whereas the cure of DMP blocked isocyanates is not [38]. 1,2,4-triazole and mixtures of 1,2,4-triazole and DMP are reported to give low curing temperatures and freedom from yellowing [39]. Concerns about possible toxic hazards with MEKO, are serving to increase emphasis on evaluating other blocking agents, such as DMP and triazole.

The largest volume coatings with blocked isocyanates are cationic electrodeposition primers (see Section 26.2). They must be stable in water indefinitely; 2-ethylhexyl alcohol blocked isocyanates give the necessary hydrolytic stability. Amine groups on the resin are neutralized with weak acids, such as lactic or formic acids. During electrodeposition, the protonated amine groups are converted into free amines at the cathode, causing the polymer to form a uniform coating on the cathode. The coatings are subsequently cured at relatively high temperatures.

Tris(alkoxycarbonylamino)triazine (TACT), a mixed methyl and butyl carbamate derivative of melamine, acts as an alcohol blocked isocyanate [40]. TACT is much more reactive than conventional alcohol blocked isocyanates, since they are activated by the triazine ring. It cross-links hydroxy-functional acrylics at 125°C in 30 minutes, comparable to (or lower than) oxime blocked isocyanates. The reaction is not catalyzed by tin compounds; it is catalyzed to a small extent by dodecylbenzene sulfonic acid.

TACT also reacts (at 125°C in 30 minutes) with epoxy-functional acrylic resins when catalyzed with tetrabutylammonium acetate; yielding N-substituted carbamates and oxazolidinones derived from the carbamates by ring closure.

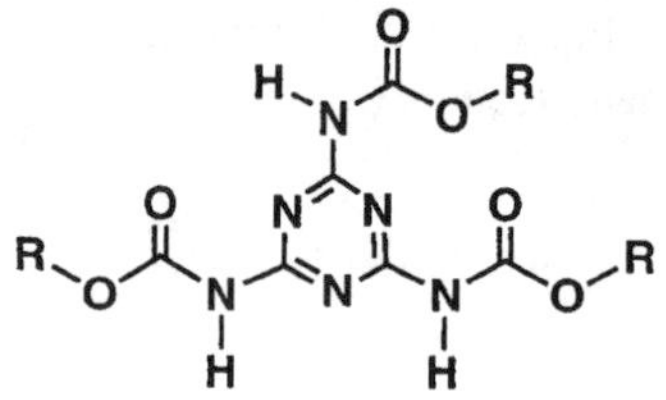

TACT (R = mixed CH$_3$, and −C$_4$H$_9$)

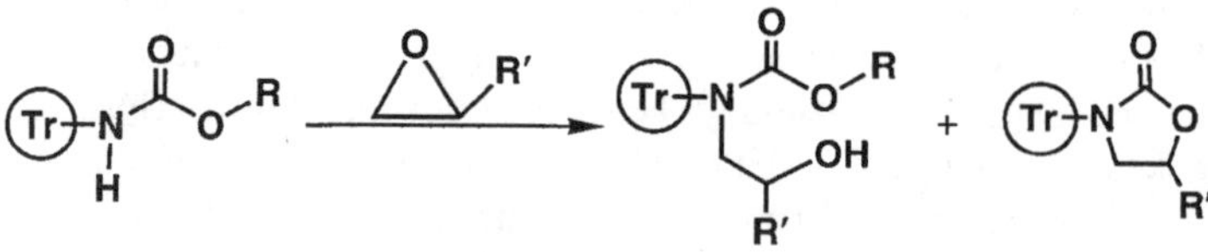

Alcohol blocked TMXDI is sufficiently reactive when catalyzed by DBTDL to be used in solvent-borne or powder coatings [41]. Lowest temperatures were found to be with methyl alcohol blocked compounds. It is suggested that the lower temperature cures resulted from the ease of diffusion and high volatility of the small methyl alcohol molecules. The low molecular weight of methyl alcohol minimizes VOC emissions.

Diethyl malonate blocked diisocyanates cross-link with polyols at low temperatures, a typical curing schedule of 120°C for 30 minutes. However, reaction of diethyl malonate blocked isocyanates with alcohols and amines does not proceed by either pathway shown in Scheme 10.4. Reaction with alcohols does not yield urethanes, but rather, transesterification occurs [42]. Correspondingly, reaction with amines yields amides, not ureas.

The transesterification proceeds fairly rapidly at 120°C; storage stable coatings can be formulated by using monofunctional alcohol in the solvent mixture [30]. The monofunctional alcohol extends the stability by competing with the polyols for reaction with the blocked isocyanate. After the coating is applied, the monofunctional alcohol evaporates, and reaction with polyol gives a cross-linked coating.

The active hydrogen of malonic ester blocked isocyanates reacts with MF resins [43]. Clear coats for automobiles have been formulated with a combination of a hydroxyfunctional acrylic resin, malonic ester blocked HDI and IPDI trimers, and an MF resin [44]. Cross-linking reactions result from reactions of isocyanates from the blocked isocyanate with hydroxyl groups, MF resin cross-linking of hydroxyl groups, and MF resin reaction with the malonic ester blocking groups.

A large scale use of blocked isocyanates is in powder coatings. (See Section 27.1.3.) The principal blocking agent is ε-caprolactam, the cyclic lactam of H$_2$N(CH$_2$)$_5$COOH. The most widely used isocyanate has been IPDI isocyanurate but it has been shown that H$_{12}$MDI reacts at lower temperatures [17]. Use of caprolactam is decreasing due to the high temperatures required for curing and oven build up that is difficult to clean. Oxime blocked isocyanates cure at lower temperature, but there is concern about possible toxic hazards. A blocked isocyanate derived from 1,2,4-triazole and a mixture of monomeric HDI and HDI isocyanurate permits relatively low temperature cure of acrylic powder coatings while giving good mechanical properties [45]. Blocking with a mixture of 1,2,4-triazole and 3,5-dimethylpyrazole is reported to give non-yellowing and low temperature cure [39]. Polyol derivatives of IPDI uretdiones are used as cross-linkers in powder coatings, some will permit curing at lower temperatures than caprolactam. The uretdione groups disassociate to isocyanates at curing temperatures providing cross-linking with no volatile blocking agent [46].

10.6. MOISTURE CURING POLYISOCYANATE COATINGS

Polyurethane moisture curing coatings cross-link under ambient conditions by reaction of isocyanates with atmospheric water. Most such coatings use isocyanate-terminated resins made from hydroxy-terminated polyesters by reacting the terminal hydroxyl groups with excess diisocyanate like oligomeric MDI or, if color retention is needed, IPDI. So the coating does not contain unreacted diisocyanate, the resin is prepared utilizing a ratio of $N{=}C{=}O$ to OH that is significantly less than 2 to 1. The coatings are stable when stored in the absence of water, but cross-link after application by reaction of the isocyanates with water to form amines, which react with another isocyanate to form substituted urea cross-links. Cure rates depend on the water content of the air; at low temperatures, higher relative humidity is required than at higher temperatures, since the relative humidity decreases with increasing temperature. At high humidity and temperature, cure is rapid, but the carbon dioxide released by the reaction of isocyanate with water can be trapped as bubbles, especially in thick films. See Ref. [47] for discussion of the effects of temperature and humidity and other application considerations. Moisture curing urethane coatings are used for applications such as floor coatings, for which abrasion resistance and hydrolytic stability are important. They are called urethane coatings even though these cross-links are urea rather than urethane groups. Since urea groups also form intermolecular hydrogen bonds, presumably, they can affect resistance to mechanical stress similarly to urethane groups.

Hydrolytic resistance of coatings is affected by several variables, including cross-link density and free volume availability, as well as the functional groups present. Model compound studies show that substituted ureas are somewhat more easily hydrolyzed under both neutral and acidic (pH 1.34) conditions than are analogous urethanes [48]. Ureas and urethanes from aliphatic isocyanates are somewhat more resistant to hydrolysis than those from aromatic isocyanates. Both urea and urethane groups are generally more stable to hydrolysis than unhindered ester groups.

Moisture curing coatings are less susceptible to error in application than 2K coatings, for which there is a chance of incorrect ratios of packages affecting the stoichiometric ratios. A disadvantage is that solvents, pigments, and other components must be essentially water free. Use of moisture cure coatings is primarily in clear, gloss coatings since it is expensive to remove adsorbed water from pigments. Molecular sieves can be used to adsorb water, but can reduce gloss. Pigmented moisture cure coatings can be made using a water scavenger such as an alkyl orthoformate or *p*-toluenesulfonylisocyanate. Such water scavengers, especially sulfonylisocyanates, must be handled with care due to toxicity. IPDI can be used as a moisture scavenger [49]; moisture cure urethane coatings using IPDI are used on military vehicles. Oxazolidines are reported to be effective moisture scavengers for aliphatic isocyanate based coatings, presumably with less toxic hazard and lower cost [50]. 4-Ethyl-2-methyl-2-(3-methylbutyl)-1,3-oxazolidine provides rapid water removal, while not adversely affecting stability or performance of a coating.

10.7. HYDROXY-TERMINATED POLYURETHANES

Diisocyanates can be reacted with diols and triols at an $N=C=O/OH$ ratio of less than 1 to make hydroxy-terminated polyurethanes, which can be cross-linked with MF resins or with other cross-linkers that react with hydroxyl groups. Urethane groups also cross-link with MF resins. (See Section 9.3.4.) Compared to polyesters, hydrolytic stability is better and coatings provide the toughness and abrasion resistance associated with urethane coatings. However, the solids at equal molecular weight and viscosity are lower due to intermolecular hydrogen bonding, and the urethane groups increase moisture absorption. Compared to hydroxy-functional acrylic resins, they offer the advantage of abrasion resistance and the possibility of using lower molecular weight resins. As discussed in Section 12.2.1, molecular weight reduction of acrylic resins made by conventional free radical initiated polymerization is limited by the problem of ensuring that at least two hydroxyl groups are present on each oligomer molecule. In hydroxy-terminated polyur-ethanes (as with polyesters), all molecules have two (or more) terminal hydroxyl groups, even at very low molecular weight.

Hydroxy-terminated polyurethanes with low molecular weight and viscosity were synthesized by reaction of diisocyanates with excess diols [51]. These *urethane diols*, as reactive diluents, permit formulation of higher solids acrylic/MF coatings with good properties. It was also suggested that they may be useful in water-reducible systems, reducing the organic solvent requirement (see Section 12.3). Gardon has shown that 1,2- and 1,3-diols react with hexamethylene diisocyanate isocyanurate selectively to give narrow molecular weight distribution urethane polyols [52]. If the diol has at least five carbon atoms (e.g., neopentyl glycol or 2-ethyl-1,3-hexanediol), the urethane polyols give coatings with excellent environmental etch resistance when cross-linked with MF resins.

10.8. WATER-BORNE POLYURETHANES

VOC emission regulations have led to work on water-borne polyurethane systems; a variety of approaches has been investigated [53,54].

One approach has been to make polyurethane dispersion resins that are predominantly linear, relatively high molecular weight polyurethane polymers dispersed in water. Such materials are latexes, but they are generally called *polyurethane dispersions* (PUD). One way to prepare such dispersions is by the *acetone process* [53]. An isocyanate-terminated polymer is made in acetone solution from a diisocyanate and a diol (or mixture of diols). This resin is chain extended by reacting with a sulfonate substituted diamine, $H_2NCH_2CH_2NHCH_2CH_2SO_3^-Na^+$. The acetone solution is diluted with water, and the acetone removed by distillation. The sulfonate salt substituent acts as a charge repulsion stabilizer for the aqueous dispersion. The average number of sulfonate groups per molecule can be adjusted by using mixtures of unsubstituted and sulfonate substituted diamines. One can incorporate a small amount of trifunctional isocyanate in making the initial isocyanate-terminated polymer. The result is a dispersed polymer with a low degree of cross-linking—low enough that the particles can still coalesce to a continuous film, but high enough that film properties are superior to those obtained with linear polymers [55]. Since urethanes hydrogen bond strongly with water, the particles are swollen with water, plasticizing the polymer thereby permitting film formation with higher T_g polymers than is possible with acrylic latexes [56].

One can also make combined acrylic/urethane (hybrid) aqueous dispersions [56]. Acrylic monomers are emulsion polymerized in the presence of an aqueous dispersion of a hydroxy-terminated polyurethane (e.g. one made with DMPA, see the next paragraph). The polyurethane stabilizes the aqueous dispersion, minimizing need for surfactant. Coalescence requires balance of the T_g of both the urethane and acrylic parts of the system. Compositions based on an IPDI/polypropylene glycol/DMPA urethane with styrene/methyl methacrylate/butyl acrylate are reported to form films at low temperatures.

Cross-linkable polyurethane dispersions are also made. For example, a hydroxy-functional polyurethane can be made using a diisocyanate, a diol, a triol and 2,2-dimethylolpropionic acid (DMPA) in a *one-step process*. The sterically hindered carboxylic acid group on DMPA is so low in reactivity with isocyanate that few amide bonds are formed. After neutralization of carboxylic acid groups with an amine, the resin is diluted with water to yield a dispersion analogous to the water-reducible acrylics discussed in Section 12.3. Commonly, N-methylpyrollidone is used as a solvent to reduce the viscosity of the resin. Such resins do not have the hydrolytic stability problems of water-reducible polyesters; also, the probability of formation of low molecular weight, cyclic, nonfunctional molecules is low. Lower viscosity polyurethane dispersions can be made by a *two-step process* [57]. First, a linear isocyanate-terminated polyurethane is made using a diisocyanate, a diol, and DMPA; then, in a second step, the terminal isocyanate groups are reacted with triols.

Dispersions can also be made by solvent-free processes. For example, isophorone diisocyanate, polycaprolactone diol, polytetrahydrofurandiol, and dimethylolpropionic acid are reacted with a NCO:OH ratio of 1.6:1 to make an isocyanate-terminated prepolymer. The prepolymer is neutralized with triethylamine and dispersed in water. The prepolymer is then chain extended by reacting with hydrazine hydrate [57]. Polyurethane dispersions can also be prepared by a nonisocyanate process [58]. An aliphatic polyester diol is reacted with a bis-(2-hydroxypropyl carbamate), a triol, and a transesterification catalyst, such as dibutyltin oxide. Carboxy functionality is then introduced by reacting some of the remaining hydroxy groups with an anhydride.

These hydroxy-functional polyurethane dispersions can be cross-linked with melamine-formaldehyde resins or blocked isocyanates. Blocked isocyanates can also be used with water-reducible anionic acrylic or polyester resins instead of an MF resin; oxime blocked isocyanates are commonly used, owing to their high reactivity. Oxime blocked aliphatic isocyanates are more stable to hydrolysis than oxime blocked aromatic isocyanates [55]. As discussed in Section 26.2, alcohol blocked isocyanates are used in cationic electro-deposition coatings.

Since isocyanates react relatively readily with water, it was assumed for many years that they could not be used directly in water-borne coatings; however, in the last few years, 2K water-borne coatings have been developed. It is relatively easy to make 2K coatings with good properties in the laboratory; however, production use is more difficult. There are several potential problems. In some systems, it is difficult to assure that uniform stoichiometric ratios are obtained throughout the film. As with solvent-borne 2K coatings, water-borne 2K coatings can be expected to have limited pot life. In solvent-borne coatings, pot life can be determined by monitoring viscosity increases. In many 2K water-borne coatings, viscosity does not change as reactions between N=C=O and OH take place, since they change the viscosity inside the aggregate particles, but not the bulk viscosity. In fact, the generation of CO_2 by the reaction of isocyanate decreases the pH, leading to shrinking of the aggregate particles, which can result in viscosity reduction. If reaction occurs to a significant extent before application, coalescence will be inhibited, and therefore, film properties may be poor. If there is too much reaction before volatiles are evaporated, complete coalescence of the dispersion particles may not occur. Since isocyanates also react with water, full reaction of the polyols requires an excess of isocyanate. Furthermore, the reaction with water results in formation of CO_2, which can lead to blistering, especially as thicker films are applied.

One approach to solving these problems is to combine conventional polyisocyanates with conventional, amine neutralized, water-reducible resins in a water-miscible solvent and then add water. A similar process, used for water-reducible coatings cross-linked with melamine-formaldehyde resins is described in Section 12.3. Conventional water-reducible acrylic resins in solvents such as 1-methoxy-2-propanol acetate have been used. Water soluble glycol ethers do not significantly interfere since they are primarily in the water phase with little chance to interact with isocyanates. Tertiary amines such as triethylamine or N-alkylmorpholines are used to neutralize the resin rather than the dimethylethanol amine widely used in MF cross-linked coatings. 2-amino-2-methyl-1-propanol is a particularly poor choice, since the primary amines are very reactive with isocyanate. The amine is added to the resin solution before the polyisocyanate. Finally, the coating is reduced to application viscosity with water. Mixing the polyisocyanate into the acrylic solution requires good agitation, and with conventional polyisocyanates, erratic results are common. It is necessary to use a high NCO/OH ratio, often above two, to compensate for nonproductive reactions of isocyanate with water; the high ratio increases cost. Since TMXDI reacts more slowly with water than other commercial isocyanates, successful use of this approach is reported with lower NCO/OH ratios [59].

Alternatively, the conventional polyisocyanate can be mixed in after reduction with water, just before spraying. Spray equipment designed to provide in line intensive mixing of the two components has been used to apply polyisocyanate 2K water-reducible OEM automotive clear coats with properties comparable to those obtained with solvent-borne 2K coatings [60]. A large fraction of the water evaporates during flash off and initial baking, reducing the extent of isocyanate reaction with water, minimizing CO_2 evolution, and permitting lower ratios of NCO/OH.

Mixing conventional polyisocyanates into the polyol after reduction with water can result in problems resulting from nonuniform stoichiometry. Hydrophilically-modified polyisocyanates made by reacting a fraction of the −NCO groups on a polyisocyanate (such as HDI or IPDI isocyanurate) with a polyglycolmonoether are more easily mixed. The modified polyisocyanate is stirred into an aqueous dispersion of a coreactant to form a heterogeneous dispersion in which the polyisocyanate and coreactant are in separate dispersed particles [61,62,63].

Although alcohols react more rapidly with isocyanates than water, in a water system, there is a very large excess of water, so a significant fraction of the isocyanate groups react with water; but. these reactions also give cross-links. To assure reaction of most of the hydroxyl groups, NCO/OH ratios of 2 or greater are recommended. Some applications, such as wood kitchen cabinets, do not require that cross-linking be complete; in such cases, lower NCO/OH ratios can be used, permitting lower cost and longer pot life [61]. Results are dependent on the process used to mix the two packages together; high, but not excessive, shear is required to obtain relatively uniform particle size with an average diameter of about 150 nm [64]. Two-component mixing spray guns are reported to give the most dependable results. Hydrophylically-modifed polyisocyanate 2K coatings give films with adequate performance for many applications, but the reduction in functionality resulting from the reaction with the polyetherpolyol results in films with inadequate properties for such applications as automotive clear coats.

REFERENCES

1. J. D. Nordstrom and A. H. Dervan, *Proc. Waterborne High-Solids Coat. Symp.*, New Orleans, 1993, p. 3.
2. C. Zweiner, L. Schmalstieg, M. Sonntag, K. Nachtkamp, amd J. Pedain, *Farbe Lack*, **97**, 1052 (1991).
3. D. A. Wicks and P. E. Yeske, *Prog. Org. Coat.*, **30**, 265 (1997).
4. G. A. Howartl, *Proc. Waterborne Higher-Solids, Powder Coat. Symp.*, New Orleans, 1996, p. 92.
5. M. Bock and R. Halpaap, *J. Coat. Techol.*, **59** (755), 131 (1987).
6. M. Sato, *J. Am. Chem. Soc.*, **82**, 3893 (1960).
7. S. W. Wong and K. C. Frisch, *J. Polym Sci. A: Polym. Chem.*, **24**, 2867; 2877 (1986).
8. K. Schwetlick and R. Noack, *J. Chem. Soc. Perkin Trans.* **2**, 395 (1995).
9. N. Hazel, I. Biggin, I. Kersey, and C. Brooks, *Proc. Waterborne High-Solids Powder Coat. Symp.*, New Orleans, 1997, p. 237.
10. J. D. Nordstrom, R. J. Barsotti, and V. L. Stolarski, *Proc. Waterborne High-Solids Powder Coat. Symp.*, New Orleans, 1997, p. 70.
11. I. Yilgor and J. E. McGrath, *J. Appl. Polym. Sci.*, **30**, 1733 (1985).
12. S. D. Seneker and T. A. Potter, *J. Coat. Technol.*, **63** (793), 19 (1991).
13. Anonymous, *Product Bull. KK-996*, King Industries, Norwalk, CT.
14. S.-G. Luo, H.-M. Tan, J.-G. Zhang. Y.-J. Wu, F.-K. Pei, and X.-H. Meng, *J. Appl. Poly. Sci.*, **65**, 1217 (1997).
15. F. W. Van der Weij, *J. Polym. Sci., A: Polym. Chem.*, **19**, 381 (1981).
16. F. Richter, R. Halpaap, and T. Engbert, U.S. Patent 5717091 (1998); F. Richter, German Patent Application, DE 19611849A1 (1997).
17. T. A. Potter and W. E. Slack, U.S. Patent 5124427, (1992).
18. R. T. Wojcik, *Polym. Mat. Sci. Eng.*, **70**, 114 (1994).
19. R. S. Dearth, H. Mertes, and P. J. Jacobs, *Prog. Org. Coat.*, **29**, 73 (1996).
20. R. Lomolder, F. Plogmann, and P. Speier, *J. Coat. Technol.*, **69** (868), 51 (1997).
21. H.-K.Ono, F. N. Jones, and S. P. Pappas, *J. Polym. Sci., C: Polym. Letters.*, **23**, 509 (1985).
22. K. Hatada, K. Ute, K.-I. Oka, and S. P. Pappas, *J. Polym. Sci., A: Polym. Chem.*, **28**, 3019 (1990).
23. V. Alexanian, R. G. Lees, and D. E. Fiori, U. S. Patent, 5254651 (1993).
24. D. E. Fiori and R. W. Dexter, *Proc. Water-Borne Higher-Solids Coat. Symp.*, New Orleans, 1986, p. 186.

25. R. T. Wojick, R. S. Blackwell, J. W. Reisch, and J. M. O'Connor, *Proc. Water-Borne Higher-Solids Coat. Symp.*, New Orleans, 1995, p. 287.

26. S. A. Jorissen, R. W. Rumer, and D. A. Wicks, *Proc. Water-Borne Higher-Solids Powder Coat. Symp.*, New Orleans, 1992, p. 182.

27. P. J. Mormile, S. Dantiki, D. Guyomard, S. Shepler, and B. M. Richards, U. S. Patent 5214086 (1993).

28. Z. W. Wicks, Jr., *Prog. Org. Coat.*, **3**, 73 (1975).

29. Z. W. Wicks, Jr., *Prog. Org. Coat.*, **9**, 3 (1981).

30. T. A. Potter, J. W. Rosthauser, and H. G. Schmelzer, *Proc. Water-Borne Higher-Solids Coat. Symp.*, New Orleans, 1986, p. 331.

31. T. Provder, *J. Coat. Technol.*, **61** (770), 33 (1989).

32. S. R. Seshadri, M. Gitlitz, and E. C. Bossert, *Proc. Waterborne High-Solids Powder Coat. Symp.*, New Orleans, 1996, p. 492.

33. S. P. Pappas and E. H. Urruti, *Proc. Water-Borne Higher-Solids Coat. Symp.*, New Orleans, 1986, p. 146.

34. J. S. Witzeman, *Prog. Org. Coat.*, **27**, 269 (1996).

35. H. R. Lucas and K. J. Wu, *J. Coat. Technol.*, **65** (820), 59 (1993).

36. M. W. Shaffer, T. A. Potter, L. D. Venham, and P. D. Schmitt, U. S. Patent 5504178 (1996).

37. R. T. Wojcik, J. M. O'Connor, H. G. Barnowski, Jr., M. J. Morgan, F. A. Stuber, R. S. Blackwell, G. H. Temme, and R. R. Wells, *Proc. Water-Borne Higher-Solids Powder Coat. Symp.*, Part 2, New Orleans, 1994, p. 474.

38. J. D. Nordstrom, *Proc. Waterborne Higher Solids Powder Coat. Symp.*, New Orleans, 1995, p. 492.

39. Th. Engbert, E. Konig, and E. Jurgens, *Farbe Lack*, **102**, 51 (1996); M. Sohoni and P. Figlioti, *Proc. Waterborne High-Solids Powder Coat. Symp.*, New Orleans, 1998, p. 267.

40. A. Essenfeld and K.-J. Wu, *Proc. Waterborne High-Solids Powder Coat. Symp.*, New Orleans, 1997, p. 246.

41. Y. Huang, G. Chu, M. Nieh, and F. N. Jones, *J. Coat. Technol.*, **67** (842), 33 (1995).

42. Z. W. Wicks, Jr. and B. W. Kostyk, *J. Coat. Technol.*, **49** (634), 77 (1977).

43. W. J. Blank, Z. A. He, E. T. Hessell, and R. A. Abramshe, *Polym. Mater. Sci. Eng.*, **77**, 391 (1997).

44. U. Rockrath, G. Wigger, and U. Poth, U. S. Patent 5516559 (1996).

45. F. M. Witte, G. Kieft, W. H. A. van den Elshout, R. Baijards, and M. Houweling, *Proc. Waterborne High-Solids Powder Coat. Symp.*, New Orleans, 1995, p. 32.

46. R. M. Guida, *Modern Paint Coat.*, July, 34 (1996).

47. G. Gardner, *J. Protective Coat. Linings*, February, 81 (1996).

48. T. M. Chapman, *J. Polym. Sci., A: Polym. Chem.*, **27**, 1983 (1989).

49. B. H. Urs, U. S. Patent 4304706 (1981).

50. G. N. Robinson, J. F. Alderman, and T. L. Johnson, *J. Coat. Technol.*, **65** (820), 51 (1993).

51. W. J. Blank, *Prog. Org. Coat.*, **20**, 235 (1992).

52. J. L. Gardon, *J. Coat. Technol.*, **65** (819), 25 (1993).

53. J. W. Rosthauser and K. Nachtkamp, "Water-Borne Polyurethanes," in K. C. Frisch and D. Klempner, Eds., *Advances Urethane Sci. Technol.*, Technomic Publ., Westport, CT, **10**, 121-162 (1987).

54. J. C. Padget, *J. Coat. Technol.*, **66** (839), 89 (1994).

55. P. L. Jansse, *J. Oil Colour Chem Assoc.*, **72**, 478 (1989).

56. R. Satguru, J. McMahon, J. C. Padget, and R. G. Coogan, *J. Coat. Technol.*, **66** (830), 47 (1994).

57. C. R. Hegedus and K. A. Kloiber, *J. Coat. Technol.*, **68** (860), 39 (1996).

58. W. J. Blank and V. J. Tramontano, *Prog. Org. Coat.*, **27**, 1 (1996).

59. D. E. Fiori, *Prog. Org. Coat.*, **32**, 65 (1997).

60. M. Dvorchak and H. Bui, *Proc. Waterborne High-Solids Powder Coat. Symp.*, New Orleans, 1998, p. 80.

61. P. B. Jacobs and P. C. Yu, *J. Coat Technol.*, **65** (822), 45 (1993).
62. M. J. Dvorchak, *J. Coat. Technol.*, **69** (866), 47 (1997).
63. C. R. Hegedus, A. G. Gilicinski, and R. J. Haney, *J. Coat. Technol.*, **68** (852), 51 (1996).
64. H. Bui, M. Dvorchak, K. Hudson, and J. Hunter, *Europ. Coat. J.*, 476 (1997).

CHAPTER 11

Epoxy and Phenolic Resins _____

Epoxy and phenolic resins are two other classes of step-growth resins important in coatings.

11.1. EPOXY RESINS

Terminology of epoxies can be confusing. Epoxy groups (also called epoxides) are three-membered cyclic ethers; in IUPAC and Chemical Abstracts nomenclature, they are called oxiranes. Most commercially important epoxy resins are derived from (chloromethyl)oxirane, more commonly known as epichlorohydrin (ECH). The resins generally contain oxiranylmethyl ethers or esters, usually called glycidyl ethers or esters. In addition to the uses discussed in this chapter, so-called epoxy esters are discussed in Section 15.8 because of their close relationship to the chemistry of alkyds; acrylated epoxy resins are discussed in Section 16.4.

| Epichlorohydrin | Glycidyl Ether | Glycidyl Ester |

11.1.1. Bisphenol A Epoxy Resins

The first epoxy resins used in coatings, and still the largest volume, are bisphenol A (BPA) epoxies made by reacting BPA with ECH. Under basic conditions, the initial reaction is formation of a BPA anion (BPA$^-$), which attacks ECH and results in the formation of a new oxirane ring with elimination of chloride anion (Cl$^-$), as shown in Scheme 11.1.

The initial product is the monoglycidyl ether of BPA (MGEBPA). Analogous reaction of the phenolic group of MGEBPA with NaOH and ECH gives the diglycidyl ether of BPA

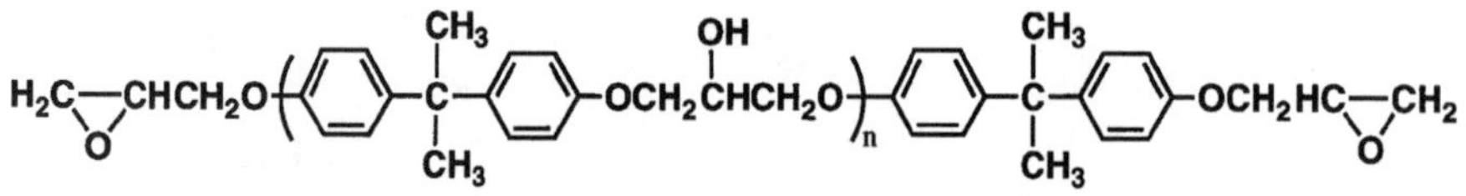

Bisphenol A Epoxy Resins

Scheme 11.1

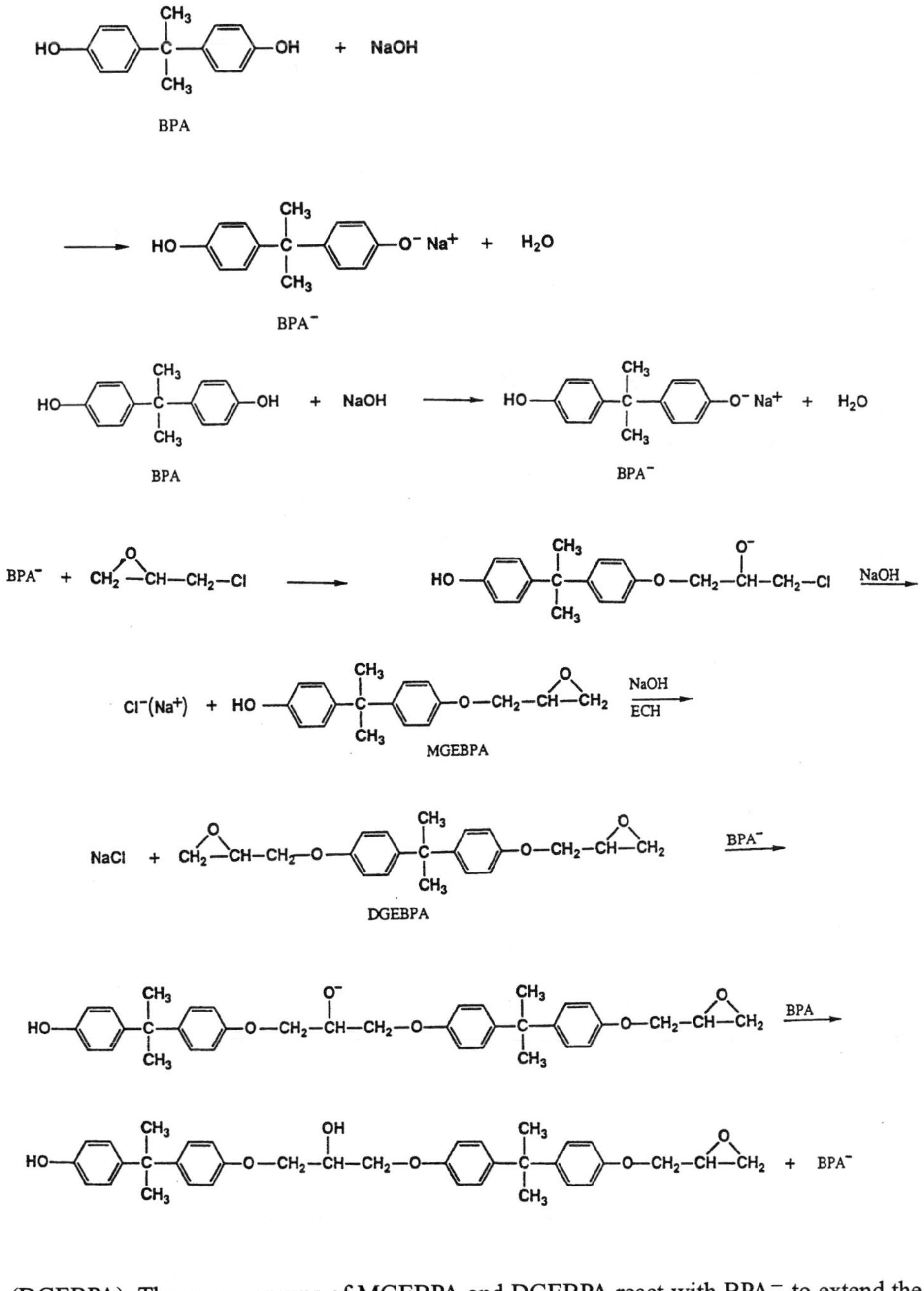

(DGEBPA). The epoxy groups of MGEBPA and DGEBPA react with BPA⁻ to extend the chain, as shown for DGEBPA; these reactions introduce alcohol groups on the backbone. Continuation of these reactions results in linear polymers, since both the BPA and ECH are difunctional. Bisphenol A epoxy resins are made with excess ECH, so the end groups are glycidyl ethers. The polymers may be represented by the following general formula, where the molar ratio of ECH to BPA determines the average n value.

Molecular weight is controlled by the ratio of ECH to BPA. With a large excess of ECH, it is possible to make a resin that is dominantly DGEBPA, that is, where $n = 0$ in the general formula. The pure $n = 0$ compound is a crystalline solid, but commercial grades are liquids with n values of 0.11 to 0.15 (so-called *standard liquid resin*). As the ratio of ECH to BPA is reduced, molecular weight and the n value of the epoxy resin increase. Viscosity also increases with molecular weight. Above an average n value of 1, the resins are amorphous solids with increasing T_g. Although the resins are said to have melting points, they do not melt in the sense that a crystalline solid does. Rather, under specified test conditions, the resins flow to some standard extent at the so-called *melting point*. The term *softening point* rather than melting point is sometimes used. Commercially available higher molecular weight resins are often designated as types 1001, 1004, 1007, and 1009. Table 11.1 gives average n values, epoxy equivalent weights (EEW), and melting points for commercial BPA epoxy resins [1]. As molecular weight increases, epoxy equivalent weight and average hydroxy functionality also increase. In some even higher molecular weight epoxy resins, the amount of epoxy groups present is so small that the resins approach being just polyfunctional alcohols, commonly called *phenoxy resins*.

Although theoretically, there should be two epoxy groups on each molecule, a variety of other end groups are present to small extents. To minor degrees, unreacted phenol and 1,2-chlorohydrin terminal groups are present; additional nonepoxy end groups result from side reactions. About 2% of the terminal groups are 1,2-glycol derivatives. These can result from a hydrolytic side reaction, probably HO^- catalyzed, of epoxy groups, as shown in Eq. 11.1. The glycol can also result from the hydrolysis of the chlorine of the epichlorohydrin to the corresponding hydroxy-substituted compound, after reaction with a phenol group, would also yield a glycol terminal group.

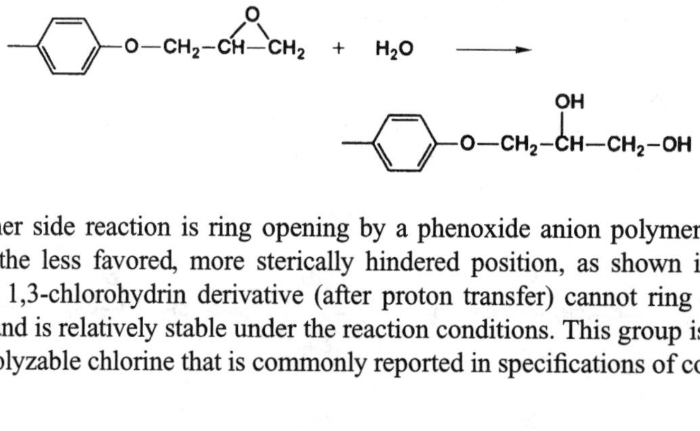

$$—\langle \rangle—O—CH_2—CH—CH_2 \quad + \quad H_2O \longrightarrow$$

$$—\langle \rangle—O—CH_2—\overset{OH}{CH}—CH_2—OH \qquad (11.1)$$

Another side reaction is ring opening by a phenoxide anion polymer end group with ECH at the less favored, more sterically hindered position, as shown in Eq. 11.2. The resulting 1,3-chlorohydrin derivative (after proton transfer) cannot ring close to give an oxirane and is relatively stable under the reaction conditions. This group is the main site of nonhydrolyzable chlorine that is commonly reported in specifications of commercial epoxy resins.

Table 11.1 Characterization of Commercial BPA Epoxy Resins

Resins	n Value	EEW	Melting Point (°C)
Standard liquid	0.13	190	Liquid
1001 Type	2	500	65–75
1004 Type	5.5	950	95–105
1007 Type	14.4	2250	125–135
1009 Type	16	3250	145–155

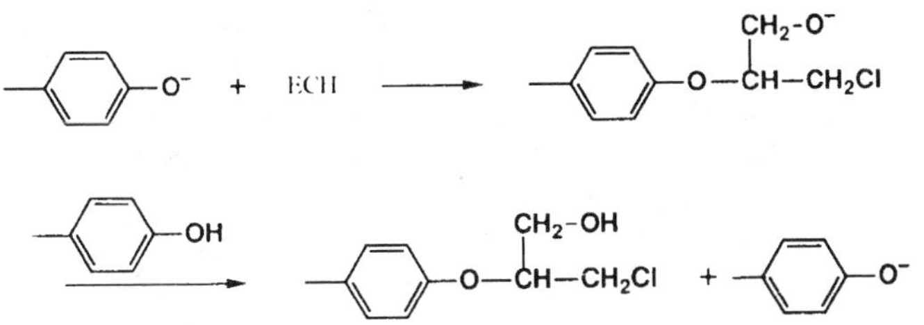

$$(11.2)$$

Instead of ring closing to give an epoxy group, the hydroxyl group of the 1,2-chlorhydrin can react with another ECH molecule to give the addition product shown in Eq. 11.3.

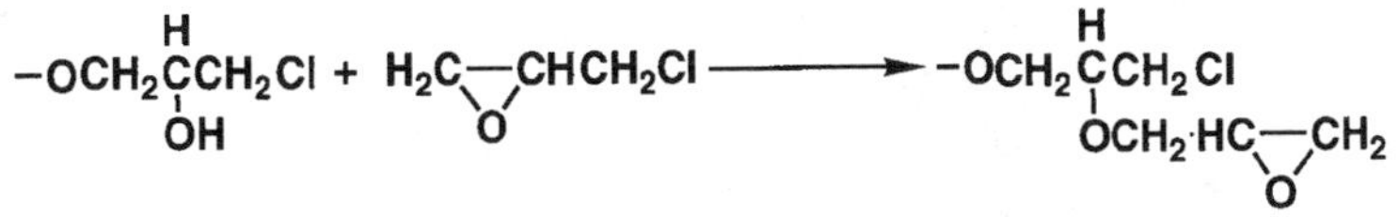

$$(11.3)$$

A hydroxyl group on the polymer backbone can also react with ECH to yield branched molecules with an epoxy functionality ($\bar{f}_n$) of three, as shown in Eq. 11.4.

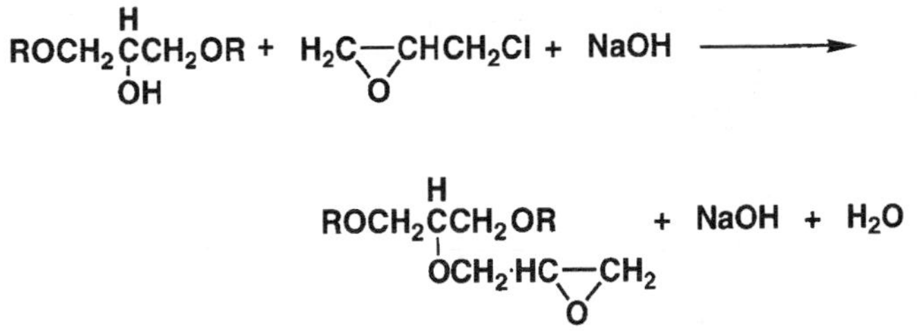

$$(11.4)$$

If there were no side reactions, the epoxy $\bar{f}_n$ of BPA epoxy resins would be 2, however, the net effect of side reactions is that commercial BPA epoxy resins generally have an $\bar{f}_n$ less than 2, commonly about 1.9. In some cases, this lower $\bar{f}_n$ can have an important effect on film properties. The presence of terminal glycol groups can give lower viscosity resins since they result from chain termination, reducing molecular weight. It has been shown that the presence of small controlled amounts of terminal glycol groups can have beneficial effects on adhesion [2].

The procedure shown in Scheme 11.1 for making BPA epoxy resins is called the *taffy process*. Stoichiometric amounts of NaOH are required, resulting in formation of a large amount of NaCl, which must be removed from the resin by washing with water. The washing step is not difficult for the standard liquid resin due to its low viscosity. However, as the ratio of ECH to BPA is decreased, higher average molecular weight products are produced, the reaction mixture becomes highly viscous, and water washing becomes difficult. Also, as molecular weight and viscosity increase, the probability of branching increases. The taffy process is now used only for resins with average n values less than 4.

Higher molecular weight resins are prepared by reacting the standard liquid epoxy resin ($n = 0.13$) with BPA in the presence of a catalyst such as ethyltriphenylphosphonium hydroxide. The catalyzed reaction of BPA with both epoxide groups of the standard resin results in a higher molecular weight resin with BPA end groups, which further react with standard resin to give epoxide end groups. The average molecular weight depends on the ratio of liquid resin to BPA. There can be variations in the resins obtained, depending on the catalyst used. This procedure for making epoxy resins is called the *advancement process*. The reaction is carried out at higher temperature; hence, the viscosity is lower, agitation is better, and there is less branching. No NaCl is produced; therefore, the product need not be washed free of salt, simplifying the manufacturing process.

Development of advanced HPLC analytical techniques has permitted improvements in process development and control. Not only can the individual oligomers ($n = 0, 1, 2, 3,$ etc.) be separated, but also the oligomers in which one or both ends have 1,3-chlorohydrin or 1,2-dihydroxy groups can be separated [3] (see Eqs. 11.1 and 11.2). Analysis of the products resulting when minor process changes are made in reaction conditions permits establishing process parameters to meet relatively narrow product specifications. Tight process control is particularly critical for epoxy resins to be used in electronic applications and powder coatings (see Chapter 27). Resins made by the taffy process consist of oligomers with $n = 0, 1, 2, 3, 4, 5,$ and so on, whereas resins made by the advancement process have largely even number n value oligomers, that is, $n = 0, 2, 4, 6,$ and so on. The predominance of oligomers with even number n values in resins from the advancement process follows from starting with the diglycicyl ether of BPA. Roughly 10 wt% of odd n molecules result from the presence of about 10% of $n = 1$ resin in the liquid resin.

Bisphenol A epoxy resins perform especially well in coatings applications in which excellent adhesion and corrosion resistance are required. A limitation of their use is poor exterior durability, primarily resulting from direct absorption of UV radiation by the aromatic ether groups, which ultimately leads to photoxidative degradation.

11.1.2. Other Epoxy Resins

A variety of other epoxy resins is available. Bisphenol F (BPF) epoxy resins (from reaction of ECH with BPF in place of BPA) have the advantage of lower viscosities at the same n value. Standard liquid BPA epoxies have viscosities on the order of 12 Pa·s at 25°C, whereas the viscosities of comparable BPF epoxies are 6 Pa·s or less.

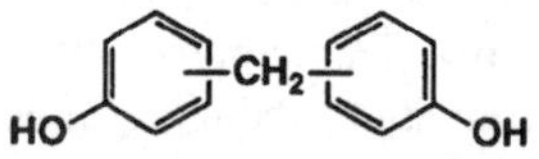

BPF

Still lower viscosities can be obtained by reacting BPA epoxies with an alcohol, such as *n*-butyl alcohol, followed by reacting the resultant hydroxyl groups with ECH, using $(CH_3)_4NCl$ as catalyst [4]. A new epoxy resin is obtained, a major component of which is 2,2'-bis[*p*-(3-*n*-butoxy-2-gylcidyloxypropyloxy)phenyl]propane. Viscosities on the order of 1 Pa·s are reported [5]. The effect of the butyl ether group is to lower the T_g and, hence, the viscosity. Cross-linked films are softer and more impact resistant than comparable films using BPA resins.

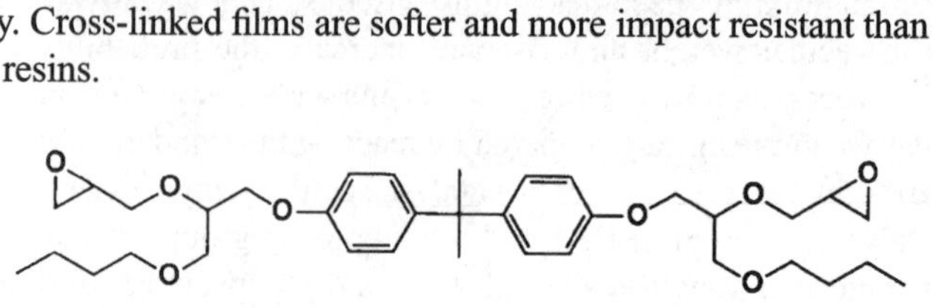

Epoxy resins with lower viscosity than BPA epoxy resins can be made by substituting a flexible diol for BPA in the advancement process to make what might be called *copolymer epoxy resins*. Examples are advancement resins made from combinations of propylene or dipropylene glycol and BPA [6]. When cross-linked with phenolic resins, these resins form films that are more flexible than films from homopolymer BPA epoxy resins, but still provide excellent adhesion. Lower T_g, lower viscosity, and more flexible epoxy resins can also be prepared from bisphenols with longer chain links between the two phenol rings, especially multiple 1,2-diether groups [7]. Experimental resins to demonstrate the idea were prepared by reacting a series of such bisphenols with DGEBPA, using ethyltriphenylphosphonium acetate/acetic acid in methyl alcohol as a catalyst.

Epoxy resins are also prepared by reaction of ECH with novolac phenolic resins. (See Section 11.6.2.) The resulting novolac epoxy resins are useful in applications in which more than two epoxy groups per molecule are desirable. Epoxy resins derived from the reaction of *o*- or *p*-cresol-formaldehyde novolacs and ECH are available, with an epoxy $\bar{f}_n$ of 2.2 to 5.5. A general structure of novolac epoxies is shown below:

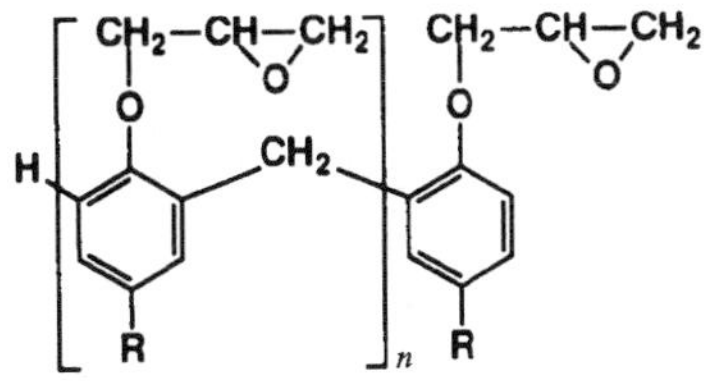

Epoxy Novolac Resin

Epoxy resins analogous to BPA resins, but with hydrogenated aromatic rings, have lower T_g and lower viscosity than do BPA resins at the same n value. Hydrogenated-BPA (HBPA) epoxy resins have better exterior durability relative to both BPA and BPF resins, resulting from the absence of the UV absorbing aromatic ether groups.

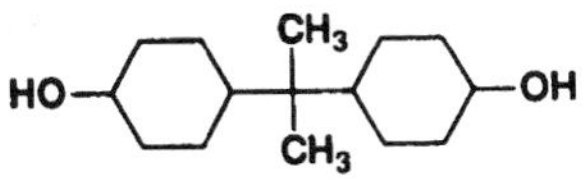

Hydrogenated-Bisphenol A Resin (HBPA)

Triglycidylisocyanurate (TGIC) is a solid trifunctional epoxy cross-linker used in powder coatings. The presence of three functional groups gives higher cross-link density than obtained with BPA epoxy resins, and photochemical stability of the cured coatings is superior. However, there is concern that use of TGIC may present toxic hazards.

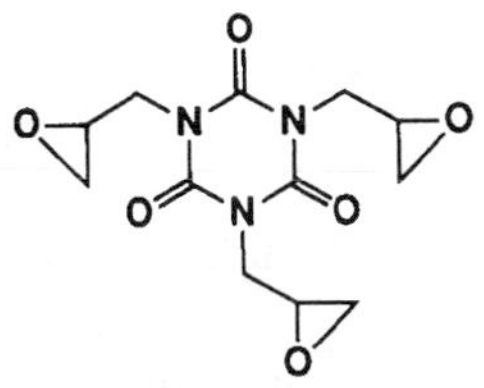

Triglycidylisocyanurate

A variety of other aliphatic epoxy products is available. Epoxidized soy and linseed oils are used in making acrylate derivatives for UV cure resins (see Section 16.4) and thermal cationic cure resins (see Section 11.3.5). Others are made by the reaction of polyols-such as glycerol, sorbitol, and polyethylene or polypropylene glycols-with ECH and a base catalyst. Such aliphatic epoxies are used as reactive diluents to reduce viscosity and give films with better exterior durability.

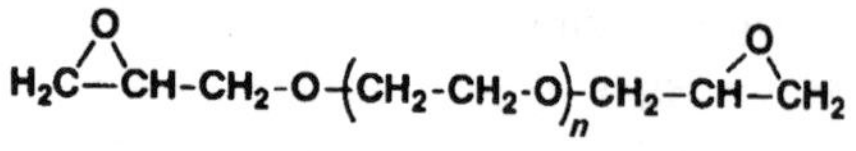

$$H_2C-CH-CH_2-O-(CH_2-CH_2-O)_n CH_2-CH-CH_2$$

Polyethylene Glycol Diglycidyl Ether

Also available are low molecular weight cycloaliphatic diepoxy compounds such as 3,4-epoxycyclohexylmethyl 3′,4′-epoxy-4-cyclohexylcarboxylate (**1**), prepared by epoxidation of the corresponding alkenes, generally with peracetic acid. Another is diglycidyl 1,2-cyclohexanedicarboxylate (**2**). Such low molecular weight epoxy derivatives are particularly useful as reactive diluents in cationic coatings (see Sections 11.3.4 and 28.1.2). They can also be used as cross-linking agents for polyols (see Section 11.3.3), carboxylic acids, and anhydrides (see Section 11.3.2).

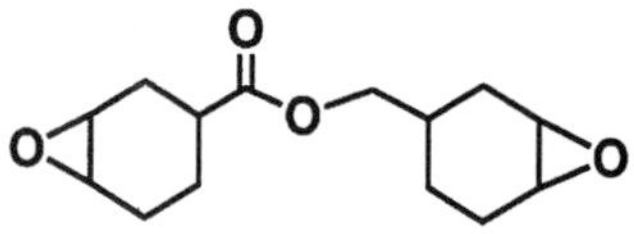

1 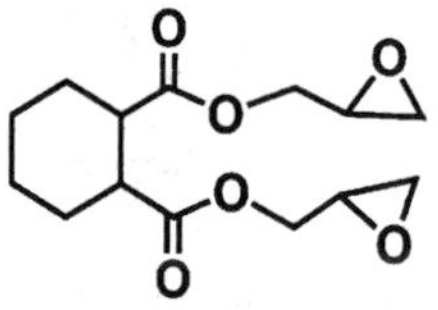   **2**

A widely used way to make epoxy-functional resins is by free radical polymerization of acrylic esters with glycidyl methacrylate (GMA) as a comonomer. By varying comonomers, GMA content, and molecular weight, a range of materials can be made. Exterior durability and acid resistance can be excellent. GMA modified acrylic resins are used in top coats for automobiles, among other applications.

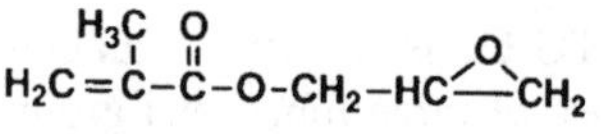

$$H_2C=C-C-O-CH_2-HC-CH_2$$

Glycidyl Methacrylate (GMA)

11.2. EPOXY-AMINE SYSTEMS

Epoxy groups react at ambient temperatures with primary amines to form secondary amines and with secondary amines to form tertiary amines. Tertiary amines react at higher temperatures to form quaternary ammonium compounds.

Reaction rates depend on epoxy and amine structure, concentration, catalysis, and media effects. Terminal epoxy groups, such as glycidyl ethers and esters, are more reactive

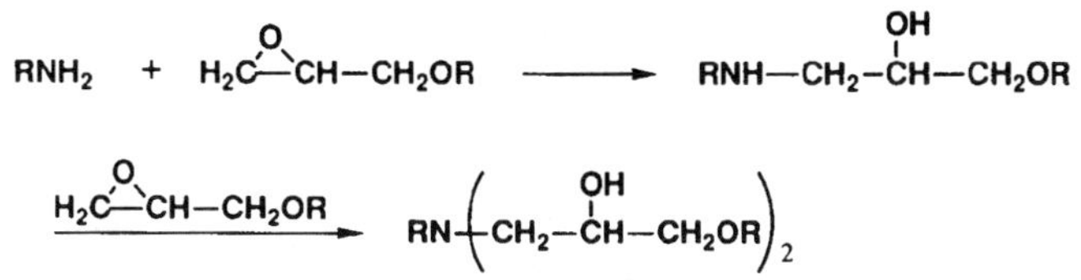

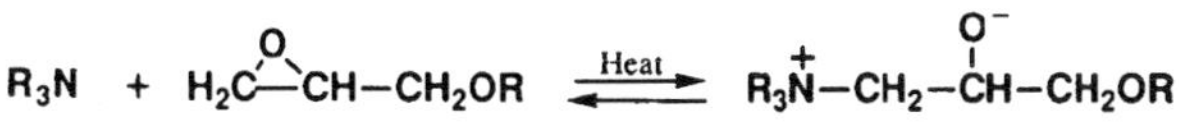

$$RNH_2 \ + \ H_2C\overset{O}{\diagdown}CH-CH_2OR \ \longrightarrow \ RNH-CH_2-\overset{OH}{\underset{|}{C}H}-CH_2OR$$

$$\underset{H_2C\overset{O}{\diagdown}CH-CH_2OR}{\xrightarrow{\hspace{3cm}}} \ RN\left(CH_2-\overset{OH}{\underset{|}{C}H}-CH_2OR\right)_2$$

$$R_3N \ + \ H_2C\overset{O}{\diagdown}CH-CH_2OR \ \underset{}{\overset{Heat}{\rightleftharpoons}} \ R_3\overset{+}{N}-CH_2-\overset{O^-}{\underset{|}{C}H}-CH_2OR$$

than internal epoxy groups [e.g., cycloaliphatic diepoxide (1)], which are more sterically hindered.

Reactivity of amines tends to increase with base strength and decrease with steric crowding. The general order of reactivity, primary > secondary ≫ tertiary amines, can be attributed to steric effects, as well as to the absence of a transferable proton in the case of tertiary amines. Cycloaliphatic amines have reduced reactivity; the second reaction of such an amine is particularly slow. Aliphatic amines are more reactive than aromatic amines, which are less basic. The reaction is catalyzed by water, alcohols, tertiary amines, and weak acids (most notably by phenols), which promote ring opening by proton complexation with the epoxide oxygen. Strong acids are not effective catalysts for the reaction. Hydrogen-bond acceptor solvents tend to reduce reaction rates, probably by complexing with hydrogen donors in competition with the epoxy group.

Scheme 11.2 shows the attack on an epoxy group by the nonbonded electron pair of a secondary amine. Attack is primarily at the less sterically hindered terminal end of the epoxy group. Scheme 11.2 also shows catalysis by a weak acid (HA) by hydrogen complexation and hydrogen donation to the epoxide oxygen, which facilitates ring opening. The catalyst is regenerated by removing the proton from the amine nitrogen, which takes on a positive charge during the reaction. The conjugate base A^- could also participate in the catalysis by assisting removal of the amine proton during the ring opening reaction. Based on this reasonable possibility, an explanation for the more effective catalysis by weak acids relative to strong acids noted earlier can be hypothesized. A strong acid (e.g., HCl) exists predominately as the amine salt ($R_2NH_2^+Cl^-$). Because of the large excess of amine reactant as compared to the catalyst, even weaker acids such as phenols are also expected to exist largely as the corresponding salts ($R_2NH_2^+ArO^-$). This suggests that the proton donor in the reaction is the same amine ion in both cases. On the other hand the proton acceptor, ArO^-, is a substantially stronger base than Cl^- because conjugate base strength increases as acid strength decreases. The proton acceptor ArO^- more effectively promotes the second reaction shown in Scheme 11.2. Although we are unaware of experimental evidence directly supporting this hypothesis in epoxy-amine reactions, the general importance of concerted weak acid-weak base catalysis is well established [8]. 2,4,6-[Tris(dimethylaminomethyl)]phenol, which has both phenolic and tertiary amine groups, is an important catalyst for epoxy-amine reactions.

Scheme 11.2

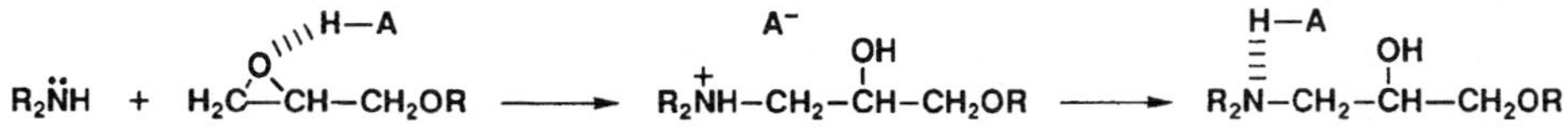

$$R_2\overset{..}{N}H \ + \ H_2C\overset{O^{\prime\prime\prime\prime}H-A}{\diagdown}CH-CH_2OR \ \longrightarrow \ R_2\overset{+}{N}H-CH_2-\overset{OH}{\underset{|}{C}H}-CH_2OR \ \xrightarrow{\hspace{1cm}} \ R_2N-CH_2-\overset{OH}{\underset{|}{C}H}-CH_2OR$$

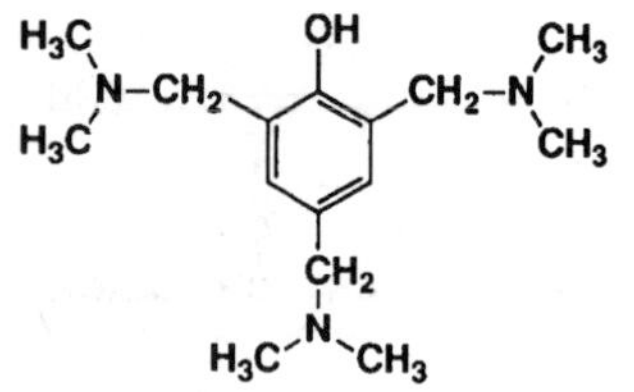

2,4,6[-tris(dimethylamino)]phenol

Epoxy-amine systems are important in several coatings applications. Epoxy-amine reactivity is too high at ambient temperatures to allow sufficient storage stability of a coating containing polyamine and polyepoxide in the same package; thus two package (2K) coatings are required. With many aliphatic amines the pot life, is limited to a few hours, and the coating generally takes about a week to cure at ambient temperature.

11.2.1. Pot Life and Cure Time Considerations

Epoxy-amine coatings are formulated to maximize pot life and minimize cure time. (See Section 2.3.2 for a discussion of kinetic limitations that prevent both long-term, one-package stability and moderate cure temperatures and of ways to obviate these limitations.) Many factors must be considered, including reactive group concentrations; the structural effects of amine, epoxy, and solvents on reaction rates; the equivalent and molecular weights; and $\bar{f}_n$ of the reactants. As the molecular weight of a BPA epoxy resin is increased, the number of equivalents per liter of epoxy groups decreases; therefore, the reaction rate is slower. Furthermore, as the molecular weight increases, the viscosity increases; thus, to formulate to the same viscosity, the amount of solvent must be increased. This decreases concentration of both the amine and epoxy groups and lengthens pot life. Unfortunately, the need to reduce VOC emissions forces formulation of higher solids coatings, which have shorter pot lives.

An approach to the problem of lengthening pot life without importantly reducing cure time is to use blocked amine cross-linkers. Ketones react with primary amines to give ketimines, which do not readily react with epoxy groups. However, ketimines hydrolyze with water to release free amine plus ketone; the reverse reaction of ketimine formation. Ketimine-epoxy moisture cure coatings are stable in a moisture-free environment, but cure after application and exposure to ambient moisture. Most commonly, methyl ethyl ketone (MEK) is used. The high volatility of MEK from thin coating films minimizes reverse reaction with amine.

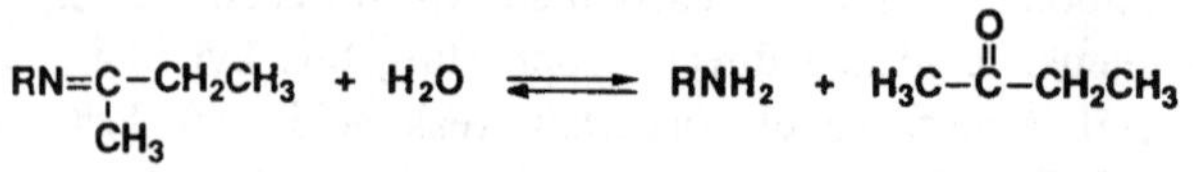

Ketimine   MEK

Ketimine-epoxy systems are stable indefinitely in the absence of water and should, therefore, permit formulation of one package coatings. Nevertheless, they are most commonly used in long pot life two package coatings; because of the difficulty of drying all of the components in a coating. Many epoxy-amine coatings are pigmented, and pigment surfaces have a layer of water on them before they are dispersed. The water remains in the coating and hydrolyzes the ketimine. While removal of water from the surface of the pigment is possible, it adds to the cost. The amount of water is not high, usually less than 1% of the pigment weight, but the molecular weight of water is so low that a little water will hydrolyze a large amount of ketimine. Solvents also contain water, and use of anhydrous solvents would also increase costs. A further difficulty is that, cure rate depends upon humidity.

An interesting approach to increasing storage stability of epoxy coatings sufficiently to permit formulation of one package coatings that can be cured at moderately elevated temperatures is to use a cross-linker that must undergo a phase change in order to react. A solid polyamide prepared from phthalic anyhydride and diethylenetriamine [DETA, $HN(CH_2CH_2NH_2)_2$] is an example [9]. When dispersed in an epoxy resin and heated above 100°C, the solid polyamide liquefies and reacts to form imides, releasing amine cross-linkers, including free DETA. Because the polyamide is a solid during storage, its reaction rate is not controlled by the kinetics of the reaction, so Arrhenius restrictions on package stability versus cure schedule (see Section 2.3.2) do not apply.

A commonly used cross-linker for epoxy powder coatings is dicyandiamide (dicy), a crystalline compound (mp 205°C) that also provides latency by insolubility. Dicyandiamide decomposes into soluble reactive products below its melting point, and cure temperatures can be reduced substantially by using as accelerators tertiary amines (e.g., benzyldimethylamine), imidazoles (e.g., 2-methylimidazole), or a substituted urea [e.g., *N*-3-(*p*-chlorophenyl)-*N'*,*N'*-dimethylurea] that releases free amine at elevated temperatures. The complex reactions of epoxies with dicy have been the subject of numerous studies [10], but are not fully understood.

Dicyandiamide (dicy)

11.2.2. Toxic and Stoichiometric Considerations

Other factors are involved in selecting amines besides reaction rates. Many amines are toxic. While they are easily handled safely in a chemical plant, toxic hazards can arise if relatively inexperienced, careless, or uninformed personnel mix and apply 2K epoxy coatings incorporating some amines. For example, diethylenetriamine (DETA) is an efficient cross-linker for epoxies, but the handling hazard is high. In general, toxic hazards are reduced by increasing molecular weight and reducing water solubility. As molecular weight increases, volatility decreases, which reduces the chances of inhaling dangerous amounts of amine. Also, as water solubility decreases and molecular weight increases, permeability through body membranes such as skin decreases, generally reducing the toxic hazard. There are, of course, exceptions to these broad generalizations. Safety data sheets supplied by the manufacturer should always be read for safe handling recommendations.

Other disadvantages of using low molecular weight, highly functional amines are their low equivalent weights and viscosities. The equivalent weight of pure DETA is 21. If DETA were used with an epoxy resin with an equivalent weight of about 500, the stoichiometric weight ratio of the two components would be about 25:1. This disparate amount would increase the difficulty of obtaining proper mixing, and result in a risk of error in mixing stoichiometric amounts in two package coatings. The difference in viscosity between the DETA and epoxy resin also makes uniform mixing difficult.

One approach to designing amine cross-linkers with higher equivalent weight and lower toxic hazard is to make so-called *amine adducts*. Standard liquid BPA epoxy ($n = 0.13$) is reacted with an excess of a multifunctional amine such as DETA. The excess amine is removed by vacuum distillation, leaving the amine-terminated adduct, as shown in the idealized structure below. The low molecular weight DETA is handled only in the chemical factory with proper precautions. Similar adducts are prepared using a variety of amines to provide adducts with a range of cure rates and pot lives.

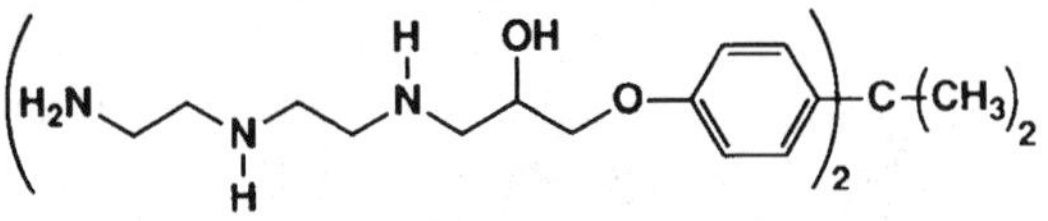

Epoxy Amine Adduct

Conventional amine curing agents are unsatisfactory when it is necessary to cure an epoxy coating at temperatures below 5°C—the cure rate is too slow. This problem was addressed by using amine Mannich bases, prepared by reacting a methylolphenol with excess polyamine [11]. Although the functionality of the amine is reduced, the presence of the phenolic hydroxyl accelerates the epoxy/amine reaction.

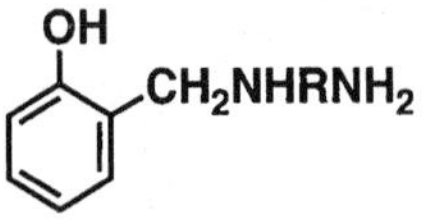

R = cycloalkyl
Mannich base

Another approach is to react a multifunctional amine (e.g., DETA) with aliphatic mono- or dicarboxylic acids to form amine-terminated amides. Dimer fatty acids are widely used; they are complex mixtures, predominantly C_{36} dicarboxylic acids, made by acid catalyzed dimerization of unsaturated C_{18} fatty acids (see Section 14.3.1). The reaction yields amine-functional polyamides. Terminology is confusing, the amides derived from dimer acids are frequently just called polyamides. The amide nitrogen groups do not react with epoxy groups; only the terminal amine groups do. (There are similar polyamides that are not amine-functional and not useful as cross-linkers for epoxy resins.) An example is made by reacting dimer acids with excess DETA. A mixture of products is obtained, including the simplest amino-amide, shown below. The amine equivalent weight is the corresponding molecular weight divided by the number of reactive NH groups (six in the polyamide shown). A range of polyfunctional amines is used in making polyamides, including DETA, triethylenetetramine (TETA), and aromatic diamines, such as *m*-phenylene diamine. Lower viscosity reactants can be obtained by reacting monocarboxylic acids and DETA; they are frequently called amido-amines.

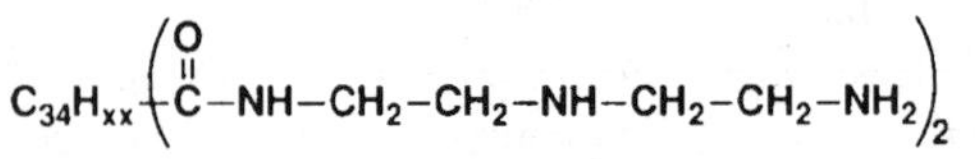

Polyamide

In the preparation of polyamides using an amine such as DETA, the amino-amide can react further by eliminating water and forming a terminal imidazoline group. The extent of this reaction varies, depending upon reaction conditions. Imidazoline formation reduces viscosity, but also reduces average functionality and cross-link density of cured films [12,13].

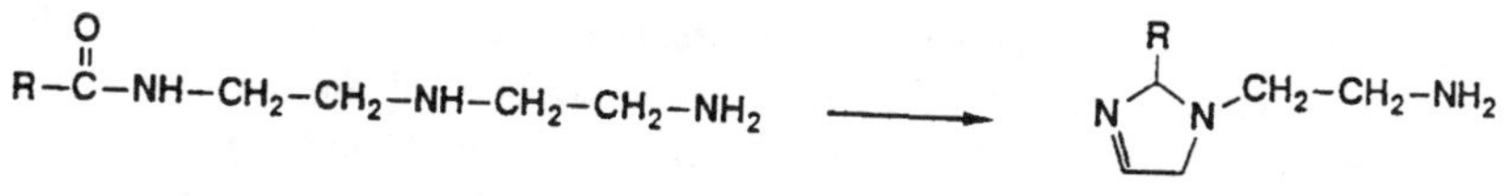

$$R-\overset{\overset{\displaystyle O}{\|}}{C}-NH-CH_2-CH_2-NH-CH_2-CH_2-NH_2 \longrightarrow$$

Imidazoline

11.2.3. Graininess and Blushing

Bisphenol A epoxy resins and polyamides are mutually soluble in the solvents used in epoxy-amine coatings, but most are not compatible in the absence of solvents. Thus, as solvent evaporates, phase separation can occur, resulting in a rough surface, called *graininess*. Graininess can be avoided by allowing the coating to stand for 30 minutes to 1 hour after the two packages are mixed. Partial reaction of the two components takes place during the waiting period. The elimination of graininess can be attributed to increased viscosity and/or the formation of reaction products that improve the compatibility of the mixture. When liquid standard resin is used, longer times are required after mixing to avoid phase separation. Proprietary amine cross-linkers are available that exhibit better compatibility, minimizing the problem [13].

Another phenomenon, called *blushing* or *scumming*, is the appearance of a greyish, greasy deposit on the surface of films, usually accompanied by incomplete surface cure. Low temperature, high-humidity conditions increase the probability of blushing. Blushing decreases gloss, increases yellowing, gives poor recoatability, and may interfere with intercoat adhesion. Blushing is said to result from formation of relatively stable carbamate salts of some amine groups on exposure to carbon dioxide and water vapor in the atmosphere [14].

$$RNH_2 + CO_2 + H_2O \longrightarrow RNHCO_2H \xrightarrow{RNH_2} RNH_3^+ \ ^-OCONHR$$

As with graininess, it is often possible to minimize blushing by mixing the epoxy and amine components 1 hour or so before application. During the waiting period, some of the most reactive amine groups (those most likely to form carbamates) react with epoxy groups, so when the film is applied after this waiting period, blushing is less likely to occur. If the waiting period is too long, viscosity increase is excessive. Amine adduct and Mannich base epoxy coatings show little if any blushing.

The term blushing is also used to describe other phenomena, for example, the whitish appearance sometimes obtained when coatings with fast evaporating solvents are sprayed on humid days.

11.2.4. T_g Considerations

It is critical to select a combination of epoxy resin and amine cross-linker that will give a final T_g that permits relatively complete reaction of the amine and epoxy groups at the temperature encountered during application. As polymerization and cross-linking proceed T_g increases. If the T_g of the final reacted coating is higher than the cure temperature, the reaction becomes very slow when a small fraction of the functional groups are unreacted. As the T_g of a homogeneous network approaches the cure temperature, $(T - T_g)$ and free volume decrease, and the reaction rate becomes limited by the mobility of the reactants, rather than by their reactivity. If the T_g approaches a value of about 40 to 50°C above the reaction temperature, reaction essentially ceases. (See Section 2.3.3.) Unreacted functional groups cause adverse effects on mechanical properties and solvent resistance. If a coating that cures well when applied to an offshore oil rig in the Caribbean is applied to the support structure of an oil rig in the North Sea, where the water temperature even in summer does not exceed 4°C, adequate curing may not occur. Water plasticizes epoxy-amine coatings, lowering T_g. A possible example of the effect of T_g is a report that, after 7 days curing at 25°C, films from butyl ether modified BPA epoxy resin (see Section 11.1.2) cross-linked with an amine adduct had better methyl alcohol resistance than films made with the same amine adduct and BPA epoxy resin [5]. There is no evident explanation for these results based on the chemical compositions. The superior methyl alcohol resistance may have resulted from a greater extent of reaction of the lower T_g butylated derivative, before mobility limitation slowed down the reaction.

11.2.5. Other Formulating Considerations

If a coating is to be applied under water, it is essential that the polyamine be insoluble in water and that solubility of water in the polyamine be minimal. Epoxy-amine coatings tend to exhibit limited solvent resistance and are particularly susceptible to attack by acidic solvents, such as acetic acid. At least in part, this sensitivity results from diffusion of acetic acid into the film, followed by formation of acetate salts with amine groups. The hydrophilic salt groups increase the solubility of water in the film, increasing water permeability and softening the film, making it more susceptible to damage. This situation is exacerbated when cross-link density is low. Bisphenol A epoxy resins have an $\bar{f}_n$ of about 1.9, so even with highly functional amine components, cross-link density is limited, especially if there is deviation from stoichiometry. Since the amines are usually polyfunctional, it is generally best to formulate with a small excess (about 10%) of amine cross-linker (about 10%) to assure that the epoxy groups are fully reacted. The problem can be alleviated by use of higher functionality epoxy resins such as novolac epoxies, which have $\bar{f}_n$ up to five. With the higher functionality and more viscous novolac epoxies, care must be exercized in selection of an amine cross-linker to ensure adequate reaction. Sometimes, blends of BPA and novolac epoxy resins are used.

Another factor to be considered when formulating epoxy-amine coatings is the effect of solvent composition on the coating. As noted earlier, hydrogen-bond acceptor solvents extend pot life. However, esters should be avoided, since they undergo aminolysis, especially with primary amines, at room temperature. Alcohols and water catalyze the reaction and also can react with epoxy groups, affecting the stability of the epoxy component package. Alcohols react very slowly with epoxy groups at room temperature. When the reactant is a monoalcohol, there is little change in viscosity. Formulators generally judge package stability by changes in viscosity, but in this case, there could be a change in epoxy functionality without a significant change in viscosity. To the extent that epoxy groups react with an alcohol, the potential for cross-linking is decreased. The $\bar{f}_n$ of a BPA epoxy resin is somewhat less than 2 and further loss of functionality could cause the final film to have inferior properties. Apparently, in many systems, there is no difficulty. For example, some epoxy resins are sold as solutions in glycol monoethers. However, in other cases, a decrease in epoxy content has been observed after storage over a period of several months.

A similar situation can arise with water. Again, reaction of an epoxy group with water consumes epoxy groups. In some cases, in which the epoxy package is pigmented with TiO_2, it has been shown that epoxy content decreases with storage time, presumably as a result of the reaction of water from the surface of the TiO_2. This reaction, as well as that with alcohols, may be catalyzed by basic and/or acidic impurities or by alumina (basic) and silica (acidic), generally present as surface treatments on TiO_2. In most cases epoxy-package pigmented coatings have been used without problems. Apparently, the effects of both alcohols and water on package stability are system dependent. In view of such reports, it is recommended that package stability be checked periodically by epoxy group analysis rather than by relying solely on monitoring viscosity changes.

11.2.6. Water-Borne Epoxy-Amine Systems

To reduce solvent content, water-borne epoxy-amine coatings have been widely investigated. One approach has been to make emulsion systems. Incorporation of emulsifying agents in either or both the amine and the epoxy package permits the addition of water during the mixing. Application solids of a water-borne coating are lower than those of an organic solvent-soluble system, but the VOC content is lower, since a substantial fraction of the volatiles is water. Some loss of epoxy groups can be expected during the pot life of the mixed coating due to the high water content. Proprietary "self-emulsifiable" epoxy resins and polyamides are available; properties approaching those of solvent-borne coatings can be achieved [15]. These are made by reacting a surfactant with the epoxy resin or polyamide.

Another approach to water-borne systems, is to use salts of amine-functional resins. The amine groups of a concentrated solution of the resin in organic solvents are neutralized with hydrochloric acid. When the solution is diluted with water, polymer aggregates, swollen with solvent and water, are formed with the amine salt groups on the outer periphery of the aggregates suspended in a continuous water phase. The behavior is analogous to that of water-reducible resins, discussed in Section 12.3. When a solution of epoxy resin in solvent is mixed into the system, it goes inside the resin aggregates. The epoxy groups are, thus, kept separate from the amine hydrochloride groups, permitting a pot life of several days. After the coating is applied, the water and solvent evaporate, leaving the amine hydrochloride and epoxy groups in the same phase. They react to yield a chlorohydrin and a free primary amine. Then, the amine can react twice with two additional epoxy groups. Since BPA epoxy resins have less than two epoxy groups per molecule and about one in three of these groups will be converted to chlorohydrins, it is desirable to use at least some novolac epoxy with $\bar{f}_n$ up to five. Then, when a third of the groups are converted to chlorohydrins, there is still an average of more than two epoxy groups per molecule to cross-link with amine groups.

A novel approach is to use a weakly acidic solvent, such as nitroalkanes [16, 17]. Nitroalkanes form salts of amines, as shown in the equation below; the salt groups stabilize epoxy-amine emulsions and allow the system to be reduced with water. Following application, the nitroalkane solvent evaporates shifting, the acid-base equilibrium to the free amine, that is, to the left in the equation. Thus, the amine-nitroalkane combination functions as a transient emulsifying agent that stabilizes the emulsion during storage and, yet, is not present to adversely effect the final film properties. Conversion of amine groups to salts prolongs pot life of the mixed composition, since polar salt groups are oriented outwards into the water phase, while the epoxy groups are in the interior of the emulsion particles.

$$R-NH_2 \; + \; R_2CH-NO_2 \; \rightleftharpoons \; R-\overset{+}{N}H_3 \quad R_2C=NO_2^-$$

Resins used in automotive primers applied by cathodic electrodeposition are prepared by the reaction of epoxy and amine resins. The use of aqueous dispersions of carboxylate salts of such resins is discussed in Section 26.2.

11.3. OTHER CROSS-LINKING AGENTS FOR EPOXY RESINS

11.3.1. Phenols

BPA epoxy resins can be cross-linked with phenolic resins; both resole and novolac phenolic resins, discussed in Sections 11.6.1 and 11.6.2, respectively, can be used. The reaction with phenols occurs predominantly at the less hindered CH_2 site of unsymmetrical epoxides (e.g., glycidyl derivatives), as shown. The reaction is acid catalyzed; pTSA and phosphoric acids are commonly used.

The phenolic hydroxyl groups of both resole and novolac phenolics react with epoxy groups. In addition, the methylol groups of the resole phenolics undergo self-condensation and probably also react with hydroxyl groups on the epoxy resin. Thus, cross-link density is higher with resole phenolics. There is also reaction with the hydroxy groups formed by the reaction with phenols. Bisphenol A epoxies or polyfunctional epoxy resins (e.g., novolac epoxies) can be used. The coatings require baking, and package stability is relatively limited. Package stability is enhanced with etherified resole resins. Increased solids and high functionality are reported using butoxymethylolated BPA as the phenolic resin [5]. For other examples, see Section 11.6.3.

Unpigmented epoxy-phenolic coatings are used as linings for beverage cans and for some types of food cans. Concern has been raised because of the possible endocrine disruption by free bisphenol A, an estrogen mimic. Studies are underway to determine whether trace amounts of BPA are extracted in food or beverage cans from BPA epoxy containing can linings [18]. Pigmented epoxy-phenolic coatings are used as high performance primers. In both cases, the major advantages are adhesion to metals, even in the presence of water, and complete resistance to hydrolysis. In both of these applications, neither the discoloration that occurs on baking, nor poor exterior durability is important.

Water-borne epoxy-phenolic emulsion coatings have been developed. Since conventional surfactants remain in the final film and detract from water resistance, water-soluble amine salts of methacrylic acid copolymers have been recommended [19]. For example, an amine salt of a copolymer of 40 : 20 : 20 : 20 methacrylic acid/methyl methacrylate/ethyl acrylate/styrene was shown to be an effective emulsifying agent. A further approach to water-borne epoxy-phenolic coatings is to prepare carboxylic acid-functional derivatives [20]. For example, a polyalkylidene phenol can be reacted with a BPA epoxy resin and then, with formaldehyde and a haloacetic acid. The product has carboxylic acid, epoxy, and methylol groups. Aqueous dispersions are prepared using a tertiary amine.

11.3.2. Carboxylic Acids and Anhydrides

Carboxylic acids are effective cross-linkers for epoxy coatings. Reference [21] provides a review of the patent literature. The reaction of a carboxylic acid and an epoxy group yields a hydroxy ester. Ring opening occurs predominantly at the less hindered CH_2 carbon, although reaction at the more hindered CH–R site is significant.

$$R'-\overset{\overset{\text{O}}{\|}}{C}-OH \;+\; H_2C\overset{\text{O}}{\diagup}CH-R \longrightarrow R'-\overset{\overset{\text{O}}{\|}}{C}-O-CH_2-\overset{\overset{\text{OH}}{|}}{CH}-R \tag{11.5}$$

The reaction has been shown to be second order in carboxylic acid. Most likely, one of the acid groups functions by nucleophilic attack at the CH_2 group, whereas the other functions as an electrophile and assists ring opening by complexation with the epoxy oxygen (see Scheme 11.2.) The second-order dependence of reaction rate on acid concentration also results in a rapidly decreasing rate with conversion, exacerbating the difficulty of achieving high conversions.

$$\text{Rate} = k[\text{epoxy}]\,[\text{RCOOH}]^2$$

The third-order dependence of the reaction (first order in epoxy and second order in carboxylic acid) also results in a small Arrhenius A value owing to the high molecular ordering of the epoxide and two carboxylic acid groups required in the transition state, reducing reactivity at all temperatures. Glycidyl methacrylate copolymers and cycloaliphatic epoxides such as (1) and (2) react more rapidly than BPA epoxies. Tertiary amines catalyze the reaction of carboxylic acids with epoxies. Triphenylphosphine is also reported to be a particularly effective catalyst. With triphenylphosphine catalysis and an excess of epoxy groups to carboxylic acid groups, coatings can be formulated that cross-link at 25°C [22].

Latent amine catalysts have been designed for powder coatings formulated with BPA epoxy resins and carboxylic acid-functional polyesters [23]. The catalysts are crystalline solid amic acids (e.g., the amic acid derived from the reaction of 3-methylphthalic anhydride and N,N-dimethylaminopropylamine). The amic acids are zwitterions, in which the tertiary amine is protonated, minimizing both their solubility in the powder coating and their catalytic activity. On heating, the amic acid melts and undergoes intramolecular cyclization to an imide with the elimination of water. The tertiary amine group on the cyclic imide catalyzes the carboxylic acid-epoxy reaction.

Reaction of hydroxyl groups with epoxies competes with that of carboxylic acid groups. When carboxylic acids are used as cross-linking agents, part of the cross-links result from the reaction of epoxies with hydroxyl groups originally present on the epoxy resin or generated in the epoxy-carboxylic acid reaction. Esterification of carboxylic acid groups with hydroxyls may also occur.

Carboxylic acid-functional acrylics can be cross-linked with BPA epoxy resins. Acrylic copolymers with pendant epoxy groups, made using glycidyl methacrylate (GMA) as a comonomer, cross-link with carboxylic acid-functional acrylate copolymers [21]. Self cross-linking acrylics can be made by incorporating both (meth)acrylic acid and GMA in the same polymer. While the reaction rate of epoxy groups with carboxylic acid groups at storage temperatures is slow, it is not zero. When self cross-linking resins are made, the time of storage stability starts when the resin is made. On the other hand, when the functional groups are on two different resins, the time of storage stability does not start until the resins are mixed to make the liquid coating.

Closely related to carboxylic acids is use of cyclic anhydrides. See Ref. [21] for a review of the patent literature. Reaction of anhydrides with epoxy resins can occur initially with the epoxy resin hydroxyl groups, yielding esters and carboxylic acids. The resulting carboxylic acid groups then react with epoxy groups. This reaction generates a new hydroxyl group, and so on. Epoxy groups can also react directly with anhydrides. Tertiary amines are generally used as catalysts; they probably function primarily by reacting with the epoxy to form a transient zwitterion, which then reacts with the anhydride.

11.3.3. Hydroxyl Groups

BPA epoxy resins are not sufficiently reactive with hydroxyl groups to be useful as cross-linkers for hydroxy-functional resins, although when BPA epoxies react with resole phenolic resins, carboxylic acids, and anhydrides, there is also reaction with hydroxyl groups. However, with proper catalysis, cycloaliphatic epoxies serve as cross-linking agents for polyols for films baked at 120°C. This reactivity has been used to formulate water-borne coatings based on caprolactone polyols and epoxycyclohexylmethyl epoxycyclohexyl carboxylate (1) with diethylammonium triflate as a blocked catalyst [24].

BPA epoxy resins can be cross-linked by reactions of their hydroxyl groups. Both MF and UF amino resins are used, cross-linking occurs mainly between the activated ether groups of the MF or UF resin and the hydroxyl groups of the epoxy resin by transetherification (see Chapter 8). Generally, amine salts or esters of pTSA or another sulfonic acid are used as latent catalysts. Polyisocyanates also cross-link the hydroxyl groups of epoxy resins. Blocked isocyanates are generally preferred because they permit one-package stability. (See Section 10.5.).

11.3.4. Mercaptans

Mercaptans (thiols, RSH) react with epoxies to yield sulfides. The reaction is strongly catalyzed by tertiary amines, which convert the mercaptan into the more highly reactive mercaptide anion RS^-.

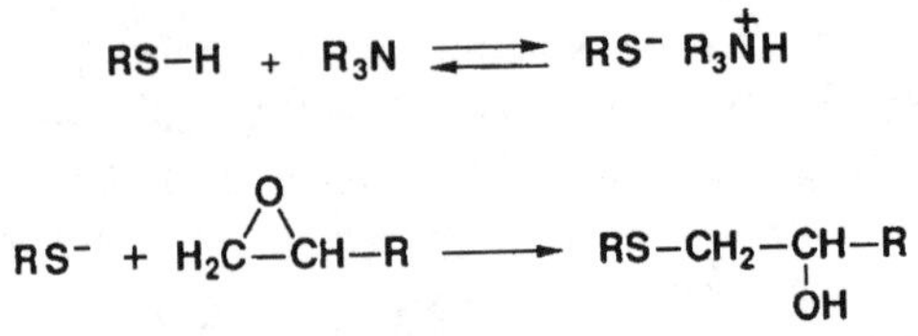

$$RS-H + R_3N \rightleftharpoons RS^- \; R_3\overset{+}{N}H$$

$$RS^- + H_2C\overset{O}{\overset{\triangle}{-}}CH-R \longrightarrow RS-CH_2-\underset{OH}{CH}-R$$

Reactivity of the mercaptide anion is sufficient for ambient temperature cure 2K coatings. Low molecular weight polysulfide rubbers are mercaptan-terminated polymers; they have been used to cross-link with BPA epoxy resins in primers for aircraft. The unpleasant odor of mercaptans is a drawback for some applications.

11.3.5. Homopolymerization

Epoxy groups undergo homopolymerization to polyethers in the presence of tertiary amines, Lewis acids, and very strong protic acids (super acids). Acid precursors are most commonly used as initiators. There are two types: blocked or latent acids that undergo thermal decomposition to give the free acid and photoinitiators that release acid on exposure to UV. UV curable epoxy coatings are discussed in Section 28.3.1. The following equation shows initiation and the first step of polymerization:

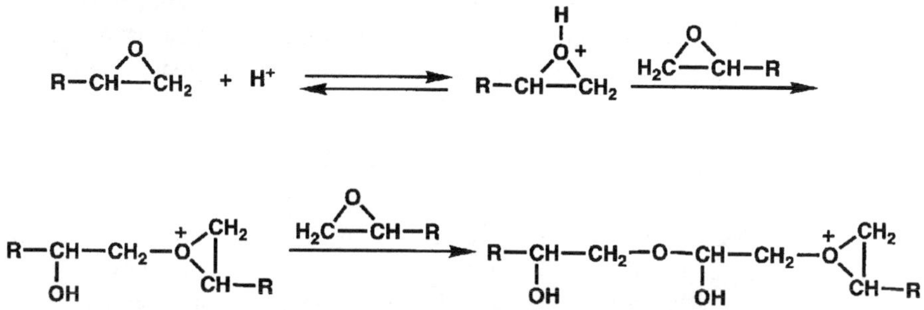

Suitable super acids are trifluoromethylsulfonic acid (triflic acid) (F_3CSO_3H), hexafluoroantimonic acid ($HSbF_6$), hexafluoroarsenic acid ($HAsF_6$), and hexafluorophosphoric acid (HPF_6). Since strong acids have weak conjugate bases, the corresponding counter ions are nonnucleophilic. Only super acids are effective for homopolymerization of epoxies. Relatively strong acids, such as HCl and pTSA, are ineffective because the conjugate bases of such acids are nucleophilic enough to add to the protonated epoxy group, preventing addition of a second epoxy group, as in the homopolymerization reaction. The result is addition to form a chlorohydrin (or sulfonate ester), as shown in Eq. (11.6), rather than polymerization.

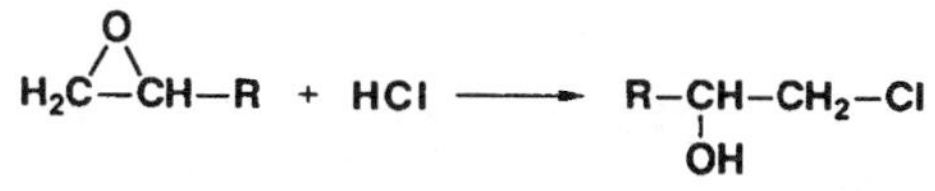

$$\text{(11.6)}$$

Homopolymerization can also be used for thermosetting coatings. α,α-Dimethylbenzylpyridinium hexafluoroantimoniate is a blocked catalyst that permits curing at 120°C of a GMA copolymer using (**1**) as a reactive diluent while retaining adequate pot life [25]. Cycloaliphatic epoxides such as (**1**) have been used along with polyols in thermal cationic cure coatings; part of the cross-linking is from homopolymerization and part from reaction with hydroxyl groups [26]. Epoxidized linseed oil can be added to increase impact resistance.

11.4. WATER-REDUCIBLE EPOXY ACRYLIC GRAFT COPOLYMERS

A large scale use of epoxy resins is to make acrylic graft copolymers for interior linings of beverage cans [27,28]. To prepare the graft copolymers, a solution of a BPA epoxy resin in a glycol ether solvent is reacted with ethyl acrylate, styrene, and methacrylic acid using benzoyl peroxide as initiator. The reaction is carried out at about 130°C, at which, both benzoyloxy and phenyl radicals are generated. These radicals can initiate polymerization, but also abstract hydrogens from the epoxy resin [29]. Abstraction of a hydrogen results in a free radical on the epoxy resin backbone, which serves as an initiating site for polymerization of the vinyl comonomers. Thus, a graft copolymer is formed with acrylic/styrene side chains substituted with carboxylic acid groups (from the methacrylic acid). The product is a mixture of epoxy/acrylic graft copolymer, nongrafted acrylic copolymer, and unreacted epoxy resin. The complex resin mixture is neutralized with an amine such as dimethylaminoethanol. Class I MF resin is added as a cross-linker, and the system is diluted with water. The result is a dispersion that is used as a spray applied coating for the interior of two piece beverage cans. Since the water-solubilizing groups are attached by C–C bonds rather than by ester groups, the resins are resistant to hydrolysis during storage. Sometimes, a latex is blended with the dispersion to reduce cost.

11.5. EPOXY RESIN PHOSPHATE ESTERS

Phosphoric acid reacts with BPA epoxy resins to generate phosphate esters. Complex reactions occur; the predominant product is the monophosphoric acid ester of the primary alcohol, but other products are also present [30]. Most epoxy groups are hydrolyzed during the reaction to give the corresponding dihydroxyl derivative. Low molecular weight *epoxy phosphates* are used as adhesion promoters. Higher molecular weight epoxy resins can also be modified by reacting with minor amounts of phosphoric acid and water. Use of phosphoric acid-modified epoxy resins in epoxy-phenolic formulations gives coatings with improved adhesion and flexibility as compared with corresponding unmodified epoxy-phenolic coatings [30]. A phosphoric acid catalyst is not needed, as it is with conventional epoxy/phenolic coatings.

11.6. PHENOLIC RESINS

Although their importance has waned, phenolic resins still have significant uses. Phenolics are made by reaction of formaldehyde with phenol and substituted phenols. The products depend on the phenol(s) used, the stoichiometric ratio of phenol to formaldehyde, and the pH during the reaction. Phenolic resins are divided into two broad classes: *resole phenolics*, which are made using alkaline catalysts and high ratios of formaldehyde to phenol, and *novolac phenolics*, which are made using acid catalysts and low ratios of formaldehyde to phenol. Phenolic resins are used on a large scale in plastics and adhesives applications. (See Ref. [31] for details of the chemistry and the wide range of applications.)

11.6.1. Resole Phenolic Resins

Under alkaline conditions, the initial reaction product of phenol and formaldehyde is a mixture of ortho and para methylolated phenols. The methylolated phenols are more reactive with formaldehyde than the unsubstituted phenol resulting in the rapid formation of 2,4-dimethylolphenol and, subsequently, 2,4,6-trimethylolphenol; the latter is the predominant product, with a large excess of formaldehyde and a relatively short reaction time. With lower ratios (but still a molar excess) of formaldehyde to phenol and longer reaction times, formation of higher molecular weight resole phenolic resins is favored.

Polymerization occurs primarily by a methylol group on one phenol reacting at the ortho or para position of another phenol to form a methylene bridge connecting the two phenols. Dibenzylether bridges connecting two phenols also form by reaction of two methylol groups with each other. With excess formaldehyde, methylol groups are present on the terminal phenol groups of resole resins. Although not shown in the general structure, some of the aromatic rings have three substituents.

Such phenol-based resole resins cross-link on heating and are used in adhesive and plastics applications. However, they are not suitable for coatings applications, primarily because their cross-link density is higher than appropriate for any coating. Furthermore, the package stability of the resins is limited. Resole phenolics useful in coatings applications are made from monosubstituted phenols and mixtures of monosubstituted phenols with phenol. The use of substituted phenols reduces the potential cross-link density. There are two broad categories of such resins: (1) those that are soluble in alcohol and other low molecular weight oxygenated solvents, commonly called *alcohol-soluble, heat-reactive phenolics*; and (2) those that are soluble in vegetable oils and are called *oil-soluble, heat-reactive phenolics*.

Resole Phenolic Resin (from *p*-cresol)
(Idealized Structure)

Alcohol-soluble, heat-reactive resole resins are prepared by reacting phenol, *o*- or *p*-cresol, and formaldehyde in the presence of a base catalyst at less than 60°C while removing water under vacuum. The catalyst is neutralized, alcohol is added, and the salt resulting from catalyst neutralization is removed by filtration. Potential cross-link density is controlled by the ratio of phenol to cresol; molecular weight is controlled by the ratio of formaldehyde to phenols and by reaction time.

Such resole phenolic resins are used in interior can coatings and tank linings. They require baking using an acid catalyst in order to cure in short times. To enhance flexibility and adhesion, they are commonly blended with low molecular weight poly(vinyl butyral) as a plasticizer. The films are resistant to swelling by oils such as encountered in canned fish and are completely resistant to hydrolysis. These resins and other heat-reactive phenolics discolor during baking, which restricts their usage to applications for which development of a yellow-brown color is permissible. These resins are also blended with epoxy resins in thermosetting coatings for applications such as primers and can coatings. (See Section 11.3.1). The absence of hydrolyzable bonds and generally excellent adhesion properties are their chief advantages.

Oil-soluble, heat-reactive phenolics are prepared by reacting a para substituted phenol, (e.g., *p*-phenylphenol, *p-t*-butylphenol, or *p*-nonylphenol) with somewhat less than 2 moles of formaldehyde per mole of substituted phenol. The resulting resole phenolics are cast from the reactor after neutralization of the catalyst. The resulting resole phenolics are solid, linear resins with terminal methylol groups. The most common use for such resins is to make varnishes with linseed oil and/or tung oil. However, since varnishes have, in large measure, been replaced by other vehicles, the consumption of these resins has declined markedly.

11.6.2. Novolac Phenolic Resins

Novolac phenolics of interest for coatings are made with acid catalysts and *o*- or *p*-substituted phenols. Molecular weight is controlled by the molar ratio of phenol and formaldehyde, which is always greater than 1. In contrast to resole phenolics, the terminal phenol groups are not metholylated, as shown in the idealized structure.

Novolac Phenolic Resin

Three types of novolac resins are used in coatings.

1. Alcohol-soluble nonheat-reactive, low molecular weight phenolics derived from *o*- or *p*-cresol. An important use for these resins is in the preparation of novolac epoxy resins by reaction with epichlorohydrin. (See Section 11.1.2.)

2. Oil-soluble, nonheat-reactive novolac phenolic resins are made using a low ratio of formaldehyde, an acid catalyst, and a substituted phenol (e.g., *p*-phenylphenol, *p-t*-butyphenol, or *p*-nonylphenol). They are used together with drying oils, particularly tung oil or tung/linseed oil mixtures, to make varnishes. Such varnishes are still used to a small extent as marine spar varnish, for which their reputation for durability maintains their position in the do-it-yourself marine yacht market. The durability of phenolic varnishes may well result, at least in part, from the antioxidant activity of the phenolic groups. (See Section 5.2.2 for a discussion of phenolic antioxidants.)

3. Rosin-modified phenolic resins are the principal type of modified phenolics still in use. Their use in coatings is limited to low cost varnishes. The largest volume use is in printing inks. For inks, the phenolic resin is prepared in the presence of rosin esters and/or zinc or calcium salts of rosin. The structures of the reaction products, which are high melting, hydrocarbon-soluble resins, are not completely known. An important example of their use is in publication gravure inks such as those used in mail-order catalogs and some magazines. They are also used, to a degree, in heat-set letterpress printing inks for magazine and paperback book applications.

11.6.3. Ether Derivatives of Phenolic Resins

The package stability of alcohol-soluble resole resins and their compatibility with epoxy resins can be improved by partial conversion of the methylol groups to ethers. The ether groups undergo exchange reactions with hydroxyl groups in the presence of acid catalysts. Allyl ethers have been used for many years along with epoxy resins in interior can coatings.

More recently, low molecular weight butyl ethers have been made available for use in cross-linking epoxy resins and other hydroxy-substituted resins, primarily by etherification and transetherification reactions [32]. A typical resin has an average of 2.2 aromatic rings per molecule. Due to its low molecular weight (320) the resin has a moderate viscosity, as supplied in butyl alcohol solution. As might be expected with such a low molecular weight resin, some free phenol remains in the resin. The reactive groups in the resins are primarily butoxymethyl groups, but there are also benzyloxy groups and some free methylol groups. Furthermore, the phenol groups can react with epoxy groups. Acid catalysts, such as phosphoric or sulfonic acids, are required. Blocked acids may be utilized to extend shelf life.

GENERAL REFERENCES

C. A. May, Ed., *Epoxy Resins—Chemistry and Technology*, Marcel Dekker, New York and Basel, 1988.

B. Ellis, Ed., *Chemistry and Technology of Epoxy Resins*, Blackie Academic & Professional, London, 1993.

REFERENCES

1. L. V. McAdams and J. A. Gannon, "Epoxy Resins," in *Encyclopedia of Polymer Science and Engineering*, 2nd. ed., Vol. 6, Wiley, New York, 1986, pp. 322–382.
2. P. S. Sheih and J. L. Massingill, *J. Coat. Technol.*, **62** (781), 25 (1990).
3. D. R. Scheuing, *J. Coat. Technol.*, **57** (723), 47 (1985).
4. M. M. Bagga, *European Patent Application* 22073 (1981).
5. E. G. Bozzi and D. Helfand, FSCT Symposium, Louisville, KY, May 1990. See also K. L. Payne and J. S. Puglisi, *J. Coat. Technol.*, **59** (752), 117 (1987).
6. J. L. Massingill, P. S. Sheih, R. C. Whiteside, D. E. Benton, and D. K. Morisse-Arnold, *J. Coat. Technol.*, **62** (781), 31 (1990).
7. R. A. Dubois and P. S. Sheih, *J. Coat. Technol.*, **64** (808), 51 (1992).
8. W. P. Jencks, *Catalysis in Chemistry and Enzymology*, McGraw-Hill, New York, 1969, pp. 199–211; W. P. Jencks, *Chem. Rev.*, **72**, 705 (1972).
9. M. Agostinho and V. Brytus, *J. Coat. Technol.*, **60** (764), 61 (1988).
10. M. D. Gilbert, N. S. Schneider, and W. J. MacKnight, *Macromolecules*, **24**, 360 (1991); M. Fedtke, F. Domaratius, K. Walter, and A. Pfitzmann, *Polym. Bull.*, **31**, 429 (1993).
11. D. J. Weinmann, K. Dangayach, and C. Smith, *J. Coat. Technol.*, **68** (863), 29 (1996).
12. V. Brytus, *J. Coat. Technol.*, **58** (740), 45 (1986).
13. R. F. Brady, Jr. and J. M. Charlesworth, *J. Coat. Technol.*, **65** (816), 81 (1993).
14. R. W. Tess, in C. A. May, Ed., *Epoxy Resins—Chemistry and Technology*, Marcel Dekker, New York and Basel, 1988, p. 743.
15. A. Wegmann, *J. Coat. Technol.*, **65** (827), 27 (1993).
16. R. Albers, *Proc. Water-Borne Higher-Solids Coat. Symp.*, New Orleans, 1983, pp. 130–143; U.S. Patent 4352898 (1982).
17. J. A. Lopez, U.S. Patent 4816502 (1989).
18. S. R. Howe, L. Borodinsky, R. S. Lyon, *J. Coat. Technol.*, **70** (877), 69 (1998); R. J. Wingender, P. Niketas, and C. K. Switala, *J. Coat. Technol.*, **70** (877), 75 (1998).
19. S. Kojima, T. Moriga, and Y. Watanabe, *J. Coat. Technol.*, **65** (818), 25 (1993).
20. P. Oberressl, T. Burkhart, D. Chambers, and A. Slock, *Proc. Waterborne High-Solids Powder Coat. Symp.* New Orleans, 1997, p. 296.

21. M. Ooka and H. Ozawa, *Prog. Org. Coat.*, **23**, 325 (1994).

22. M. D. Shalati, J. R. Babjak, R. M. Harris, and W. P. Yang, *Proc. Intl. Conf. Coat. Sci. Technol.*, Athens, 1990, p. 525.

23. S. P. Pappas, V. D. Kuntz, and B. C. Pappas, *J. Coat. Technol.*, **63** (796), 39 (1991).

24. R. F. Eaton and K. T. Lamb, *J. Coat. Technol.*, **68** (860), 49 (1996).

25. S. Nakano and T. Endo, *Prog. Org. Coat.*, **28**, 143 (1996).

26. R. F. Eaton, *Polym. Mater. Sci. Eng.*, **77**, 381 (1997).

27. J. T. K. Woo, V. Ting, J. Evans, R. Marcinko, G. Carlson, and C. Ortiz, *J. Coat. Technol.*, **54** (689), 41 (1982).

28. J. T. K. Woo and A. Toman, *Polym. Mater. Sci. Eng.*, **65**, 323 (1991).

29. J. T. K. Woo and A. Toman, *Prog. Org. Coat.*, **21**, 371 (1993).

30. J. L. Massingill, *J. Coat. Technol.*, **63** (797), 47 (1991).

31. P. W. Kopf, "Phenolic Resins," in *Encyclopedia of Polymer Science and Engineering*, 2nd. ed., Vol. 11, Wiley, New York, 1988, pp. 45–95.

32. Monsanto Chemical Co. (now Solutia Inc.), *Santolink EP 560*, Springfield, MA, 1990.

CHAPTER 12

Acrylic Resins

Acrylic resins are used as the primary binder in a wide variety of industrial coatings. Their major advantages are photostability and resistance to hydrolysis.

12.1 THERMOPLASTIC ACRYLIC RESINS

Thermoplastic acrylic (TPA) polymers have many excellent properties, especially exterior durability, but their use has declined because they require large amounts of solvent to reduce viscosity low enough for application. Lacquers based on TPA polymers were used for automotive topcoats from the 1950s to the 1970s. Their high solvent levels make possible brilliant metallic colors since they permit orientation of aluminum flake pigment in the films parallel to the surface (see Section 29.1.2). Acrylic lacquers are still used to a degree in refinishing automobiles (see Section 32.3) and for other specialized coatings, but as VOC regulations become more restrictive, they, too, will probably be phased out.

Solution thermoplastic acrylic polymers are prepared by chain-growth polymerization, as discussed in Section 2.2.1. They are copolymers of various methacrylic monomers; with a T_g of over 70°C. Monomers are selected on the basis of cost and effect on properties, notably, outdoor durability and T_g. Methyl methacrylate (MMA), styrene (S), and n-butyl acrylate (BA), are often used. The acrylic copolymers are used with plasticizers. Control of molecular weight in the synthesis is critical. Film strength increases with molecular weight, although above $\bar{M}_w$ of about 90,000, the change in properties with increasing molecular weight is small. The upper end of molecular weight is limited because solutions of acrylic polymers with $\bar{M}_w$ greater than about 100,000 exhibit *cobwebbing* on spraying; rather than atomizing to small droplets from the orifice of a spray gun, the lacquer comes out as threads. As molecular weight increases, viscosity of solutions increases and the solids, which can be spray applied decreases. Viscosity is particularly affected by the very high molecular weight fractions; therefore, it is critical to minimize this fraction by controlling molecular weight distribution ($\bar{M}_w/\bar{M}_n$) to a narrow range. (See Section 2.2.1.) Commercial TPAs have $\bar{M}_w$ of 80,000 to 90,000 and $\bar{M}_w/\bar{M}_n$ of 2.1 to 2.3. The solids of TPA lacquers at application viscosity are in the range of 11 to 13 NVV (nonvolatile volume).

12.2. THERMOSETTING ACRYLIC RESINS

Solids can be increased by using thermosetting acrylic resins (TSA). Since they are designed to react chemically after application, they can have lower molecular weights. The postreaction leads to a polymer network, which is ideally one interconnected molecule with a very high molecular weight and, hence, potentially good film properties. The cross-linked film is not soluble in any solvent. The term *thermosetting acrylic* seems to imply that the resins cross-link by themselves; however, most TSAs are resins bearing functional groups that are reacted with a different functional polymer or cross-linker. However, there are examples of self cross-linking acrylics.

12.2.1. Hydroxy-functional acrylics

Hydroxy-functional acrylic resins are copolymers of nonfunctional monomers, often MMA, S, and BA, with a hydroxy-functional monomer. They are cross-linked with either melamine-formaldehyde (MF) resins (see Chapter 9) or with polyisocyanates (see Chapter 10). Solvent-borne hydroxy-functional TSAs for conventional-solids coatings were first developed in the 1950s. Their $\bar{M}_n$s are usually 10,000 to 20,000 with $\bar{M}_w/\bar{M}_n$ of 2.3 to 3.3. Thermosetting acrylics are prepared by free radical initiated chain-growth polymerization under monomer-starved conditions, as described in Section 2.2.1. Polymerization of conventional TSAs is easier to control than that of TPAs, since there is not as critical effect of molecular weight on application and film properties. Due to their lower molecular weight, one can use less expensive aromatic hydrocarbon solvents for the polymerization medium. However, the lower molecular weight of TSA polymers means that there are more end groups per unit weight. Therefore, the structure of the end groups can have greater effects on performance than in the case of TPA polymers. Reference [1] provides a review of reactions in free radical polymerizations that can lead to various end groups. Azo initiators are used predominantly, since they cause relatively few side reactions and form end groups with minimal photochemical reactivity. Certain peroxy initiators also meet these requirements, but others leave end groups that reduce exterior durability. Chain transfer to solvent can also result in end groups that reduce exterior durability. For example, terminal groups resulting from chain transfer to a ketone solvent, such as methyl *n*-amyl ketone (MAK), can have a deleterious effect [2].

Hydroxyl groups are often introduced by using a hydroxy-functional comonomer such as 2-hydroxyethyl methacrylate (HEMA). If a lower T_g reactive monomer is desired, 2-hydroxyethyl acrylate (HEA) can be used, but its greater toxic hazard increases handling costs; it is often preferable to reduce T_g by increasing the proportion of low T_g nonreactive monomers such as *n*-butyl acrylate. Commercial grades of HEMA (and HEA) contain appreciable amounts of diester, ethylene glycol dimethacrylate (EGDMA) in the case of HEMA. Separation of the mono- and diesters is not easy since their boiling points are similar. When a relatively small fraction of hydroxy-functional monomer is used, small amounts of diester cause some branching, which does not seem to be harmful. However, larger amounts cause higher molecular weight, broader molecular weight distribution, and even gelation.

2-Hydroxypropyl methacrylate (HPMA) can also be used. It costs less than HEMA and generally has a lower diester content. It is a mixture of isomers in which the secondary alcohol predominates. Since secondary alcohols are less reactive than primary alcohols, TSAs with HPMA substituted for HEMA require higher baking temperatures, longer baking times, or more catalyst to cure to a given cross-link density with MF resins. For example, baking temperatures may have to be increased by 10 to 20°C. Isocyanate cross-linkers also react slower with HPMA-containing TSAs, but there is the accompanying advantage of longer pot life. Use of HPMA in place of HEMA may reduce resistance to photoxidation as a result of the hydrogens on tertiary carbons, which are more easily abstractable. An alternative method for introducing hydroxy functionality is to react a carboxylic acid-functional copolymer made using methacrylic (or acrylic) acid (MAA or AA) with an oxirane such as propylene oxide. Raw material cost is lower with propylene oxide than with HPMA, but exacting process control is required. Another approach to making hydroxy-functional acrylics is to react an AA or MAA copolymer with a glycidyl ester such as glycidyl versatate [3]. An example of a hydroxy-functional TSA is an MMA/S/BA/HEMA/MAA copolymer with a weight ratio of $50:15:20:14:1$, corresponding to a mole ratio of $54.3:15.6:16.9:11.7:1.5$. Typical molecular weights are $\bar{M}_w$ of about 35,000 and $\bar{M}_n$ of about 15,000 ($\bar{M}_w/\bar{M}_n = 2.3$). With the given ratio of comonomers, the $\bar{P}_w$ is about 320 and the $\bar{P}_n$ about 140. For stoichiometric comparisons, one uses the number average $\bar{M}_n$. This resin has a hydroxyl equivalent weight of slightly over 900, with a number average functionality $\bar{f}_n$ of about 16 hydroxyl groups per polymer molecule. The small amount of carboxylic acid-functional monomer (MAA) is introduced to reduce the probability of pigment flocculation in a liquid coating.

The nonfunctional monomers (MMA, S, and BA, in this case) are selected on the basis of their effect on T_g, exterior durability, and cost. This monomer combination provides good exterior durability with a relatively high T_g at a moderate cost, as is appropriate for an automotive top coat. This resin also has a relatively high level of hydroxy-functional monomer, appropriate when relatively high cross-link density (XLD) films are desired. For applications requiring more flexible films, such as coil coatings or exterior can coatings, one would use a lower HEMA content. By adjusting T_g and functionality, TSAs can be designed for a wide range of end uses. Countless variations can be prepared.

Thermosetting resins and cross-linkers must be selected or designed for use as a system. The appropriate composition of hydroxy-functional TSAs depends on the cross-linker to be used. A TSA resin appropriate for a TSA-MF coating would not necessarily be appropriate to use with a particular polyisocyanate cross-linker. Depending on the polyisocyanate used with a TSA, the T_g of the final cross-linked films could be higher or lower than would be obtained by cross-linking the same TSA with an MF resin. Thermosetting acrylic resins for polyisocyanate cross-linking tend to have lower T_g and lower $\bar{f}_n$ than TSAs for cross-linking with MF resins in order to compensate for the effect of the intermolecular hydrogen bonding between urethane groups in the cross-linked films (see Section 10.4).

Further increase in solids became necessary to meet lower VOC emission requirements. During the 1970s, major efforts were launched to make high solids solution acrylic resins. It may seem that this problem is simple; why not just reduce $\bar{M}_n$ from 15,000 to 1500? Actually, the problem is complex; as molecular weight goes down, greater care must be exercised in every aspect of polymerization, coating formulation, and application. In polymerization, the amount of non- or monofunctional resin must be kept to a very low fraction. Due to the larger number of chain ends, the effect of chain transfer to solvents such as ketones, which can introduce photoreactive end groups, is potentially more serious. More care is needed in establishing the stoichiometric ratio of cross-linker, and, as noted in Section 9.3.1.4, the cure window becomes narrower. Controlling film defects during application becomes more difficult as the solids content becomes higher. (See Chapter 23.) It is particularly difficult to avoid sagging of spray applied baking coatings.

An example of a TSA resin has the following composition: S/MMA/BA/HEA; weight ratio $15:15:40:30$; $\bar{M}_w$, 5200; $\bar{P}_w$, 54; $\bar{M}_n$, 2300; $\bar{P}_n$, 20; $\bar{P}_w / \bar{P}_n = 2.7$; equivalent weight, 400; $\bar{f}_n = 5.7$ [4]. It was made for studying MF cross-linking of fairly high solids coatings. The resin was prepared at 65 NVW in methyl amyl ketone (MAK). In comparing this TSA with the conventional TSA described earlier in this chapter, note that $\bar{M}_n$ has been reduced by a factor of 6.5, whereas $\bar{f}_n$ is only reduced by a factor of 2.8. This difference results from the higher content of hydroxy-functional monomer required in the high solids resin. To attain about the same XLD in the final film, the low molecular weight resin must undergo more reactions. Although not reported, application solids of coatings formulated with this resin would be around 45 NVV. This solids content corresponds to that used in high solids metallic automotive top coats. However, the solids are still too low to meet EPA regulations for some current end uses, and even stricter regulations are anticipated.

Free radical polymerization inherently limits the extent to which molecular weight of a TSA can be reduced. Satisfactory performance requires that a high fraction of the molecules have at least two hydroxyl groups, but this becomes statistically less probable as molecular weight is reduced. The problem can be illustrated by comparing the conventional TSA ($\bar{M}_n = 15,000$, $\bar{P}_n = 140$) described in Section 12.2 with a hypothetical high solids TSA having the same monomer ratio with $\bar{M}_n = 1070(\bar{P}_n = 10)$, which is about the level needed to formulate with VOCs around 300 g L^{-1}. The conventional resin has an $\bar{f}_n$ of 16 hydroxyl groups. While individual molecules have more or less than 16 hydroxyl groups, statistically the number of molecules with fewer than 2 hydroxyl groups is very low, and virtually all the molecules are capable of cross-linking. In contrast, $\bar{f}_n$ of this high solids TSA would be only about 1.2, and a very large fraction of the molecules could not be cross-linked. Molecules with no hydroxyl groups would either volatilize or remain in the film as plasticizers, to the detriment of film properties. Molecules with one hydroxyl group would terminate cross-linking reactions, leaving *loose ends* in the coating. The theory of elasticity predicts that loose ends seriously diminish the mechanical properties of a network. Experimental experience confirms that loose ends have a substantial effect on initial film properties [5,6]. For example, each percent by weight of monohydroxy oligomer in a TSA may reduce T_g of the TSA-MF enamel by about 1°C [5].

The problem results from both molecular weight and functional group distribution. Figure 2.2 (Chapter 2) shows the distribution of molecules of differing degrees of polymerization of a series of resins with $\bar{P}_n = 12$ and $\bar{P}_w/\bar{P}_n = 1.07$, 1.5, and 3.0. The theoretical minimum polydispersity that can be obtained by anionic polymerization is 1.07, and the theoretical minimum achievable by free radical initiated polymerization is 1.5. The difference is explained by different kinetic characteristics; with anionic polymerization, the rate of propagation can be slower than the rate of initiation, while with free radical polymerization, the rate of propagation is generally faster. It is difficult to approach the theoretical minima in practice; $\bar{P}_w/\bar{P}_n = 2.5$ is typical of reasonably well controlled free radical polymerizations. The other factor involved is sequence length distribution, which is the distribution of sequences of different lengths of nonfunctional monomers separating functional monomers. If these sequence lengths are short compared to the chain length of the low molecular weight parts of the resin, multiple functional groups will be incorporated in the molecules. But, if the sequence lengths of nonfunctional monomers are long in comparison to chain lengths, molecules with one or no functional monomer units can be expected.

Statistical methods have been used to calculate the proportions of nonfunctional molecules that would be formed during random copolymerization of monomer mixtures with differing monomer ratios to different molecular weights and molecular weight distributions [7,8]. Due to the assumptions involved, the calculations are approximate. An example of such results for S/BA/HEA (30:50:20 wt%) copolymers of varying $\bar{P}_n$ is given in Table 12.1 [8]. For $\bar{P}_n = 9.5$ ($\bar{M}_n = 1125$) about 36% of all molecules, corresponding to 13% of the weight, is nonfunctional. (The mol% and wt% differ because of the molecular weight distribution, low molecular weight molecules are more likely to be nonfunctional than the higher molecular weight ones.) The weight fraction of monofunctional oligomer is more difficult to calculate but would be expected to exceed that of nonfunctional oligomer. At the higher HEA content (30%) of the example given previously with a $\bar{P}_n = 20$, the proportion of monofunctional oligomer molecules would be minimized. But, if one wished to have volume solids of about 70 NVV, one would need to have a $\bar{P}_n$ of about 10 with a narrow distribution of molecular weight. The calculations show that the functional monomer content would have to be very high—so high that the XLD of the cross-linked film would be too high for acceptable film properties.

Viscosity of acrylic resins can be reduced by changing the proportions of comonomers to reduce T_g. However, this approach also reduces T_g of the cross-linked film, other factors being equal. The effect of reduced T_g on film properties can be offset to an extent by increasing cross-link density, but if carried too far, this approach leads to an unsatisfactory combination of properties. Methacrylate esters of relatively rigid alcohols, such as 3,3,5-trimethylcyclohexyl [9] and isobornyl [10] methacrylates, with a compact shape can combine relatively low viscosity and high T_g can be used as partial replacements for MMA and S.

Molecular weight is reduced by using chain transfer agents, but some chain transfer agents can have an adverse effect on exterior durability [6]. Use of a chain-transfer agent with a functional group is an interesting approach. For example, when 2-mercaptoethanol is used as a chain-transfer agent, initiating free radicals bearing a hydroxyl group are formed. These lead to a high fraction of molecules with a hydroxyl group on one end, reducing the fractions of non- and monofunctional molecules in the resin [11]. Improved film properties are obtained, but the unpleasant odor of mercaptoethanol causes handling problems and may leave residual odor in the resin.

High levels of initiator with close control of variation of temperature and concentrations during polymerization can minimize polydispersity. While there have been no basic studies published, it appears that with high initiator concentration, polydispersity is narrower than theory predicts. Proprietary acrylic resins for general purpose white coatings with

Table 12.1. Percentage of Nonfunctional Molecules Statistically Predicted for S/BA/HEA (30:50:20 by wt) Copolymers

$\bar{P}_n$	$\bar{M}_n$	Mol%	Wt% Nonfunctional
36.8	4357	15	1.8
19.2	2273	24	5.8
9.5	1125	36	12.8

Source: See Ref. [8].

moderately demanding durability requirements have been reported with sprayable viscosities at about 54 to 56 NVV (70–72 NVW) [12]. Another commercial resin with $\bar{M}_n$ of 1,300 and $\bar{M}_w/\bar{M}_n = 1.7$ can be formulated with Class I MF resins in white coatings at about 77 NVW [13]. These resins provide hard, chemical resistant films with many good properties, although the films tend to be brittle.

Initiator choice can be critical. Benzoyl peroxide gives wide distributions due to chain transfer to polymer resulting from hydrogen abstraction. Azo initiators such as AIBN give less branching; AIBN is also generally preferred over benzoyl peroxide because exterior durability of films is superior. Initiators based on *t*-amyl peroxides, such as ethyl 3,3-di(*t*-amylperoxy)butyrate, are reported to give narrow distributions [14]. Other workers recommend initiators derived from *t*-butyl peroxide, such as *t*-butyl peracetate [15].

Group transfer polymerization has been used to make acrylic copolymers with narrow molecular weight distributions [16,17]. A rapid, reversible complexation of small concentrations of enolate anions with silyl ketene acetals is proposed to explain the living nature of these polymerizations. The overall reactions are proposed to be:

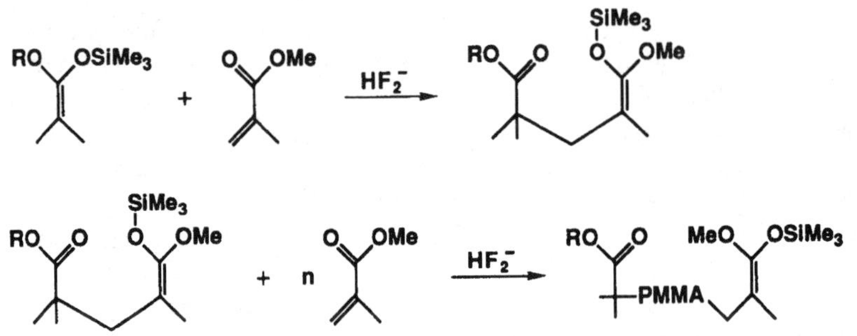

The polymerization permits close control of the structure of acrylic polymers; $\bar{M}_w/\bar{M}_n$ can be reduced to 1.2 or less, distribution of functional groups among molecules can be narrowed, functional groups can be placed at specific locations within each molecule, and multiarmed star polymers can be produced. While resins prepared in this manner have not been successfully commercialized for binders, group-transfer polymerization has been used to produce dispersing agents for pigments (see Section 20.1.3).

Another approach to increasing solids is to blend acrylic polyols with other low viscosity polyols, such as polyesters. Some reduction in VOC can be achieved in this way. The film properties, such as exterior durability and chemical resistance, with such blends are often not quite as good as those of straight TSA-MF coatings, but they may be adequate for many uses, and some highly weather resistant coatings have been reported [18].

12.2.2. Acrylics with Other Functional Groups

Carboxylic acid-functional acrylic resins are made using acrylic or methacrylic acid as comonomers; they are cross-linked with epoxy resins. (See Section 11.3.2.) Note that close quality control of acrylic acid is required because it can dimerize.

Acrylics with amide groups can be used in several ways. Acrylamide copolymers can be cross-linked with MF resins by reaction with the amide group; the curing temperature required is higher than for hydroxy-functional resins. Alkoxymethyl derivatives of acrylamide copolymers can be made by reacting the amide groups with formaldehyde, followed by etherification, analogous to the preparation of amino resins, as described in Section 9.4.4. These resins can be used as cross-linkers for hydroxy-functional TSAs. Alternatively one can make alkoxymethyl derivatives of monomeric acrylamides for use as comonomers. For example, *N*-(isobutoxymethyl)methacrylamide has been used as a comonomer along with hydroxy-functional comonomers to make copolymers that are "self cross-linking" [19]. Acrylamide-HEMA copolymers with nonreactive comonomers can also be reacted with formaldehyde and then with an alcohol to make a self cross-linking resin.

Epoxy-functional acrylics are made using glycidyl methacrylate (GMA) as a comonomer (see Section 11.3.2). Such resins have been recommended for powder clear coats for automobiles [20-22] (see Section 27.1.4). They can be cross-linked with dicarboxylic acids, such as dodecanoic acid [20,21] or with carboxylic acid-functional acrylic resins [22].

Isocyanato-functional acrylics can be prepared by copolymerizing isopropenyldimethylbenzylisocyanate (TMI) with acrylates (see Section 10.3.2) [23]; they can be cross-linked with polyols or hydroxy-functional acrylic resins. They can also be reacted with hydroxypropylcarbamate to yield carbamate-functional acrylic resins [24,25]. Carbamate-functional acrylics can be cross-linked with Class I MF resins to give films with better environmental etch resistance than MF cross-linked hydroxy-functional acrylics while retaining the advantage of mar resistance (see Section 9.3.4). Carbamate-functional acrylics can also be prepared by reacting acrylic resins with urea [25]. They can also be prepared by ester interchange between a hydroxyl group on a hydroxy-functional acrylic and the carbamate from propylene glycol monomethyl ether [26].

Trialkoxysilyl-functional acrylics can be prepared using a trialkoxysilylalkyl methacrylate as a comonomer [27]. Clear coats made with them are cross-linked by moisture in the air. Properties are said to be excellent; see Section 16.5.4 for further discussion.

12.3. WATER-REDUCIBLE THERMOSETTING ACRYLIC RESINS

Another way to reduce VOC emissions from TSA coatings is to make TSA resins that can be diluted (reduced) with water. Such resins are sometimes called "water-soluble resins," but this terminology is misleading since they are not soluble in water. Rather, solutions of amine salts of these resins in organic solvents can be diluted with water to form reasonably stable dispersions of polymer aggregates swollen by solvent and water. To minimize confusion, we have chosen to use the following terminology: *water-borne* is used broadly for all coatings with aqueous media, including latex coatings; *water-reducible coatings* is used only for water-borne coatings based on resins having hydrophilic groups in most or all molecules, which excludes latexes; and *water-soluble* is limited specifically to substances that are soluble in water.

A typical water-reducible acrylic resin is a copolymer of MMA/BA/HEMA/AA in a weight ratio of $60:22.2:10:7.8$ prepared by free radical initiated polymerization using an azo initiator [28]. The polymerization is carried out at high solids (generally 70 NVW or higher) in a water-miscible solvent. Glycol ethers, such as 1-(*n*-propoxy)-2-propanol or 2-butoxyethanol and butyl alcohols are the most widely used solvents. Except for the solvent and the higher proportion of acrylic acid, such resins are similar to conventional solids TSAs described in Section 12.2 with $\bar{M}_w$ and $\bar{M}_n$ of about 35,000 and 15,000, respectively.

Water-reducible acrylic resins typically have acid numbers of 40 to 60. (Acid number is determined by titration and is defined as mg of KOH required to neutralize 1 g of resin solids. Equivalent weight equals 56,100/acid number.) Residual monomer is removed by distilling off a small fraction of the solvent, and the resin is stored as a concentrated solution. The first step in preparing a coating is to add an amine, such as 2-(dimethyl-amino)ethanol (DMAE). As discussed later, less than the theoretical amount of amine required to neutralize all of the carboxylic acid groups is commonly used. The ratio used is called the *extent (of) neutralization* (EN). For example, if 75% of the theoretical amount were used, the EN would be 75. Other coating components (pigments, MF resin, sulfonic acid catalyst) are dispersed or dissolved in this solution and before application the coating is diluted with water.

The change in viscosity with dilution of water-reducible resins is abnormal. An example is given in Figure 12.1 showing log viscosity as a function of concentration for a model resin [29]. The height of the peak in the dilution curve is dependent on the particular resin and formulation, and the systems are highly shear thinning in the peak region. Another abnormality of water-reducible resins is that their pH is over 7 (commonly 8.5 to 9.5), even though less than the theoretical amount of amine necessary to neutralize the carboxylic acid is used. For comparison purposes, a dilution curve for the same resin using an organic solvent (*t*-butyl alcohol in this case) is included in the figure, as is a typical dilution curve of a latex.

The morphology of water-reducible TSAs has been studied fairly extensively [29–31]. In Figure 12.1, the water-reducible resin is a 54% solution of a 90 : 10 butyl methacrylate (BMA)/AA copolymer in *t*-butyl alcohol at 75 EN with DMAE. The solvent dilution curve is fairly linear, as is typical of the log viscosity against concentration relationship of most resin solutions in good solvents. The water dilution curve shows the abnormal response that is typical of water-reducible resins. During dilution with water, two kinds of changes are occurring simultaneously: The concentration is reduced, and the ratio of solvent to water is also reduced.

In the first stages of dilution, viscosity drops more rapidly with water than with solvent. It is hypothesized that the viscosity is high before dilution because of association of ion pairs on different molecules. Water associates strongly with ion pairs, separating most of the intermolecular ion pairs and causing rapid reduction in viscosity. However, as dilution with water continues, the viscosity levels off and then increases, passing through a maximum. Still further dilution causes a very steep drop of viscosity. At application viscosity of about 0.1 Pa·s, such systems typically have solids in the range of 20 to 30 NVW. Thus the application solids of coatings made with such resins is low; however, since a major part of the volatile components is water, VOC emissions are still low.

The amine salt of the resin is soluble in the *t*-butyl alcohol and in *t*-butyl alcohol-water solutions in which the ratio of solvent to water is high. But, as additional water is added, the ratio of solvent/water decreases to the point where some of the molecules are no longer soluble in the mixed solvent. These molecules do not precipitate in a separate macrophase, but the nonpolar segments of various molecules associate with each other to form aggregates. The predominantly nonpolar parts of the molecules are in the interior of the aggregates, and the highly polar carboxylic acid salt groups are on their periphery. As dilution continues, more and more molecules join in aggregates. Since the solvent is soluble in the resin, some dissolves in the aggregates, swelling their volume. Also, water associates with the salt groups, and some dissolves in the solvent inside the aggregates, further swelling them. As the aggregates form, the system changes from a solution to a

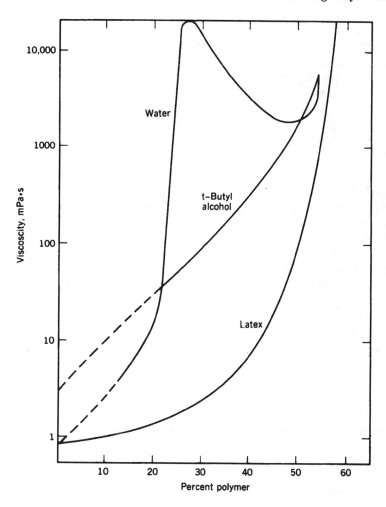

Figure 12.1. Viscosity dependence on concentration for a 10 mol% acrylic acid copolymer, 75 EN with DMAE, dissolved in *t*-butyl alcohol and then diluted with water. Also shown are curves for dilution of the same resin with *t*-butyl alcohol and a typical viscosity-solids latex dilution with water. (From Ref. [29], with permission.)

dispersion of aggregates in a continuous phase. As the number and volume of aggregates grow, the volume fraction of internal phase increases and the aggregates get more and more closely packed, leading to the increase in viscosity. (See Section 3.5 for discussion of the factors controlling the viscosity of liquids containing dispersed phases.) At the maximum viscosity, the material is predominantly a dispersion of highly swollen aggregates in water containing some solvent. The high viscosity can lead to difficulty in stirring. On still further dilution, the viscosity drops rapidly. This drop results from two factors. There is a dilution effect, that is, the decrease in volume fraction internal phase. However, the drop is even steeper than would result from this effect alone. The balance of the decrease is due to a decrease in the swelling of the aggregates. The partitioning of solvent and water between the aggregates and the continuous phase changes throughout the dilution process, and as more water is added, more solvent moves to the continuous phase, shrinking the swollen aggregates.

Viscosity of the solutions in the organic solvent is Newtonian, as it is when only small amounts of water have been added. However, in the mid-range around the peak of the water dilution curve, the dispersions show a high degree of shear thinning [30]. The dilute systems show little or no shear thinning. This behavior can be explained on the basis that, in the shear thinning stage, the swollen aggregate particles (the internal phase) are not rigid. When shear is applied, the particles distort, decreasing the shape factor, increasing the packing factor, and, as a result, decreasing the viscosity as a function of increasing shear rate. With further dilution, some of the organic solvent is extracted from the aggregates into the continuous phase, and the aggregates become smaller and less easily distorted. In the more dilute stage, the viscosity is lower, so the shear stress exerted on the aggregates at a given shear rate is less, reducing the probability of distortion. Thus, the flow properties are Newtonian or only slightly shear thinning at application viscosities.

The abnormal pH effect noted earlier can be explained in a way consistent with this picture of the morphology of the system [32]. When one neutralizes a simple carboxylic acid, like acetic acid, with 75% of the theoretical equivalent amount of an amine like DMAE, the pH is about 5.5. However, the situation with the carboxylic acid groups on the polymer chain is different. The acrylic acid groups are relatively randomly spaced along the polymer chain. In some cases, there are carboxylic acid groups near each other; in other cases, there are single acrylic acid groups separated from others on both sides by several hydrophobic ester monomer units. As aggregation occurs during dilution, many of the carboxylic acid groups are near the surface, where all, or almost all, are neutralized by the water-soluble amine. However, geometric factors require that many of the carboxylic acid groups be in the interior of the aggregates; those most widely separated from other carboxylic acid groups by hydrophobic monomer residues are presumably most likely to be "hidden" in the interior of aggregates. 2-(Dimethylamino)ethanol partitions among the continuous (water/solvent) phase, the surface region of the aggregates, and the interior of the aggregates. Since DMAE and DMAE salts are highly soluble in water, it is expected that DMAE concentrates in the first two regions, leaving a fraction of carboxylic acid groups unneutralized in the interior of aggregates. Even when there is only 75% of the amine necessary to neutralize all of the carboxylic acid groups, there is more than enough to neutralize those carboxylic acid groups at or near the surface of the aggregates. Part of the amine is in the continuous phase, resulting in a basic pH reading. Since the interaction is between a weak base and a weak acid, the change of pH with amine addition is slow. Due to this insensitivity, pH is not appropriate as a quality control specification.

This description of the morphology of a water-reducible system accounts for its behavior, but does not provide direct evidence for the presence of aggregates. Evidence was found by studying the effects of changing solvent to water ratios at a constant resin to amine concentration. A BMA/AA copolymer (mol ratio 84 : 16) at 75 EN with DMAE was "dissolved" in a series of solvent to water solutions at the same concentration of 21.2% of resin plus DMAE [30]. Viscosity of this series of dispersions goes through a peak at 30 : 70 solvent to water. The samples were then subjected to ultrafiltration. No resin was retained on the filter in the case of samples with a ratio of 80 : 20 solvent to water up to pure solvent. As the solvent ratio decreased from 80 : 20 to 30 : 70, the amount of resin retained increased until at 30 : 70, essentially all of the resin was retained by the ultrafilter. All of the resin in samples with still less solvent was also retained. The samples were also examined by phase contrast microscopy, which revealed the presence of particles in all those samples where resin was retained by the ultrafilter.

Many variables affect the morphology of these systems. The shape of the viscosity-concentration (dilution) curve varies. In some cases, there is a very high viscosity peak-higher than the viscosity of the original undiluted material; in other cases, there may be only a shoulder in the dilution curve. Dependence of viscosity on molecular weight depends on solvent structure and the ratio of solvent to water, in addition to concentration [30]. The effect of molecular weight on viscosity is different at different stages of dilution. Log viscosity of the all solvent solution varies with approximately the square root of the molecular weight. Viscosity in the peak area of the dilution curve is very dependent on molecular weight. As molecular weight increases, the viscosity of the interior of the aggregate particles is higher, so they are more difficult to distort; hence, the viscosity of the whole system does not decrease as much at any given shear rate. Very high peak viscosity leads to difficulty in dilution. It is necessary to limit the molecular weight so that the coating can be thoroughly agitated throughout the dilution cycle with the available mixing equipment. The viscosity of systems diluted with water to application viscosity is independent of molecular weight. This advantage of these resins permits application of coatings made with a resin having a molecular weight in the same range as those used in conventional solvent-borne TSA coatings, but with VOC contents equivalent to relatively high solids coatings.

The dilution behavior of TSAs depends on their carboxylic acid content. The effect of varying the mole percent of acrylic acid in a series of BA/AA copolymers from 10 to 50% is shown in Figure 12.2 [29]. With 50% acrylic acid, the salts exhibit a viscosity dependence on dilution behavior approaching that of solution systems. As carboxylic acid content is reduced, the abnormal rheological properties become more pronounced. Notice that the concentration at viscosities near those required for application are highest for the lowest acid content systems. This is an important reason that resins of this type are designed with the lowest carboxylic acid content (acid number 40 to 60) that still provides a stable dispersion at application viscosity. Low acid numbers favor formation of relatively unswollen aggregates, keeping viscosity down, but if the acid numbers are too low, the system separates into macrophases instead of forming a stable microphase dispersion.

The required acid content is lower for resins with increasing hydroxyl group content. While salts of carboxylic acid groups are much more hydrophilic than hydroxyl groups, hydroxyl groups are sufficiently hydrophilic to affect the solubility in water-solvent blends, as indicated by the water solubility of the homopolymer of 2-hydroxyethyl acrylate. Therefore, the minimum required acid content decreases as the content of hydroxyl groups increases.

Melamine-formaldehyde resins are most commonly used as cross-linkers for the coatings. Class I and Class II monomeric, methylolated MF resins are miscible in these systems. They cross-link the TSA by reacting with hydroxyl and carboxyl groups. The reaction with OH groups is faster, and the resultant ether bonds are more stable to hydrolysis than the ester bonds resulting from cross-linking with COOH groups. Thus, while the cross-link density of the final film depends on the total functionality of COOH and OH groups, it is generally desirable to attain as much of the cross-linking as possible through OH groups and to adjust cross-link density by varying the $\bar{f}_n$ of OH. A further advantage of using the lowest possible level of COOH is that less amine is required. Amines are relatively expensive and must be included as part of the VOC emissions. See Section 10.8 for use of polyisocyanates as cross-linkers.

Various types and amounts of amine can be used [29,33,34]. Generally, less than the stoichiometric amount of amine is used. The lower the amine content, the lower the viscosity of the fully diluted systems, that is, the higher the solids at a fixed application viscosity. Figure 12.3 shows viscosity response to dilution as a function of extent

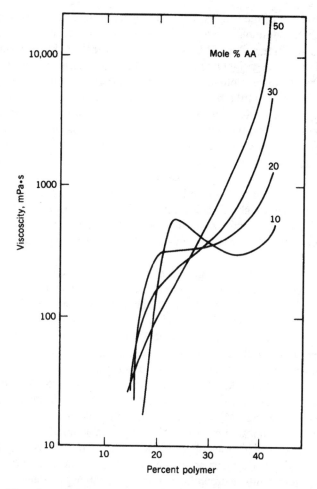

Figure 12.2. Viscosity variations during water dilutions of 42 NVW copolymers having 50, 30, 20, and 10 mol% AA; each at 75 EN with DMAE. (From Ref. [29], with permission.)

neutralization of a 90 : 10 BMA/AA resin with DMAE. For any resin-amine combination, there is a minimum amount of amine required to give a stable dispersion at application viscosity, that is, to prevent macrophase separation. In the example shown in Figure 12.3, the 50 EN sample could not be fully diluted without macrophase separation. In a similar experiment using an 80 : 20 BMA/AA resin, 50 EN with DMAE gave a stable dispersion even when diluted until the viscosity was below 0.1 Pa·s. Viscosity at application viscosity changes rapidly with concentration, and it is possible to overshoot a desired reduction. The viscosity is also sensitive to amine content, so if a coating has been reduced with too much water, resulting in too low a viscosity, the viscosity frequently can be brought back up by addition of small amounts of amine.

Another variable is the structure of the neutralizing amine. While there may be some effect of base strength (less amine may be required with increasing base strength), the principal variable seems to be the water solubility of the amine. The amount of amine

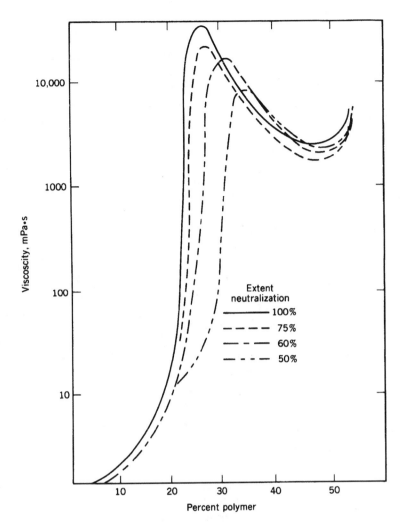

Figure 12.3. Viscosity as a function of weight percent resin when neutralized with varying levels of DMAE. Resin has 10 mol% AA; dilution started at 54 wt% solids. (From Ref. [32], with permission.)

required for stable dispersions increases in the order DMAE < triethylamine (TEA) < tripropylamine (TPA). Hydroxyl-substituted amines are the most widely used, but morpholine derivatives such as N-ethylmorpholine (NEM) are also effective.

It is important in selecting amines, to consider not only the effect on dispersion stability, but also the effect on package stability and curing of the coating [32]. The amine improves the package stability by minimizing reactions of MF resins during storage. If Class II MF resins are to be used, the amine must be tertiary. If, as is more common, Class I MF resins are used, primary or secondary amines can also be used.

2-(Dimethylamino)ethanol, while widely used, has a disadvantage. It has been shown that transesterification occurs between ester groups on the acrylic resin and the hydroxyl of the amino alcohol, thereby covalently bonding part of the amine to the resin [33]. Since bound amine cannot volatilize during application and baking, it inhibits the acid catalyzed cross-linking reactions. N-Ethylmorpholine, which has essentially the same boiling point as DMAE, permits faster curing, since it cannot undergo transesterification. 2-Amino-2-methyl-1-propanol (AMP) can be used with Class I MF resins. It provides stable dispersions and, in spite of its higher boiling point, gives coatings that cure more rapidly than even those neutralized with NEM. 2-Amino-2-methyl-1-propanol has a hydroxyl group that can transesterify, but also a primary amine that can undergo aminolysis reactions with ester groups on the resin [33]. By reacting at both sites, AMP can act as a cross-linking agent, supplementing cross-linking with the MF resin. Furthermore, AMP has been shown to cyclize, to an oxazolidine under curing conditions, reducing basicity and increasing cure rate [34]. The difference in curing response with DMAE and AMP has also been demonstrated by dynamic mechanical analysis [35].

Another factor involved in the choice of amine is the effect on wrinkling [33]. When using TEA as the neutralizing amine, it is common to get low gloss coatings. The low gloss results from the development of a fine wrinkle pattern on the surface during curing of the film, which is caused by faster curing of the surface layer than the interior of the film. When the interior of the film does cure, it shrinks, causing the immobilized surface of the film to wrinkle. (See Section 23.6 for further discussion of wrinkling.) This phenomenon occurs when amine is essentially absent near the surface of the film, but still present in the lower layers of the film; the MF resin readily cross-links the acrylic at the surface, while the reaction is still inhibited in the lower layers.

The probability of wrinkling increases as film thickness increases because there is more likely to be a differential in amine content as film thickness increases. The probability of wrinkling is also affected by amine structure. A series of amines gave decreasing probability of wrinkling in the order TEA > DMAE > NEM [33]. TEA is a relatively strong base (pK_a of conjugate acid $= 10.9$), and hence, diffuses slowly through a film containing COOH groups; it also has a relatively low boiling point (90°C), so on reaching the surface, it volatilizes rapidly. This combination of factors leads to a differential amine content even with relatively thin films. NEM, which is a weaker base (pK_a of conjugate acid $= 7.8$) and has a higher boiling point (139°C), can diffuse relatively more rapidly through the film, but volatilizes more slowly from the surface. The result is a more nearly uniform amine gradient and more uniform timing of curing of surface and lower layers of a film. DMAE has an intermediate base strength ($pK_a = 9.3$) and a relatively high boiling point (134°C), and exhibits intermediate behavior between that of TEA and NEM.

REFERENCES

1. R. P.-T. Chung and D. H. Solomon, *Prog. Org. Coat.*, **21**, 227 (1992).
2. J. L. Gerlock, D. F. Mielewski, D. R. Bauer, and K. R. Carduner, *Macromolecules*, **21**, 1604 (1988).

3. R. W. Ryan and M. B. Walt, *Proc. Waterborne High-Solids Powder Coat. Symp.*, New Orleans, 1994, p. 786.
4. L. W. Hill and K. Kozlowski, *J. Coat. Technol.*, **59** (751), 63 (1987).
5. T. Nakamichi and K. Shibato, *J. Jpn. Soc. Colour Mater.*, **59**, 592 (1986).
6. H. J. Spinelli, *Org. Coat. Appl. Polym. Sci.*, **47**, 529 (1982).
7. L. W. Hill and Z. W. Wicks, Jr., *Prog. Org. Coat.*, **10**, 55 (1982).
8. K. O'Driscoll, *J. Coat. Technol.*, **57** (705), 57 (1983).
9. K. J. H. Kruithof and H. J. W. van den Haak, *J. Coat, Technol.*, **62** (790), 47 (1990).
10. A. J. Wright, *Eur. Coat. J.*, 696 (1996).
11. R. A. Gray, *J. Coat. Technol.*, **57** (728), 83 (1985).
12. J. E. Nylund and S. Pruskowski, *Resin Review*, **39**, 17 (1989).
13. *Technical Bulletin on Joncryl 500, PSN 0-104-12/83*, S.C. Johnson & Son, Inc., Racine, WI, 1983.
14. V. R. Kamath and J. D. Sargent, Jr., *J. Coat. Technol.*, **59** (746), 51 (1987).
15. G. G. Myers, *J. Coat. Technol.*, **67** (841), 31 (1995).
16. D. Y. Sogah, W. H. Hertler, O. W. Webster, and G. M. Cohen, *Macromolecules*, **20**, 1473 (1987).
17. J. A. Simms and H. J. Spinelli, *J. Coat. Technol.*, **59** (752), 125 (1987).
18. K. Okadaa, K. Yamaguchi, and H. Takeda, *Prog. Org. Coat.*, **34**, 169 (1998).
19. J. Reitberg, J. H. van Roon, and R. van der Linde, *Proc. Intl. Conf. Org. Coat. Sci. Tech.*, Athens, 1989, p. 209.
20. M. Ooka and H. Ozawa, *Prog. Org. Coat.*, **23**, 325 (1994).
21. J. C. Kenny, T. Ueno, and K. Tsutsui, *J. Coat. Technol.*, **68** (855), 35 (1996).
22. T. Agawa and E. D. Dumain, *Proc. Waterborne High-Solids Powder Coat. Symp.*, New Orleans, 1997, p. 342.
23. A. Vazken, R. G. Lees, and D. E. Fiori, U. S. Patent 5254651 (1993).
24. J. W. Rehfuss and D. L. St. Aubin, U.S. Patent 5356669 (1994).
25. J. W. Rehfuss and D. L. St. Aubin, U.S. Patent 5605965 (1997).
26. S. V. Barancyk, C. A. Verardi, and W. A. Humphrey, International Patent Application, WO 95/29947 (1995).
27. M. J. Chen, F. D. Osterholtz, E. R. Pohl, P. E. Ramdatt, A. Chaves, and V. Bennett, *J. Coat. Technol.*, **69**, (870), 43 (1997).
28. B. C. Watson and Z. W. Wicks, Jr., *J. Coat. Technol.*, **55** (698), 59 (1983).
29. L. W. Hill and B. M. Richards, *J. Coat. Technol.*, **51** (654), 59 (1979).
30. Z. W. Wicks, Jr., E. A. Anderson, and W. J. Culhane, *J. Coat. Technol.*, **54** (688), 57 (1982).
31. R. Armat, S. G. Bike, G. Chu, and F. N. Jones, *J. Appl. Polym. Sci.*, **60**, 1927 (1996).
32. L. W. Hill and Z. W. Wicks, Jr., *Prog. Org. Coat.*, **8**, 161 (1980).
33. Z. W. Wicks, Jr. and G. F. Chen, *J. Coat. Technol.*, **50** (638), 39 (1978).
34. P. E. Ferrell, J. J. Gummeson, L. W. Hill, and L. J. Truesdell-Snider, *J. Coat Technol.*, **67** (851), 63 (1995).
35. L. W. Hill, P. E. Ferrell, and J. J. Gummeson, in T. Provder, M. A. Winnik, and M. W. Urban, Eds., *Film Formation in Waterborne Coatings*, American Chemical Society, Washington, DC, 1996, p. 235.

CHAPTER 13

Polyester Resins

In fibers and plastics, *polyester* is applied to high molecular weight, partially crystalline, linear, thermoplastic polymeric esters of a short-chain diol and terephthalic acid. In coatings, it is used to describe a different sort of material, most coating polyesters have relatively low molecular weights and are amorphous, branched, and ultimately cross-linked. It is applied only to certain polyesters, those prepared from polyols and polybasic acids. Alkyds are also polyesters in the chemical sense of the word, but they are not called polyesters in coatings. As discussed in Chapter 15, alkyds are prepared from polyols and dibasic acids, and, in addition, monobasic acids, usually derived from vegetable oils. Coating polyesters are sometimes called *oil-free polyesters* or *oil-free alkyds* to distinguish them from alkyds.

The general principles of step-growth polymerization of polyesters are discussed in Section 2.2.2. The relatively low molecular weight polyesters used in coatings are usually made from mixtures of diols, triols, and dibasic acids. Most commonly, excess polyol is used; hence, the polyesters are *hydroxy-terminated polyesters*. They are most commonly cross-linked with melamine-formaldehyde (MF) resins or polyisocyanates. Resins with excess carboxylic acid are also made; these *carboxylic acid-terminated polyesters* are cross-linked with epoxy resins, MF resins, or 2-hydroxyalkylamides. *Water-reducible polyesters* are synthesized with both terminal hydroxyl and carboxylic acid groups; they are usually cross-linked with MF resins.

Branched polyesters are made from monomer mixtures that include one or more monomers having functionality of $F > 2$. Kinetic analysis of such polymerizations is complex, but a few generalizations can be pointed out. As the proportion of monomer of $F > 2$ increases, $\bar{M}_n$, $\bar{M}_w/\bar{M}_n$, and number average functionality $\bar{f}_n$ all increase. The average $\bar{F}$ must be controlled to avoid gelation at high extents of reaction.

In general terms, thermosetting polyesters give coatings with better adhesion to metal substrates and better impact resistance than thermosetting acrylics (TSAs) (see Section 12.2). On the other hand, TSAs give coatings with superior water resistance and exterior durability. The differences result primarily from the presence of the ester linkages in the backbone of polyesters-linkages that impart flexibility, but are vulnerable to hydrolysis. Thus, polyesters tend to be used for one-coat coatings on metal, and TSAs tend to be used in applications for which exterior durability and moisture resistance are particularly important, often with primers that provide excellent adhesion of the coating system to the metal. Many exceptions to these generalizations can be found.

13.1. HYDROXY-TERMINATED POLYESTERS FOR CONVENTIONAL SOLIDS COATINGS

Most hydroxy-terminated polyesters are made by co-esterifying four types of monomers: two polyols (a diol and a triol) and two diacids (an aliphatic dibasic acid or its anhydride and an aromatic dicarboxylic acid or its anhydride). Tribasic and cycloaliphatic acids are used to some extent. The ratio of moles of dibasic acid to polyol must be less than 1 to give terminal hydroxyl groups and avoid gelation. Molecular weight is controlled by this ratio; the smaller the ratio, the lower the molecular weight. The molecular weight distribution, $\bar{M}_n$, and $\bar{f}_n$ are all controlled by the diol to triol ratio; $\bar{f}_n$ is critical because it affects the potential cross-link density after cure. With the same mole ratio of dibasic acid to total polyol, an increase in triol component increases the average number of hydroxyl groups per molecule, decreases hydroxy equivalent weight, broadens molecular weight distribution, and increases cross-link density of a fully cross-linked film. The ratio of aromatic to aliphatic dibasic acids is the principal factor controlling the T_g of the resin, but differences in polyol structure also affect T_g.

The monomers are esterified at 220 to 240°C while water is removed. Organotin compounds, titanium orthoesters, or zinc acetate are frequently used as catalysts. Strong protonic acids also catalyze esterification, but can cause side reactions and discoloration. The reaction is continued until a high conversion has been reached. Acid numbers of resins to be cross-linked with MF resins are generally 5–10 mg KOH per g of resin solids; those for use with polyisocyanates are commonly esterified until the acid number is less than 2. Polyesters made from polyfunctional reactants can cross-link during polymerization, causing gelation. Since the probability of gelation increases with conversion and high conversions are generally desirable, an equation relating formula composition, extent reaction, and probability of gelation would be useful. Progress has been made in predicting gelation in simple systems in which the reactivity of all hydroxyl groups and carboxylic acid groups are equal, but even with these "simple" systems, the equations give only fair predictions because they do not take into consideration the formation of cyclic esters. See Ref. [1] or [2] for discussions of such equations. In the case of polyesters used in coatings, the situation is far from simple; the reactivity of the various functional groups commonly varies substantially, and side reactions occur. Therefore, the textbook gelation equations are of little value for polyesters used in coatings. If the mole ratio of dibasic acid to polyol exceeds, or even closely approaches 1, gelation may occur before esterification is essentially complete. This is seldom a problem in making hydroxy-functional polyesters for coatings, but it can become critical in synthesis of alkyds (see Section 15.1).

Misev [3] has published equations that permit calculation of ratios of polyols and polybasic acids to use to make polyesters with a desired $\bar{M}_n$ and $\bar{f}_n$, at a desired extent reaction of the acid groups. The equations are complex and require iterative computer solution. The equations are simplified if the choice of raw materials is limited, as it generally is, to diols, triols, and dibasic acids, and if essentially all of the acid groups are to be esterified. In commercial practice, polyesters are cooked until the acid number is reduced to about 5, a level that corresponds to the esterification of about 99% of the carboxylic acid groups. Therefore, an assumption of complete esterification is reasonable.

In these simplified equations N_{p2} represents the number of moles of diol; N_{p3}, the number of moles of triol; and N_a, the number of moles of dibasic acid. M_{p2} represents the molecular weight of the diol; M_{p3}, the triol; and M_a, the dibasic acid. If more than one dibasic acid (or other component) is used, then additional terms would be added to the equation for each additional component. Using these equations, one can calculate how much diol, triol, and dibasic acid should be used in an initial laboratory cook for a desired $\bar{M}_n$ and $\bar{f}_n$.

In making hydroxy-terminated polyesters, the number of moles of polyol is one more than of dibasic acid, as shown in Eqs. 13.1 and 13.2:.

$$N_{p2} + N_{p3} = N_a + 1 \tag{13.1}$$

or

$$N_a = N_{p2} + N_{p3} - 1 \tag{13.2}$$

The relationship between the average functionality and the moles of reactants is expressed by Eq. 13.3:

$$\bar{f}_n = 2N_{p2} + 3N_{p3} - 2N_a \tag{13.3}$$

Substituting the value of N_a from Eq. 13.2 into Eq. 13.3 gives Eqs. 13.4 and 13.5:

$$\bar{f}_n = N_{p3} + 2, \tag{13.4}$$

or

$$N_{p3} = \bar{f}_n - 2 \tag{13.5}$$

When the esterification is carried out 2 mole of water (total molecular weight $= 36$) are split out for each mole of dibasic acid, resulting in Eq. 13.6 for $\bar{M}_n$. If an anhydride is used, only 1 mole of water ($M = 18$) is split out for each mole of anhydride, and the appropriate adjustment in Eq. 13.6 is needed.

$$\bar{M}_n = N_{p2}M_{p2} + N_{p3}M_{p3} + N_a(M_a - 36) \tag{13.6}$$

Substituting in Eq. 13.6 for N_a using Eq. 13.2 and for N_{p3} using Eq. 13.5 gives Eq. 13.7, which relates $\bar{M}_n$ to $\bar{f}_n$.

$$\bar{M}_n = N_{p2}M_{p2} + (\bar{f}_n - 2)M_{p3} + (N_{p2} + \bar{f}_n - 3)(M_a - 36) \tag{13.7}$$

A wide range of polyesters has been used commercially. In conventional polyester—MF coatings, polyesters with $\bar{M}_n$ on the order of 2000 to 6000, $\bar{M}_w/\bar{M}_n$ on the order of 2.5 to 4,

Table 13.1. Starting Formulation for a Conventional Polyester Resin

Raw Material	Weight	Moles	Equivalents
NPG	2050	19	38
TMP	270	3	9
IPA	1740	10.5	21
AA	1530	10.5	21

and with $\bar{f}_n$ of 4 to 10 hydroxyl groups are usual. Table 13.1 gives an example of an initial laboratory formulation calculated for a polyester with an $\bar{M}_n$ of 5000 and an $\bar{f}_n$ of 5 using Eqs. 13.1–13.7. Neopentyl glycol is NPG, trimethylolpropane is TMP, isophthalic acid is IPA, and adipic acid is AA. Structural formulas for these compounds are given in Sections 13.1.1 and 13.1.2.

A formulation such as that in Table 13.1 does not take into consideration all of the complexities of polyesterification; thus, it can only be expected to provide a starting point for experimentation. A common complication is partial loss of monomer from the reactor by volatilization. The diol is usually the most volatile component, so it is generally necessary to use some excess diol. The amount of excess diol depends upon the particular reactor and conditions; the efficiency of separating water and diol, the rate of flow of inert gas, the temperature of reaction, and so forth. A formula established in the laboratory has to be adjusted when production is scaled up in a pilot plant, again when it is scaled up in a production reactor, and yet again when production is shifted from one reactor to another.

Esterification is a reversible reaction. Consequently, the rate of reaction depends on the rate of removal of the water, particularly near the end of the process. A few percent of a reflux solvent, such as xylene, may be added to accelerate removal of water and to assist in return of volatilized diol to the reactor. Alternatively, an inert gas purge can be used in the later stages of the process to help remove the last of the water. Polyols can undergo self-condensation to form polyethers, which also produces water. Ester groups that form early in the process may hydrolyze or transesterify during the process and re-form many times, leading to a mixture of kinetic and thermodynamic control of the structure of the final product.

13.1.1. Selection of Polyols

Polyols are selected on the basis of cost, rate of esterification, stability during high temperature processing (minimal decomposition and discoloration), ease of separation from water during processing, viscosity of esters at equal molecular weight and function-ality, effect on T_g, rate of cross-linking with MF resins (or other cross-linkers), and hydrolytic stability of their esters. Obviously, compromises are necessary. The following polyols are representative of the many available:

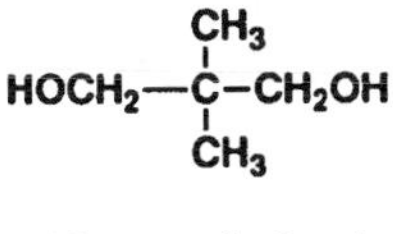

Neopentyl glycol
NPG

Cyclohexanedimethanol
CHDM

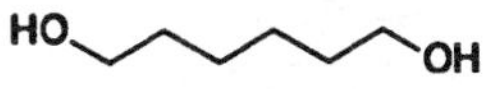

1.6-Hexanediol
HD

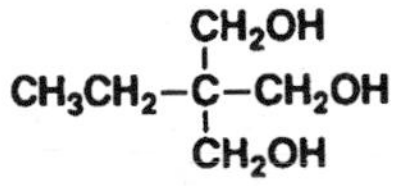

Trimethylolpropane
TMP

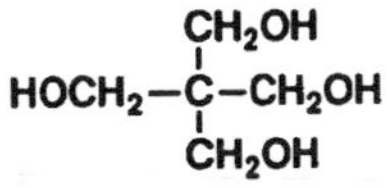

Pentaerythritol
PE

The most widely used diol is probably NPG, and the most widely used triol is TMP. CHDM esterifies significantly more rapidly than NPG, and TMPD more slowly [4]. It is a fair, but not always accurate, generalization that the slower the rate of esterification, the greater the resistance of the ester to hydrolysis. The hydrolytic stability of esters of NPG and TMP is better than those from less hindered glycols like ethylene glycol or propylene glycol. Neopentyl glycol (bp 213°C) is volatile at esterification temperatures, so well designed fractionating and condensing equipment are needed to permit the removal of water with a minimum loss of NPG during processing at 220 to 240°C. Most diols start decomposing in the presence of strong acids at temperatures above about 200°C, so strong acid esterification catalysts should be avoided. Organotin compounds and titanium orthoesters are examples of appropriate catalysts.

While the kinetics and mechanisms of esterification and hydrolysis reactions have been extensively studied from a fundamental point of view, the literature contains little basic data dealing with polyols of most interest in coatings resins. Turpin [5] has proposed that a variant of Newman's "rule of six" may be useful in predicting the relative hydrolytic stability of esters. Newman's rule states that if one numbers the atoms in the structure of an ester starting with the carbonyl oxygen as 1, the larger the number of substituents in the 6 positions, the better the hydrolytic stability. Turpin suggested that substituents in the 7 position might also affect the stability and proposed an empirical steric factor:

$$\text{Steric factor} = 4(\text{No. 6 position atoms}) + (\text{No. 7 position atoms})$$

For example for an NPG ester,

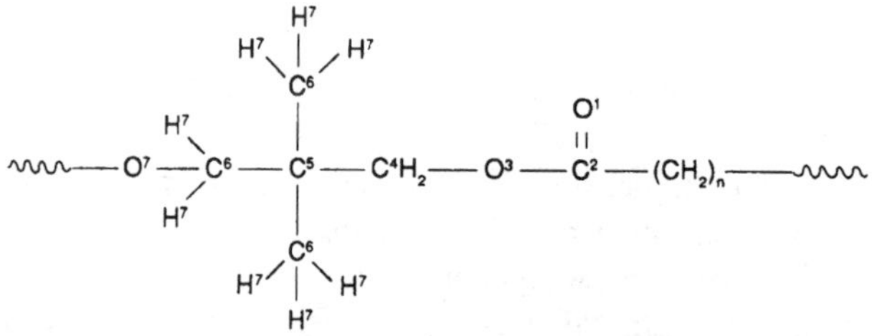

$$
\begin{aligned}
3 \text{ No. 6 atoms} \times 4 &= 12 \\
9 \text{ No. 7 atoms} \quad\quad &= \underline{9} \\
\text{Total} &= \overline{21}
\end{aligned}
$$

The value of 21 for NPG esters is the same for TMP esters, but for ethylene glycol and 1,6-hexanediol esters, the steric factors are only 13 and 15, respectively. Turpin's steric factor seems to correlate reasonably well with experience and with the limited published data for esters of acyclic alcohols. Turpin also pointed out that an *anchimeric*, or *neighboring group*, effect may be involved. For example, the location of the hydroxyl group in a half ester of ethylene glycol may promote the hydrolysis of its neighboring ester through a cyclic transition state and contribute to the poorer hydrolytic stability of ethylene glycol esters as compared to those of 1,6-hexanediol [5].

Esters of cycloaliphatic diols, such as cyclohexanedimethanol (CHDM), seem more resistant to hydrolysis than predicted by Turpin's steric factor. A comparison of results of testing films of coatings made with a series of polyesters from several polyols has been published [6]. Coatings based on CHDM polyesters cross-linked with MF resins showed the best test results for properties related to hydrolytic stability. In contrast to general behavior, CHDM also esterifies more rapidly than NPG [4]. It has also been shown that CHDM polyesters give cured films with MF resins more rapidly than NPG based polyesters [7].

13.1.2. Selection of Polyacids

Most polyesters are made using a mixture of aromatic and aliphatic diacids. The ratio of aromatic to aliphatic acid is a major factor that controls the T_g of the resin.

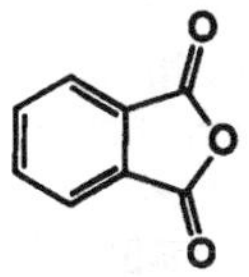

Phthalic Anhydride
PA

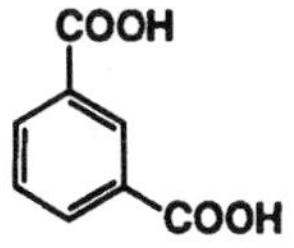

Isophthalic acid
IPA

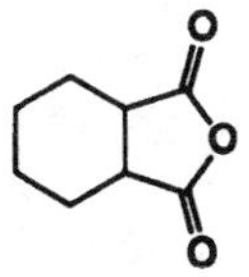

Hexahydrophthalic anhydride
HHPA

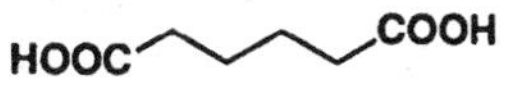

Adipic acid
AA

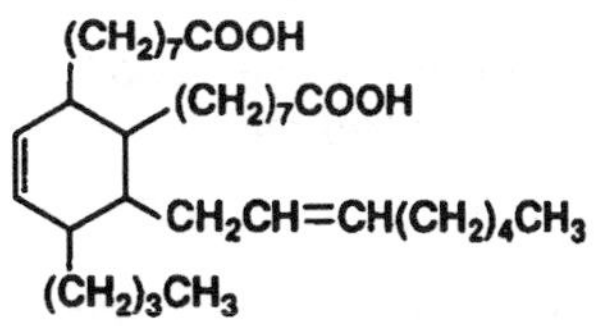

Dimer fatty acid
(One isomer of mixture)

Azelaic acid
AZA

Aromatic acid esters hydrolyze more slowly than aliphatic esters, unless there is an anchimeric effect. Isophthalic acid (IPA) is the predominant aromatic acid used even though phthalic anhydride (PA) can be processed at lower cost. The preference for IPA is based on superior exterior durability of coatings made with IPA polyesters; this has been attributed to greater hydrolytic stability of isophthalic esters. Half esters of (ortho)phthalic acid are more easily hydrolyzed than half esters of isophthalic acid in the pH range of 4 to 8 [8]. This difference results from the anchimeric effect of the carboxylic acid group of the phthalic acid half ester in a location, which increases the catalytic effect of the acid on the hydrolysis of the ester. During exterior exposure, resistance to hydrolysis in contact with rain water, which has a pH in the range of 4 to 6, is most important. Under these conditions, isophthalic acid polyesters are more resistant than phthalic acid polyesters. On the other hand, mono- and diesters of isophthalic acid hydrolyze more rapidly than those of phthalic acid under alkaline conditions [9,10].

A processing cost advantage of phthalic anhydride results from its lower melting point (131°C); it can be handled as a molten liquid, it is readily soluble in the reaction mixture, and its reactive anhydride structure rapidly forms monoesters at about 160°C. Isophthalic acid (mp > 300°C) is more difficult to process because it dissolves more slowly in the reaction medium, and lower reactivity slows the processing. Due to longer time at high temperatures, volatilization losses and acid catalyzed side reactions (e.g., etherification or dehydration of polyols; see Section 15.6.2) are more serious problems with IPA than with PA.

Adipic acid is probably the most widely used aliphatic dibasic acid. Succinic and glutaric acids give esters with inferior hydrolytic stability, presumably due to the anchimeric effect [5]. Longer chain acids such as azelaic and sebacic acids may give somewhat better hydrolytic stability and greater reduction of T_g (on a molar basis) as compared with adipic acid, but are more expensive. Dimer acids, derived from dimerization of drying oil fatty acids and, hence, predominantly C_{36} acids, are widely used. Fatty acid dimerization is discussed in Section 14.3.1. Dimer acids are relatively inexpensive and are available in various grades. The high quality grades of dimer acids normally used in polyesters have little monobasic or tribasic acid contaminants and have been hydrogenated to saturate any double bonds, which could be a weak point for photodegradation. Dimer acid-isophthalic polyesters impart an excellent balance of properties to MF cross-linked coil coatings, providing high resistance to marring combined with excellent resistance to cracking during fabrication of coated metal.

Cycloaliphatic dibasic acids or anhydrides can be used. Hexahydrophthalic anhydride (HHPA) is said to give exterior durability comparable to IPA. It has been speculated that HHPA half esters may be less susceptible to anchimeric effects than phthalic acid half esters. The esters may isomerize to the *trans* isomers during processing. 1,4-Cyclohexanedicarboxylic acid (CHDA) provides a good balance of hardness, flexibility, and stain resistance. CHDA esters are much more stable to hydrolysis at pH 8.9 than IPA esters, but somewhat less stable to hydrolysis at pH 4.5 [9,10].

13.2. POLYESTER RESINS FOR HIGH SOLIDS COATINGS

Low viscosity polyesters are required for reduction of VOC emissions; viscosity of concentrated polyester solutions depends on several variables. Molecular weight and molecular weight distribution are two important factors. The number of functional groups per molecule also affects the viscosity; increasing the number of hydroxyl groups (and to an even greater extent, the number of carboxylic acid groups) increases the probability of intermolecular hydrogen bonding, in turn increasing viscosity. The hydrogen-bond effect can be minimized, but not eliminated, by using hydrogen-bond acceptor solvents such as ketones. The T_g of the resin also affects viscosity at a given concentration. For this reason, reducing the ratio of aromatic to aliphatic dibasic acids and using acyclic polyols rather than cyclic polyols, such as CHDM [6,7], give lower viscosities, but there is a lower limit of T_g below which desired film properties cannot be attained.

Polyesters have significant advantages over acrylics in high solids coatings. In contrast to the difficulty of making low molecular weight TSAs with at least two hydroxyl groups on essentially all molecules, the synthesis of polyesters with two or more hydroxyl groups on essentially all of the molecules is relatively easy. With the low mole ratios of dibasic acids to polyols (2:3 is typical) required to make low molecular weight resins, the probability of cyclization reactions to form nonfunctional materials is small. Virtually all of the final reactions result in terminal hydroxyl groups. While lowest viscosity is obtained with difunctional resins, such resins also require the most careful formulation and curing to achieve good film properties. If a hydroxyl is left unreacted on a conventional polyester with an average of five or so hydroxyl groups per molecule, the effect on film properties would probably be relatively small. However, failing to react all of the hydroxyl groups on a resin with only two hydroxyl groups per molecule would magnify the effect. A compromise is required; commonly, an $\bar{f}_n$ of between two and three hydroxyl groups per molecule is used.

The major factor controlling molecular weight ($\bar{M}_n$) of a polyester is the mole ratio of dibasic acid to polyol. A ratio on the order of 2:3 is typical. Loss of polyol during production of a polyester results in a higher molecular weight than predicted from the starting ratio. Condensers are designed to remove water rapidly, while returning a high fraction of NPG (or other polyol), but none is completely efficient. It is necessary to add some extra glycol to compensate for losses. Since production equipment and conditions vary, the final decision on the mole ratio to load must be based on experience with the particular reactor and with the particular formula. The amount of glycol being lost can be estimated by checking the refractive index of the water being removed from the reaction.

Some low molecular weight components of a resin may volatilize when a thin film is baked. Such loss has to be counted as part of the VOC emissions, together with solvent loss. Belote and Blount studied this volatility both in the absence of cross-linking agent and in its presence [11]. They made a series of model resins from NPG and a 1 : 1 mole mixture of AA and IPA; the polyacid to polyol ratio was varied from 1 : 2 to 1 : 1.15. The presence of the cross-linker (Class I MF resin) reduced volatile loss. They concluded that an $\bar{M}_n$ of about 800 to 1000 was optimum for achieving lowest VOC with resins having normal molecular weight distributions. At $\bar{M}_n$ around 600, so much low molecular weight fraction volatilized that total VOC was higher than at $\bar{M}_n$ of 800, *even though the amount of solvent required was lower*. A broad molecular weight distribution not only increases volatile losses because of the low molecular weight material, but also gives a higher viscosity because of the disproportionate effect on viscosity of the high molecular weight fraction. Lowest viscosity results from making as narrow molecular weight distribution as possible.

Use of 2,2,4-trimethyl-1,3-pentanediol (TMPD) in combination with a two-stage addition of the triol TMP during synthesis of a polyester, gives lower viscosity and improved film properties [12]. Use of TMPD, an unsymmetrical diol, presumably gives narrower molecular weight distribution because of the differential reactivity of the two hydroxyl groups. A drawback is the need for somewhat higher baking temperatures for coatings, presumably a result of the low reactivity of the hindered terminal hydroxyl groups derived from TMPD. Three of the resin formulations used in this study are given in Table 13.2. Addition of the TMP in two stages resulted in a reduction in $\bar{M}_n$ from 1500 to 1100 (compare Resins 1 and 2) with a corresponding reduction in viscosity. Further reduction was achieved by adjusting the monomer ratios (Resin 3).

A difunctional polyester with a narrow molecular weight distribution and a reported $\bar{M}_n$ of 425 is commercially available [13]. This resin is presumably manufactured by preparing

Table 13.2. High Solids Polyester Formulations

Raw Material	Equivalents		
	Resin 1	Resin 2	Resin 3
TMPD	11.96	11.96	12.24
TMP	1.72	0.86	0.88
IPA	4.56	4.56	4.38
AA	4.56	4.56	4.38
Second Stage			
TMP	–	0.86	0.88
Resin $\bar{M}_n$	1500	1075	1000
Coating NVW	75.6	76.5	77.6

a polyester with excess CHDM and then distilling off the low molecular weight fraction with a vacuum thin-film evaporator. Low viscosity polyester diols and triols prepared by reacting caprolactone with a diol or a combination of diol and triol are also available [14]. These types of low $\bar{M}_n$ resins seldom provide satisfactory coatings when used by themselves, but they are useful in blends to increase the solids content of somewhat higher molecular weight polyester or acrylic based coatings. They are often called *reactive diluents*.

Linear polyesters made by transesterification of 1,4-butanediol with a mixture of glutarate, adipate, and azaleate methyl esters give low viscosity hydroxy-terminated polyesters when $\bar{M}_n$ 680 [15]. They gave solvent-free two package coatings with hexamethylene diisocyanate trimer. However, with Class I MF resins, baked films were soft due to low T_g.

13.3. CARBOXYLIC ACID-TERMINATED POLYESTER RESINS

Carboxylic acid-terminated polyesters are copolymers of a diol and a molar excess of dibasic acid. A modest amount of a tri- or polyfunctional monomer, most commonly trimellitic anhydride (TMA), is almost always included to increase $\bar{f}_n$. Carboxylic acid-functional polyesters can be cross-linked with MF resins, but in most cases are cross-linked with epoxy resins or 2-hydroxyalkylamides, as discussed in Sections 11.3.1 and 16.6.1, respectively. With high ratios of high melting point diacids like IPA, it is usually preferable to use a two-step process. The slow dissolution and high melting point of IPA prevents clean reaction when it is used in excess. First, one prepares a hydroxy-functional polyester with the IPA and then reacts the terminal hydroxyl groups with a lower melting acid or anhydride. Careful control is needed to assure reproducible products, since varying degrees of transesterifcation occur in the second stage. Final product structure is governed both by kinetic and thermodynamic factors; thus, resin properties are affected by reaction time and other process variables.

13.4. WATER-REDUCIBLE POLYESTER RESINS

Polyesters are also made for use in water-borne coatings with reduced VOC. Most such polyesters have both hydroxyls and carboxylic acids as terminal groups. As with other water-reducible resins, acid numbers in the range of 35 to 60 are required to give amine salt solutions in solvent that can be diluted with water to give reasonably stable dispersions of aggregates of resin molecules swollen with water and solvent. These resins show abnormal viscosity dilution curves similar to those described for water-reducible acrylic resins in Section 12.3.

In theory, one could prepare such water-reducible polyesters by simply stopping the esterification of a combination of diol, triol, AA, and IPA at the desired acid number; in practice, it is almost impossible to do so reproducibly. Rather, it is necessary to find combinations of polybasic acids that have sufficiently different reactivity to permit the control of the ratio of unreacted hydroxyl and carboxyl groups. The most widely used method has been to prepare a hydroxy-functional resin and then add enough TMA to esterify a fraction of the hydroxyl groups, generating two carboxyl groups at each site. Advantage is taken of the higher reactivity of the anhydride groups of the TMA at 180°C,

Table 13.3. Water-Reducible Polyester Formulation

Materials	Weight	Mole Ratio	Equivalent Ratio
NPG	685	1.0	1.0
Adipic Acid	192	0.2	0.2
TMA	84	0.067	0.1
IPA (85:15 IPA to TPA)	655	0.6	0.6

React at 235°C until the acid number is 16-18; cool to 180°C, then add:

TMA	84	0.067	0.1

React at 180°C until the acid number is 40-45; cool to 160°C, and thin to 80% solids with diethylene glycol monobutyl ether.

relative to the carboxylic acid groups formed in the reaction. Table 13.3 gives an example of a formulation for preparing a water-reducible polyester [16].

This approach has been used on a large scale, but it has disadvantages. Most seriously, the ester linkages in the resin are vulnerable to hydrolysis during storage of the formulated coating. The ester group of partially esterified TMA is particularly subject to hydrolysis because of the anchimeric effect of the adjacent carboxylic acid group. Hydrolysis of the TMA partial ester results in removal of the solubilizing carboxyl groups and, thereby, destabilizes the resin dispersion. The film properties may also be adversely affected. In addition, the use of primary alcohols as a solvent, as recommended in Table 13.3, has been found to be undesirable. Primary alcohols have been found to esterify carboxylic acid groups and transesterify ester groups during the thinning at 160°C and, at a slow, but appreciable, rate, during storage of the resin. This problem can be minimized by using a secondary alcohol as solvent, along with careful control of process temperatures. 1,4-Cyclohexanedicarboxylic acid esters are much more stable to hydrolysis under basic conditions than IPA [10]. Water-reducible polyesters are used in applications for which good storage stability and hydrolytic stability are not important, such as industrial coatings with a fast turnover.

An alternative approach that provides somewhat better, but still limited, hydrolytic stability is to use 2,2-dimethylolpropionic acid as one of the diol components. The carboxylic acid group of this monomer is highly hindered by being located on a tertiary carbon. The resulting differential reactivity makes it possible to esterify the hydroxyl groups while leaving most of the acid groups unreacted. While they are too hindered to esterify readily, the acid groups still readily form salts.

Hydrolytic stability is also affected by the choice of polyol. In addition to the steric effect discussed earlier, it has been shown that polyols with low water solubility give polyesters that are more stable to hydrolysis under basic conditions than those with higher water solubility, presumably because the polymers are more hydrophobic [10]. For example, polyesters made with 2-butyl-2-ethyl-1,3-propanediol have much greater stability to hydrolysis than NPG esters.

The problem of hydrolysis can be minimized by making powdered solid polyesters. An example of such a solid polyester is made from purified isophthalic acid, adipic acid, neopentyl glycol, cyclohexanedimethanol, hydrogenated bisphenol A, and trimellitic anhydride [17]. The resin is powdered and stored until a coating is to be made; then, it is stirred into a hot aqueous solution of dimethylethanolamine to form a dispersion.

Another possible problem with water-reducible polyesters results from intramolecular reaction of terminal hydroxyl and carboxylic acid groups to form some low molecular weight nonfunctional cyclic polyesters. When coatings are baked, a small amount of the cyclic esters can volatilize out of the coating and gradually accumulate in cool spots in the oven. Eventually, sufficient resin can accumulate to drip down on products going through the ovens, marring their finish. Since the amounts are small, dripping may only start after weeks or months of operation of the coating line.

Water-thinnable polyester coatings have been formulated with low molecular weight oligomeric hydroxy-terminated polyesters [15]. Up to about 20% of water dissolves in a polyester-Class I MF resin binder, reducing the viscosity to about half. This permits making solvent free coatings. Further work is needed but the idea is appealing not only because of the low VOC from the lack of need of cosolvent, but also because no amine is needed. Absence of amine should reduce the problem of hydrolysis.

13.5. POLYESTER RESINS FOR POWDER COATINGS

Polyester resins for powder coating are brittle solids with a relatively high T_g (50–60°C), so the powder coating does not sinter (partially fuse) during storage. These requirements are met with terephthalic acid (TPA) and NPG as the principal monomers. Smaller amounts of other monomers are added to increase $\bar{f}_n$ and to reduce T_g to the desired level. The relatively high T_g makes it possible to prepare hard, tough films with relatively low cross-link density. Widely used commercial products are amorphous, not crystalline [18]. Both hydroxy- and carboxy-terminated polyesters are used. The former are most commonly cross-linked with blocked isocyanates (see Section 10.5) and the latter with epoxy resins (see Chapter 11). Other cross-linkers include 2-hydroxyalkylamides (see Section 16.6.1) and tetramethoxymethylglycoluril (see Section 9.4.3). (See Chapter 27 for discussion of powder coatings.)

1,4-Cyclohexanedicarboxylic acid (CHDA) has been suggested as a replacement for isophthalic acid [19]. CHDA polyesters have lower T_g and lower melt viscosities; if the T_g is too low, NPG can be fully or partially replaced with hydrogenated bisphenol A to provide for storage stability. Processes involving direct esterification of TPA are used, but since TPA has a very high melting point, it is common to make the hydroxy-terminated polyester by transesterification of dimethyl terephthalate. An appropriate transesterification catalyst is tetraisopropyl titanate. If desired, the hydroxy-terminated TPA polyester can then be reacted with other polyacids to form a carboxylic acid-terminated product in a second stage.

REFERENCES

1. G. W. Odian, *Principles of Polymerization*, 3rd. ed., Wiley, New York, 1991, pp. 40–125.
2. H. G. Elias, *An Introduction to Polymer Science*, VCH, Weinheim, New York, 1997, pp. 120–142.
3. T. A. Misev, *J. Coat. Technol.*, **61** (772), 49 (1989).
4. D. E. Van Sickle, M. A. Taylor, and L. A. Bass, *Polym. Mater. Sci. Eng.*, **76**, 288 (1997).
5. E. T. Turpin, *J. Paint Technol.*, **47** (602), 40 (1975).

6. D. J. Golob, T. A. Odom, Jr., and R. W. Whitson, *Polym. Mater. Sci. Eng.*, **63**, 826 (1990).

7. G. Chu and F. N. Jones, *J. Coat. Technol.*, **65** (819), 43 (1993).

8. M. L. Bender, F. Chloupek, and M. C. Neveu, *J. Am. Chem. Soc.*, **80**, 5384 (1958).

9. E. Kivinen and E. Tommila, *Suomen Kemistilehti*, **14B**, 7 (1941); G. V. Rao and N. Venkatasu-bramanian, *Aust. J. Chem.*, **24**, 201 (1971); A. Cambon and R. Jullien, *Bull. Soc. Chim. France*, 2003 (1973).

10. T. E. Jones and J. M. McCarthy, *J. Coat. Technol.*, **67** (844), 57 (1995).

11. S. N. Belote and W. N. Blount, *J. Coat. Technol.*, **53** (681), 33 (1981).

12. J. D. Hood, W. W. Blount, and J. T. Sade, *Proc. Water-Borne Higher-Solids Coat. Symp.*, New Orleans, 1986, p. 14.

13. L. J. Calbo, *Proc. Water-Borne Higher-Solids Coat. Symp.*, New Orleans, 1986, p. 356.

14. Union Carbide Corp., Specialty Polymers and Composites Div., *Tech. Bull., TONE Polyols*, 1986.

15. F. N. Jones, *J. Coat. Technol.*, **68** (852), 25 (1996).

16. Amoco Chemicals Corp., *Tech. Bull. TMA-109e*, 1984.

17. R. Engelhardt, *Proc. Waterborne Higher-Solids Powder Coat. Symp.*, New Orleans, 1996, p. 408.

18. F. N. Jones and G. Teng, unpublished results.

19. L. K. Johnson and W. T. Sade, *J. Coat. Technol.*, **65** (826), 19 (1993).

CHAPTER 14

Drying Oils

Among the oldest binders for paints are *drying oils*, which are liquid vegetable or fish oils that react with oxygen to form solid films. They have been used since prehistoric times; in the 19th and early 20th centuries, binders of most paints were drying oils. Their use has decreased; however, they still have applications. Most importantly, they are raw materials for other binders such as alkyd resins, epoxy esters, and uralkyds. (See Chapter 15.) These resins can be considered to be synthetic drying oils; an understanding of drying oil chemistry is a necessary foundation for understanding them. The chapter only summarizes the information available; see the general references for further details.

14.1. COMPOSITION OF NATURAL OILS

Naturally occurring oils are *triglycerides*, triesters of glycerol and fatty acids. Some triglycerides are drying oils, but many are not. The reactivity of drying oils with oxygen results from the presence of diallylic groups (i.e., two double bonds separated by methylene groups, $-CH=CHCH_2CH=CH-$) or conjugated double bonds. Esters of many different fatty acids occur in nature. Fatty acids with 18 carbon atoms are most common; those found in oils that are most important in coatings follow. The letters c and t represent *cis* and *trans* orientation, respectively and the numbers designate the position of the first carbon of the double bond.

Common Name	Structure
Stearic acid	$CH_3(CH_2)_{16}COOH$
Palmitic acid	$CH_3(CH_2)_{14}COOH$
Oleic acid	$CH_3(CH_2)_7CH=CH(CH_2)_7COOH$ (9c)
Linoleic acid	$CH_3(CH_2)_4CH=CHCH_2CH=CH(CH_2)_7COOH$ (9c 12c)
Linolenic acid	$CH_3CH_2CH=CHCH_2CH=CHCH_2CH=CH(CH_2)_7COOH$ (9c 12c 15c)
Pinolenic acid	$CH_3(CH_2)_4CH=CHCH_2CH=CHCH_2CH_2CH=CH(CH_2)_3COOH$ (5c 9c 12c)
Ricinoleic acid	$CH_3(CH_2)_5\overset{\displaystyle OH}{C}HCH_2CH=CH(CH_2)_7COOH$ (9c)
α-Eleostearic acid	$CH_3(CH_2)_3CH=CHCH=CHCH=CH(CH_2)_7COOH$ (9t 11c 13t)

The oils are mixtures of mixed triglycerides with different fatty acids distributed among the triglyceride molecules. An example of one of the many triglycerides in linseed oil follows:

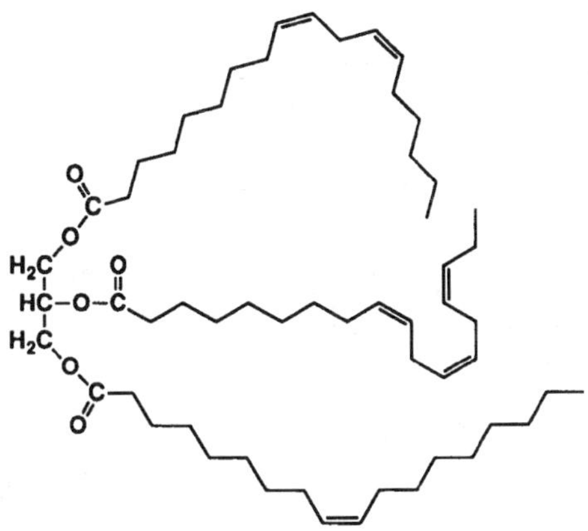

A Triglyceride

Separation of the tens or hundreds of different triglyceride molecules present in an oil borders on the impossible. Therefore, oils are characterized by high performance liquid chromatography (HPLC) or gas chromatography (GC) of the methyl esters obtained by transesterification [1]. Typical fatty acid contents of some oils are given in Table 14.1.

Table 14.1. Typical Fatty Acid Compositions of Selected Oils

	Fatty Acids				
Oil	Saturated[a]	Oleic	Linoleic	Linolenic	Other
Linseed	10	22	16	52	
Safflower	11	13	75	1	
Soybean	15	25	51	9	
Sunflower, MN	13	26	61	trace	
Sunflower, TX	11	51	38	trace	
Tung	5	8	4	3	80[b]
Tall oil fatty acids[c]	8	46	41[d]	3	2[e]
Tall oil fatty acids[f]	2.5	30	45	1	14[g]
Castor	3	7	3		87[h]
Coconut	91	7	2		

[a] Saturated fatty acids are mainly mixtures of stearic (C_{18}) and palmitic (C_{16}) acids; coconut oil also contains C_8, C_{10}, C_{12}, and C_{14} saturated fatty acids.
[b] α-Eleostearic acid.
[c] North American origin.
[d] Linoleic plus geometric and conjugated isomers.
[e] Rosin.
[f] European origin.
[g] Pinolenic acid.
[h] Ricinoleic acid.
Sources: See Refs. 2–6.

Compositions of the oils vary, sometimes quite widely, with variations in plant strain, climate, soil, and other growth conditions. Differences in sunflower oils are especially large, as illustrated in Table 14.1 for sunflower oils from Minnesota and Texas. In general, oils derived from seeds grown in colder climates contain larger fractions of more highly unsaturated fatty acids as triglyceride esters.

Included in Table 14.1 is an important kind of fatty acid, called *tall oil fatty acids* (TOFA). The word tall is the Swedish word for pine. Tall oil fatty acids are obtained as a byproduct of the sulfate pulping process for making paper. Oils present in the trees are saponified to soaps in the process. Tall oil fatty acids are obtained by acidification, followed by fractional distillation to separate the fatty acids from rosin acids and other byproducts also present. These acids have a range of compositions, especially between acids of North American and European origin as shown. They are often chemically modified to give conjugated bonds (see Section 14.3.3).

Animal oils and fats are also triglycerides, but the only animal oils used to any degree in coatings are refined fish oils. They contain triglycerides of a wide spectrum of fatty acids, including C_{18} to C_{26} fatty acids with up to five nonconjugated double bonds separated by single methylene groups.

14.2. AUTOXIDATION AND CROSS-LINKING

Drying, semidrying, and nondrying oils are often defined based on their *iodine value*, that is, grams of iodine required to saturate the double bonds of 100 g of an oil. Some authors classify oils as follows: drying oils, iodine value greater than 140; semidrying oils, iodine value, 125–140; and nondrying oils, iodine value less than 125 [7]. Although iodine values can serve as satisfactory quality control specifications, they are not useful and can be misleading in defining a drying oil or for predicting reactivity.

14.2.1. Nonconjugated Drying Oils

A useful empirical relationship is that nonconjugated oils that have a *drying index* greater than 70 are drying oils [8]. The drying index is calculated as follows:

$$\text{Drying index} = (\% \text{ linoleic acid}) + 2(\% \text{ linolenic acid})$$

Using this formula with the data from Table 14.1, the drying index of linseed oil is 120, it is a drying oil; the drying index of soybean oil is 69, it is a semidrying oil. The active group initiating drying is the diallylic group ($-\text{CH}=\text{CHCH}_2\text{CH}=\text{CH}-$). Linoleic and linolenic acids have one and two diallylic groups per molecule, respectively. Criteria for predicting drying capacity can be stated more generally: Drying is related to the average number of diallylic groups per molecule. If this number is greater than about 2.2, the oil is a drying oil; if it is moderately below 2.2, the oil is a semidrying oil; there is no sharp dividing line between semidrying oils and nondrying oils. These statements apply to synthetic drying oils as well as to natural oils. Since diallylic groups are the sites for cross-linking, it is convenient to relate the average number of such groups per molecule to the number average functionality $\bar{f}_n$ of the triglyceride or synthetic drying oil. We do so to facilitate the ensuing discussion, although it is possible that some of the sites are involved in more than one cross-linking reaction.

As previously indicated, the reactivity of nonconjugated drying oils is related to the average number of diallylic groups per molecule. The methylene groups are activated by their allylic relationship to two double bonds and are much more reactive than methylene groups allylic to only one double bond. Evidence for these assertions is provided by the relative rates of autoxidation of synthetic triolein (glyceryl trioleate), trilinolein, and trilinolenein, which are $1:120:330$, respectively [9]. The number of diallylic groups ($\bar{f}_n$) of the three triglycerides are 0, 3, and 6, respectively; the theoretical iodine values are 86, 173, and 262. The autoxidation rates are more closely related to the number of diallylic groups between double bonds, $\bar{f}_n$, than to iodine values, which are proportional to the average number of double bonds per molecule. Based on the data in Table 14.1, the $\bar{f}_n$ of the linseed oil is 3.6; it is a drying oil. The $\bar{f}_n$ of soybean oil is 2.07; it is a semidrying oil. The higher the $\bar{f}_n$ of a drying oil, the more rapidly a solvent-resistant, cross-linked film forms on exposure to air.

The reactions taking place during drying are complex. Early studies of the chemistry of drying were done without benefit of modern analytical instrumentation, which has recently been brought to bear on this complex problem [5,10–13]. Cross-linked films form from linseed oil in the following stages: (1) an induction period during which naturally present antioxidants (mainly tocopherols) are consumed, (2) a period of rapid oxygen uptake with a weight gain of about 10% (FTIR shows an increase in hydroperoxides and appearance of conjugated dienes during this stage), and (3) a complex sequence of autocatalytic reactions in which hydroperoxides are consumed and cross-linked film is formed. In one study, steps 1, 2, and 3 were far along in 4, 10, and 50 hours, respectively, when catalyzed by a *drier* (see Section 14.2.2) [10]. Cleavage reactions to form low molecular weight byproducts also occur during the latter stages of film formation. Slow continuing cleavage and cross-linking reactions through the lifetime of the film lead to embrittlement, discoloration, and slow formation of volatile byproducts. Oils with significant quantities of fatty acids with three double bonds, such as linolenic acid, discolor to a particularly marked degree.

The following scheme illustrates some of the many reactions that occur during cross-linking. Initially, naturally present hydroperoxides decompose to form free radicals:

$$ROOH \rightarrow RO\cdot + HO\cdot$$

At first, these highly reactive free radicals react mainly with the antioxidant, but as the antioxidant is consumed, the free radicals react with other compounds. Hydrogens on methylene groups between double bonds are particularly susceptible to abstraction, yielding the resonance stabilized free radical (**1**):

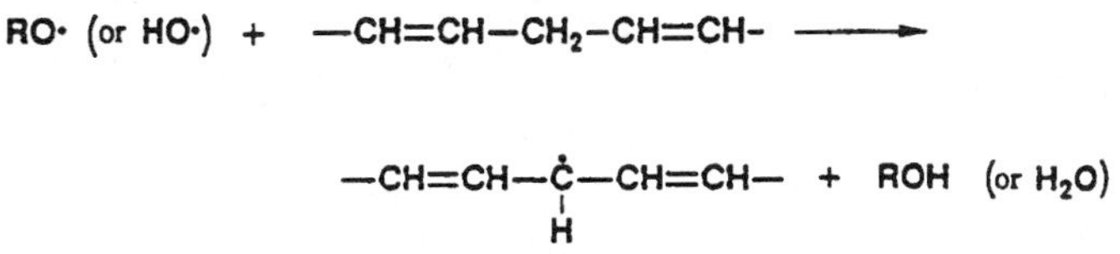

$$RO\cdot \text{ (or } HO\cdot) + \quad -CH=CH-CH_2-CH=CH- \quad \longrightarrow$$

$$-CH=CH-\overset{|}{\underset{H}{\dot{C}}}-CH=CH- \quad + \quad ROH \text{ (or } H_2O)$$

$$\mathbf{1}$$

Free radical (**1**) exists as three resonance contributors and reacts with oxygen to give predominantly a conjugated peroxy free radical such as (**2**):

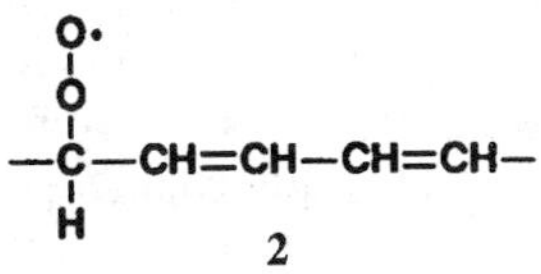

$$\overset{O\cdot}{\underset{|}{\overset{|}{O}}} \\ -\overset{|}{\underset{H}{C}}-CH=CH-CH=CH-$$

$$\mathbf{2}$$

The peroxy free radicals can abstract hydrogens from other methylene groups between double bonds to form additional hydroperoxides and generate free radicals like (**1**). Thus, a chain reaction is established, resulting in autoxidation. At least part of the cross-linking occurs by radical-radical combination reactions forming C–C, ether, and peroxide bonds. These reactions correspond to termination by combination reactions in free-radical chain-growth polymerization. (See Section 2.2.1.)

$$R\cdot \; + \; R\cdot \; \longrightarrow \; R{-}R$$
$$RO\cdot \; + \; R\cdot \; \longrightarrow \; R{-}O{-}R$$
$$RO\cdot \; + \; RO\cdot \; \longrightarrow \; RO{-}OR$$

Reactions analogous to the addition step in chain-growth polymerization could also produce cross-links. For example, the residual conjugated double bonds after intermediate (**2**) has cross-linked may undergo addition reactions:

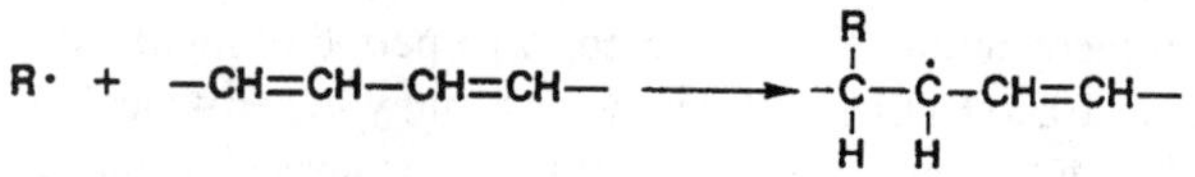

$$R\cdot \; + \; {-}CH{=}CH{-}CH{=}CH{-} \; \longrightarrow \; \underset{\overset{|}{H}}{\overset{\overset{R}{|}}{C}}{-}\underset{\overset{|}{H}}{\dot{C}}{-}CH{=}CH{-}$$

3

Such reactions could yield C–C or C–O linkages depending on the structure of R·. Subsequently, free radical (**3**) can rearrange, add oxygen to form a peroxy free radical, abstract a hydrogen from a methylene between double bonds, combine with another free radical, or add to a conjugated double bond system.

Studies of the reactions of ethyl linoleate with oxygen in the presence of driers (see Section 14.2.2) by ^{1}H and ^{13}C NMR showed that the predominant cross-linking reactions were those that formed ether and peroxy cross-links [11,12]. Mass spectroscopic studies showed that only about 5% of the cross-links were new C–C bonds [11]. Substantial levels of epoxy groups were detected in the reaction mixture, rising to a maximum in about 5 days and virtually disappearing in 100 days; it is suggested that epoxy groups may react with carboxyl groups to form ester cross-links. Rearrangement and cleavage of hydroperoxides to aldehydes and ketones, among other products, lead to low molecular weight byproducts. The characteristic odor of oil and alkyd paints during drying is attributable to such volatile byproducts, as well as to the odor of organic solvents.

Undesirable odor has been a factor motivating replacement of oil and alkyds in paints with latex, particularly for interior applications. The reactions leading to these odors have been extensively studied in connection with flavor changes of vegetable cooking oils [13]. Aldehydes have been shown to be major byproducts from the catalyzed autoxidation of methyl oleate, linoleate, and linolenate, as well as from curing of drying oil-modified alkyd resins [11,14]. It has also been shown that C_9 acid esters remain in the nonvolatile reaction mixture [14].

Dried films, especially of oils with 3 double bonds, yellow with aging. The yellow color bleaches significantly when exposed to light; hence, yellowing is most severe when films are covered, such as by a picture hanging on a wall. The reactions leading to color are complex and are not fully understood. Yellowing has been shown to result from incorporation of nitrogen compounds and is markedly increased by exposure to ammonia. It has been proposed that ammonia reacts with 1,4-diketones formed in autoxidation to yield pyrroles, which oxidize to yield highly colored products [15].

14.2.2. Catalysis of Autoxidation and Cross-Linking

The rates at which uncatalyzed nonconjugated drying oils dry are slow. Many years ago, it was found that metal salts (*driers*) catalyze drying. The most widely used driers are oil-soluble cobalt, manganese, lead, zirconium, and calcium salts of octanoic or naphthenic acids. Salts of other metals, including rare earths, are also used. Cobalt and manganese salts, so-called *top driers* or *surface driers*, primarily catalyze drying at the film surface. Lead and zirconium salts catalyze drying throughout the film and are called *through driers*. Calcium salts show little, if any, activity alone, but may reduce the amount of other driers needed. The surface-drying catalysis by cobalt and manganese salts results from the catalysis of hydroperoxide decomposition:

$$Co^{2+} + ROOH \longrightarrow RO\cdot + OH^- + Co^{3+}$$

$$Co^{3+} + ROOH \longrightarrow ROO\cdot + H^+ + Co^{2+}$$

The net result is formation of water and free radicals that undergo reactions such as given above. Note that cobalt cycles between the two oxidation states. The activity of through driers has not been adequately explained.

Combinations of metal salts are almost always used. Mixtures of lead with cobalt and/or manganese are particularly effective, but, as a result of toxicity control regulations, lead driers can no longer be used in consumer paints sold in interstate commerce in the United States. Combinations of cobalt and/or manganese with zirconium, frequently with calcium, are commonly used. Calcium does not undergo redox reactions; it has been suggested that it may promote drying by preferentially adsorbing on pigment surfaces, minimizing adsorption of active driers. The amounts of driers needed are system specific. Their use should be kept to the minimum possible level, since they not only catalyze drying, but also catalyze the reactions that cause postdrying embrittlement, discoloration, and cleavage.

14.2.3. Conjugated Drying Oils

Oils containing conjugated double bonds, such as tung oil, dry more rapidly than any nonconjugated drying oil. Free radical polymerization of the conjugated diene systems can lead to chain-growth polymerization, rather than just a combination of free radicals to form cross-links. High degrees of polymerization are unlikely because of the high concentration of abstractable hydrogens acting as chain-transfer agents. However, the free radicals formed by chain transfer also yield cross-links. In general, the water and alkali resistance of films derived from conjugated oils are superior, presumably because more of the cross-links are stable carbon-carbon bonds. However, since the α-eleostearic acid in tung oil has three double bonds, discoloration on baking and aging is severe.

14.3. SYNTHETIC AND MODIFIED DRYING OILS

Several types of chemical modification of drying oils are practiced. In addition to those discussed here, drying oil-modified alkyds, epoxy esters, and uralkyds are discussed in Chapter 15.

14.3.1. Heat Bodied Oils, Blown Oils, and Dimer Acids

Both nonconjugated and conjugated drying oils can be thermally polymerized by heating under an inert atmosphere to form *bodied oils*. Bodied oils have higher viscosities and are used in oil paints to improve application and performance characteristics. Process temperatures may be as high as 300–320°C for nonconjugated oils and 225–240°C for conjugated oils, although the reactions occur at an appreciable rate at somewhat lower temperatures [3]. At least in part, bodying may result from thermal decomposition of hydroperoxides, always present in natural oils, to yield free radicals, resulting in a limited degree of cross-linking. It has also been shown that thermal rearrangement to conjugated systems occurs followed by Diels-Alder reactions, which lead to formation of dimers [16]. Since tung oil has a high concentration of conjugated double bonds, it undergoes thermal polymerization more rapidly than nonconjugated oils (e.g., linseed oil). One must be careful to control the heating of tung oil or the polymerization will lead to gelation.

Viscosity of drying oils can also be increased by passing air through oil at relatively moderate temperatures, 140-150°C, to produce *blown oils* [3]. Presumably, reactions similar to those involved in cross-linking cause autoxidative oligomerization of the oil.

Polyunsaturated acids dimerize or oligomerize by heat treatment; the reactions are acid catalyzed. For example, the doubly unsaturated fatty acids of TOFA can be dimerized or oligomerized followed by removal of the residual monobasic acids by distillation. Under the high reaction temperatures, some decarboxylation occurs. Decarboxylation can be minimized by heating under pressure in the presence of a small percentage of water and activated clay. The products obtained are called *dimer acids*. They are predominantly C_{36} dicarboxylic acids, with small fractions of monocarboxylic acid and some C_{54} tricarboxylic acids. For some uses, it is desirable to eliminate residual double bonds by hydrogenation. Dimer acids are used to make polyesters (see Section 13.1.2) and polyamides (see Section 11.2.2).

14.3.2. Varnishes

The drying rate of drying oils can be increased by dissolving a solid resin in the oil and diluting with a hydrocarbon solvent. Such a solution is called a *varnish*. The solid resin serves to increase the T_g of the solvent-free film so that film hardness is achieved more rapidly. The rate of cross-linking does not increase so the time required for the film to become solvent resistant is not shortened. Any high melting thermoplastic resin soluble in drying oil will serve the purpose. The higher the "melt point" of the resin, the greater the effect on drying time. Naturally occurring resins, such as congo, copal, damar, and kauri resins, synthetic resins, such as ester gum (glyceryl esters of rosin), phenolic resins (see Section 11.7), and coumarone-indene resins, have been used.

In varnish manufacture, the drying oil (usually linseed oil, tung oil, or mixtures of the two) and the resin are cooked together to high temperature to obtain a homogeneous solution of the proper viscosity. The varnish is then thinned with hydrocarbon solvents to application viscosity. During cooking, some dimerization or oligomerization of the drying oil occurs; in some cases, reaction between the oil and resin has been demonstrated. Varnishes were widely used in the 19th and early 20th centuries, but have been almost completely replaced by a variety of other products, especially alkyds, epoxy esters, and uralkyds. The term varnish has come to be used more generally for transparent coatings, even though few of them today are varnishes in the original meaning of the word.

14.3.3. Synthetic Conjugated Oils

Tung oil dries rapidly, but is expensive and its films discolor rapidly due to the presence of three double bonds. These shortcomings led to efforts to synthesize conjugated oils, especially those containing esters of fatty acids with two conjugated double bonds. One approach is to dehydrate castor oil using acid catalysis. A major component (87%) of the fatty acid content of castor oil triglycerides is ricinoleic acid, 12-hydroxy-(Z)-9-octadecenoic acid, which dehydrates to mixed geometric isomers of 9,11-conjugated and 9,12-nonconjugated fatty acid esters. *Dehydrated castor oil* dries relatively rapidly at room temperature, but on further exposure to air, the surface becomes tacky. This *aftertack* has been attributed to the presence of various geometric isomers formed during the dehydration. Dehydrated castor oil and its fatty acids are mainly used to prepare alkyds and epoxy esters for baking coatings, in which aftertack does not occur.

Nonconjugated oils can be partially isomerized to conjugated oils by heating with a variety of catalysts, mostly alkaline hydroxides. Some *cis-trans* isomerization occurs in the process but the extent is apparently not great enough to result in aftertack. A similar process can be used to partially conjugate double bonds of TOFAs. Synthesis of conjugated fatty acids by treatment of oils at high temperature with aqueous alkali hydroxides accomplishes isomerization and saponification simultaneously [17]. The principal use of such conjugated oils and fatty acids has been to make alkyds and epoxy esters.

14.3.4. Esters of Higher Functionality Polyols

As mentioned in Section 14.2.1, the time required for nonconjugated oils to form a solvent-resistant, cross-linked film decreases as the average number of diallylic groups $\bar{f}_n$ increases. When oil-derived fatty acids are reacted with polyols with more than three hydroxyl groups per molecule, the number of cross-linking sites per molecule increases relative to the corresponding natural triglyceride oil. While soybean oil is a semidrying oil since $\bar{f}_n$ is 2.07, the pentaerythritol (PE) tetraester of soybean fatty acids has an $\bar{f}_n$ of 2.76 and is a drying oil. The PE ester of linseed fatty acids has an $\bar{f}_n$ of about 5 and gives dry, solvent-resistant films more rapidly than linseed oil. Still faster drying rates can be achieved with still higher functionality polyols, such as di- and tripentaerythritol. Alkyds, epoxy esters, and uralkyds (see Chapter 15) made with fatty acids from such oils as soybean and linseed oils can be considered as higher functionality synthetic drying oils.

14.3.5. Maleated Oils

Both conjugated and nonconjugated oils react with maleic anhydride to form adducts. Conjugated oils, such as dehydrated castor oil, react at moderate temperatures by a Diels-Alder reaction:

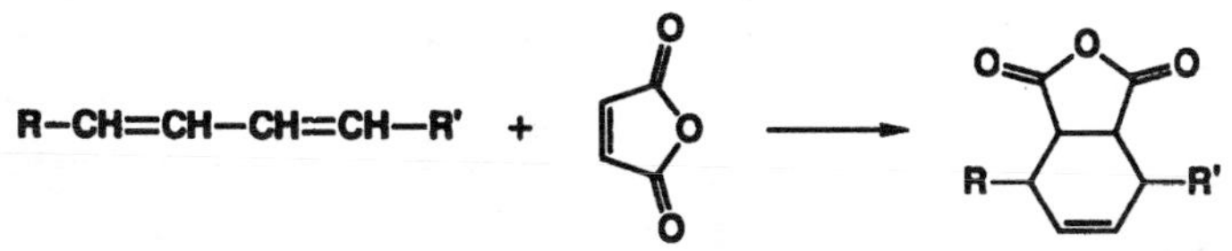

Nonconjugated oils, such as soybean and linseed oils, require higher temperatures (>100°C) and form a variety of adduct structures. Model compound studies using methyl linoleate demonstrate that maleic anhydride undergoes an ene reaction to give succinyl anhydride adducts. The ene reaction results in conjugated bonds, as shown in general structure (**4**) subsequent Diels-Alder reaction with a second maleic anhydride gives a dianhydride, as shown in general structure (**5**) [18].

The products of these reactions, termed *maleated oils*, or sometimes, *maleinized oils*, react with polyols to give moderate molecular weight derivatives that dry faster than the unmodified drying oils. For example, maleated soybean oil esterified with glycerol dries at a rate comparable to that of a bodied linseed oil with a similar viscosity.

Maleated linseed oil can be made water-reducible by hydrolysis with aqueous ammonia to convert the anhydride groups to the ammonium salts of the diacid. Such products have not found significant commercial use, but as described in Sections 15.3 and 15.8, the process is used to make water-reducible alkyds and epoxy esters.

14.3.6. Vinyl-Modified Oils

Both conjugated and nonconjugated drying oils react in the presence of a free radical initiator with such unsaturated monomers as styrene, vinyltoluene, and acrylic esters. High degrees of chain transfer result in the formation of a variety of products, for example, low molecular weight homopolymer of the monomer, short-chain graft copolymers, and dimerized drying oil molecules. The reaction of drying oils with such monomers is not commercially important, but the same principle is used to make modified alkyds. (See Section 15.4.) Linseed oil modified with cyclopentadiene has found fairly sizeable commercial use. This product is made by heating a mixture of linseed oil and dicyclopentadiene above 170°C, the temperature above which the reverse Diels-Alder reaction liberates monomeric cyclopentadiene at an appreciable rate. The product is inexpensive and dries faster than linseed oil. However, its odor and dark color limit its applications.

GENERAL REFERENCES

F. L. Fox, *Oils for Organic Coatings*, Federation of Societies for Coatings Technology, Blue Bell, PA, 1965.

A. E. Rheineck and R. O. Austin, "Drying Oils," in R. R. Myers and J. S. Long, Eds., *Treatise on Coatings*, Vol. I, No. 2, Marcel Dekker, New York, 1968, pp. 181–248.

Z. W. Wicks, Jr., "Drying Oils," in *Kirk-Othmer Encyc. Chem. Technol.*, 4th ed., **8**, 519–532 (1993).

REFERENCES

1. R. G. Ackman, *Prog. Chem. Fats Other Lipids*, **12**, 165 (1972); G. R. Khan and F. Scheinmann, *Prog. Chem. Fats Other Lipids*, **15**, 343 (1977); J. W. King, E. C. Adams, and B. A. Bidlingmeyer, *J. Liq. Chromatography.*, **5**, 275 (1982).

2. Z. W. Wicks, Jr., *Kirk-Othmer Encyc. Chem. Technol.*, 4th ed., **8** (1993), p. 521.

3. M. W. Formo, "Drying Oils," in D. Swern, Ed., *Industrial Oil and Fat Products*, Wiley, New York, Vol. I (1979), pp. 177–232 and 687–816, Vol. II (1982), pp. 343–406.

4. F. L. Fox, *Oils for Organic Coatings*, Federation of Societies for Coatings Technology, Blue Bell, PA, 1965.

5. N. A. Porter, L. S. Lehman, B. A. Weber, and K. J. Smith, *J. Am. Chem. Soc.*, **103**, 6447 (1981).

6. K. S. Ennor and J. Oxley, *J. Oil Colour Chem. Assoc.*, **50**, 577 (1967); Anon., *Lipid Technol.*, **6** (5), 110 (1994).

7. A. E. Rheineck and R. O. Austin, "Drying Oils," in R. R. Myers and J. S. Long, Eds., *Treatise on Coatings*, Vol. 1, No. 2, Marcel Dekker, New York, 1968, pp. 181–248.

8. J. H. Greaves, *Oil Colour Trades J.*, **113**, 949 (948).

9. J. R. Chipault, E. E. Nickell, and W. O. Lundberg, *Off. Digest*, **23**, 740 (1951).

10. J. H. Hartshorn, *J. Coat. Technol.*, **54** (687), 53 (1982).

11. W. J. Muizebelt, J. J. Donkerbroek, M. W. F. Nielen, J. B. Hussem, and M. E. F. Biermond, *J. Coat. Technol.*, **70** (876), 83 (1998).

12. N. A. R. Falla, *J. Coat. Technol.*, **64** (815), 55 (1992).

13. E. N. Frankel, *Prog. Lipid. Res.*, **19**, 1 (1980).

14. R. A. Hancock, N. J. Leeves, and P. F. Nicks, *Prog. Org. Coat.*, **17**, 321, 337 (1989).

15. T. L. T. Robey and S. M. Rybicka, *Paint Research Station Tech. Paper No. 217*, Vol. 13 (1), 2 (1962).

16. D. H. Wheeler and J. White, *J. Am. Chem. Soc.*, **44**, 298 (1967).

17. T. F. Bradley and G. H. Richardson, *Ind. Eng. Chem.*, **34**, 237 (1942).

18. A. E. Rheineck and T. H. Khoe, *Fette Seifen Anstrichm.*, **71**, 644 (1969).

CHAPTER 15 _____

Alkyd Resins _____

While no longer the largest volume vehicles in coatings, alkyds still are of major importance. Alkyds are prepared from polyols, dibasic acids, and fatty acids. They are polyesters, but in the coatings field the term polyester is reserved for "oil-free polyesters," discussed in Chapter 13. The term *alkyd* is derived from *alc*ohol and ac*id*. Alkyds tend to be lower in cost than most other vehicles and tend to give coatings that exhibit fewer film defects during application. (See Chapter 23.) However, durability of alkyd films, especially outdoors, tends to be poorer than films from acrylics, polyesters, and polyurethanes.

There are many types of alkyds. One classification is into *oxidizing* and *nonoxidizing* types. Oxidizing alkyds cross-link by the same mechanism as drying oils, as discussed in Chapter 14. Nonoxidizing alkyds are used as polymeric plasticizers or as hydroxy-functional resins, which are cross-linked by melamine-formaldehyde (MF), or urea-formaldehyde (UF) resins or by isocyanate cross-linkers. A second classification is based on the ratio of monobasic fatty acids to dibasic acids utilized in their preparation. The terminology used was adapted from terminology used to classify varnishes. Varnishes with high ratios of oil to resin were called long oil varnishes; those with a lower ratio, medium oil varnishes; and those with an even lower ratio, short oil varnishes. Oil length of an alkyd is calculated by dividing the amount of "oil" in the final alkyd by the total weight of the alkyd solids, expressed as a percentage, as shown in Eq. 15.1. The amount of oil is defined as the triglyceride equivalent to the amount of fatty acids in the alkyd. The 1.04 factor in Eq. 15.2 converts the weight of fatty acids to the corresponding weight of triglyceride oil. Alkyds with oil lengths greater than 60 are *long oil alkyds*; those with oil lengths from 40 to 60, *medium oil alkyds*, and those with oil lengths less than 40, *short oil alkyds*. There is some variation in the dividing lines between these classes in the literature.

$$\text{Oil Length} = \frac{\text{Weight of "Oil"}}{\text{Weight of Alkyd} - \text{Water Evolved}} \times 100 \qquad (15.1)$$

$$\text{Oil Length} = \frac{1.04 \times \text{Weight of Fatty Acids}}{\text{Weight of Alkyd} - \text{Water Evolved}} \times 100 \qquad (15.2)$$

Another classification is *unmodified* or *modified alkyds*. Modified alkyds contain other monomers in addition to polyols, polybasic acids, and fatty acids. Examples are *styrenated alkyds* (see Section 15.4) and *silicone alkyds* (see Section 16.5.3).

268

15.1. OXIDIZING ALKYDS

Oxidizing alkyds can be considered as synthetic drying oils. They are polyesters of one or more polyols, one or more dibasic acids, and fatty acids from one or more drying or semidrying oils. The most commonly used polyol is glycerol, the most commonly used dibasic acid is phthalic anhydride (PA), and a widely used oil is soybean oil. Let us consider a simple, idealized example of the alkyd prepared from 1 mol of PA, 2 mol of glycerol, and 4 mol of soybean fatty acids. Using the fatty acid composition data for soybean oil from Table 14.1, we can calculate that this alkyd would have an $\bar{f}_n$ of 2.76 [i.e., it would have an average of 2.76 activated diallylic groups ($-CH=CHCH_2CH=CH-$) per molecule (see Sections 14.2.1 and 14.3.4 for discussion of $\bar{f}_n$)] and, therefore, would dry to a solid film. The alkyd would form a solvent resistant film in about the same time as a pentaerythritol (PE) ester of soybean fatty acids (see Section 14.3.4), since they have the same $\bar{f}_n$. However, the alkyd would form a tack-free film faster because the rigid aromatic rings from PA increases the T_g of the film.

If the mole ratio of PA to glycerol were 2 to 3, 5 mol of soybean fatty acid could be esterified to yield an alkyd with an $\bar{f}_n$ of 3.45. This alkyd would cross-link more rapidly than the 1 : 2 : 4 mole ratio alkyd and would also form tack-free films even faster because the ratio of aromatic rings to long aliphatic chains would be 2 : 5 instead of 1 : 4. As the ratio of PA to glycerol is increased further, the average functionality for autoxidation increases and the T_g after solvent evaporation increases because of the increasing ratio of aromatic to long aliphatic chains. For both reasons, films dry faster.

A theoretical alkyd prepared from 1 mol each of glycerol, PA, and fatty acid would have an oil length of about 60. A highly idealized structure of such an alkyd is shown below. However, if one were to try to prepare such an alkyd, the resin would gel prior to complete reaction. Gelation would result from reaction of a sufficient number of trifunctional glycerol molecules with three difunctional PA molecules to form cross-linked polymer molecules, swollen with partially reacted components. Gelation can be avoided by using a sufficient excess of glycerol to reduce the extent of cross-linking. When the reaction is carried to near completion with excess glycerol, there are few unreacted carboxylic acid groups, but many unreacted hydroxyl groups.

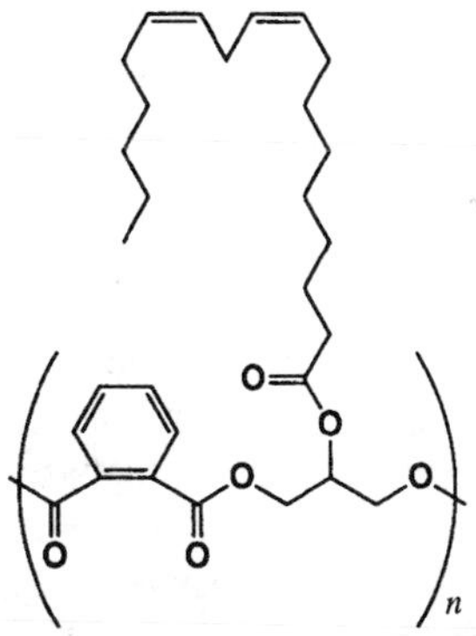

Idealized structure of an alkyd made from 1 mol of PA, 1 mol of glycerol, and 1 mol of linoleic acid.

There have been many attempts, none fully successful, to calculate the ratios of functional groups and the extent of reaction that can be reached without encountering gelation. The problem is complex. The reactivity of the hydroxyl groups can be different;

for example, glycerol contains both primary and secondary alcohol groups. Under esterification conditions, polyol molecules can self-condense to form ethers and, in some cases, can dehydrate to form volatile aldehydes (see Section 15.6.2). Reactivity of the carboxylic acids also varies. The rate of formation of the first ester from a cyclic anhydride is more rapid than formation of the second ester. Aliphatic acids esterify more rapidly than aromatic acids. Polyunsaturated fatty acids and their esters can dimerize or oligomerize to form cross-links. Of the many papers in the field, that by Blackinton recognizes the complexities best [1]. In addition to the above complexities, particular emphasis is placed on the extent of formation of cyclic compounds by intramolecular esterification reactions.

Misev has developed equations that permit calculation of ratios of ingredients theoretically needed to prepare an alkyd of any desired oil length, number average molecular weight, and hydroxy content [2]. Just like in other equations, the important effect of dimerization of fatty acids is not included as a factor in Misev's equations. In practice, alkyd resin formulators have found that the **mole** ratio of dibasic acid to polyol should be less than 1 to avoid gelation. How much less than 1 depends on many variables. Composition variables are discussed in Sections 15.1.1 through 15.1.3; the effect of variables in reaction conditions is discussed in Section 15.6.

For medium oil alkyds, the ratio of dibasic acid to polyol is not generally changed much relative to alkyds with an oil length of about 60, but the fatty acid content is reduced to the extent desired. This results in a larger excess of hydroxyl groups in the final alkyd. It is commonly said that as the oil length of an oxidizing alkyd is reduced below 60, the drying time decreases to a minimum at an oil length of about 50. However, this conventional wisdom must be viewed cautiously. The ratio of aromatic rings to aliphatic chains continues to increase, increasing T_g after the solvent evaporates from the film tending to shorten the time to form a tack-free film. However, at the same molecular weight, the number of fatty acid ester groups per molecule decreases as the oil length decreases below 60, since more hydroxyl groups are left unesterified. Therefore, the time required to achieve sufficient cross-linking for solvent resistance increases.

Long oil alkyds are soluble in aliphatic hydrocarbon solvents. As the oil length decreases, mixtures of aliphatic and aromatic solvents are required, and oil lengths below about 50 require aromatic solvents, which are more expensive than aliphatics. The viscosity of solutions of long oil alkyds, especially of those with oil lengths below 65, is higher in aliphatic than in aromatic solvents; in medium oil alkyds, which require mixtures of aliphatic and aromatic solvents, viscosity decreases as the proportion of aromatic solvents increases. In former days, and to some extent still today, it was considered desirable to use a solvent mixture that gave the highest possible viscosity; then, at application viscosity, the solids were lower and the raw material cost per unit volume was less. Accordingly, alkyds were designed to have high dilutability with aliphatic solvents. This was false economy, but it was a common practice and is still being practiced to some extent. Increasingly, the emphasis is on reducing VOC, so the question becomes how to design alkyds with low solvent requirements rather than high dilutability potential. Furthermore, the aromatic solvents are on the HAP list. High solids alkyds are discussed in Section 15.2.

15.1.1. Monobasic Acid Selection

Drying alkyds can be made with fatty acids from semidrying oils, since the $\bar{f}_n$ can be well above 2.2. For alkyds made by the monoglyceride process (see Section 15.6.1), soybean oil is used in the largest volume. Soybean oil is economical and supplies are dependable because it is a large scale agricultural commodity; alkyd production takes only a few percent of the world supply. For alkyds made by the fatty acid process (see Section 15.6.1), tall oil fatty acids (TOFA) are more economical than soybean fatty acids. Both soybean oil and TOFA contain roughly 40–60% linoleic acid and significant amounts of linolenic acid. (See Table 14.1.) White coatings containing linolenic acid esters gradually turn yellow. Premium cost "nonyellowing" alkyds are made with safflower or sunflower oils, which are high in linoleic acid, but contain very little linolenic acid.

Applications in which fast drying and high cross-link density are important require alkyds made with drying oils. The rate of oxidative cross-linking is affected by the functionality of the drying oils used. At the same oil length and molecular weight, the time required to achieve a specific degree of cross-linking decreases as the average number of diallylic groups ($\bar{f}_n$) increases. Linseed long oil alkyds therefore cross-link more rapidly than soybean long oil alkyds. The effect is especially large in very long oil alkyds and less noticeable in alkyds with oil lengths around 60, where $\bar{f}_n$ is very high even with soybean oil and the effect of further increase in functionality by using linseed oil is small. Because of the large fraction of esters of fatty acids with three double bonds in linseed alkyds, their color and color retention is poorer than that of soybean alkyds. Tung oil based alkyds, because of the high proportion of esters with three conjugated double bonds, dry still faster. Tung oil alkyds also exhibit a high degree of yellowing. Dehydrated castor alkyds have fairly good color retention, since they contain only a small proportion of esters of fatty acids with three double bonds; they are used primarily in baking coatings.

As discussed in Section 14.3.1, drying oils and drying oil fatty acids undergo dimerization at elevated temperatures. Dimerization occurs concurrently with esterification during alkyd synthesis; it generates difunctional acids, increasing the mole ratio of dibasic acids to polyol. The rate of dimerization is faster with drying oils having a higher average number of diallylic groups per molecule and with those having conjugated double bonds. Thus, the molecular weight and, therefore, the viscosity of an alkyd made with the same ratio of reactants depends on the fatty acid composition. The higher the degree of unsaturation, the higher the viscosity due to the greater extent of dimerization. Linseed alkyds have higher viscosities than soybean alkyds made with the same monomer ratios under the same conditions. The effect is particularly marked with tung oil. It is difficult to prepare straight tung alkyds because of the risk of gelation; commonly, mixed linseed-tung alkyds are used when high oxidative cross-linking functionality is desired.

A critical factor involved in the choice of fatty acid is cost. Drying oils are agricultural products and, hence, tend to be volatile in price. By far, the major use of vegetable oils is for foods. Depending upon relative prices, one drying oil is often substituted for another in certain alkyds. By adjusting for functionality differences, substitutions can frequently be made without significant changes in properties.

Fatty acids are not the only monobasic acids used in making alkyds. Benzoic acid is also used, especially to esterify some of the excess hydroxyl groups remaining in the preparation of medium oil alkyds. The benzoic acid increases the ratio of aromatic to aliphatic chains in the alkyd, thus contributing to a higher T_g of the solvent-free alkyd and more rapid formation of a tack-free film. At the same time, the reduction in the free hydroxyl content may somewhat reduce water sensitivity of the dried films. Rosin can also be used in the same fashion. Although rosin is not an aromatic acid, its polynuclear ring structure is rigid enough to increase T_g. If the critical requirement in drying is rapid development of solvent resistance, such benzoic acid and rosin modifications do not serve the purpose; they only reduce tack-free time. Frequently, benzoic acid-modified alkyds are called *chain-stopped* alkyds. The implication of the terminology is that the benzoic acid stops chain growth. This is not the case; the benzoic acid simply esterifies hydroxyl groups that would not have been esterified if the benzoic acid were absent. The effect on degree of polymerization is negligible.

15.1.2. Polyol Selection

Glycerol is the most widely used polyol because it is present in naturally occurring oils from which alkyds are commonly synthesized. (See Section 15.6.1.) The next most widely used polyol is pentaerythritol (PE). In order to avoid gelation, the tetrafunctionality of PE must be taken into account when replacing glycerol with PE. If the substitution is made on a mole basis, rather than an equivalent basis, chances for gelation are minimized. As mentioned earlier, the ratio of moles of dibasic acid to polyol should be less than 1, and generally, a slightly lower mole ratio is required with PE than with glycerol. At the same mole ratio of dibasic acid to polyol, more moles of fatty acid can be esterified with PE. Hence, in long oil alkyds, the average functionality for cross-linking is higher, and the time to reach a given degree of solvent resistance is shorter for a PE alkyd as compared to a glycerol alkyd. Due to this difference, one must be careful in comparing oil lengths of glycerol and PE alkyds.

$$CH_2OH$$
$$HOCH_2-\overset{\displaystyle CH_2OH}{\underset{\displaystyle CH_2OH}{C}}-CH_2OH$$

Pentaerythritol
PE

$$HOCH_2-\overset{\displaystyle CH_2OH}{\underset{\displaystyle CH_2OH}{C}}-CH_2-O-CH_2-\overset{\displaystyle CH_2OH}{\underset{\displaystyle CH_2OH}{C}}-CH_2OH$$

Dipentaerythritol
DiPE

$$HOCH_2-\overset{\displaystyle CH_2OH}{\underset{\displaystyle CH_2OH}{C}}-CH_2-O-CH_2-\overset{\displaystyle CH_2OH}{\underset{\displaystyle CH_2OH}{C}}-CH_2-O-CH_2-\overset{\displaystyle CH_2OH}{\underset{\displaystyle CH_2OH}{C}}-CH_2OH$$

Tripentaerythritol
TriPE

When PE is synthesized, di- and tripentaerythritol are byproducts, and commercial PE contains some of these higher polyols. Consequently, care must be exercised in changing sources of PE, since the amount of the higher polyols may differ. Because of the very high functionality, diPE and triPE ($F = 6$ and 8, respectively) are useful in making fast drying low molecular weight alkyds. (See Section 15.2.)

To reduce cost, it is sometimes desirable to use mixtures of PE and ethylene or propylene glycol. A 1 : 1 mole ratio of tetra- and difunctional polyols gives an average functionality of three, corresponding to glycerol. The corresponding alkyds can be expected to be similar, but not identical. Trimethylolpropane (TMP) can also be used, but the rate of esterification is slower than with glycerol. Although all of TMP's alcohol groups are primary, they are somewhat sterically hindered by the neopentyl structure [3]. Trimethylolpropane, however, gives a narrower molecular weight distribution, which provides alkyds with a somewhat lower viscosity than the comparable glycerol-based alkyd. A kinetic study demonstrated that esterification of one or two of the hydroxyl groups of TMP has little effect on the rate constant for esterification of the third hydroxyl group [4]. It can be speculated that PE behaves similarly.

15.1.3. Dibasic Acid Selection

Dibasic acids used to prepare alkyds are usually aromatic. Their rigid aromatic rings increase the T_g of the resin. Cycloaliphatic anhydrides, such as hexahydrophthalic anhydride, are also used. While they are not as rigid as aromatic rings, the cycloaliphatic rings also increase T_g.

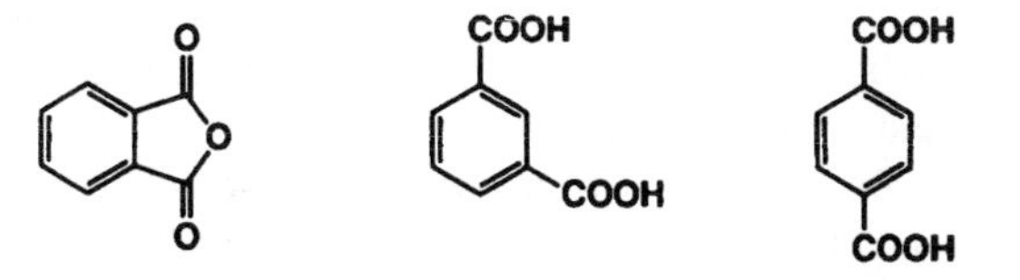

| Phthalic Anhydride | Isophthalic acid | Terephthalic acid |
| PA | IPA | TPA |

By far, the most widely used dibasic acid is PA. It has the advantage that the first esterification reaction proceeds rapidly by opening the anhydride ring. The amount of water evolved is lower, which also reduces reaction time. The relatively low melting point (the pure compound melts at 131°C) is desirable, since the crystals melt and dissolve readily in the reaction mixture. In large scale manufacturing, molten PA is used, which reduces packaging, shipping, and handling costs.

The next most widely used dibasic acid is isophthalic acid (IPA). As discussed in Section 13.1.2, esters of IPA are more resistant to hydrolysis than are those of PA in the pH range of 4 to 8, the most important range for exterior durability. On the other hand, under alkaline conditions esters of phthalic acid are more resistant to hydrolysis than isophthalic esters. The raw material cost for IPA is not particularly different from PA (even after adjusting for the extra mole of water that is lost), but the manufacturing cost is higher. The high melting point of IPA (330°C) leads to problems getting it to dissolve in the reaction mixture so that it can react. High temperatures are required for longer times than with PA; hence more dimerization of fatty acids occurs with TPA resulting in higher viscosity. The longer time at higher temperature also leads to greater extents of side reactions of the polyol components, which are discussed in Section 15.6.2 [5]. Thus, when substituting IPA for PA, one must use a lower mole ratio of IPA to polyol in order to make an alkyd of similar viscosity.

15.2. HIGH SOLIDS OXIDIZING ALKYDS

The need to minimize VOC emissions has led to efforts to increase solids content of alkyd resin coatings. Since xylene is on the HAP list, its use is being reduced. Some increase in solids can be realized by a change of solvents. Aliphatic (and to a somewhat lesser degree, aromatic) hydrocarbon solvents promote intermolecular hydrogen bonding, especially between carboxylic acids, but also between hydroxyl groups, thereby increasing viscosity. Use of at least some hydrogen-bond acceptor solvent, such as an ester, or hydrogen-bond acceptor-donor solvent such as an alcohol, gives a significant reduction in viscosity at equal solids.

An approach to increasing solids is to decrease molecular weight, which is easily accomplished by decreasing the dibasic acid to polyol ratio and going to longer oil length alkyds. However, making a significant reduction in VOC by this route requires having an alkyd with lower functionality for cross-linking and a lower ratio of aromatic to aliphatic chains. Both changes increase the time for drying. The effect of longer oil length on functionality can be minimized by using drying oils with higher average functionality. Use of oils containing linolenic or α-eleostearic acid is limited by their tendency to discolor. One can use safflower oil, which has a higher linoleic acid content, and less linolenic acid than soybean oil. Proprietary fatty acids with 78% linoleic acid are commercially available. Increasing the concentration of driers (see Section 14.2.2) accelerates drying, but also accelerates yellowing and embrittlement.

Solids can be increased by making resins with narrower molecular weight distributions. For example, one can add a transesterification catalyst near the end of the alkyd cook; this gives more uniform molecular weight and a lower viscosity product. To study the effect of molecular weight distribution, model alkyds with very narrow molecular weight distribution were synthesized by using dicyclohexylcarbodiimide, which allows low temperature esterification [3]. With the same ratio of reactants, the $\bar{M}_n$ and polydispersity were lower than the conventional alkyd control. These differences resulted from less dimerization through reactions of the double bond systems of the fatty acids and avoidance of self-etherification of polyol in the low temperature preparation. It was found that the solids could be 2–10% higher than with the conventionally prepared alkyd of the same raw material composition. The model alkyds dried more rapidly, but their film properties, especially impact resistance, were inferior to those obtained with control resins with the usual broad molecular weight distribution [6]. Conventionally prepared TMP alkyds had lower molecular weights and viscosities than the glycerol alkyds. This difference may result from less self-etherification of TMP as compared to glycerol.

High solids alkyds for baking applications have been made using tripentaerythritol. The high functionality obtained using this polyol ($F = 8$) gives alkyds that cross-link as rapidly as shorter oil length, higher viscosity glycerol alkyds [7]. However, for air dry applications, the lower aromatic to aliphatic ratio lengthens the tack-free time. Presumably, progress could be made using a high functionality polyol with some combination of phthalic and benzoic acid, together with fatty acids with as high functionality fatty acids as possible. The cost of such an alkyd would be high.

Another approach to high solids alkyds is to use *reactive diluents* in place of part of the solvent. The idea is to have a component of lower molecular weight and much lower viscosity than the alkyd resin, which reacts with the alkyd during drying, so it is not part of the VOC emissions. At least three types of reactive diluents have been used to formulate high solids alkyd coatings.

Polyfunctional methacrylate monomers (e.g., trimethylolpropane trimethacrylate) have been used in force dry coatings (coatings designed to be cured in the range of 60 to 80°C). Although polymerization of these monomers is inhibited by oxygen and there is a substantial degree of termination by chain transfer in the coreaction of such polyfunctional monomers, cross-linking still occurs. Polyfunctional acrylates can be used at ambient temperatures [8], but some people exhibit skin irritation on repeated exposure to polyfunctional acrylates.

A second example is use of dicyclopentadienyloxyethyl methacrylate as a reactive diluent [9]. It is difunctional, due to the easily abstractable allylic hydrogen on the dicyclopentadiene ring structure and the methacrylate double bond. The compound coreacts with drying oil groups in the alkyd.

Dieyclopentadienyloxyethyl methacrylate

A third example is the use of mixed acrylic and drying oil fatty acid amides of hexa(aminomethoxymethyl)melamine [10,11]. Such reactive diluents contain high functionalities of $>NCH_2NHCOCH=CH_2$ and $>NCH_2NHCOC_{17}H_x$ moieties and promote fast drying.

Using optimized resins and, in some cases, reactive diluents, good quality air dry and baking alkyd coatings can be formulated with VOC levels of 280 to 350 g/L of coating. A 250 g/L level is attainable only with some sacrifice of application and film properties; still lower limits of permissible VOC are projected.

15.3. WATER-REDUCIBLE ALKYDS

As with almost all other resin classes, work has been done to make alkyd resins for coatings that can be reduced with water. One approach that has been more extensively studied in Europe than in the United States is the use of alkyd emulsions [12,13] (see Section 25.3 for further discussion). The emulsions are stabilized with surfactants and can be prepared with little, if any, volatile solvent. It is common to add a few percent of an alkyd-surfactant blend to latex paints to improve adhesion to chalky surfaces (see Section 31.1) and, in some cases, to improve adhesion to metals (see Section 32.1.3). It is important to use alkyds that are as resistant as possible to hydrolysis. Hybrid alkyd-acrylic latexes have been prepared by dissolving an oxidizing alkyd in the monomers used in emulsion polymerization, yielding a latex with an alkyd grafted on the acrylic polymer [14,15].

Another approach has been to make alkyds with an acid number in the range of 50 using secondary alcohols or ether alcohols as solvents. The acid groups are neutralized with ammonia or an amine. The resultant solution can be diluted with water to form a dispersion of solvent swollen aggregates in water. Such resins are analogous to the water-reducible acrylics discussed in Section 12.3, and their behavior during water dilution is similar. Note that the use of primary alcohol solvents must be avoided because they can more readily transesterify with the alkyd during resin production and storage, leading to reduction in molecular weight and $\bar{f}_n$[16].

Hydrolytic stability can be a problem with water-reducible alkyds. If the carboxylic acid groups are half esters of phthalic or trimellitic acid, the hydrolytic stability will be poor and probably inadequate for paints that require a shelf life of more than a few months. Due to the anchimeric effect of the neighboring carboxylic acid group, such esters are relatively easily hydrolyzed. As hydrolysis occurs, the solubilizing acid salt is detached from the resin molecules, and the aqueous dispersion loses stability. A more satisfactory way to introduce free carboxylic acid groups is by reacting a completed alkyd with maleic anhydride. Part of the maleic anhydride adds to the unsaturated fatty acid esters as discussed in Section 14.3.5. The anhydride groups are then hydrolyzed with amine and water to give the desired carboxylate salt groups, which are attached to resin molecules with C–C bonds and cannot be hydrolyzed off. There is still a hydrolytic stability problem with the alkyd backbone, but hydrolysis does not result in destabilization of the dispersion. Similarly acrylated fatty acids can be used to synthesize water-reducible alkyds with improved hydrolytic stability [17].

After the film is applied, the water, solvent, and amine evaporate, and the film cross-links by autoxidation. Since there are a fairly large number of residual carboxylic acid groups left in the cross-linked binder, the water resistance and particularly the alkali resistance of the films are reduced, but are still satisfactory for some applications [18]. *Early water resistance* can be a problem if, for example, a freshly painted surface is rained on before all the amine has evaporated from the film. Commonly, ammonia is used as the neutralizing amine because it is assumed that ammonia volatilizes faster than any other amine. This assumption is not necessarily so; if the T_g of the alkyd film is sufficiently high before all of the amine has volatilized, loss of amine becomes controlled by diffusion rate. The rate of diffusion of amine through the carboxylic acid-functional film is affected by the base strength of the amine. A less basic amine, such as morpholine, may leave the film before ammonia even though its volatility is considerably lower.

15.4. STYRENATED ALKYDS

Oxidizing alkyds can be modified by reaction with vinyl monomers. The most widely used monomers are styrene, vinyl toluene, and methyl methacrylate, but essentially any vinyl monomer can be reacted in the presence of an alkyd to give a modified alkyd. Methyl methacrylate imparts better heat resistance than styrene but at higher cost.

In making styrenated alkyds, an oxidizing alkyd is prepared in the usual way and cooled to about 130°C in the reactor; then styrene and a free radical initiator such as benzoyl peroxide are added. The resulting free radical chain process leads to a variety of reactions, including formation of low molecular weight homopolymer of styrene, grafting of polystyrene onto the alkyd, and dimerization of alkyd molecules. The reaction is generally carried out at about 130°C, which favors decomposition of benzoyl peroxide to form phenyl free radicals; phenyl radicals have a greater tendency to abstract hydrogen, which favors grafting. After the reaction is complete, the resin is diluted with solvent. The ratio of alkyd to styrene can be varied over a wide range; commonly 50% alkyd and 50% styrene is used. The ratio of aromatic rings to aliphatic chains is greatly increased, and as a result, the T_g of styrenated alkyds is higher and tack-free time is shorter. Styrenated alkyds give a "dry" film in 1 hour or less versus 4 to 6 hours for the counterpart nonstyrenated alkyd. However, the average functionality for oxidative cross-linking is reduced, not just by dilution with styrene, but also because the free radical reactions involved in the styrenation consume some activated methylene groups. As a result, the time required to develop solvent resistance is longer than for the counterpart alkyd. The fast drying and low cost make styrenated alkyds very attractive for some applications, but in other cases, the longer time required for cross-linking is more critical, in which case styrenated alkyds are not appropriate.

Styrenated alkyd vehicles are often used for air dry primers. One must be careful to apply top coat almost immediately or not until after the film has had ample time to cross-link. During the intermediate time interval, application of top coat is likely to cause nonuniform swelling of the primer, leading to what is called *lifting* of the primer. The result of lifting is the development of wrinkled areas in the surface of the dried film. End users who are accustomed to using alkyd primers, which do not give a hard film until a significant degree of cross-linking has occurred, are particularly likely to encounter problems of lifting if they switch to styrenated alkyd primers.

15.5. NONOXIDIZING ALKYDS

Certain low molecular weight short-medium and short oil alkyds are compatible with such polymers as nitrocellulose and thermoplastic polyacrylates. Therefore, such alkyds can be used as plasticizers for these polymers. They have the advantage over monomeric plasticizers (e.g., dibutyl or dioctyl phthalate) that they do not volatilize appreciably when films are baked. It is generally not desirable to use oxidizing alkyds, which would cross-link and lead to embrittlement of the films, especially on exterior exposure. Therefore, nondrying oil fatty acids (or oils) are used in the preparation of alkyds for such applications. For exterior acrylic lacquers, pelargonic acid (n-$C_8H_{17}COOH$) alkyds combine excellent resistance to photodegradation with good compatibility with the thermoplastic acrylic resins. An interesting sidelight on terminology is that these pelargonic alkyds have been called polyesters rather than alkyds because the word polyester connotes higher quality than the word alkyd. Castor oil derived alkyds are particularly appropriate for nitrocellulose lacquers for interior applications, since the hydroxyl groups on the ricinoleic acid promote compatibility.

All alkyds, particularly short-medium oil and short oil alkyds, are made with a large excess of hydroxyl groups to avoid gelation. These hydroxyl groups can be cross-linked with MF resins or with polyisocyanates. In some cases, relatively small amounts of MF resin are used to supplement the cross-linking during baking of medium oil oxidizing alkyds. To achieve compatibility, butylated MF resins are used. Such coatings provide somewhat better durability and faster curing than alkyd resins alone, with little increase in cost. The important advantage of relative freedom from film defects common to alkyd coatings can be retained. (See Chapter 23.) However, the high levels of unsaturation remaining in the cured films reduce resistance to discoloration on overbake and exterior exposure and cause loss of gloss and embrittlement on exterior exposure. These difficulties can be reduced by using nondrying oils with minimal levels of unsaturated fatty acids. Coconut oil has been widely used; its performance can be further enhanced by hydrogenation of the small amount of unsaturated acids present in it.

Since isophthalic (IPA) esters are more stable to hydrolysis in the pH range of 4 to 8 than phthalate esters, the highest performance exterior alkyd-MF enamels use nonoxidizing IPA alkyds. Exterior durability of such coatings is satisfactory for automobile top coats with opaque pigmentation. The films have an appearance of greater "depth" than acrylic-MF coatings. The films are perceived to be thicker than films acrylic-MF coatings of comparable thickness and pigmentation. However, for many applications, alkyd-MF coatings have been replaced with acrylic-MF or polyester-MF coatings to improve the overall balance of film properties.

15.6. SYNTHETIC PROCEDURES FOR ALKYD RESINS

Various synthetic procedures, each with many variables, are used to produce alkyd resins; the general references and Refs. [19] and [20] provide useful reviews of manufacturing procedures. Alkyds can be made directly from oils or by using free fatty acids as raw materials.

15.6.1. Synthesis from Oils or Fatty Acids

Monoglyceride process. In the case of glycerol alkyds, it would be absurd to first saponify an oil to obtain fatty acids and glycerol, and then reesterify the same groups in a different combination. Rather, the oil is first reacted with sufficient glycerol to give the total desired glycerol content, including the glycerol in the oil. Since PA is not soluble in the oil, but is soluble in the glycerol, transesterification of oil with glycerol must be carried out as a separate step before the PA is added; otherwise glyceryl phthalate gel particles would form early in the process. This two-stage procedure is often called the monoglyceride process. The transesterification reaction is run at 230 to 250°C in the presence of a catalyst; many catalysts have been used. Before the strict regulation of lead in coatings, litharge (PbO) was widely used; the residual transesterification catalyst also acted as a drier. Examples of catalysts now used in the United States are tetraisopropyl titanate, lithium hydroxide, and lithium ricinoleate. The reaction is run under an inert atmosphere such as CO_2 or N_2 to minimize discoloration and dimerization of drying oils.

While the process is called the monoglyceride process, the transesterification reaction actually results in a mixture of unreacted glycerol, monoglycerides, diglycerides, and unconverted drying oil. The composition depends on the ratio of glycerol to oil and on catalyst, time, and temperature. In general, the reaction is not taken to equilibrium. At some relatively arbitrary point, the PA is added, beginning the second stage. The viscosity and properties of the alkyd can be affected by the extent of reaction before the PA addition. While many tests have been devised to evaluate the extent of transesterification, none is very general because the starting ratio of glycerol to oil varies over a considerable range, depending on the oil length of the alkyd being made. (In calculating the mole ratio of dibasic acid to polyol, the glycerol already esterified in the oil must also be counted.) A useful empirical test is to follow the solubility of molten PA in the reaction mixture. This test has the advantage that it is directly related to a major requirement that must be met. In the first stage, it is common to transesterify the oil with less expensive pentaerythritol to obtain mixed partial esters. The second stage, esterification of the "monoglyceride" with PA, is carried out at a temperature of 220 to 255°C.

Fatty Acid Process. It is often desirable to base an alkyd on a polyol (e.g., PE) other than glycerol. In this case, fatty acids must be used instead of oils, and the process can be performed in a single step with reduced time in the reactor. Any drying, semidrying, or nondrying oil can be saponified to yield fatty acids, but the cost of separating fatty acids from the reaction mixture increases the cost of the alkyd. A more economical alternative is to use TOFA, which have the advantage that they are produced as fatty acids. Tall oil fatty acid composition is fairly similar to that of soybean fatty acids. (See Table 14.10.) Specially refined tall oils with higher linoleic acid content are available, as are other grades that have been treated with alkaline catalysts to isomerize the double bonds partially to conjugated structures. Generally, when fatty acids are used, the polyol, fatty acids, and dibasic acid are all added at the start of the reaction, and the esterifcation of both aliphatic and aromatic acids is carried out simultaneously in the range of 220 to 255°C.

15.6.2. Process Variations

Esterification is a reversible reaction; therefore, an important factor affecting the rate of esterification is the rate of removal of water from the reactor. Most alkyds are produced using a reflux solvent, such as xylene, to promote the removal of water by azeotroping. Since the reaction is run at a temperature far above the boiling point of xylene, less than 5% of xylene is used. The amount is dependent on the reactor and is set empirically such that there is enough to reflux vigorously, but not so much as to cause flooding of the condenser. Some of the xylene is distilled off along with the water; water is separated and xylene is returned to the reactor. A schematic diagram of equipment for alkyd preparation using refluxing solvents is given in Figure 15.1. The presence of solvent is desirable for other reasons: vapor serves as an inert atmosphere, reducing the amount of inert gas

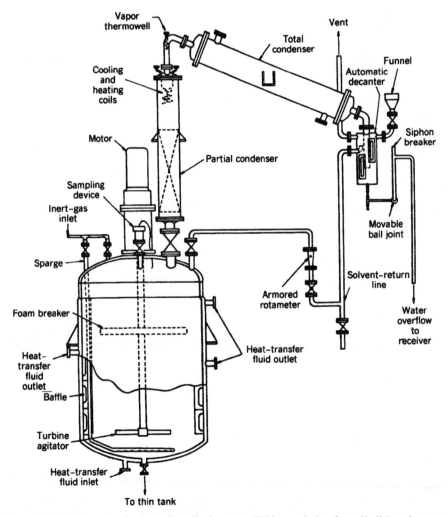

Figure 15.1. Schematic diagram of an alkyd reactor. (With permission from Shell Development Co., US 2643272, 1947, US 3763699, 1956; C. P. Van Dejk and F. J. F. Van der Plas GB 765765, 1955; GB 790166, 1956.)

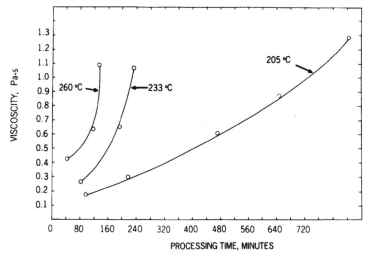

Figure 15.2. Effect of temperature and time on viscosity. (Adapted from Ref. [20], with permission.)

needed, and the solvent serves to avoid accumulation of sublimed solid monomers, mainly PA, in the reflux condenser.

Reaction time is affected by reaction temperature. Figure 15.2 shows the relation among temperature, time, and viscosity in preparing an alkyd. Figure 15.3 shows the relationship of the variables with acid number (i.e., mg KOH required to titrate the acid in 1 g resin solids). There are economic advantages to short reaction times. Operating costs are reduced, and the shorter times permit more batches of alkyd to be produced in a year, increasing capacity without capital investment in more reactors. Therefore, it is desirable to operate at as high a temperature as possible without risking gelation.

A critical aspect of alkyd synthesis is deciding when the reaction is completed. Disappearance of carboxylic acid is followed by titration, and increase in molecular weight is followed by viscosity. Determination of acid number and viscosity both take some time. Meanwhile, in the reactor, the reaction is continuing. After it is decided that the extent of reaction is sufficient, the reaction mixture must be "dropped" into a larger tank containing solvent. When a 40,000-L batch of alkyd is being made, a significant time is required to get the resin out of the reactor into the reducing tank; meanwhile, the reaction is continuing. The decision to start dropping the batch must be made so that the acid number and viscosity of the batch will be right after the continuing reaction that occurs between the time of sampling, determination of acid number and viscosity, and discharging of the reactor. The time for these determinations becomes the rate controlling step in production. If they can be done rapidly enough, the reaction can be carried out at 240°C or even higher without overshooting the target acid number and viscosity. On the other hand, if the control tests are done slowly, it may be necessary to run the reaction at only 220°C, which may require 2 hours or more of additional reaction time (see Figs. 15.2 and 15.3). Automatic titration instruments permit rapid determination of acid number, so the usual limit on time required is viscosity determination. While attempts have been made to use viscosity of the resin at reaction temperature to monitor change in molecular weight, the dependence of viscosity on molecular weight at that high temperature is not sensitive enough to be very useful. The viscosity must be determined on a solution at some lower

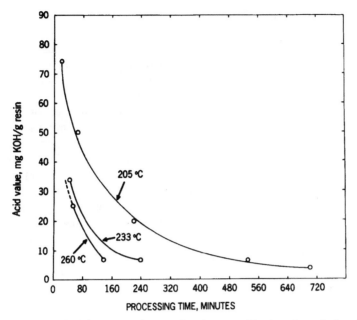

Figure 15.3. Effect of temperature and processing time on acid value of a typical medium oil linseed alkyd. (Adapted from Ref. [20], with permission.)

standard temperature. Since viscosity depends strongly on solution concentration and temperature, these variables must be carefully controlled.

In alkyd production, viscosity is commonly determined using Gardner bubble tubes as discussed in Section 3.3.4. The cook is continued until the viscosity is high enough so that by the time the resin batch is dropped into the solvent and the batch cooled, its viscosity will be what is called for in the specification. This means starting to discharge the reactor when the test sample is at some lower viscosity. It is not possible to generalize how large this difference should be; it depends on the specific alkyd composition, the temperature at which the reaction is being run, the time required to do the determination, the time required to empty the reactor, and so on. Viscosities can be determined more rapidly using a cone and plate viscometer (see Section 3.3.2) than with bubble tubes; the very small sample required for a cone and plate viscometer can be cooled and equilibrated at the measurement temperature more quickly.

Many variables affect the acid number and viscosity of alkyds. One is the ratio of reactants: The closer the ratio of moles of dibasic acid to polyol approaches 1, the higher the molecular weight of the backbone of the resin, but also the greater the likelihood of gelation. A useful rule of thumb for a starting point is to use a mole ratio of 0.95. The final ratio is determined by adjustments such that the combination of acid number and solution viscosity come out at the desired levels. The greater the ratio of hydroxyl groups to carboxylic acid groups, the faster the acid groups are reduced to a low level. The degree of completion of the reaction is an important factor controlling the viscosity, as well as the acid number. It is usually desirable to have a low acid number, typically in the range of 5 to 10.

The composition of the fatty acids is a major factor affecting the viscosity, and compositions of an oil or grade of TOFA can be expected to vary somewhat from lot to lot. Dimerization and oligomerization of the unsaturated fatty acids occur in the same temperature range at which the esterification is carried out. (See Section 14.3.1.) Fatty acids with conjugated double bonds dimerize more rapidly than those with nonconjugated bonds, and dimerization rates increase with the level of unsaturation. At the same ratio of phthalic to polyol to fatty acids, alkyds of the same acid number and solution concentration will increase in viscosity in the order soybean < linseed < tung.

Some volatilization of polyol, PA, and fatty acids out of the reactor will occur, depending on the design of the reactor, the rate of reflux of the azeotroping solvent, the rate of inert gas flow, and the reaction temperature, among other variables; the amount and ratio of these losses affect the viscosity at the standard acid number. The exact ratio of reactants must be established in the reactor that is actually used for synthesis. Since gelation can occur if the ratio of dibasic acid to polyol is too high, it is better not to put all the PA into the reactor in the beginning. If the viscosity is too low when the acid number is getting down near the standard, more PA can easily be added. The amount of PA held back can be reduced as experience is gained cooking a particular alkyd in a particular reactor.

Side reactions can affect the viscosity-acid number relationship. Glycerol and other polyols form ethers to some degree during the reaction. Glycerol can also form toxic acrolein by successive dehydrations.

$$\underset{\substack{OH\\|}}{HOCH_2-CH-CH_2OH} \longrightarrow \underset{\substack{OH\qquad\qquad OH\\|\qquad\qquad\quad|}}{HOCH_2-CH-CH_2-O-CH_2-CH-CH_2OH}$$

$$\underset{\substack{OH\\|}}{HOCH_2-CH-CH_2OH} \longrightarrow \underset{\substack{O\\\|}}{H_2C=CH-C-H}$$

When these reactions occur, the mole ratio of dibasic acid to polyol increases and the number of hydroxyl groups decreases; therefore, at the same acid number, the molecular weight will be higher. Excessively high viscosity and even gelation can result. Ether formation is catalyzed by strong protonic acids; therefore, it is desirable to avoid them as catalysts for the esterification. Monobutyltin oxide has been used as an esterification catalyst; presumably, it does not significantly catalyze ether formation. As noted earlier, PE and TMP seem less vulnerable than glycerol to undesirable side reactions such as ether formation, and glycerol is the only polyol that can decompose to form acrolein. A hydroxyl group on one end of a growing polyester chain can react with a carboxylic acid group on another end of the same molecule, leading to ring formation. Transesterification of chain linkages can have the same result. Since cyclization reactions reduce chain length, their net effect is to reduce viscosity.

Many alkyd resins have broad, uneven molecular weight distributions. (See Figure 2.3.) Kumanotani has shown that even modest changes in reaction conditions can cause large differences in molecular weight distribution, which can have significant effects on final film properties [21]. In many alkyds, very small gel particles, microgels, are formed. Kumanotani has shown that these microgels play an important role in giving greater strength properties to final films. Process changes that may make the alkyd more uniform may be undesirable. For example, allowing glycerolysis to approach equilibrium before addition of PA and using transesterification catalysts in the final stages of esterification both favor narrower molecular weight distributions and lower viscosities, but films made from the more uniform alkyds may exhibit inferior mechanical properties.

15.7. URALKYDS

Uralkyds are also called *urethane alkyds* or *urethane oils*. They are alkyd resins in which a diisocyanate, usually toluene diisocyanate (TDI), has fully or partly replaced the PA usually used in the preparation of alkyds. One transesterifies a drying oil with a polyol such as glycerol or pentaerythritol to make a "monoglyceride" (see Section 15.6.1) and reacts it with some PA (if desired) and then with somewhat less diisocyanate than the equivalent amount of N=C=O based on the free OH content. To assure that no N=C=O groups remain unreacted, methyl alcohol is added at the end of the process. Just like alkyds, uralkyds dry faster than the drying oil from which they were made, since they have a higher average functionality (more activated diallylic groups per average molecule). The rigidity of the TDI aromatic rings also speeds up the drying by increasing the T_g of the resin.

Two principal advantages of uralkyd over alkyd coatings are superior abrasion resistance and resistance to hydrolysis. Disadvantages are inferior color retention (when TDI is used) of the films, higher viscosity of resin solutions at the same percent solids, and higher cost. Uralkyds made with aliphatic diisocyanates have better color retention, but are more expensive and have lower T_g. The largest use of uralkyds is in architectural coatings. Many so-called varnishes sold to the consumer today are based on uralkyds; they are not really varnishes in the original sense of the word. (See Section 14.3.) They are used as transparent coatings for furniture, woodwork, and floors; applications in which good abrasion resistance is important. Since they are generally made with TDI, they tend to turn yellow and then light brown with age; yellowing is acceptable in clear varnishes, but would be a substantial drawback in light colored pigmented paints.

Water-reducible uralkyds can be made by reacting an isocyanate-terminated polyurethane made with dimethylolpropionic acid (see Section 10.8) with a "monoglyceride" of a drying oil.

15.8 EPOXY ESTERS

BPA epoxy resins can be converted to what are commonly called *epoxy esters* by reacting with fatty acids. Drying or semidrying oil fatty acids are used so that the products cross-link by autoxidation. The epoxy groups undergo a ring-opening reaction with carboxylic acids to generate an ester and a hydroxyl group. (See Section 11.3.2.) These hydroxyl groups, as well as the hydroxyl groups originally present on the epoxy resin, can esterify with fatty acids. They are generally made by starting with a low molecular weight epoxy resin (i.e., the standard liquid resin, $n = 0.13$) and extending with BPA by the advancement process (see Section 11.1.1) to the desired molecular weight. Off-specification epoxy resin is often used to reduce cost. The fatty acids are added to the molten, hot resin, and the esterification reaction is continued until the acid number is low, usually less than 7 mg of KOH per gram of resin. In the esterification reaction with fatty acids, the average number of sites for reaction is the n value, corresponding to the number of hydroxyl groups on the resin, plus twice the number of epoxy groups. The esterification is carried out at high temperatures (220–240°C). The rate of esterification slows as the concentration of hydroxyl groups diminishes, and side reactions occur, especially dimerization of the drying oil fatty acids (or their esters). It is not practical to esterify more than about 90% of the potential hydroxyl groups, including those from ring opening the epoxy groups. The lower useful limit of the extent of esterification is about 50%, this is required to ensure sufficient fatty acid groups for oxidative cross-linking.

Tall oil fatty acids are commonly used because of their low cost. Linseed fatty acids give faster cross-linking coatings due to higher average functionality. However, their viscosity is higher because of the greater extent of dimerization during esterification, and their cost is higher. For still faster cross-linking, part of the linseed fatty acids can be replaced with tung fatty acids, but the viscosity and cost are still higher. The color of epoxy esters from linseed and linseed-tung fatty acids is darker than the tall oil esters. Dehydrated castor oil fatty acids give faster curing epoxy esters for baked coatings. The rate of formation of a dry film from epoxy esters depends on two factors: the average number of diallylic groups $\bar{f}_n$ and the ratio of aromatic rings to long aliphatic chains. The $\bar{f}_n$ can be maximized by using higher molecular weight BPA epoxy resin and by using enough fatty acid to react with a large fraction of the epoxy and hydroxyl groups. The ratio of aromatic rings to fatty acids can be maximized by using high molecular weight epoxy resin and esterifying a smaller fraction of epoxy and hydroxyl groups.

Epoxy esters are used in coatings in which adhesion to metal is important. While the reasons are not completely understood, it is common for epoxy coatings, including epoxy esters, to have good adhesion to metals and to retain adhesion after exposure of the coated metal to high humidity, a critical factor in corrosion protection. A distinct advantage of epoxy esters over alkyd resins is their greater resistance to hydrolysis and saponification. The backbone of alkyds is held together with esters from PA and the polyol, whereas in epoxy esters, the backbone is held together with C–C and ether bonds. Of course, the fatty acids are bonded to the backbone with ester groups in both cases, but the fraction of polymer bonds in a dry film subject to hydrolysis is substantially lower in the case of epoxy esters. On the other hand, exterior durability of epoxy ester coatings is poor, as is the case with all films made with BPA epoxy resins. As a result of these advantages and disadvantages, the major uses for epoxy resins are in primers for metal and in can coatings, such as for crowns (bottle caps), in which the important requirements are adhesion and hydrolytic stability. In baking primers, it is sometimes desirable to supplement the cross-linking through oxidation by including a small amount of MF resin in the formulation to cross-link with part of the free hydroxyl groups on the epoxy ester.

Epoxy ester resins with good exterior durability (better than alkyds) can be prepared by reacting epoxy-functional acrylic copolymers (made with glycidyl methacrylate) with fatty acids. The product is an acrylic resin with multiple fatty acid ester side chains. By appropriate selection of acrylate ester comonomers and molecular weight, the T_g of the resin can be designed so that a tack-free film is obtained by solvent evaporation; then the coating cross-links by autoxidation. For an application like repainting an automobile at ambient temperatures, the cross-linking can proceed relatively slowly and need not be catalyzed by metal salt driers. The rate of cross-linking is slower without driers, but exterior durability is better.

Epoxy esters can also be made water-reducible. The most widely used water-reducible epoxy esters have been made by reacting maleic anhydride with epoxy esters prepared from dehydrated castor oil fatty acids. Subsequent addition of a tertiary amine, such as 2-(dimethylamino)ethanol, in water results in ring opening of the anhydride to give amine salts. Like other water-reducible resins (see Section 12.3), these resins are not soluble in water, but form a dispersion of resin aggregates swollen with water and solvent in an aqueous continuous phase. The hydrolytic stability of these epoxy esters is better than corresponding alkyds, and sufficient for use in electrodeposition primers (see Section 26.1) until anionic primers were replaced by cationic primers. Water-reducible epoxy esters are still used in spray applied baking primers and primer-surfacers. They are also used in dip coating primers in which nonflammability is an advantage. Their performance equals that of solvent-soluble epoxy ester primers.

GENERAL REFERENCES

Anonymous, *The Chemistry and Processing of Alkyd Resins*, Monsanto Chemical Co., (now Solutia, Inc.), 1962.

F. N. Jones, "Alkyd Resins," in *Ullman's Encyclopedia of Industrial Chemistry*, 6th ed. (1st Intl. ed.), Vol. A-1, VCH Verlagsgesellschaft mbH, Weinhein, Germany, 1985, pp. 409–423.

K. F. Lin, "Alkyd Resins" in *Kirk–Othmer Encyclopedia of Chemical Technology*, 4th ed., Vol. 2, Wiley, New York, 1992, pp. 53–85.

REFERENCES

1. R. J. Blackinton, *J. Paint Technol.*, **39** (513), 606 (1967).
2. T. A. Misev, *Prog. Org. Coat.*, **21**, 79 (1992).
3. S. L. Kangas and F. N. Jones, *J. Coat. Technol.*, **59** (744), 89 (1987).
4. R. Bacaloglu, M. Maties, C. Csunderlik, L. Cotaraca, A. Moraru, J. Gros, and N. Marcu, *Angew. Makromol. Chem.*, **164**, 1 (1988).
5. R. Brown, H. Ashjian, and W. Levine, *Off. Digest*, **33**, 539 (1961).
6. S. L. Kangas and F. N. Jones, *J. Coat. Technol.*, **59** (744) 99, (1987).
7. P. Kass and Z. W. Wicks, Jr., U.S. Patent 2577770 (1951).
8. E. Levine, *Proc. Water-Borne Higher-Solids Coat. Symp.*, New Orleans, 1977, p. 155.
9. D. B. Larson and W. D. Emmons, *J. Coat. Technol.*, **55** (702), 49 (1983).
10. *Tech. Bull., Resimene AM-300 and AM-325*, Monsanto Chemical Co., (now Solutia, Inc.), January 1986.
11. W. F. Strazik, J. O. Santer, and J. R. LeBlanc, U.S. Patent 4293461, (1981).
12. G. Osterberg, M. Hulden, B. Bergenstahl, and K. Holmberg, *Prog. Org. Coat.*, **24**, 281 (1994).
13. G. Ostberg and B. Bergenstahl, *J. Coat. Technol.*, **68** (858), 39 (1996).
14. T. Nabuurs, R. A. Baijards, and A. L. Germna, *Prog. Org. Coat.*, **27**, 163 (1996).
15. J. W. Gooch, S. T. Wang, F. J. Schork, and G. W. Poehlein, *Proc. Waterborne, High Solids, Powder Coat. Symp.*, New Orleans, 1997, p. 366.
16. C. J. Bouboulis, *Proc. Water-Borne Higher-Solids Coat. Symp.*, New Orleans, 1982, p. 18.
17. B. Zuchert and H. Biemann, *Farg och Lack Scandinavia*, (2) 9 (1993); W. Weger, *Fitture e Vernici*, **B66** (9), 25 (1990).
18. R. Hurley and F. Buona, *J. Coat. Technol.*, **54** (694), 55 (1982).
19. J. Kaska and F. Lesek, *Prog. Org. Coat.*, **19**, 283 (1991).
20. Anonymous, *The Chemistry and Processing of Alkyd Resins*, Monsanto Chemical Co. (now Solutia, Inc.), 1962.
21. J. Kumanotani, H. Hironori, and H. Masuda, *Adv. Org. Coat. Sci. Tech. Ser.*, **6**, 35 (1984).

CHAPTER 16 _____

Other Resins and Cross-Linkers ___

Chapters 8–15 describe the types of resins and cross-linkers used in the largest volume in coatings. In this chapter, other commercially important materials and selected promising developments are discussed.

16.1. HALOGENATED POLYMERS

Halogenated polymers have desirable properties for certain applications. For example, their water permeability is low; some are used in top coats for corrosion protection. Some are sufficiently soluble in polyolefin plastics that they are used in tie coats to provide adhesion for top coats.

16.1.1. Solution Grade Thermoplastic Vinyl Chloride Copolymers

Solution grade vinyl chloride resins are copolymers of vinyl chloride and vinyl acetate frequently with a small amount of a third monomer to impart special properties. The vinyl acetate lowers T_g and broadens the range of applicable solvents. For example, a terpolymer of vinyl chloride, vinyl acetate, and maleic acid with a weight ratio of 86 : 13 : 1 (mole ratio 81 : 17 : 1) and an $\bar{M}_w$ on the order of 75,000 has been used in interior coatings for beverage cans. The maleic acid promotes adhesion; other solution vinyl resins are made with hydroxy functionality. Use of solution vinyl resins has declined because of the low solids required for application, about 10 to 12 nonvolatile volume solids (NVV).

Vinyl chloride copolymers require stabilization to prevent thermal and photochemical degradation. As discussed in Section 5.3, the polymers undergo dehydrochlorination in an autocatalytic chain reaction. A variety of stabilizing agents are used, including organotin esters such as dibutyl tin dilaurate; barium, cadmium, and strontium soaps; maleates; and oxirane compounds.

16.1.2. Vinyl Chloride Dispersion Copolymers

Dispersion grade vinyl chloride copolymers with high molecular weights can be used in high solids coatings. They are prepared by suspension polymerization as particles with an

286

average diameter of a few micrometers. In this process, a solution of an initiator in the monomer mixture is added with vigorous stirring to a hot dilute solution of poly(vinyl alcohol) (PVA) in water. The monomers are dispersed into small droplets inside of which the polymerization is initiated and completed, resulting in the formation of particles. A monomer-soluble initiator is used in suspension polymerization, in contrast to the water-soluble initiators used in emulsion polymerization (see Chapter 8). The PVA stabilizes the suspension by minimizing coalescence of the particles when the temperature of the reaction is above the T_g. Since the PVA has many easily abstractable hydrogens, graft side chains grow on the PVA. After polymerization is complete, the polymer particles are separated by filtration, washed, and dried. The molecular weight of polymers made by suspension polymerization depends primarily on initiator structure and on the combination of reaction temperature and initiator concentration. Commonly, $\bar{M}_w$ is on the order of 100,000.

In plastics, dispersion grade vinyl chloride copolymers are used in plastisols. Plastisols are polymer particles dispersed in a plasticizer. Since the T_g of the polymer is well above room temperature, and also because the polymer is partially crystalline, the polymer does not dissolve in the plasticizer at an appreciable rate at room temperature. When the plastisol is heated above the polymer T_g and above the melting point of the crystalline regions in the polymer, the polymer dissolves in the plasticizer and the particles coalesce to a molten state. When cooled, the product is a plastic consisting of a homogeneous solution of polymer and plasticizer. The viscosity of plastisols is generally too high for application as coatings. The viscosity can be lowered by addition of solvent. Solvents must be chosen that dissolve the plasticizer without significantly swelling the polymer particles. Application viscosities can generally be achieved with 20% solvent or less. These systems are properly called *organosols*, but they are also frequently called *plastisols*, even though there is volatile solvent present.

16.1.3. Fluorinated Polymers

Polytetrafluoroethylene [PTFE, $-(CF_2CF_2)_n-$] has the greatest exterior durability and heat resistance of any polymer used in coatings. However, PTFE is insoluble in solvents, and its fusion temperature is so high that its coating uses are limited to applications in which the substrate can withstand high temperatures [1]. For example, aqueous dispersions of PTFE are used to coat the interior of chemical processing equipment and cookware [2]. After application, the polymer particles are sintered at temperatures as high as 425°C. Like all highly fluorinated compounds PTFE has such a low surface free energy that it is not wet by either water or oils; hence, it provides a "nonstick" cooking surface.

Since its fusion temperature is somewhat lower, poly(vinylidene fluoride), [PVDF, $-(CH_2CF_2)_n-$] can be used in additional applications. PVDF is used in coil coatings as a plastisol-like dispersion in a solution of acrylic resin [1,3]. The fusion temperature of the films is reported to be 245°C. The exterior durability is outstanding, but the cost is high and only low gloss coatings are possible. Copolymers of vinylidene flouride (VDF) are also finding use in powder coatings.

Fluorinated copolymers with functional groups such as hydroxyl groups can be cross-linked after application. Copolymers of VDF with a hydroxy-functional monomer cross-linked with a polyisocyanate give coatings with superior wet adhesion and corrosion as compared with PVDF homopolymer [4]. Perfluoroalkyl acrylate esters have been copolymerized with hydroxyethyl methacrylate (HEMA) to form solvent-soluble resins;

however, the monomers are expensive. Halofluoroethylene (CF_2=CFX)/vinyl ether copolymers have been used on steel building panels and, to a limited extent, in clear coats for automobiles [1]. Vinyl ethers and CF_2=CFX form alternating polymers. Functional groups can be introduced by copolymerizing hydroxy- and/or carboxy-substituted vinyl ether comonomers. The T_g is controlled by the ratio of fluorinated monomer to vinyl ether monomers and by the chain length of the alkyl group on the vinyl ether. Copolymers with hydroxyl groups can be cross-linked with MF resins or poly-isocyanates. Gloss retention on exterior exposure is reported to be intermediate between poly(vinylidene fluoride) and acrylic coatings.

16.1.4. Chlorinated Rubber and Ethylene/Vinyl Acetate Copolymers

Chlorinated rubber is used in top coats for heavy duty maintenance paints because of the barrier properties that result from its low water permeability. It is also used in tie coats on polyolefin plastics. Chlorinated rubber is compatible with some alkyd resins and is used to impart fast drying qualities to alkyd-based traffic paints. Like PVC, chlorinated rubber dehydrochlorinates and requires stabilizers similar to those used with PVC. Some metal salts, especially iron, tend to promote degradation of chlorinated rubber, so it degrades when applied over rusty steel [5].

Chlorinated rubber is prepared from natural rubber that has been milled to reduce its molecular weight. A solution of milled rubber in CCl_4 is reacted with chlorine or another chlorinating reactant. The reactions are complex: Addition to the double bonds, substitution, and cyclization reactions all occur. To eliminate most of the double bonds from the rubber, the final product contains 65–68 wt% chlorine. Excess free chlorine and HCl are removed by washing; the polymer is precipitated by addition of alcohol, filtered, and dried. Various molecular weight grades are manufactured. Polymer strength increases with molecular weight, but the viscosity of solutions also increases. The formulator makes an appropriate compromise between coating solids and film properties for a particular application.

Environmental regulations in the United States have increased the cost of producing chlorinated rubber to prohibitive levels; production has been discontinued and only imported material is available. Chlorinated ethylene/vinyl acetate copolymers have been developed that can be used to replace chlorinated rubber in at least some applications [6]. Since the initial resin does not contain double bonds, the chlorine content need not be as high as in chlorinated rubber; grades with 52, 55, and 58% chlorine are available. The higher the chlorine content, the better the barrier properties but the lower the flexibility and impact resistance. Performance is said to be comparable to chlorinated rubber, with the advantage of better storage stability.

16.2. CELLULOSE DERIVATIVES SOLUBLE IN ORGANIC SOLVENTS

Cellulose is a naturally occurring polymer of repeating anhydroglucose units; it is insoluble in water and organic solvents. Derivatives of cellulose soluble in water and others soluble in organic solvents can be made by reactions of the hydroxyl groups.

Many soluble cellulose derivatives have been prepared; those of most importance in coatings are nitrocellulose (NC) and cellulose acetobutyrate (CAB). Ethyl cellulose, cellulose acetate, and cellulose propionate have been used to some extent. Water-soluble

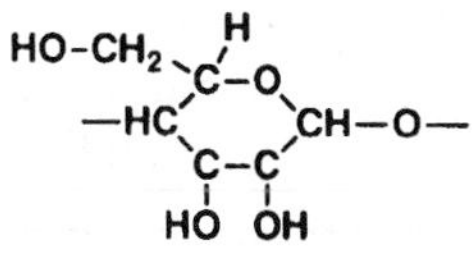

HO-CH₂

Anhydroglucose Unit

cellulose derivatives are used to modify the flow properties of latex paints and are discussed in Chapter 31.

16.2.1. Nitrocellulose

Nitrocellulose (NC) is another example of poor nomenclature, since it is not really a nitro compound; it is a nitrate ester and should be called cellulose nitrate. Three types of NC are available with differing extents of esterification. Their compositions are given in Table 16.1. An isomeric dinitroanhydroglucose unit of nitrocellulose is:

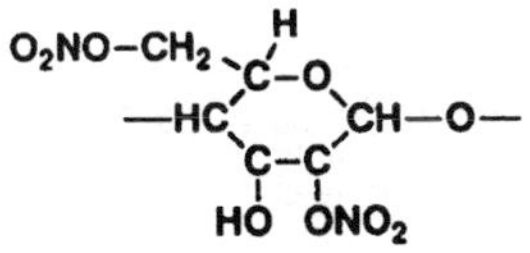

O_2NO-CH_2

Dinitroanhydroglucose Unit

The SS type of nitrocellulose is alcohol-soluble and is used in flexographic printing inks but has few uses in coatings. The RS type is soluble in esters and ketones and is generally used in coatings. The explosive type is included in Table 16.1 to emphasize the hazards of nitrocellulose. The RS type used in coatings is not explosive, but is highly flammable. To reduce the handling hazard, nitrocellulose is shipped wet with a nonsolvent. In most cases, RS type nitrocellulose is shipped at 70 NVW, wet with ethyl or isopropyl alcohol. While RS type nitrocellulose is not soluble in alcohol, it is soluble in mixtures of ketones and esters with alcohols and hydrocarbons. (See Section 17.2.)

Nitrocellulose is made by reacting cellulose with nitric acid. Small amounts of water are present, and some hydrolysis of the cellulose occurs, reducing the molecular weight. Several molecular weight grades are available. The higher the molecular weight, the greater the polymer strength, and the lower the solids of coatings that can be made. Viscosity of nitrocellulose solutions is tested with a falling ball viscometer, in which the time for a ball to fall a specified distance through a nitrocellulose solution of standard concentration is measured. The various molecular weight grades are designated as 1/4, 1/2, 1 second, and so forth. The times are proportional to the time in the falling ball test; the higher the number of seconds, the higher the molecular weight.

Table 16.1. Composition of Nitrocellulose Types

Type	Nitrogen (%)	–ONO₂ groups per anydroglucose unit
SS (Spirit Soluble)	10.7–11.2	1.9–2.0
RS (Regular Solubility)	11.8–12.3	2.15–2.25
Explosive	12.3–13.5	2.25–2.5

Nitrocellulose was used in lacquers for automobile topcoats before acrylic lacquers, which have better exterior durability, were developed. They continued to be used for some years in refinish lacquers for repairing automobiles because they polished more easily and gave a deeper gloss appearance than room temperature dried acrylic lacquers. Nitrocellulose lacquer primers continue to be used to some degree for refinishing automobiles. The principal remaining use of NC lacquers is in wood finishing. (See Section 32.1.) These lacquers have relatively low solids but continue to be used to a significant, if decreasing, extent because they enhance the appearance of wood grain to a greater extent than any other coating. Other desirable characteristics of NC lacquers for wood finishing are rapid drying, which permits handling and shipment soon after finishing, and easy damage repair, since the films are thermoplastic. Increasingly stringent VOC emission regulations can be expected to force further reductions in use of nitrocellulose.

16.2.2. Cellulose Acetobutyrate

A variety of organic esters of cellulose are used in fibers, plastics, and, to a degree, in coatings. The only one still used to a significant extent in coatings is a mixed acetate, butyrate ester (CAB). CAB lacquers are used to some extent; they exhibit lighter color and better color retention with reduced handling risk relative to NC lacquers. Their principal use has been as a component in acrylic automotive coatings. CABs help control flow and, particularly, facilitate orientation of aluminum flakes parallel to the surface of the film immediately during application. Orientation of the flakes in the final film enhances the appearance of metallic finishes. (See Ref. [7] and Chapter 29.)

Several types of CAB are available with different ratios of acetate to butyrate to unesterified hydroxyl groups and with differing molecular weights. The ratio affects the solubility and compatibility of the polymer. The larger the ratio of butyrate ester, the larger the amount of hydrocarbon that can be tolerated in a solvent mixture. Also, as the butyrate content increases, the T_g tends to be lower. Compatibility with acrylic resins is dependent on both the CAB and the acrylic polymer. An example of a CAB grade used in acrylic lacquers has an average of 2.2 acetate groups, 0.6 butyrate groups, and 0.2 unreacted hydroxyl groups per anhydroglucose unit.

16.3. UNSATURATED POLYESTER RESINS

Unsaturated polyester resins are maleic acid-containing polyesters dissolved in styrene. They contain unsaturated maleate [Z-(cis) configuration] and fumarate [E-(trans) configuration] groups in the backbone. These resins are used on a large scale in glass-reinforced plastics. They are used on a smaller scale in some coating applications. With a free radical initiator, a styrene solution of unsaturated polyester is converted into a cross-linked polymer. The product is a complex mixture derived from copolymerization of monomer with the maleate or fumarate double bonds, graft copolymerization, and homopolymerization of styrene.

A variety of diacids and diols can be used to make unsaturated polyesters; the most common are phthalic anhydride (PA), maleic anhydride (MA), and propylene glycol (or a mixture of propylene glycol and propylene oxide). These are the lowest cost raw materials, furthermore, the initial reaction can be carried out at moderate temperatures, and a minimum amount of water is formed. Cross-link density of the cured product is affected by the ratio of MA to PA; mole percentages in the range of 15 to 40% of maleic anhydride are commonly used. During esterification, some of the maleate ester groups isomerize to fumarate ester groups. The fraction isomerized depends on processing time and temperature and can also vary depending on the other components of the polyester. Fumarate groups are more reactive during cross-linking than maleate groups. Reaction conditions during esterification must be carefully controlled to ensure that the extent of isomerization is reproducible from batch to batch. Use of fumaric acid instead of maleic anhydride results in a higher fumarate diester content; such polyesters generally give harder plastics than those made with maleic anhydride. More flexible plastics can be obtained by substituting aliphatic diacid (e.g., adipic acid) for part of the PA to reduce T_g. While other glycols can be used, economics favor use of propylene glycol (or propylene oxide). The properties of the cured materials are also affected by the ratio of polyester to styrene; typically, a 70 : 30 ratio is used.

The resin/styrene solution is cross-linked using free-radical initiators. Peroxides, such as benzoyl peroxide, are used at temperatures of 70 to 100°C. Many applications require cross-linking at ambient temperatures. In these cases, an initiator, such as methyl ethyl ketone (MEK) peroxide, and promoters, such as a mixture of dimethylaniline and cobalt naphthenate, are added to the resin-styrene solution just before use. The combination reacts rapidly at room temperature to generate free radicals, which initiate cross-linking. MEK peroxide is a complex mixture in which the major components contain both peroxide and hydroperoxide groups [8], as shown. The cobalt salt acts as a redox catalyst for the decomposition of the peroxide and hydroperoxide groups into free radicals; dimethylaniline further promotes that reaction.

$$\underset{\underset{CH_3}{|}}{\overset{\overset{OOH}{|}}{C_2H_5-C-OOH}} \qquad \underset{\underset{CH_3}{|}}{\overset{\overset{OOH}{|}}{C_2H_5-C}}-O-O-\underset{\underset{CH_3}{|}}{\overset{\overset{OOH}{|}}{C}}-C_2H_5$$

Major components of MEK Hydroperoxide

A plastics application sometimes classified as a coating is gel coats. Gel coats are pigmented unsaturated polyester-styrene coatings (with initiator and promoter); they are sprayed on the inside of a mold surface. The exposed (back) surface of the gel coat is then sprayed with glass fiber-loaded unsaturated polyester-styrene compound and then covered with plastic film. After the entire composition cross-links and is taken from the mold, the surface is the gel coat that cured in the absence of oxygen. Many glass-reinforced plastic objects, ranging from prefabricated shower stalls to boat hulls, are made this way. In the case of boat hulls, hydrolytic stability is particularly important so that the surface retains gloss after outdoor exposure and immersion in water. Unsaturated polyesters made using neopentyl glycol, MA, and isophthalic acid provide better gloss retention than those made from propylene glycol and PA, but cost is higher.

The free radical polymerization is oxygen inhibited. This is not a problem for many plastics applications, since the reaction is commonly carried out in a mold so that the resin surface is not exposed to air. In most coating applications, the top surface is exposed to the air and will remain sticky after the polymerization is complete below the surface. A further problem is the volatility of styrene. These problems can be minimized by incorporating some insoluble semicrystalline paraffin wax in the formulation. After the coating is applied, the low surface tension wax particles preferentially come to the surface. The wax layer reduces the rate of styrene loss and also reduces oxygen concentration in direct contact with the polymerizing coating, thereby minimizing the difficulty of surface cure. The wax layer, however, results in a relatively uneven, low gloss surface, suitable for some applications but not for others.

Unsaturated polyesters can be used in UV cure coatings. (See Chapter 28.) A photoinitiator is used that generates free radicals on exposure to UV radiation. High intensity radiation sources are used, which generate very large numbers of free radicals sufficiently rapidly at the surface, so that the oxygen in the air at the surface is depleted (by reaction with radicals) and polymerization can proceed. Acrylate systems (see Section 16.4) are more widely used than styrene/unsaturated polyesters.

Oxygen inhibition can be minimized using coreactants that have allyl groups with styrene-unsaturated polyesters [9]. A free radical abstracts a hydrogen atom from the methylene group activated by both the adjacent double bond and the ether oxygen. The resulting free radicals from this reaction then react with oxygen to form peroxy free radicals, which in turn abstract hydrogen from the activated methylene groups to form hydroperoxides. The reactions consume some of the oxygen at the surface and generate new hydroperoxide in a chain reaction; as a result, oxygen inhibition is reduced.

Use of allyl ether reactions has been extended to water-borne unsaturated polyester resins. A low molecular weight ester is prepared by reacting 2 mol of maleic anhydride with 1 mol of a mixture of low molecular weight polyalkylene glycols and diols. The resulting partial ester is further esterified with 2 mol of trimethylolpropane diallyl ether [10]. The polyalkylene glycol ester segments are efficient emulsifying agents, so the polyester can be emulsified in water for application. Films can be cured either with hydroperoxide-cobalt initiators or by photoinitiators and UV radiation.

16.4. (METH)ACRYLATED OLIGOMERS

Thermosetting resins can be made with (meth)acrylate double bonds as cross-linkable groups. Such resins are used in radiation cured coatings. (See Section 28.1.4.) They can also be used in ambient cure coatings and at somewhat elevated temperatures (force-dry) with free radical initiators. Both acrylate and methacrylate polymerizations are air inhibited, particularly methacrylate systems, which generally require higher cure temperatures to minimize the effect of oxygen inhibition.

Acrylated oligomers can also be cross-linked with polyfunctional primary amines by a Michael addition reaction, as illustrated for monofunctional reactants. This reaction is fast, so practical systems are based on blocked amines, commonly ketimines [11]. When a coating film is exposed to atmospheric moisture, the ketimine reacts with water to release the free primary amine.

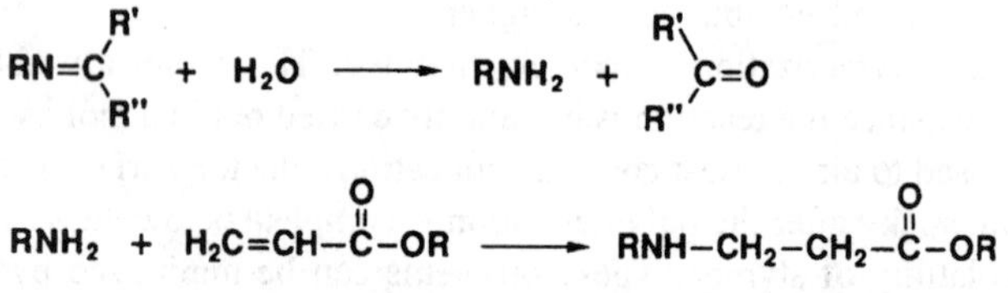

$$RN=C\genfrac{}{}{0pt}{}{R'}{R''} + H_2O \longrightarrow RNH_2 + \genfrac{}{}{0pt}{}{R'}{R''}C=O$$

$$RNH_2 + H_2C=CH-\overset{O}{\underset{\|}{C}}-OR \longrightarrow RNH-CH_2-CH_2-\overset{O}{\underset{\|}{C}}-OR$$

Acrylated oligomers can also be cross-linked by a Michael reaction with acetoacetylated resins and their enamine derivatives, as discussed in Section 16.6.2.

(Meth)acrylated oligomers are prepared from a variety of starting oligomers. The properties of films made from the (meth)acrylated oligomers are affected by the average number of acrylic double bonds per molecule and by the structure of the "core" oligomer. For example, acrylated urethane oligomers tend to give coatings with a good combination of hardness and elasticity, and epoxy resin derivatives tend to give coatings with good toughness, chemical resistance, and adhesion. Any polyol or hydroxy-terminated oligomer (HO–R–OH) can be reacted with excess diisocyanate (OCN–R′–NCO) to yield an isocyanate-terminated oligomer. This oligomer can then be reacted with hydroxyethyl acrylate at ambient or moderately elevated temperature to yield an acrylated urethane oligomer. Hydroxyethyl methacrylate yields the analogous methacrylated oligomer.

$$HO-R-OH \ + \ 2 \ O{=}N{=}C{-}R'{-}N{=}C{=}O \ \longrightarrow \ \left[O{=}N{=}C{-}R'{-}\overset{\overset{O}{\parallel}}{C}{-}O \right]_2 R$$

$$1$$

$$1 \ + \ 2 \ HO-CH_2-CH_2-O-\overset{\overset{O}{\parallel}}{C}-CH{=}CH_2 \ \longrightarrow$$

$$H_2C{=}CH{-}\overset{\overset{O}{\parallel}}{C}{-}O{-}CH_2{-}CH_2{-}O{-}\overset{\overset{O}{\parallel}}{C}{-}NH{-}R'{-}NH{-}\overset{\overset{O}{\parallel}}{C}{-}O{-}R{-}O{-}\overset{\overset{O}{\parallel}}{C}{-}NH{-}R'{-}NH{-}$$

$$-\overset{\overset{O}{\parallel}}{C}{-}O{-}CH_2{-}CH_2{-}O{-}\overset{\overset{O}{\parallel}}{C}{-}CH{=}CH_2$$

Acrylated Urethane Oligomer

Another route is to react oxirane groups of epoxy resins with (meth)acrylic acid. The ring-opening reaction yields the acrylic ester and a hydroxyl group:

$$CH_2{=}CH{-}\overset{\overset{O}{\parallel}}{C}{-}O{-}CH_2{-}\overset{\overset{OH}{|}}{CH}{-}CH_2{-}O{-}\!\!\left\langle\bigcirc\right\rangle\!\!{-}\overset{\overset{CH_3}{|}}{\underset{\underset{CH_3}{|}}{C}}{-}\!\!\left\langle\bigcirc\right\rangle\!\!{-}O{-}CH_2{-}\overset{\overset{OH}{|}}{CH}{-}CH_2{-}O{-}\overset{\overset{O}{\parallel}}{C}{-}CH{=}CH_2$$

Acrylated Epoxy Oligomer

Various catalysts (e.g., triphenylphosphine) are used so that the reaction is carried out at as low a temperature as possible. Care is required to avoid polymerization of the acrylic acid or esters during the process. Inhibitors are added to trap free radicals. Some inhibitors, notably phenolic antioxidants, are effective only in the presence of oxygen, so the reaction is commonly carried out under a mixed inert gas, air atmosphere. Variations in reaction conditions and catalyst composition can result in significant differences in the product. The most widely used epoxy resin is the standard liquid BPA epoxy resin ($n = 0.13$) to yield predominantly the acrylated diglycidyl ether of BPA. Epoxidized soybean or linseed oil also react with acrylic acid to give lower T_g oligomers with higher functionality.

Acrylated melamine-formaldehyde (MF) resins have been prepared by reacting etherified MF resins with acrylamide [12]. These resins have the potential to cure in two ways: by UV curing through the acrylate double bonds and thermally by residual alkoxymethylol groups on the MF resin. It is reported that UV curing followed by thermal curing provides films with increased hardness, improved stain resistance, and durability.

16.5. ORGANIC SILICON DERIVATIVES

Three classes of organic silicon derivatives are used in coatings: silicones, reactive silanes, and orthosilicates.

16.5.1. Silicones

Silicones, more properly called *polysiloxanes*, are polymers with backbones consisting of –[Si(R)$_2$–O]– repeating units. Polysiloxanes are prepared from chlorosilanes. The principal commercial monomers are the following methyl and phenyl substituted silanes.

Me$_3$SiCl	Trimethylchlorosilane
Ph$_2$SiCl$_2$	Diphenyldichlorosilane
PhSi(Me)Cl$_2$	Phenylmethyldichlorosilane
Me$_2$SiCl$_2$	Dimethyldichlorosilane
PhSiCl$_3$	Phenyltrichlorosilane
MeSiCl$_3$	Methyltrichlorosilane

Chlorosilanes react with water to form silanols that, in turn, can condense to form siloxanes. For example, dimethyldichlorosilane reacts with water to form a polysiloxane, often represented as a linear polysiloxane:

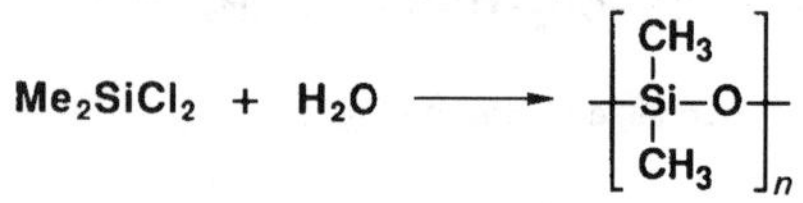

$$Me_2SiCl_2 \ + \ H_2O \longrightarrow \left[\begin{array}{c} CH_3 \\ | \\ Si-O \\ | \\ CH_3 \end{array} \right]_n$$

Structures of the polymers can be more complex than represented by this linear structure. Siloxanes are more apt than hydrocarbons to form large ring cyclic structures, such as (Me$_2$SiO)$_4$. Such cyclic compounds are intermediates in the polymerization and are present to varying degrees in the finished product.

Oligomers are made using monochlorosilanes as chain-terminating groups; molecular weight is reduced in proportion to the monochlorosilane to dichlorosilane ratio. The oligomeric products are called *silicone fluids* or *silicone oils*. As discussed in Chapter 23, small amounts of silicone fluids are used as additives in coatings. Poly(dimethylsiloxane) fluids reduce surface tension because the ease of rotation around the Si–O–Si bonds leads to rapid orientation of methyl groups at the surface. The methyl groups give a surface with very low surface tension. Low molecular weight silicone fluids can be used as VOC exempt solvents.

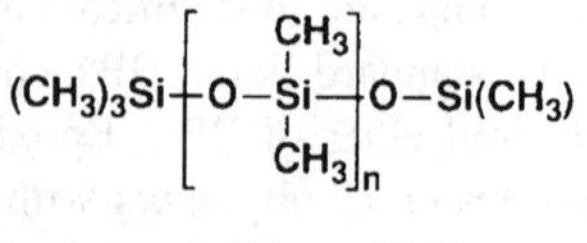

$$(CH_3)_3Si \left[\begin{array}{c} CH_3 \\ | \\ O-Si \\ | \\ CH_3 \end{array} \right]_n O-Si(CH_3)$$

A Silicone Fluid

Poly(dimethylsiloxanes) have limited compatibility with many coating resins. Chemically modified silicone fluids with broader ranges of compatibility have been described [13]. Examples are polysiloxane/polyether block copolymers, polyether copolymers that also have some longer alkyl chains on the siloxane groups, and poly(dimethylsiloxanes) modified with esters and aryl-substituted alkyl groups.

16.5.2. Silicone Resins for Coatings

Trichlorosilanes are used to provide chain branching. They can be copolymerized with dichlorosilanes to make silicone rubbers. While silicone rubbers are expensive, they have unique and useful properties; they exhibit excellent resistance to oxidation, accompanied by retention of flexibility at low temperatures. They have a low T_g as a result of the ease of rotation around the Si–O–Si bonds. Chain branching is not the only possible outcome when trichlorosilanes are copolymerized. They can also react with water to form three-dimensional cluster compounds (silesquioxanes) exemplified by the cubic octamer $(C_6H_5)_8Si_8O_{12}$.

Polymerization of a mixture of mono-, di-, and trichlorosilanes results in a silicone resin with some unreacted hydroxyl groups. The initial reaction is carried out by hydrolyzing a solution of the monomers in a water-immiscible solvent. The silicone resin is separated from the acidic water phase, and residual acid is removed. A stable solution of a silicone resin is obtained if the last water is removed from the reaction mixture at relatively low temperatures in the absence of catalyst. The $\bar{M}_n$ of such resins is generally in the range of 700–5000. Different copolymerization processes yield different proportions of linear, branched, cyclic, and three-dimensional cluster structures; the composition affects properties.

Silicone resins are baked to form a cross-linked film. The cross-linking process is reversible. A typical cure schedule is 1 hour at 225°C. The time and temperature can be reduced with catalysts such as zinc octanoate. Since the cross-linking reaction is reversible, silicone films are sensitive to water, especially under basic conditions. Ammonia and amines are especially destructive to such films. Evidence has been presented that aminolysis occurs selectively at Si groups with three oxygens (i.e., at cross-linking sites), probably resulting from the greater electrophilicity of such Si groups with three electronegative oxygen atoms [14].

The molecular weight and viscosity of silicone resins depend on the proportion of monochloro-, dichloro-, and trichlorosilane monomers and on the process. The ultimate cross-link density depends on the fraction of trichlorosilane in the recipe and on the fraction of this monomer that forms branched, rather than cyclic or cluster structures, during the process.

Most silicone resins are copolymers of methyl and phenyl substituted monomers; other alkyl monomers are sometimes included. Properties depend on the phenyl to methyl ratio, as summarized in Table 16.2 [15]. The rate of the cross-linking reaction is faster with high-methyl substituted silicone resins. Consequently, a larger amount of material is lost during heat curing of high-phenyl resins since low molecular weight components evaporate in the oven to a larger degree before polymerizing. The faster cure rate of the high-methyl substituted silicones and the longer package stability of the high-phenyl substituted silicones follow from the same factor. The greater UV resistance of high-methyl relative to high-phenyl substituted silicone resins gives greater gloss retention during exterior exposure. The exterior durability of well-formulated silicone coatings is better than that of acrylic-MF, polyester-MF, or urethane coatings and approaches that of highly fluorinated polymers.

Table 16.2. Properties of High-Methyl versus High-Phenyl Silicone Resins

High-Methyl	High-Phenyl
Lower weight loss in curing	Greater weight loss in curing
Faster cure rate	Longer package stability
Higher UV stability	Higher heat stability
Lower temperature flexibility	Greater oxidation resistance

Source: Adapted from Ref. [15].

On the other hand, high-phenyl silicones are superior to high-methyl silicones for applications requiring high temperature resistance, and far superior to other organic coatings except certain fluoropolymers. The thermal stability of high-methylsilicones is much greater than that of silicones with longer alkyl groups. The half-lives at 250°C, based on weight loss studies, are reported to exceed 100,000 hours for phenylsilicone films compared with over 10,000 hours for methylsilicone films and only 2 hours for propylsilicone films [15]. The useful life of a methylsilicone film at 350°C is given as 1000 hours; for comparison a "polyester" film (presumably cross-linked with a MF resin) has a useful life of 1000 hours at 223°C.

When silicones are thermally decomposed, the ultimate product is silicon dioxide, which, though brittle, can serve as a temperature-resistant coating binder. For example, chimney paints are made from silicone resins pigmented with aluminum flake for use at over 500°C for years. At the high service temperature, the organic substituents burn off, leaving behind a film of the aluminum pigment in a matrix of silicone dioxide, possibly with some aluminum silicate-essentially, glass. While it is brittle, this film continues to provide protection if it is not mechanically damaged. See Ref. [15] for discussion of coatings with varying heat resistance.

As expected from the difference noted earlier in the effects of aromatic and aliphatic substituents, coatings from methylsilicone resins have low temperature flexibility superior to those from phenylsilicones and to most other organic coatings. They also combine the properties of being repellant to liquid water, but permeable to water vapor.

High solids silicone resins have been made available that cure either directly using zinc octonoate catalysts or by cross-linking with trialkoxysilanes using a titanate catalyst [16]. Water-borne silicone resins have also been developed. One approach is to emulsify a silicone resin in water using an emulsifying agent [17].

16.5.3. Silicone-Modified Resins

The high cost of pure silicone coatings and the long, high temperature cures can be reduced by combining silicones with other coating resins. The earliest approach was simply to add a silicone resin to an alkyd resin in the reactor at the end of the alkyd cook. While some covalent bonds between silicone resin and alkyd might form, probably most of the silicone resin simply dissolves in the alkyd. Exterior durability of silicone-modified alkyd coatings is significantly better than unmodified alkyd coatings. The improvement in durability is roughly proportional to the amount of added silicone resin; 30% silicone resin is a common degree of modification. Silicone resins designed for this purpose may contain higher alkyl, as well as methyl and phenyl, groups to improve compatibility. Alkyd coatings modified with high-phenylsilicone resins are reported to have greater thermoplasticity, faster air drying, and higher solubility than high-methylsilicone-modified alkyds. These results can be understood when we consider that the higher rigidity of the aromatic rings leads to a "solid" film at an earlier stage of cross-linking. Less cross-linking in the phenylsilicone-modified coatings makes them more thermoplastic and soluble. Further improvements in exterior durability are obtained by coreacting a silicone intermediate during synthesis of the alkyd. Such intermediates react readily with free hydroxyl groups. Silicone-modified alkyds are used mainly in outdoor air dry coatings for which application is expensive (e.g., in a top coat for steel petroleum storage tanks).

Silicone-modified polyester and acrylic resins are used in baking coatings, especially in coil coatings for metal siding. See Ref. [15] for examples of formulations and preparation of several examples of silicone-modified resins. Silicone-modified epoxy resins have also been made [18]. Water-borne silicone resins can be prepared from water-reducible acrylic and polyester resins. Also, acrylic latexes prepared with hydroxyethyl (meth)acrylate as a comonomer can be modified with silicone intermediates [16].

Silicone-modified resins are made by reacting silicone intermediates with hydroxy-functional resins; in some cases, the silicone intermediates have silanol (Si–OH) groups. These can undergo cocondensation with hydroxyl groups on the resin to be modified, as well as self-condensation with other Si–OH groups. The ratio of the two reactions is controlled by catalyst choice. The best catalysts for promoting the desirable cocondensation between the resin and the silicone intermediate are titanates such as tetraisopropyl- or tetraisobutyltitanate. In model compound studies, tetraisopropyltitanate was shown to favor cocondensation over self-condensation by 3.4 : 1 as compared to a 0.23 : 1 ratio with no catalyst. Cocondensation between a hydroxy-functional silicone intermediate and acrylic polyol is shown below. Self-condensation between silicone intermediates can also occur resulting in some Si–O–Si cross-links. Excessive co- or self-condensation of a polyfunctional silicone intermediate and a polyfunctional resin can result in gelation.

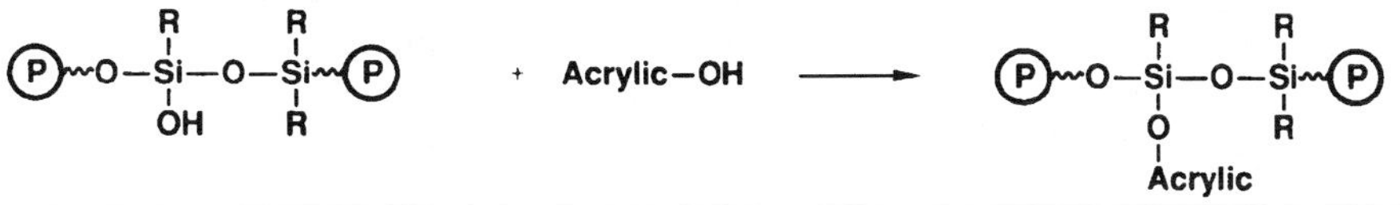

It is easier to control the reaction when the reactive functional groups on the silicone intermediates are silylmethoxy (Si–OMe) rather than silanol (Si–OH), groups. A methoxylated silicone intermediate has been described that has a weight average molecular weight of 470 and an equivalent weight of 155 [15]. An idealized structure of the intermediate is:

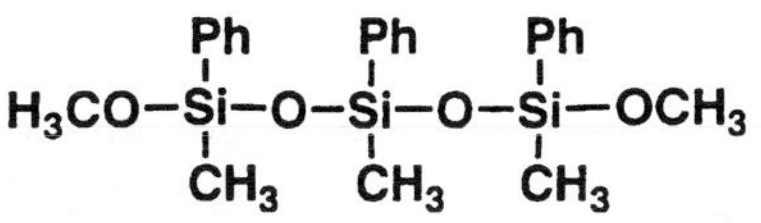

A solution of the polyester or acrylic is heated with the desired amount of silicone intermediate and the titanate catalyst at 140°C until a predetermined viscosity is reached.

The reaction is relatively slow at 140°C; higher temperatures reduce reaction time and, hence cost, but increase the risk of overshooting the desired viscosity, as well as the risk of gelation. Typical silicone-modified polyesters and acrylics have 30–50 wt% silicone. Modification with less than about 25% silicone provides little improvement in exterior durability. Exterior durability is enhanced with increasing silicone content above 30%, but cost also increases.

The same reactions occur during cross-linking after the coating is applied. Usually, zinc octanoate is used as the catalyst for cross-linking the coating, since titanate esters are hydrolyzed by the water brought into the coating by pigments. The principal applications are in coil coatings (see Section 29.4), for which a typical cure schedule is 90 seconds at peak metal temperatures of 300°C. Such coatings tend to soften when exposed to high humidity for prolonged periods, probably resulting from hydrolysis of the cross-links, which, as discussed above, have three oxygen atoms bound to a single silicon atom. The softening is called reversion. The film hardens again if the ambient humidity decreases, but the film is subject to physical damage if scraped while it is soft. To minimize this problem, a small amount of MF resin can be added as a secondary cross-linker.

16.5.4. Reactive Silanes

Reactive silane is the term used to identify compounds having a substituted alkyl group terminated with a trialkoxysilyl group. They have been used in coatings in several different ways; two review papers are Refs. [19] and [20]. Several reactive silanes are commerically available, including: 3-aminopropyltriethoxysilane, *N*-(2-aminoethyl)-3-aminopropyl-trimethoxysilane, 3-glycidyloxypropyltrimethoxysilane, and 3-methacryloxypropyltri-methoxysilane.

Trialkoxysilyl groups can react directly, or indirectly in the presence of water, with hydroxyl groups; the other functional group can participate in the cross-linking reaction with the resins in a coating. For example, the trialkoxysilyl group on 3-aminopropyl-triethoxysilane can react with hydroxyl groups on the surface of glass, while the amine group can react with epoxy groups in an epoxy resin. The interactions with surfaces are complex [19]. The relative rates of hydroylsis of the alkoxysilyl groups to silanol groups, self-condensation reactions, and reactions with substrate hydroxyl groups vary with pH, water concentration, alcohol concentration, reactivity of substrate hydroxyl groups, and rates of reverse reactions. In the case of amino-functional silanes, the situation can be further complicated by the competing possiblity of adsorption of amine groups on the substrate. An idealized picture of the interaction of 3-aminopropyltrimethoxysilane with the surface of glass and an epoxy resin is shown in Scheme 3 in Section 6.4.3. Before the advent of reactive silanes, it was difficult to achieve adhesion to glass that would resist displacement with water. Use of reactive silanes has made possible coatings and sealants with excellent adhesion to glass in the presence of water. While water can hydrolyze some of the linkages, the reaction is reversible. With multiple attachments, adhesion is not lost. (See Section 6.4.3 for further discussion).

Reactive silanes also form stable condensation products with other oxides, such as those of aluminum, zirconium, tin, titanium, and nickel; less stable bonds are formed with oxides of carbon, iron, and boron [20]. Reactive silane additives are used not only as adhesion promoters, but also to treat the surface of oxide pigments to improve dispersion properties.

Bis(trisilylalkoxy)alkanes are being investigated to treat the surface of steel to increase adhesion [21]. Clean steel is rinsed with water, and then the wet steel is dipped first in an aqueous solution of bis(trimethoxysilyl)ethane (BTSE) and then in an aqueous solution of a reactive silane. The BTSE reacts with water and hydroxyl groups on the steel and silanols from other molecules of BTSE to give a water-resistant anchor to the steel. The reactive silane reacts with other silanol groups from the BTSE, and the reactive group can react with a coating binder. There are many variables in the treatment; excellent results have been obtained in laboratory work and commercial use is anticipated.

Resins with multiple trialkoxysilyl groups can be used as binders for moisture-cure coatings. For example, an isocyanate-terminated resin can be reacted with 3-aminopropyl-triethoxysilane to give a resin with terminal triethoxysilyl groups. Coatings made using such resins cross-link to a polymer network after application and exposure to humid air. Multifunctional isocyanates may be used to prepare resins with increased cross-link density. The original isocyanate-terminated resin would also moisture cure, but an advantage of the triethoxysilyl resin is that part of the solvent can be ethyl alcohol. The ethyl alcohol permits a reasonable pot life in the presence of water, such as comes into a coating from pigment surfaces. If free isocyanate groups are present, the coating is not stable in the presence of either ethyl alcohol or water. The trialkoxysilyl approach also avoids the formation of CO_2, which can lead to film imperfections such as pinholing, in moisture cure urethanes. Trialkoxysilyl acrylic resins can be made by copolymerizing a trialkoxysilylalkyl methacrylate with other acrylic monomers [22,23,24]. Coatings based on such resins cure on exposure to atmospheric moisture; the reaction is catalyzed with organotin compounds or organic acids. The resulting coatings are reported to have excellent exterior durability, resistance to environmental etching and marring, and adhesion to aluminum. They are candidates for use in automobile refinish clear coats.

Automotive clear coats are being made by combining trialkoxysilylalkyl-functional and hydroxy-functional acrylic resins with MF resins or blocked isocyanates [24]. No detailed studies of the cross-linking reactions have been published, but they presumably include hydrolysis and condensation of the trialkoxysilyl groups to form siloxane cross-links, reaction of hydroxyl groups with MF resins or blocked isocyanates, and transetherification of the hydroxyl groups with the trialkoxysilyl groups. The resulting coatings are reported to have excellent exterior durability and very good resistance to environmental etching and marring. Another advantage is that formaldehyde emissions are reduced because lower levels of MF resin are used.

Trialkoxysilylated acrylic and vinyl acetate latexes can be prepared using 3-methacryloxypropyltriisobutoxysilane as a comonomer in emulsion polymerization [25]. The isobutoxysilyl derivatives are used, since they are more resistant to hydrolysis at polymerization temperatures. They are stable latexes that do not cross-link until water evaporates from a film; possibly, the alkoxysilyl groups hydrolyze during storage, but the silanols formed do not condense because of the large excess of water present until after a film is formed. Coatings made from them are reported to have superior adhesion, as well as high chemical, solvent, and mar resistance. Carboxylic acid-functional acrylic latexes can be cross-linked using β-(3,4-epoxycyclohexyl)ethyltriethoxysilane [26].

16.5.5. Orthosilicates

Tetraethylorthosilicate, $Si(OEt)_4$, is used to make binders for zinc-rich primers, which are widely used primers for corrosion protection of steel (see Sections 7.4.3 and 32.3). An ethyl alcohol solution of $Si(OEt)_4$ is partially polymerized by the addition of a small amount of water-just enough to increase the molecular weight sufficiently to give the desired viscosity. The following equation shows dimer formation, higher ratios of water produce oligomers:

$$Si(OC_2H_5)_4 \ + \ H_2O \ \rightleftharpoons \ (C_2H_5O)_3Si{-}O{-}Si(OC_2H_5)_3 \ + \ C_2H_5OH$$

Zinc powder pigment is dispersed in the resulting oligomer solution. Note that the reaction is reversible; the ethyl alcohol solvent retards polymerization by small amounts of water in the coating formulation during storage. When the coating is applied and the alcohol evaporates, water is absorbed from the atmosphere, and the cross-linking reaction continues to completion at ambient temperatures. While the cross-linked binder is predominantly polysilicic acid, zinc salts form by reaction with the zinc hydroxide and carbonate always present on the surface of zinc metal pigments. Ferric ions from the steel surface may also be incorporated in the matrix. For water-borne zinc-rich primers, aqueous solutions of potassium, sodium, and/or lithium silicate are being used.

16.6. OTHER APPROACHES TO CROSS-LINKING

16.6.1. 2-Hydroxyalkylamide Cross-Linkers

Esterification of most alcohols with carboxylic acids is too slow for practical use as a cross-linking reaction. However, 2-hydroxyalkylamides undergo esterification reactions more rapidly than simple alcohols. Polyfunctional 2-hydroxyalkylamides (e.g., the tetra-functional hydroxyalkylamide derived from aminolysis of dimethyl adipate with diisopro-panolamine) can serve as cross-linkers for carboxylic acid-functional acrylic or polyester resins [27].

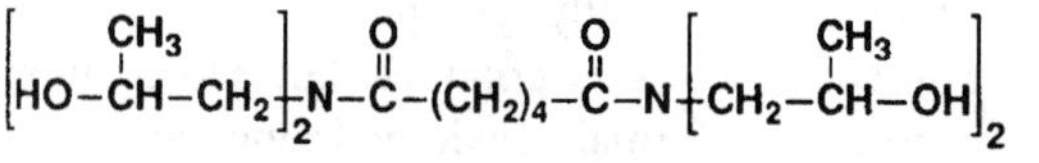

$$\left[\begin{matrix} CH_3 \\ HO-CH-CH_2 \end{matrix}\right]_2 N-\overset{O}{\overset{\|}{C}}-(CH_2)_4-\overset{O}{\overset{\|}{C}}-N\left[\begin{matrix} CH_3 \\ CH_2-CH-OH \end{matrix}\right]_2$$

Hydroxyalkylamide Cross-Linker

The properties of coatings obtained by cross-linking carboxylic acid-functional acrylic resins with hydroxyalkylamides compare favorably with those obtained using MF resins as cross-linkers with the same resins. An advantage relative to MF cross-linkers is the absence of formaldehyde, which is emitted in low concentrations when MF based coatings are baked. A disadvantage is that high baking temperatures are required; the lowest reported satisfactory bake is 150°C for 30 minutes.

Hydroxyalkylamides are soluble in both water and common coating solvents and, hence, are useful as cross-linkers in either water-borne or solvent-borne coatings. Tetra-N,N,N',N'-(2-hydroxyethyl)adipamide is a solid well suited for use in powder coatings [28] (see Section 27.1.3). However, some discoloration on baking can occur, [29] presumably, due to formation of some free amine groups by thermal rearrangement of the hydroxylalkylamides.

The cross-linking reaction exhibits unique features compared to general esterification reactions. It is not catalyzed by acid; also, 2-hydroxyalkylamides with secondary hy-droxyls esterify more rapidly than a corresponding primary alcohol derivative. Aromatic carboxylic acids esterify faster than aliphatic carboxylic acids, the reverse of other alcohols. In addition to the faster reaction of 2-hydroxyalkylamides with aromatic acids, the saponification resistance of the products is greater than that of the corresponding esters from aliphatic acids. Mechanistic studies of the hydroxyalkylamide-carboxylic acid esterification reaction, as well as related reactions, have been reported. An explanation for the results, involving participation of the amide group and intermediate formation of oxazolinium groups, has been advanced which is consistent with the experimental observations [30]. Direct evidence for formation of oxazolinium intermediates has been obtained [29].

It has also been shown that 2-hydroxyalkylamides react with epoxies. At elevated temperatures, the amide rearranges to the aminoester, and the resulting amine group then reacts with the epoxy [31].

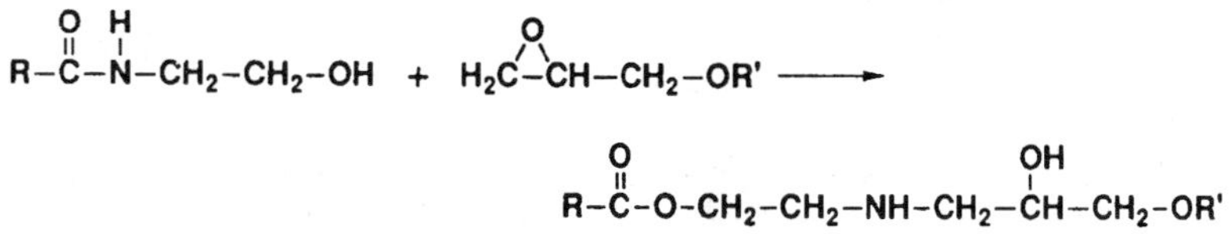

$$R-\overset{\overset{\displaystyle O}{\|}}{C}-\overset{\overset{\displaystyle H}{|}}{N}-CH_2-CH_2-OH \;+\; H_2\overset{\overset{\displaystyle O}{\diagup\!\diagdown}}{C}-CH-CH_2-OR' \longrightarrow$$

$$R-\overset{\overset{\displaystyle O}{\|}}{C}-O-CH_2-CH_2-NH-CH_2-\overset{\overset{\displaystyle OH}{|}}{C}H-CH_2-OR'$$

16.6.2. Acetoacetate Cross-Linking Systems

The chemistry of β-keto esters such as acetoacetic esters has been widely studied. Acetoacetic esters exist to a significant degree in the tautomeric enol form.

$$H_3C-\overset{\overset{\displaystyle O}{\|}}{C}-CH_2-\overset{\overset{\displaystyle O}{\|}}{C}-OR \;\longleftrightarrow\; H_3C-\overset{\overset{\displaystyle OH}{|}}{C}=CH-\overset{\overset{\displaystyle O}{\|}}{C}-OR$$

They undergo a wide range of reactions, some of which can be adapted for coatings purposes. Several methods are available to prepare resins bearing acetoacetate groups. Acetoacetoxy-functional acrylic solution resins can be made by copolymerizing aceto-acetoxyethyl methacrylate (AAEM) with other acrylate monomers [32,33]. One can also use AAEM to make acrylic latexes [34]. However, the long-term stability of acetoacetic ester polymers in aqueous media has been questioned. Hydroxy-functional resins can be reacted with diketene or transesterified with methyl acetoacetate to form acetoacetylated resins. Replacement of the hydroxyl group with the less polar acetoacetate group leads to a reduction in viscosity, permitting higher solids at the same application viscosity.

A variety of coreactants with acetoacetylated resins have been investigated. MF resins react with acetoacetate groups in the presence of an acid catalyst, although somewhat less rapidly than with hydroxyl groups [32]. Film properties of the resulting coatings using acetoacetylated resins are reported to be comparable with those made from hydroxy-functional resins. There are indications of improved wet adhesion, perhaps resulting from chelating interactions with the surface of the steel.

Isocyanates also react with acetoacetate groups; the cure rate is slower than with hydroxyl groups, but pot life is longer. Film properties are reported to be similar to those of conventional polyurethanes [32].

Polyacrylates (e.g., trimethylolpropane triacrylate) undergo Michael reactions with acetoacetate groups at ambient temperatures in the presence of strongly basic catalysts such as tetramethylguanidine (TMG) [34]. Formic acid has been reported to be an effective

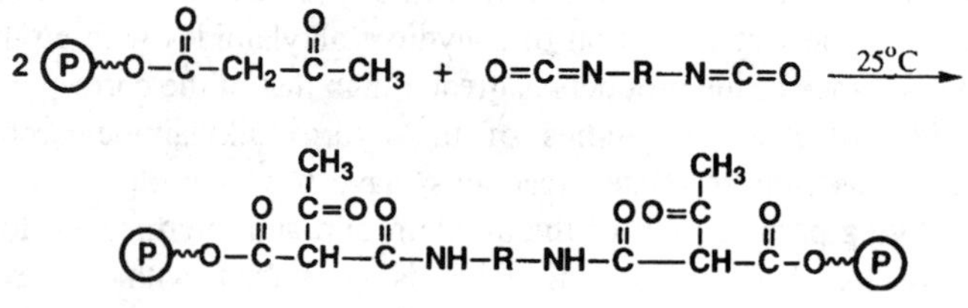

$$2 \; \textcircled{P} \text{\raisebox{0pt}{$\sim$}} O-\overset{O}{\overset{||}{C}}-CH_2-\overset{O}{\overset{||}{C}}-CH_3 \;+\; O=C=N-R-N=C=O \;\xrightarrow{25^\circ C}$$

$$\textcircled{P}\text{\raisebox{0pt}{$\sim$}}O-\overset{O}{\overset{||}{C}}-\overset{\overset{\displaystyle CH_3}{|}}{\underset{}{C}H}\overset{\overset{\displaystyle C=O}{|}}{-}\overset{O}{\overset{||}{C}}-NH-R-NH-\overset{O}{\overset{||}{C}}\overset{\overset{\displaystyle O=C}{|}}{-}\overset{\overset{\displaystyle CH_3}{|}}{C}H-\overset{O}{\overset{||}{C}}-O\text{\raisebox{0pt}{$\sim$}}\textcircled{P}$$

volatile blocking agent to extend pot life when 1,8-diazabicyclo[5,4,0]undec-7-ene is used as a catalyst [33].

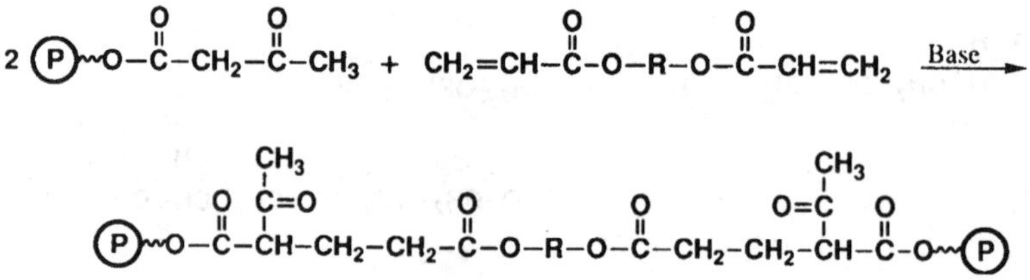

$$2 \; \textcircled{P}\text{\raisebox{0pt}{$\sim$}}O-\overset{O}{\overset{||}{C}}-CH_2-\overset{O}{\overset{||}{C}}-CH_3 \;+\; CH_2{=}CH-\overset{O}{\overset{||}{C}}-O-R-O-\overset{O}{\overset{||}{C}}-CH{=}CH_2 \;\xrightarrow{\text{Base}}$$

$$\textcircled{P}\text{\raisebox{0pt}{$\sim$}}O-\overset{O}{\overset{||}{C}}-\overset{\overset{\displaystyle CH_3}{|}}{\underset{}{C}H}\overset{\overset{\displaystyle C=O}{|}}{-}CH_2-CH_2-\overset{O}{\overset{||}{C}}-O-R-O-\overset{O}{\overset{||}{C}}-CH_2-CH_2-\overset{\overset{\displaystyle O=C}{|}}{C}H-\overset{O}{\overset{||}{C}}-O\text{\raisebox{0pt}{$\sim$}}\textcircled{P}$$

Amines react rapidly with acetoacetic esters; hence, polyamines can be used as cross-linkers for acetoacetate-functional latexes [34]. In solution coatings, the reaction is so fast at ambient temperatures that pot life is limited. This problem can be minimized by blocking the amine groups with ketones. The resultant ketimine hydrolyzes in the presence of water, permitting the cross-linking reaction to proceed [35]. The cross-links are tautomeric ketimine-eneamine groups that are thought to interact strongly with metal surfaces [36]. A ketimine-acetoacetate cross-linking primer is reported to give excellent adhesion and corrosion resistance when applied to an aircraft grade aluminum alloy with a chromate-free pretreatment.

$$\overset{O}{\overset{||}{RO\,C}}\;CH_2\overset{\overset{\displaystyle CH_3}{|}}{C}{=}NR' \;\longleftrightarrow\; \overset{O}{\overset{||}{RO\,C}}\;CH{=}\overset{\overset{\displaystyle CH_3}{|}}{C}NHR'$$

Ketimine   Enamine

Ketimine-eneamine tautomers can also serve as Michael reactant cross-linkers with polyacrylates [37]. Acetoacetate pendant resins also can be cross-linked by reaction with aldimines at room temperature [38].

16.6.3. Polyaziridine Cross-Linkers

Aziridine, the nitrogen three-membered ring counterpart of oxirane, and its derivatives have been studied for many years; in some cases polyaziridines have been used as cross-linkers. The common name of aziridine is ethylene imine. Ethylene imine is highly toxic and may be carcinogenic. Polyaziridines are skin irritants, and some individuals may become sensitized. Mutagenicity of polyaziridines is controversial, however, dilution by coating vehicles reduces their possible toxic effects [39]. Ethylene imine is even more reactive with acids than ethylene oxide. In the presence of relatively strong acids, it polymerizes very rapidly to yield poly(ethylene imine), $(-CH_2CH_2NH-)_n$.

Among the many reactions of aziridines, the one of greatest interest in coatings applications is the reaction of a polyaziridine with a polyfuntional carboxylic acid to form 2-aminoester cross-links. Some 2-aminoesters can spontaneously rearrange to the corresponding 2-hydroxyamides, but this reaction does not break the cross-link.

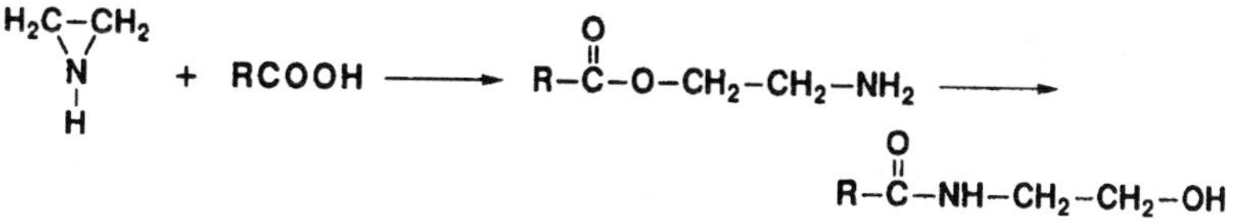

$$H_2C-CH_2 \atop \underset{H}{\overset{\diagup}{N}} + RCOOH \longrightarrow R-\overset{O}{\overset{\|}{C}}-O-CH_2-CH_2-NH_2 \longrightarrow$$

$$R-\overset{O}{\overset{\|}{C}}-NH-CH_2-CH_2-OH$$

A variety of polyfunctional aziridines have been investigated. An example is the trifunctional Michael addition product of 3 mol of aziridine to 1 mol of trimethylolpropane triacrylate.

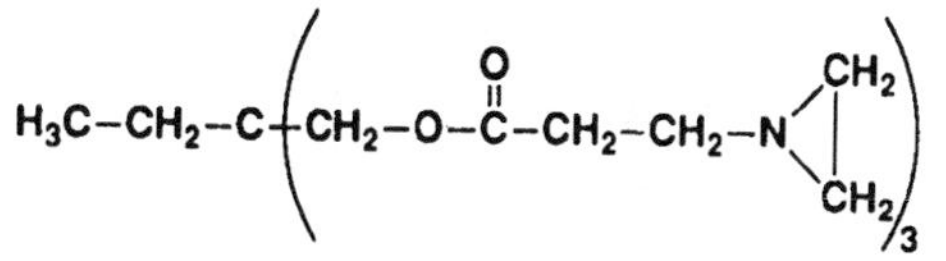

$$H_3C-CH_2-C\left(CH_2-O-\overset{O}{\overset{\|}{C}}-CH_2-CH_2-N\overset{\diagup CH_2}{\diagdown CH_2}\right)_3$$

Methyl aziridine (propylene imine) is also used. The main use of such polyaziridines is to cross-link with carboxylic acid groups on latexes and water-borne polyurethanes. Reaction with the carboxylic acid is much faster than the reaction of the aziridine groups with water, but the reaction rate with water is such that pot lives are 48–72 hours [39]. With water, the aziridine hydrolyzes to an amino alcohol, but it is said that there is no indication that the hydrolyzed aziridine adversely affects film properties. Additional crosslinker can be added to restore reactivity. In view of the potential toxic hazards, manufacturer's recommendations for safe handling should be carefully followed.

16.5.4. Polycarbodiimide Cross-Linkers

Carbodiimides react with carboxylic acids, and react slowly enough with water so that they can be used in water-borne systems. The product of the reaction with a carboxylic acid is an N-acylurea.

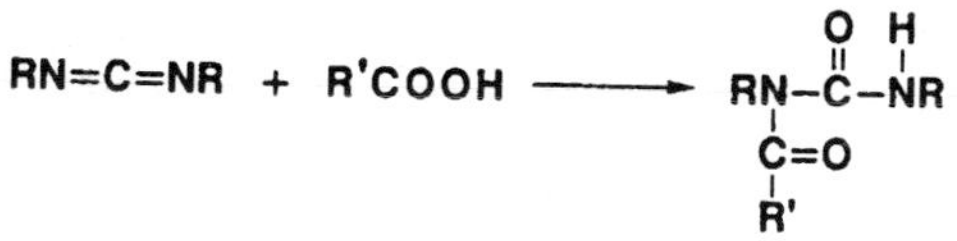

$$RN=C=NR + R'COOH \longrightarrow RN-\overset{O}{\overset{\|}{C}}-\overset{H}{\overset{|}{N}}R$$
$$\overset{|}{C}=O$$
$$\overset{|}{R'}$$

Multifunctional carbodiimides are available to cross-link carboxylic acid-functional resins, including aqueous polyurethane dispersions and latexes [40]. Cross-linking occurs within several days at ambient temperature and much faster with heat [41]. In one study with latexes, curing conditions ranged from 60°C to 127°C for 5–30 minutes and the higher temperatures gave better films [40]. Apparently, film properties depend on physical film formation as well as on the extent of chemical cross-linking.

16.6.5. Transesterification

Hydroxy-functional resins can cross-link by transesterification reactions with ester groups on another molecule releasing monofunctional alcohol. To be useful in coatings, a catalyst is required. Orthotitanate catalysts are generally destroyed by water, and strong base catalysts, such as sodium methoxide, are too toxic and difficult to handle to be of practical use in coatings. Epoxy nucleophile catalyzed transesterification (ENCAT) has been shown to give good cure under normal bake conditions [42]. Catalytic amounts of the epoxy nucleophile are formed by the reversible reaction of a weaker nucleophile. While the weak nucleophile does not catalyze transesterification, it can form enough epoxy nucleophile to be an effective catalyst.

$$
\text{R}_4\text{N}^+ \ \ ^-\overset{\text{O}}{\overset{\|}{\text{O}}}\text{CCH}_3 + \text{R}'\text{OCH}_2\overset{\text{O}}{\overset{\diagdown}{\text{CH}}}-\text{CH}_2 \ \rightleftharpoons \ \text{R}'\text{OCH}_2\overset{\text{O}^-}{\underset{\,}{\text{CH}}}\text{CH}_2\overset{\text{O}}{\overset{\|}{\text{O}}}\text{CCH}_3 \ \ \text{R}_4\text{N}^+
$$

In one example, the catalyst was a combination of liquid BPA epoxy resin and benzyltrimethylammonium acetate [42]. Transesterification was most rapid with primary alcohols and esters of aromatic acids. While hydroxy-functional acrylic resins self cross-link, it was found that better properties, especially impact resistance, were obtained using a combination of hydroxy-functional acrylics and polyesters. Strong acids, such as sulfonic acids, catalyze sufficiently to cross-link acrylic resins with carboxyl and hydroxyl functionality in 30 minutes at 140–150°C [43].

REFERENCES

1. S. Munekata, *Prog. Org. Coat.*, **16**, 113, (1988).
2. K. Batzar, *Proc. Intl. Conf. Org. Coat.*, Athens, (1989) p. 13.
3. J. E. Gaske, *Coil Coatings*, Federation of Societies for Coatings Technology, Blue Bell, PA, 1987.
4. A. Barbucci, E. Pedroni, J. L. Perillon, and G. Cerisola, *Prog. Org. Coat.*, **29**, 7 (1996).
5. M. Morcillo, J. Simancas, J. L. G. Fierro, S. Feliu, Jr., and J. C. Galvan, *Prog. Org. Coat.*, **21**, 315 (1993).
6. Anonymous, *Hypalon CP Chlorinated Polyolefins*, Technical Bulletin, Du Pont Company, Wilmington, DE, 1989.
7. B. N. McBane, *Automotive Coatings*, Federation of Societies for Coatings Technology, Blue Bell, PA, 1987.
8. C. S. Sheppard, "Peroxy Compounds," in *Encyclopedia of Polymer Science and Engineering*, 2nd. ed., Vol. 11, Wiley, New York, 1988, pp. 1–21.
9. H.-J. Traenckner and H. U. Pohl, *Angew. Makromol. Chem.*, **108**, 61 (1982).
10. M. J. Dvorchak and B. H. Riberi, *J. Coat. Technol.*, **64** (808), 43 (1992).
11. A. Noomen, *Prog. Org. Coat.*, **17**, 27 (1989).
12. J. J. Gummeson, *J. Coat. Technol.*, **62** (785), 43 (1990).
13. F. Fink, W. Heilen, R. Berger, and J. Adams, *J. Coat. Technol.*, **62** (791), 47 (1990).
14. Y.-C. Hsiao, L. W. Hill, and S. P. Pappas, *J. Appl. Polym. Sci.*, **19**, 2817 (1975). See also S. P. Pappas and R. L. Just, *J. Polym. Sci. Polym Ed.*, **18**, 527, (1980).
15. W. A. Finzel and H. L. Vincent, *Silicones in Coatings*, Federation of Societies for Coatings Technology, Blue Bell, PA, 1996.
16. W. A. Finzel, *J. Coat. Technol.*, **64** (809), 47 (1992).
17. T. Laubender, *Proc. Intl. Conf. Water-Borne Coat.*, Milan, 1992, Paper No. 21.
18. R. A. Ryntz, V. E. Gunn, H. Zou, Y. L. Duan, H. X. Xiao, and K. C. Frisch, *J. Coat. Technol.*, **64** (813), 83 (1992).
19. V. A. Ogarev and S. L. Selector, *Prog. Org. Coat.*, **21**, 135 (1992).
20. G. L. Witucki, *J. Coat. Technol.*, **65** (822), 57 (1993).

21. W. J. van Ooij and T. Child, *Chemtech*, February, 26 (1998).

22. M. Ooka and H. Ozawa, *Prog. Org. Coat.*, **23**, 325 (1993).

23. H. Furukawa, Y. Kato, N. Ando, M. Inoue, Y. K. Lee, I. Hazan, and H. Omura, *Prog. Org. Coat.*, **24**, 81 (1994).

24. J. D. Nordstrom, *Proc. Water-borne, Higher-Solids, Powder Coat. Symp.*, New Orleans, 1995, p. 192.

25. M. J. Chen, A. Chavez, F. D. Osterholtz, E. R. Pohl, and W. B. Herdle, *Surface Coat. Intl.*, **79**, 539 (1996); M. J. Chen, F. D. Osterholtz, E. R. Pohl, P. E. Ramdatt, A. Chaves, and V. Bennett, *J. Coat. Technol.*, **69** (870), 43 (1997).

26. M. J. Chen, F. D. Osterholtz, A. Chaves, P. R. Ramdatt, and B. A. Waldman, *J. Coat. Technol.*, **69** (875), 49 (1997).

27. J. Lomax and G. F. Swift, *J. Coat. Technol.*, **50** (643), 49 (1978).

28. A. Mercurio, *Proc. Intl. Conf. Org. Coat*, Athens, Greece, 1990, p. 235.

29. D. Stanssens, R. Hermanns, and H. Wories, *Prog. Org. Coat.*, **22**, 379 (1993).

30. Z. W. Wicks, Jr., M. R. Appelt, and J. C. Soleim, *J. Coat. Technol.*, **57** (726), 51 (1985).

31. F. N. Jones and I.-C. Lin, *J. Appl. Polym. Sci.*, **29**, 3213 (1984).

32. F. D. Rector, W. W. Blount, and D. R. Leonard, *J. Coat. Technol.*, **61** (771), 31 (1989). J. S. Witzeman, W. D. Nottingham, and F. D. Rector, *J. Coat. Technol.*, **62** (789), 101 (1990).

33. T. Li and J. C. Graham, *J. Coat. Technol.*, **65** (821), 64 (1993).

34. P. J. A. Geurnik, L. van Dalen, L. G. J. van der Ven, and R. R. Lamping, *Prog. Org. Coat.*, **27**, 73 (1996).

35. R. J. Clemens and F. D. Rector, *J. Coat. Technol.*, **61** (770), 83 (1989).

36. K. H. Zabel, R. E. Boomgaard, G. E. Thompson, S. Turgoose, and H. A. Braun, *Prog. Org. Coat.*, **34**, 236 (1998).

37. W. Schubert, *Prog. Org. Coat.*, **22**, 357 (1993).

38. N. Chen, J. N. Drescher, R. K. Pinschmidt, T. M. Santosusso, and C.-F. Tien, *Proc. Waterborne, High-Solids, Powder Coat Symp.*, New Orleans, 1996, p. 198.

39. G. Pollano, *Polym. Mater. Sci. Eng.*, **77**, 383 (1997).

40. J. W. Taylor and D. R. Bassett, "The Application of Carbodiimide Chemistry to Coatings," in J. E. Glass, Ed., *Technology for Waterborne Coatings*, American Chemical Society, Symp. Ser. No. 663, Washington, DC, 1997, p. 137.

41. "Ucarlink® Crosslinkers," *Technical Bulletin UC-350*, Union Carbide Corp., Danbury, CT, 1994.

42. G. P. Craun, *J. Coat. Technol.*, **67** (841), 23 (1995); G. P. Craun, C.-Y. Kuo, and C. M. Neag, *Prog. Org. Coat.*, **29**, 55 (1996).

43. G. Chu, F. N. Jones, R. Armat, and S. G. Bike, *Polym. Mater. Sci. Eng.*, **73**, 235 (1995).

CHAPTER 17

Solvents

Most coatings contain volatile material that evaporates during application and film formation. The volatile components reduce viscosity for application and control viscosity changes during application and film formation. Selection of volatile components affects popping, sagging, and leveling and can affect adhesion, corrosion protection, and exterior durability. Too often, formulators do not consider the effects solvent choice can have on coating performance. In some cases, the volatile material must be a solvent for the resins in a formulation; in other cases, nonsolvent volatile components are desired. Usually, any volatile organic material is called a solvent whether it dissolves the resin or not.

As discussed in Sections 17.7 and 17.8, almost all solvents are classified by the U.S. Environmental Protection Agency (EPA) as photochemically reactive *volatile organic compounds* (VOCs), and their use has been regulated to reduce air pollution since the 1970's. In 1990, the U.S. Congress listed certain common solvents as *hazardous air pollutants* (HAPs), limiting their usage.

17.1. SOLVENT COMPOSITION

A variety of organic compounds and mixtures are used as solvents. They can be classified in three broad categories: weak hydrogen-bonding, hydrogen-bond acceptor, and hydrogen-bond donor-acceptor solvents.

Weak hydrogen-bonding solvents are aliphatic and aromatic hydrocarbons. Commercial aliphatic solvents are mixtures of straight chain, branched chain and alicyclic hydrocarbons. They vary in volatility and solvency. Varnish makers and painters (VM&P) naphthas are aliphatic solvents with high volatility. Mineral spirits are slower evaporating aliphatic hydrocarbons; special grades with low aromatic content and hence, less odor are available. An advantage of aliphatics is low cost, especially on a volume basis since their densities as well as their price per unit weight are low. Aromatic hydrocarbon solvents are more expensive than aliphatic ones, but dissolve a broader range of resins. Toluene, xylene, and high flash aromatic naphthas have been widely used. Commercial xylene is a mixture of isomeric xylenes and ethyl benzene. Higher boiling mixtures of aromatic hydrocarbons (high flash aromatic naphthas) are predominantly mixed alkyl and dialkyl benzenes with 3 to 5 carbons in the substituent groups. Use of benzene is prohibited because of toxicity. Toluene and xylene have been used on a large scale; however, they are listed as HAPs. (See Section 17.8.)

Esters and ketones are hydrogen-bond acceptor solvents. Ketones are generally less expensive than esters with corresponding vapor pressures. The cost differential is particularly marked on a volume basis since ketones have lower densities. Use of methyl ethyl ketone (MEK) and methyl isobutyl ketone (MIBK) is being reduced because they are on the HAP list; use of acetone is increasing because it has been "*delisted*," that is, no longer included as a volatile organic compound by the EPA. Use of esters is increasing. Slow evaporating esters such as 1-methoxy-2-propyl acetate and 2-butoxyethyl acetate are often preferred over slow evaporating ketones like isophorone and methyl *n*-amyl ketone because of odor. Esters of 2-ethoxyethanol were widely used at one time but are suspected of long-term adverse health effects. Esters should not be used as solvents for resins bearing primary or secondary amine groups because of aminolysis reactions, which convert amines to amides. Nitroparaffins such as a 2-nitropropane are highly polar hydrogen-bond accepting solvents. Their high polarity results in increased electrical conductivity. This is useful for adjusting solvent combinations for electrostatic spraying. (See Section 22.2.3.)

Alcohols are used as strong hydrogen-bond donor-acceptor solvents. The most widely used volatile alcohols are methyl, ethyl, isopropyl, *n*-butyl, *sec*-butyl, and isobutyl alcohols. Most latex paints contain a slow evaporating, water-soluble solvent, such as ethylene or propylene glycol that does not dissolve in the polymer particles. (See Section 31.1.) In water-reducible acrylic and polyester resin coatings, low volatility ether alcohols, such as 1-propoxypropan-2-ol, 2-butoxyethanol, or the monobutyl ether of diethylene glycol, are used.

17.2. SOLUBILITY

In the early days of the paint industry, there was little problem selecting solvents-almost all resins used dissolved in hydrocarbons. An exception was shellac, which dissolved in ethyl alcohol. Early in the 20th century, nitrocellulose was introduced, and solvent selection became a greater challenge. RS type nitrocellulose (see Section 16.2.1) is soluble in esters and ketones, but not in hydrocarbons or alcohols. However, nitrocellulose does dissolve in a mixture of ketones or esters and hydrocarbons, which reduces cost. If ethyl alcohol is included in the solvent mixture, a higher ratio of hydrocarbons can be used, further reducing cost. Esters and ketones were classified as *true solvents*, hydrocarbons as *diluents*, and alcohols as *latent solvents*.

To control drying rate, under various conditions, mixtures of solvents are required. At a minimum, nitrocellulose lacquer solvent blends contain two esters or ketones, two hydrocarbons, and an alcohol. As many as 10 solvents are blended to control drying and solubility through the drying cycle at the lowest volume cost. Solvent mixtures must be selected to maintain solubility throughout solvent evaporation from films. If the slowest evaporating solvent were a hydrocarbon, nitrocellulose would precipitate before all the solvent was gone to give an uneven film with poor appearance and physical properties.

17.2.1. Solubility Parameters

After 1930, more and more types of resins were adopted for coatings, and empirical selection of solvents and solvent mixtures became more difficult. The general rule of "like dissolves like" was broadened by the experience with nitrocellulose that mixtures of weak hydrogen-bonding hydrocarbons with strong hydrogen-bonding alcohols give solvency similar to medium hydrogen-bonding esters and ketones. In the 1950s, Burrell initiated studies to develop a more scientific basis for selecting solvents and formulating solvent mixtures [1]. He turned to the work of Hildebrand, who analyzed miscibility of small molecule organic compounds in terms of thermodynamics [2]. Hildebrand showed that the tendency of a pair of chemicals to mix spontaneously can be described in terms of Gibb's free energy equation:

$$\Delta G_m = \Delta H_m - T\Delta S_m$$

Miscibility is thermodynamically favored when the change in free energy of mixing, ΔG_m, is negative. Hildebrand pointed out that the change in entropy ΔS_m is usually positive, since a solution is usually less ordered after mixing than before. (There are exceptions when strong interactions among the different molecules increase the order of mixtures relative to the separate materials, resulting in a negative entropy change.) In most cases, ΔS_m is positive and the entropy change favors mixing. Hildebrand focused on ΔH_m, the enthalpy of mixing, since it usually determines the sign of ΔG_m. Change in ΔH_m is in turn related to change in the energy of mixing ΔE_m; R is the gas constant and T is temperature (Kelvin):

$$\Delta E_m = \Delta H_m - RT$$

Liquids have intermolecular attractive forces strong enough to hold the molecules together; otherwise, they would be gases. The forces can be measured by determining the energy needed to vaporize a liquid at that temperature. Results are expressed in terms of the molar energy of vaporization divided by the molar volume V. Hildebrand reasoned that the energy required to separate molecules during mixing is related to cohesive energy density $\Delta E_v/V$. He expressed the change in ΔE_m of ideal solvents by the following equation, where V_m is the average molar volume and ϕ_1 and ϕ_2 are the volume fractions of two components.

$$\Delta E_m = V_m\phi_1\phi_2\left[(\Delta E_v/V)_1^{1/2} - (\Delta E_v/V)_2^{1/2}\right]^2$$

The square root of cohesive energy density of a solvent was defined as the *solubility parameter* δ. A pair of solvents are miscible if the differences in solubility parameters approach zero. Then, the ΔE_m of the pair approaches zero, the ΔH_m is small, and the free energy of mixing is controlled by ΔS_m. When ΔS_m is positive, the solvents are miscible. ΔE_v and V_m vary with temperature; therefore, δ varies with temperature. Most tables of solubility parameters give the values at 25°C, although commonly, no temperature is specified.

There is risk of confusion about units. The older units of solubility parameter were (cal cm^{-3})$^{1/2}$, sometimes designated as hildebrands, h. In the SI system, the proper units are (MPa)$^{1/2}$; 1(cal cm^{-3})$^{1/2}$ = 0.488(MPa)$^{1/2}$. The SI units have not been widely adopted and (cal cm^{-3})$^{1/2}$ are still used, although frequently not stated.

When data are not available for energy of vaporization, solubility parameters can be estimated by using empirical equations relating boiling points, vapor pressure data, or surface tension data [3]. They can also be estimated by summation of Small's molar attraction constants, G using the following equation. where ρ is density and M is molecular weight:

$$\delta = \left(\frac{\rho}{M}\right) \Sigma G = \left(\frac{1}{V}\right) \Sigma G$$

Selected Small's constants are listed in SI units in Table 17.1. As with solubility parameters, the units for Small's constants are commonly not specified and are frequently the old units. Hoy [4] used Small's constants to calculate solubility parameters of 640 compounds [4].

Both Small's constants and solubility parameters of hydrogen bond donor-acceptor molecules vary with the environment. The value for the alcohol hydroxyl group included in Table 17.1 depends on other groups in the solvent with which the hydroxyl group might hydrogen bond and on the polarity of other components in mixed solutions. Water is the most extreme case. While specific values of a solubility parameter of water are sometimes encountered, the values vary widely, since they are very dependent on the medium. Their usefulness is limited to comparing systems with similar compositions.

Burrell tried to apply solubility parameters to prediction of solubility of resins [1]. The volatility of resins is so low that ΔE_v at 25°C cannot be determined. Burrell got around this by determining solubility of a resin in a series of solvents with known solubility parameters. The solubility parameter range of a resin was taken as the range of solubility parameters of the solvents that would dissolve the resin. He found that in many cases, resins were not soluble in all the solvents with solubility parameters that fell within the solubility parameter range determined for the resin. Most of the cases in which solubility parameters did not work had solvents and resins with markedly different potential for hydrogen bonding. He divided solvents into three classes: poor, medium, and high hydrogen bonding. He then determined solubility of the resins in a series of each class of solvents. The resulting ranges of solubility parameters for a resin in the three groups of solvents permitted fairly good predictions of solubility. Burrell found that one can predict with some confidence whether a solvent mixture would dissolve a resin. One can calculate the weighted average solubility parameter of a mixture known to dissolve a resin using the following relationship, where the x values are the mole fraction of solvents in the mixture:

$$\delta_{\text{mix}} = \frac{x_1 v_1 \delta_1 + x_2 v_2 \delta_2 + x_3 v_3 \delta_3 + \cdots}{x_1 v_1 + x_2 v_2 + x_3 v_3 + \cdots}$$

Table 17.1. **Small's Molar Attraction Constants, $(\text{MPa})^{1/2}\ \text{cm}^3/\text{mol}$ at 25°C**

Hydrocarbon Groups	G	Other Groups	G
–CH$_3$	284	O (ethers)	236
–CH$_2$–	270	O (oxiranes)	361
–CH–	176	Cl	420
		CO (ketones)	539
=CH–	249	COO (esters)	668
=CH$_2$	259	OH (see text)	463
Phenyl	1400		
Phenylene	1370		

Since molar volumes for most solvents are fairly similar, a simplifying approximation is to use volume fractions to calculate average solubility parameter.

$$\delta_{mix} = \phi_1\delta_1 + \phi_2\delta_2 + \phi_3\delta_3 + \cdots$$

One does a similar calculation to determine the average degree of hydrogen bonding. Using this procedure, it was almost always possible to predict whether an alternative solvent mixture would also dissolve the resin. Next, Burrell applied the idea to prediction of solvents or solvent mixtures for new resins using Small's molar constants to estimate the solubility parameter of the new resin. Gram equivalent weight of the average repeating unit of the resin was used for molecular weight in Small's equation. Results were fair; in a majority, but not in all cases, the resins were soluble in solvents and solvent mixtures with solubility parameters similar to those calculated for the resin, provided the need for similar levels of hydrogen bonding was taken into consideration.

A variety of refinements and extensions of Burrell's techniques have been proposed. Lieberman mapped solubility of polymers in a grid in which the axes are solubility parameter and hydrogen-bonding index [5]. The plot predicts that any solvent or solvent blend with solubility parameter and hydrogen-bonding index values that fall within the boundaries of the plot will dissolve the resin. Such plots can be useful, but their preparation requires substantial experimental effort [6].

17.2.2. Three-Dimensional Solubility Parameters

Other investigators considered this two-dimensional approach to be too simplistic and proposed three-dimensional systems. The system of Hansen came to be most widely accepted [7]. Hansen reasoned that since there are three types of interactive forces between molecules, there should be three types of solubility parameters: dispersion, δ_d; polar, δ_p; and hydrogen bond, δ_h. The total solubility parameter was arbitrarily set equal to the square root of the sum of the squares of the partial solubility parameters:

$$\delta = (\delta_d{}^2 + \delta_p{}^2 + \delta_h{}^2)^{1/2}$$

For mixed solvents, one can calculate a weighted average of the three partial solubility parameters:

$$\delta_d(\text{blend}) = (\phi\delta_d)_1 + (\phi\delta_d)_2 + \cdots + (\phi\delta_d)_n$$
$$\delta_p(\text{blend}) = (\phi\delta_p)_1 + (\phi\delta_p)_2 + \cdots + (\phi\delta_p)_n$$
$$\delta_h(\text{blend}) = (\phi\delta_h)_1 + (\phi\delta_h)_2 + \cdots + (\phi\delta_h)_n$$

Three-dimensional solubility parameters can be determined or calculated by a variety of methods. Although values are commonly given as three figures, it should not be inferred that three figures are significant. Table 17.2 gives solubility parameters for some representative solvents selected from tables in Ref. [7] and the Polymer Handbook [8].

Hansen determined three-dimensional solubility parameters for a group of resins by experimentally testing solubilities of 34 polymers in 90 solvents [7]. As an alternative to this laborious procedure, Hoy calculated three-dimensional solubility parameters of resins by a method analogous to use of Small's constants[9]. The values determined or calculated

Table 17.2. Three-Dimensional Solubility Parameters (MPa)$^{1/2}$

Solvent	Total	δ_d	δ_p	δ_h
n-Hexane	14.9	14.9	0	0
Toluene	18.2	18.0	1.4	2.0
o-Xylene	18.0	17.0	1.4	3.1
Acetone	19.9	15.5	10.4	7.0
Methyl ethyl ketone	19.0	16.0	9.0	5.1
Methyl isobutyl ketone	17.0	15.3	6.1	4.1
Isophorone	19.8	16.6	8.2	7.4
Ethyl Acetate	18.2	15.8	5.3	7.2
Isobutyl acetate	16.8	15.1	3.7	6.3
n-Butyl acetate	17.4	15.8	3.7	6.3
Methyl alcohol	29.7	15.1	12.3	22.3
Ethyl alcohol	26.6	15.8	8.8	19.4
Isopropyl alcohol	23.5	16.4	6.1	16.4
n-Butyl alcohol	23.1	16.0	5.7	15.8
2-Butoxyethanol	20.9	16.0	5.1	12.3

by various methods are frequently not the same; tables of (presumably) self-consistent values have been published and are available in computer databases. Three-dimensional solubility parameters are based on thermodynamic laws, but several assumptions and arbitrary choices are involved in their derivation and use. Assertions that they are "theoretically sound" can be misleading [3]. They are best regarded as an empirical method that has proven useful for finding alternative solvent mixtures with similar solvency characteristics.

Attempts have been made to apply solubility parameters to other problems, but the theoretical foundations are shaky and actual results have been erratic. For example, they do not accurately predict solubility of all polymers, and use of solubility parameters to predict solubility of one polymer in another (*compatibility*) frequently gives erroneous predictions. It has been increasingly recognized that three-dimensional solubility parameters are an oversimplification of the complex factors involved in solubility. The difficulties result from two probably interrelated factors. First, their application has assumed that entropy changes can be neglected. Second, the hydrogen-bond solubility parameter combines the effects of donors and acceptors. Important effects of entropy changes are particularly likely to occur in systems in which hydrogen bonding is important. Hansen has recognized the problem and in some cases suggests giving increased weight to the hydrogen bond parameter [7].

The difficulties increase as molecular weight increases. As molecules get larger, interaction between solvent and polymer molecules must become greater to overcome intermolecular polymer-polymer interactions. Intramolecular interactions may also play a role. Estimates of solubility parameters of polymers by use of group attraction constants do not take molecular weight into consideration.

People first working with polymers may be surprised by some of the solubility effects. They are accustomed to upper limits of solubility; for example, 36.1 g of NaCl dissolves in 100 g of water at 25°C. In general, there is no upper limit to the solubility of a polymer. If a small amount is entirely soluble, it is quite safe to conclude that any larger amount will be soluble in that solvent. However, it is common for there to be a lower limit of solubility.

A high concentration of polymer may be soluble in some solvent(s), but on dilution, part of the polymer may precipitate. This phenomenon can be used to fractionate polymers by dilution. The fractions precipitating first are generally the highest molecular weight components. As dilution is continued further, progressively lower molecular weight fractions are precipitated. In some cases, the fractionation is based on the polarity of parts of the resin. Many alkyds are soluble at high concentrations in aliphatic solvents, but partially precipitate on dilution. The first fractions precipitating are high molecular weight molecules with a larger than average number of hydroxyl and/or carboxylic acid groups. One can think of the situation in reverse: Some solvent is soluble in the polymer (resin), and all of the polymer is soluble in the combination of polymer and solvent. As more solvent is added, solvency of the system changes, and parts of the polymer are not soluble in the more dilute polymer solution. The precipitate is highly swollen with solvent.

Hoy showed that glycol ethers can change apparent polarity to assume the polar nature of their surrounding environment [4]. In polar solvents, they behave as polar solutes. However, in nonpolar solvents, they undergo intramolecular bonding or intermolecular hydrogen bonding, to form dimers. In view of these results, it is not surprising that it is difficult to predict interactions of many coating resins with various solvents because the resins are often polydisperse, polyfunctional materials with multiple hydrogen-bond donor and acceptor sites.

17.2.3. Other Solubility Theories

Numerous refinements of solubility parameters have been proposed. Some have involved adding a fourth parameter, for example by dividing δ_h into hydrogen-bond donor and acceptor terms. Huyskens and Haulait-Pirson proposed equations that reflect changes in entropy and attempt to consider the differences resulting from both hydrogen-bond acceptor and donor groups [10]. They provide somewhat better predictions of the solubilities of a limited number of resin-solvent combinations. However, their equations still do not give perfect predictions, and only the relatively simple cases of poly(vinyl acetate), poly(methyl methacrylate), and poly(ethyl methacrylate) were studied. Their paper explains the shortcomings of the three-dimensional system, but falls short of giving a broadly applicable alternative.

A major series of studies involving cooperative work in several laboratories led to the development of what a technical news story called a "universal solubility equation" [11]. The researchers themselves "believe this to be somewhat of an overstatement" [12]. The complexity of polymer solubility can be seen from the fact that this universal equation has 13 factors in a 5 term equation and still does not take into consideration several important situations. Among the limitations, critical to coatings, is that the equation has thus far been applied mainly to monofunctional solutes.

17.2.4. Practical Considerations

Meanwhile, what can formulators do? Solvent changes are frequent, motivated by new toxic hazard information, changes in relative cost, new regulations, temporary shortages, and so on. This is the type of problem for which solubility parameters are most useful, recognizing that the results may be imperfect. One can use solvent computer programs with data banks that include solubility parameters, evaporation rates, cost, density, etc. This permits calculation of alternative solvent mixtures with solubility characteristics that

Table 17.3. Solvent Blends, Weight Percents

Component	1	2	3
MEK	9.9	6.1	14.7
MIBK	29.7	32.8	19.7
Xylene	21.9	25.2	24.3
Toluene	20.1	18.4	18.2
n-Butyl alcohol	13.5	17.5	17.3
2-Ethoxyethyl acetate	4.9		
1-Methoxy-2-propyl acetate			5.7
Property			
Solubility parameter, δ^a	19	19	19
δ_d	18	18	18
δ_p	5.7	5.7	5.3
δ_h	5.7	5.1	5.3
Relative evaporation rate	1.16	1.17	1.14
Cost[b]	0.70	0.68	0.68

[a] $(MPa)^{1/2}$.
[b] At the time of study, in dollars per kg.

do not change substantially during evaporation, while minimizing cost per unit volume. Several solvent suppliers have such computer programs. The result of such a calculation is given in Table 17.3 in which solubility parameter units are converted to $(MPa)^{1/2}$ and rounded off [13]. Solvent blend 1 contained 2-ethoxyethyl acetate. which, because of health concerns, had to be removed from the formulation. Blends 2 and 3 were calculated to have close to equal average solubility parameters and evaporation rates. Blend 2 was unsatisfactory because the slowest evaporating solvent, xylene, was not a solvent for the resin. Blend 3 was a better starting point since the slowest evaporating solvent, 1-methoxy-2-propyl acetate, was a true solvent for the resin. Relative evaporation rates are also given in Table 17.3. Relative evaporation rates of mixed solvents must be viewed with caution, see Section 17.3.3.

A second type problem faced by formulators is to predict solvents and solvent combinations for a new resin. Solubility parameters may be useful, but in most cases, qualitative application of the idea that like dissolves like is just as effective for finding a starting point and takes less time. The problem of selecting solvents for new resins has become easier because, in order to increase solids, lower molecular weight resins are being used. A wider range of solvents dissolves a resin as the resin's molecular weight gets lower.

17.3. SOLVENT EVAPORATION RATES

During application and film formation, the volatile material evaporates out of the coating. The rate at which evaporation occurs affects not only time required to convert a coating to a dry film, but also appearance and physical properties of the final film. Like so many topics in the coatings field, evaporation rate appears simple, but turns out to be complex.

17.3.1. Evaporation of Single Solvents

The rate of evaporation of a solvent is affected by four variables: temperature, vapor pressure, surface to volume ratio, and rate of air flow over the surface. The rate of evaporation of water is also affected by relative humidity.

The important temperature is the temperature at the surface. While this temperature may initially be that of the surrounding air, it decreases as solvent evaporates. While the surface is cooled by evaporation, it is warmed by thermal diffusion from within the sample and its surroundings. Depending on the circumstances, thermal diffusion may occur rapidly so that surface temperature does not fall much during evaporation, or slowly, resulting in a sharp drop in surface temperature. The cooling effect is largest for those solvents and situations in which the solvent evaporates most rapidly. Also, the higher the heat of vaporization of the solvent, the greater the extent of temperature drop if all other variables are equal.

Vapor pressure of the solvent at the temperature(s) at which evaporation occurs is the important vapor pressure. A common error is to assume that boiling points [the temperatures at which solvents have a vapor pressure of 101.3×10^2 kPa (1 mm Hg $= 0.1333$ kPa; 1 atm $= 101.3$ kPa)] are directly related to vapor pressures at other temperatures. However, boiling points are poor indicators. For example, benzene has a boiling point of 80°C and ethyl alcohol has a boiling point of 78°C, but at 25°C, their respective vapor pressures are 1.3 and 0.79 kPa. Consequently, benzene evaporates more rapidly than ethyl alcohol at 25°C under the same conditions. Similarly at 25°C, n-butyl acetate (bp 126°C) evaporates more rapidly than n-butyl alcohol (bp 118°C).

The effect of the ratio of surface area to volume results because solvent evaporation occurs at the solvent-air interface. If 10 g of solvent are spread out over an area of 100 cm^2, it evaporates more rapidly than if the surface area is 1 cm^2. Solvent evaporates more slowly from a coating in an open can than after application as a film. When a coating is applied by a spray gun, it is atomized to small particles as it comes out of the orifice of the gun. The ratio of surface area to volume is very high, and the rate of solvent loss is very high. Thus, a major fraction of solvent evaporates after the spray droplets leave the orifice of the spray gun and before they arrive at the surface to which the coating is being applied. As coating films dry, because the ratio of surface to volume affects evaporation rate, the fraction of solvent present in a 50 μm film after a given time is greater than that remaining in a 25 μm film. The concentration of the resin solution and the viscosity increase more slowly when a coating is applied in thicker films.

The rate of air flow over the surface is important because the rate of evaporation depends on the partial pressure of the solvent vapor in the air at the air-solvent interface. If vaporized solvent molecules are not carried away from the surface quickly, the partial pressure of solvent builds up and evaporation is suppressed. Air flow rates vary substantially, depending on the application method, and therefore, the solvents used in a coating must be selected for the particular application conditions. For example, spraying a coating with an air spray gun results in significantly more loss of solvent than when the coating is sprayed from an airless gun because of the greater air flow over the surface of the droplets by the former method. (See Sections 22.4.1 and 22.4.2.) The rate of solvent loss from a freshly coated surface depends on the rate of air flow through the spray booth. If the same coating is applied to the outside and the inside of a pipe, solvent evaporates more rapidly from the outside unless the inside is ventilated. Air flow effects cause nonuniform evaporation from coated objects; solvent evaporates more rapidly near the edges of a coated panel than from its center.

Relative humidity has little effect on the evaporation rates of most solvents, however, it has a major effect on the evaporation rate of water. The higher the relative humidity, the more slowly water evaporates when all other conditions are equal. As relative humidity approaches 100, the rate of evaporation of water approaches zero. Relative humidity decreases as air is warmed; sometimes it is feasible to partly compensate for the effects of humidity on drying by modestly increasing the air temperature, since warming the air both increases the vapor pressure of the water and decreases the relative humidity.

17.3.2. Relative Evaporation Rates

As mentioned in Section 17.2, the introduction of nitrocellulose lacquers led to the need to formulate complex solvent combinations. One of the important criteria involved in solvent selection was evaporation rate. From the standpoint of solvent selection, vapor pressure is an important variable that affects evaporation rate. However, it is difficult to look at a set of vapor pressure data and judge how much more rapidly one solvent will evaporate than another. Procedures for measuring absolute evaporation rates were developed, but because of the difficulties of controlling air and heat flows in different types of apparatus it became common to determine *relative evaporation rates*. Rates of evaporation of other solvents were related to the evaporation rate of *n*-butyl acetate, which at the time was the standard nitrocellulose solvent. Relative evaporation rate E is defined by Eq. 17.1, where t_{90} is the time for 90 wt% of a sample to evaporate in a given type of apparatus under controlled conditions:

$$E = \frac{t_{90}(n\text{-butyl acetate})}{t_{90}(\text{test solvent})} \qquad (17.1)$$

Using Eq. 17.1, the relative evaporation rate of *n*-butyl acetate is 1 by definition. Some authors express E as a percentage figure, corresponding to 100 for *n*-butyl acetate. In either case, the higher the E value, the faster the relative evaporation rate. One should be sure which reference point is being used; there is at least one paper in the literature in which an author used data from two different tables without realizing that he was mixing data that differed by a factor of 100; his conclusions were absurd.

Determination of relative evaporation rates requires measurement under carefully controlled, standardized conditions. The Shell Thin Film Evaporometer is an example of an apparatus. A top loading balance is placed in an enclosure designed to minimize variation in air flow over the surface of the balance pan. A flow of 25°C air with a relative humidity of less than 5% at a rate of 21 L min^{-1} is maintained. A sample of 0.70 mL solvent is dispensed onto a piece of filter paper on the balance pan, and the time required to lose 90% of the sample weight is determined. Rates determined this way are volume-based relative evaporation rates. While volume-based evaporation rates are the most common in the literature, some experimental procedures use a fixed weight, rather than a fixed volume of solvent. Care must be taken not to mix data obtained by the different methods. One should also avoid comparing relative evaporation rates determined using different instruments because different evaporation conditions can change not only the absolute evaporation rates, but also the relative evaporation rates.

A study by Rocklin illustrates the effects of changes in conditions on relative evaporation rates [14]. He compared the relative evaporation rates of 66 solvents measured by the standard procedure (evaporation from filter paper) with rates measured by evaporation directly from the flat aluminum pan of the balance. Table 17.4 gives his

Table 17.4. Volume-Based Relative Evaporation Rates at 25°C

Solvent	E_{paper}	E_{metal}	Ratio
n-Pentane	12	38	0.32
Acetone	5.7	10	0.55
Ethyl acetate	4.0	6.0	0.67
Methyl ethyl ketone	3.9	5.3	0.74
n-Heptane	3.6	4.3	0.83
Toluene	2.0	2.1	0.92
Ethyl alcohol	1.7	2.6	0.65
Methyl isobutyl ketone	1.7	1.7	1.0
Isobutyl acetate	1.5	1.5	1.0
n-Butyl acetate	1	1	1
sec-Butyl alcohol	0.93	1.2	0.81
m-Xylene	0.71	0.71	1.0
n-Butyl alcohol	0.44	0.48	0.92
2-Ethoxyethanol	0.37	0.38	0.98
Water	0.31	0.56	0.56
Methyl n-amyl ketone	0.34	0.35	0.96
2-Ethoxyethyl acetate	0.20	0.19	1.1
n-Decane	0.18	0.16	1.1
2-Butoxyethanol	0.077	0.073	1.1
Isophorone	0.023	0.026	1.0

data for several solvents. Note that both sets of data are relative to n-butyl acetate. As can be seen by comparing the E_{paper}/E_{metal} ratios, significant differences exist between relative evaporation rates when the solvents evaporate from filter paper and the corresponding relative rates when those same solvents evaporate from a smooth metal surface in the same instrument at the same settings.

The most notable differences between E_{paper} and E_{metal} occur with faster evaporating solvents and with water and alcohols. In the case of evaporation from filter paper, the ratio of surface area to volume is much higher than in the case of the evaporation from the metal. Therefore, at first solvents evaporate more rapidly from filter paper, resulting in a sharp drop in temperature, which leads to a decrease in vapor pressure that slows evaporation. In the case of evaporation from a metal surface, the surface area is much smaller and the thermal conductivity is much higher. Both factors minimize the temperature drop, leading to less decrease in vapor pressure and evaporation rate. The difference is greater with fast evaporating solvents such as n-pentane and acetone, than with n-butyl acetate which evaporates more slowly, resulting in their low ratios of E_{paper} to E_{metal}. The experimental temperature was 25°C but that is the air temperature. What controls the rates of evaporation and, hence, the relative rates is the actual temperature(s) of the surfaces.

The E_{paper} to E_{metal} ratios are also low in the cases of water and alcohols. This results from the greater extent of hydrogen-bond interaction of the hydroxyl groups with the very large surface area of cellulose of the paper as compared with the smooth aluminum surface. This retards evaporation of water and alcohols relative to n-butyl acetate. Note that relative to n-butyl acetate, sec-butyl alcohol evaporates more rapidly from the smooth metal surface and more slowly from the filter paper.

When formulating baking coatings for spray application, it is common to use a mixture of fast and very slow evaporating solvents. A significant fraction of the fast evaporating solvent evaporates before the spray droplets reach the object being coated, raising viscosity and reducing the tendency of the coating to *sag*, while the slow evaporating solvent keeps the viscosity low enough to promote *leveling* and to minimize the probability of *popping* when the coated object is put into a baking oven. (See Chapter 23 for further discussion of sagging, leveling, and popping.) In selecting slow evaporating solvents, formulators have generally used tables of relative evaporation rates at 25°C and boiling points, which are available for a large number of solvents. Evaporation rate data have only been published for a limited number of slow evaporating solvents over the range of 75°C to 150°C [15]. The rates were determined using a thermogravimetric analyzer (TGA) isothermally at a series of furnace temperatures. The rate relationships can be quite different at different temperatures. For example, at 25°C, the evaporation rate for a commercial mixture of the dimethyl esters of succinic, glutaric, and adipic acids is about five times slower than for isophorone. However, at 150°C, their evaporation rates are approximately equal.

Which evaporation rate data are "correct"? All of them are "correct"; they depend on the particular circumstances under which they were determined. However, we do not apply coatings to filter paper (or to aluminum balance pans or to TGA pans) nor do we dry them in a Shell Evaporometer (or a TGA furnace). The rates at which solvents evaporate in actual use depend on the particular situation. Solvents evaporate more rapidly from coatings applied by a spray gun than if they are applied by roller coating. The type of spray gun can make a substantial difference. Other significant variables include the rate of air flow through the spray booth, the shape and mass of the object coated, the film thickness being applied, the flash-off time before entering an oven, and the way heat in the oven is zoned.

17.3.3. Evaporation of Mixed Solvents

Evaporation of solvent blends, rather than pure solvents, adds further complications. In *ideal* homogeneous solutions, vapor pressure is governed by Raoult's law, which predicts that the vapor pressure P_i of the ith component of the solution is reduced from the vapor pressure of the pure liquid P_i^o in proportion to its mole fraction x_I:

$$P_i = x_i P_i^o$$

Since vapor pressures of the different solvents are different, the composition of the solvent that evaporates is different than the composition of the solvent blend. Therefore, the partial vapor pressures change continuously as solvent evaporates from a mixture. Raoult's law provides a good approximation for many combinations of solvents, especially when structures are similar and intermolecular interactions are minimal. However, other solvent mixtures are *nonideal* as a result of interaction effects. These effects change as ratios of solvents change. The vapor pressure P_{total} of any mixture of miscible solvents can be calculated by the following equation, where χ represents an empirical adjustment factor for the interaction effects, usually given the more dignified-sounding title of an activity coefficient.

$$P_{total} = P_1 + P_2 + \cdots + P_i = \chi_1 P_1^0 x_1 + \chi_2 P_2^0 x_2 + \cdots + \chi_i P_i^0 x_i$$

Activity coefficients for many solvents have been evaluated and incorporated into computer programs that can calculate partial vapor pressures of each solvent in a mixture throughout its evaporation. One such program is called UNIFAC [16]. Results of such calculations are often presented as the partial vapor pressures at each 10% interval through the evaporation. Several assumptions are involved in calculating vapor pressures, and in order to relate vapor pressure to evaporation rate, it is assumed that all other factors (temperature, surface to volume ratio and air flow) affecting evaporation rate are fixed. Thus, the results of such calculations are only approximations. However, in view of the uncertainties of relative evaporation rates, a high level of accuracy is not needed. An extensive review paper discusses the evaporation of solvent blends [17].

Relative evaporation rates of mixed solvent compositions can be determined experimentally in an evaporometer. Relative evaporation rate of a mixed solvent E_T, is calculated from volume fractions c, activity coefficients a, and relative evaporation rates E, of the individual solvents [18]:

$$E_T = (caE)_1 + (caE)_2 + \cdots + (caE)_n$$

Such calculated E_T values are of dubious accuracy: the composition changes over time, so the value of E_T changes. The difference between experimental and calculated relative evaporation rates is especially large when comparing a solvent mixture with a narrow range of E values with another with a wide range of E values. For example, one could calculate E_T for a mixture of two solvents with high and low E values that would be equal to the E value of a single solvent with an intermediate E value. However, it is evident that the mixed solvent would actually have a smaller relative evaporation rate than the single solvent since after the fast evaporating solvent was gone, the slow evaporating solvent would evaporate more slowly than the single intermediate E value solvent.

This description makes it sound almost impossible to formulate a satisfactory solvent mixture for a coating. However, formulation does not start in a vacuum of knowledge. Experience with other formulations that proved useful under somewhat similar circumstances provides guidance. Using such formulations together with any reasonable table of relative evaporation rates, one can make a first attempt at a new formulation. The coating is then applied under the particular circumstances, and adjustments made as needed. The final adjustment is almost always made by an experienced person in the user's factory with production spray guns and operating conditions. The use of relative evaporation rate tables assists development of a formulation that is in the ball park of what is needed; because final, in-factory adjustments are needed anyway, it may not matter much which relative evaporation rate table was used.

Evaporation rates of water-organic solvent mixtures require special consideration for at least four reasons. First, strong interactions often cause deviations from Raoult's law. Second, relative humidity (RH) affects the evaporation rate of water, but has little or no effect on the evaporation rates of organic solvents. Third, azeotropic effects may occur. Fourth, the heat capacity and heat of vaporization of water are unusually high.

The humidity effect is illustrated by the fact that the relative evaporation rate E of water at 0 to 5% RH and an air temperature of 25°C is 0.31, but at 100% RH it is 0. If a solution of 2-butoxyethanol ($E = 0.077$) in water evaporates at low RH, water evaporates more rapidly, and the remaining solution becomes enriched in 2-butoxyethanol. At high RH, 2-butoxyethanol evaporates more rapidly, and the remaining solution becomes enriched in water. At some intermediate RH, the relative evaporation rates of water and 2-butoxyethanol are equal. When an aqueous solution evaporates at this RH, the composition of the remaining solution is constant. This RH has been called the *critical relative humidity*

(CRH) [19]. The CRH for 2-butoxyethanol solutions in water is estimated at about 80%. If the relative evaporation rate of the solvent were larger than that of water, even at 0 to 5% RH, there would be no CRH, since the solvent would evaporate more rapidly than water at any RH. At the other end of the scale, if the relative evaporation rate of the solvent were very low, the CRH would approach 100%.

Azeotropic behavior is likely to occur with water solutions. Rocklin has studied the role of azeotropy in speeding up water/solvent evaporation in humid air [20]. He reports development of a computer model, the AQUEVAP program, that permits calculation of the fastest evaporating water-solvent blends at various RH values. For example, at 40% RH, the time required for evaporation of 90% of a 20 wt% solution of 2-butoxyethanol in water is 1820 seconds compared with 2290 for water alone. Thus, coevaporation of water and 2-butoxyethanol accelerates evaporation.

The high heat capacity and heat of vaporization of water can also affect the evaporation rates of water and water-solvent blends in an oven. For example, the times for 99% weight loss of 2-butoxyethanol (bp 171°C), water, and a 26 : 74 blend of 2-butoxyethanol/water in a TGA when room temperature samples were put into the furnace at 150°C were 2, 2.6, and 2.5 minutes, respectively [21]. While the air temperature in the furnace was 150°C, the samples took some time to heat up. The higher heat of vaporization of water (2260 J g^{-1} at its boiling point) compared to 2-butoxyethanol (373 J g^{-1} at its boiling point) slowed the rate of heating of the water and water-solvent blend enough to more than offset the expected evaporation rates based on boiling points or E values.

17.3.4. Evaporation of Solvents from Coating Films

Except in high solids coatings (see Section 17.3.5.), the resin or other coating components have little effect on initial rate of solvent evaporation when coating films are applied. Initial rates of evaporation of solvent from resin solutions are, within experimental error, the same as or close to the rates of evaporation of the solvents alone under the same conditions. Such observations are not inconsistent with Raoult's law, which predicts that dissolved resins will have little effect on vapor pressure because of their high molecular weights. However, as solvent loss from a coating continues, a stage is reached at which the rate of evaporation slows sharply. As the viscosity of the remaining coating increases, availability of free volume decreases and the rate of solvent loss becomes dependent on the rate of diffusion of solvent through the film to the surface, rather than on the rate of evaporation from the surface. The solids level at which the transition from evaporation rate control to diffusion rate control occurs varies widely, but is often in the 40–60 NVV range.

Hansen described the situation as first stage and second stage losses of solvent [22]. In the first stage, the rate is governed by the factors that govern evaporation of solvent mixtures-vapor pressure, surface temperature, air flow over the surface, and surface to volume ratio; at this stage, there is a first-power dependence on wet film thickness. After a transitional stage, evaporation slows and the rate of solvent loss becomes dependent on the rate of diffusion of the solvent molecules through the film. During this stage, the evaporation rate depends on the square of film thickness. In line with the changes in viscosity, the first stage is sometimes called the *wet stage* and the second stage the *dry stage*. An example of a plot of weight of solution remaining as a function of time for evaporation of MIBK from a soluble vinyl chloride copolymer resin solution is shown in Figure 17.1 [23].

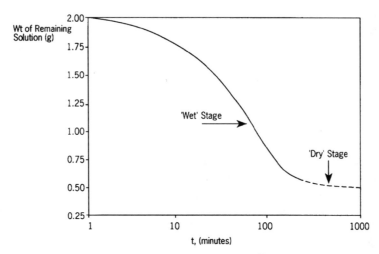

Figure 17.1. Two-stage release of solvent; MIBK in Vinylite® VYHH at 23°C, initially at 20 wt% polymer. (From Ref. [23], with permission.)

During the second stage, diffusion rate is primarily controlled by free volume availability. That is, the solvent molecules move through the film by jumping from free volume hole to free volume hole. The most important factor controlling free volume availability is $(T - T_g)$. If solvent evaporation is occurring at a temperature well above the T_g of the solvent-free resin, diffusion rate will not restrict the rate of solvent evaporation at any stage in drying. If the T_g of the resin is above the temperature at which drying is occurring, solvent evaporation will become diffusion rate controlled as $(T - T_g)$ becomes small. No experimental data are available as to the values of $(T - T_g)$ when diffusion rate control becomes important; it is probable that such values are system dependent.

When the second stage is reached, the rate of evaporation becomes dependent on how fast the solvent molecules get to the surface, not on their vapor pressure. As solvent loss continues, concentration of the remaining resin solution increases, and therefore, T_g increases and the rate of diffusion slows further. If the T_g of the resin is sufficiently higher than the temperature of the film, the rate of solvent loss will, in time, approach zero. Years after films have been formed, there will still be residual solvent left in the film. If solvent must be essentially completely removed in a reasonable time, the film must be baked at a temperature above the T_g of the solvent-free resin. It is a common mistake for novice polymer chemists to think that the rate of solvent loss from polymer samples at temperatures that are below the T_g of the blend can be increased by drying under a vacuum. Since the rate of solvent loss has reached a stage at which it is independent of vapor pressure of the solvent, it is also independent of atmospheric pressure above the sample.

Solvent evaporation is affected by the addition of a plasticizer. The concentration of solvent at which solvent loss becomes diffusion controlled decreases as plasticizer concentration increases. The amount of solvent retained after some time interval decreases as plasticizer concentration increases. Perhaps, less obviously, in the case of mixed-solvent systems, the ratio of slow to fast evaporating solvent retained in the film increases as plasticizer concentration increases.

There have been many attempts to quantify the diffusion of solvents from coating films during solvent evaporation, but success has been limited [24,25]. Some of the problems are

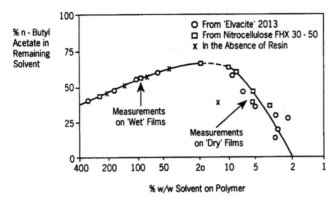

Figure 17.2. Changes in remaining solvent concentration during wet and dry stages of solvent evaporation at 23°C from films of an acrylic resin (Elvacite® 2013) and nitrocellulose, each initially in 60:40 IBAc to BAc. (From Ref. [23], with permission.)

discussed in Ref. [26] and possible approaches to modeling the diffusion through scaling analysis are suggested. As a broad generalization, small molecules tend to evaporate faster than large molecules in both the first and second stages because small molecules tend to be more volatile in the first stage and to diffuse faster in the second stage. However, other factors are involved. One is molecular configuration. Since the molecules must jump from hole to hole, the smaller the size of the solvent molecule, the greater its chance of finding sufficiently large holes. Even though its relative evaporation rate is higher, cyclohexane is retained in films to a greater degree than toluene because cyclohexane is bulkier.

Figure 17.2 shows the results of an experiment illustrating the effect of cross-sectional size. Nitrocellulose and acrylic lacquers were made at 20 wt% solids in 60 : 40 isobutyl acetate (IBAc) and n-butyl acetate (BAc), and the ratios of the two solvents remaining in the films were monitored. Like other branched-linear isomeric pairs, the relative evaporation rate of branched IBAC is higher than that of linear BAC. Thus, the ratio of IBAc to BAc remaining in the film decreased during the first, wet stage of solvent loss, reaching a minimum of about 35 : 65. During the second, dry stage, the linear compound diffused more rapidly, so the ratio of IBAc to BAc remaining in the film increased and approached 90 : 10 [23]. Equations have been developed that model effect of solvent size on diffusion based on free volume of polymers [27].

In air dry coatings, the T_g often increases to a level above ambient temperature, while significant amounts of solvent remain in the film. When this happens, solvent evaporation becomes extremely slow, and detectable amounts of volatile solvents may remain in the film for years. Residual solvent has been shown to affect film properties such as corrosion (see Section 7.3.1.1.) and moisture resistance adversely. Thus, selection of solvents that have minimal detrimental effects may be a factor in optimizing long-term film performance.

17.3.5. Evaporation of Solvents from High Solids Coatings

Generally, it is more difficult to control sagging (see Section 23.3) of spray applied high solids coatings than of their lower solids counterparts. While other factors may be involved, there is considerable evidence that high solids formulations lose much less solvent than conventional coatings while atomized droplets are traveling between a spray gun and the object being coated [28,29]. As a result, there is less increase in viscosity and, therefore, a greater tendency to sag. Explanations for this decreased solvent loss do not appear to be adequate. One factor is the colligative effect of the lower mole fractions of solvent(s) in a high solids coatings. A sample calculation illustrates this difference. If the vehicle of a conventional coating consists of a 30% by weight solids solution of a resin with $\bar{M}_n$ of 15,000 in 70% of a solvent mixture with $\bar{M}_n$ of 100, the mole fraction of solvent is 0.997. However, if a high solids coating has 70% solids of a resin with $\bar{M}_n$ of 1000 and 30% of a solvent mixture with $\bar{M}_n$ of 100, the mole fraction of solvent is 0.811. While this difference is in the direction of slowing solvent evaporation from the high solids coating, it does not seem large enough to account for the large differences in solvent loss that have been reported.

It was suggested, without experimental evidence, that high solids coatings may undergo transition from first-stage to second-stage solvent loss with relatively little solvent loss as compared to conventional coatings [30]. The reasoning was that the T_g of the solution in a high solids coatings could change more rapidly with concentration, hence, reach a stage of free volume limitation of solvent loss after only a little loss of solvent. Consistent with this hypothesis, Ellis found that high solids polyesters were formulated at concentrations that were already above the transition concentration where solvent loss rate becomes diffusion controlled [31]. He also found that the transition points occur at higher solids with linear molecules such as n-octane versus isooctane, and n-butyl acetate as compared to isobutyl acetate. Further research is needed to fully understand the reasons for slower loss of solvents from high solids coatings; meanwhile, slower solvent loss can be a serious problem for the application of high solids coatings. More detailed discussion of sagging and sag control in high solids coatings appears in Sections 23.3 and 24.2.2.

17.3.6. Volatile Loss from Water-Borne Coatings

For solvent-borne coatings, the formulator has available a variety of solvents having a wide range of properties and volatility. But, for water-borne coatings the major volatile component is water, which comes with only one vapor pressure-temperature curve, one heat of vaporization, and so forth. Furthermore, the relative humidity at the time of application and drying of the coatings can have a major effect on rates of volatile losses from water-borne coatings. Formulators use limited levels of organic solvents to modify evaporation rates; however, future regulations can be expected to reduce the levels permitted.

A major class of water-borne coatings is based on water-reducible resins. As described in Section 12.3, these resins are not soluble in water. A solution of a salt of the resin in an alcohol or an ether alcohol is diluted with water. During dilution, the resin forms aggregates that are swollen with solvent and water, dispersed in a continuous aqueous phase that also contains some solvent. The relative evaporation rate of water is affected by the relative humidity, so with solvents like 2-butoxyethanol, there is a CRH, as described in Section 17.3.3. It has been found that the CRH is different for coatings than for water-solvent blends without resin. For example, CRH is 65% for 10.6 wt% (based on volatile components) 2-butoxyethanol in a coating, in contrast to 80% calculated for the same concentration of a 2-butoxyethanol-water blend without resin [32]. A possible explanation of the difference in CRH is that the concentration of 2-butoxyethanol in the continuous water phase is below the average and the solvent-water ratio inside the aggregates is above the average. This type of distribution and its effect on the composition of vapor above a model aggregate system using a more volatile solvent, t-butyl alcohol, has been demonstrated [33]. Such effects can be critical in controlling sagging and, particularly, popping of such water-borne coatings. (See Sections 23.3, 23.7, and 25.1.)

The other major class of water-borne coatings uses latexes as the principal binder. Evaporation of water from a drying latex paint film resembles first-stage drying throughout most of the process; it is controlled by temperature, humidity, evaporative cooling, and rate of air flow over the surface [34]. After most of the water has left, evaporation slows as a result of coalescence of a surface layer through which water must diffuse. In latex paints that are to be applied by brush or roller, it is desirable to retard the development of a partially coalesced surface layer to permit lapping of wet paint on wet paint. This generally requires the presence of some slow evaporating solvent, such as ethylene or propylene glycol. The presence of these solvents does not affect the initial rate of water loss, but does slow the development of a surface skin [35]. The presence of such a water-soluble solvent also facilitates the loss of coalescing solvent. Use of both types of solvents is discussed further in Chapter 31.

17.4. VISCOSITY EFFECTS

Solvent selection can have a major effect on viscosity of resin solutions. This results from two factors: the viscosity of the solvent itself and the effect of solvent-resin interactions. The direct effect of solvent viscosity on solution viscosity can be seen in equations that relate viscosity and resin concentration. Even the simplest relationship, which is valid over only a narrow range of concentration of solutions with viscosities in a range of about 0.1 to 10 Pa · s, shows this dependence on solvent viscosity:

$$\ln \eta_{\text{solution}} = \ln \eta_{\text{solvent}} + K(\text{Conc})$$

For example, a difference of viscosity of 0.2 mPa · s between the viscosities of two solvents (1.0 mPa · s and 1.2 mPa · s) may appear trivial, but can cause a difference of 2000 mPa · s in the viscosity of a 50 wt % solution of a resin (10 Pa · s vs. 12 Pa · s) [36]. The relationship between concentration and viscosity of resin solutions is discussed in more detail in Section 3.4.3.

One must be careful in comparing viscosities of solutions. There are data in the literature that make comparisons based on weight relationships, volume relationships, weight of solvent per volume of coating, and weight of resin per volume of solution. One must decide what type of comparison is appropriate for a particular purpose. Most commonly, weight ratios are used when considering the viscosity of resin solutions, but they can be misleading because air pollution regulations are based on weight of solvent per volume of coating. If one compares the viscosity of resin solutions with different solvents on a weight basis, one might make the wrong choice in trying to minimize weight of solvent per volume of coating. Since flow is generally related to volume considerations, use of volume fraction comparisons might seem desirable, but polymer solutions are seldom ideal, creating uncertainty about the significance of volume fraction numbers.

An example of the effect of solvent choice on viscosity of an acrylic solution containing 400 g L^{-1} of solution of the various solvents is given in Table 17.5 [37]. For comparison purposes, the densities and viscosities of the solvents alone are also given. The data in Table 17.5 illustrate the substantial effect that solvent selection can have on solution viscosity of a given polymer. The weight concentrations of these solutions are not constant;

Table 17.5. Viscosities at 25°C of Solutions of a High Solids Acrylic; Solvent Concentration 400 g L⁻¹ Solution

Solvent	Solvent Viscosity (mPa·s)	Solvent Density (g mL⁻¹)	Solution Viscosity (mPa·s)
Methyl propyl ketone	0.68	0.805	80
Methyl isobutyl ketone	0.55	0.802	110
Ethyl acetate	0.46	0.894	121
Methyl amyl ketone	0.77	0.814	147
n-Butyl acetate	0.71	0.883	202
Toluene	0.55	0.877	290
Isobutyl isobutyrate	0.83	0.851	367

it would have been interesting to compare also the viscosities of equal weight ratio solutions. The data suggest that solvents with low density and low viscosity, as predicted by theory, tend to give low solution viscosity. However, there are other factors involved.

Another factor that is sometimes important is the effect of solvent/solvent interactions on viscosity of solvent mixtures. In general, these effects are small except for mixtures that include alcohols or water. When relatively small (generally < 40%) amounts of alcohols are mixed with other solvents, they do not increase the viscosity of the solution proportionally to their relatively high neat viscosities [38]. This results from reduced multiple hydrogen-bond interactions between the alcohol molecules in the presence of other solvents. The effects are larger and less obviously predictable in the case of solutions of solvents in water.

Large viscosity effects can result from solvent-resin interactions. To distinguish these effects from the effects of solvent viscosity discussed previously, it is common to compare relative viscosities (η/η_s) of solutions. The effects of interactions are complex and not fully understood in even relatively simple systems. At least two factors play a role. First, most resins used in coatings have polar and hydrogen-bonding substituents, such as hydroxyl groups or carboxylic acid groups, that tend to associate with polar groups on other molecules, often substantially increasing viscosity. Solvents that can prevent or minimize such interactions by interacting with the polar groups tend to reduce viscosity. Polar solvents with single hydrogen-bond acceptor sites are effective in minimizing relative viscosity; examples are ketones, ethers, and esters. Water can also serve to reduce viscosity of oligomers with hydrogen bond interactions. It has been shown that up to 20% (depending on the formulation) of water can dissolve in solvent-free coatings and that adding water, usually 5 to 15%, reduces viscosity by 40 to 60% [39].

A second factor is the effect of a given solvent on the hydrodynamic volume of the individual resin molecules with their closely associated solvent molecules. If the interaction between resin and solvent molecules is strong, the chains of the resin molecules become extended, and their hydrodynamic volume increases. If the interaction is not so strong, the molecules contract, and the hydrodynamic volume becomes smaller. Relative viscosity tends to be directly related to hydrodynamic volume. However, if solvent-resin interaction is so weak that there can be resin-resin interaction rather than only resin-solvent interaction, clusters of resin molecules form and relative viscosity and viscosity increase. In the majority of coating applications, and especially high solids coatings, which use resins of low molecular weight, resin-resin interactions are the stronger of the two effects, although hydrodynamic volume effects can be significant. Thus, most high solids formulations include hydrogen-bond acceptor solvents. The situation with mixed solvents is more complex. Only one paper has been published that attempts to study the viscosity effects with solvent mixtures. Erickson and Garner [40] found in a study involving a limited number of systems, that relative viscosity was dominated by the effect of the solvent that interacted most strongly with the resin.

A word of caution: Relative viscosities can be useful for understanding the factors affecting viscosity, but the viscosities important in applying coatings are not relative viscosities. A solvent giving a comparatively high viscosity solution could give a relatively low relative viscosity solution. Viscosity of resin solutions is discussed further in Section 3.4.

Solvent selection can also affect storage stability. Inclusion of alcohols in formulations containing melamine-formaldehyde resins can increase storage life by minimizing cross-linking during storage. (See Section 9.3.3.) Polyesters can transesterify with primary alcohols during storage; this can especially be a problem with water-reducible polyesters. (See Section 13.4.) Alcohols can react slowly with epoxy resins during storage. Hydrogen-bond acceptor solvents can extend the pot life of epoxy-amine coatings. (See Section 11.2.5.) Alcohols react with isocyanates, urethane grade esters and ketones that contain no more than traces of alcohols or water are available. Reaction rates of isocyanates with hydroxyl groups are reduced in hydrogen bond acceptor solvents, extending pot life. (See Section 10.4 and Ref. [41].)

17.5. FLAMMABILITY

Fires and explosions of flammable coating solvents can inflict terrible burns. Many tragic accidents have occurred in which people were killed or severely burned. Sadly, these accidents could have been prevented. Most solvents used in coatings are flammable. Care should be exercised when working with solvents in the laboratory, in the coatings factory, and in end use environments. Flammability depends on structure and vapor pressure. Generally, there is both an upper and a lower level of vapor concentration that limit flammability or explosion. If the partial pressure is low enough, not enough energy is released during burning to maintain the vapor-air mixture above the ignition temperature of the system. If the partial pressure of the solvent is high enough, there is insufficient oxygen for explosion or fire. Tables of upper and lower explosive limits of many solvents are available. A full container of solvent may present less of a fire risk than a recently emptied container; the vapor phase in the former might have the solvent at concentrations above the upper explosive limit, while the "empty" container may have a concentration in the explosive range.

The most common cause for fires in coating factories has been static electricity. Solvent flowing out of one tank and into another tank by gravity can pick up enough electrostatic charge to cause a spark, which can set off a fire or explosion. To avoid such charge accumulation, all tanks, pipes, and so forth used in handling solvents and solvent-containing mixtures should be electrically grounded at all times. A second cause is sparking of electrical motors or faulty electrical connections. Factories and laboratories should be equipped with explosion proof electrical installations. One should never try to bypass these installations. And, of course, smoking is a potential ignition source.

There are two main types of flammability tests: *open cup* and *closed cup*, both measure a *flash point*, the minimum temperature at which solvent can be ignited by a hot wire. ASTM specifies standard conditions for both tests. Generally, open-cup testers give results more appropriate for indicating degree of hazard of a mixture when exposed to air, as during a spill. The closed-cup flash point more nearly describes the fire hazard of a liquid enclosed in a container. U.S. Department of Transportation regulations for shipment of flammable liquids are based on closed-cup tests. Transportation costs can be substantially affected by flash points of the material being shipped. A closed-cup flash point is lower than an open-cup flash point. Solvents used in architectural paints should have closed cup flash points over 38°C. Extensive tables of flash points of solvents are available. Different sources sometimes disagree on flash point values. Variations are not surprising for solvents such as naphthas and mineral spirits that are variable mixtures, but variations are also found with single component solvents such as *n*-butyl acetate. The ASTM methods are said to be accurate to ±2.5°C. Reproducibility is poorer at temperatures below 0°C.

Many coatings contain a mixture of solvents. It is safest to determine flash points of mixtures experimentally. A discussion of the factors affecting flash points, including molecular interactions in blends, is given by Ellis [42]. It has been reported that good predictions of closed cup flash points for mixed solvents can be made by a UNIFAC computer program that requires only flash points and molecular structures of the pure components [43]. The best results were reported using UNIFAC group interaction parameters derived from flash points of binary solvent mixtures. Calculation of useful approximations of flash points of mixed solvents with simpler equations has been reported [44].

Risk of fire or explosion can be eliminated by meeting either of two conditions. If solvent vapor concentrations in air are kept well outside of the concentration range within which ignition is possible or if all sources of ignition are eliminated, there can be no fire. Unfortunately, many accidents have occurred because neither of these conditions were satisfied. Because it is difficult to be certain that either of these conditions can be met in all circumstances, prudent practice dictates that all possible steps should be taken to meet both conditions, providing redundancy. The importance of good ventilation cannot be over-emphasized, especially since the vapors of relatively dense solvents may stratify in stagnant air, which tends to be concentrated in the lower part of the workspace.

17.6. OTHER PHYSICAL PROPERTIES

Density can be an important variable. It can have a major effect on cost. Most solvents are sold on a weight basis but the critical factor and, correspondingly, critical cost in almost all cases in the coatings field is the cost per unit volume. Most U.S. air pollution regulations are based on weight of solvent per unit volume of coating (see Section 17.9) which also favors use of low density solvents in formulations.

Conductivity can affect solvent choice. As discussed in Section 22.2.3, use of electrostatic spray guns requires control of the conductivity of the coating. In general, formulations that have appreciable, but low, conductivity work best. The conductivity of hydrocarbon solvents is too low to permit pickup of adequate electrostatic charge. Alcohols, nitroparaffins, and small amounts of amines are common solvents or additives to increase conductivity to a desired range. The conductivity of water-borne coatings can pose problems for electrostatic application, such as the need to insulate the spray apparatus and relatively fast loss of charge from spray droplets. Additives can minimize the latter problem. For example, inclusion of glycol ethers improves sprayability of water-borne coatings, apparently by reducing surface conductivity of water. Presumably, the alkyl groups of the ethers orient quickly to the surface of the droplets.

Surface tension (see Section 23.1) can be another important factor influencing solvent selection. Solvent can affect the surface tension of coatings, which, in turn, can have important effects on the flow behavior of coatings during application as discussed at length in Chapter 23. Solvent selection can also be a factor affecting the development of surface tension differentials across the surface of a drying film during application and film formation; these differentials can affect flow behavior [45,46]. Since surface tensions depend on temperature and concentration of resins in solution, solvent volatility can have a large effect on the development of surface tension differentials. For a coating to wet a substrate, the surface tension of the coating must be lower than that of the substrate. Solvent selection can be a factor particularlly in aqueous systems. While surface tension of aqueous systems can be reduced by adding surfactant, it is often more desirable to accomplish the same purpose with a solvent such as 2-butoxyethanol. Solvent evaporates, while surfactant leaves residues in the final film that may be detrimental to properties such as adhesion and humidity resistance.

17.7. TOXIC HAZARDS

In considering the toxic risks of volatile solvents, the extent of exposure must be considered in combination with toxicity data. All solvents are toxic at some level of exposure. Obviously, one should avoid ingesting them. The hazard from skin contact can be controlled by wearing protective clothing. Generally, the greatest potential risk comes from inhalation.

Three general types of toxicity data are important. Acute toxicity data indicates the level of intake in single doses that can be injurious or lethal, this kind of information can be particularly important in cases of accidental ingestion or spills. A second type of toxicity data concerns the level of exposure that is safe when people are going to be exposed 8 hours a day for long periods of time. This kind of data is used, for example, to set the upper concentration limit for solvent in a spray booth. Third, is risk of exposure over periods of years to low levels of materials that can increase health risks such as cancer. When it is found, usually by animal tests, that a solvent may be carcinogenic, very low levels of permissible exposure are set. The levels are frequently too low to be controlled by economically feasible methods. The effect is that carcinogenic solvents are banned. For example, benzene has not been used in coatings for many years for this reason. While most solvents used in large quantities in coatings have been tested and are thought to be noncarcinogenic, it is prudent for the user to be aware of current knowledge of the materials he/she is using and to minimize inhalation and contact with all solvents.

Extensive tables of all three types of data are available. In formulating coatings, it is also necessary to consider the clientele that will use them. While coatings sold to retail consumers are carefully labeled to include application cautions, one must assume that many people will not read the labels. When selling to a large corporation, it is reasonable to assume that the Material Safety Data Sheets will be read and appropriate practices will be established. But when selling to small industrial customers such as automobile repair shops, one cannot assume that the customer will pay attention to precautions. It may be ethical to sell a coating with a somewhat toxic solvent to one class of customer but unethical, although perhaps legal, to sell it to another class of customer. A common difficulty is to know what the level of exposure will be. Reference [47] describes an approach to assessing possible exposures when retail consumers apply coatings in a room.

In 1990 the U.S. Congress listed hazardous air pollutants (HAP) for which use is to be reduced [48]. Among those of importance in the coatings field are methyl ethyl ketone, methyl isobutyl ketone, *n*-hexane, toluene, xylenes, methyl alcohol, ethylene glycol, and ethers of ethylene glycol. The U.S. EPA's Hazardous Air Pollutants Strategic Implementation Plan describes regulatory efforts [49]. The first step was a voluntary program aimed at reducing emissions of 17 chemicals, including MEK, MIBK, toluene, and xylene, by 50% (of 1988 levels) by 1995. Mandatory HAP limits are included in EPA's Unified Air Toxics Regulations, issued for all major categories of coatings users in 1995–1999; for an example, see Ref. [50]. Compliance is required within three years of the issue date. A group of solvent producers has petitioned for removal of 2-butoxy ethanol, MEK and MIBK from the HAP list. The present HAP list motivates replacement of listed solvents with solvents not on the HAP list. Such substitutions may do little or nothing to improve air quality if the unlisted solvent is as hazardous as the listed one.

17.8. ATMOSPHERIC PHOTOCHEMICAL EFFECTS

Since the 1950s, it has been realized that the presence of organic compounds in the atmosphere can lead to serious air pollution problems. Terminology can be confusing; in the older literature, such compounds are referred to as "hydrocarbons," meaning any organic compound, not just unsubstituted hydrocarbons. More recently, they are called *volatile organic compounds* (VOC). The term *reactive organic* gases (ROG) is also used. Particularly in Europe, solvents are classified by their *photochemical ozone creation potential* (POCP). Three end effects of VOC emissions into the atmosphere are important: formation of eye irritants, particulates, and toxic oxidants, especially ozone.

While all of these factors are important, the most critical for coatings is ozone. While ozone is a naturally occurring component of the atmosphere, it is toxic to plants and animals. When you go up in the pine forests in the Rocky Mountains and smell the wonderful "fresh air," the odor is ozone. The pine trees emit substantial quantities of VOC into the atmosphere, the UV level increases with altitude, and as a result, ozone generation is high. Plants and animals, including humans, evolved in the presence of some ozone and can tolerate its presence up to a point. However, with the rapid growth of VOC emissions from man-made sources since 1900, ozone levels on many days of the year in many parts of the world, especially in and around cities, have exceeded the levels that many plants can withstand and endanger human health, especially of susceptible individuals.

By far the largest source of man-made VOC emissions comes from transportation: auto and truck tailpipe emissions, along with fuel leakage during distribution. The second largest source is coatings; in 1995, coatings and adhesives accounted for 11.6% of man-made VOC emissions [51]. There is not general agreement on the level of ozone that is considered safe. From 1978 to 1997 the U.S. National Ambient Air Quality Standard for ozone allowed no more than a one-hour period per year (averaged over three years) when the concentration of ozone in the air exceeded 0.12 ppm. Many urban areas in the U.S. have had trouble meeting this standard. In 1997, the EPA reduced the standard to 0.08 ppm. To attain this standard, the 3-year average of the fourth-highest daily maximum

8-hour average of continuous ambient air monitoring data over each year must not exceed 0.08 ppm [52]. It is estimated that 47 of the 50 states have some areas that do not comply with the new standard. See Ref. [52] for further information on ozone regulations.

Photochemical reactions in the atmosphere are complex and dependent on many variables in addition to the amount and structure of VOCs, especially the concentrations of various nitrogen oxides. Full discussion of the reactions is beyond the scope of this introductory text. References [53–55] provide brief and more detailed reviews, respectively, on the state of the science. Probably, the principal pathways leading to the generation of ozone are by way of hydrogen abstraction from the VOC compounds. Some of the reactions that have been proposed to explain ozone generation are shown in Scheme 17.1, in which RH represents a VOC compound:

Scheme 17.1

$$RH + \cdot OH \rightarrow H_2O + R\cdot$$
$$R\cdot + O_2 \rightarrow ROO\cdot$$
$$ROO\cdot + NO \rightarrow RO\cdot + NO_2$$
$$NO_2 + h\nu \rightarrow NO + O$$
$$O_2 + O \rightarrow O_3$$

Note that nitrogen oxides (NO_x) participate in the ozone forming reaction sequence shown in Scheme 17.1. A 1991 study by the National Research Council concluded that, in many areas of the United States, substantial reductions of ozone can be accomplished only by reducing NO_x levels "in addition to or instead of reducing VOC" [56]. Coatings emit almost no NO_x; the main sources are electrical generation and transportation.

An important variable in the amount of ozone generated from VOCs is the ease of abstraction of hydrogens from reactive organic gases by free radicals, such as the hydroxyl free radicals shown in the first equation of Scheme 17.1. Data are available on the rate constants for a wide variety of organic compounds [55]. In general terms, compounds with hydrogens on carbon atoms alpha to amines or ethers, hydrogens on tertiary carbon atoms, allylic hydrogens, and benzylic hydrogens are examples of easily abstractable hydrogens. (See Section 5.1.)

In early investigations of the effects of VOC on air pollution, organic compounds were divided by rabbit eye irritation tests into photochemically active compounds of high and low reactivity. It was proposed that if the emission of the highly reactive compounds could be limited, the less reactive ones could dissipate and avoid high local concentration of pollutants. This led to the establishment of a definition of *photochemically reactive solvents* in *Rule 66* of the Los Angeles Air Control District. After some years of experience, it was realized that most organic compounds are photoreactive and that the extent of dissipation in the atmosphere after local emission had been overestimated. Changing from highly reactive compounds to less reactive ones might diminish the air pollution effects near the scene of emission in exchange for increasing air pollution downwind from the emission site. Furthermore, the coatings industry objected to having to use different solvent combinations in different parts of the country. This situation led to the conclusion that it would be best to limit the emission of almost all organic compounds into the atmosphere.

Current U.S. regulations treat almost all solvents useful in coatings (except water, acetone, CO_2, and certain silicone fluids and fluorinated solvents) as equally undesirable.

Removal of methyl acetate and *t*-butyl acetate from the list has been requested. Within this context, regulators have recognized that some coatings require higher VOC levels than others for adequate performance. The EPA assessed the most advanced technology for each end use and established different maximum VOC guidelines for major applications. States and localities may, and sometimes do, enforce more rigorous guidelines. During most of the 1990s, the EPA guidelines ranged from 0.23 to 0.52 kg L^{-1} (1.9 to 4.3 lb gal^{-1}) for most major industrial coating operations [57]. Tighter EPA guidelines are expected by the year 2000.

In establishing future regulations, there is a difference of opinion as to whether all solvents should be considered as equally undesirable in the atmosphere, as they are now. The present approach is simpler to enforce. However, it may well be that using less reactive solvents to replace more reactive ones would be advantageous by allowing at least some opportunity for dissipation in the atmosphere to minimize the probability of local excess ozone concentrations. In Europe, some regulations are based on the POCP of individual solvents. Reference [55] provides a list of POCP values and examples of reformulation of solvents to minimize POCP emissions. The POCP values have been calculated by computer modeling, using reaction kinetic data to assess changes in ozone formation and checked in smog chambers [53]. Values for about 200 VOCs are available. Xylene and related aromatic hydrocarbons have particularly high values.

An ingenious approach to VOC reduction is use of supercritical carbon dioxide as a component in a solvent mixture [58]. The critical temperature and pressure of CO_2 are 31.3°C and 7.4 MPa (72.9 atm), respectively. Below that temperature and above that pressure, CO_2 is a supercritical fluid. It has been found that under these conditions, solvency properties of CO_2 are similar to aromatic hydrocarbons. The concept is to ship the coating in a concentrated form; then the high solids coating and supercritical CO_2 are metered into a proportioning spray gun in such a ratio as to reduce the viscosity to the level needed for proper atomization. Airless spray guns are used; it has been found that the rapid evaporation of the CO_2 as the coating leaves the orifice of the spray gun assists atomization. (See Section 22.2.5.) VOC emission reductions of 50% or more have been reported.

The units used in most U.S. regulations are weight of solvent per unit volume of coating as applied, excluding water (grams of VOC per liter as applied minus water). This unusual definition follows from the conventions that air pollution data are based on weights of pollutants and that coatings are sold and applied on a volume basis. (Film thickness is a volume measurement, since it is equivalent to film thickness per unit area.) The limitations are based on VOC content of the coating "as applied." If, as is very common, the user is to add solvent to the coating as it is shipped by the coating manufacturer, the VOC is based on the coating after this solvent has been added. The exclusion of water is to prevent circumventing regulations by developing coatings with low solids that are diluted with water, but still have high VOC contents relative to the dry film thickness that has to be applied.

VOC emissions can be substantially affected by the *transfer efficiency* in spray applying coatings. When a coating is sprayed, only a part of the coating is actually applied to the object being coated. Transfer efficiency represents the percentage of the coating used that is actually applied to the product. As the transfer efficiency increases, the VOC emissions decrease, since less coating is used. Transfer efficiency depends on many variables, particularly the type of spray equipment utilized (see Section 22.2). In some cases, regulations have been established requiring certain kinds of spray equipment or setting a lower limit on transfer efficiency, such as 65%.

Besides reducing the VOC content of coatings and increasing transfer efficiency, there are two other broad approaches to minimizing VOC emissions: solvent recovery and incineration. In some cases, it is feasible to recover the solvent used in coatings. One can pass the air flowing out of a drying chamber through activated carbon beds where the solvent is adsorbed. After the surface of the carbon is saturated with solvent, the material is heated to distill and recover the solvent. Alternatively, the solvent-laden air can be passed through condensers cooled with liquid nitrogen; the nitrogen that is vaporized is fed into the drying chamber, reducing the oxygen content and, hence, permitting higher concentrations of solvent without exceeding the lower explosive limit. Solvent recovery is desirable when it is feasible, but feasibility is limited by low solvent concentration in the air stream needed to stay safely below the lower explosive limits. In the case of effluent air from spray booths, it must be below concentrations that could have toxic hazards; the solvent concentration is generally too low to permit economic recovery of the solvent.

VOC emissions can also be minimized by incineration. The effluent solvent-laden air stream is heated in the presence of a catalyst to a temperature high enough to burn the solvent. As with solvent recovery, this approach is feasible only when solvent concentrations are relatively high. Incineration has been found to be particularly applicable in coil coating (see Section 29.4). In this case, most of the solvent is released in the baking oven. Part of the effluent air from the baking oven is recirculated back into the oven; the amount of such recirculation is limited so that the solvent content does not approach the lower explosive limit. The balance of the effluent air is fed to the gas burners that heat the oven. The solvent in the air is burned along with the gas; the fuel value of the solvent reduces the gas requirement. Thus, VOC emission is minimized and the fuel value of the solvent is used. Since the 1970s, VOC emissions from typical auto assembly plants have been reduced by 70–80% by combining reduced use of VOCs, improved transfer efficiency, and incineration [59].

17.8.1. Determination of VOC

The amount of VOC emitted by a coating is not easily determined. Solvent can be retained in films for very long periods of time. In latex paints, coalescing solvents are used that are only slowly released from the coating. In cross-linking coatings, volatile byproducts may be generated by the reaction. For example, MF cross-linking leads to the evolution of a molecule of volatile alcohol for each cocondensation reaction and in self-condensation reactions, there can be emission of alcohol, formaldehyde, and methylal. The amount released depends on curing conditions and the amount of catalyst used. On the other hand, when slow evaporating glycol ether solvents are used in an MF cross-linking system, it is probable that some of the glycol ether transetherifies with the MF resin and is not emitted from the film. Amines used in "solubilizing" water-reducible coatings may volatilize to different extents, depending on conditions and amine structure [60,61]. With high solids two package coatings, the amount of volatile material can be affected by many variables, most obviously by the time interval between mixing the two packages and application. Very high solids coatings use low molecular weight oligomers, particularly when baked, some of the oligomer may volatilize. Thus, in many cases, only approximations of potential VOC emissions can be calculated, even when the formulation of a coating is known.

Experimental determination of VOC is not straightforward. The amount of VOC released depends on conditions under which the coating is used. Time, temperature, film thickness, air flow over the surface, and, in some cases, the amount of catalyst are among the variables that affect the results obtained. While it would seem that determination under the conditions of actual use would be most appropriate, this is not easy to do. For air dry coatings, the time required for the determination would be very long. For baking coatings, there can be differences in conditions for use of the same coating. It is generally agreed that it would be desirable to have a standard method for determining VOC. However, there is little agreement as to what that standard method should be; in view of the effect of application variables, it is doubtful if a single appropriate standard method will ever be developed.

Methods for determination of VOC and other useful information were collected by Brezinski in 1993 [57]. While these methods remain current as of 1998, improvements are needed. For example, a round robin study of ASTM D-2369-89 (the standard method for measuring volatile content) gave results 5–34% higher than expected, with large standard deviations for two component coatings; results with one component coatings were more reasonable [62]. An updated ASTM method, D-2369-95, uses similar procedures; its reproducibility ranges from ± 3.4 to 5.3%, with little difference between one- and two-component coatings. The standard method for determining VOC is ASTM D-3960-96; in a round robin test of solvent-borne automotive coatings its reproducibility was 2.96%.

Measuring VOC of water-borne coatings is made more complex by the need to determine water content. A round robin test of water-borne automotive coatings by ASTM D-3960-96 had a reproducibility of 9.75%, not surprising in view of the poor reproducibility of water analysis methods. A standard gas chromatography method, ASTM D-3792-91, is only reproducible to 7.5%. Three rather complicated Karl Fischer procedures (ASTM D-4017-96a) are reproducible to 4.2 to 7.4%. Research on improved water analysis methods is underway. A modified Karl Fischer method in which the water in a coating is azeotropically distilled before titration was more accurate and convenient [63].

In solvent-borne coatings, VOC is calculated by the following equation, in which NVW is the weight solids of the coating, determined under specified conditions, and ρ_1 is the density of the coating in g mL^{-1}. The factor 10 serves to adjust the VOC units to grams of solvent per liter of coating:

$$VOC = 10(100 - NVW)\rho_1$$

For units of kg L^{-1}, one divides by 100 instead of multiplying by 10. In some U.S. regulations, the units are pounds per gallon; multiplication by 8.345×10^{-3} converts these units into g L^{-1}. The VOC of water-borne coatings can be calculated using the following equation, where 0.997 is the density of water at 25°C:

$$VOC = \frac{\rho_1(100 - NVW) - \rho_1(\%H_2O)}{100 - \rho_1(\%H_2O)/0.997} \times 1000$$

An alternative to VOC analysis is to calculate VOC based on the formulation of the coating. This calculation requires knowledge of the solvent content of all coating components and assumptions about what fraction of the solvents are actually emitted and about how much additional VOC is produced by chemical reactions such as cross-linking. Even so, calculated VOC values may be more reliable than measured values in many cases, especially for water-borne coatings with very low VOC, in which the error of water analysis is magnified.

Some regulators would prefer that the units be kilograms per liter of applied coatings solids. This may be a desirable objective, but requires determination of the thickness and density of the applied cured film. ASTM Method D-2697 includes a procedure for using Archimedes' liquid volume displacement principle to determine the density of the film cured on a metal disk for 60 minutes at 110°C. The accuracy of the values is limited by the same factors discussed previously.

GENERAL REFERENCE

W. H. Ellis, *Solvents*, Federation of Societies for Coatings Technology, Blue Bell, PA, 1986.

REFERENCES

1. H. Burrell, *Off. Digest*, **27**, 726 (1955).
2. J. Hildebrand, *J. Am. Chem. Soc.*, **38**, 1452 (1916).
3. T. C. Patton, *Paint Flow and Pigment Dispersion*, 2nd. ed., Wiley-Interscience, New York, 1979, pp. 306–310.
4. K. L. Hoy, *J. Paint Technol.*, **42** (541), 76 (1970).
5. E. P. Lieberman, *Off. Digest*, **34** (444), 30 (1962).
6. W. H. Ellis, *Solvents*, Federation of Societies for Coatings Technology, Blue Bell, PA, 1986.
7. C. M. Hansen, "Solubility Parameters," in J. V. Koleske, Ed., *Paint & Coatings Testing Manual*, 14th ed., ASTM, Philadelphia, PA, 1995, pp. 383–404.
8. J. Brandrup and E. H. Immergut, *Polymer Handbook*, 3rd. ed., Wiley, New York, 1989, pp. VII/540–543.
9. K. L. Hoy, *Tables of Solubility Parameters*, Union Carbide Corp., Chemicals & Plastics, R&D Dept., Charleston, WV, 1969.
10. P. L. Huyskens and M. C. Haulait-Pirson, *J. Coat. Technol.*, **57** (724), 57 (1985).
11. R. Rawls, *Chem. Eng. News*, March 18, 20 (1985).
12. M. J. Kamlet, R. M. Doherty, J.-L. M. Abboud, M. H. Abraham, and R. W. Taft, *Chemtech*, September, 566 (1986).
13. ARCO Chemical Co., *ARCOCOMP Computer Solvent Selector Program*, Newton Square, PA, 1987.
14. A. L. Rocklin, *J. Coat. Technol.*, **48** (622), 45 (1976).
15. H. L. Jackson, *J. Coat. Technol.*, **58** (741), 87 (1986).
16. S. Skjold-Jorgenson, B. Kolbe, J. Gmehling, and P. Rasmussen, *Ind. Eng. Chem. Prod. Res. Dev.*, **18**, 714 (1979).
17. T. Yoshida, *Prog. Org. Coat.*, **1**, 72 (1972).
18. T. C. Patton, *Paint Flow and Pigment Dispersion*, 2nd. ed., Wiley-Interscience, New York, 1979, p. 340.
19. P. W. Dillon. *J. Coat. Technol.*, **49** (634), 38 (1977).
20. A. L. Rocklin, *J. Coat. Technol.*, **58** (732), 61 (1986).
21. B. C. Watson and Z. W. Wicks, Jr., *J. Coat. Technol.*, **55** (698), 59 (1983).
22. C. M. Hansen, *Ind. Eng. Chem. Prod. Res. Dev..*, **9**, 282 (1970).
23. D. J. Newman and C. J. Nunn, *Prog. Org. Coat.*, **3**, 221 (1975).
24. C. M. Hansen, *J. Oil Colour Chem. Assoc.*, **51**, 27 (1968).
25. R. A. Waggoner and F. D. Blum, *J. Coat. Technol.*, **61** (768), 51 (1989).
26. R. C. Lasky, E. J. Kramer, and C.-Y. Hui, *Polymer*, **29**, 673 (1988).
27. J. S. Vrentas, C. M. Vrentas, and N. Faridi, *Macromolecules*, **29**, 3272 (1996).
28. S. H. Wu, *J. Appl. Polym Sci..*, **22**, 2769 (1978).

29. D. R. Bauer and L. M. Briggs, *J. Coat. Technol.*, **56** (716), 87 (1984).
30. L. W. Hill and Z. W. Wicks, Jr., *Prog. Org. Coat.*, **10**, 55 (1982).
31. W. H. Ellis, *J. Coat. Technol.*, **53** (696), 63 (1983).
32. L. B. Brandenburger and L. W. Hill, *J. Coat. Technol.*, **51** (659), 57 (1979).
33. Z. W. Wicks, Jr., E. A. Anderson, and W. J. Culhane, *J. Coat. Technol.*, **54** (688), 57 (1982).
34. S. G. Croll, *J. Coat. Technol.*, **59** (751), 81 (1987).
35. D. A. Sullivan, *J. Paint Technol.*, **47** (610), 60 (1975).
36. T. C. Patton, *Paint Flow and Pigment Dispersion*, 2nd. ed., Wiley-Interscience, New York, 1979, p. 109.
37. G. F. Sprinkle, Jr., *Modern Paint & Coatings*, April, 44 (1983).
38. A. L. Rocklin and G. D. Edwards, *J. Coat. Technol.*, **48** (620), 68 (1976).
39. F. N. Jones, *J. Coat. Technol.*, **68** (852), 25 (1996).
40. J. R. Erickson and A. W. Garner, *Org. Coat. Plast. Chem. Prepr.*, **37** (1), 447 (1977).
41. N. Hazel, I. Biggin, I. Kersey, and C. Brooks, *Proc. Waterborne High-Solids Powder Coat. Symp.*, New Orleans, 1997, p. 237.
42. W. H. Ellis, *J. Coat. Technol.*, **48** (614), 45 (1976).
43. D. T. Wu, S. Lonsinger, and J. A. Klein, *FATIPEC Congress Book*, Vol. IV, 1988, p. 227.
44. J. L. McGovern, *J. Coat. Technol.*, **64** (810), 33, 39 (1992).
45. F. J. Hahn, *J. Paint Technol.*, **45** (562), 58 (1971).
46. W. S. Overdiep, *Prog. Org. Coat.*, **14**, 159 (1986).
47. R. L. Smith, L. J. Culver, and S. L. Hillman, *J. Coat. Technol.*, **59** (747), 21 (1987).
48. J. J. Brezinski, in J. V. Koleske, Ed., *Paint and Coating Testing Manual*, ASTM, Philadelphia, 1995, pp. 3–14.
49. Anon. (U.S. EPA), *Hazardous Air Pollutants Strategic Implementation Plan*, http://www.epa.gov/ttn/uatw/happlan.html (1997).
50. Anon. (U.S. EPA), *Final Air Toxics Regulation for Wood Furniture Manufacturing Operations*, http://www.epa.gov/ttn/utaw/fswood.html (1995).
51. Anon (U.S. EPA), *National Air Pollutant Emission Trends, 1900–1995*. EPA-454/R-96-007, Research Triangle Park, NC.
52. Anon. (U.S. EPA), *Final Revisions to the Ozone and Particulate Matter Air Quality Standards*, http://www.epa.gov/oar/oaqps/ozpmbro/current.htm and http://www.epa.gov/ai . . . /greenbk/ozone1hr/dec97/o3std.html.
53. I. D. Dobson, *Prog. Org. Coat.*, **24**, 55 (1994).
54. J. H. Seinfeld, *Science*, **243**, 745 (1989).
55. R. Atkinson, *Atmospheric Environment*, **24A**, 1 (1990).
56. Anon. (National Research Council), *Rethinking the Ozone Problem in Urban and Regional Air Pollution*, National Academy Press, Washington, 1991.
57. J. J. Brezinski, Ed., *Manual on Determination of Volatile Compounds in Paints, Inks, and Related Coating Products*, ASTM Manual Series: **MNL 4**, 2nd. ed. ASTM, Philadelphia, PA, 1993.
58. K. A. Nielsen, J. N. Argyropoulos, D. C. Busby, and J. J. Lear, *Polym Mater. Sci. Eng.* **70**, 170 (1990).
59. E. A. Praschan, *ASTM Standardization News*, October, 24 (1995).
60. Z. W. Wicks, Jr. and G. F. Chen, *J. Coat. Technol.*, **50** (638), 39 (1978).
61. L. W. Hill, P. E. Ferrell, and J. J. Gummeson, in T. Provder, M. A. Winnik, and M. W. Urban, Eds., *Film Formation in Waterborne Coatings*, American Chemical Society, Washington, DC, 1996, p. 235.
62. B. Ancona, R. Alexander, J. Aviles, J. Bove, R. Granata, P. Kuzma, P. Peterson, C. Pohan, D. Schober, and N. Shearer, *J. Coat. Technol.*, **65** (820), 45 (1993).
63. V. C. Jenkins, J. C. Reilly, B. Sypowicz, and M. T. Wills, *J. Coat. Technol.*, **67** (841), 53 (1995).

CHAPTER 18 _____

Color and Appearance _____

Color and the interrelated topic of gloss are important to the decorative aspects of the use of coatings and, sometimes, to the functional aspects of their use. We have all dealt with color since we were babies, but most people have little understanding of color. Many technical people think of it as an aspect of physics dealing with the distribution of visible light. While that is a factor, color is a psychophysical phenomenon. The difficulty of understanding color can be seen by considering the most rigorous definition of color that has been prepared: Color is that characteristic of light by which an observer may distinguish between two structure-free fields of view of the same size and shape. In effect, it says that color is what is left to distinguish between two objects when all the other variables are removed. Not a very satisfying definition.

Color has three components: an observer, a light source, and an object. (The single exception is when the light source is the object being viewed.) There is no color on an uninhabited island. This is not just a semantic statement; color requires an observer. There is no color in the absence of light; in a completely darkened room there is no color, not because you cannot see it, but because it is not there. There must be an object; if you look out the window of a spaceship without looking at a planet or star, there is no color—there is an observer, there is light, but there is no object.

Another major factor affecting appearance is surface roughness. If a surface is very smooth, it has a high gloss; if it is rough on a scale below the ability of the eye to resolve the roughness, it has a low gloss. If, however, the roughness can be resolved visually, a film may exhibit brush marks, orange peel, texture, and so forth. Furthermore, there can be a combination of small scale and larger scale roughness, so films can, for example, have a low gloss and brushmarks or high gloss and orange peel. The eye can resolve irregularities in surface smoothness of approximately 25 μm, depending on the distance from the object. Adding to the complications, color and gloss interact, changing either changes the other.

18.1. LIGHT

Light is electromagnetic radiation to which our eyes are sensitive. The range of visible wavelengths varies somewhat between individuals, but in most cases is 390 through 770 nm. The sensitivity of the eye varies as a function of wavelength, see Figure 18.1) [1].

Figure 18.1 also shows the response of photomultiplier tubes and silicon photodiodes. The eye is relatively insensitive to the shorter and longer ends of the range. The response of our eyes depends on the distribution of wavelengths of light emitted by the light source. In the case of monochromatic light, the colors we see range from violet through blue, green, yellow, and red as we look at monochromatic light sources of increasing wavelengths. We see different colors as the ratios of wavelengths in polychromatic light sources change. If we look at a light source with nearly equal content of all wavelengths, we see white.

Sunlight is considered the standard light, but sunlight varies depending on the time of day, latitude, season, cloudiness, and so forth. When work toward understanding color began, light from overcast north sky (in the Northern Hemisphere) was accepted as standard. Based on many measurements of energy distribution, a light source designated as D_{65} was adopted as a standard related to average daylight. The mathematical description of a light source is called an *illuminant*. A graph of standard illuminant D_{65} is shown in Figure 18.2 [2]. Tables giving energy distributions as a function of wavelength for various bandwidths from 1 nm to 20 nm wide are available [3].

The energy distribution from tungsten lights is different. Another standard light source, A, is a carefully specified tungsten light operated under specified conditions. A graph of illuminant A is also shown in Figure 18.2, and tables of its energy distributions are available. Fluorescent lights are another type of light source, and many types are available. As shown in Figure 18.3, they exhibit a continuous energy distribution with peaks at a series of wavelengths. Even if the underlying continuous spectrum were the same as D_{65}, the peaks would lead to changes in colors when these lights are used as a light source in comparison with daylight.

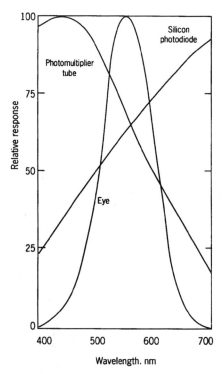

Figure 18.1. Sensitivity of the eye, photomultiplier tube, and silicon photodiode as a function of wavelength. (From Ref. [1], with permission.)

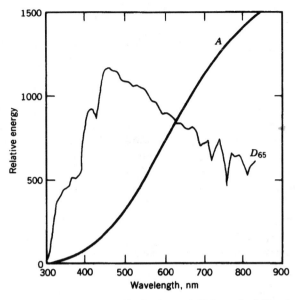

Figure 18.2. Relative spectral power distributions of CIE standard illuminants A and D_{65}. (From Ref. [2], with permission.)

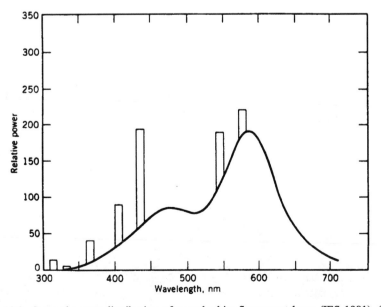

Figure 18.3. Spectral power distribution of a cool white fluorescent lamp (IES 1981). (From Ref. [1], with permission.)

18.2. LIGHT–OBJECT INTERACTIONS

An important factor that affects color is the interaction between the light and the object being viewed.

18.2.1. Surface Reflection

When a light beam is directed at a surface, some light is reflected at the surface and some passes into the object. As shown in Figure 18.4, if the surface is optically smooth, light is reflected at the same angle (r) as the angle of incidence (i). This kind of reflectance is called *specular* (mirror-like) *reflectance*. By convention, an angle of incidence normal to the surface is designated as $0°$ and the grazing angle is $90°$. The fraction of light reflected (R) varies with angle of incidence and difference in refractive index (n) between the two phases. If there is no difference in index of refraction, no light is reflected at the interface; as the difference increases, the fraction of light reflected increases. For angles of incidence near $0°$, the fraction reflected can be calculated using the following equations given in both the general form, Eq. 18.1a, and where the first medium is air ($n_1 = 1$) Eq. 18.1b:

$$R = \left(\frac{n_2 - n_1}{n_2 + n_1}\right)^2 \tag{18.1a}$$

$$R = \left(\frac{n - 1}{n + 1}\right)^2 \tag{18.1b}$$

Most resins have refractive indexes of about 1.5; Eq. 18.1b shows that approximately 4% of incident light is reflected when the angle of incidence is near $0°$. The dependence of

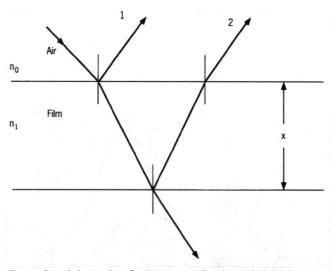

Figure 18.4. External and internal reflection and refraction of light by a nonabsorbing film (refractive index, n_1, thickness, x) with optically smooth parallel surfaces. (From Ref. [4], with permission.)

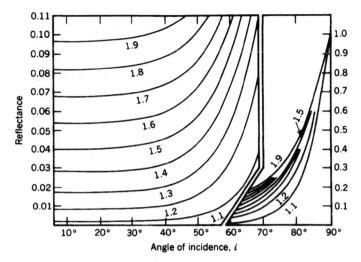

Figure 18.5. Fraction of light reflected from a smooth surface as a function of angle of incidence i with various differences in refractive index. (From Ref. [5], with permission.)

reflection on angle of incidence is illustrated in Figure 18.5 [5]; reflectance approaches 100% as the angle of incidence approaches 90°. The family of curves in Fig. 18.5 illustrates the combined effects of refractive index difference and angle of incidence.

Light not reflected at the surface enters the object. When a light beam enters an object it is refracted; that is, it is bent. The angle of refraction varies with the ratio of the refractive indexes of the two media as shown by Eq. 18.2.

$$\sin r = \frac{n_1}{n_2} \sin i \tag{18.2}$$

Figure 18.4 shows that when light passes from air into a nonabsorbing plastic film with parallel, optically smooth surfaces and an index of refraction of 1.5, the angle of refraction is smaller than the angle of incidence. If the angle of incidence is near 0°, the fraction reflected at the first surface is 0.04, and 0.96 is transmitted into the film. If there is no absorption, 0.96 reaches the second film/air interface. There, 0.96×0.96 of the original light is transmitted into the air on the other side of the film, being refracted so that the angle of refraction is equal to the original angle of incidence. There is also reflection at the second surface, and 0.04×0.96 is reflected back towards the first surface. Again, there is reflection of 4% and transmission of 96% at the back of the first surface. As a result, $(0.96)^2 \times (0.04)$ of the original light is transmitted out of the film at an angle equal to the original angle of incidence, and $(0.96) \times (0.04)^2$ is reflected back into the film. This bouncing back and forth continues, if there is no absorption, until the total transmission equals 0.92 and total reflectance equals 0.08.

When a light beam passes from a medium of higher refractive index to one of lower refractive index, the angle of the beam increases. If the angle of incidence is high enough, all light is reflected back and none is transmitted out. If the refractive indexes are 1.5 and 1, the critical angle (at which all the light is reflected) is calculated to be 41.8°. Total reflection is the basis for optical fibers. If light enters the end of a nonabsorbing fiber and in which are no bends such that the angle of incidence with the inner surface of the fiber is less than its critical angle, all of the light will follow the fiber, even all the way across an ocean.

If the thickness of a film is small (60–250 nm), interference colors will be seen. The intensity of some wavelengths of reflected light are enhanced, and others are transmitted selectively. With a refractive index of 1.5 in air, at a film thickness of 66 nm, bluish white color is reflected and yellowish white is transmitted; as thickness is increased, other colors are preferentially reflected. The colors seen are affected by the angle of incidence of the light.

18.2.2. Absorption Effects

In almost all coatings, the color we observe is affected by differential absorption of various wavelengths of light. Colorants, dyes, pigments, and, to a degree, some resins, absorb some wavelengths of light more strongly than others. These absorptions are controlled by the chemical structures of the colorants. First, we consider the absorption effects in transparent systems, that is, systems in which the colorant is in solution or is so finely divided that it does not significantly reflect light at the colorant-resin interface. The extent of absorption depends on chemical composition, wavelength, particle size, optical path length (film thickness), concentration, and media-colorant interactions.

Each colorant has an absorption spectrum that controls the absorption of various wavelengths of light. Chemists generally speak of molar absorptivities, ε, in units of L mol^{-1} pathlength^{-1}, corres-ponding to the absorbance per molar concentration of a substance. Physicists generally speak of absorption coefficients, K, in units of pathlength^{-1} mass^{-1}, corresponding to the absorbance per unit mass of the system.

The smaller the particle size, the greater the fraction of light absorbed by the same quantity of a colorant. The highest molar absorptivities are exhibited by individual molecules in solution. In the case of pigments, the smaller the particle size of a pigment, the greater the absorption. The longer the path followed by a beam of light through a medium containing absorbers, the greater the degree of absorption. In transmission of a beam of light at $0°$, the optical path equals the film thickness. At any other angle, the optical path length is greater than the film thickness. If, in passing through a unit path length, one-half of the light of a particular wavelength is absorbed and one-half is transmitted (ignoring surface reflection), on passing through two units of path length, three-fourths is absorbed and one-fourth is transmitted. Mathematically this relationship is expressed by an exponential equation, as shown in Eqs. 18.3a and 18.3b, in which X is path length, I is intensity of light transmitted, and I_0 is the original light intensity. [Chemists commonly use base 10 (Eq. 18.3a), and physicists base e (Eq. 18.3b).]

$$I/I_0 = 10^{-\varepsilon X} \tag{18.3a}$$

$$I/I_0 = e^{-KX} \tag{18.3b}$$

Ideally, the same relationship holds when the concentration of the colorant in the medium is changed, as shown in Eq. 18.4:

$$I/I_0 = 10^{-\varepsilon CX} \tag{18.4}$$

Equation 18.4 holds only over a limited range of concentrations; the width of the range is system dependent. When absorbers are in solution, there are less likely to be intermolecular interactions between molecules in dilute solutions than in more concentrated solutions. In the case of pigment dispersions, concentration effects are likely to be larger. Another complication is possible interactions with the media. In the case of solutions, a change in solvents can sometimes lead to association of molecules effectively increasing particle size and hence reducing absorption. There can also be effects of hydrogen bonding between a soluble dye molecule and different solvents. This changes the structure, leading to a change in absorption spectrum. In dispersions of pigments, changes in the medium, such as dilution with solvent, can lead to flocculation (agglomeration) of pigment particles; this causes a larger particle size and, hence, reduced absorption.

Figure 18.6 shows transmission spectra of idealized reddish magentas. Spectra *a* and *b* result from transmission of light through the same transparent coating of path length X and $2X$, respectively. (Surface reflection effects are ignored.) The fraction of light of all wavelengths transmitted through the thicker coating, shown in spectrum *b*, is less. But also, the relative transmissions of the blue (B), green (G) and red (R) parts of the spectra are different. The color seen in case *b* would be a redder purple than in case *a*. The composition is the same but the shade of the color is affected by path length. The same type of change takes place if the concentration is doubled at the same film thickness. This is an idealization of a real situation. One of the inks in four color process printing that permits printing of multicolored pictures is a reddish magenta. The film thickness of magenta ink affects the color of the magenta print and, hence, the combined color picture.

18.2.3. Scattering

Scattering is another phenomenon that can occur during the passage of light through a film. If there are small particles dispersed in the film that have a different refractive index than the medium, light is reflected and diffracted at the interfaces between the particles and the medium. The physics involved in scattering is complex and beyond the scope of this brief presentation, but, the results can be stated in a simplified manner. When a beam of light passes through a film containing nonabsorbing particles, the light is internally reflected in all directions, so it changes from a beam to diffuse illumination inside the film. Light that reaches the back of the top surface at angles greater than the critical angle is reflected back into the film; part of the light reaching this surface at angles less than the critical angle leaves the film. If the film is thick enough, no light can pass completely

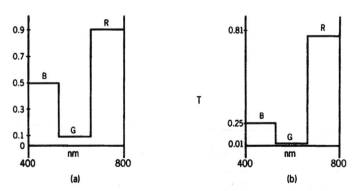

Figure 18.6. Transmission spectra of idealized magentas: (*a*) path length $= x$; (*b*) path length $= 2x$.

through it; all the light is reflected back out of the top surface. The reflection, however, is not just at the specular angle, but at all angles, that is, this light is diffusely reflected even though the top surface of the film is optically smooth. The extent of light scattering by nonabsorbing particles depends on: refractive index difference between particles and medium, particle size, film thickness, and particle concentration.

The greater the difference in refractive index, the greater the degree of light scattering. As shown in Figure 18.7, the degree of scattering increases steeply as the difference in refractive index increases. The degree of scattering is the same whether the particles have a higher or lower refractive index than the media. For example, water droplets in air (fog) scatter light as efficiently as air droplets of the same size at the same concentration in water (foam). An ideal white pigment would absorb no light and have a very high refractive index, so the difference in refractive index from that of binders would be large. Rutile TiO_2 comes close to meeting these specifications; it has an average refractive index of 2.73, but it absorbs some light below about 420 nm. Another crystal type of TiO_2, anatase, absorbs less light, but its refractive index is lower, 2.55. The smaller index of refraction difference leads to less efficient scattering.

Scattering is affected by particle size. Figure 18.8 shows the relationship between scattering coefficient and particle size for 560 nm light for rutile TiO_2 in a resin with a refractive index of about 1.5. As particle size decreases, scattering coefficient increases until a maximum is reached at 0.19 µm diameter, when the particle size decreases further, scattering coefficient drops off steeply [6]. Commercial TiO_2 has a range of particle sizes. Since efficiency drops off more rapidly on the small diameter size side of the maximum, TiO_2 pigments are produced with an average particle diameter somewhat over 0.2 µm. See Ref. [7] for calculations showing the effect of particle size distribution and particle agglomeration on scattering. The particle diameter with maximum scattering depends on refractive index difference. For example, as also shown in Figure 18.8, the scattering coefficient of calcium carbonate ($n = 1.57$) goes through a maximum at about 1.7 µm (note the use of different scales). As expected from the small difference in refractive index scattering by even optimum particle size is low.

The extent of light scattering is affected by film thickness. If no absorption occurs, the light that is not reflected back out of the top surface is transmitted, unless the film thickness is great enough so that all the light is reflected back. Scattering is also affected by

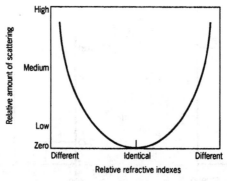

Figure 18.7. Scattering as a function of refractive index difference; particles have higher refractive indexes than the media on the right hand part of the curve and lower values on the left. (From Ref. [1], with permission.)

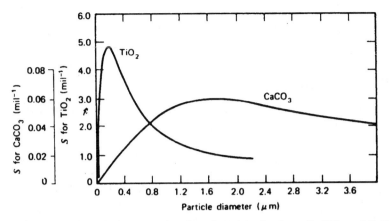

Figure 18.8. Scattering coefficients as a function of particle size for rutile TiO_2 and $CaCO_3$. (From Ref. [6], with permission.)

concentration. At low concentrations of particles such as rutile TiO_2, transmission through a film (neglecting surface reflection) follows a relationship, shown in Eq. 18.5, similar to that for absorption:

$$I/I_0 = e^{-ScX} \tag{18.5}$$

However, as concentration increases, scattering efficiency decreases, in the case of rutile TiO_2 pigmented coatings, the efficiency decreases so much that the fraction of light transmitted increases and that reflected decreases. The result can be seen in a plot of scattering coefficient S as a function of pigment volume concentration (PVC) of rutile TiO_2 in dry films, shown in Figure 18.9 [6]. PVC is defined as the volume percent pigment in a **dry** film of coating. In commercial practice, the cost effectiveness of pigmenting with TiO_2 drops off sufficiently at PVCs above about 18% that it is not generally economically sound to use a higher PVC of TiO_2. The optimum value varies somewhat from system to

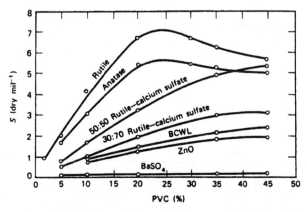

Figure 18.9. Experimental plots of scattering coefficients versus PVC for selected pigments in a dry acrylic lacquer. BCWL is basic carbonate of white lead. (From Ref. [6], with permission.)

system; it depends on the stability of the dispersion and on the TiO_2 content of the particular TiO_2 pigment, which may vary from grade to grade by more than 10%. (See Section 19.1.1.)

18.2.4. Multiple Interaction Effects

The effects of surface reflection, absorption, and scattering are interdependent. Light sources generally are not narrow beams, but relatively broad beams or diffuse sources. Surfaces are not often optically smooth; in some cases, we strive for surface roughness. In coatings, we seldom deal with pigments or pigment combinations that only absorb or only scatter light; commonly, both occur simultaneously. We are seldom interested in free films, but rather films on a substrate so that the reflectance at the bottom surface of a film is not the same as that of the top surface of the film. The eye does not distinguish between the light reaching it that has been reflected from the top surface, from within the film, or from the bottom surface of a film. The combined light from all three sources is integrated when the eye reacts to it.

As shown in Figure 18.5, surface reflectance increases as the angle of incidence increases. If a panel of a high gloss (smooth surface) blue paint illuminated with a beam of light is observed at any angle other than the specular angle, one sees a comparatively dark blue color, resulting from the diffuse reflectance back out of the film. On the other hand, if the angle of incidence is small and one observes at the specular angle, the color is a lighter blue because one sees a mixture of that same light plus some "white" light reflected from the surface. If the angle of incidence approaches 90°, one sees a very light blue color at the specular angle because the observed light contains a high fraction of surface reflected light and only a little light from within the film. If the illumination is diffuse instead of a beam, the color is a lighter blue from most angles of observation because one sees a higher fraction of surface reflected light. If the blue paint is low gloss (rough surface) with exactly the same composition as in the previous example, a lighter blue color is seen under diffuse illumination at most viewing angles, as compared with the high gloss paint. This results from a greater ratio of surface reflected light reaching the eye. The color is darkened at most viewing angles by wetting the surface with water, since the water fills in the surface roughness.

The effects of concentration and film thickness on coatings that absorb or weakly scatter light without rescattering the primary scattered light can be modeled by simple equations. However, interactions resulting from both absorption and multiple scattering together are more complex, and equations that model reflection and transmission are more complex. Figure 18.10 gives a set of Kubelka-Munk equations that model reflectance, R_1 and transmittance T_1 of a translucent or opaque film or slab of material of thickness X containing light absorbers and scatterers. The observer is viewing the sample in air, while the absorbers and scatterers are embedded in a medium such as plastic or resin.

The Kubelka-Munk reflectance R and transmittance T of the film in the absence of a refractive index boundary include the effects of absorption K, scattering S, and film thickness X. The boundary reflections from both sides of the top surface r_1, and the inner and bottom sides of the film, r, as well as the corresponding boundary transmittances t_1 and t influence the reflectance R_1 and the transmittance T_1 seen by the observer viewing the sample in an air medium. Two boundary reflectances r_1 and r are required because light normally incident on the film from air has a reflectance value of 0.04, while diffuse light incident on the boundary from inside the resin has a reflectance value of 0.596 for a

$$R_1 = r_1 + \frac{t_1 t[R(1 - rR) + rT^2]}{(1 - r_1 R)^2 - r^2 T^2}$$

$$T_1 = \frac{t_1 tT}{(1 - r_1 R)^2 - r^2 T^2}$$

$$t_1 = (1 - r_1) \qquad t = (1 - r)$$

$$R = \frac{\sinh bSX}{a \sinh bSX + b \cosh bSX}$$

$$T = \frac{b}{a \sinh bSX + b \cosh bSX}$$

$$a = 1 + \frac{K}{S}; \qquad b = (a^2 + 1)^{1/2}$$

For completely opaque slabs or films, simplified equations can be used.

$$\frac{K}{S} = \frac{(1 - R_\infty)^2}{2R_\infty}; \qquad R_\infty \text{ is the reflectance at complete hiding}$$

Surface reflection must be corrected for using Saunderson's Equation:

$$R_1 = r_1 + \frac{t_1 tR_\infty}{(1 - rR_\infty)}$$

Assumptions:

1. The equations apply to a single wavelength at a time.
2. K and S are the same throughout the film.
3. Pigment particles are randomly oriented.
4. The internal radiation flux is perfectly diffuse.
5. Edge effects are ignored.

Figure 18.10. Kubelka-Munk equations and assumptions.

material of refractive index about 1.5. The differences between R_1 and R, as well as those between T_1 and T, in the case of typical plastics and resins are very significant. Other Kubelka-Munk equations are available that permit calculations when the reflectances at the inner side of the and bottom sides of the film are different [8]. As discussed in Section 18.9.2, simplifying assumptions permit use of these equations in calculations for color matching; an example in which the films have complete hiding (see Section 18.3) is also shown in Figure 18.10.

18.3. HIDING

Color is affected by reflectance back through the film of light reaching the substrate to which the coating is applied. It is common to compare the colors of a coating applied over a striped black and white substrate. If one can see the pattern of the stripes through the coating, the coating is said to exhibit poor *hiding*. The difference results from the reflection of light reaching the white stripes compared with absorption of light reaching the black stripes. The effect is as if some black pigment was put in the coating above the black stripes. If all light entering the film is absorbed or scattered back out of the film prior to reaching the substrate, there is no effect of substrate on color; hiding is complete.

Hiding is a complex phenomenon and is affected by many factors. Hiding increases as film thickness increases. Low hiding coatings require thicker films; that is, coverage decreases; the area covered (hidden) by a liter of coating is less and the cost is higher. Hiding increases as the efficiency of light scattering increases, that is, hiding is affected by the refractive index differences, particle sizes, and concentrations of scattering pigments present. Hiding increases as absorption increases. Black pigments, which have high absorption coefficients for all wavelengths, are particularly effective. Colorants, also increase hiding but not as much as black. Surface roughness increases hiding; a larger part of the light is reflected at the top surface reducing the differences of reflection resulting from differences in the substrate to which a coating is applied.

An important factor affecting hiding, sometimes forgotten in testing coatings, is uniformity of film thickness. Application of coatings commonly results in nonuniform thickness. Coatings are generally formulated to level, that is, to flow after application to make the film thickness more uniform. (See Section 23.2.) Leveling is often incomplete, however, and there can be effects on hiding. Consider a poor leveling coating where an average film thickness of 50 μm of dry coating has been applied, but there are brushmarks remaining so adjacent to each other are lines with film thicknesses of 65 μm and 35 μm. If the hiding at 50 μm is just adequate, there will not be good hiding at 35 μm. The difference in color is emphasized by the thin layers of coating being right next to the thick layers. The contrast results in rating hiding by the uneven film as poorer than that of a uniform 35 μm film of the same coating. Further complicating hiding is variation of substrates over which the coating is applied. Hiding of a white coating applied over a white surface might be rated as excellent, but hiding of the same coating over a black surface might be rated poor.

There are quality control tests that compare hiding of batches of the same or similar coatings, but no test is available that can provide an absolute measure of hiding [9]. The only way to establish covering power of a coating is to apply it to an appropriate large surface with a film thickness that gives adequate hiding and then calculate the actual coverage in $m^2\ L^{-1}$.

In some cases, hiding is not desired, an example is a coating for metallized plastic Christmas tree ornaments. One wants transparent red, green, blue, etc. coatings. Preparation of a transparent coating requires that there is no light scattering within the film; therefore, the particle size of colorant particles must be very small.

18.4. METALLIC AND INTERFERENCE COLORS

Metallic coatings are widely used on automobiles. They are made with transparent colorants together with *nonleafing* aluminum pigment (see Section 19.2.5). These metallic coatings exhibit unusual shifts in color as a function of viewing angle. As noted in Section 18.2.4, regular high gloss paints exhibit dark colors when a panel is looked at from relatively small viewing (normal) angles and light when a panel is observed from large angles of view. Metallic coatings are lighter in color when viewed near the normal angle (the *face color*) and darker when viewed from a larger angle (the *flop color*). It is desirable to achieve a high degree of *color flop*, that is, a large difference between face and flop colors. To do so requires the surface be smooth (high gloss), that there be no light scattering from the resin or color pigment dispersion, and that the aluminum flake particles be aligned parallel to the surface of the film.

As can be seen in the idealized diagram in Figure 18.11, when an observer looks at a metallic film from an angle near the normal, the path length of light through the film is short because it is reflected back by the aluminum. Therefore, there is little chance for absorption by pigment, and the color is light. On the other hand, from a greater viewing angle, the observer sees light that has been reflected back and forth within the film so that the path length is longer and the color is darker. The film in which the aluminum flakes are oriented must be transparent. If there is light scattering within the film, the optical path length of the light viewed at large angles will be reduced and flop will be less. See Section 29.1.2 for discussion of possible mechanisms of orientation of aluminum during film formation.

Pigments that produce colors by interference are also used in automotive coatings. The first such pigments used were *pearlescent* pigments, which exhibit a pearly luster with iridescent colors. Pearlescent pigments are mica flakes on which thin films of titanium dioxide or iron oxide have been deposited, serving to give interference reflection of light striking the pigment surface. At some spots on the flake surface, some wavelengths of light are strongly reflected and others are transmitted. At other spots, where the film thickness of the oxide coating on the mica is different, different wavelengths are reflected and transmitted. The result is a mother-of-pearl effect. The hues of the coatings vary with angles of illumination and viewing. They can be used alone or in combination with other colored pigments and/or aluminum pigments to give unusual styling effects. See Ref. [4] for discussion of pearlescence.

Recently, another type of interference color pigment has become available and is beginning to be used in automobile coatings [10]. The pigments are flakes with thicknesses on the order of 1 μm of composite materials, with a center layer of opaque metal sandwiched between two clear layers; then on both surfaces, there are layers of metal so thin that they are semitransparent. The thin layers have uniform film thicknesses and cause selective reflection and absorption of various wavelengths of light. Depending on the thicknesses, various color pigments can be made. Color is also affected by the angle of

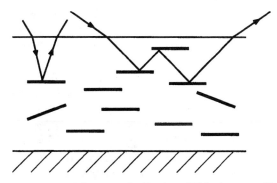

Figure 18.11. Idealized diagram of reflection of light in a metallic coating.

illumination and viewing, since the path length of light through the layers depends on the angles of illumination and viewing. Furthermore they can be used in formulations with conventional transparent pigment dispersions. Orientation parallel to the surface is required, using the same techniques as used in metallic colors. The multicolor effects are being introduced to the automotive industry and other applications that have used metallic coatings.

18.5. THE OBSERVER

The third critical aspect of color is the observer, both the eye and the brain are involved. The retina of the eye has two types of photodetectors, rods and cones. These detectors are stimulated by light photons to send signals through the optic nerves to the brain. Rods are sensitive at low levels of illumination to all wavelengths of light and become insensitive as the level of illumination increases. Cones are insensitive at low levels of illumination but are sensitive at higher levels. The response of cones is wavelength dependent. Cones have three ranges of sensitivity that overlap, one peaks in the blue region of the spectrum, another in the green, and the third in the red region. These overlapping ranges lead to complex signals to the brain, which integrates the signals so that we can see thousands of different colors. At low levels of illumination, we see only with rods, and vision is limited to shades of gray. At intermediate levels of illumination, there are responses from both rods and cones; hence, one does not see bright colors, only grayish colors. At higher levels of illumination, the rods are inactive and we see a full range of colors from the responses of cones.

The mechanism of color vision is complex and only partly understood and is beyond the scope of this introductory text. There are variations from person to person in the degree of response to the three ranges of sensitivity. As a result, colors seen by different observers are not identical. Usually, these differences are small, but in some cases they are large. In extreme cases, some people are *color blind*. There are different types of color blindness; the most common is red-green color blindness.

To specify color and predict responses to mixtures of colorants, a mathematical model of a *standard human observer* was established by an international committee of experts known by the acronym of its French name as CIE. Figure 18.12 shows a graph of three functions that model color vision. At any given wavelength, the standard observer has the same response to the ratio of the values of $\bar{x}$, $\bar{y}$, and $\bar{z}$ at that wavelength as to monochromatic light of that wavelength. Tables of these *CIE Color Matching Functions* are available as a function of wavelength with bandwidths of 1, 10, and 20 nm [3].

Within the range of higher illumination, the eye adapts to changes in levels of illumination. For example, if an area of white is surrounded by black, the white looks whiter than if the black were not present. The eye adapts to the level of light reflected by the combined black and white and responds more to the white in the presence of the black. Similar effects occur when two strongly colored fields of view are adjacent to each other. Yellow surrounded by blue-green looks more orange than if the blue-green were absent. In general, if one looks at color chips when selecting a coating, the color is different than if that coating is applied to a large surface because of the effect of the surrounding area on the color. Many other effects result from such interactions between the eyes and combinations of colors on a surface. (See Ref. [5] for further discussion.)

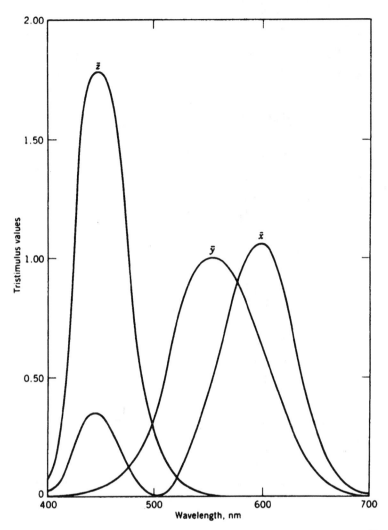

Figure 18.12. CIE color matching functions, $\bar{x}$, $\bar{y}$, and $\bar{z}$, for equal energy spectra. (From Ref. [1], with permission.)

18.6. INTERACTIONS OF LIGHT SOURCE, OBJECT, AND OBSERVER

Color depends on the interaction of three factors: light source, object, and observer. If any factor changes, the color changes. If we observe an object under a light source with the energy distribution of illuminant A and shift to a different illuminant, for example, illuminant C (similar to D_{65}), the color changes. Light source A has relatively lower emission in the blue end of the spectrum and relatively higher emission in the red end. The light reaching the eye when the object is illuminated with source A has more red light and less blue light than the light reflected from the same object illuminated with source C. The color is different. The situation is illustrated in spectral sets a and b in Figure 18.13, which shows the different responses as product spectra of light source x object x observer for an

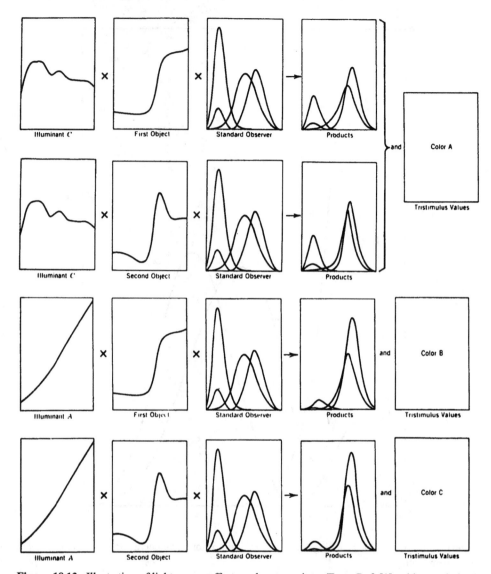

Figure 18.13. Illustration of light source effects and metamerism. (From Ref. [1], with permission.)

object with illuminants *A* and *C*. The only way to be sure that a pair of coatings will match under all illuminations is to use colorants with identical chemical composition and physical states.

If the chemical composition and physical state of the colorants in two coatings are the same, their reflectance spectra are identical, and the coatings match under any light source. It is possible for two materials with different colorant compositions and, hence, different reflectance spectra to result in the same color under a certain light source. However, such a pair will not match under light sources with different energy distributions. This phenomenon is called *metamerism*. In a spectral match, the two panels change color with a new light source, but it is the same change in both cases. In a metameric pair, the color is the same with one light source; the colors of both panels also change when the light source is changed, but the extent of change is different between the two panels. This situation is illustrated in Figure 18.13, which shows responses of a second object with illuminants C (spectral set c) and A (spectral set d). The first and second objects (a metameric pair) are identical under illuminant C, but their colors are different under illuminant A.

18.7. COLOR SYSTEMS

The human eye can discriminate thousands of colors. However, it is difficult for a person to tell another person what colors he/she sees. A color description such as "light grayish blue-green" illustrates this difficulty. Many systems have been developed to permit definitive identification of colors. There are two types of systems: one that uses color samples in specific arrangements and one that identifies colors mathematically. All color systems have at least three dimensions to include all the possible colors.

The most widely used visual color system in the United States is the *Munsell Color System* with carefully prepared and selected color chips are classified in a three-dimensional system. The dimensions of the Munsell System are called *hue*, *value*, and *chroma*. Hue refers to the dimension of color described as blue, blue-green, green, greenish-yellow, yellow, yellowish-red, red, reddish-purple, purple, purplish-blue, and back to blue. The value dimension characterizes the lightness of a color as compared to a series of gray samples, a *gray scale*. The value 0 is assigned to pure black and the value 10 to pure white. A light blue has a high value, while a dark blue of the same hue has a low value. The chroma dimension refers to the difference between a color and a gray of the same value and hue. A bright red has a high chroma, whereas a grayish red of the same hue and value has a low chroma.

The color chips in the Munsell System are prepared so that there are equal visual differences between all pairs of adjacent chips. The chips are systematically labeled; for example, a chip labeled G5/6 is a green with a value of 5 and a chroma of 6. One can look at a set of Munsell chips and know the color that someone meant when they made such a designation. There are two limitations to this statement. The light source must be specified. Chip G5/6 gives a different color under source A than under source D_{65}. Secondly, surface roughness affects color, so comparisons have to be made at equal gloss levels. Two sets of Munsell chips are available: one with high gloss and the other with low gloss. Significant errors can result in comparing either of these with other materials that are semigloss. In Europe, a different color chip system, the Natural Color System, is most widely used [11].

The mathematical color system is the *CIE Color System*. It is based on mathematical descriptions of light sources, objects, and a standard observer. Light sources are specified by their relative energy distributions, objects are specified by their reflectance (or transmission) spectra, and the observer is specified by the CIE standard human observer tables. For color analysis, the light reflected (or transmitted) from (or through) an object is measured with a spectrophotometer. Since in most cases, the reflection is diffuse, it is essential to use a spectrophotometer with an integrating sphere so that all of the light being reflected is sampled, not just that at some narrow angle. For the most accurate data, reflectance measurements are taken at each wavelength and the values over the range of 380 to 770 nm would be used in the summation. For most purposes, the accuracy is sufficient using 16 measurements at 20 nm intervals from 400 through 700 nm.

To identify the color resulting from the interaction of a light source, an object, and a standard observer, one uses the data for these three dimensions to calculate the *tristimulus values X, Y,* and *Z* using Eqs.18.6*a, b,* and *c.*

$$X = \sum_{380}^{770} \bar{x}_\lambda E_\lambda R_\lambda \qquad\qquad (18.6a)$$

$$Y = \sum_{380}^{770} \bar{y}_\lambda E_\lambda R_\lambda \qquad\qquad (18.6b)$$

$$Z = \sum_{380}^{770} \bar{z}_\lambda E_\lambda R_\lambda \qquad\qquad (18.6c)$$

Tristimulus values are different for the same object and the same observer when the energy distribution E of a different illuminant is used. This is as it should be, since we know that colors change as the light source changes. The tristimulus values uniquely and unequivocally define colors. For example, $X = 14.13$, $Y = 14.20$, and $Z = 51.11$ is a definitive description of a color, but what color? Unfortunately, even experts often cannot say by looking at the numbers. This set of tristimulus values is for a blue but few people could look at them and tell you it is a blue, much less whether it is a grayish-blue and approaches being a purplish-blue, which it happens to be.

The X and Y tristimulus values can be converted to *chromaticity values x* and *y* by normalization as shown in Eq. 18.7.

$$x = \frac{X}{X + Y + Z} \qquad y = \frac{Y}{X + Y + Z} \qquad\qquad (18.7)$$

If one has a metameric pair of coated panels, the tristimulus values and the chromaticity values are the same with the light source under which the panels match. They are not the same, however, if calculated with the energy distributions from another light source. When the light source changes, the X, Y, and Z and the x and y values of both panels change, but the changes are to different degrees.

The chromaticity values of each wavelength of the spectrum can be calculated and plotted against each other to form the *CIE spectrum locus*; see Figure 18.14. The ends of the locus are connected by a straight line called the *purple line*. There are no purples in the spectrum; in CIE color space, hues of purple lie along this line. As shown, the plot can be divided into color areas, so one can look at the x and y values and have a reasonable idea of the shade of a color. The third dimension rises vertically from the plot; it is the Y tristimulus axis, *luminance*; $Y = 100$ (or 1, in some conventions) at the x, y values of the light source. The Y at the spectrum locus line approaches 0. At x, y points between the spectrum locus and the point of the source, Y is always less than 100. As the value of Y gets larger, the gamut of possible colors narrows, as can be seen in Figure 18.15.

If one draws a line from the point of the source through the point of a sample extended out to the spectrum locus, the wavelength at the intercept is called the *dominant wavelength* of the color. This dimension corresponds to the hue dimension in the Munsell System, but the scale is different. If this extrapolation intercepts the purple line, then the line is extrapolated in the opposite direction, and the intercept with the spectrum locus is called the *complementary dominant wavelength*. If one divides the distance from the

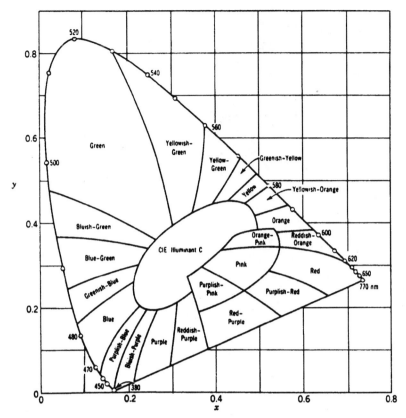

Figure 18.14. CIE chromaticity diagram showing the location of various hues. (From Ref. [1], with permission.)

source point to the sample point by the total distance from the source point to the spectrum locus (some workers express it as a percentage), one obtains the *purity*. Purity is the same dimension as chroma in the Munsell System, but the scale is different. The vertical Y dimension is a gray scale, as in the Munsell value dimension, but again the scale is different.

Figure 18.15 shows a topographical diagram of three-dimensional CIE Color Space with illuminant C. All real colors that can be seen with light source C fall inside this color space. In the Munsell system, the boundaries are limited not by reality of color, but by the color purity of the available pigments to make the reference chips. As shown by in Figures 18.14 and 18.15, CIE color space is not visually uniform. For example, a small difference in x and y in the blue part of the color space represents a substantial difference in color, whereas in the green part of color space, the same difference in x and y values represents little difference in color. In this sense, Munsell color space is preferable, since the differences are visually uniform. However, color calculations of the types needed for instrumental color matching are feasible with CIE color space, but not with Munsell space. If the comparisons are to be visual, one commonly uses the Munsell system. If the comparisons are to be mathematical and include all possible color space, one uses the CIE system.

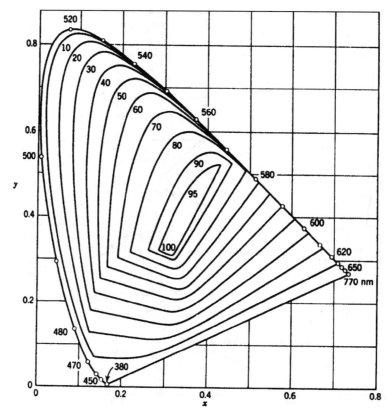

Figure 18.15. Topographical diagram of three-dimensional CIE color space with illuminant C. (From Ref. [1], with permission.)

There have been many attempts to mathematically transform CIE color space to a visually equal color space; then differences could be used as meaningful measures for specification purposes. Some progress has been made, and color differences (ΔE) can be calculated using the CIE 1976 $L^*a^*b^*$ equations. L^* represents lightness, a^* redness-greenness, and b^* yellowness-blueness. These quantities are calculated from tristimulus values by the following equations:

$$L^* = 116(Y/Y_n) - 16$$

$$a^* = 500[f(X/X_n) - f(Y/Y_n)]$$

$$b^* = 200[f(Y/Y_n) - f(Z/Z_n)]$$

In these equations, X, Y, and Z are the tristimulus values of the sample; X_n, Y_n, and Z_n are the tristimulus values of the light source; $f(Y/Y_n) - (Y/Y_n)^{1/3}$ for values of $(Y/Y_n > 0.008856$; $f(Y/Y/_n) = 7.787(Y/Y_n) + 16/116$ for values of (Y/Y_n) equal to or less than 0.008856; and the functions $f(X/X_n)$ and $f(Z/Z_n)$ are similarly defined.

The equation for CIELAB color difference is:

$$\Delta E(L^*a^*b^*) = [(\Delta L^*)^2 + (\Delta a^*)^2 + (\Delta b^*)^2]^{1/2}$$

These equations still do not represent fully uniform color space. If specifications are written for a line of colors specifying a fixed $\pm$ range for ΔE for the whole series, the requirements will be more stringent for some colors than for others. Even if there were color difference equations available that were visually uniform, there would still be a difficulty using them for specifications. By using such a specification, the color would be permitted to vary equally in any direction from the central standard. However, it is common for people to be more concerned about deviations in one direction in color space as compared to other directions. For example, there is commonly a greater tolerance for whites to be off in the blue direction than for them to be off in the yellow direction.

18.8. COLOR MIXING

There are two types of color mixing: *additive* and *subtractive*. In additive mixing, the primary colors are red, green, and blue. Additive mixing is involved in theatrical stage use of overlapping colored spot lights, as well as in color television, where three colors of dots (red, green, and blue) are projected near each other on the screen. The lights from nearby dots are "added" when we look at them to give colors, which depend on the ratio of the three colors in nearby dots. In additive color mixing, equal amounts of blue and green light give blue-green (*cyan*); similarly, blue and red light give purple (*magenta*); green and red light give *yellow*; and equal addition of all colors gives white light. With appropriate light

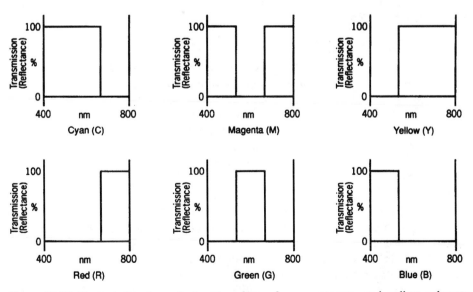

Figure 18.16. Transmission (or reflectance) spectra of cyan, magenta, and yellow colorants, together with their complementary colors.

sources, all color lights can be made. Note that we see as yellow the combination of all wavelengths in the green, yellow, orange, and red segments of the spectrum.

In almost all cases, however, color mixing encountered in coatings is not additive but subtractive mixing. We use colorants that absorb (subtract) some of the light of some wavelengths from white light. If we add a second colored pigment to a coating containing colored pigment, we subtract more; we cannot add back any intensity of wavelengths by using a further colorant. The primary colors in subtractive mixing are cyan, magenta, and yellow. If we mix equal quantities of an ideal cyan and an ideal magenta, the result is blue. That is, the cyan absorbs red and the magenta absorbs green, so blue remains. Similarly, cyan and yellow yield green; magenta and yellow yield red; and a mixture of equal amounts of all three ideal absorbing colorants absorbs all light and yields black. Idealized transmission (or reflectance) spectra of the primary colors, which illustrate their complementary nature, are provided in Figure 18.16.

18.9. COLOR MATCHING

A large fraction of pigmented coatings is color matched. The customer originally chose some color he or she liked for a refrigerator, automobile, or other product, and a coating formulator was given a sample of a material with this color and was asked to match the color using a coating formulation appropriate for the end use. After the formulator establishes the appropriate ratio of the appropriate pigments and after the customer approves the color match, then the factory must color match batch after batch of the coating to meet this standard.

18.9.1. Information Requirements

Before starting the initial laboratory color match, the color matcher needs a substantial amount of information:

1. *Metamerism.* Is a spectral (nonmetameric) match possible? That is, can exactly the same colored pigments (including white and black) be used to establish the match as were used to make the customer's sample? If not, the customer must realize that any match will be metameric; that is, the colors will match under some light source, but not under others. For example, if the sample is a dyed fabric, the color of the textile cannot be exactly matched under all lights by any coating, since the colorants cannot be identical. If, as another example, the customer has been using a coating made with one or more pigments containing lead compounds and wishes to have a lead-free coating, only a metameric match is possible.

2. *Light sources.* If the match is to be metameric, the customer and the supplier must agree on the light source(s) under which the color is to be evaluated. If there is to be more than one light source, a decision should be made as to whether it is more desirable to have a close match under one light source without regard to how far off the match might be under other light sources, or to have a fair color match under several light sources, but not a close match under any individual light source.

3. *Gloss and texture.* The color of a coating depends on its gloss and texture. Some of the light reaching the eye of an observer is reflected from the surface of the film and some from within the film. The color seen by the observer depends on the ratio of the two types

of reflected light. At most angles of viewing, more light is reflected from the surface of a low gloss coating than from the surface of a high gloss coating. It is impossible to match the colors of a low gloss and high gloss coating at all angles of viewing. There must be agreement as to the gloss of the coating, and if the gloss of the standard is different from the gloss desired for the new coating, the angles of illumination and viewing must be agreed on. It is impossible to make even a metameric match of the color of a fabric sample with a paint at all angles of viewing since both the colorants and the surface textures have to be different. When paint manufacturers advertise on TV that their paint stores can do this, they are engaging in misleading advertising.

4. *Color properties.* Colorants have to be chosen that permit formulation of a coating that can meet the performance requirements. Need the coating have exterior durability, resistance to solvents, resistance to chemicals like acids and bases, resistance to heat, meet some regulation for possible toxicity, and so on?

5. *Film thickness and substrate.* Since in many cases, the coating will not completely hide the substrate, the color of the substrate will affect the color of the coating applied to it and the extent of that effect will vary with the film thickness. This variable is particularly important in applications like can coatings and coil coatings for which relatively thin coatings are applied (<25 µm). A thin coating that was color matched over a gray primer will not match over a red primer. A coating set up for one coat application on aluminum will not match the color standard if it is applied instead over steel.

6. *Baking schedule.* Since the color of many resins and some pigments are affected by heating, particularly at high temperatures, the color of a coating can be affected by the time and temperature of baking. It is also necessary to specify what the color requirements are for overbaking.

7. *Cost.* The color matcher should know the cost limitations. There is no sense in making an excellent color match with pigments that are too expensive for a particular application.

8. *Tolerance.* How close a color match is needed? Coatings for exterior siding or automobile top coats require very close color matches. For many others, close matching is unnecessary, but some customers, in their ignorance, set tight tolerance limits. Overly tight tolerances raise cost without performance benefits. For coatings that are going to be produced over time with many repeat batches, the most appropriate way to set color tolerances is to have an agreed on set of limit panels. For example, for a deep yellow coating, these would be the limits of greenness and redness, brightness and darkness, and high and low chroma. Since panels can change with age, spectrophotometric measurements should be made of the standard and the limit panels, and CIE tristimulus values calculated. The customer would then agree to accept any batch of coating with a color within this volume of color space. As mentioned earlier, while ΔE color differences, calculated from the 1976 CIELAB Color Difference Equations, have been used to set color tolerances, they have limitations. It is undesirable to use them as a general statement of tolerance for a series of colors. Furthermore, as noted previously, they permit variation equally in any direction from the standard color, which is frequently undesirable.

18.9.2. Color Matching Procedures

There are two approaches to matching colors: visual and instrumental. In visual color matching, experienced color matchers look at the sample to be matched, and from their experience select a combination of pigment dispersions that they think will permit matching the color. A formulation is mixed and the coating applied; since gloss affects color, first the gloss must be compared with the standard and adjusted if needed. The color matcher looks at the resulting color in comparison to the standard and decides what further addition of either one or more of the same pigment dispersions, or perhaps dispersions of different pigments, are needed to adjust the color match.

Spectrophotometric curves of the original standard sample can be used in an analytical mode to help identify component colorants in the standard, thereby simplifying pigment selection. Organic pigments can often be identified by examining the absorption spectra of solutions of colorants from a sample of coating [12]. The process continues until a satisfactory combination of pigment dispersions has been selected. Color matching must be based on dry films. It can be useful to compare samples of wet coatings to estimate progress towards a match, but since color changes considerably during application, film formation, and drying, decisions have to be based on dry films prepared under conditions approximating the way in which the coating will actually be used. When matching a new color, even a highly experienced color matcher seldom matches a color in less than 3 *hits*, and commonly, it takes several more hits, sometimes requiring changes in one or more of the original choices of pigments.

The laboratory color matcher not only has the responsibility of selecting the colorants and their ratios to make a color match, but he/she must also attempt to make the color match so that it will be as easy as possible to produce efficiently batch after batch in the factory. It is not possible for pigment manufacturers to produce successive batches of color pigments that are exactly the same color and color strength. Furthermore, there will be batch to batch variation in dispersions made from these pigments. The formulation should be set up so that it can be adjusted in the factory to permit matching the color standard in spite of these variations. It is desirable to use four pigments (counting black and white, but not inerts) to make the original match. This provides the four degrees of freedom necessary to move in any direction in three-dimensional color space. Sometimes use of single pigments cannot be avoided, but it is undesirable especially in applications in which close color matches are needed. Sometimes, more than 4 pigments are needed but this adds to the complications of production and should be avoided, if possible. The factory should not make any changes in the pigments in the formulation because the result will be a metameric match.

Visual color matching is a highly skilled craft requiring years of apprenticeship to master. It is still practiced, but the trend is toward computerized instrumental color matching. Instrumental color databases, along with computer programs, can be used to select colorants and their ratio both for original color matches in the laboratory and to provide information as to the amount of the different pigment dispersions to be added in the factory to match production batches. Establishing such a program requires a major effort. The database must be carefully set up. The pigments to be used must be made into dispersions, and multiple single color coatings with different concentrations of color pigment dispersion in a series of appropriate white coatings must be made and applied. The reflectance values must be measured at 16 wavelengths; for critical matches, values at 35 wavelengths may be needed. See Ref. [11] for further discussion of creation of a pigment database.

Complete discussion of computer color matching is beyond the scope of this introductory text. See Refs. [11] and [13] (and the references they give) for reviews of computer color matching and more detailed discussion. Solid color coatings at complete hiding can be accurately color matched by current computer color matching software. The basis for these programs is the Kubelka-Munk theory outlined in Figure 18.10. These programs attempt to achieve a tristimulus match between the sample and standard or object to be color matched. The mathematics of this operation is complicated by the nonlinear nature of the relation between the colorant concentrations in the sample and the tristimulus values of the sample and standard.

Typical color measurements consist of either 16 or 35 reflectance measurements at a set of wavelengths in the visible region. To utilize this data for color matching, the reflectances must be corrected for boundary reflections by means of Saunderson's equation to give Kubelka-Munk reflectances, see Figure 18.10. These reflectances can be used to calculate the ratio K/S for the sample to be matched at each wavelength. The K and S values are assumed to be linear functions of the pigment concentrations chosen to match the sample. In the case of a four pigment match:

$$K = k_1 c_1 + k_2 c_2 + k_3 c_3 + k_4 c_4$$

$$S = s_1 c_1 + s_2 c_2 + s_3 c_3 + s_4 c_4$$

and

$$[(K/S)s_1 - k_1]c_1 + [(K/S)s_2 - k_2]c_2 + \cdots + [(K/S)s_4 - k_4]c_4 = 0$$

for each of the 16 (35) wavelengths, since the measured K/S) ratio, k_i and s_i values for the individual pigments are different at each wavelength. In addition, the pigment concentrations are usually expressed as percents to satisfy the equation:

$$c_1 + c_2 + c_3 + c_4 = 100$$

It would seem easy to select three wavelengths to give three equations that combined with the last equation would give four linear equations with four unknowns that on solution would give the required pigment concentrations to match the sample. Unfortunately, this is not the case in practice. Even using all 16 (35) K/S equations and using various estimates such as tristimulus weighted averaging or least squares solutions of the determined set of 17 (36) equations in 4 unknowns does not give, in most cases, a close tristimulus color match. Allen devised a method that is used by most commercial software to obtain a tristimulus match [14]. Starting from an initial estimate of pigment concentrations obtained using the K/S ratios and tristimulus weighting, these concentrations are iteratively refined to approximate a tristimulus match. The iterative refinement process consists of setting up and solving the following set of equations:

$$\Delta X = \frac{\delta X}{\delta c_1} \Delta c_1 + \frac{\delta X}{\delta c_2} \Delta c_2 + \frac{\delta X}{\delta c_3} \Delta c_3 + \frac{\delta X}{\delta c_4} \Delta c_4$$

$$\Delta Y = \frac{\delta Y}{\delta c_1} \Delta c_1 + \frac{\delta Y}{\delta c_2} \Delta c_2 + \frac{\delta Y}{\delta c_3} \Delta c_3 + \frac{\delta Y}{\delta c_4} \Delta c_4$$

$$\Delta Z = \frac{\delta Z}{\delta c_1} \Delta c_1 + \frac{\delta Z}{\delta c_2} \Delta c_2 + \frac{\delta Z}{\delta c_3} \Delta c_3 + \frac{\delta Z}{\delta c_4} \Delta c_4$$

$$0 = \Delta c_1 + \Delta c_2 + \Delta c_3 + \Delta c_3 + \Delta c_4$$

The ΔX value is the difference between the measured X value of the object to be matched and the X value of the match calculated from the initial concentrations. The partial derivatives of X with respect to concentration are calculated from the initial match concentrations. The values for Y and Z are calculated in the same fashion. The details of these calculations can be found in the cited references. The resulting four equations are solved for the four unknown Δc_i values, and a new set of match concentrations are calculated by adding the Δc_i values to the initial estimate of c_i values used to set up these equations. The process is repeated until a tristimulus match or no further improvement in the match is obtained. Usually, the best tristimulus match with the four chosen pigments is obtained in less than three iterations.

If the four colorants chosen are close to or identical with the colorants in the object to be matched, a nonmetameric match results. If the degree of metamerism as judged by the variation in color difference between the object and match for two or more illuminants is unacceptable, different four pigment combinations can be tried until an acceptable match is attained.

Matching of metallic colors has been difficult to computerize because the colors have to match at multiple angles. The reflectances of the samples and all of the colorants in the database must be determined at multiple angles. Matching colors with pearlescent flakes is even further complicated by changes in hue as a function of angle, and computer programs have yet to be developed. Measurement of metallic and pearlescent coatings is the topic of ongoing research between instrument manufacturers, coating suppliers, and users with the ASTM; Ref. [15] summarizes the approaches.

Other information on the colorants, such as cost, properties, regulatory restrictions, and so forth, can be built into the database. The computer can then calculate a series of alternative colorant combinations selected for lowest cost, least degree of metamerism, excluding pigments with inadequate exterior durability, and so forth. The formulator can then choose the most appropriate combination for the particular color matching assignment. The formulator then makes up an experimental batch of paint, applies it to the appropriate substrate, bakes or dries it, and measures the reflectance values. The result is seldom a satisfactory color match on the first attempt. Reflectance values of the first hit are instrumentally compared with those of the standard, and the computer calculates the amounts of the various colorants needed so that the color will match. The number of hits required to make color matches has been found to be significantly less using computer color matching than the number required by even experienced visual color matchers.

The same program is used in production of factory batches of colored coatings. The colorants to be used have already been selected and must not be changed. The laboratory formulation is used for the initial mix, but a fraction of each of the colorants is held back. (Otherwise, if the factory batch of some pigment dispersion were stronger than standard, too much of that colorant would be put in.) The batch is mixed, a panel coated, and its reflectance measured. The data is then used to calculate what additions have to be made to the batch to obtain a color match. If necessary, the process is repeated.

The savings in time and, hence, in cost using computer color matching can be large, but the cost of establishing and maintaining the database is substantial. Since the colors of pigment dispersions vary from batch to batch of pigment and of pigment dispersion, measurements usually have to be made of each batch for comparison with the standard database values. Computer programs can be written to correct for small differences in the colors from batch to batch.

18.10. GLOSS

Gloss is a complex phenomenon. Individuals frequently disagree on gloss difference. Partly because of the difficulty of visual assessment, progress in developing useful mathematical treatments or measurements of gloss has been limited. Unfortunately, many people assume that measurements made every day by hundreds of technicians are more meaningful than they really are.

People generally do not have a clear definition of gloss, partly because there are several types of gloss. *Specular* (mirror-like) *gloss* is the type most often considered in coatings. A high gloss surface reflects a large fraction of the light that is reflected from the surface at the specular angle, that is, the angle of reflected light equal to the angle of the incident light beam. Lower gloss surfaces reflect a larger fraction at nonspecular angles. When considering gloss, people visually compare the amount of light reflected at the specular angle with the amounts reflected at other angles. If the contrast in reflection is high, gloss is said to be high. Note that gloss is *not* as seems to be commonly assumed, directly related to the fraction of light reflected from the surface. The fraction of light reflected at a surface increases as the angle of illumination increases as shown in Figure 18.5. At most angles of illumination, surface reflection from a low gloss surface is greater than from a high gloss surface. Surface reflection at the specular angle increases as the refractive index of an object increases, as also shown in Figure 18.5.

If a surface is rough on a micro scale, the angle of incidence of a beam of light is not the same as the geometric angle of the surface with the light beam. Light is reflected at specular angles between the light beam and individual rough facets of a surface. If a surface has many small facets oriented at all possible angles, a beam of light is reflected in all directions. Such a surface is a *diffuse reflector* and has a low gloss; it is called *matte* or *flat*. Flat is unfortunate terminology, since a perfectly smooth (flat) surface gives high gloss, whereas a microscopically rough surface gives what we call a flat. At intermediate surface roughness, the gloss is intermediate and is called *semigloss, eggshell, satin*, or a variety of other terms.

A phenomenon related to specular gloss is *distinctness-of-image* (DOI) gloss. A perfect specular reflector is a perfect mirror with an image that exactly mimics the original. If a surface has perfect diffuse reflection, no mirror image can be seen. At intermediate stages, the image is more and more blurred as the ratio of specular to diffuse reflection decreases. Furthermore, larger surface irregularities lead to distortion of the image. Commonly, one sees some degree of both blurring and distortion.

Sheen, as the term is used in the coatings industry, refers to reflection of light when a low gloss coating is viewed from near a grazing angle. A high gloss coating reflects a high fraction of light whose angle is near grazing. Reflection from most low gloss surfaces is low at a grazing angle. A low gloss coating is said to have a high sheen if there is significant reflection at a grazing angle. The effect is easy to see, but difficult to describe; there is not a glare, (as when a glossy surface is illuminated and viewed at a grazing angle), but a "soft," relatively high reflection.

Luster is another type of gloss effect. (To illustrate the problem of defining gloss, the dictionary definition of luster is: gloss, sheen.) Luster is directional gloss. For example, some woven fabrics are much glossier when viewed parallel to the warp than parallel to the woof. Such fabrics exhibit variations in gloss looking at folds in a draped fabric. If the contrasts are great, the fabric is said to have a high luster. Similar effects can sometimes be observed in textured coatings.

Haze can be said to be a form of gloss. When light enters a hazy film, it is scattered to some degree, causing some diffuse reflection to reach the eye of the observer. The result is similar to having some of the light reflected at nonspecular angles at the surface. The contrast between the fractions of light reflected at specular and nonspecular angles is reduced. In pigmented coatings, it may be difficult to distinguish visually between gloss reduction resulting from haze and from scattering of light by a pigment. It is always desirable for a formulator to make a batch of coating with no pigment to check the clarity of a dried, cross-linked, pigment-free film.

Bloom gives a similar effect to haze. If a liquid component of a coating film is not soluble in the resin binder, it can separate from the body of the film in small droplets. These can come to the surface, making it uneven, diffusing light beams striking the surface and reducing gloss. In contrast to haze, bloom can be wiped off a surface with a cloth damp with solvent for the blooming material; commonly, bloom reappears.

18.10.1. Variables in Specular Gloss

Some of the variables of specular gloss can be visualized by use of schematic diagrams based on hypothetical measurements by an idealized goniophotometer. A goniophotometer is an instrument in which a beam of light can be directed at a surface at any angle of incidence and can measure the amount of light reflected at any angle in the same plane as the incident beam. An idealized instrument is shown schematically in Figure 18.17. Figure 18.17(*a*) shows a cross section through the plane of the beam of light. Figure 18.17(*b*) shows a view looking down on the instrument. The drawing shows that the only light detected by the photometer is that reflected in the plane of the beam of light. No instrument can measure light reflected at all angles in all planes. Also, there is no way to build an instrument that would measure light reflected at 90°. In this idealized instrument, the beam of light is narrow enough so that only a point on the surface is illuminated, and the photometer can detect a reflected beam with a beam width approaching zero.

All light reflected from a perfect mirror is at the specular angle and in the same plane as the beam of light. From a perfect diffuse reflector, light is reflected at all angles in all planes. Line S_0 in Figure 18.18 shows photocurrent as a function of angle of viewing from a perfect diffuse reflector illuminated at 45°, as measured by an ideal goniophotometer. At first glance, the results do not seem rational, but it must be recalled that the photometer can measure all the light reflected at 0° and only a fraction approaching zero of the light reflected in all planes at 90°.

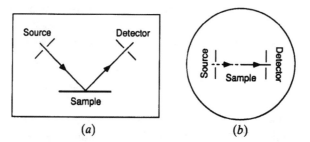

(a) *(b)*

Figure 18.17. Simplified schematic representation of a goniophotometer.

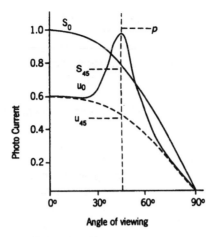

Figure 18.18. Schematic representation of the photocurrent reflected as a function of angle of viewing. (From Ref. [16], with permission).

Also shown in Fig. 18.18 is the idealized response curve u for a semigloss white coating. Comparison of these two curves illustrates three factors important in considering specular gloss. First is the *relative brightness A* of the sample compared to that of the diffuse reflector. Here A is the ratio of light intensities in the perpendicular direction u_0/S_0. Second is *height-of-gloss*, h which is calculated from the *peak height p* by Eq. 18.8.

$$h = \frac{P - u_{45}}{S_{45}} = \frac{p}{S_{45}} - \frac{A}{100} \tag{18.8}$$

Third is the *distinctness-of-image* aspect of gloss δ, as defined by Eq. 18.9, in which P is the total area under the sample curve, U is the area of diffuse background reflection from the sample, and S is the area under the standard diffuse reflector curve:

$$\delta = \frac{h}{F} \qquad F = \frac{P - U}{S} \tag{18.9}$$

If A is large, an observer will consider the gloss of a sample to be lower than that of a sample with a lower A. If h is small, the gloss of a panel will be lower than that of a sample with a higher h. It is instructive to compare a white gloss coating with a black gloss coating. A is larger and h is smaller for the white coating because white scatters light that enters a film, resulting in a high extent of diffuse reflection. On the other hand, a black coating absorbs almost all the light entering its film, so there is little diffuse reflection from within the film. If the surfaces of the two films are equally smooth, surface reflection is equal, but the black coating will be seen as glossier even though it reflects less of the incident light. It is impossible to make a white coating with as high a gloss as that of a high gloss black one. The possible gloss of colored coatings is intermediate; the darker the color, the higher the potential gloss. We saw in Section 18.2.4 that gloss affects color; now we see that color affects gloss. If the value of δ is high, the surface acts as a mirror, giving a clear image. As the value of δ decreases, there is more and more blurring of the specular image.

One must be careful when comparing these values with observer evaluation of gloss. As has already been stated different observers rate the gloss of pairs of panels differently. Intensity of reflection (peak height) is not a major factor in an observer's evaluation of gloss; the principal factors are the contrast between reflection at the specular angle and the distinctness of image. Gloss is also affected by the distance between the observer and an object. If the observer is close enough to visually resolve irregularities on the surface, he or she will say that a surface is a rough, high gloss surface. On the other hand, if the observer is far enough away from the same surface so he or she cannot visually resolve the irregularities, he or she will say it is a smooth, low gloss surface.

The major factor controlling gloss of coatings is pigmentation. As a film shrinks with volatile loss, pigment particles can cause irregularities in the surface. Roughness of the surface varies with the ratio of pigment volume concentration (PVC) to the critical pigment volume concentration (CPVC) in the dry film. (See Section 21.1.) Reference [17] discusses the effect of pigmentation on gloss as solvent evaporates.

In solvent-borne, high gloss coatings, the pigment concentration in the top micrometer or so of a dry film contains little, if any, pigment. Thus, the first few percent of pigment volume have little effect on gloss. This unpigmented layer results from the motions within a film as solvent evaporates. Initially, convection currents are set up within the film, and both resin solution and dispersed pigment particles move freely. As solvent evaporates, viscosity of the film increases, and movement of pigment particles is slowed. Movement of resin solution continues longer so the top surface contains little pigment. Application of a clear top coat over a pigmented base coat can markedly increase gloss.

As PVC increases, the amount of pigment at the surface increases, decreasing gloss. Particle size of the pigment affects gloss; if aggregates of pigment are not broken up in the dispersion process, gloss will be lower. Flocculated pigment systems have a lower CPVC, so at the same PVC there will usually be a greater likelihood of low gloss. However, since large particles stop moving before small ones, flocculated particles will stop moving sooner than well stabilized ones, which can lead to increased gloss in low PVC coatings. References [18] and [19] discuss effects of pigment particle size and clear layer thickness on specular gloss.

In some coatings (e.g., furniture lacquer), it is desirable to have a low gloss, but still a high degree of transparency. This is accomplished using a small quantity of very fine particle size silicon dioxide as a pigment. The combination of small particle size and low refractive index difference results in minimal light scattering as long as concentration is low. When solvent evaporates from such a lacquer, the SiO_2 particles keep moving until the viscosity of the surface of the film becomes high. The result is a higher than average concentration of pigment in the top of the film, reducing gloss at relatively low PVC.

Poor leveling can reduce gloss. If the irregularities are large, as is commonly the case with brush marks, the surface is seen as a glossy, but wavy surface. However, if the irregularities are small, as is commonly the case with orange peel, gloss will be rated as lower. An irregular surface can result from poor leveling over a smooth surface, but can also be the result of applying a coating over a rough substrate. It has been shown that smoothness and gloss are affected by the extent of roughness of a substrate, film thickness, and viscosity [20]. See Section 23.2 for further discussion of leveling.

Wrinkling affects gloss. If the surface of a film cross-links before the lower layers of the film, wrinkling is likely to occur. (See Section 23.6.) When the lower layers cross-link, shrinkage occurs, causing the top of the film to pucker up in wrinkles. If the size of the wrinkles is small enough, gloss will be low, but looked at in a microscope, the surface can be seen to be an irregular glossy surface.

Since the fraction of light reflected at a surface increases as refractive index increases, gloss of high gloss coatings tends to increase as refractive index of a coating increases, since the contrast between reflection at the specular angle and other angles is greater. Refractive index differences between different binders are small, so while detectable, this effect is generally small compared to surface roughness [17,18]. Refractive indexes of pigment can also have an effect. Metals have very high refractive indexes, so metallic coatings tend to exhibit high gloss.

Latex coatings generally exhibit lower gloss than solvent-borne coatings at equal levels of pigmentation in the dry film. There are several reasons for this:

1. Latex coatings have both resin and pigment particles as dispersed phases. During drying of a latex paint film, there is not the same opportunity for separation to give a pigment-free thin layer at the top of the film as in a solvent coating. Latexes with smaller particle size give somewhat higher gloss films than larger particle size latexes. Some difference in segregation may occur, since large size particles stop moving first as a film dries.

2. Pigment-free dry films of many latexes are not transparent; they are hazy, reducing gloss. The haze is due to the presence of dispersants and water-soluble polymers that are not completely soluble in the latex polymer.

3. Fairly commonly, surfactants bloom to the surface of a latex coating, reducing gloss.

4. It is generally more difficult to achieve good leveling with latex coatings. Surface roughness such as brush marks reduce gloss. The effect of brush marks depends on how close an observer is to the surface. If one is close enough to visually resolve the brush marks, the coating can look like a wavy, gloss surface. From a greater distance, the gloss appears to be low. Factors affecting leveling of latex coatings, as well as other factors affecting their gloss are discussed in Section 31.3.

Gloss can change during the life of an applied film. In some cases, the surface of the film embrittles and then cracks as the film expands and contracts. Generally, this mechanical failure is progressive, and after initial loss of gloss, there is film erosion. In other cases, especially clear coats, erosion occurs first, and loss of gloss is only evident after erosion is deep enough to cause protrusion of pigment particles of the base coat. Erosion of binder in pigmented films can proceed to a stage in which pigment particles are freed from binder on the surface and rub off easily; this phenomenon is called *chalking*. Chalky surfaces have dramatically lower gloss, and, therefore, the color changes to a lighter color. Loss of gloss can also result from loss of volatile components after a film is exposed; this causes film shrinkage and increases surface roughness [21]. Reference [21] gives an excellent review of durability and gloss.

In some cases, the gloss of a low gloss surface increases with use. It is fairly common for the gloss of flat wall paints to increase if rubbed, for example, near a light switch. This increase in gloss is called *burnishing*; means to minimize burnishing are discussed in Section 31.2.

18.10.2. Gloss Measurement

No fully satisfactory method for measuring gloss is available, and no satisfactory rating scale for visual observation of gloss has been developed. While all people will agree as to which film is glossier if the gloss difference is large; they frequently disagree in ranking if the difference is small. Even the same observer experiences difficulty in consistently ranking a series of panels if the differences in gloss are small.

Caution must be exercised in using instruments that measure gloss. In Section 18.10.1, we discussed an idealized goniophotometer. In practice, one has to accept less than this ideal. The beam diameter cannot actually approach zero because there must be sufficient light intensity reflected from all angles to give a measurable response on available photodetectors. In real instruments, there is a light source shining on a slit aperture at a fixed distance from a sample surface. The reflected light goes through another slit aperture at a fixed distance to a photodetector. In research instruments, both angles of illumination and viewing can be varied independently. These instruments are expensive and relatively difficult to maintain. They are used for research and, for calibrating standards for less sophisticated instruments.

Specular gloss meters are widely used, but correspondence between meter readings and visual comparisons is limited. Such instruments give significantly different readings with differences in intensity of reflected light, but observers are relatively insensitive to such differences. Furthermore, the aperture of the slit in a gloss meter is about 2° whereas the limit of resolution of a human eye is about 0.0005° of arc [17]. A gloss meter is, therefore, less sensitive to distinctness-of-image than the eye (when the observer is close to the object). The distance between the aperture and a panel is fixed in a gloss meter, whereas a person can view a panel from any distance.

The most widely used gloss meters, also called reflectometers, are simplified goniophotometers in which one measures a response only at the specular angle. Those most commonly used in the coatings industry can make measurements when the angles of incidence and viewing are 20°, 60°, and 85°. A schematic drawing is shown in Figure 18.19. The first step for using a gloss meter is to calibrate the instrument with two standards: one with high gloss and the other with a lower gloss. If the second standard does not give the standard reading after the instrument is set with the first standard, something is wrong; most commonly, one or both of the standards is dirty or scratched. Other possible problems include panel misalignment, deterioration of the light source, or a malfunction of the photometer. One must use the standard that has been calibrated at the selected angle. Black and white standards are available. As evident from the discussion in Section 18.10.1, reflection at the specular angle is not the same from a white and a black standard with equal surface roughness. It is not feasible to have standards of all different colors; white

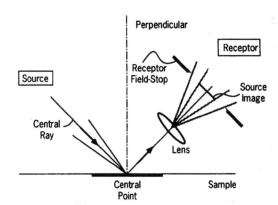

Figure 18.19. Schematic drawing of a gloss meter. (From Ref. [16], with permission.)

standards are used for light colors and black standards for dark colors. (One should always report which set of standards is used.)

In normal practice, one first measures at 60°. If the reading obtained is over 70, readings should be made at 20° rather than at 60° since the precision is higher nearer the midpoint of the meter reading. It is common to read low gloss panels at both 60° and 85°. Readings at 85° may have a relationship to sheen. It is essential to report the angle at which readings are made. Multiple readings should be taken and analyzed statistically. This reduces the probability that local surface irregularities or dirt particles are affecting the meter reading. It has been said that slight brush marks do not affect readings if the direction of the brush marks is parallel to the plane of incidence. If results are to be compared between two laboratories, it is essential to check the compatability of their instruments. This is best done by measuring at least three black and three white standards on each instrument. Readings are reproducible on carefully calibrated instruments to ±3%, but it is better to express error in terms of units, such as ±2 gloss units [22]. This is a high percent of error in the low gloss range.

There is some confusion as to what the numbers mean. Commonly, though undesirably, they are expressed as percents. It is better to call them gloss units or just meter readings. They are *not*, as some people seem to believe, the percent of light reflected at the surface. They are closer to being the percent of light reflected at that angle compared to the reading that would be obtained if a perfectly smooth surface were measured. As noted previously, the total reflection from a black matte surface is much higher at most angles of illumination and viewing than that from a high gloss black surface. The point can be emphasized by comparing the meter readings of the same panels at 20° and 60°. As shown in Figure 18.20, the meter readings are lower for the same panel when the setting is 20°, as compared to 60°.

With sufficient care, the instruments can be used for quality control comparisons of lots of the same or very similar coatings and for following loss of gloss on aging. For specification purposes other than quality control, specular glossmeters are not appropriate, and one must rely on standard visual panels. The customer selects three panels of each

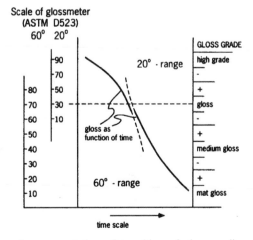

Figure 18.20. Schematic representation of transition of gloss readings according to Freier [23]. (From, Ref. [16], with permission.)

color, one with the desired gloss and the others representing the upper and lower acceptable limits.

Figure 18.20 illustrates a procedure to follow loss of gloss when a high gloss panel is exposed to exterior weathering. Initial readings were taken at 20°; then after some exposure, at both 20° and 60°, then for panels which have shown considerable loss of gloss only at 60°. Different scale units in the 20° and 60° readings (left) provide for a continuous slope at the transition point. The right scale designates gloss grade as observed visually.

Distinctness-of-image (DOI) meters as shown in Figure 18.21, rely on using the sample as a mirror. The reflection of a grid on the surface of the panel is compared visually to a set of photographic standards ranging from a nearly perfect mirror reflection to a blurred image in which the grid cannot be detected. One reports the comparison of the degree of blurring and also a qualitative statement about distortion. Correspondence with visual assessment in the high gloss range is better than with specular gloss meters, but DOI tends to be insensitive to small differences in the low gloss range. Instruments are available to make comparisons based on optical density, rather than relying on visual comparisons.

New instruments are being made with linear diode array detectors, which permit multiple measurements of light reflected at small angle increments without using an aperture in front of the detector [24]. Computerized instruments make multiple measurements of reflectance from small areas (approximately 100 μm diameter) over a 10 cm^2 area. This permits separation of the reflection from micro roughness and macro roughness, thus giving a numerical rating for gloss and a separate measurement of macro roughness such as orange peel or texture.

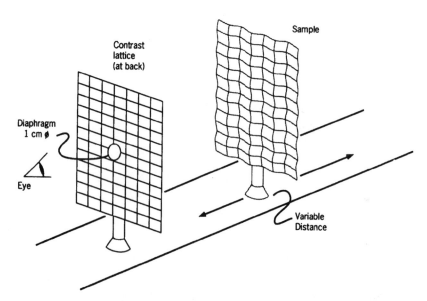

Figure 18.21. Schematic representation of a DOI meter. (From Ref. [16], with permission.)

GENERAL REFERENCES

D. B. Judd and G. Wysecki, *Color in Business, Science, and Industry*, 3rd. ed., Wiley, New York, 1975.

F. W. Billmeyer, Jr. and M. Saltzman, *Principles of Color Technology*, 2nd. ed., Wiley, New York, 1981.

G. Wysecki and W. S. Stiles, *Color Science*, 2nd. ed., Wiley-Interscience, New York, 1982.

Anonymous, *ASTM Standards on Color and Appearance Measurement*, 3rd., ed., ASTM, Philadelphia, 1991.

P. E. Pierce and R. T. Marcus, *Color and Appearance*, Federation of Societies for Coatings Technology, Blue Bell, PA, 1994.

U. Zorll, *Prog. Org. Coat.*, **1**, 113 (1972).

R. S. Hunter and R. W. Harold, *The Measurement of Appearance*, 2nd. ed., John Wiley & Sons, New York, 1987.

H. K. Hammond III and G. Kigle-Boeckler, "Gloss," in J. V. Koleske, Ed., *Paint and Coating Testing Manual*, 14th ed., ASTM, Philadelphia, 1995, pp. 470–480.

REFERENCES

1. F. W. Billmeyer, Jr. and M. Saltzman, *Principles of Color Technology*, 2nd. ed., Wiley, New York, 1981.
2. F. W. Billmeyer, Jr., "Color," in *Encyclopedia of Chemical Technology*, 3rd. ed., Vol. 6, Wiley, New York, 1979, pp. 523–548.
3. G. Wyszecki and W. S. Stiles, *Color Science*, 2nd. ed., Wiley-Interscience, New York, 1982.
4. L. M. Greenstein, "Nacreous (Pearlescent) Pigments and Interference Pigments," in P. A. Lewis, Ed., *Pigment Handbook*, 2nd. ed., Vol. 1., Wiley-Interscience, New York, 1988, pp. 829–858.
5. D. B. Judd and G. Wysecki, *Color in Business, Science, and Industry*, 3rd. ed., Wiley, New York, 1975.
6. P. B. Mitton, "Opacity, Hiding Power, and Tinting Strength," in T. C. Patton, Ed., *Pigment Handbook*, Vol. 3, Wiley, New York, 1973, pp. 289–339.
7. D. P. Fields, R. J. Buchacek, and J. G. Dickinson, *Surface Coat. Intl.*, **76**, 87 (1993).
8. Z. W. Wicks, Jr. and W. Kuhhirt, *J. Paint Technol.*, **47** (610), 49 (1975).
9. E. Cremer, *Prog. Org. Chem.*, **9**, 241 (1981).
10. F. J. Droll, *Paint Coat. Ind.*, February, 54 (1998).
11. P. E. Pierce and R. T. Marcus, *Color and Appearance*, Federation of Societies for Coatings Technology, Blue Bell, PA, 1994.
12. R. Kumar, F. W. Billmeyer, Jr., and M. Saltzman, *J. Coat. Technol.*, **57** (720), 49 (1985).
13. D. C. Rich. *J. Coat. Technol.*, **67** (840), 53 (1995).
14. E. Allen, *J. Opt. Soc. Amer.*, **64**, 993 (1974).
15. A. B. J. Rodrigues, *ASTM Standardization News*, October, 98 (1995).
16. U. Zorll, *Prog. Org. Coat.*, **1**, 113 (1972).
17. J. H. Braun, *J. Coat. Technol.*, **63** (799), 43 (1991).
18. L. A. Simpson, *Prog. Org. Coat.*, **6**, 1 (1978).
19. J. H. Braun and D. P. Fields, *J. Coat. Technol.*, **66** (828), 93 (1994).
20. P.-A. P. Ngo, G. D. Cheever, R. A. Ottaviani, and T. Malinski, *J. Coat. Technol.*, **65** (821), 29 (1993).
21. J. H. Braun and D. P. Cobranchi, *J. Coat. Technol.*, **67** (851), 55 (1995).
22. S. Huey, *Off. Digest*, **36**, 344 (1964).
23. H.-J. Freier, *Farbe Lack*, **73**, 316 (1967).
24. K. B. Smith, *Surface Coat. Intl.*, **80**, 573 (1997).

CHAPTER 19

Pigments

Pigments are insoluble, fine particle size materials used in coatings for one or more of five reasons: to provide color, to hide substrates, to modify the application properties of a coating, to modify the performance properties of films, and/or to reduce cost. Pigments are divided into four broad classes: white, color, inert, and functional pigments. Pigments are insoluble materials used as colloidal dispersions. Dyes are soluble colored substances; they are used only in specialized coatings such as stains for wood furniture. (See Section 30.1.) Some pigments are called *lakes*. The original meaning of lake was a dye that had been converted into a pigment by irreversible adsorption on some insoluble powder. The term lake is now sometimes used when a colored pigment is blended with an inert pigment; when the pigment is essentially all color pigment, it is sometimes called a *toner*.

Particle size affects color strength, transparency or opacity, exterior durability, solvent resistance, and other properties of pigments. For any given pigment, the manufacturer selects the most appropriate compromise for the particle size and designs the process to produce that average particle size consistently. Manufacture of most pigments involves precipitation from water; the process conditions determine the particle size. Many pigments are surface treated during or after precipitation. The precipitated pigment is filtered and the filter cake is dried. During drying, pigment particles become cemented together in aggregates. The coating manufacturer generally receives pigment from the pigment manufacturer as a dry powder of aggregates and must disperse these aggregates to break them up to their original particle size and make a stable dispersion. (See Chapter 20.)

19.1. WHITE PIGMENTS

A large fraction of coatings contain white pigment. White pigments are used not only in white coatings, but also in a substantial fraction of other pigmented coatings to give lighter colors than would be obtained using color pigments alone. Furthermore, many color pigments give transparent films, and the white pigment provides a major part of the hiding power of the coating. As discussed in Section 18.2.3, the ideal white pigment would absorb no visible light and would have a high scattering coefficient. Since a major factor controlling scattering efficiency is the difference in refractive index between the pigment

370

and the binder, refractive index is a critical property of a white pigment. Reference [1] is a monograph on white pigments.

19.1.1. Titanium Dioxide

The most important white pigment used in coatings is TiO_2. The technology of TiO_2 pigments is reviewed in Refs. [1] and [2]. The factors affecting the hiding power of TiO_2 are discussed in Refs. [1] and [3]. Two different crystal types are used: *rutile* and *anatase*. Rutile is used in larger volume primarily because it gives about 20% greater hiding power than anatase; the average refractive index of rutile is 2.73, compared to 2.55 for anatase. On the other hand, rutile absorbs some violet light whereas anatase absorbs almost no light; see Fig. 19.1. While a coating pigmented with only rutile looks white, it looks yellowish alongside a coating pigmented with anatase.

The color of rutile TiO_2 coatings can be adjusted by tinting. If one adds a small amount of a dispersion of a violet pigment, such as carbazole violet, the white coating obtained is less yellow. The violet pigment strongly absorbs wavelengths other than violet and reduces the difference in reflection of the violet wavelengths as compared to the others. This makes the coating a light gray, but unless there is a higher reflectance white present for comparison, it is seen as white, not gray. At the same time, there is the added advantage that hiding power is increased due to the additional light absorption. Phthalocyanine blue is also used as a tinting pigment. Phthalocyanine blue does not give as "white a white" as carbazole violet, but is less expensive and gives a slightly greater improvement in hiding. The most common tinting pigment used in whites is carbon black. Carbon black is less expensive than either violet or blue pigments, and gives a greater increment to hiding. Since it absorbs all wavelengths of light, it does not reduce the yellowness of the white, but provides a lower cost method of increasing hiding.

In white pigmented UV cure coatings, the lower absorption of near UV radiation by anatase TiO_2 is an advantage over rutile, since there is less interference with absorption of

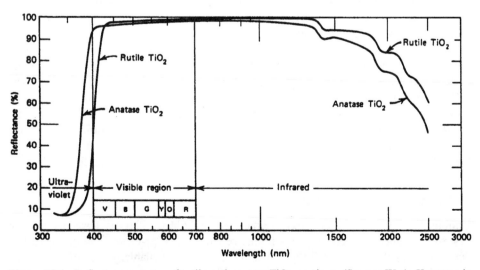

Figure 19.1. Reflectance spectra of rutile and anatase TiO_2 coatings. (Source: W. A. Kampton in Ref. [6], 1st ed., Vol. I, p. 2, with permission.)

UV radiation by photoinitiators. Another difference between rutile and anatase TiO$_2$ is in their photoreactivity. Anatase pigmented coating films fail more quickly on outdoor exposure by *chalking* than rutile pigmented films. Chalking is the formation of loose pigment particles on the exposed coating surface resulting from the erosion of binder by photodegradation. (See Section 5.2.4.) Chalking is usually undesirable, although *self-cleaning* white house paints are deliberately formulated with anatase TiO$_2$ so that they chalk readily. When the surface of the film erodes away, dirt accumulation is also removed so that the paint surface stays whiter.

In most exterior applications, chalking must be minimized, and rutile TiO$_2$ is used. However, while rutile is less photoreactive than anatase, it is still sufficiently reactive to reduce exterior durability. The problem is minimized by treating the surface of the TiO$_2$ particles during pigment manufacture [1,2]. The most common surface treatments are silica (SiO$_2$) and/or alumina (Al$_2$O$_3$); a variety of other metal oxides are also used with silica and/or alumina. A typical surface treatment for exterior grades of rutile is about 6 wt% silica and 4% alumina. An ideal surface treatment would be a continuous shell on all TiO$_2$ particles to prevent contact between the TiO$_2$ and the binder. Perfection has not been achieved, but the best chalk-resistant grades of rutile with a shell of SiO$_2$ show little acceleration of degradation. The best treated anatase pigments accelerate chalking more than treated rutiles, but much less than the untreated anatase pigment. Since TiO$_2$, especially rutile, strongly absorbs UV (see Fig. 19.1), it competes with the binder in a coating for UV absorption. If chalk-resistant grades are used, the reduced UV absorption by the binder tends to enhance exterior durability.

Surface treatment affects other properties of the pigments besides chalking. Alumina treatments improve dispersibility in solvent-borne coatings. However, alumina can partly neutralize acid catalysts, such as those used with coatings containing Class I MF resins, reducing cure rates. (See Section 9.3.1.) Specially treated TiO$_2$ pigments permit production of more stable dispersions in water-borne coatings. TiO$_2$ dispersions in water, called *slurries*, are used in the manufacture of latex paints. (See Section 20.3.) The slurries are manufactured by the TiO$_2$ producer and shipped in tank cars or trucks to replace dry pigment. The price on a pigment basis is about the same, but the slurry is a pumpable liquid and aggregation of pigment particles is minimized, so there are substantial savings in material handling and dispersion costs.

A variety of TiO$_2$ pigments is offered by each TiO$_2$ manufacturer. Some are designed for maximum exterior durability, some to provide maximum gloss, some are more easily dispersed in water systems, and others are more easily dispersed in solvent systems. The TiO$_2$ content of the various types of pigments varies from as high as 98% to as low as 75%. Generally, the hiding varies with TiO$_2$ content, so the amount of TiO$_2$ needed for hiding varies with pigment type. Formulators need to be aware of the differences among the types. Coating performance problems and economic waste have resulted from use of the wrong type of TiO$_2$ and from indiscriminate substitution of one type for another.

19.1.2. Other White Pigments

Until the late 1930s, white lead (basic lead carbonate, 2PbCO$_3$·Pb(OH)$_2$) was widely used as a white pigment. Its relatively low refractive index, 1.94, gives low hiding power, less than one-tenth that of rutile TiO$_2$. When TiO$_2$ became available, usage of white lead dropped rapidly. White lead is somewhat soluble in water and presents a toxic hazard. Many children suffer from lead poisoning attributed to eating paint containing white lead that peels from the walls and woodwork in older buildings. As a result, the lead content of any paint sold to retail consumers through interstate commerce in the United States is limited to 0.06% of the dry weight.

Zinc oxide, ZnO, was formerly used as a white pigment, but because of its low refractive index, 2.02, it cannot compete for hiding power with TiO_2. As discussed in Section 19.4, ZnO is used in exterior house paints as a fungicide and in some can linings as a sulfide scavenger. Zinc oxide should not be used in primers \because it is somewhat water soluble and can, therefore, induce blistering by osmotic pressure when water vapor penetrates through the top coats into the primer coat. Other white pigments, important at one time, but largely obsolete, are zinc sulfide (ZnS), with a refractive index of 2.37, and lithopone, $ZnS/BaSO_4$.

Small air bubbles scatter light and are another white "pigment." In some cases, cost reductions are possible by taking advantage of the scattering by air bubbles in coating films. The refractive index of air, 1, gives a refractive index difference of about 0.5 in a typical binder. While this difference is small compared with the difference of 1.25 between binder and rutile TiO_2, it still provides some light scattering and, hence, hiding. A common way of incorporating air bubbles in a coating film is by use of such high levels of pigmentation that after the solvent evaporates, there is not enough resin to adsorb on all the pigment surfaces and fill all the interstices between the pigment particles. The result is air voids, which increase hiding. Such coatings are said to be formulated at pigment volume levels above the *critical pigment volume concentration* (CPVC). (See Chapter 21.) Since interfaces between TiO_2 and air are developed, the very large refractive index difference of 1.75 is especially effective in increasing hiding. However, the voids make the film porous and reduce the protective properties of the films. In ceiling paints, advantage is taken of the added hiding, since requirements for other properties are not as important.

Particles containing air bubbles provide added hiding from the air-resin interface without imparting porosity. For example, high T_g latexes containing water trapped within the particles are available and finding increasing usage in latex paints as a partial replacement for TiO_2 [4]. The latexes are prepared by sequential emulsion polymerization [5]. Initially a low T_g latex copolymer of acrylic acid and acrylic esters is prepared; these particles are swollen with water, especially at alkaline pHs. These core particles are encapsulated in shells of high T_g cross-linked polymer such as a styrene/divinylbenzene copolymer. When the paint is applied and the water evaporates, the conventional latex particles of the binder coalesce to form the film, but the high T_g latex particles do not coalesce. Water diffuses out of the particles, leaving air bubbles within the particles, which add to the hiding.

19.2. COLOR PIGMENTS

A wide variety of color pigments is used in coatings. References [6] and [7] provide detailed discussions of their chemistry, properties, economics, and uses. Reference [8] gives a more condensed coverage of organic pigments. First, it is appropriate to summarize the considerations involved for selecting color pigments for a particular coating application.

1. *Color.* The first criterion is the color of the pigment. Pigment suppliers provide technical bulletins that contain color chips showing the color obtainable with each

pigment. Generally, there are two or three color chips for each pigment. The *mass tone* is displayed by a color chip in which the pigment is used as the sole pigment in the coating. Then, there are one or two additional color chips showing the color obtained when the pigment is used together with TiO$_2$ in different ratios; these are called *tint colors*. In some cases, color chips showing the effect with added aluminum flake pigment (see Section 19.2.5) are also provided.

2. *Color strength.* Some colorants are strong and others weak corresponding to the absorption coefficients and particle sizes of the pigments. It may well be more economical to use an expensive (i.e., high cost per unit volume), strong pigment than a weak, lower cost pigment.

3. *Opacity or transparency.* Depending on the end use, it may be desirable to use a pigment that increases hiding by both scattering and absorbing radiation, or it may be important to select pigments that scatter little, if any, light in the coating film so that a transparent color can be obtained.

4. *Ease of dispersion.* Some pigments are more easily dispersed than others. If everything else were equal, one would select the most easily dispersed pigment. Many pigments are surface treated by the manufacturers to enhance their ease of dispersion. (See Ref. [9] for a review.) Many types of surface treatment have been used, including adsorption on the surface of a molecule with similar structure to the pigment, but with polar groups to enhance dispersibility; monolayers of polymers; and, more recently, layers of inorganic oxides on the surface of organic pigments [10]. Pigment dispersion is a complex phenomenon involving several factors; it is discussed in detail in Chapter 20.

5. *Exterior durability.* Some pigments are more sensitive to photodegradation than others; photodegradation leads to loss of color or change of hue. Pigment manufacturers provide data that can be useful for screening pigments that may be appropriate for use outdoors. However, there can be significant variation in exterior durability, depending on the combination of pigment and resin. Therefore, for critical applications, the durability of specific formulations must be determined. (See Chapter 5 for discussion of and testing for exterior durability.)

6. *Heat resistance.* Coatings that are baked or exposed to high service temperatures require heat resistant pigments. A few pigments (e.g., yellow iron oxides) undergo chemical changes with heat; a more common problem is that some organic pigments slowly sublime at elevated temperatures.

7. *Chemical resistance.* In use, many coatings are exposed to chemicals, most often to acids and bases, and the pigments must resist color change under those conditions. For example, automobiles are exposed to acid rain, and home laundry machines are exposed to alkaline detergents.

8. *Water solubility.* In most applications, pigments with any significant water solubility are best avoided because they leach out of a film, resulting in loss of properties. The presence of water-soluble substances in a primer can lead to blistering of the film when water permeates through it. In latex paints, water-soluble materials can reduce the stability of the dispersion of the latex particles. The stability of colloidal aqueous dispersions can be particularly affected by the presence of low concentrations of polyvalent salts. On the other hand, pigments that help protect against corrosion by passivation (see Section 19.4) must be somewhat soluble in water to be effective. (See Section 7.4.2.)

9. *Solvent solubility.* Partial solubility of pigments in some solvents can be a problem. For example, if one is making a red coating for a bicycle that is to have a white stripe on

the red coating, the red pigment must not dissolve in solvents used in the white striping coating. If the red pigment is somewhat soluble, the stripes will become pink. Such pigments are said to *bleed*.

10. *Moisture content*. Most "dry" pigments adsorb small to moderate amounts of water. The adsorbed water can result in serious problems with water-reactive binders such as polyisocyanates.

11. *Toxic and environmental hazards*. Filter face masks should always be worn when handling dry powder pigments; it is dangerous to inhale fine-particle dusts, even if they are chemically inert. In most cases, there is little or no hazard once the pigment is incorporated into a coating. However, toxicity even in a coating film can be important. For example, as mentioned previously, white lead is no longer permitted in architectural coatings in the United States. Lead pigments are also prohibited in coatings for children's toys and furniture. Zinc chromate pigments are known to be human carcinogens. In some countries, their use in coatings is prohibited; regulations can be expected to become increasingly restrictive. Another factor that can affect pigment choice is the increasing cost of disposal of materials containing heavy metals.

12. *Cost*. Cost is important, but as noted earlier, one cannot tell which is the least expensive pigment by just looking at the price per unit weight. The critical question is: How much will the final coating cost on a volume basis, and this is affected by color strength and density.

19.2.1. Yellow and Orange Pigments

Inorganic Yellows and Oranges. Iron oxide yellows, FeO(OH), are low chroma brownish-yellow pigments. They give opaque films with good hiding and high exterior durability; chemical and solvent resistance are excellent. The pigments are generally easily dispersed and are comparatively inexpensive. When heated above 150°C, they gradually change color to a low chroma red because they dehydrate to form iron oxide red, Fe_2O_3. Most iron oxide yellows are made synthetically, but some natural ore ochre pigments are used. In some cases, particularly with the natural pigments, the presence of soluble iron and other metal salts can affect the stability of coatings, which cure by a free radical mechanism. Extremely fine particle size iron oxide pigments are also available; they are used when transparency is required. This effect of particle size on opacity is an excellent example of the relationship between light scattering and particle size. (See Section 18.2.3.)

Chrome yellow pigments are bright, high chroma yellows. Medium chrome yellow pigments are predominantly lead chromate ($PbCrO_4$). Greenish yellow colorants, called primrose yellow and lemon yellow, are cocrystals of lead chromate with lead sulfate. Redder yellows, that is chrome oranges, are cocrystals of lead chromate with PbO. Still redder oranges, molybdate oranges, are cocrystals of lead chromate with lead molybdate ($PbMoO_4$) and lead sulfate. Chrome yellows are relatively low cost, although higher than iron oxide, at least partly because their density is higher. Chrome yellows discolor on exterior exposure to form lower chroma yellows but their exterior durability is adequate for many outdoor uses. They are bleed and heat resistant. Due to their lead content, their use is not permitted in consumer paints in the United States. Their use in industrial applications is declining because of concern about use of lead compounds. Their major current use is in yellow traffic striping paint. Use of chromates for this purpose has been banned in many

European countries and in some U.S. states. Monoarylide yellows are their predominant replacement.

Titanium yellows are made by introducing other metal ions into the lattice of anatase TiO_2 crystals, followed by calcining to convert to the rutile crystal structure. Greenish yellow shades are based on introducing antimony and nickel; reddish yellow shade grades contain antimony and chromium. Resistance to exterior exposure, chemicals, heat, and solvents is excellent. However, only relatively weak yellow colors can be produced, so cost is high, and the range of colors that can be made is restricted.

Organic Oranges and Yellows. Chemical structures of representative pigments are shown in Figure 19.2. The code designations in this and subsequent tables are from the Color Index system developed by the Society of Dyers and Colourists in the United Kingdom and the Association of Textile Chemists and Colorists in the United States. The P in the code stands for pigment, the next letter designates the hue, and the number is chronologically assigned.

Diarylide yellows are bisazo pigments derived from 3,3′-dichlorbenzidene, for example, PY 13, as shown in Figure 19.2. They have high color strengths and high chroma. The hue and photostability are controlled by the number, positions, and structure of substituent groups on the aromatic ring of the anilide portion of the molecule. Even the most photochemically stable diarylide yellow pigments fade on exterior exposure, especially when used in tints. On the other hand, their solvent, heat, and chemical resistances are excellent. Due to their high strength and low density, their cost is relatively low. They are used in interior coatings for which a bright yellow color is needed, in tints, and in such

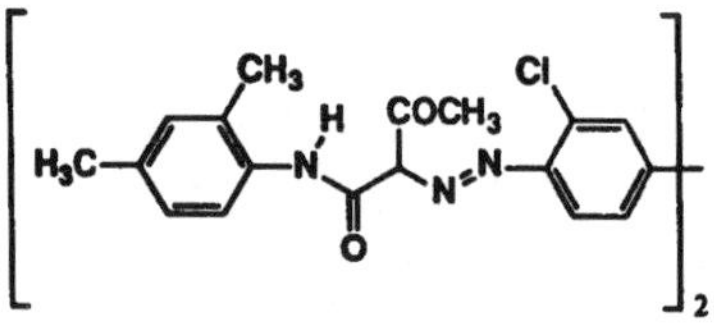

A diarylide yellow
PY 13

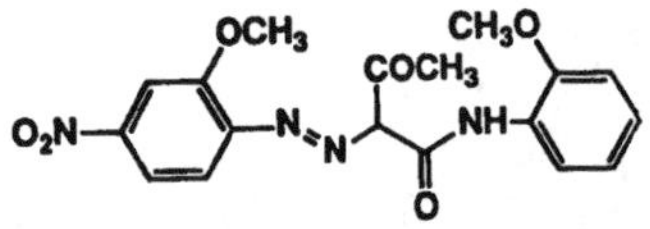

A monoarylide yellow
PY 74

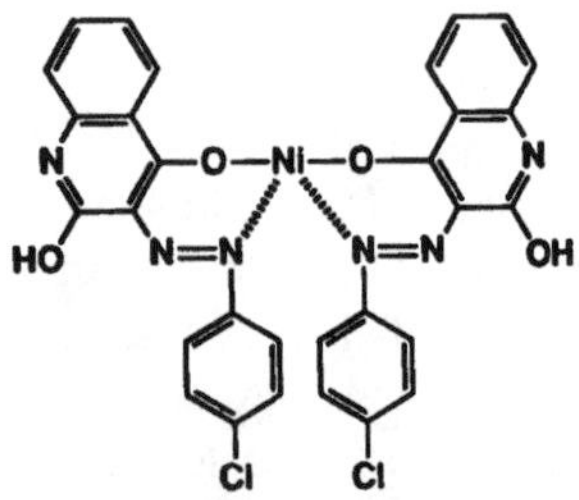

Nickel azo yellow
PG 10

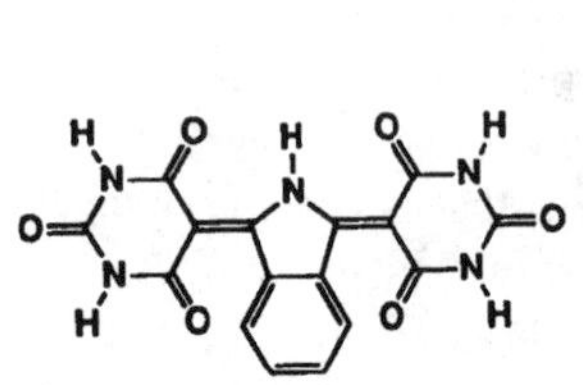

Isoindoline yellow
PY 139

Figure 19.2. Examples of organic yellow pigments.

applications as coatings for pencils. Grades with excellent transparency can be made. Diarylide yellows are the major yellows used in printing inks, especially in yellow inks used in four-color process printing.

Monoarylide (monoazo) yellow pigments, such as PY 74, also have high chromas. They have poor bleed resistance and sublime when exposed to high temperatures. However, their lightfastness is better than diarylide yellows, although still inferior to inorganic yellow pigments. Some opaque grades have sufficient light resistance for use in outdoor coatings. They are replacing chrome yellows in traffic paints.

Nickel azo yellow is a relatively weak, very greenish yellow; in fact, as can be seen from the designation PG 10 in Figure 19.2, it is classified as a green. Nickel azo yellow exhibits excellent exterior durability and heat resistance. It gives transparent films and is used predominantly in automotive metallic coatings. There have been some situations in which bleeding into stripes has been reported, so specific applications must be checked for bleeding.

Vat yellow pigments, as exemplified by isoindoline yellow PY 139, give transparency with excellent exterior durability and heat and solvent resistance. They are expensive and used only when their outstanding properties are required (e.g., in automotive metallic coatings).

Benzimidazolone orange pigments offer excellent light fastness and resistance to heat and solvent. They are used as replacements for molybdate orange.

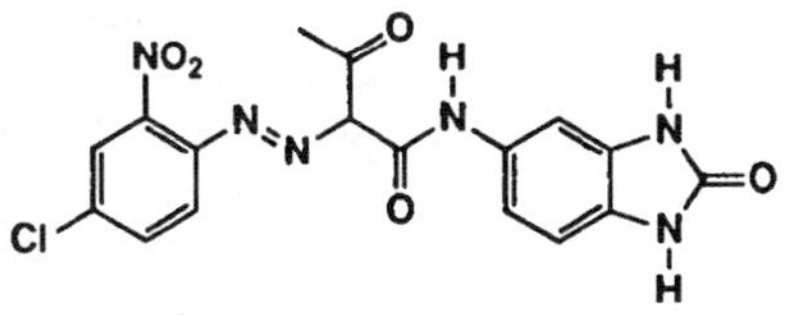

Benzimidazolone Orange
PO 36

19.2.2. Red Pigments

Inorganic Reds. Iron oxide (Fe_2O_3) gives the familiar barn red color. It is a low chroma red with excellent properties and low cost. In contrast to iron oxide yellows, iron oxide reds are thermally stable. When their particle size is optimal for scattering, they provide a high degree of hiding. There are also very fine particle grades available that provide transparent films. The excellent exterior durability makes the transparent grades suitable for use with aluminum in metallic automotive top coats.

Organic Reds. Toluidine red pigment, PR 3, is a moderate cost, bright red azo pigment with high color strength, good exterior durability in deep colors, good chemical resistance, and adequate heat resistance to permit use in baking enamels. As shown in Figure 19.3, it is an azo derivative of β-naphthol. Toluidine red is soluble in some solvents and gives coatings that are likely to bleed.

Bleed resistance with azo pigments can be achieved by the presence of carboxylic acid salts. For example, 2-hydroxy-3-naphthoic acid (BON) can be coupled with diazo compounds. Permanent Red 2B is an example of such bleed resistant high chroma red azo pigments. It is available as the calcium, barium, or manganese salt. The somewhat higher cost, manganese salt shows better exterior durability than the calcium or barium salts. A variety of related azo pigments that have somewhat different shades is also

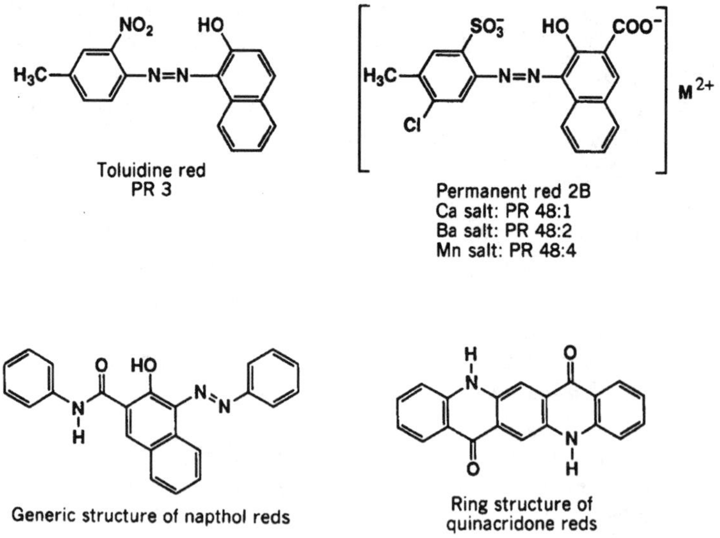

Toluidine red
PR 3

Permanent red 2B
Ca salt: PR 48:1
Ba salt: PR 48:2
Mn salt: PR 48:4

Generic structure of napthol reds

Ring structure of
quinacridone reds

Figure 19.3. Examples of organic red pigments.

available. This general class is the largest volume of organic red pigments used in coatings and inks. However, many are sensitive to bases and are not suitable for some latex paints.

Naphthol reds are a large family of azo pigments with various ring substituents (Cl, OCH_3, NO_2, etc.) in the generic structure shown in Figure 19.3. They are more resistant to bases, soap, and acid than the permanent reds, and also have fairly good exterior durability and solvent resistance.

Quinacridone pigments are nonbleeding, heat and chemical resistant, and give outstanding exterior durability, even in light shades. However, their cost is high. Depending on substitution and crystal form, a variety of orange, maroon, scarlet, magenta, and violet colors are available. Large particle size grades are used when opaque pigments are needed, and fine particle size grades are available for use in metallic automotive top coats. Surface treatments are applied to increase resistance of their dispersions to flocculation [11]. Many other high performance red pigments are available; see Refs. [6] and [7].

19.2.3. Blue and Green Pigments

Inorganic Blues and Greens. Iron blue, ferric ammonium ferrocyanide, $FeNH_4Fe(CN)_6$, is an intense reddish shade blue with fairly good properties. Since the 1930s, it has been increasingly supplanted by phthalocyanine blues, which have greater color strength. Cocrystals of various ratios of iron blue and chrome yellow are called chrome greens. The popularity of these pigments has decreased because of the high color strength of phthalocyanine green; they can no longer be used in architectural coatings in the United States because of their lead content.

Organic Blues and Greens. The principal blue and green pigments are copper phthalocyanine (CPC) pigments, commonly called phthalo blue and phthalo green (Figure 19.4). They exhibit outstanding exterior durability, bleed and chemical resistance,

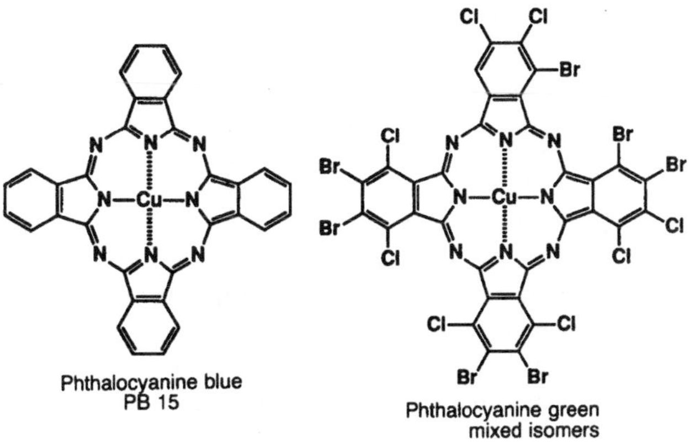

Figure 19.4. Representative phthalocyanine pigments.

are heat stable, and have high tinting strengths. Although their cost per pound is fairly high, the high tinting strength and quite low density mean their use cost is moderate.

Phthalo blues are available commercially in three crystal forms: alpha, beta, and the seldom-used epsilon. The beta form has a relatively green shade of blue and is stable. The alpha form is redder, but not as stable; in some cases, when some grades of the alpha form are used, there can be serious problems of change of color and strength during storage of coatings or during baking. More stable alpha form pigments are available; these incorporate various additives that stabilize the crystal form and minimize problems with flocculation of dispersions. Some grades of phthalocyanine blue are slightly chlorinated; these have greener blue shades.

Phthalo greens are made by halogenating copper phthalocyanine to produce mixtures of isomers in which many of the 16 hydrogens of CPC have been replaced with chlorine or mixtures of chlorine and bromine; see Figure 19.4. Phthalo greens with 13-15 chlorines but no bromine have blue-green shades, with the lower chlorine content pigments being the bluest greens. Partial replacement of chlorine with bromine shifts the color toward a yellow-green shade. The yellowest shades have a high ratio of bromine to chlorine.

19.2.4. Black Pigments

Almost all black pigments used in coatings are carbon blacks. They absorb UV radiation as well as light, and with most binders, black colors are the most stable on exterior exposure. Carbon blacks are made by a variety of processes of partial combustion and/or cracking of petroleum products or natural gas. Depending on the process, the particle size and, therefore, the degree of *jetness* (intensity of blackness) varies. High color channel blacks have the smallest particle sizes, with diameters of 5 to 15 nm, and have the highest jetness. They are used when intensely black, glossy coatings are desired. Furnace blacks are lower in cost, give less jet blacks, and have larger particle sizes. Various grades are available with average diameters of 50 to 200 nm. Lampblacks have still larger particle size, on the order of 0.5 μm, and have lower color strengths than other carbon blacks. They are primarily used in making gray coatings. They are preferable to high color blacks for this purpose; if a small excess of high color black is added by mistake, it may be necessary to increase the batch size 50% or more to make up for the mistake. The shade of gray obtained changes much more slowly with the addition of lampblack, making color matching easier.

While all carbon blacks are predominantly elemental carbon arranged in polynuclear six-member rings, the chemical structure of the surfaces varies depending on the raw material and process used in their manufacture. The surfaces are generally quite polar and in some cases are acidic. There are also variations in the porosity. Black coatings can present difficulties in formulating coatings, especially when channel blacks are used. Due to the small particle size, the ratio of surface area to volume of the pigment particles is very high. The large surface area leads to adsorption of high ratios of resin on the pigment particles—commonly, many times their volume of polymer-substantially increasing their volume and, hence, giving high viscosity at relatively low pigment loading levels. Due to the polarity of the surface and the large surface area, they can selectively adsorb polar additives such as catalysts from a coating formulation. For example, in drying oil and oxidizing alkyd formulations, the metal salt driers slowly adsorb on carbon black so that the coating dries more and more slowly as it ages.

Acetylene black pigment increases the electrical conductivity of films. Conductivity is required for coatings such as primers for plastic parts that are to be coated by electrostatic spray. (See Section 30.3.2.)

19.2.5. Metallic and Interference Pigments

The most important metallic pigments are aluminum flake pigments. They are produced by milling finely divided aluminum metal suspended in mineral spirits in steel ball mills to make thin flakes. A variety of particle size pigments is available. There are two major classes used in coatings: *leafing* and *nonleafing* pigments.

Leafing aluminum pigments are surface treated (e.g., with stearic acid) so that they have a low surface tension. When a coating containing leafing aluminum is applied, as a result of their low surface tension, flakes orient at the surface. This gives a bright metallic appearance and acts as a barrier to permeation of oxygen and water vapor through the film. As a result of barrier development, leafing aluminum pigments are used in top coats for corrosion protection of steel structures. (See Section 7.3.3.)

Nonleafing aluminum pigments have higher surface tensions and do not come to the surface of the coating. They are used in metallic top coats for automobiles. As discussed in Section 18.4, formulations using nonleafing aluminum are designed to maximize the fraction of flakes that are oriented parallel to the film surface. Films made with transparent color pigments and nonleafing aluminum change depth of color and shade with the angle of viewing. Gold colored aluminum flakes can be made by vapor deposition of iron carbonyl on the surface of aluminum flakes followed by oxidation to generate a thin layer of iron oxide [12]. Nonleafing aluminum pigments are surface treated to minimize their sensitivity to acid in the environment. The aluminum pigments used in water-reducible coatings are treated to minimize reaction of water with the aluminum (which generates hydrogen). In early work it was necessary to mix in the aluminum paste just before application. Various treatments have been developed to permit formulation of premixed coatings [13]; an example is treatment with a low acid number styrene/maleic acid copolymer.

Bronze, nickel, and stainless steel pigments with platelets similar to those of aluminum pigments are also available, but are used in lower volume. Bronze alloy flakes have "gold" colors; depending on the alloy composition, shades from greenish yellow to reddish golds are available. Generally, they are surface treated, so they leaf. Due to the presence of copper, bronze alloy pigments change color on outdoor exposure to a blotchy, muddy green appearance.

Interference pigments are flake pigments that lead to color by interference. (See Sections 18.2.1 and 18.4.) Pearlescent pigments are platelets of mica having thin surface treatment layers of TiO_2 or Fe_2O_3, which serve to give interference reflection of light striking the pigment surfaces [14,15]. As a result, at some spots on the surface of the platelets, some wavelengths of light are strongly reflected and others are transmitted. At other spots, where the film thickness of the treatment layer is different, different wavelengths are reflected and transmitted. The result is a mother-of-pearl effect. These pigments are used in automotive coatings to give an effect related to that obtained with aluminum flake, but with addition of hue changes with angle.

A new type of pigment that gives colors as a result of interference effects has recently become available [16]. The pigment particles are thin (around 1 μm in thickness) flakes composed of a thin layer of reflective metal sandwiched between two layers of clear plastic, and with very thin top layers of a semi opaque very thin metal coating on each surface. As a result of interference reflection (see Section 18.3), certain wavelengths of light are reflected and others absorbed. The color obtained depends on the film thicknesses and the angles of illumination and viewing. As with other flake pigments, these flakes must be oriented parallel to the surface of the coating.

19.3. INERT PIGMENTS

Inert pigments absorb little, if any, light and have refractive indexes close enough to those of binders that they give little light scattering when used as pigments. Several synonymous terms are used: *inert pigments*, *inerts*, *fillers*, and *extenders*. Commonly, but not always, they are inexpensive and reduce coat of a coating. The principal function of most inert pigments is often to occupy volume in a film. Other functions are to adjust the rheological properties of fluid coatings and the gloss and mechanical properties of films. As discussed in Chapter 21, many film properties are controlled by the volume of pigment in the film. References [6] and [17] give detailed discussions of the multitude of inert pigments available.

Calcium carbonate ($CaCO_3$) pigments are widely used inerts. The lowest cost grades are ground limestone or the mixed calcium magnesium carbonate ore, dolomite. Synthetic calcium carbonate pigments are more expensive, but they are whiter. Pigments are available with a variety of average particle sizes. In some applications, the reactivity of calcium carbonate with acids makes carbonate pigments undesirable; especially in exterior paints, degradation of film properties may be accelerated by acid rain. Calcium carbonate pigments should not generally be used in exterior latex paints. Water and carbon dioxide permeate through the film of a latex paint; some calcium carbonate reacts to give calcium bicarbonate, which is water soluble and permeates back out of the film. On the film surface, the water evaporates and the reaction reverses, leaving a *frosting* of insoluble calcium carbonate deposited on the film. Frosting is especially noticeable on dark color paints.

A wide range of clays (aluminum silicates) is used as inert pigments. They are available in various particle size ranges. Cost is frequently related to whiteness. As mentioned in Section 19.4, bentonite and attapulgite clays are used to modify viscosity of coatings. Mica (aluminum potassium silicate) has a platelet structure and can be useful in reducing permeability of films to oxygen and water vapor when it orients parallel to the surface of the film.

Magnesium silicate minerals are also used as inert pigments. Talcs of various crystal structures affect the film strength of coatings differently. Some talcs are platey and reduce vapor permeability; others are fibrous and may be particularly effective in film reinforcing. Asbestos is a very fibrous magnesium silicate that is no longer used because it causes lung cancer when fibers are inhaled.

Silicon dioxide pigments are an important group of inert pigments. Ground natural silicon dioxide is used in a variety of particle sizes. An unusual example of an SiO_2 pigment is diatomaceous earth, also called fuller's earth. It is composed of fossil skeletons of diatoms. The large ratio of surface area to volume affects film properties. (See Section 21.2.) Very fine particle size synthetic silicon dioxide pigments are used to reduce gloss of clear coatings and to impart shear thinning flow properties to coatings, as briefly discussed in the next section. They are expensive.

Barytes (barium sulfate) has been used as an inert pigment, especially in automotive primer formulations. It is said to provide a "harder" primer. The density of barytes is high (4.5), roughly twice that of most other inert pigments. In at least some cases, it is possible that use of barytes is based on weight comparisons with other inert pigments, rather than volume comparisons as should be done. While barytes is not expensive on a weight basis, it is more expensive than most other fillers on a volume basis.

While most inert pigments are inorganic minerals, organic materials can also be used as inert pigments. For example, powdered polypropylene is insoluble and acts as an inert pigment. High T_g latexes such a polystyrene latex can be used as an inert pigment in latex paints. Synthetic fibers, such as aramid fibers, have been shown to be effective for increasing the mechanical strength of coating films.

19.4. FUNCTIONAL PIGMENTS

Functional pigments are used to modify the application characteristics, appearance, or film properties of coatings. An important example is corrosion-inhibiting pigments. Complex zinc chromate pigments, red lead, zinc phosphate, and many others are used in primers to inhibit corrosion of steel by passivation of anodic areas. In contrast to other pigments, they must be somewhat soluble in water in order to function. Zinc metal pigment provides corrosion protection by cathodic action and is used in a class of primers known as zinc-rich primers. Discussion of these pigments is found in Sections 7.4.2 and 32.1.3. References [6] and [18] provide considerable information.

Flatting pigments are a type of functional pigments used to reduce gloss. While low gloss is often attained by formulating coatings with a high volume ratio of pigment in the dry film, sometimes, this is not a desirable approach. For example, in lacquers for wood furniture, it is essential to have the transparency of the dry film as complete as possible so that the beauty of the wood grain is not concealed. Fine particle size silicon dioxide is widely used in such coatings as a flatting pigment. As discussed in Section 18.10.1, during solvent evaporation from the coating film, convection currents carry the fine particles to the surface so that the pigmentation at the surface of the dry film is high enough to reduce gloss. The combination of low total pigmentation, small particle size, and small refractive index difference gives little reduction in transparency. Powdered polypropylene has been used in a somewhat analogous fashion.

Some pigments are used as biocides. For example, zinc oxide is used as a fungicide. Barnacles, algae, and other organisms can grow on the outer hulls of ships, to minimize this growth; *antifouling coatings* are used. Cuprous oxide and organotin pigments have been used in antifouling coatings; although their use is now constrained by environmental concerns. (See Section 32.2.)

Zinc oxide is used in linings for cans used for packing vegetables, such as corn, that evolve some hydrogen sulfide during cooking. The ZnO reacts with the H_2S to form white zinc sulfide, preventing black stains that result from the formation of tin sulfide by reaction between the H_2S and the tin oxide layer on the tin surface of the tin-plated cans.

Antimony oxide (Sb_2O_3) is a white pigment with a refractive index too low (2.18) and a cost too high to use simply for hiding. It is used in fire retardant coatings. When a combination of Sb_2O_3 and a chlorinated or brominated polymer is heated to high temperature, combustion byproducts are generated that suppress flame propagation.

Another class of functional pigments is viscosity modifiers. In general terms, these pigments increase the low shear viscosity of coatings either to inhibit pigment settling during storage of the coating and/or to reduce sagging after application of the coating. Important examples used in solvent-borne coatings are quaternary ammonium salt treated bentonite clays [19] and fine particle size silicon dioxide. Attapulgite clay is used in water-borne systems to reduce settling and syneresis (the separation of a layer of pigment and latex free liquid at the surface of cans of latex paint). The effect of such pigments on flow is discussed in Section 3.5.

GENERAL REFERENCES

T. C. Patton, Ed., *Pigment Handbook*, 3 Vols., Wiley-Interscience, New York, 1973; P. A. Lewis, Ed., 2nd. ed., Vol. I, Wiley-Interscience, New York, 1988.

J. H. Braun, *White Pigments*, Federation of Societies for Coatings Technology, Blue Bell, PA, 1995.

W. Herbst and K. Hunger, *Industrial Organic Pigments*, VCH, New York, 1997.

REFERENCES

1. J. H. Braun, *White Pigments*, Federation of Societies for Coatings Technology, Blue Bell, PA, 1995.
2. J. H. Braun, A. Baidans, and R. E. Marganski, *Prog. Org. Coat.*, **20**, 105 (1992).
3. D. P. Fields, R. J. Buchacek, and J. G. Dickinson, *Surface Coat. Intl.*, **76**, 87 (1993).
4. D. M. Fasano, *J. Coat. Technol.*, **59** (752), 109 (1987).
5. J. W. Vanderhoff, J. M. Park, and M. S. El-Aasser, *Polym. Mater. Sci. Eng.*, **64**, 345 (1991).
6. T. C. Patton, Ed., *Pigment Handbook*, 3 Vols., Wiley-Interscience, New York, 1973. P. A. Lewis, Ed., 2nd ed., Vol. I, Wiley-Interscience, New York, 1988.
7. W. Herbst and K. Hunger, *Industrial Organic Pigments*, VCH, New York, 1997.
8. P. A. Lewis, *Organic Pigments*, Federation of Societies for Coatings Technology, Blue Bell, PA, 1995.
9. B. G. Hays, *Am. Inkmaker*, June, 28 (1984); October, 13 (1986); November, 28 (1990).

10. P. Bugnon, *Prog. Org. Coat.*, **29**, 39 (1996).
11. E. E. Jaffe, C. D. Campbell, S. B. Hendi, and F. Babler, *J. Coat. Technol.*, **66** (832), 47 (1994).
12. R. Iden, *J. Coat. Technol.*, **67** (843), 57 (1995).
13. B. Muller, *J. Coat. Technol.*, **67** (846), 59 (1995).
14. L. M. Greenstein, "Nacreous (Pearlescent) Pigments and Interference Pigments," in P. A. Lewis, Ed., *Pigment Handbook*, Vol. I, Wiley-Interscience, New York, 1988, pp. 829–858.
15. D. P. Chapman, *J. Coat. Technol.*, **68** (862), 19 (1996).
16. F. J. Droll, *Paint Coat. Industry*, **14** (2), 54 (1998).
17. D. H. Solomon and D. G. Hawthorne, *Chemistry of Pigments and Fillers*, Wiley-Interscience, New York, 1983.
18. A. Smith, *Inorganic Primer Pigments*, Federation of Societies for Coatings Technology, Blue Bell, PA, 1989.
19. S. J. Kemnetz, A. L. Still, C. A. Cody, and R. Schwindt, *J. Coat. Technol.*, **61** (776), 47 (1989).

CHAPTER 20

Pigment Dispersion

As described in Chapter 19, pigments are manufactured with the particle size distribution that gives the best compromise of properties, but the particles often become cemented together into aggregates during processing. Breaking these aggregates and forming stable dispersions of optimally sized pigment particles is a critical process in the manufacture of coatings. Making dispersions involves three aspects: (1) *wetting*, (2) *separation*, and (3) *stabilization*. Most authors agree that there are three aspects to dispersion but different terms are used, sometimes with conflicting meanings. Be careful reading papers to know how an author is using the terms.

20.1. DISPERSIONS IN ORGANIC MEDIA

Because of effects of the high surface tension and polarity of water, consideration of pigment dispersion is separated in two sections. This section discusses dispersion in organic media, usually solutions of resins in organic solvent. Dispersion in aqueous media is treated in Section 20.2.

20.1.1. Wetting

Wetting is displacement of air, and sometimes of water or other contaminants, from the surface of pigment particles and aggregates by the vehicle. It is an essential requirement for pigment dispersion. Wetting requires that the surface tension of the vehicle be lower than the surface free energy of the pigment. In organic media this is the case for all inorganic and most organic pigments. If a pigment has an especially low surface free energy, then it is necessary to use a medium with even lower surface tension. There can be important differences in the rate of wetting. When a dry pigment is added to a vehicle to make a mill base, it tends to clump up in clusters of pigment aggregates. For wetting to occur, the vehicle must penetrate through these clusters and into the pigment aggregates. The rate of wetting is dominantly controlled by the viscosity of the vehicle; lower viscosity leads to more rapid wetting.

20.1.2. Separation

Processes are designed to separate pigment aggregates into individual crystals without grinding crystals to smaller particle size; it is generally undesirable to decrease the crystal size. Many different types of machinery are used to carry out the separation stage. Some important examples are described in Section 20.5. Dispersion machines work by applying a shear stress to the aggregates suspended in the vehicle. If the aggregates are easily separated, the machinery only needs to be able to exert a comparatively small shear stress. If aggregates require a relatively large force for separation, then machinery that can apply a higher shear stress is required. Pigment manufacturers have been increasingly successful in processing and surface treating pigments so that their aggregates are relatively easily separated.

Recall from Section 3.1, that shear stress is equal to shear rate times the viscosity of the mill base. The available shear rate for any particular dispersion equipment is set by machine design. The formulator must select appropriate dispersion machinery that can transfer sufficient shear stress to the aggregates and formulate a mill base for the most efficient use of the selected machinery. For the fastest rate of separation of aggregates, the mill base should have as high a viscosity as the equipment can handle efficiently. Then, the highest shear stress is exerted on the pigment aggregates, and separation is accomplished in minimum time. The engineering theory and equations modeling the forces for separation are discussed in an excellent series of papers by Winkler and coworkers [1].

20.1.3. Stabilization

Wetting and separation are important steps in making pigment dispersions, but it is seldom a problem to carry out these two stages in solvent systems. On the other hand, stabilization can frequently be a serious problem and is usually the key to making good pigment dispersions. If the dispersion is not stabilized, the pigment particles will be attracted to each other and undergo *flocculation*. Flocculation is a type of aggregation, but the aggregates formed are not cemented together like the aggregates in the dry pigment powder. Although substantial shear stress is required to separate the original aggregates, flocculation can be reversed by applying relatively low levels of shear stress.

Flocculation is almost always undesirable. With light scattering pigments, the larger particle size resulting from flocculation reduces scattering and, therefore, reduces hiding. With color pigments, the larger particle size reduces light absorption and, thus, color strength. Larger size floccules in the final film affect gloss. Flocculation of pigments, including inert pigments, can change critical pigment volume concentration (CPVC) and, thus, affect properties of the coating films (see Section 21.2). Flocculated dispersions are shear thinning and have higher viscosities at low shear rates than well-stabilized dispersions. Flocculated dispersions do have the advantage that any settling forms soft pigment-bearing sediments that are easily stirred back to uniformity. However, settling problems can usually be minimized by addition of thixotropic additives such as treated clay or gels without the adverse results of flocculation.

There are two mechanisms for stabilization: *charge repulsion* and *entropic repulsion*. (As discussed in Section 8.1.1, some authors prefer to use the terms osmotic repulsion or steric repulsion instead of entropic repulsion). In charge repulsion, particles with like electrostatic charges repel each other. Calcium and zinc soaps can sometimes be effective pigment dispersants, by absorbing on the pigment surfaces, leading to ionic charges on the surface that stabilize dispersions. However, in most cases, entropic repulsion is the more important stabilizing mechanism. Charge repulsion is more important in aqueous dispersions, as discussed in Section 20.3.

Entropic repulsion is a term used to describe the repelling effect of layers of adsorbed material on the surface of the particles of a dispersion that prevents the particles from getting close enough together for flocculation to occur. In many dispersions of pigments in organic media, the adsorbed layer consists of resin molecules swollen with solvent. The particles are in rapid, random (Brownian) motion. As they approach each other, their adsorbed layers become crowded; there is a reduction in the number of possible conformations of molecules of resin and associated solvent in the adsorbed layers. The resulting decrease in disorder constitutes a reduction of entropy. Reduction of entropy corresponds to an increase in energy and requires force; hence, resistance to reduction of entropy leads to repulsion. Similarly, compression of the layers could lead to a more ordered system by squeezing out solvent. The accompanying reduction in entropy would again lead to repulsion.

Much of our understanding of entropic stabilization of pigment dispersions comes from the seminal work of Rehacek [2]. He devised an experimental technique to determine the thickness and composition of the adsorbed layer on the surface of a pigment dispersed in a resin solution. First, he made a series of solutions of the resin in a solvent with different concentrations, c_1. He then dispersed a known amount of pigment in each of these solutions. Samples of the pigment dispersions were centrifuged until a pigment-free layer formed. The concentration of resin, c_2, in this supernatant layer was determined. He plotted $(c_1 - c_2)/P$, where P is g pigment, for each sample against c_2. (Unfortunately, he labeled $(c_1 - c_2)/P$ as "adsorption." It would have made it easier to understand his papers, if he had labeled this axis "apparent adsorption".)

A schematic representation of typical data is presented in Figure 20.1; in all cases, the shape of the curve is the same. Rehacek extrapolated the straight line portion of the curve to intercepts with both axes. The intercept with the c_2 axis represents the concentration of resin in the adsorbed layer on the pigment surface. The intercept with the $(c_1 - c_2)/P$ axis represents mg resin solids adsorbed per g pigment. It can be proven mathematically that these two values remain constant at all points on the linear portion of the curve. At the intercept with the c_2 axis, apparent adsorption is zero; this does not mean that no resin is adsorbed. It means that the concentration of resin in the adsorbed layer is the same as the concentration in the initial resin solution, c_1, and in the resin solution after pigment was dispersed in the resin solution, c_2. Thus, the amount of resin and solvent in the adsorbed layer is established. The surface area of a gram of pigment can be determined by nitrogen adsorption. Using the densities of the pigment and the resin solution, one can calculate the thickness of the adsorbed layer of resin and solvent in the linear range of the plot.

In Rehacek's work, as well as that of others [3,4], it has been found that if the adsorbed layer thickness of resin plus solvent is less than 9–10 nm, the dispersion is not stable and flocculation occurs. This adsorbed layer thickness is an average layer thickness; some places are thinner and others are thicker. Similar studies in mixed solvent systems with no resin present show that the adsorbed layer thickness of a combination of solvents is 0.6–0.8 nm and not sufficient to stabilize against flocculation [4]. With monofunctional surfactants, the adsorbed layer can be thinner and still protect against flocculation. In one case, McKay [5] has shown that an adsorbed layer thickness of 4.5 nm of surfactant and associated solvent was adequate. In contrast to the adsorbed layer of resin, which is nonuniform in thickness, the surfactant layer is comparatively uniform, so it does not have

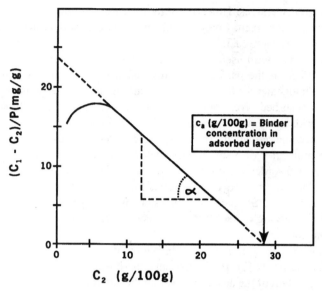

Figure 20.1. Schematic plot of $(c_1 - c_2)/P$ versus c_2 in studies of adsorption of resin and solvent on pigment surfaces. (Adapted from Ref. [2], with permission.)

to be as thick to provide stabilization. The advantages and disadvantages of resin-solvent versus surfactant-solvent stabilization are discussed later.

Referring back to Figure 20.1, it is of interest to understand why the plot deviates from linearity at lower values of c_2. The values of $(c_1 - c_2)/P$ are lower at low values of c_2. In other words, there is less apparent absorption of resin per unit area of pigment surface. This can be explained by competition between resin and solvent adsorption that depends on both the relative affinity of resin and solvent molecules for the pigment surface and the concentration of resin. If the concentration of resin is high enough, the resin "wins" and there is a complete adsorbed resin layer swollen with solvent. However, at lower concentrations, both solvent and resin are adsorbed on the particle surface, so the average layer thickness is insufficient to prevent flocculation.

Rehacek observed three other differences in behavior of dispersions with c_2 values above and below the start of the linear section of the curve. The low shear viscosities of the dispersions below the critical concentration were higher than those above it, separation of pigment during centrifugation was more rapid, and the bulk of the centrifugate formed was greater. This behavior indicates that the system is flocculated below the critical concentration. Viscosity at low shear rates increases and the system becomes shear thinning when flocculation occurs, as discusses in Section 3.5. Also, in a flocculated system under low shear conditions, particle sizes are larger than in a nonflocculated system; this causes more rapid settling in centrifugation. Furthermore, the floccules occupy more volume than a stable dispersion of the same amount of pigment, since continuous phase is trapped inside the floccules; this leads to the bulkier layer of sediment.

What controls the thickness of the adsorbed layer? In the case of a surfactant with its polar end adsorbed on the surface of a polar pigment, the length of the nonpolar aliphatic chain is the primary factor. In the case of resins with several adsorption sites, the largest single factor is probably molecular weight. For example, Saarnak showed that the adsorbed layer thickness on TiO_2 dispersed in a series of BPA epoxy resins in MEK, increased from 7 to 25 nm as the molecular weight of the epoxy resin increased [3]. With the lowest molecular weight resin, the layer thickness of 7 nm was insufficient to prevent flocculation. Dispersions in solutions of the higher molecular weight epoxy resins were stable.

Adsorbed layer thickness is also affected by the pigment surface. Relatively less polar organic pigments are generally more likely to give significant differences in adsorbed layer thickness with different resin-solvent combinations than are more polar inorganic pigments. However, even with inorganic pigments, significant differences can be encountered. For example, a TiO_2 surface treated with alumina forms a more stable dispersion than a TiO_2 surface treated with silica in the same long oil alkyd solution [6]. The authors proposed that the adsorbed layer is more compact on the silica-treated TiO_2. Commonly, the interactions between pigment surfaces and adsorbed molecules are considered to be hydrogen-bond interactions, but some authors prefer to interpret the interactions as acid-base interactions [7,8].

The spacing and number of functional groups along the resin chain affect layer thickness. As an extreme example, a linear aliphatic chain resin with a polar group on every other carbon atom would be expected to adsorb strongly on the surface of a polar pigment like TiO_2. At equilibrium, adsorption of single molecules with interaction of successive polar groups with the pigment surface would be favored, resulting in a thin adsorbed layer. However, if the resin has only occasional polar groups along the chains, at equilibrium, the longer segments between polar groups give loops and tails of resin swollen with solvent projecting out from the pigment surface, hence, a thicker adsorbed layer.

Layer thickness is also affected by solvent-resin interaction. If the loops and tails interact more strongly with the solvent, there are more solvent molecules in the layer, the average conformation of the resin is more extended, and the layer thickness is greater. Resin molecules that have multiple adsorbing groups have an advantage in competition with solvent molecules, but if the solvent interacts strongly with the pigment surface and the resin only interacts weakly, the more numerous solvent molecules will "win" the contest. For example, it has been found that toluene favors adsorption of macromolecules such as nitrocellulose, polyurethanes, and phenoxy resins on magnetic iron oxide pigment particles as compared to tetrahydrofuran [8]. Addition of solvent to a stable pigment dispersion can, in some circumstances, lead to flocculation. If the ratio of resin to solvent is just sufficient to allow adequate adsorption of resin to stabilize the dispersion, addition of more of the same solvent can shift the equilibrium, displacing part of the resin and reducing the average adsorbed layer thickness below the critical level for stabilization, resulting in flocculation. The dispersion is said to have been subjected to *solvent shock*.

For most conventional solvent-borne coatings, the resin used as binder in the coating can stabilize the pigment dispersion. Most conventional alkyds, polyesters, and thermosetting acrylics stabilize dispersions of most pigments. If there is a problem, the two most common changes to give adequate stability are to increase the molecular weight or the number of hydroxyl, amide, carboxylic acid, or other polar groups on at least that part of the resin to be used in the mill base. It has been shown that the highest molecular weight components of resins are selectively adsorbed [7].

If a low concentration of monofunctional surfactant is used, the solvent is likely to out compete most surfactant molecules due to the larger number of solvent molecules, even though the surfactant is more strongly adsorbed. This can be offset by increasing the concentration of surfactant molecules to shift the equilibrium favor of surfactant. However, doing so leaves excess surfactant in the final film, probably reducing performance properties. Thus, in organic media, monofunctional surfactants are not generally a desirable choice for stabilization. There are combinations of pigment/resin-solvent for which stability against flocculation cannot be achieved and an additive such as a surfactant is required. Additives should be used only as a last resort. They may be effective in stabilizing the pigment dispersion, but they may also interfere with some other critical property. For example, adhesion to metals may be reduced by added surfactant. A commonly used additive for stabilization is lecithin, a naturally occurring choline ester of phosphoglycerides. It is quite strongly adsorbed on the surface of many pigments, and therefore, lesser amounts are required than of most surfactants.

It is frequently difficult to make stable pigment dispersions in high solids coatings. Increasing the solids of organic solution coatings requires decreasing the molecular weight of the resins and reducing the number of functional groups per molecule. The low molecular weight results in thinner adsorbed layers of resin and associated solvent molecules. The reduced number of functional groups per resin molecule decreases the probability of the adsorption of resin molecules; there is a greater probability of solvent adsorption being favored and hence a greater likelihood of flocculation.

There is a further problem of adsorbed layer thickness on pigment surfaces in high solids coatings. In conventional coatings, the volume of pigment in the wet coating of even highly pigmented coatings is quite low, so differences in adsorbed layer thickness do not make a large difference in the viscosity of the wet coating (provided the dispersion is stable against flocculation). However, in high solids coatings, especially highly pigmented high solids coatings, the adsorbed layer thickness can have a significant effect on the viscosity. The effect of layer thickness can be seen by considering a modification of the Mooney equation shown in Eq. 20.1 (see Section 3.5) taking into account that the internal phase volume due to the adsorbed layer, V_a, in addition to that due to the pigment, V_p, is included in V_i. Using Eq. 20.1, model calculations were made to illustrate the effect of adsorbed layer thickness on viscosity [9].

$$\ln \eta = \ln \eta_e + \frac{2.5(V_p + V_a)}{1 - \frac{(V_p + V_a)}{\phi}} \tag{20.1}$$

Figure 20.2 shows the calculated dependence of the viscosity of coatings with 70 NVV as a function of the pigment volume concentration in dry films to be prepared from the coatings. The possible level of pigmentation is more limited with the thicker adsorbed layer than with the thinner one. The following principal assumptions were made in making the calculations: $V_i = 1.2V_p$ and $V_i = 2V_p$ (these values are equivalent to an 8 nm and a 25 nm adsorbed layer, respectively, on a 200 nm pigment); $\phi = 0.65$; solvent: $\rho = 0.8$ and $\eta = 0.4$ mPa·s; and oligomer: $\rho = 1.1$, and $\eta = 40$ mPa·s at 70 NVV, 4×10^5 mPa·s at 100 NVV. The viscosity dependence was assumed to follow Eq. 20.2, in which w_r is weight fraction of resin:

$$\ln \eta = \ln \eta_s + \frac{w_r}{0.963 + 0.763w_r} \tag{20.2}$$

Surfactants have been designed that are so strongly adsorbed on a pigment that they are preferentially adsorbed even in the presence of a large amount of solvent; then, little excess

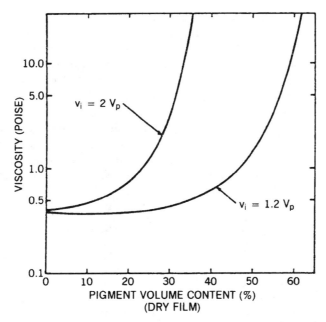

Figure 20.2. Calculations of the effect of PVC on the viscosity of two formulations, both at 70 NVV, but differing in thickness of the layer of adsorbed polymer solution. See text for assumptions. (From Ref. [9], with permission.)

over the amount sufficient to saturate the pigment surface area is needed to stabilize a dispersion. For example, phthalocyanine blue modified by covalently attaching long aliphatic side chains has been used as a surfactant with phthalocyanine blue pigment; the phthalocyanine end of the molecules of surfactant in effect joins the crystal structure of the surface of the pigment particles so that little, if any, is in solution [5]. The average adsorbed layer thickness needed to protect against flocculation was shown to be about 4.5 nm. Specific, proprietary surfactants have been designed for other pigments.

In many high solids coatings, it is not possible to make stable dispersions using the resins of the coating. This has lead to the design of special dispersing aids called *hyperdispersants*, or *superdispersants* [10–13]. Jakubauskas describes the design parameters of such dispersants [10]. He concluded that the most effective class of dispersant has a polar end with several functional groups and a less polar tail of sufficient length to provide for a surface layer that at least 10 nm thick. Also see Ref. [13] for a further review of the use of hyperdispersants. Examples of such dispersants include polycaprolactone-polyol-polyethyleneimine block copolymers [10]; polycaprolactone capped with toluene diisocyanate post-reacted with triethylenetetramine [10]; acrylic resins made by group transfer polymerization starting with methyl methacrylate, followed by glycidyl methacrylate, post-reacted with polyamines or polycarboxylic acids [11]; low molecular weight polyesters from polyhydroxystearic acid [12]; and proprietary hyperdispersants [13]. In some cases, it is desirable to treat the surface of a pigment so that there are "anchors" on the surface to interact with the dispersant [14,15].

A single dispersant that would anchor firmly to all pigments and be compatible with all types of coatings would be desirable. Then a single series of dispersions could be used in all products, substantially reducing costs. This ideal is not attainable, but part of the potential savings can be realized by using broadly effective dispersants to reduce the number of series of dispersions. Commonly, dispersants custom tailored for a single type of coating optimize properties.

Another approach to stabilization is to bond long chains covalently to the surface of pigments. For example, the surface of SiO_2 pigment particles has been reacted with trialkoxysilanes with long chain alkyl substituents [16]. Titanate orthoesters with three ethyl groups and one alkyl group with a long alkyl chain reduce the viscosities of dispersions of some inert pigments. It may be that the ethyl groups exchange with hydroxyl groups on the pigment surfaces. In many cases, pigment manufacturers offer pigments that have been surface treated to provide stable dispersions even when the pigment is dispersed in solvent. See Refs. [15] and [17] for reviews of surface treatment of pigments.

20.2. FORMULATION OF MILL BASES

The combination of resin (and/or dispersant), solvent, and pigment, used to make a pigment dispersion is called a *mill base*. The formulator must design a mill base for dispersing a pigment in the most appropriate dispersion equipment at optimum efficiency. Pigment dispersion machinery is the most expensive machinery in a coatings factory, from both the standpoints of capital and operating costs. It is, therefore, important to maximize the amount of pigment dispersed per unit time. Higher pigment loading means more efficient production; high loadings are possible when the viscosity of the vehicle (solvent plus resin) used in the mill base is low. Low viscosity also gives faster wetting. A properly stabilized pigment dispersion exhibits Newtonian flow, and its viscosity follows the Mooney equation as shown in Eq. 20.3:

$$\ln \eta = \ln \eta_e + \frac{K_E V_i}{1 - \dfrac{V_i}{\phi}} \qquad (20.3)$$

The volume of pigment (internal phase) can be maximized by using the lowest possible viscosity (η_e) vehicle. Solvent alone gives low viscosity, fast pigment wetting, and high pigment content, but solvent alone cannot stabilize a dispersion against flocculation. Therefore, it necessary to include some resin (or hyperdispersant) in the mill base. For maximum pigment loading, it is desirable to use the minimum concentration of resin solution that provides stability. The Rehacek procedure could be used to determine the minimum concentration that still gives a point on the linear section of the curve. However, the procedure is time consuming. Many years ago, Fred Daniel devised a simpler and faster, although less accurate, method [18].

20.2.1. Daniel Flow Point Method

The *Daniel flow point method* is a powerful tool for formulating mill bases efficiently, especially for dispersions to be made in ball and sand mills and related types of equipment [18]. It provides an estimate of the most appropriate resin concentration to use with a particular pigment. One makes a series of solutions with different resin concentrations.

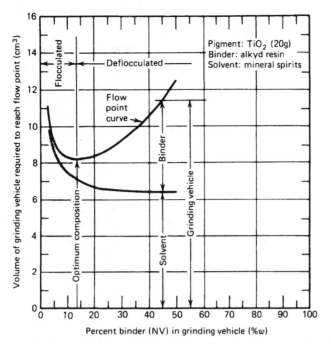

Figure 20.3. Daniel flow point plot: mL of solutions of an alkyd resin in mineral spirits per 20 g TiO$_2$ as a function of NVW resin in solution. (From Ref. [18], with permission.)

Then one determines the amount of each solution that must be added to a weighed amount of pigment so that when it is dispersed by rubbing vigorously with a spatula on a flat glass plate, the resulting dispersion has a viscosity just low enough to flow readily off the spatula. The spatula is used as both a dispersion machine and a viscometer. Since the clearance between the spatula and the glass plate is small, the shear rate is high; a spatula is a fairly good dispersing machine. One plots the volume of each solution needed for flow against the concentration of that solution. This plot is an isoviscosity plot; that is, the viscosity of each of the dispersions is approximately the same. (The actual low shear rate viscosity is approximately 10 Pa·s.) An example is shown in Figure 20.3; in addition to the volume of resin solution, the volume of solvent in the resin solution is also plotted.

Any resin-solvent-pigment combination that can give a stable dispersion shows a minimum in the curve. Let us consider the significance of the minimum. To the right of the minimum, the viscosity of the external phase is increasing because the concentration of resin is increasing. As can be seen from Eq. 20.3, if η_e increases and the viscosity stays constant, then V_i must decrease; that is, the amount of vehicle required per unit of pigment must increase. To the left of the minimum, the amount of vehicle per unit of pigment is also increasing, even though η_e is decreasing. The concentration of resin is insufficient to stabilize the dispersion. The pigment dispersion flocculates increasingly as the concentration of resin in the solutions decreases and the viscosity increases more and more steeply, so additional resin solution must be added to reach the isoviscosity level. The minimum point corresponds to the minimum concentration of that resin in that solvent that permits preparation of a stable dispersion of that pigment. Since the determination is not highly accurate, it is usual to start experimental dispersions with a somewhat higher resin concentration. If one adds more solvent to a stable dispersion with near the minimum resin concentration, the dispersion will flocculate. In some cases, no minimum is found, the amount of resin solution required per unit of pigment keeps decreasing as the concentration increases. This behavior signifies that a stable dispersion of that pigment cannot be made with that combination of resin and solvent.

Patton gives examples of data on the Daniel flow point method [18]. He also gives information expressed in volume units of pigment as well as the more conventional weight units.

20.2.2. Oil Absorption Values

In the early days of coating formulating, it was observed that the weight of pigment that could be incorporated with linseed oil in a mill base varied greatly from pigment to pigment. To simplify formulating mill bases to approximately equal viscosity, *oil absorption values* were determined. Linseed oil was slowly added to a weighed amount of pigment while rubbing with a spatula. Initially, balls of pigment wet with oil formed. When just enough oil had been added and worked into the pigment so that one coherent mass formed, the end point was reached. The oil absorption value is calculated as the number of pounds (g) of linseed oil required to reach the endpoint with 100 lb (100 g) of pigment. At the endpoint, there is just enough linseed oil to adsorb on the surface of all the pigment particles and fill the interstices between close-packed particles.

Oil absorption values for different pigments vary over a wide range. The smaller the particle size, the higher the oil absorption. A small particle size pigment has more surface area; therefore, a larger amount of linseed oil is adsorbed on the surface. In some cases, such as some grades of carbon black, the pigment particles are porous. Some oil penetrates into these pores, increasing the amount of oil required and, hence, the oil absorption value. Diatomaceous earth (see Section 18.3) has a very high surface area and, hence, very high oil absorption values. Pigment density also has a major effect. High density pigments require less weight of oil to adsorb on the surface of a unit weight of pigment and to fill the interstices; hence, they have lower oil absorptions. Other vehicles than linseed oil give similar oil absorption values. Therefore, data obtained with linseed oil can be used in formulating mill bases with any vehicle that gives stable dispersions. Pigment suppliers provide oil absorption values for their pigments. The oil absorption value permits one to set a starting point for pigment content when formulating a mill base.

The precision of oil absorption determinations by the spatula method is not high. Operators working in different laboratories commonly report values only within $\pm 15\%$. It has been said that with experience, deviations by a single operator can be reduced to ± 2–3%. In spite of these error ranges, it is common to see oil absorption values given with three supposedly significant figures. Improved accuracy and precision can be achieved by using a mixing rheometer (see Section 3.3.2), such as a Brabender Plastometer, for carrying out the determination [19]. The mixer is loaded with a known amount of linseed oil pigment is added slowly. The mixer imparts the necessary shear to separate pigment aggregates. The power required to turn the blades is recorded. As the amount of pigment is increased, the power requirement increases, and when the oil absorption endpoint is passed, the dispersion mass breaks up into chunks, leading to erratic readings. The values obtained are more reproducible and are generally a few percent higher than those obtained by the spatula method. The relationship between oil absorption of pigments and the critical pigment volume concentration of films containing those pigments is discussed in Section 21.3.

20.3. DISPERSIONS IN AQUEOUS MEDIA

Dispersion of pigments in aqueous media involves the same factors as in organic media: wetting, separation, and stabilization. However, the unique properties of water add extra complications. First, the surface tension of water is high, so there is more likely to be a problem in wetting the surface of pigment particles. Secondly, in some cases, water interacts strongly with the surface of pigments; therefore, the functional groups on the stabilizers have to interact more strongly with the pigment surface to compete with water. Furthermore, many applications of aqueous dispersions are in latex paints, so the systems have to be designed such that stabilization of the latex dispersion and the pigment dispersion do not adversely affect each other.

Inorganic pigments such as TiO_2, iron oxides, and most inert pigments have highly polar surfaces, so there is no problem with wetting them with water. The surfaces of inorganic pigments interact strongly with water, but the adsorbed layer of water does not by itself stabilize against flocculation. Most organic pigments, however, require the use of a surfactant to wet the surfaces. Some organic pigments are surface treated with adherent layers of inorganic oxides to provide a polar surface that is more easily wet by water [15].

In contrast to dispersions in organic media, stabilization by charge repulsion can be a major mechanism in aqueous media. (See Section 8.1.1.) The stability of the dispersions depends on pH, since pH affects surface charges. For any combination of pigment, dispersing agent, and water, there is a pH at which the surface charge is zero; this pH is called the *isoelectric point* (*iep*). At *iep*, there is no charge repulsion; above *iep*, the surface is negatively charged; and below *iep*, it is positively charged. The stability of dispersions is at a minimum at *iep* ± 1 pH unit [20]. The *iep* value for pigments varies, for example, 4.8 for kaolin clay and 9 for $CaCO_3$.

Surface treatments can have important effects on the stability of aqueous pigment dispersions. Special treated TiO_2 pigments have been developed for water-borne coatings. Substantial differences in dispersion stability can result from differences in the composition and completeness of the surface treatment [21]. In some cases, the amount of surface treatment is such that the TiO_2 content is as low as 75% of the pigment weight. Since hiding is related to the actual TiO_2 content, larger amounts of such highly treated pigments are needed to obtain equivalent hiding.

A broad study of stabilization of aqueous TiO_2 dispersions by anionic and nonionic surfactants concluded that a high molecular weight nonionic copolymer provided the greatest resistance to flocculation both in the dispersion and during drying of a gloss latex paint film [22]. Since stabilization resulted primarily from entropic stabilization, not charge repulsion, it was not affected by changes in pH.

Slurries of up to 80wt% TiO_2 in water are used on a large scale with substantial cost savings. (See Section 19.1.1.) The dispersions are stabilized with carboxylic acid-functional dispersants; pH is controlled by adding amine. The formulations are established to permit minimal flocculation in order to retard settling. Bactericide is added to inhibit microbial growth.

Most latex paint formulations contain several pigments and several surfactants. The *iep* of the various pigments are different, which complicates the problem of charge stabilization. Commonly, mixtures of surfactants are used. Anionic surfactants are frequently used as one component. Polymeric anionic surfactants (such as salts of acrylic copolymers in which acrylic acid and hydroxyethyl acrylate are used as comonomers) provide salt groups for strong adsorption on the polar surface of the pigment and hydroxyl groups for interaction with the aqueous phase; nonpolar intermediate sections add adsorbed layer thickness. Polymeric surfactants are less likely to lead to performance problems in the final films than monomeric surfactants. Nonionic surfactants are frequently used along with an anionic surfactant. It is common to also add potassium tripolyphosphate, the basicity of which may assure that the pH is above the *iep* of all pigments. Note that potassium tripolyphosphate is used in paints, not the sodium salt used in laundry detergents. The potassium salt is less likely to deposit on the surface of a film as scum after being leached out of a dry paint film by water. Large paint producers develop proprietary surfactants for their major product lines.

Organic pigments, unless surface treated, generally have surface free energies lower than the surface tension of water. Therefore, surfactants are needed to reduce the surface tension of the water to permit the wetting, that is essential for pigment dispersion. Either anionic or nonionic surfactants can be used. In the case of anionic surfactants, the dominant mode of stabilization is probably charge repulsion. In the case of nonionic surfactants, the relatively long polyether alcohol end is oriented out into the water and associated with multiple water molecules; stabilization is probably predominantly by entropic repulsion. Block copolymers with hydrophobic and hydrophilic segments made by group transfer polymerization have been recommended for stabilization of aqueous dispersions of a range of organic pigments [23]. The block copolymers also stabilize dispersions of latex particles.

In many cases, paint manufacturers make only white paints and supply separate color dispersions and formulations for the paint stores to mix with the white paints to make a large number of colored paints for their customers. Many latex paint manufacturers buy, rather than manufacture, their color pigment dispersions. The wetting and stabilization methods used are proprietary.

Additives can sometimes affect dispersion stability. For example, a hydrocarbon solvent used in a latex paint as a defoamer was shown to cause flocculation [24]. The paint formula had several different surfactants. It was found that the order in which the surfactants were added controlled whether or not flocculation occurred.

There is need for more research on the factors leading to stability of dispersions in latex paints. There is reasonable understanding of the stabilization of single pigment systems. However, paints contain several pigments, commonly with both polar and nonpolar surface pigments, and one or more latexes. (See Chapter 31.) Furthermore, the formulations contain one or more water-soluble polymers and/or associative thickeners (see Section 31.3) to adjust the rheological properties of the paint. As a result, the selection of appropriate dispersing agent combinations is done largely on an empirical basis. As in any situation in which empirical knowledge is the key to success, it is desirable to build a database of combinations that work and do not work to facilitate formulating with a new pigment combination or latex.

20.4. DISPERSION EQUIPMENT AND PROCESSES

A variety of equipment is used to make pigment dispersions. An important difference between different machines is in the level of shear stress that they can exert on the pigment aggregates. Easily separated pigment aggregates can be dispersed in low shear stress equipment whereas some pigment aggregates require very high shear stress for separation. Discussion of all types of such machinery and their operation is beyond the scope of this text. Patton provides more discussion of some types and detailed engineering information is available from machinery manufacturers [18]. Reference [1] deals more fundamentally with engineering aspects of some dispersion methods. We discuss a few important types with emphasis on their advantages and disadvantages. A formulator must become familiar with dispersion equipment available in his or her company's factories so that he or she can design formulations appropriate for production with that machinery.

From a processing point of view, there are three stages in making and using pigment dispersions: (1) premixing, that is, stirring the dry pigment into the vehicle and eliminating lumps; (2) imparting sufficient shear stress to separate the pigment aggregates; and (3) let down, that is, combining the pigment dispersion with the balance of the ingredients to make a coating. Some machines can carry out only the second step, some can do two of the three, and others can do all three.

Some pigment dispersions are made for specific batches of a coating; after the pigment dispersion is made, it is let down with the other components to make the final coating. Other pigment dispersions are made for use in several related types of coatings or as *tinting pastes* for color matching a variety of coatings. To minimize inventory of tinting pastes, one tries to select vehicles for them that are compatible with a range of coatings made by that company.

20.4.1. High-Speed Disk (HSD) Dispersers

High-speed disk (HSD) dispersers consist of a shaft with a disk that rotates at high speed in a vertical cylindrical tank. High-speed disk dispersers are also called high speed impellers. A schematic drawing of a typical disk is shown in Figure 20.4. Increasingly, the disks are made from engineering plastics that provide greater abrasion resistance than steel. The shearing action takes place by the differential laminar flow rates streaming out from the edge of the disk, typically rotates with a peripheral speed of 4000 to 5000 feet per minute (fpm). Sometimes, the speed of an HSD is expressed in revolutions per minute (rpm), but the important speed is the peripheral speed, which depends on the radius of the disk. The shear stress developed is relatively low, and therefore, HSD machines are appropriate only for relatively easily separated pigments. Obtaining predominantly laminar flow requires some minimum viscosity depending on the dimensions; it is usually somewhat over 3 Pa·s. The higher the viscosity, the greater the shear stress exerted on the aggregates and, therefore, the faster the process. The viscosity should be set such that the motor driving the shaft is run at peak power.

For maximum efficiency, pigment loading should be maximized. In solvent-based coatings, this can be done by using a vehicle solution with as low a resin concentration as

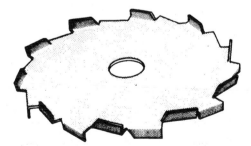

Figure 20.4.   Schematic drawing of a high speed impeller disk. (From Ref. [18], with permission.)

provides stabilization against flocculation. The Daniel flow point method can be used to estimate the resin-solvent ratio. Then, the ratio of this vehicle to the pigment is set to achieve a viscosity at high shear rate such that the motor draws peak amperage. In dispersions in water for latex paints, water-soluble polymers are commonly included in the pigment dispersion stage of making a latex paint to increase viscosity during dispersion.

A diagram of flow in using an HSD is shown in Figure 20.5. Centrifugal force leads to flow up the sides of the tank. If the mill base is Newtonian and the dimensions and operating conditions are appropriate, the whole charge becomes intimately mixed and all portions of the mill base pass repeatedly through the zone of highest shear, near the edges of the disk. However, if the mill base is shear thinning, the viscosity is high at the upper edge of the material on the sides of the tank where shear rate is low, resulting in hang up and incomplete mixing. Some mill bases contain a pigment designed to make the final coating shear thinning. The resulting hang up problem can be minimized by dispersing all of the pigments except the one that gives the shear thinning effect. After the balance of the pigments has been separated, the final pigment is slowly added. A desirable approach to minimizing the hang up problem is to design the dispersing tank with a slow-speed scraping blade that travels around the upper inside of the tank while the high-speed disk is spinning in the center of the tank.

HSD machines are commonly used for premix, dispersion, and let down operations. One initially loads the vehicle components, mixing at a low rpm, and then slowly load the dry pigment near the shaft. After the pigment is loaded, the speed is turned up to about 5,000 fpm for the dispersion stage. In a properly formulated mill base, the dispersion stage requires about 15 minutes. The speed is reduced and the let down is carried out with the rest of the components in the formulation. When producing latex paints, the latex cannot be present during the dispersion stage, since most latexes coagulate when exposed to high shear. The latex is added in the let down stage at low speed.

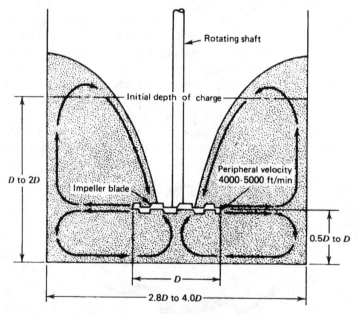

Figure 20.5. Diagram of a high-speed impeller disperser showing correct positioning of disk and optimum dimension ratios. (From Ref. [18], with permission.)

Compared to other dispersion equipment, HSD machines generally have the lowest capital and operating costs. No separate premix is required, and the let down can be carried out in the same tank. Cleaning to change from one color to another is relatively easy. Solvent loss can be kept to a minimum by using a covered tank. The major limitation is that the shear stress imparted to pigment aggregates is relatively low, so the equipment can only be used for relatively easily separated pigments. Pigment manufacturers have made substantial progress in making pigments with easily separated aggregates. Laboratory disk dispersers are available, and results correlate well with production machines.

20.4.2. Ball Mills

A *ball*, or *pebble*, *mill* is a cylindrical container, mounted horizontally and partially filled with balls or pebbles. The mill base components are added to the mill, and it is rotated at a rate such that the balls are lifted up one side and then roll in a cascade to the lower side, as shown schematically in Figure 20.6. Intensive shear is imparted to pigment aggregates when the balls roll over each other with a relatively thin layer of mill base between them.

Two classes of such mills are used. In steel ball mills, the balls and lining of the cylindrical mill are steel. In pebble mills, the balls are ceramic and the mill lining is porcelain. In early days, the balls were large round pebbles, hence the term pebble mills. The term ball mill is used by some to mean either class of mill and by others to mean only steel ball mills. We use ball mill in the general sense and steel ball mill and pebble mill to specify each particular class. Steel balls have the advantage of higher density and therefore, shearing is greater and milling times can be shorter. There is always some ball wear. If one disperses TiO_2 in a steel ball mill, one obtains a gray, rather than a white. (Dispersions for primers are sometimes run in steel ball mills, since they are commonly gray anyhow.) Pebble mills are used when discoloration is a disadvantage.

Ball mills operate most efficiently when their diameter is large so that the length of the cascade is long. The efficiency of operation is dependent on the loading of the mill. A mill should be loaded about half full of balls; this gives the longest cascade. The mill base volume should be just over the volume needed to cover the balls when the mill is at rest. If excess mill base is used, the time required for satisfactory separation increases. If the balls are all the same diameter spheres, the volume of balls in a half full mill is approximately 32% of the total volume of the mill. The volume occupied by the mill base when loaded for

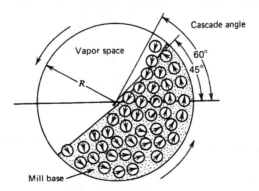

Figure 20.6. Schematic drawing of the cascading pattern in a ball mill. (From Ref. [18], with permission.)

optimum efficiency is a little over 18%. If the balls are not uniform in diameter, the packing factor is higher and there is less space for mill base.

Operation of a ball mill requires care in setting the rpm of the mill. If the rpm is too low, the balls are not carried high enough. If it is too high, the balls are carried past the 60° angle shown in Figure 20.6, and some fall rather than cascade. This reduces shear, can lead to ball breakage, and is more likely to lead to breaking the ultimate crystals of the pigment rather than just shearing the aggregates. If the rpm is still higher, the balls centrifuge and little dispersion action is exerted. Experienced mill operators can tell whether the rpm setting is proper by the sound. Milling efficiency is also affected by the viscosity of the mill base. If the viscosity is too low, ball wear is high. If the viscosity is too high, the balls roll more slowly and efficiency is reduced. Optimum viscosity is dependent on ball size and density; the larger the ball size and the higher the density of the balls, the higher the optimum viscosity. Viscosities in the neighborhood of 1 Pa · s are commonly used.

The Daniel flow point method (see Section 20.2.1) was originally developed to help formulate mill bases for ball mills. It permits loading the maximum amount of pigment at the viscosity required. Proper mill base formulating and use of proper volumes of balls and mill bases can make large differences in the time required for satisfactory separation. The time required depends on how easily the pigment aggregates are separated. The minimum time is usually on the order of 6 to 8 hours. Even difficult pigments should require no more than 24 hours. Sometimes one hears of 72 or more hours being required for dispersion. This almost always indicates poor formulation or mill loading.

Although the capital cost of ball mills is relatively high, operating cost is low. No premixing is required, and the mill can run unattended. In some cases, let down is carried out in the mill, but it is usually done separately. There is no volatile loss. Ball mills can be used to disperse all but the most difficult to separate pigments. They are difficult to clean and, therefore, most appropriate for making batch after batch of the same dispersion. The dispersion can be made, the batch emptied out, and the next batch started without cleaning the mill. If a series of colors is to be made in the same mill, one should start with the lightest color and work toward the darkest color. Another difficulty is that batch sizes are limited by the mills available, since the mills must be properly loaded.

When the mill base is dumped at the end of a batch, a significant amount remains in the mill and must be rinsed; the rinse material is then added to the batch. This rinsing should not be done with solvent alone. If the mill base has been properly formulated with only a little higher concentration of resin solution than indicated by the minimum point on the Daniel flow point curve, the addition of solvent will flocculate the dispersion. Rinsing should be done with a resin solution at least as concentrated as that used in the mill base.

There is no directly comparable laboratory mill available. This is because the efficiency of a ball mill is so dependent on its diameter. Production mills range from 1.25 to 2.5 m (4 to 8 ft) in diameter. Some laboratories use so-called jar mills. They are usually less than 30 cm (1 ft) in diameter and are usually rolled at a much less than ideal rpm. Correlation with production operations is usually poor. *Quickee mills* are more appropriate and much faster for making laboratory batches of dispersions roughly similar to production dispersions in a ball mill. A steel container is filled a little over half full with 30 mm steel balls and enough mill base to somewhat more than cover the balls. The container is then shaken on a paint shaker of the type used in paint stores. Easily separated pigments require 5 to 10 min. Difficult to separate pigments may take up to an hour of shaking. Jars partly filled with glass beads, sand, or ceramic media can be used as quickee mills in cases for which discoloration from the steel balls is excessive.

20.4.3. Media Mills

Media mills were invented to get around batch size limitations of ball mills. In the original mills, the media was sand that had an average diameter of about 0.7 mm, and the mills were called *sand mills*. It is common to use the term sand mill to include all media mills. Sand is still used but more frequently, small ceramic balls are used; some times the mill is then called a *bead mill*. Small steel balls are also used; then the mill is commonly called a *shot mill* because the steel balls originally used were the shot used in shotgun shells. Shot mills are generally used to disperse carbon black or other dark pigments.

A schematic drawing of a media mill is shown in Figure 20.7. There is a high-speed rotor in the center, and the space between the rotor and the cylinder is partially filled with sand or other particles. A premixed mill base is pumped into the bottom of the mill. It flows up through the mill, being exposed to shear as it passes between the rapidly moving particles, and out through a screen that keeps the media in the mill. Horizontal mills are also available. A typical media mill operates with a peripheral rotor speed of 10 m s^{-1}. This impels the particles at high speed so that even though their size is small, the shear rate between them is high, and, of course, in passing through the mill, a pigment aggregate passes between many pairs of particles. Mill bases should be formulated using slightly higher resin concentrations than the minimum in the Daniel flow point curve. The higher the viscosity of the mill base, the longer the average residence time in the mill and, hence, the greater the degree of shear. For easily separated pigments, it is desirable to use lower viscosity to achieve greater flow rate through the mill. Viscosities range from 0.3 to 1.5 Pa · s. For more difficult to separate pigments, two or three passes through a mill (or a battery of mills in series) may be required. Mills have been designed that automatically recycle part of the mill base. The mill shown in Figure 20.7 has a vertical rotor, but use of horizontal media mills is increasing.

Capital investment and operating costs of media mills are relatively low. Premixing is required and let down must be done separately. Batch size is flexible. Cleaning is a problem; generally, certain mills are reserved for similar colors to minimize cleaning. Media mills are effective for separating pigment aggregates, provided that the aggregates are small compared to the particles. Laboratory media mills are available that give quite good correlation with production mills. Instead of pumping in the bottom of the mill, they are designed to pour the premix into the top of the mill. Horizontal laboratory mills are also available. The use of media mills and related equipment is probably second only to HSD machines.

20.4.4. Three-Roll and Two-Roll Mills

Although formerly widely used to make dispersions for coatings, the use of *three-roll mills* is much more limited today. Figure 20.8 shows a schematic drawing of a three-roll mill. Aggregate separation results from the shear developed as the mill base passes through the nip between each pair of rolls. The viscosity of the mill bases is higher than for the other methods discussed thus far − 5 to 10 Pa · s or even higher. Since the mill base is exposed on the rolls, solvents must have low vapor pressures to minimize evaporation.

Three-roll mills have comparatively high capital and operating costs. Skilled operators are required. Premixes are required and let down is a separate operation. On the positive side, shear rates are high so difficult to separate pigment aggregates can be processed, batch size is versatile, and cleanup is relatively simple. Production usage today is limited

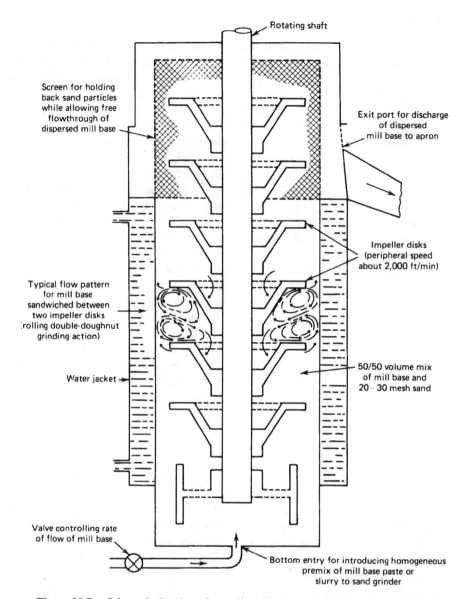

Rotating shaft

Screen for holding
back sand particles
while allowing free
flowthrough of
dispersed mill base

Exit port for discharge
of dispersed
mill base to apron

Impeller disks
(peripheral speed
about 2,000 ft/min)

Typical flow pattern
for mill base
sandwiched between
two impeller disks
(rolling double-doughnut
grinding action)

50/50 volume mix
of mill base and
20 - 30 mesh sand

Water jacket →

Valve controlling rate
of flow of mill base

Bottom entry for introducing homogeneous
premix of mill base paste or
slurry to sand grinder

Figure 20.7.   Schematic drawing of a media mill. (From Ref. [18], with permission.)

almost entirely to small batches of dispersions with no solvent, or only low volatility solvent, to make high viscosity dispersions. Three-roll mills are convenient for laboratory use.

Two-roll mills exert even greater shear rates than three-roll mills. They are usually used with solvent-free high molecular weight polymer-pigment systems. They can separate even the most difficult aggregates. They are most commonly used for dispersing very expensive pigments. The high capital and operating costs are justified when the economic value of achieving the last 10–20% of potential color yield means substantial difference in product

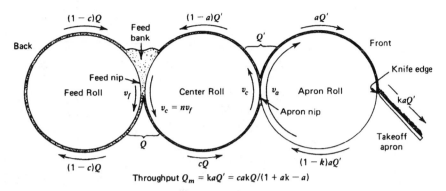

Figure 20.8.   Schematic diagram of a three-roll mill. (From Ref. [18], with permission.)

cost. Two-roll mills are particularly appropriate when the dispersion is to be used in a transparent coating. This means that essentially all of the pigment aggregates should be broken down to ultimate particle size to eliminate (or at least minimize) light scattering. Another use is for dispersing certain carbon black pigments when the desired jetness is only attained with virtually complete separation.

Two-roll mills have very high capital and operating costs. Volatile loss is complete. Premixes are required, and further processing of the dispersion is required to make a liquid dispersion that can be incorporated into a coating. Few coatings companies operate two-roll mills; most purchase pigment dispersions of the type for which two-roll mills are appropriate from companies that specialize in making pigment dispersions.

20.4.5. Extruders

Extruders are increasingly used for pigment dispersion. Pigment dispersion for almost all powder coatings is done in extruders (see Section 27.4.1); also, some high viscosity liquid dispersions are processed in extruders. An extruder has a screw feeding through a cylinder and pressing the material out the end through a die. The barrel of the extruder can be operated over a wide temperature range. The shear impressed in the screw is very high, and any pigment aggregates can be broken down in an extruder. Difficult to disperse pigments require longer residence times. In the case of solids, the product is usually chopped into small pieces and then pulverized either for use as a powder or for ease of dissolving.

20.5. EVALUATION OF DEGREE OF DISPERSION

Assessment of *degree of dispersion* is a critical need for establishing original formulations and optimizing processing methods as well as for quality control. Differences in degree of dispersion come from two factors: incompleteness of separation of the original aggregates into individual crystals and flocculation after separation. Frequently, the coatings industry does a poor job in this critical evaluation.

For white and colored pigments, the most effective evaluation method is by determination of *tinting strength* in comparison to a standard. For a white dispersion, one weighs out a small sample of the dispersion and mixes into it a small weighed amount of a standard color dispersion; as an illustration blue. Also, one weighs out a standard white sample with the same ratio of the blue standard. After thorough mixing, one puts a small amount of each tint mixture adjacent to the other on a piece of white paper and draws down both samples with a stiff flat-ended spatula so that the edges of the two samples touch each other. One can then compare the color of the two drawdowns. If the batch is a darker blue than the standard, the tinting strength of the batch of white is low, meaning that it is not equally dispersed. To test a color dispersion, say blue, one carries out the same procedure but uses the standard white in both samples and the standard blue and the batch of blue.

With dispersions of pigments that are to be used in applications like automotive metallic coatings, transparency is a critical requirement. Transparency can be tested by drawing down a standard and a batch side by side on a glass plate. The degree of haze or difference from the standard can be evaluated visually or measured instrumentally.

One can check for flocculation by pouring some of the tint mix made for the draw down onto a piece of tin plate. One the, rubs the wet coating with a forefinger. If the color changes, the dispersion is flocculated. For example, if a mix of blue with white becomes bluer where it is rubbed, the blue pigment dispersion is flocculated. If the rubbed portion becomes a lighter blue, the white is flocculated. Flocculation can also be detected by examining flow of the dispersion. Well-stabilized dispersions have Newtonian flow properties. If a dispersion is shear thinning (and does not contain a component designed to make it shear thinning), it is flocculated.

A further method of assessing pigment dispersion is by settling or centrifugation experiments. The rate of settling is governed by particle size and difference in density of the dispersed phase from the medium. A well-separated, well-stabilized dispersion settles or centrifuges slowly, but when settling is complete the amount of sediment is small. A well-separated but poorly-stabilized dispersion settles quickly to a bulky sediment. The floccules settle more quickly because of their large size and they form a bulky sediment because continuous phase is trapped within the floccules. Flocculated sediment is readily stirred, or shaken back to a uniform suspension, in contrast to the sediment formed from the nonflocculated dispersion. If the pigment settles or centrifuges relatively quickly to a compact layer, the separation step is incomplete. The larger aggregates settle more quickly but the volume is not large, because less internal phase is trapped in the sediment than with a flocculated dispersion. Settling or centrifugation tests provide qualitative or semiquantitative information sufficient for development work and quality control purposes. For research purposes, one can determine more quantitative data; the degree of flocculation can be calculated from rates of centrifugation [5].

One can also examine the dispersion with a microscope; one must use caution in preparing the samples for examination. In general, it is necessary to dilute the sample. If the sample is diluted with solvent, there is a possibility of flocculation. In such a case, one may report that the dispersion is flocculated when actually the flocculation was a result of sample preparation. Electron microscope studies of the surfaces of etched dry coating films can be useful for assessing variations in dispersion [25].

A procedure called the *flocculation gradient technique* is probably the most rapid and accurate method for quantitative study of the degree of pigment dispersion in both liquid coatings and dry films. The method was originally developed by Balfour and Hird to study TiO_2 dispersions [26]. The extent of scattering of 2500 nm infrared radiation by a film as a function of film thickness is measured. (Particles, generally, scatter longer wavelength radiation more effectively than they do visible light.) There must be a significant difference in the refractive index at 2500 nm between the pigment and the vehicle. A plot of backscatter against film thickness gives a straight line whose gradient increases with increasing flocculation. The technique has proven useful in evaluation of flocculation of pigments [27]. Infrared scattering measurements are also useful in the determination of the degree of flocculation in liquid dispersion samples. Other examples of the use of infrared backscattering measurements are given in Refs. [8] and [28].

The most widely used method of testing for *fineness of grind* in the coatings industry is a drawdown gauge, commonly the *Hegman gauge*. A sketch of one such gauge is shown in Figure 20.9. A sample of the dispersion is placed on the steel block before the zero reading and drawn down by a steel bar scraper. One then lifts the block up and quickly looks across the drawdown sample to see at which graduation one can start to see particles projecting or streaks caused by particles being dragged along. It is said that the higher the scale reading, the "better" the dispersion.

Unfortunately, the device is not capable of measuring degree of dispersion. First, we should remember that a major problem in making satisfactory dispersions is avoiding flocculation. However, the gauge cannot detect flocculation at all, since the drawing down step breaks up any floccules. Next, the particle sizes of properly dispersed pigments are small compared to the depth of the groove on the gauge. The depth on some gauges ranges from 0 to 10 mil (250 μm) in graduation units of 1.25 mil (approx. 30 μm). The gauge shown in Figure 20.9 ranges from 4 mil (100 μm) to 0 in steps of 0.5 mil (12.5 μm). TiO_2 pigment particles have an average size of about 0.23 μm, about 2 orders of magnitude smaller than the graduation steps of the gauge. Even aggregates of 10^4 to 10^5 particles could escape detection. Many color pigment particles are even smaller and carbon black particles can be as small as 5 nm. Some inert pigment particles are as large as 2 or 3 μm, still an order of magnitude smaller than the groove depths. Obviously, the gauge cannot test whether or not all or most of the particles are less than the groove depths. Blakely showed that in TiO_2 dispersions only approximately 0.1% of the total pigmentation of a coating was responsible for an unacceptable fineness of grind rating [29].

Why are such devices used? They do give some idea of whether the big aggregates are getting broken up, and they do give some idea of the presence of dirt particles. The determinations are fast, taking about a half minute. A tinting strength determination by an experienced person requires 2 or 3 minutes, but after spending this time, one has an assessment of the dispersion instead of a close to meaningless number. In the 19th century,

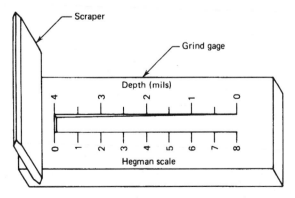

Figure 20.9. Sketch of a grind gauge and scraper for measuring "fineness of grind" of pigment dispersions. (From Ref. [18], with permission.)

when very coarse, difficult to separate pigments were all that was available, the gauge might have had some value. There will be no excuse for using it in the 21st century.

GENERAL REFERENCES

T. C. Patton, *Paint Flow and Pigment Dispersion*, 2nd. ed., Wiley-Interscience, New York, 1979.
G. D. Parfitt, *Dispersions of Powders in Liquids*, 3rd. ed., Applied Science Publishers, London, 1981.

REFERENCES

1. J. Winkler, E. Klinke, and L. Dulog, *J. Coat. Technol.*, **59** (754), 35 (1987); J. Winkler, E. Klinke, M. N. Sathyanarayana, and L. Dulog, *J. Coat. Technol.*, **59** (754), 45 (1987); J. Winkler and L. Dulog, *J. Coat. Technol.*, **59** (754), 55 (1987).
2. K. Rehacek, *Ind. Eng. Chem. Prod. Res. Dev.*, **15**, 75 (1976).
3. A. Saarnak, *J. Oil Colour Chem. Assoc.*, **62**, 455 (1979).
4. L. Dulog and O. Schnitz, *Proc. FATIPEC Congress*, Vol. II (1984) p. 409.
5. R. B. McKay, *Proc. Intl. Conf. Org. Coat. Technol.*, Athens, 1980, p. 499.
6. A. Brisson and A. Haber, *J. Coat. Technol.*, **63** (794), 59 (1991).
7. J. Lara and H. P. Schreiber, *J. Coat. Technol.*, **63** (801), 81 (1991).
8. S. Dasgupta, *Prog. Org. Coat.*, **19**, 123 (1991).
9. L. W. Hill and Z. W. Wicks, Jr., *Prog. Org. Coat.*, **10**, 55 (1982).
10. H. L. Jakubauskas, *J. Coat. Technol.*, **58** (736), 71 (1986).
11. J. A. Sims and H. J. Spinelli, *J. Coat. Technol.*, **59** (752), 125 (1987).
12. J. D. Schofield and J. Toole, *Polym. Paint Colour J.*, December 10/24, 170 (1980).
13. J. D. Schofield, *J. Oil Colour Chem. Assoc.*, **74**, 204 (1991).
14. E. E. Jaffe, C. D. Campbell, S. B. Hendi, and F. Babler, *J. Coat. Technol.*, **66** (832), 47 (1994).
15. P. Bugnon, *Prog. Org. Coat.*, **29**, 39 (1996).
16. K. Hamann and R. Laible, *Proc. FATIPEC Congress* (1978), p. 17.
17. J. Schroeder, *Prog. Org. Coat.*, **16**, 3 (1988).
18. T. C. Patton, *Paint Flow and Pigment Dispersion*, 2nd. ed., Wiley-Interscience, New York, 1979.
19. T. K. Hay, *J. Paint Technol.*, **46** (591), 44 (1974).
20. W. H. Morrison, Jr., *J. Coat. Technol.*, **57** (721), 55 (1985).
21. T. Losoi, *J. Coat. Technol.*, **61** (776), 57 (1989).
22. J. Clayton, *Surface Coat. Intl.*, **94**, 414, (1997).
23. H. J. Spinelli, *Prog. Org. Coat.*, **27**, 255 (1996).
24. R. E. Smith, *J. Coat. Technol.*, **60** (761), 61 (1988).
25. A. Brisson, G. L'Esperance, and M. Caron, *J. Coat. Technol.*, **63** (801), 111 (1991).
26. J. G. Balfour and M. J. Hird, *J. Oil Colour Chem. Assoc.*, **58**, 331 (1975).
27. J. E. Hall, R. Benoit, R. Bordeleau, and R. Rowland, *J. Coat. Technol.*, **60** (756), 49 (1988).
28. J. E. Hall, R. Bordeleau, and A. Brisson, *J. Coat. Technol.*, **61** (770), 73 (1989).
29. R. R. Blakely, *Proc. FATIPEC Congress* (1972) p. 187.

CHAPTER 21

Pigment Volume Relationships ___

Traditionally, coatings formulators have worked with weight relationships, but volume relationships are generally of more fundamental importance and practical significance. While there had been a few previous isolated examples of recognition of the importance of volume considerations in performance of coatings, credit for full realization of this importance belongs to Asbeck and Van Loo [1]. They looked at a series of performance variables as a function of the *pigment volume concentration* (*PVC*), that is, the volume percent of pigment in a *dry* film. (Some authors call PVC pigment volume content.) The term PVC should never be used to specify the volume of pigment in a wet coating. This has been done occasionally in the literature and has been responsible for serious misinterpretations. Although PVC is usually expressed as a percentage value, some authors express PVC as pigment volume fraction in equations without bothering to tell the reader.

Asbeck and Van Loo observed that many properties of films change abruptly at some PVC as PVC is increased in a series of formulations. They designated the PVC at which these changes occurred as the *critical pigment volume concentration*, CPVC. They also defined CPVC as that PVC where there is just sufficient binder to provide a complete adsorbed layer on the pigment surfaces and to fill all of the interstices between the particles in a close-packed system. Below CPVC, the pigment particles are not close-packed and binder occupies the "excess" volume in the film. Above CPVC, the pigment particles are close packed, but there is not enough binder to occupy all of the volume between the particles resulting in voids in the film. Slightly above CPVC, the voids are air bubbles in the film, but as PVC increases, the voids interconnect and film porosity increases sharply.

Bierwagen et al. have emphasized that when films are prepared from coatings with PVC near CPVC, there may not be uniform distribution of pigment through the dry film, so parts of the film may locally be above CPVC and other parts below CPVC [2]. Also, some properties start to change as soon as PVC increases to the extent that there are air voids in films, and other properties change when the PVC is sufficiently greater than CPVC that the film begins to be porous. Coatings with flocculated pigment clusters result in films with nonuniform distribution of pigment particles, and therefore, CPVC with flocculated pigment particles is lower than the CPVC with the same pigment combination that is not flocculated.

CPVC with different pigments varies over a wide range, at least from 18 to 68. (See Section 21.2 for discussion of factors affecting CPVC.) The precision and accuracy of CPVC values vary, depending on the method of determination. (See Section 21.3.)

21.1. RELATIONSHIPS BETWEEN FILM PROPERTIES AND PVC

As PVC is increased in a series of coatings made with the same pigments and binders, density increases to a maximum when PVC equals CPVC and then decreases, as shown schematically in Figure 21.1(a). The increase reflects that pigments, with few exceptions, have higher densities than binders. However, above CPVC, the lower density of air reduces the film density.

As shown schematically in Figure 21.1(b), tensile strength generally increases with PVC to a maximum at CPVC but then decreases above CPVC. Below CPVC, the pigment particles serve as reinforcing particles and increase the strength. It can be considered that polymer molecules adsorb on the surface of multiple pigment particles, providing the equivalent of cross-links. Accordingly, more force is required to break this physical network as pigment level increases. However, above CPVC, air voids weaken the film; abrasion and scrub resistances of films drop above CPVC.

Plots of adhesion as a function of PVC are sometimes reported, as shown in Figure 21.1(c). It appears that adhesion goes through a maximum at CPVC, but the relationship may not be that simple. In adhesion testing, failure can be cohesive or adhesive. (See Section 6.6.) In cohesive failure, the coating film breaks; in adhesive failure, there is separation at the interface between coating and substrate. In many, if not all, cases, in which results such as shown in Figure 21.1(c) are obtained, the failures at low PVC are cohesive failures, as are those above CPVC; only those failures near CPVC, if any, are adhesive failures. Such results reflect changes in cohesive tensile strength of the film, not changes in adhesion between substrate and coating. If the cohesive strength of a film is lower than its adhesive strength, failure does not provide any information about adhesive strength.

Stain resistance decreases above CPVC, since staining liquids can penetrate into pores, leaving color behind that is difficult to remove. Porosity also affects other properties. If one applies a single coat of a coating with PVC above CPVC to steel and exposes the panel to humidity, rapid rusting can occur, since the pores permit water and oxygen to get to the

MECHANICAL PROPERTIES

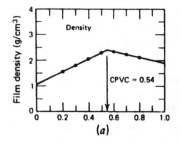 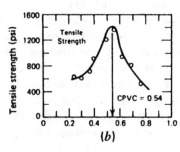 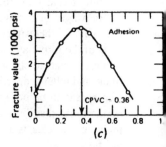

(a)   (b)   (c)

Figure 21.1. Relationship between PVC (volume fractions) and (a) density, (b) tensile strength, and (c) "adhesion." (From Ref. [3], with permission.)

surface of the steel with little interference. On the other hand, if one applies an alkyd-based coating with PVC above CPVC to a wood substrate, one is less likely to get blistering than with a similar coating with PVC below CPVC. When water gets into the wood behind the coating, it can escape through pores in an alkyd coating when PVC is above CPVC but not when PVC is below CPVC.

Gloss is related to PVC. In general, unpigmented films have high gloss. The initial few percent of pigment has little effect on gloss, but above a PVC of 6 to 9, gloss drops until PVC approaches CPVC, as shown in Figure 21.2. (See Section 18.3.)

It is almost always desirable to make primers with a high PVC, since the rougher, low gloss surface gives better intercoat adhesion than a smooth, glossy surface. It is sometimes desirable to design a primer with PVC greater than CPVC. Adhesion of a top coat to such a primer is enhanced by mechanical interlocking resulting from penetration of vehicle from the top coat into pores of the primer. Many of the pores in the primer are filled with binder from the top coat, which decreases the PVC of the top coat resulting in loss of gloss of the top coat. Such a primer is said to have poor *enamel hold out*. The primer PVC should be only enough higher than CPVC to provide adhesion to minimize loss of gloss of the top coat.

PVC (see Section 18.3) also affects hiding; as pigmentation increases, hiding generally increases. Initially, hiding increases rapidly, but then hiding tends to level off. In the case of rutile TiO_2, as shown in Figure 18.9, hiding goes through a maximum, gradually decreases with further increase in PVC, and then increases above CPVC. This increase in hiding above CPVC results from air voids left in the film when PVC is above CPVC. The refractive index of air (1.0) is less than that of the binder (approximately 1.5), so there is light scattering by the air interfaces in addition to interfaces between pigment and binder. The effect becomes large as interfaces between air and pigment increase with increasing PVC. For example, if rutile TiO_2, refractive index of 2.73, is in the formulation, the difference between the refractive index of TiO_2 and air (1.73) is larger than between TiO_2 and binder (1.23).

Owing to the high cost of TiO_2, coatings are not generally formulated with a PVC of TiO_2 greater than about 18% since incremental hiding at higher PVC is not cost efficient. (This value is dependent on the actual TiO_2 content of the TiO_2 pigment and on the stability of the TiO_2 dispersion.) Rather than using large quantities of TiO_2 in high PVC coatings, lower-cost inert pigments are used to occupy additional volume. The effect is illustrated in Figure 21.3. The scattering efficiency versus PVC data are based on a series

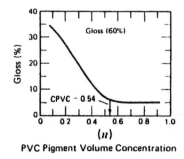

Figure 21.2. Relationship between gloss and PVC (volume fractions). (From Ref. [3], with permission.)

of paints made with four TiO_2 pigments that have differing oil absorption values and a clay pigment (Lorite). Maximums in scattering efficiencies (below CPVC) occur at about 25 PVC (the PVC is that of the combination of TiO_2 and inert pigment). Above CPVC, there is an increase in scattering coefficient, and hence, in hiding. Note that the CPVC generally increases as the oil absorption of the TiO_2 decreases, but scattering efficiency increases with oil absorption. These changes presumably reflect decreasing particle size as oil absorption increases. The scattering efficiency of TiO_2 is also affected by the particle size of the inert pigments used with it [5]. Inclusion of some inert pigment with a particle size smaller than that of TiO_2 (i.e., less than 0.2 μm) increases the efficiency by acting as a so-called *spacer* for the TiO_2 particles.

The increase in hiding above CPVC can be useful. For example, the hiding of white ceiling paints can be improved by formulating above CPVC. This permits hiding with one coat, which is particularly desirable in ceiling paints. The stain resistance and scrub resistance of the paint are inferior to similar paints with PVC less than CPVC, but stain and scrub resistance are not important in ceiling paints. The high PVC paints have the added advantage of lower cost.

Tinting strengths of white coatings increase as the PVC of a series of coatings is increased beyond CPVC. The air voids above CPVC increase light scattering so that a colored paint dries with a lighter color than one with the same amount of color pigment but with a PVC below CPVC.

Changes in volume relationships almost always control physical properties of films. Thus, there is a major advantage to formulating coatings on the basis of volume rather than weight relationships. For any particular application, there is a ratio of PVC to CPVC most appropriate for the combination of properties needed for that application. Once this ratio has been established, changes in pigment combinations for that application should be made such that this PVC/CPVC ratio is maintained. This important concept is developed in

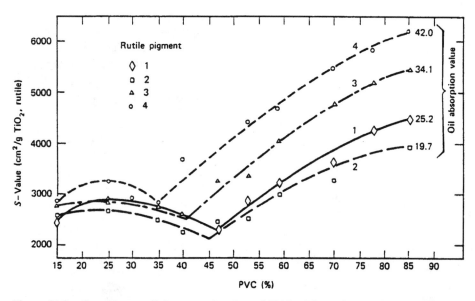

Figure 21.3. Scattering coefficient as a function of PVC of flat paints made with different oil absorption grades of rutile TiO_2 and Lorite inert pigment. (From Ref. [4], with permission.)

detail by Bierwagen [6]. He also emphasizes that one must exercize particular care when formulating with PVC near CPVC, since relatively small changes in pigment ratios or in packing, flocculation, or film formation can substantially affect film properties [7].

While it is desirable to have accurate determinations of CPVC available, even reasonable estimates can be useful, since they permit one to focus a series of experiments within a range approximating the desired PVC to CPVC ratio. The decision as to the proper pigment loading of a particular formulation should be based on actual experimental data, not on the theoretically best PVC value. Even without CPVC data, the concept is of value to the formulator. It is critical to recognize that performance properties vary with volume, not with weight relationships, and that CPVC increases with increasing particle size distribution. Qualitative use of oil absorption and density values for individual pigments, together with recognition that CPVC increases when mixtures of pigments with different particle size distributions are used, permit one to start formulating in a reasonable range of compositions. On the other hand, using weight relationships, one is working blindly.

21.2. FACTORS CONTROLLING CPVC

There are large variations in CPVC, depending on the pigment or pigment combination in a coating and the extent, if any, of pigment flocculation. With the same pigment composition, the smaller the particle size, the lower the CPVC. The ratio of surface area to volume is greater for smaller particle size pigments; hence, a higher fraction of binder is adsorbed on the surface of the smaller pigment particles and the volume of pigment in a close-packed final film is smaller. One can consider this effect from the standpoint of a modification of the Mooney equation, shown in Eq. 21.1. The volume fraction of internal phase is the sum of the volume of pigment V_p and the volume of the adsorbed layer V_a. In a solvent-free system, V_p equals PVC and V_p is equal to CPVC when $V_p + V_a = \phi$. Thus, if V_a is larger, the V_p corresponding to CPVC is smaller. Note that at CPVC, viscosity of a solvent-free system approaches infinity.

$$\ln \eta = \ln \eta_e + \frac{K_E(V_p + V_a)}{1 - \frac{(V_c + V_a)}{\phi}} \tag{21.1}$$

CPVC depends on particle size distribution; the broader the distribution, the higher the CPVC. As discussed in Section 3.5, broader particle size distribution of spherical, dispersed-phase systems increases packing factor. In low gloss coatings, the least expensive component of the dry film is inert pigment; to minimize cost, it is desirable to maximize inert pigment content. Since properties should remain constant, one does not want to change the PVC to CPVC ratio; the lowest cost systems, therefore, are those with the highest CPVC. This means minimizing the amount of very fine particle size pigment, but at the same time maximizing particle size distribution. Compromises are needed, but broad particle size distribution is advantageous.

Pigment dispersion affects CPVC; CPVC of films from coatings in which the pigment is flocculated are lower than CPVC from corresponding coatings with nonflocculated pigment. Films prepared from coatings with flocculated pigment clusters have less uniform distribution of pigment, and hence, are more likely to have portions where there are local high concentrations of pigment. Solvent containing resin is trapped inside clusters of pigment particles. When the coating dries, the solvent diffuses out of the resin solution trapped in the floccules, leaving behind insufficient binder to fill the spaces. In one example, it is reported that CPVC decreased from 43 to 28 with increasing flocculation [8]. Asbeck suggests using the term *UCPVC, ultimate CPVC*, to designate CPVC with a nonflocculated pigment combination [8]. It seems to us, better to recognize that flocculation is one of many variables that affect CPVC, especially since it may be difficult to determine UCPVC experimentally.

One might expect CPVC to be dependent on binder composition, since the composition affects the thickness of adsorbed layers; thicker adsorbed layers might be expected to lead to lower CPVC. Actually, CPVC with a given pigment or pigment combination seems to be essentially independent of the binder composition (except in latex paints, which are discussed in Section 21.4). It may be that the differences are too small to detect with the relatively imprecise methods for determining CPVC. It may also be that the shrinkage forces encountered when films are formed press the particles against each other to the extent that only some minimal layer of binder is left between particles, regardless of the original adsorbed layer thickness. No basic studies of this phenomenon have been reported.

21.3. DETERMINATION OF CPVC

Critical pigment volume concentration has been determined by many different procedures [6,9]. In many cases, the precision-that is, the reproducibility of the value-is relatively poor, and the accuracy-that is nearness to the true value-is also sometimes poor or indeterminate. Bierwagen has emphasized that in the formation of films with PVC at or near CPVC, local fluctuations within a film of volume fractions of binder and pigment are possible [2,7]. Thus, there could be parts of a film with PVC > CPVC, while the average composition might be less than CPVC. In view of these uncertainties, one must be careful in assessing the importance of small differences in PVC and CPVC values.

Many changes in film properties have been used as a means of determining CPVC; tinting strength is one of the most widely used. A series of white paints with increasing PVC are prepared and tinted with the same ratio of color to white pigment. Above CPVC, the white tinting strength of the coating increases due to the "white" air bubbles above CPVC. It is suggested that the values of CPVC obtained by this and other optical methods are somewhat low, due to nonuniform distribution of pigment, resulting in parts of a film being above CPVC while others are below [2]. The technique is most easily applicable to white paints, but can also be applied to colored paints.

Another method is determination of density as a function of PVC. Since the density of most pigments is higher than those of binders and the density of air is lower than those of binders, density maximizes at CPVC. Due to nonuniform distribution of pigment in the films, it is suggested that the density method tends to give high CPVC values [2]. The CPVC can be determined by filtering a coating and measuring the volume of the pigment filter cake. Asbeck recommends a specially designed filter that he calls a *CPVC cell* [8].

The CPVC for a pigment or pigment combination can be calculated from oil absorption (OA) (see Section 20.2.2), as shown in Eq. 21.2, provided the OA value is based on either a nonflocculated dispersion or determined at a sufficiently high shear rate that any floccules are separated. The definitions of both OA and CPVC are based on close-packed systems with just sufficient binder to adsorb on the pigment surfaces and fill the interstices between the pigment particles. Oil absorption is expressed as g of linseed oil per 100 g pigment; CPVC is expressed as mL of pigment per 100 mL of film. ρ is the density of the pigment(s), and 93.5 is 100 times the density of linseed oil. (Both OA and CPVC are expressed as percentages, not as fractions.)

$$CPVC = \frac{1}{1 + \dfrac{(OA)(\rho)}{93.5}} \qquad (21.2)$$

The significance of the interrelationship depends on the observation that oil absorption and CPVC are approximately independent of the binder, provided the pigment particles are not flocculated. Since the accuracy of calculated CPVC depends on the accuracy of OA determinations, OA values determined by a mixing rheometer, such as a Brabender Plastometer (see Section 3.3.2), are preferable to values determined by the spatula rub up method [10]. Such an intensive mixer provides greater shear rate for separation of pigment aggregates than spatula rubbing; and also the data points are taken while the dispersion is under high shear, so even if the binder does not stabilize the pigment against flocculation, the volume fractions still represent a nonflocculated system. Effects of variation in procedure for determination of OA, including the use of other liquids besides linseed oil, are reported in Ref. [11]. The discussion of OA determination in this paper is interesting and useful; however, the conclusions drawn by the author about CPVC are erroneous. The author failed to recognize that when the pigment dispersions were diluted with solvent, the pigment flocculated.

Since many OA values are not accurate, Asbeck recommends against such calculations [8], but many workers have found the calculations useful. Although the CPVC for individual pigments and for specific combinations of pigments can be calculated from oil absorptions, CPVC values of pigment combinations cannot be calculated from these values alone, since differences in particle size distribution with pigment combinations affect the packing factor. The OA values for each combination of pigments can be determined experimentally. A variety of equations have been developed to calculate CPVC from data on individual pigments [9]. The most successful equations use OA values, densities, and average particle sizes of the individual pigments [12,13]. The equations assume that the particles are spheres, a fair assumption for many, but not all, pigments. Calculated values correspond reasonably well to experimentally determined CPVC values.

21.4. CPVC OF LATEX PAINTS

There is considerable controversy about the applicability of CPVC to latex paints. Some maintain that one cannot apply the concept. Others maintain that there is no fundamental difference in the CPVC concept between solvent-borne and latex paints. Still others maintain that the concept is useful in latex paints, but that CPVC must be thought of differently in latex paints than in solvent-based paints. Bierwagen emphasizes that experimental errors in determining CPVC of latex paints are even greater than with solvent-borne paints, so one must use caution in reaching conclusions [14]. Reference [15] includes a review of the literature.

In a study of the effect of PVC on hiding of latex paints, it was concluded that CPVC was lower in latex paint than in solvent-based paint made with the same pigment composition [16]. Patton has recommended that the term *latex CPVC*, LCPVC, be used to distinguish from CPVC of solvent-borne paints. Although CPVC is approximately independent of the binder in solvent-borne paints, LCPVC varies with the latex and some other components of latex paints. It has been found that LCPVC increases as the particle size of the latex decreases. Also, LCPVC increases as the T_g of the latex polymer decreases, and a coalescing agent increases LCPVC. Since LCPVC is smaller than CPVC, the ratio of volume percent of binder in the film of a solvent based paint, V_s, to that of a latex based paint, V_1, will always be less than 1 with the same pigment combination. This ratio, shown in Eq. 21.3, has been called the *binder index e* [16].

$$e = \frac{V_s}{V_1} \tag{21.3}$$

It has been proposed and demonstrated with a limited number of examples that this ratio is independent of the pigment combination. It follows that if one knows the binder index for a latex, one can calculate the LCPVC for paints made with that latex and other pigments from the CPVC values calculated from oil absorptions (or, for mixtures, by calculation from OA, density, and particle size).

The difference between CPVCs of latex- and solvent-based paints results from the difference in film formation. To illustrate, let us consider a highly idealized comparison. Let us assume that all the pigment particles in the solvent-based paint are spheres with the same diameter and that the CPVC is 50. In the latex paint, we will use the same pigment together with spherical latex particles that have the same diameter as the pigment particles. In the solvent-based paint, there is a layer of resin swollen with solvent on the surface of all the pigment particles. Ideally, when a solvent-borne paint with PVC = CPVC is applied, the solvent evaporates and the "resin-coated" pigment particles arrange themselves in a random close-packed order with binder filling in all the spaces between the pigment particles. At the same ratio of pigment to binder as in the solvent-based paint, the idealized latex paint would contain equal numbers of latex particles and pigment particles. When we apply a layer of the latex paint, the water evaporates and we get a close-packed system of spheres. However. some of the spheres are pigment particles and some are latex particles. There would not be a uniform arrangement of alternating latex and pigment particles in a three-dimensional lattice; rather, there would be a statistical distribution of particles. In some areas there would be clusters of pigment particles, and in other areas there would be clusters of latex particles. As the film forms, the latex particles coalesce, flowing around the pigment particles. However, the viscosity of the coalesced binder is high, and it is difficult for the polymer to penetrate into the center of clusters of pigment particles. As film formation proceeds, the water left inside the pigment clusters diffuses out of the film, leaving behind voids. Although the PVC of 50 in the solvent-based paint equaled CPVC, the same PVC of 50 in the latex paint results in a film with PVC > LCPVC.

The probability of having clusters of pigment particles can be decreased by increasing the number of latex particles until void-free films are obtained, but this necessarily reduces PVC. If the T_g of the polymer is lower, the viscosity of the polymer at the same temperature is lower, and the distance that the latex can penetrate into clusters of pigment particles increases. Thus, the LCPVC of paint made with lower T_g latex is higher, although still lower than the CPVC of a solvent-based paint. Analogously, reduction of viscosity of the polymer by coalescing agent increases LCPVC. It follows that the higher the temperature during film formation, the higher the LCPVC should be, although no experimental test of this hypothesis has been published. It also follows, although again untested, that LCPVC should depend on time. Although the viscosity of the polymer is high and inhibits flow between the particles, the viscosity is not infinite, and perhaps in time, voids would fill.

If a smaller particle size latex is used with the same ratio of pigment to binder volume, the number of latex particles in our idealized paint would be larger than the number of pigment particles. Now the probability of clusters of pigment particles forming would be reduced. As a result, one would expect, as is found experimentally, that LCPVC increases with decreasing particle size of the latex. Particle size distribution of a latex affects its packing factor [18], which would presumably affect LCPVC. A quantitative study of the effect of latex particle size on CPVC using a series of monodisperse vinyl acetate/butyl acrylate latexes with TiO_2 pigment showed that CPVC depended on the ratio of the number of latex and pigment particles and the ratio of their diameters [15].

Not only does larger particle size latex decrease LCPVC, flocculation of the latex particles would be expected to decrease LCPVC. If one formulates a latex paint so that the PVC is slightly less than LCPVC with an unflocculated latex, but the latex flocculates, the PVC would be greater than LCPVC. There has been no data published on the effect of latex flocculation on LCPVC. A simulation program using particle size distributions of latex and pigment and a measure of the deformability of latex particles has been developed to predict the CPVC of simple latex paints [19]. There is need for further research on pigment volume relationships in latex paints.

GENERAL REFERENCE

G. P. Bierwagen and T. K. Hay, *Prog. Org. Coat.*, **3**, 281 (1975).

REFERENCES

1. W. K. Asbeck and M. Van Loo, *Ind. Eng. Chem.*, **41**, 1470 (1949).
2. R. S. Fishman, D. A. Kurtze, and G. P. Bierwagen, *Prog. Org. Coat.*, **21**, 387 (1993).
3. T. C. Patton, *Paint Flow and Pigment Dispersion*, 2nd. ed., Wiley-Interscience, New York, 1979, p. 172.
4. P. B. Mitton, in T. C. Patton, Ed., *Pigment Handbook*, Vol. 3, Wiley-Interscience, New York, 1973, p. 321.
5. J. Temperley, M. J. Westwood, M. R. Hornby, and L. A. Simpson, *J. Coat. Technol.*, **64** (809), 33 (1992).
6. G. P. Bierwagen and T. K. Hay, *Prog. Org. Coat.*, **3**, 281 (1975).
7. G. P. Bierwagen, *J. Coat. Technol.*, **64** (806), 71 (1992).
8. W. K. Asbeck, *J. Coat. Technol.*, **64** (806), 47 (1992).
9. R. W. Braunshausen, Jr., R. A. Baltrus, and L. De Bolt, *J. Coat. Technol.*, **64** (810), 51 (1992).
10. T. K. Hay, *J. Paint Technol.*, **46** (591), 44 (1974).
11. H. F. Huisman, *J. Coat. Technol.*, **56** (712), 65 (1984).
12. G. P. Bierwagen, *J. Paint Technol.*, **44** (574), 46 (1972).
13. C. R. Hegedus and A. T. Eng, *J. Coat. Technol.*, **60** (767), 77 (1988).
14. G. P. Bierwagen and D. C. Rich, *Prog. Org. Coat.*, **11**, 339 (1983).

15. G. del Rio and A. Rudin, *Prog. Org. Coat.*, **28**, 259 (1996).

16. F. Anwari, B. J. Carlozzo, K. Choksi, M. Chosa, M. DiLorenzo, C. J. Knauss, J. McCarthy, P. Rozick, P. M. Slifko, and J. C. Weaver, *J. Coat. Technol.*, **62** (786), 43 (1990).

17. T. C. Patton, *Paint Flow and Pigment Dispersion*, 2nd. ed., Wiley-Interscience, New York, 1979, p. 192.

18. K. L. Hoy and R. H. Peterson, *J. Coat. Technol.*, **64** (806), 59 (1992).

19. G. T. Nolan and P. E. Kavanaugh, *J. Coat. Technol.*, **42** (850), 37 (1995).

CHAPTER 22

Application Methods

Many methods are used to apply coatings; several of the most important methods are covered in this chapter. Further information and discussion of additional application methods can be found in the general reference at the end of the chapter. Many factors affect the choice of method to be used for a particular application, including: capital costs, operating costs, film thickness, appearance requirements, and the structure of the object to be coated. Reduction of VOC emissions is a driving force in development of new modifications of application methods. The amount of solvent required and the possibility of recovering the solvent are affected by the application method. Electrodeposition and application of powder coatings are discussed in Chapters 26 and 27, respectively.

22.1. BRUSHES, PADS, AND HAND ROLLERS

Brushes, pads, and hand rollers are frequently used for application of architectural paints. Although the same paints can usually be applied by spray gun, few do-it-yourselfers use spray guns; on the other hand, professional painters use spray guns whenever possible to save time.

22.1.1. Brush and Pad Application

A variety of brushes is available: narrow and wide, long handled and short handled, nylon, polyester, and hog bristle [1]. Hog bristles are appropriate for solvent-borne paints, but not for water. Nylon bristles are appropriate for water-borne paints, but are swollen by some solvents. Polyester bristle brushes can be used with either. Brushes have in common a large number of bristles that hold paint in the spaces between the bristles. When the paint is applied, pressure forces paint out from between the bristles. The forward motion of the brush splits the layer of paint so that part is applied to the surface and part remains on the brush. Open-cell polyurethane foam "brushes" are also used.

Viscosity characteristics are critical in using brushes. Pickup of paint on the brush is controlled by paint viscosity at a relatively low shear rate, around 15–30 s^{-1}, the shear rate of dipping and removing a brush from a paint can. If the viscosity is too high, too much paint is brought out of the container with the brush; if the viscosity is too low, too little paint is on the brush. Ease of brushing requires a low viscosity. The shear rate between the brush and the substrate is relatively high, in the range of 5000–20,000 s^{-1} [2]. The viscosity at high shear rate controls the ease of brushing; high viscosity leads to high "brush drag." Film thickness applied is affected by viscosity at high shear rate; applied film thickness increases with increasing viscosity. For most applications, a viscosity at high shear rate of 0.1 to 0.3 Pa·s is appropriate. (See Chapter 31 for further discussion.) Solvents with relatively slow evaporation rates must be used to slow the increase in viscosity of the paint on the brush.

When paint is applied by brush, the surface of the wet film has furrows called *brush marks*. These furrows do not result from the individual bristles of the brush, as is evident from comparing the size and number of brush marks with the bristles of the brush. When polyurethane foam "brushes" are used, brush marks still result, even though the brush has no bristles. Brush marks result from splitting the wet film between the brush and the substrate as the brush is applying the paint. Whenever layers of liquids are split, the surfaces are initially irregular. These irregularities are made into lines by the linear movement of the brush. It is desirable to formulate a coating such that the brush marks flow out before the film dries. As discussed in Sections 23.2 and 23.3, low viscosity promotes leveling, but increases the probability of sagging. Thixotropic flow properties are generally desirable because they delay the increase in viscosity after brushing, permitting a compromise between leveling and sagging. Formulation problems are particularly difficult in the case of latex gloss paints, as discussed in Section 31.3.

Pad applicators are also used in the do-it-yourself market [1]. The most common type of pad consists of a sheet of nylon pile fabric attached to a foam pad that is attached to a flat plastic plate with a handle. For low-viscosity coatings like stains and varnishes, a lambs-wool pad is used. Pads have a number of advantages as compared to brushes. Pads hold more paint than a similar width brush and can apply paint up to twice as fast. In general, pad application leaves a smoother layer than brush application. Extension handles can be used with pads, reducing the need for moving ladders. Pads, especially refills, are less expensive than brushes. On the other hand, pads require use of trays, which results in some paint loss and solvent evaporation. Cleanup of pads is more difficult than brushes.

22.1.2. Hand Roller Application

Hand rollers are the fastest method of hand application and are widely used in applying architectural paints to walls and ceilings. A variety of rollers and roll coverings are available [1]. There are rollers with built-in wells to minimize need for dipping in the tray; there are also power-filled rollers.

Viscosity requirements are similar to those for brush application. When paint is applied by roller, there is also film splitting. As the roller moves, the film is stretched; flow is in an extensional mode (see Section 3.6.3) within the film. Due to imbalances of pressures, the film breaks at different times, leaving a ribbed surface. Some of the film extends into fibers still farther from the nip. As the roller moves, fibers are drawn longer and longer and eventually break. After breaking, the ends of the fibers are pulled by surface tension to minimize surface area of the applied coating, but if there is not sufficient time at low viscosity for leveling to occur, track marks are left in the coating. A further complication arises with longer fibers that may break in two places; as a result, droplets of loose paint fly off to land on the painter or the floor. This is called *spattering*. If spattering could be eliminated, painting speed could be increased, and time spent on masking and laying drop cloths could be reduced. However, all current formulations spatter, some much more than others. The phenomenon is not completely understood. Glass [3,4] has pointed out that as the fibers are drawn out, extensional flow, rather than shear flow, is involved. (See Section 5.6.3.) It follows that extensional viscosity, rather than shear viscosity, affects development of fibers from roller application and, in turn, the roughness of the film and the degree of spattering. Spattering can occur with any kind of paint, but is particularly difficult to minimize in latex paints. Formulation variables affecting spattering of latex paints are discussed in Section 31.2.

22.2. SPRAY APPLICATION

Spraying is a widely used method for applying paints and coatings in architectural and especially in industrial applications. Spraying is much faster than application by brush or hand roller. Spraying is used on flat surfaces, but is particularly applicable to coating irregularly shaped articles. Many different kinds of equipment are used for spraying; all atomize the liquid coating into droplets. Droplet size depends on the type of spray gun and coating; variables include air and fluid pressure, fluid flow, surface tension, viscosity, and, in the case of electrostatic application, voltage. The choice of spray system is affected by capital cost considerations, efficiency of paint utilization, labor costs, size and shape of objects to be coated, among other variables. Coating formulations must be established for the particular spray equipment and conditions.

The principal disadvantage is the inefficiency of application, since only a fraction of the spray particles are deposited on the object being sprayed. One must mask areas where coating is not desired, and there is a likelihood of contaminating the area with spray dust. Some of the droplets approach the surface and *bounce back*, carried by eddy currents of air. The higher the pressure, the higher the forward velocity of the air and, as a result, the greater the percentage of bounce back. Some droplets miss the object being coated; this is called *overspray*. Some droplets may fall out of the spray pattern under the force of gravity. *Fall out* is higher with large spray patterns and longer distances between the spray gun and the surface being sprayed. The sum of all this waste determines *transfer efficiency*; transfer efficiency is defined as the percent of coating solids leaving the gun that is actually deposited on the coated product.

Spray dust can cause many problems. If spray dust lands on the surface of a wet coating dirt contamination results. As discussed in Section 23.4, if the surface tension of the spray droplets is different than that of the wet coated surface cratering will result. Minimizing spray dust by choice of application equipment, ventilation, and careful maintenance and cleaning of equipment is vital.

Transfer efficiency is an important cost factor, since low transfer efficiency means more coating must be used for the same surface area. Also, high transfer efficiency reduces VOC emissions because less coating is used. Transfer efficiency is affected by many variables. The size and shape of the product being coated is a major variable. Transfer efficiency when spraying a chain link fence is low, and transfer efficiency when spraying a large wall is high. Less obvious variables are conveyor line speed and the way objects being coated are hung on a conveyor line. The spray method has a major effect on transfer efficiency, whether the application is manual or automated; whether air, airless, or rotary spray guns are used; and whether or not an electrostatic system is used. For manual systems, the skill of the sprayers is very important. For automated systems, system design is critical. Some

Table 22.1. Typical Baseline Transfer Efficiencies

Type of Spray Gun	Transfer Efficiency (%)
Air	25
Airless	40
Air-assisted airless	50
High volume, low pressure air	65
Electrostatic air	60–85
Electrostatic rotary	65–94

Source: Reference [5].

factors involved are distance between the gun and the surface, angle of the spray gun, stroke speed uniformity, extent and uniformity of overlapping, and precision of triggering.

To compare transfer efficiency of different spray methods, the EPA along with spray gun and coatings manufacturers, has developed a standard procedure for comparing *baseline transfer efficiency*, adopted by the ASTM as method D5009-96. Typical percentages are given in Table 22.1; actual efficiencies are affected by many variables besides the spray method. For example, transfer efficiency in coating a chain link fence is low at best, and transfer efficiency in coating large wall expanses is high in any case.

Very reactive coatings with short pot lives, such as some two-package polyurethane coatings, cannot be sprayed using conventional spray guns; dual feed guns are required. In the most common type of equipment, the two components are metered into a small, efficient mixing chamber just before the spray gun orifice. The average dwell time in the mixing chamber is a fraction of a second, and the average residence time in the gun is not much longer. The equipment is designed so that when spraying is interrupted, the mixing chamber and the gun are automatically flushed with solvent to prevent clogging. Frequent checking and maintenance are required to assure that the proper ratio of components is being applied. In other types of equipment, the components are sprayed from two orifices; mixing occurs after atomization so that the coating can not gel in the equipment.

22.2.1. Air Spray Guns

Air spray guns cause atomization of the coating by fine streams of compressed air. It is the oldest spray method and is still used. Figure 22.1 shows a cross section of an air spray gun. A stream of coating is driven through the nozzle orifice by relatively low pressure, 10–50 kPa (1.5–7 psi), or, in other types of guns, by the suction caused by the rapid air flow past the outside of the orifice. The stream of coating coming out of the orifice is atomized into small droplets by fine streams of compressed air at pressures of 250–500 kPa (35–70 psi). The degree of atomization is controlled by: (1) viscosity of the coating (the higher the viscosity at the high shear rate encountered going through the orifice, the larger the particle size), (2) air pressure (higher air pressure, smaller particle size), (3) diameter of the orifice (smaller orifice, smaller particles), (4) pressure forcing or pulling the coating through the orifice (higher pressure, smaller size), and (5) surface tension (lower surface tension, smaller the particle size). The outer jets of compressed air shown in the figure adjust the shape of the stream of atomized particles coming from the gun. If these jets were not present, a cross section of the stream would be roughly circular. Generally, an elliptical

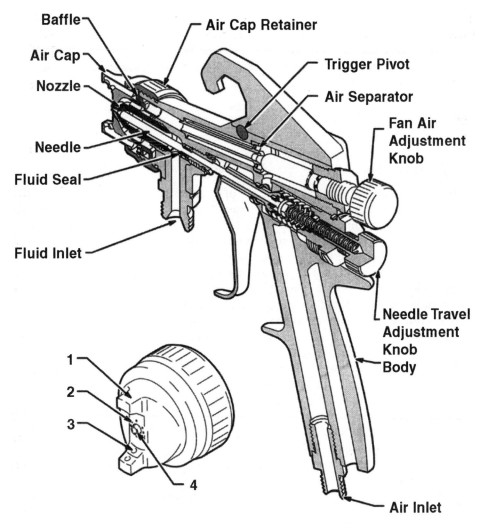

Figure 22.1. Schematic cross section of an air spray gun (Delta Spray[TM], Graco, Inc.). Schematic of spray gun nozzle: 1. Wings or horns; 2. Angular converging holes; 3. Side-port holes; and 4. Annular ring around the fluid tip. (Courtesy of Graco, Inc., Minneapolis, MN.)

cross-sectional pattern permits more efficient application; the pattern is a flattened cone, often called a *fan*. Guns can be hand held or attached to robots.

Air spray guns are less expensive than other types of guns. Atomization can be finer than with other spray guns. The system is versatile, and any sprayable object can be sprayed with air spray. The level of control can be high if the operators are skilled. However, as noted in Table 22.1, transfer efficiency is the lowest of all the spray methods.

A substantial improvement in transfer efficiency of air spraying has been made by use of *high volume, low pressure* (HVLP) air guns. These guns are designed to operate at lower air pressures, 20–70 kPa (3–10 psi), but with higher air volumes and large, unrestricted air passages to handle a large volume of air. Because of the low pressures, bounce back is reduced and transfer efficiencies of 65% or even higher can be achieved. High volume, low pressure spray guns are used increasingly in automotive repair shops. California South Coast VOC regulations require 65% transfer efficiency and air pressures of 70 kPa or less. An alternative approach to HVLP guns is low volume, low pressure (LVLP) guns. In LVLP guns, the air pressure is also less than 70 kPa, but the air volume is reduced by mixing the air and coating inside the gun.

22.2.2. Airless Spray Guns

With airless spray guns, coating is forced out of an orifice at high pressure, 5 to 35 MPa. The coating comes out of the orifice as a "sheet." As the sheet extends in moving away from the orifice, flow instabilities cause the formation of ligaments, followed by further disintegration into droplets [6]. Atomization is controlled by the relative velocity between the sheet and the contiguous air (higher relative viscosity, smaller droplets), viscosity (higher viscosity, larger particle size), pressure (higher pressure, smaller particle size), and surface tension (lower surface tension, smaller particle size). The shape of the fan or pattern of the spray is controlled by the orifice size and shape. Air-assisted, airless spray guns are also available; the atomization is airless but there are external jets to help shape the fan pattern, confining the smaller droplets within the spray pattern [7]. Both hand held and robot airless guns are available.

The size of droplets from airless spray guns is larger than those from air spray guns,: 70–150 μm, as compared with 20–50 μm [8]. Airless guns give a so-called *fishtail spray*; that is, there is a relatively sharp edge to the spray droplet fan with quite uniform droplet distribution within the fan. In contrast, the fan from air guns is *feathered* at the edge; that is, the number of droplets decreases at the edge of the fan, with some being quite widely spaced. As a result of these differences, one can generally achieve more uniform film thickness with air spray than with airless spray; air-assisted airless application gives intermediate results.

Coatings can be applied more rapidly by airless than by air guns, permitting more rapid production. However, as the application rate increases, the likelihood of applying excessive coating thickness also increases, particularly when coating objects with complex shapes. Excess coating thickness is not only wasteful, but may also lead to sagging. Since there is not a stream of compressed air accompanying the particles and because the droplet size is generally larger, there is less solvent evaporation from the atomized particles from airless guns than from air guns. Solvents with higher relative evaporation rates are generally required in formulating coatings for airless spray application.

The absence of the air stream in airless guns reduces the problem of spraying into closed recesses of irregularly shaped objects. On the other hand, spraying down a recessed section that is open on the opposite end is easier with an air gun, since the air stream helps to carry the particles along. Airless spray equipment can cause problems with some water-borne coatings. More air is dissolved in the water under the high pressure; it comes out in bubbles when the pressure is released leaving the gun. Entrapment of bubbles in the film can result in pinholing. (See Section 24.3.)

An aerosol coating container is a type of airless spray unit. A liquefied gas, commonly propane, supplies the pressure to force the coating out of the orifice. Since the pressure is relatively low, the viscosity of the coating must be low to achieve proper atomization.

22.2.3. Electrostatic Spraying

As shown in Table 22.1, transfer efficiency can be substantially higher with electrostatic spray units. In the simplest case, a wire is built into the orifice of the spray gun. An electric charge on the order of 50–125 kV is impressed on the wire. At the fine end of the wire, an electric discharge leads to ionization of the air. As atomized coating particles pass through this zone of ionized air; they pick up a negative charge. The object to be coated is electrically grounded. When the coating particles approach the grounded surface of the object, the differential in charge attracts the particles to the surface. Taking the example of a chain link fence, an increased fraction of coating deposits on the metal fence even to the extent that particles that had passed through the holes are attracted back to the back side of the fence. This *wrap around* effect permits coating both sides of the fence by spraying from only one side; there is a lower, but still fairly high, overspray loss even with electrostatic spray. With objects like automobiles or appliances, overspray losses can be reduced by over 50%, resulting in transfer efficiencies around 80%.

Different types of devices can be used for electrostatic spraying. Figures 22.2 and 22.3 show bell- and disk-type spray units, so-called rotary atomizers. In both cases, coating is pumped through a tube leading to the middle of the disk or bell; the unit is rotated at a fairly high speed on the order of 900 rpm or higher, depending on the diameter; and the coating flows out to the edge of the unit. As coating is thrown off the edge of the unit, it is atomized and then passes through an electrostatic field that charges the droplets. Droplet size is controlled by viscosity and the peripheral speed of the unit. Very high speed bells, up to 60,000 rpm, permit application of coatings with viscosities as high as 1.5–2 Pa·s compared to 0.05–0.15 Pa·s used with conventional speed disks and bells as well as spray guns. This ability to handle higher viscosity coatings permits use of less solvent. However, it has been reported that application of pigmented coatings by high speed electrostatic bells tends to give films with lower gloss [9]. It is suggested that the centrifuging effect leads to differences in pigment content among atomized particles, leading to formation of uneven films. Uniformity of application and transfer efficiency can often be improved by

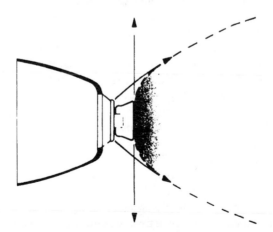

The Bell: The geometry of the spray pattern is varied according to the object to be painted by adjusting an annular compressed air shroud

Figure 22.2. Bell electrostatic spray equipment. (From Ref. [1]. with permission.)

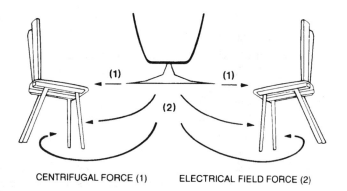

CENTRIFUGAL FORCE (1)   ELECTRICAL FIELD FORCE (2)

The Disc: The combined centrifugal and electrostatic forces of the ultra-high speed disc provide an extremely uniform finish and excellent penetration. Transfer efficiencies are high

Figure 22.3. Disk electrostatic spray equipment. (From Ref. [1], with permission.)

compressed air shaping of the cloud of atomized coating, analogously to the effect of the shaping air in conventional air guns.

Low overspray losses and good wrap around with electrostatic spray depend on the charge pickup by the atomized particles, which is controlled by conductivity of the coating. If conductivity is too low, the particles do not pick up sufficient charge from the ionized air. This is most likely to be a problem with coatings that have only hydrocarbon solvents, especially aliphatic hydrocarbons. Nitroparaffin or alcohol solvents can be substituted for a portion of the hydrocarbon solvents with good results. For coatings with free carboxylic acid groups on the resin, addition of a small amount of a tertiary amine like triethylamine increases conductivity sufficiently to give good charging. If the conductivity is too high, there is increased danger of electrical shorting. Generally, conductivity is measured, but the results are expressed as resistivity; optimum resistivity varies with equipment and the coating operation, ranging from 0.05 to 20 megaohms.

Conductivity of water-borne coatings is higher than solvent-borne coatings; resistivities of the order of 0.01 megaohms are reported [10]. Addition of slow evaporating water-miscible solvents with nonpolar ends, such as 2-butoxyethanol, somewhat reduces surface conductivity. However, spray equipment must be specially designed so that the coating line is electrically isolated; otherwise the charges will be dissipated and the atomized droplets will not be charged. Furthermore, the hazard of shocks would be high. While the whole application line can be electrically isolated, this is expensive and there is still hazard of electric shocks. Alternatively, just the atomizer and a short hose connected to a "voltage blocking device" can be charged [11]. The electrical isolation provided permits effective charging of the atomized droplets and reduces hazard of shocks at substantially lower cost than isolating the entire application line.

Electrostatic spray application is not without difficulties. The substrate must be electrically conductive so that the necessary charge differential can be set up by grounding the object to be sprayed. It also can be difficult to get coating into recessed areas even with airless electrostatic spray application due to a *Faraday cage effect*. A Faraday cage results from the pattern of field lines between the electrode on the gun and the grounded object. The strong electrical field induced by the difference in voltage establishes field lines that the atomized particles follow between the gun and the object. However, areas surrounded by grounded metal, such as inside corners of a steel case, are shielded from the electric fields by the metal; a Faraday cage is established and few particles enter such a shielded area.

Electrostatic spray is the most important method for applying powder coatings. See Section 27.5.1 for discussion of spraying powder coatings.

22.2.4. Hot Spray

Since the viscosity of coatings must be relatively low, generally, 0.05 to 0.15 Pa · s, for proper atomization, the solids, especially for lacquers made with relatively high molecular weight resins, must be quite low. An approach for increasing solids is use of *hot spray*, also called *temperature conditioned spray*. A hot spraying unit is designed with a heat exchanger to heat the coating to temperatures from 38 to 65°C. The unit is designed so that the coating is recirculated when the gun is turned off, even temporarily. At the elevated temperature, the viscosity is reduced sufficiently to permit a significant increase in solids. While these systems were originally designed for use with quite low solids coatings like lacquers, they are useful for high solids coatings. For example, it is reported that the solids of a top coat for appliances can be increased from 55 to 65% using a heater before a disk electrostatic spray gun [2]. The viscosity of high solids coatings generally decreases more sharply with increasing temperature than conventional coatings, which is a desirable characteristic for hot spray application. A further advantage is that the temperature drops between leaving the spray gun orifice and arrival on the work, leading to a correspondingly large viscosity increase. This characteristic is desirable for reducing sagging that can be serious with high solids coatings. (See Section 23.3.) Another advantage is that temperature control eliminates the effects of variation in ambient temperature on viscosity of the coating.

22.2.5. Supercritical Fluid Spray

A new approach to spray application is *supercritical fluid spraying* [8,12,13]. The supercritical fluid of choice is carbon dioxide with a critical temperature of 31.3°C and a critical pressure of 7.4 MPa. In the supercritical state, carbon dioxide exhibits solvency characteristics similar to hydrocarbons, but it is not counted as VOC. A dual feed gun is used with a low solvent content coating as one feed and supercritical CO_2 as the other. The temperature must be controlled and the pressure must be above 7.4 MPa. The process has been used with a variety of solvent-borne coatings for many applications [12]. VOC is reduced by 30–90% without changing the molecular weight of the resins. It has also been demonstrated that water-borne coatings can be applied using the process [12].

The high pressure means that airless guns must be used. The supercritical fluid spray system minimizes some of the problems of utilizing airless spray with conventional coatings [13]. When the coating leaves the orifice of the spray gun, the very rapid vaporization of the CO_2 breaks up the atomized droplets, reducing particle size and resulting in narrower particle size distribution. The droplet size is comparable to that obtained with air spray guns and significantly smaller than obtained with airless guns. Furthermore, the fan pattern is more similar to the feathered pattern obtained with air guns, rather than the usual pattern for airless guns, discussed in Section 22.2.2. Loss of the CO_2 is complete before the droplets reach the surface, so the viscosity of the applied film is relatively high, minimizing sagging. Transfer efficiency is also improved, reducing over-spray and waste disposal problems.

22.2.6. Formulation Considerations for Spray-Applied Coatings

Formulation of solvent mixtures for spray-applied coatings and effective use of spray guns requires taking into consideration the large ratio of surface area to volume of the atomized coating droplets and the flow of air over the surface of those droplets. As discussed in Section 17.3, these factors affect the rate of solvent evaporation. When lacquers are applied by spray, over half the solvent in the coating may evaporate between the orifice of the spray gun and the surface of the object. If the solvent mixture is balanced properly and the spray gun is used properly, a relatively smooth, sag-free film can be obtained. If either factor is not proper, sagging may occur when not enough solvent is evaporated or rough surfaces from dry spray. In extreme cases of poor spraying, both sagging and rough surfaces can occur on the same object. As discussed in Sections 23.2 and 23.3, low viscosity of the coating after it arrives on the surface generally facilitates leveling, but it also increases sagging.

Proper control requires a careful balance of solvent evaporation rate with the particular spray equipment and procedure. The greater the distance from the spray gun orifice to the work, the greater the fraction of solvent lost. Coatings are formulated to work best at a specific distance between the gun and the surface. This distance should be kept as constant as possible throughout the spraying operation. The problem can be illustrated by the result of a person spraying lacquer on a flat vertical surface by holding one's arm in a constant position and bending one's wrist to spray a wider area of the surface. When the gun is aimed perpendicular to the substrate surface the distance is at a minimum. When the wrist is bent to the furthest degree, the distance is at a maximum. If the solvent mixture is properly balanced for an intermediate distance, the lacquer sprayed perpendicularly would be likely to sag; the lacquer film at the extreme distance would tend to show poor leveling.

The rate of solvent loss is affected by the degree of atomization. If the average particle size is smaller, the surface to volume ratio is higher and the extent of solvent loss is greater. The rate of solvent evaporation is also affected by air flow over the surface of the droplets. More solvent evaporates with air guns than with airless guns or spinning bells. The rate of air flow through the spray booth can affect the degree of solvent loss. Temperature in the spray booth can be an important factor; during hot weather, it is common to change the solvent mixture to slow the rate of evaporation. When formulating and testing a solvent mixture for a coating in the laboratory, the same type of spray gun at approximately the same distance from the work that will be encountered in the customer's factory should be used. A final adjustment must be made in the customer's plant under regular production conditions. If those conditions change, the solvent mixture has to be changed. This is one reason why industrial coatings are almost always shipped at higher concentrations than the customer will actually use. This permits modifications of both solvent levels and solvent composition (by changing the reducing solvent) to accommodate temperature changes in the spray booth and other variables.

Viscosity of the coating must be adjusted to obtain appropriate atomization for the spray gun being used. The critical viscosity for atomization is that at the high shear rates (10^3 to 10^6 s^{-1} [2]) encountered as the coating passes through the orifice of the gun. Shear thinning coatings with relatively high viscosity at low shear rate can be sprayed successfully. The upper limitation on viscosity at low shear rate is the need to have a satisfactory flow rate of coating through the tubing to the spray gun, which varies from gun to gun. Architectural and maintenance paints, which must not sag significantly when thick films are applied to large wall expanses, are shear thinning. Generally, the same coating used for brush application can be used for spray; in the case of latex paints with some spray equipment, the viscosity at low shear rate may have to be reduced by diluting with a small amount of water.

Many conventional industrial coatings exhibit Newtonian flow; however, most high solids coatings and water-borne coatings are shear thinning. Viscosity for application is generally checked with an efflux cup. (See Section 3.3.5.) However, efflux cups should be used with care, since they do not detect shear thinning or thixotropic flow properties. They should only be used for control purposes. The proper way to establish viscosity for production spraying is by using the production spray gun under the conditions of use. Then, having found the proper degree of thinning with solvent, one can establish an efflux cup time as the standard for that coating for use in that gun under those circumstances.

Many water-reducible and high solids coatings are shear thinning after reduction to spray viscosity, which means that to give proper atomization, the efflux cup time of the reduced water-borne coating will be longer than with most solvent-borne coatings. Atomization is controlled by viscosity at high shear rate and efflux cup time is controlled by viscosity at a lower shear rate. As discussed in Section 23.3, sagging of some spray-applied high solids coatings cannot be controlled by adjustment of solvent evaporation rates. In such cases, the coating has to be formulated so that it is thixotropic.

In applying water-borne coatings by spray, air bubbles can be entrapped in the coating film. If they persist until the viscosity of the film surface becomes too high for the bubbles to escape, pinholing or popping will occur. The troublesome air voids are tiny (~ 10 μm). The problem can be particularly severe in airless spray application, since more air dissolves in the coating due to the high pressure. When the pressure is released as the coating leaves the spray gun, bubbles of air form in the droplets. While larger air bubbles can escape by rising to the surface and breaking before the viscosity at the surface increases, microform bubbles only dissipate by dissolution of the air in the coating followed by diffusion through the film [14]. In air spray, HVLP spray, and air-assisted airless spray, the problem can be minimized by using CO_2 as the driving gas instead of air. The improvement results from the higher solubility of CO_2 in a water-borne coating as compared to air.

22.2.7. Overspray Disposal

In industrial production, the advantage of decreased overspray is not just savings in the cost of the lost coating and lower VOC emissions. Any overspray has to be trapped so that it does not contaminate the surrounding area. Water-washed spray booths (a spray booth in which the wall behind the work being sprayed is a continuous waterfall that is recirculated) are widely used. The overspray is collected as a sludge. While it is sometimes possible to rework the sludge into low-grade coatings, generally the sludge must be disposed in approved hazardous solid waste disposal landfills. Such disposal has become expensive; increased transfer efficiency with electrostatic systems can reduce waste disposal cost.

Although water-wash spray booths work efficiently with solvent-borne coatings, the separation of sludge is less efficient when some water-borne coatings are sprayed. The overspray does not completely coagulate when it strikes the water as it does with solvent-borne coatings. This makes separation more difficult and can limit recirculation of the water. It has been found that froth flotation methods adapted from ore recovery processes permit relatively rapid separation of the sludge [15]. It has also been reported that separation of sludge from water-borne coatings can be improved by addition to the water tank of an emulsion of a melamine-formaldehyde resin for detackification along with water-soluble cationic and/or nonionic acrylamide polymers for coating flocculation [16].

22.3. DIP AND FLOW COATING

Dip coating can be an efficient procedure for applying coatings; it offers both relatively low capital cost equipment and low labor requirements. The principle is simple: the object is dipped into a tank of coating and pulled out; excess coating drains back into the dip tank. In practice, satisfactory coating application by dipping is more complex. While excess coating is draining off the object, a gradation of film thickness develops; the coating thickness at the top of the object is thinner than at the bottom. During draining, solvent is evaporating. The differences in film thickness can be minimized by controlling the rate of withdrawal of the object from the dip tank and the rate of evaporation of the solvent. If the object is withdrawn slowly enough and the solvent evaporates rapidly enough, film thickness approaching uniformity on vertical flat panels can be achieved. In actual production, the rate of withdrawal is usually faster than optimum so there is some thickness differential between the top and bottom of the coating.

Care must be exercised in selecting or changing solvents for dip coating because of flammability hazards and changes in viscosity of the dip tank that result from evaporation of solvent. Changes in viscosity result in changes in film thickness, which increases with increasing viscosity. Achieving consistent results requires keeping the viscosity of the coating constant, which becomes more difficult as the volatility of the solvent is increased. Solvent is added to replace solvent lost from the tank.

Successful use of dipping in production lines requires that the coating be very stable. Viscosity can increase not only by loss of solvent, but also by chemical reactions of coating components. Only a small fraction of the coating in the tank is removed each time an object is dipped. Fresh coating is added to make up for these removals, but the coating in the tank is a mixture of old and newly added coating. Some of the original charge of coating is present for a long time. If cross-linking reactions occur in the bath, viscosity increases. While the viscosity can be reduced by adding more solvent, this reduces the solids so dry film thickness is reduced. The extent of such reactions must be minimal. When oxidizing alkyds are the vehicle for a dip coating, oxidation must be avoided, since this leads to cross-linking. Stabilization requires an antioxidant, but the antioxidant must be sufficiently volatile that it escapes during the early stages of the baking cycle; otherwise, it inhibits cross-linking of the dry film. On the other hand, it must not be so volatile that it is rapidly lost from the dip tank. Isoeugenol is an example of an antioxidant for alkyds in dip coatings.

An advantage of dip coating is that all surfaces are coated, not just the outer surfaces accessible to spray. However, there are difficulties in dipping irregularly shaped objects. Coating may be held in pockets or depressions giving pools of coating that do not drain. To minimize this problem, the point(s) at which the objects hang on the hooks of the conveyor line that carries it into and out of the tank must be carefully designed and selected. Objects to be dip coated must be designed with drain holes that minimize pooling, but do not interfere with the performance or appearance of the product. Lower edges, especially lower corners, build up high film thickness. Buildup can be reduced by electrostatic *detearing*, that is, by passing the object over a highly charged electrode that causes a charge concentration at such points, causing drops to be pulled off.

Water-borne coatings are supplanting solvent-borne dip coatings in many applications. They reduce flammability hazards and VOC emissions. Electrodeposition of coatings is done by dipping, as discussed in Chapter 26.

Flow coating and dip coating are related methods. Objects to be flow coated are carried hung on a conveyor through an enclosure in which streams of coating are squirted on the object from all sides. The excess coating material runs off and is recirculated through the system. An advantage is that the volume of coating required to fill the lines of a flow coater is less than required for a dip tank for objects of the same size. This reduces inventory cost and increases coating turnover, hence reducing somewhat the problem of bath stability. Flow coaters are designed so that the atmosphere in the enclosed area is maintained in a solvent saturated condition. In this way, evaporation of solvents is minimized until flow and leveling have been obtained. There is still a gradation of film thickness from the top to the bottom of the object, but usually less than that from dip coating. Highly automated flow coating lines have been used for applying coatings to major appliances. The design permits more rapid line speeds than with conventional dip tanks.

22.4. ROLL COATING

Roll coating is widely used and is efficient, but is only applicable to uniform, flat, or cylindrical surfaces. Relatively slow evaporating solvents must be used to avoid viscosity buildup on the rolls of the coater. The pot life of the coating must be relatively long, since the rate of turnover of coating through the system is relatively low. There are many types of roll coating procedures; the two most common are *direct roll coating* and *reverse roll coating*.

In direct roll coating, the stock passes between two rollers, an applicator roller and a backup roller, that are rotated in opposite directions. The rollers pull the material being coated between them as illustrated in Figure 22.4. Direct roll coating is used for coating sheet stock and sometimes for coil stock. In direct roll coating, the applicator rollers are generally covered

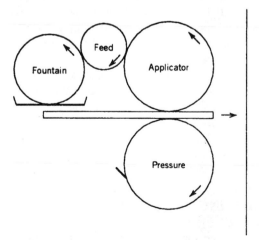

Figure 22.4. Direct roll coater. (From Ref. [1], with permission.)

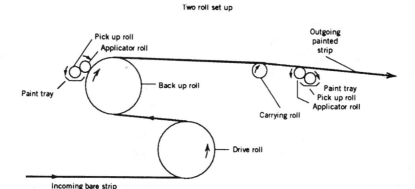

Figure 22.5. Reverse roll coater for coating both sides of coil stock. (From Ref. [1], with permission.)

with a relatively hard polyurethane elastomer. Coating is fed to the applicator roll by a smaller feed, or doctor, roll that is, in turn, fed by a pickup roll. The pickup roll runs partially immersed in a tray, called a fountain, containing the coating. Film thickness is controlled by the clearance between the feed and applicator rolls and by the viscosity.

Several variations on direct roll coating use different types of applicator rolls. The applicator rolls can have sections cut out so that they do not coat the entire substrate surface. The applicator roller can be engraved with small recessed cells over its whole surface. The cells are filled with coating, the surface is scraped clean with a doctor blade, and the coating remaining in the cells is transferred to the substrate being coated. Such so-called *precision coaters* apply a controlled amount of coating only to the areas of the substrate that contact the roller. In another variation, the applicator roller is a brush roller; it is used to apply thick coatings to relatively rough surfaces.

In direct roll coating, as the coated material comes out of the nip between the rollers, the wet layer of coating is split between the roller and the substrate. As a result, the coating has a ribbed surface as it emerges from the coater-a phenomenon called *roll tracking*. The coating must be designed to level so that this tracked appearance disappears, or at least becomes less obvious, before the coating stops flowing. To minimize solvent loss on the rolls and to keep the viscosity low to promote leveling, slow evaporating solvents are used and when possible, coatings are formulated so that they exhibit Newtonian flow. If the coating exhibits thixotropy, it is desirable to have the rate of recovery of the low shear viscosity be as slow as possible.

In reverse roll coating, the two rollers rotate in the same direction and the material being coated must, therefore, be pulled through the nip between the two rolls, as shown in Figure 22.5. It is generally not feasible to coat sheets by reverse roll coating, but the process is widely used for coating coil stock. Reverse roll coating has the advantage that the coating is applied by wiping rather than film splitting. A smoother film is formed, and the problems of leveling are minimized.

22.5. CURTAIN COATING

Curtain coating is widely used for coating flat sheets of substrate material such as wall panels. A coating is pumped through a slot in the coating head so that it flows out as a

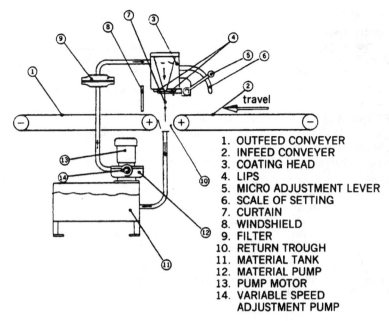

Figure 22.6. Schematic diagram of a curtain coater. (From Ref. [1], with permission.)

1. OUTFEED CONVEYER
2. INFEED CONVEYER
3. COATING HEAD
4. LIPS
5. MICRO ADJUSTMENT LEVER
6. SCALE OF SETTING
7. CURTAIN
8. WINDSHIELD
9. FILTER
10. RETURN TROUGH
11. MATERIAL TANK
12. MATERIAL PUMP
13. PUMP MOTOR
14. VARIABLE SPEED
    ADJUSTMENT PUMP

continuous curtain of coating. The material to be coated is carried under the curtain on a conveyor belt. The curtain must be wider than the substrate being coated to avoid edge effects on the film thickness. A recirculating system returns the overflow from the sides and from between sheets to the coating reservoir. A schematic diagram is shown in Figure 22.6.

The width of the slot, the pumping pressure, the viscosity of the coating, and the rate of passage of the substrate being coated control film thickness. The faster the line runs, the thinner the coating. Where applicable, curtain coating is an excellent method. No film splitting is involved, so the film that is laid down is essentially smooth. Film thickness can be very uniform. The coating must be very stable, and solvent added to make up for loss by evaporation. If, as explained in Section 23.4, a particle of low surface tension lands on the flowing curtain, surface tension differential driven flow can lead to a hole in the curtain that in turn leads to a gap in the coating being applied to the substrate. As the surface tension of the coating is increased, particles in the air are more likely to have lower surface tensions than the coating, and the probability of holes is increased.

GENERAL REFERENCE

S. B. Levinson, *Application of Paints and Coatings*, Federation of Societies for Coatings Technology, Blue Bell, PA, 1988.

REFERENCES

1. S. B. Levinson, *Application of Paints and Coatings*, Federation of Societies for Coatings Technology, Blue Bell, PA, 1988.

2. C. K. Schoff, *Rheology*, Federation of Societies for Coatings Technology, Blue Bell, PA, 1991.
3. J. E. Glass, *J. Coat. Technol.*, **50** (641), 72 (1978).
4. R. H. Fernando and J. E. Glass, *J. Rheology*, **32**, 199 (1988).
5. J. Adams, "Spray Applications Processes for Environmental Compliance," FSCT Symposium, Louisville, KY, May 1990.
6. A. H. Lefebvre, *Atomization and Sprays*, Hemisphere Publishing Corporation, New York, 1989, pp. 59–61.
7. M. G. Easton, *J. Oil Colour Chem. Assoc.*, **66**, 366 (1983).
8. K. A. Nielsen, D. C. Busby, C. W. Glancy, K. L. Hoy, A. C. Kuo, and C. Lee, *Polym. Mater. Sci. Eng.*, **63**, 996 (1990).
9. K. Tachi, C. Okuda, and K. Yamada, *J. Coat. Technol.*, **62** (791), 19 (1990).
10. M. J. Diana, *Products Finishingd*, July, 54 (1992).
11. R. D. Konieczynski, *J. Coat. Technol.*, **67** (847), 81 (1995).
12. K. A. Nielsen, J. N. Argyropoulos, D. C. Busby, D. J. Dickson, and C. S. Lee, *Proc. Water-Borne Higher-Solids Powder Coat. Symp.*, New Orleans, 1995, p. 151.
13. D. W. Senser, J. C. Colwell, and R. M. Smith, *Proc. Water-Borne Higher-Solids Powder Coat. Symp.*, New Orleans, 1995, p. 161.
14. M. S. Gebhard and L. E. Scriven, *J. Coat. Technol.*, **66** (830), 27 (1994).
15. E. W. Fuchs, G. S. Dobby, and R. T. Woodhams, *J. Coat. Technol.*, **60** (767), 89 (1988).
16. S. F. Kia, D. N. Rai, M. A. Shaw, G. Ryan. and W. Collins, *J. Coat. Technol.*, **63** (798), 55 (1991).

CHAPTER 23 _____

Film Defects _____

Obvious defects can result from incomplete coverage during application of a coatng, resulting in thin spots or holes, often called *skips* or *holidays*. Many other kinds of defects and imperfections can develop in a film during or after application. In this chapter, we deal with the most important defects and, to the extent possible, discuss the causes of the defects and approaches for eliminating or minimizing their occurrence. Unfortunately, nomenclature for many defects is not uniform. Reference [1] provides definitions for coating terms including those for film defects.

23.1. SURFACE TENSION

Many defects are related to *surface tension* phenomena. Surface tension occurs because the forces at an interface of a liquid differ from those within the liquid due to the unsymmetrical force distributions on the surface molecules. The surface molecules possess higher free energy, equivalent to the energy per unit area required to remove the surface layer of molecules. The dimensions of surface tension are force exerted in the surface perpendicular to a line; SI units are newtons per meter or millinewtons per meter ($mN\ m^{-1}$). [Older units, still commonly used, are dynes per centimeter ($1\ mN\ m^{-1} = 1$ dyne cm^{-1}).] Similar surface orientation effects are present in solids, and they have surface free energies, expressed in units of free energy per unit area, millijoules per square meter ($mJ\ m^{-2}$) that are numerically and dimensionally equal to $mN\ m^{-1}$. Frequently, people speak of surface tension of solids; while not formally correct, since the values are identical, errors do not result. Reference [2] is an excellent discussion of surface and interfacial properties, including data on many polymers.

Surface forces work to decrease the surface free energy of liquids and solids. Surface tension works to draw liquids into spheres, since a sphere encloses a minimum ratio of surface area to volume. In a spaceship, liquid droplets assume a spherical shape; on earth, the force of gravity distorts the spheres. For the same reason, surface tension drives flow of an uneven liquid surface towards becoming a smooth surface. The smooth surface has less interfacial area with air than the rough surface; hence, there is a reduction in surface free energy as the surface becomes smoother.

Segments of molecules that minimize surface tension tend to orient at the surface. The lowest surface tension results from perfluoroalkyl groups at the surface. The next lowest is methyl groups. Poly(dimethylsiloxane) has a low surface tension; the very flexible easily rotatable backbone of siloxane bonds permits orientation of a large population of methyl groups at the surface. The surface tension of linear aliphatic hydrocarbons increases as chain length increases, reflecting the larger ratio of methylene to methyl groups. Progressively higher surface tensions result from aliphatic chains, aromatic rings, esters and ketones, and alcohols. As the chain length of aliphatic esters, ketones, and alcohols increases, the surface tension increases because methylene groups give higher surface tensions than methyl groups. Water has the highest surface tension of the volatile components used in coatings.

If more than one type of molecule is present, the segments of those molecules that lead to the lowest surface free energy come to the air-liquid interface. This occurs rapidly in low-viscosity liquids and more slowly in amorphous polymers. Addition of small amounts of surfactants to water reduces surface tension because the hydrocarbon chains go to the surface. Addition of poly(dimethylsiloxane) to an organic solvent solution reduces surface tension due to orientation of the methyl groups at the surface.

When a liquid is agitated, the molecules at the surface are mixed in with the rest of the liquid. When the agitation is stopped, reorientation to give the lowest surface tension occurs, but the equilibrium surface composition is not reestablished immediately. When coating films are applied, they are subjected to considerable agitation. As Bierwagen has pointed out, the surface tension of importance in governing some aspects of coating behavior may not be the *equilibrium surface tension*, but rather a *dynamic surface tension* [3,4]. The time to regain equilibrium after agitation has ceased varies widely, depending on the composition, viscosity, temperature, and probably other factors [5]. There has not been adequate quantitative study of the rates with different systems of importance in the coatings field. Qualitatively, one can say that equilibrium is established most rapidly when the coating contains small, flexible molecules and when there are large differences in the polarity of components in the system as in water-borne coatings. Reaching equilibrium takes longer when the molecules with the lowest potential surface tension groups are polymers. But, if the polymers have moderate molecular weight and flexible backbones, they can apparently reach the surface relatively rapidly. Poly(dimethylsiloxane) mentioned previously is an example. Low molecular weight octyl acrylate copolymers are used as additives to reduce surface tension of coating films as they are forming. In water-borne coatings, it has been shown that different surfactants differ in their rates of reaching equilibrium surface tension [5]. Surface tension increases with decreasing temperature, and solvents have lower surface tensions than resins. Therefore, surface tension increases as solvent evaporates from a film of resin solution, owing to both the change in concentration and temperature.

If two liquids with different surface tensions are in contact with each other, the liquid of low surface tension flows to cover the liquid with higher surface tension, since this results in a lower overall surface free energy. Such flow is a *surface tension differential driven flow*; some authors prefer the term *surface tension gradient driven flow*. This type flow has been observed for millennia, but Carlo Marangoni, a 19th century Italian physicist, is credited with providing a scientific understanding of the phenomenon [6]. An example of the Marangoni effect is the flow that occurs when a clean glass containing wine is tipped, wetting the side of the glass with the liquid, and then returned to an upright position. Liquid then flows up the side of the glass, forming a bead of greater film thickness along the upper edge of the wetted area. When the amount of liquid collecting in the bead becomes large, droplets flow back down the side of the glass. These "tears of wine" have been known since biblical times, but why does the phenomenon occur? Ethyl alcohol has a higher relative evaporation rate and lower surface tension than water, evaporation occurs most rapidly along the edge of the layer of wine on the side of the glass. This leads to reduction in alcohol concentration and, hence, higher surface tension along the edge as compared to the bulk of the liquid. To minimize surface free energy, the low surface tension (higher alcohol concentration) liquid in the bottom of the glass flows up to cover the higher surface tension liquid at the edge. Evaporation continues at the edge, reducing the alcohol concentration and leading to a continuation of flow of lower surface tension liquid up the glass. The surface tension differential is also affected by the temperature change; as alcohol evaporates, temperature decreases, further increasing the differential in surface tension.

In summary, two types of flow result from surface tension effects. Surface tension driven flows occur to minimize surface area of a liquid. Surface tension differential driven flows occur to cover a liquid or other surface of higher surface tension with a liquid of lower surface tension.

23.2. LEVELING

Most methods of application of coatings initially lead to formation of a rough, wet film. It is usually desirable for appearance and performance to have the irregularities level out. The most widely studied leveling problem has been leveling of brush marks. While a person unacquainted with the field might first think that leveling results from gravitational effects, this is clearly not, to a significant degree, the case. If gravity were a significant factor, paints applied to ceilings should level more poorly than paints applied to floors, but that is not the case. Based on studies of flow of mineral oil, Orchard proposed that the driving force for leveling is surface tension and established mathematical models for the variables that he proposed would control the rate of leveling [7]. Orchard's treatment has been widely applied to flow of coatings. Patton [8] illustrated the model for an idealized cross section of a wet film exhibiting brush marks that follow a sine wave profile, as shown in Figure 23.1 [8]. He gives several forms of the Orchard equation and shows their derivations. A convenient form relating the change in amplitude of the sine wave to time is shown in Eq. 23.1, where: $a_0 =$ initial amplitude (cm), $a_t =$ amplitude at time t

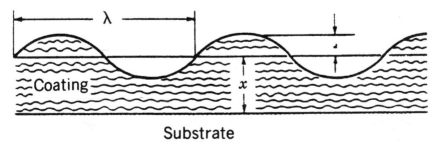

Figure 23.1. Schematic diagram of a cross section of brush marks. (From Ref. [8], with permission.)

(cm), $x =$ average coating thickness (cm), $\lambda =$ wave length (cm), $\gamma =$ surface tension (mN m^{-1}), $\eta =$ viscosity (Pa $\cdot$ s), $t =$ time (s).

$$\ln \frac{a_0}{a_t} = \frac{5.3 \gamma x^3}{\lambda^4} \cdot \frac{dt}{\eta} \tag{23.1}$$

Most rapid leveling occurs when the wavelength is small, viscosity low, surface tension high, and film thick. The formulator has little or no control over most of the variables. Wavelength is determined by application conditions; in brushing, it increases as pressure on the brush increases and as the thickness of the coating increases. High surface tension increases the rate of leveling; however, the formulator is limited in optimizing this factor, since high surface tension can lead to other defects. Thicker films promote leveling, however; increasing film thickness increases cost of coating and increases the probability of sagging (see Section 23.2) on vertical surfaces. The principal means of control left is viscosity.

The Orchard model provides satisfactory correlation between experimental data and predictions when the liquid film has Newtonian flow properties and sufficiently low volatility such that viscosity does not change during the experimental observations. In most cases, viscosity changes during the time in which leveling occurs. As solvent evaporates, viscosity increases. Furthermore, if the system is thixotropic, the viscosity is reduced by the high shear rate exerted during brushing and subsequently increases with time at the low shear rate involved in leveling. Another potential shortcoming in the Orchard treatment is the assumption of constant surface tension.

Overdiep devised methods of observation that permitted following the location of the ridges and valleys [9,10]. He found for two alkyd coatings that brush marks leveled to an essentially smooth film, but then ridges grew where there had been valleys and valleys formed where there had been ridges. Although surface tension can and does cause a ridged film to level, it cannot cause a level film to become ridged because that creates more surface area. Overdiep proposed that surface tension differential is the major driving force for emergence of the new ridges. As can be seen in Figure 23.1, the wet film thickness in the valleys of the brush marks is less than in the ridges. When the same amount of solvent evaporates per unit area of surface, the fraction of solvent that evaporates in the valleys is larger than in the ridges. As a result, concentration of resin solution in the valleys becomes higher than in the ridges and the surface tension in the valleys is higher than on the ridges. Following the Marangoni effect, coating flows from the ridges into the valleys. In other words, Overdiep proposed, and was able to demonstrate experimentally, that with volatile solvents, the primary driving force for leveling is not surface tension, but surface tension differential. In some cases, these differentials lead to overshooting of the smoothest stage and cause the growth of ridges. The extent of the flow driven by surface tension differential depends on the rate of evaporation of the solvent.

Kojima and Moriga studied solvent evaporation and leveling of water-reducible coatings and showed that the forces driving leveling depend on the solvent in the formulation [11]. With a fast evaporating solvent, such as isobutyl alcohol, surface tension increased during drying promoting leveling by surface tension differential forces, but with a slow evaporating solvent such as ethylene glycol monohexyl ether, surface tension decreased causing an adverse effect on leveling. Equations have been developed that model the drying process through the changes in surface tension differentials and changes in viscosity during solvent evaporation [12].

Overdiep was particularly interested in what happens with uneven coating films over a rough substrate [9]. He reasoned that surface tension driven flow might give the smoothest film. However, as illustrated in Figure 23.2(a), this may be undesirable because protection in thin areas might be limited. On the other hand, as shown in Figure 23.2(b), surface tension differential driven flow would tend to yield equal film thickness, with the surface of the film following the roughness of the substrate rather than being level. Overdiep suggests it might be best to adjust a coating so that both types of flow are significant to achieve a compromise with reasonable film smoothness without places where film thickness is very thin, as shown diagrammatically in Figure 23.2(c). The balance could be controlled by the volatility of the solvent; with very low volatility, leveling would be surface tension driven; with relatively high volatility, the major effect would be surface tension differential; with intermediate volatility, both phenomena could be important.

Neither the Orchard equation nor Overdiep's work takes into account the effect of thixotropy. Cohu and Magnin developed equations that predict the effect of thixotropy on leveling of nonevaporative coatings [13].

Another result of poor leveling is reduction in gloss. A study of the effect of leveling on gloss showed that three factors are important: extent of roughness of the substrate, film thickness, and viscosity [14]. The addition of fluorocarbon surfactant did not significantly change leveling in these experiments; presumably, surface tension differentials did not develop in the drying of the films.

In spray application, surface roughness consists of bumps surrounded by valleys rather than ridges and valleys. Since the effect is somewhat reminiscent of the appearance of orange skins, it is called *orange peel*. Figure 23.3 shows a photograph of an orange peel surface. The bumps are generally much larger than spray droplets. Most commonly, orange peel is encountered when spraying coatings, that have solvents with high evaporation rates. It is common for people to conclude that due to the fast evaporation, the viscosity of the

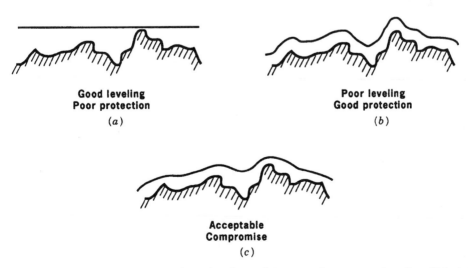

Good leveling
Poor protection
(a)

Poor leveling
Good protection
(b)

Acceptable
Compromise
(c)

Figure 23.2. Alternate leveling (a–c) results after applying a coating to a rough surface. (Adapted from Ref. [9], with permission.)

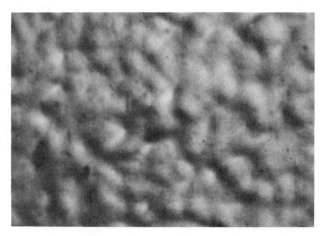

Figure 23.3. Typical orange peel pattern (15X). (From Ref. [16], with permission.)

coating on the substrate builds up so rapidly that the leveling is poor; in some situations, that is probably the case. However, in the late 1940s, it was found that leveling of sprayed lacquer films could frequently be improved by addition of very small amounts of silicone fluid. Contrary to the common wisdom that all leveling is surface tension driven, that is, promoted by high surface tension, here was a case in which adding a material known to reduce surface tension substantially improved leveling.

Hahn provided an explanation for the phenomenon [15]. When one sprays a lacquer, initially the surface is fairly smooth, and if you watch, orange peel grows. Hahn proposed that the growth of orange peel results from a surface tension differential driven flow. The last atomized spray particles to arrive on the wet lacquer surface have traveled for a longer distance between the spray gun and the surface; hence, have lost more solvent, have a higher resin concentration and, therefore, a higher surface tension than the main bulk of the wet film. The lower surface tension wet lacquer flows up the sides of these last particles to minimize overall surface free energy; that is, surface tension differential driven flow grows the orange peel. If one adds a surface tension depressant, such as a silicone fluid that orients very rapidly to the surface; the surface tension of the wet lacquer surface and the surface tension of the last atomized particles to arrive are uniformly low, so there is no differential to promote growth of orange peel. Octyl acrylate copolymer additives can also give an overall low surface tension and minimize orange peel growth.

It would be of interest to see what effect silicone fluid or octyl acrylate copolymers would have on leveling of brush applied coatings. To the extent that the leveling is surface tension differential driven, as found by Overdiep, the leveling should be made poorer, not improved. The results of such an experiment are not given in the literature, but such additives are not used to promote leveling in brush applied coatings, only in spray applied coatings.

Electrostatically sprayed coatings are likely to show surface roughness greater than nonelectrostatically sprayed coatings. As a result, the final passes in the case of automotive top coats are generally applied without electrostatic charging, even though the major part of the top coat may be applied using electrostatic spray to reduce overspray. It has been speculated that the greater surface roughness obtained with electrostatic spray results from arrival of the last charged particles on a coated surface that is quite well electrically insulated from the ground. These later arrivals may retain their charges sufficiently long to repel each other and thereby reduce the opportunity for leveling. It has also been suggested that when coatings are applied by high-speed bell electrostatic spray guns, differentials in the pigment concentration within the spray droplets may result from the centrifugal forces [17]. These pigment concentration differentials lead to rougher surfaces and reduction in gloss of the final films.

Leveling problems are particularly severe with latex paints. Latex paints, in general, exhibit a greater degree of shear thinning and more rapid recovery of viscosity after exposure to high shear rates than paints made with solutions of resins in organic solvents. Due to their higher dispersed phase content, the viscosity of latex paints changes more rapidly with loss of volatile materials than the viscosity of solvent-borne paints. No experimental work has been reported on the relative importance of surface tension and surface tension differentials in leveling of latex paints, however; it seems probable that the leveling is primarily surface tension driven. The surface tension of water is high, but the presence of surfactants imparts low surface tension to latex paints. Furthermore, it is probable that this low surface tension is established rapidly after the agitation of application stops. Perhaps more importantly, the surface tension is uniformly low, since it is almost unchanged as water evaporates. Thus, the generally poor leveling of latex paints may result in part from the absence of surface tension differentials to promote leveling. The low surface tension may not provide adequate driving force for leveling in a film whose viscosity increases rapidly with time. The problems of leveling of latex paints are discussed further in Section 31.3. The leveling of powder coatings is discussed in Section 27.3.

23.3. SAGGING

When a wet coating is applied to a vertical surface, the force of gravity causes it to flow downward (*sagging*). Differences in film thickness at various places lead to differing degrees of sagging, resulting in curtains, or drapes, of coating. Patton gives Eq. 23.2, in which: x is initial film thickness (cm); V_s is volume of coating that has sagged after time; ρ is density (g cm^{-1}); g is gravitational constant (s cm^{-2}); t is time (s); and η is viscosity (Pa · s), showing the variables that affect the volume of coating that sags as a function of time [18].

$$V_s = \frac{x^3 \rho g t}{300\eta} \tag{23.2}$$

The driving force of sagging is gravity. Density of the coating is a factor; in some cases high density inert pigments can be avoided, but generally, the formulator has little latitude to control density. Thick films should be avoided, but hiding generally dictates some minimum film thickness. Therefore, viscosity is the major variable available for controlling sagging. The tendency to sag can be evaluated by observing the behavior of films applied under conditions simulating field use. However, this does not provide a numerical basis for evaluating the extent of sagging. Various tests have been developed. The most commonly used test is a *sag-index blade*, a straight edge applicator blade with a series of 1/4-in gaps of different depths at 1/16-in intervals across the blade [20]. Using it, a drawdown, which is a series of stripes of coating of various thickness, is made on a chart, and immediately, the chart paper is placed in a vertical position. Sag resistance is rated by observing the thickest stripe that does not sag down to the next stripe. For research purposes, Overdiep has developed a more sophisticated method, the sag balance, described in Ref. [19]. He also developed equations for sagging that take into consideration the changes in viscosity after application.

In spray applied solvent solution coatings, sagging can generally be minimized while achieving adequate leveling by a combination of proper use of the spray gun and control of the rate of evaporation of solvents. The goal is to manipulate viscosity so that it is initially low for leveling, but builds up before severe sagging occurs. In brush and hand roller applied coatings where slow evaporating solvents are used, thixotropic systems permit leveling to occur before the viscosity recovers, but for which the recovery of viscosity occurs soon enough that sagging is not serious, are used. Latex paints, which are almost always thixotropic, are less likely to exhibit sagging than are solvent solution paints.

Sagging can be a serious problem with high solids coatings, especially when spray applied. Although other factors may be involved, a major cause of sagging is that substantially less solvent is lost during spraying (i.e., after the droplets leave the gun and before they arrive at the substrate surface) of high solids coatings [21,22]. This low loss of solvent leads to less increase in viscosity of a high solids coating as compared to a conventional coating; the result is greater likelihood of sagging. The reasons for the lower solvent loss have not been clearly established; there are three possible explanations. First, perhaps because of their higher surface tension, high solids coatings atomize to give larger particle size droplets than conventional coatings. The lower ratio of surface area to volume would lead to lower solvent loss. However, one should be able to adjust spray equipment and conditions to obtain equivalent atomization. A second possibility is a colligative effect on solvent evaporation. Resins in high solids coatings are lower in molecular weight, and their concentrations are higher. For both these reasons, the ratio of the number of solvent molecules to resin molecules is lower in the case of high solids coatings than for conventional coatings. This undoubtedly leads to a decrease in rate of solvent loss; see Section 17.3.5 for a model calculation. However, it seems doubtful that this difference could account entirely for the large differences in solvent loss, such as reported by Wu [21]. A third theory is that the stage of diffusion control of the rate of solvent loss is reached after less loss of solvent from high solids than from conventional coatings; thus solvent evaporation from spray droplets of high solids coatings is markedly reduced [23,24]. As discussed in Section 17.3.4, at later stages of solvent evaporation from a film, the rate of diffusion of solvent molecules to the surface becomes the factor limiting the rate of evaporation of solvent. Solvents with linear rather than branched backbones are desirable since they can diffuse faster; see Section 17.3.4.

As discussed in Section 22.2.4, hot spraying can help control sagging. When the coating cools on striking the object, the viscosity increase reduces sagging. Use of carbon dioxide under supercritical conditions is particularly helpful in controlling sagging, since the CO_2 flashes off almost instantaneously when the coating leaves the orifice of the spray gun, thereby increasing viscosity. (See Section 22.2.5.) High-speed electrostatic bell application permits application of coatings at higher viscosity, which also helps control sagging. (See Section 22.2.3.)

However, sagging of many high solids coatings cannot be adequately controlled by adjustment of solvent composition of the coating and application variables. It is then necessary to make the systems thixotropic. For example, dispersions of fine particle size silicon dioxide, precipitated silicon dioxide, bentonite clay treated with a quaternary ammonium compound, or polyamide gels can be added to impart thixotropy. One tries to formulate so that recovery to high viscosity is slow enough to permit reasonable leveling and rapid enough to control sagging. However, such agents increase the high shear viscosity somewhat, and require higher solvent levels. They also tend to lower gloss and are frequently not effective at higher temperatures.

The problem of sagging in high solids automotive metallic coatings can be particularly severe. (See Section 29.1.2.) Even a small degree of sagging, which might not be noticeable in a white coating, is very evident in a metallic coating, since it affects the orientation of the metal flakes. Use of SiO_2 to impart thixotropy is undesirable, since even the low scattering efficiency of SiO_2 is enough to reduce color flop in the coatings. Acrylic microgels have been developed that impart thixotropic flow using the swollen gel particles [25]. In the final film, the index of refraction of the polymer from the microgel is nearly identical with that of the cross-linked acrylic binder polymer so that light scattering does not interfere with color flop. The effect of the gel particles depends on interaction with the low molecular weight acrylic resin; Reference [26] discusses the rheological properties of the systems. There can also be an improvement in the strength of the final film when microgels are incorporated [27].

Another problem that can be encountered with high solids coatings is *oven sagging*, also called *hot sag* [23]. The coating appears to be fine until it is put into the oven; then sagging occurs. Oven sagging results from the strong temperature dependence of the viscosity of high solids coatings. As compared to conventional coatings, there is a much steeper drop in viscosity as the coated product enters the hot oven that promotes sagging. Oven sagging can be somewhat controlled by zoning the oven. A lower temperature in the initial zone allows more time for solvent loss, and perhaps, some cross-linking, so viscosity increases as a result of the higher solids or higher molecular weight before the film is subjected to high temperature.

Water-reducible coatings are less likely to give sagging problems than high solids coatings, but there are circumstances when they exhibit delayed sagging. The viscosity of these coatings is very dependent on the ratio of water to solvent as well as solids content. (See Section 12.3.) As the water and solvent evaporate, the residual water to solvent ratio can sometimes decrease, leading to lower viscosity in spite of the higher solids content, and sagging can result. Such behavior can depend on the relative humidity during the flash off period after spraying. It has been found with a water-reducible acrylic enamel that sagging occurred above, but not below, a critical relative humidity [28]. (See Sections 17.3.3 and 17.3.6 for the definition and discussion of critical relative humidity.)

23.4. CRAWLING, CRATERING, AND RELATED DEFECTS

If a coating that has a relatively high surface tension is applied to a substrate that has a comparatively low surface free energy, the coating will not wet the substrate. The mechanical forces involved in the application may spread the coating on the substrate surface, but since the surface is not wetted, surface tension forces tend to draw the liquid coating toward a spherical shape. Meanwhile, solvent is evaporating, and therefore, viscosity is increasing so that before the coating can pull up into spheres, the viscosity is high enough that flow essentially stops. The result is an uneven film thickness with areas having little, if any, coating adjoining areas of excessive film thickness. This behavior is

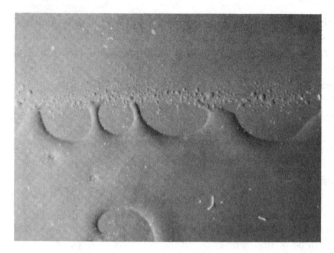

Figure 23.4. Crawling of a topcoat applied over a low surface energy primer (7X). (From Ref. [16], with permission.)

commonly called *crawling*; it is also called *retraction*. Figure 23.4 shows an example of crawling. For water-borne coatings, it has been shown that crawling can depend on the rate of establishment of equilibrium surface tension with different surfactants [5].

Crawling can result from applying a coating to steel with oil contamination on the surface. It is especially common in coating plastics. In some cases, crawling results from failure to completely remove a mold release agent from a plastic molded part. Application of a high surface tension top coat to a low surface tension primer can also lead to crawling. If a coating contains silicone fluids or fluorocarbon surfactants, there is likely to be crawling when a subsequent coat is applied. If one handles a primer surface with bare hands and then applies a relatively high surface tension top coat, it is likely that the top coat will draw away from the oils left behind in fingerprints. This type of crawling, which copies a pattern of low surface tension areas on the substrate, has been called *telegraphing*. Care is needed with the term telegraphing since this is only one of several phenomena called telegraphing.

Crawling can also result from the presence in the coating of surfactant-type molecules that can orient rapidly on a highly polar substrate surface. Even though the surface tension of the coating is lower than the surface free energy of the substrate, it could be higher than the surface free energy of the substrate after a surfactant in the coating orients on the substrate surface. This can occur if the polar group of the surfactant associated with the substrate and the long nonpolar end becomes the surface that the coating must wet. If one adds excess silicone fluid to a coating to correct a problem like orange peel, small droplets of insoluble fractions of the poly(dimethylsiloxane) can migrate to the substrate surface and spread on it, leaving a new substrate surface that the coating cannot wet, resulting in crawling and/or cratering. A little silicone fluid can solve some defect problems but even a small excess can cause what might be a worse problem. It has been reported that higher molecular weight fractions of poly(methylsiloxane) are insoluble in many coating formulations [29]. Modified silicone fluids, such as polysiloxane-polyether block copolymers, have been developed that are compatible with a wider variety of coatings and are less likely to cause undesirable side effects. The effect of a series of additives on crawling and other film defects has been reported [30].

High solids coatings may have higher surface tensions than conventional coatings. To achieve high solids, lower molecular weight resins with lower equivalent weights must be used, which means that the concentration of polar functional groups such as hydroxyl groups is higher, and hence, surface tension is generally higher. Also, the solvents that give the lowest viscosity coatings are likely to be relatively high surface tension solvents. Therefore, there is a greater likelihood of crawling problems with high solids coatings.

Cratering is the appearance of small round defects that look somewhat like volcanic craters on the surface of coatings. A schematic drawing and a photograph are shown in Figures 23.5 and 23.6, respectively. Cratering should not be confused with popping, which is discussed in Section 23.7. Cratering results from a small particle or droplet of low surface tension contaminant, which is on the substrate, in the coating, or lands on the wet surface of a freshly applied film [15]. Some of the low surface tension material dissolves in the adjacent film, creating a localized surface tension differential. As a result of the Marangoni effect, this low surface tension part of the film flows away from the particle to try to cover the surrounding higher surface tension liquid coating. Since as the flow occurs, solvent evaporates, the differential in surface tension increases, and flow continues. However, loss of solvent causes an increase in viscosity, which impedes flow, leading to formation of a characteristic crest around the pit of the crater. Commonly, as shown in Figure 23.5, a particle of contaminant can be seen in the center of a crater.

Craters must be avoided in applying coatings. The user applying the coating should try to minimize the probability of low surface tension contaminants arriving on the wet

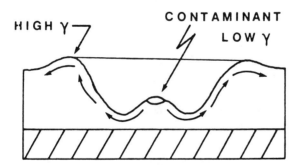

Figure 23.5. Schematic diagram of a crater. (From Ref. [16], with permission.)

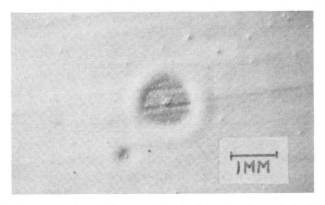

Figure 23.6. A typical crater. (From Ref. [16], with permission.)

coating surface. For example, the spraying of lubricating oils or silicone fluids on or near the conveyor carrying freshly coated parts is an almost sure way to cause craters. However, in most factories, the presence of some contaminating particles cannot be avoided; therefore, the formulator must design coatings that minimize the probability of cratering. Lower surface tension coatings are less likely to form craters, since a smaller number of contaminating particles will have an even lower surface tension. Alkyd coatings have low surface tensions and seldom give cratering problems (or crawling problems either). In general, polyester coatings are more likely to give cratering problems than acrylic coatings, which tend to have lower surface tensions. High solids coatings, because of their generally higher surface tensions, are more likely to give cratering problems than conventional coatings. Some water-borne coatings are also vulnerable to cratering. In a number of cases, particles from cosmetics and similar products worn by production workers have caused cratering of automotive coatings. Powder coatings are quite vulnerable to cratering, as discussed in Chapter 27.

Additives can be used to minimize cratering. Small amounts of silicone fluid generally eliminate cratering, but, as noted before, caution is required in selecting the amount and type of silicone fluid used in order to avoid crawling or recoat adhesion problems. Octyl acrylate copolymer additives usually reduce cratering. These additives operate by reducing the surface tension to a level lower than most contaminants that could cause cratering. If the entire surface has a uniformly low surface tension, there will be no surface tension differential driven flow. A comparison of the effects of a range of additives on the control of defects such as cratering has been reported [31,32]. Side effects such as reduction of gloss and loss of adhesion of coatings applied over the coatings are discussed.

There are many other examples of film defects that result from surface tension differential driven flows. In coating tin plate sheets, the coating is applied by roller and the coated sheets are passed on to warm wickets that carry the sheets approximately vertically through an oven. In some cases, one can see a pattern of the wicket as a thin area on the final coated sheet. The heat transfer to the sheet is fastest where it is leaning against the metal wicket. The surface tension of the liquid coating on the opposite side drops locally because of the higher temperature. This lower surface tension material flows toward the higher surface tension surrounding coating, leaving an area of thin coating. This defect has also been called telegraphing. Similarly, on large plastic parts with thicker reinforced areas on the back of the part, telegraphing through to the top coat(s) can be encountered.

In spraying flat sheets, one can get an effect called *picture framing*, or *fat edge*; the coating is thickest at the edges and just in from the edge, the coating is thinner than average. The contrast in hiding of the substrate can make the differences in film thickness very evident. Solvent evaporates most rapidly from the coating near the edge, where the air flow is greatest. This leads to an increase in resin concentration at the edge and to a lower temperature. Both factors increase the surface tension there, causing the lower surface tension coating adjacent to the edge to flow out to the edge to cover the higher surface tension coating.

Surface tension differential driven flow can also result when overspray from spraying a coating lands on the wet surface of a different coating. If the overspray has lower surface tension than the wet surface, cratering occurs. If the overspray has high surface tension compared to the wet film, local orange peeling results.

In applying coatings by curtain coating (see Section 22.6), the curtain of coating must remain intact. If a particle or droplet of contaminant of lower surface tension than the coating lands on the surface of the flowing curtain, surface tension differential driven flow will cause a thin area in the curtain, which can cause a hole in the curtain. When this part of the curtain is deposited on the panel being coated, an uncoated area results. The problem is minimized by using coatings of the lowest possible surface tension. Since the curtain is flowing, dynamic surface tension is the important quantity. Bierwagen has discussed this phenomenon [4].

23.5. FLOATING AND FLOODING; HAMMER FINISHES

Two related defects result from uneven distribution of pigment in a film as it is drying; floating and flooding. Unfortunately, terminology is not uniform; some people call both phenomena floating. We follow the more common terminology by which floating describes a mottled effect and flooding is used when the surface color is uniform, but darker or lighter in color than should result from the pigment combination used.

Floating is most evident in coatings pigmented with at least two pigments. For example, a light blue gloss enamel panel can show a mottled pattern of darker blue lines on a lighter blue background. The pattern tends to be hexagonal, but seldom perfectly so. Alternatively, with a different light blue coating, the color pattern might be reversed: The lines could be light blue with the background a darker blue. These effects result from pigment segregations that occur as a result of convection current flows driven by surface tension differentials while a film is drying. Rapid loss of solvent from a film during drying leads to considerable turbulence. Convection patterns are established whereby coating material flows up from lower layers of the film and circulates back down into the film. As the fresh material flows across the surface before it turns back down, solvent evaporates, concentration increases, temperature drops, and surface tension increases. The resultant surface tension differential drives continuation of the convection current. The flow patterns are roughly circular, but as they expand, they encounter other flow patterns and the convection currents are compressed. If the system is quite regular, a pattern of hexagonal *Bénard cells* is established. The cells are named after a 17th-century French scientist who pointed out the commonness of hexagonal flow patterns in nature. As solvent evaporation continues, viscosity increases and it becomes more difficult for the pigment particles to move. The smallest particle size, lowest density particles continue moving longest and the largest particle size, highest density particles stop moving sooner. The segregated pattern of floating results.

Floating is particularly likely to occur if one pigment is flocculated and the other is a nonflocculated dispersion of fine particle size. The fine particle size pigment keeps moving longest and is trapped where the convection current turns back into the film at the border between adjacent cells. The border between the cells has a higher concentration of the finer particle size material whereas the center of the cells is more concentrated in the coarser pigment. If, in the example of the light blue coating, the white pigment is flocculated and the blue is not, one will find darker blue lines on a lighter blue background. If, on the other hand, the blue is flocculated and not the white, there will be lighter blue lines on a darker blue background. Figure 23.7 shows a schematic diagram of convection patterns in Bénard cell formation. Floating can also occur in single pigment coatings, resulting in uneven pigmentation of the surface, which reduces gloss.

Floating can be reduced by properly stabilizing pigment dispersions so that neither pigment is flocculated. However, one can get floating even without flocculation. This can result from the use of pigments with very different particle sizes and densities. An example

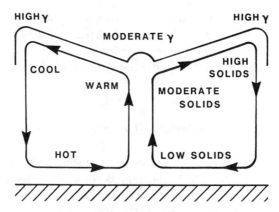

Figure 23.7. Schematic diagram of Bénard cell formation. (From Ref. [16], with permission.)

is use of fine particle size high-color carbon black with titanium dioxide to make a gray coating. Not only is the particle size of the TiO_2 several times that of the carbon black, but the TiO_2 also has about fourfold higher density. A much larger particle size, weaker black, such as lamp black, can be used to make a gray with a lower probability of floating.

Floating is reduced if slower evaporating solvents are used. Surface tension differentials are then less likely to be established, and Marangoni flow and floating do not occur. If a coating exhibits floating, the formulator should first reformulate to eliminate the problem; most commonly, selective pigment flocculation is the cause and should be corrected. Sometimes the flocculation is caused by the solvents used to reduce the coating to application viscosity, in which case solvent change eliminates the problem. In other cases, a change in the resin or dispersing agent used to stabilize the pigment dispersion is required. (See Section 20.1.3.) Second choice is to use an additive. As with other flow phenomena driven by surface tension differentials, floating can be prevented by adding a small amount of a silicone fluid. To avoid crawling, the minimum amount of silicone must be used. The probability of problems can be reduced by using a very dilute solution of the silicone fluid in solvent. Since slow evaporating solvents also can help reduce floating, it is common to use slow evaporating solvents to make additive solutions.

The term *flooding* is applied when the color of the surface is uniform, but different than should have been obtained from the pigment combination involved. For example, one might have a uniform gray coating, but a darker gray than that expected from the ratio of black to white pigments. The most troubling part of flooding is that the extent of flooding can vary with the conditions encountered during application, leading to different colors on articles coated with the same coating. Flooding results from surface enrichment by one or more of the pigments in the coating [16,33]. The stratification is thought to occur as a result of different rates of pigment settling within the film, which are caused by differences in pigment density and size or flocculation of one of the pigments. Flooding is accentuated by thick films, low vehicle viscosity, and low evaporation rate solvents—any factor that tends to keep the film at low viscosity longer and allow more pigment settling. The remedies are to avoid flocculation and low density fine particle size pigments, and if possible, use faster evaporating solvents and higher viscosity vehicles. Floating, flooding, and related color defects are discussed in Ref. [33].

While floating is usually undesirable, ingenious coatings formulators have taken advantage of the problem by purposely inducing floating to make attractive coatings.

The coatings are called *hammer finishes* because they look like the pattern one would get by striking a piece of metal with a ball peen hammer. Hammer finishes were once used on a large scale, especially for coating cast iron components, for which it was desirable to hide the surface roughness. Such coatings contain large particle size, nonleafing aluminum pigment and dispersions of transparent, fine particle size pigments, commonly phthalocyanine blue. One way of getting a hammer effect is to spray a metallic blue coating and then spray a small amount of solvent on the wet film. Surface tension is lowest where drops of solvent land, and surface tension differential driven convection flow patterns are set up, leading to floating where the lines have more blue and the centers of the patterns have more aluminum with less blue. There are *self-hammer* coatings, formulated to give a hammer finish pattern without need for a spatter spray of solvent. Fast evaporating solvents are used with a resin, such as a styrenated alkyd, resin that gives fast drying. Use of hammer finishes has decreased as smooth plastic molded parts replaced rough metal castings in many end uses.

23.6. WRINKLING; WRINKLE FINISHES

The term *wrinkling* refers to the surface of a coating that looks shriveled or wrinkled into many small hills and valleys. In some cases, a wrinkle pattern is so fine that to the unaided eye, the film appears to have low gloss rather than to look wrinkled. However, under magnification, the surface can be seen to be glossy but wrinkled. In other cases, the wrinkle patterns are broad or bold and are readily visible to the naked eye. Wrinkling results when the surface of a film becomes high in viscosity while the bottom of the film is still relatively fluid. It can result from rapid solvent loss from the surface, followed by later solvent loss from the lower layers. It can also result from more rapid cross-linking at the surface of the film than in the lower layers of the film. Subsequent solvent loss or cure in the lower layers results in shrinkage, which pulls the surface layer into a wrinkled pattern. Wrinkling is more apt to occur with thick films than with thin films because the possibility of different reaction rates and differential solvent loss within the film increases with thickness.

The earliest examples of wrinkling were with drying oil films, especially if all or part of the oil was tung oil and cobalt salts were used as the only drier. Tung oil cross-links relatively rapidly when exposed to oxygen from air, and cobalt salts are active catalysts for the autoxidation reaction, but are poor through driers (see Section 14.2.2). These factors favor differential surface cure; which results in wrinkling of the surface layer. Depending on the ratio of tung oil to other drying oils or, in the case of alkyd systems, the oil length of the alkyd, and the ratio of cobalt to driers such as lead (or zirconium) salts that promote through dry, the wrinkle pattern can be fine or bold. While in many cases, wrinkling is undesirable, ingenious coating formulators turned the disadvantage into an advantage and for many years, *wrinkle finishes* were sold on a large scale for applications such as office equipment. Like hammer finishes, wrinkle finishes covered uneven cast metal parts. Their usage has dropped since plastic molded parts have supplanted many metal castings.

Today, wrinkling is usually an undesired defect. It is most often encountered in improperly formulated or applied MF cross-linked coatings in which amines are used to neutralize acidity in the coating formulation and/or amine blocked sulfonic acid catalysts are used for package stability. The probability of wrinkling in such coatings increases as the volatility of the amine increases. For example, triethylamine leads to wrinkling under conditions in which dimethylaminoethanol does not. Increasing catalyst concentration tends to increase wrinkling, as does greater film thickness [34].

Another situation in which wrinkling can occur is in UV curing of pigmented acrylate coatings with free radical photoinitiators. (See Section 28.1.3.) High concentrations of photoinitiator are required to compete with absorption by the pigment. Penetration of UV through the film is reduced by absorption by the pigment as well as by the photoinitiator. There is rapid cross-linking at the surface and slower cross-linking in the lower layers of the film, resulting in wrinkling. Wrinkling is likely to be more severe if the curing is done in an inert atmosphere rather than in air. In the latter case, the cure differential is reduced by oxygen inhibition of surface cure. UV curing by cationic polymerization, which is not air inhibited, is even more prone to surface wrinkling.

23.7. BLISTERING AND POPPING

Blistering is the formation of bubbles near the surface of a film; *popping* is the formation of broken bubbles at the surface of a film that do not flow out. Both effects can occur when the surface viscosity increases to a high level while volatile material remains in the lower parts of the film. If the viscosity is very high at the surface, the bubbles of solvent rise to the surface, but do not break, blistering instead. Popping occurs when the viscosity at the surface of a film increases sufficiently that solvent bubbles break, but the coating does not flow out before further viscosity increase prevents leveling. If the "pops" are very small, the effect is sometimes called *pinholing*.

Popping and blistering result from rapid loss of solvent at the surface of a film during initial flash off. The surface develops a high viscosity relative to that of the solvent rich lower layers of the film. When the coated object is put into an oven, solvent volatilizes in the lower layers of the film, creating bubbles that do not readily pass through the high-viscosity surface. As the temperature increases further, the bubbles expand, finally bursting through the top layer resulting in popping. The viscosity of the film meanwhile has increased enough so that the coating cannot flow together to heal the eruption. Popping can also result from, or at least be made worse by, entrapment of air bubbles in a coating. If the surface of the film has a high viscosity, the bubbles may remain in the film until the coating goes into the oven. The air expands with higher temperature, and the bubbles may burst through the surface. Air bubbles are especially likely to be entrapped during spray and hand roller application of water-borne coatings. (See Section 23.8.) Blistering and popping can result from solvent that remains in primer coats or base coats when the top coat is applied. In coating plastics, solvents can dissolve in the plastic and then cause blistering or popping when a coating is applied over the plastic and then baked. Another potential cause of popping is evolution of volatile byproducts of cross-linking after the surface viscosity has increased to the extent that the bubbles of volatile material cannot readily escape through the surface.

The probability of popping increases with film thickness, since there is a greater chance of developing a differential in solvent content as film thickness increases. A means of evaluating the relative likelihood that a series of coatings will show popping is to determine the maximum film thickness of each coating that can be applied without popping when the films are prepared, flashed off, and baked under standardized conditions [35]. This thickness is called the *critical film thickness for popping*. Popping can be minimized by spraying more slowly in more passes, by longer flash-off times before the

Table 23.1. Critical Film Thickness for Popping

Copolymer T_g (°C)	Critical Dry Film Thickness (μm)	
	Water	Solvent
−28	50	120
−13	30	>70, <95
−8	20	>70, <95
14	10	55
32	5	25

object is put into the oven, and by zoning the oven so that the first stages are relatively low in temperature. The probability of popping can also be reduced by having a slow evaporating, good solvent in the solvent mixture. This tends to keep the surface viscosity low enough for bubbles to pass through and heal before the viscosity at the surface becomes too high.

Popping can be particularly severe with water-reducible baking enamels, as shown in Table 23.1. The pairs of enamels were identical except that in one set, they were reduced for application with solvent and in the other set, with water [35]. As can be seen, the critical film thickness for popping was consistently lower for the water-reduced coatings. The data in Table 23.1 also illustrate another variable that affects the probability of popping. Critical film thickness for popping in these coatings decreases as the T_g of the acrylic resin in the coating increases. This is true in both the solvent-reduced and water-reduced coatings, but the effect is particularly large in the water-reduced compositions.

There are probably many reasons for the greater difficulty of controlling popping of water-reducible coatings. A variety of solvents with different evaporation rates is available for adjusting formulations of solvent-borne coatings, but water has only one vapor pressure/temperature curve, and its curve is steeper than for any organic solvent. Water can be retained by forming relatively strong hydrogen bonds with polar groups on resin molecules at room temperature; these hydrogen bonds break at higher temperature releasing the water. The heat of vaporization of water, 2260 Jg^{-1}, is higher than that of organic solvents—373 Jg^{-1} for 2-butoxyethanol as an example. This higher heat of vaporization slows the rate of temperature increase of films of water-reduced coatings in an oven, further increasing the probability of popping [33]. (See Section 17.3.6.)

In contrast to increased probability of popping with higher T_g water-reducible coatings, popping is more likely to occur with lower T_g latex polymers. Coalescence of the surface before the water has completely evaporated is more likely with a lower T_g latex.

23.8. FOAMING

During manufacture and application, a coating is subjected to agitation and mixing with air, creating the opportunity for foam formation. Incorporation of foam in a coating can lead to pinholing or popping. The problem can be particularly severe when applying water-borne coatings by spray, especially airless spray, or by hand rollers. Formation of a foam involves the generation of a large amount of surface area; it follows, therefore, that the lower the surface tension, the less the energy required to generate a given amount of foam. However, foam bubbles in pure low viscosity liquids are not stable and break almost instantaneously; there must be something present to stabilize the foam. Although water has a high surface tension and therefore might not generate bubbles easily, bubbles in water are easier to stabilize since a wider variety of components can be put in water that rapidly migrate to the surface of a bubble to stabilize it. A surfactant not only reduces the surface tension of water, facilitating foam formation, but also migrates to the surface of the droplets to give an oriented surface layer with a high viscosity, stabilizing the foam bubbles. In formulating a latex paint, an important criterion in selecting surfactants or water-soluble polymers as thickeners is their effect on foam stabilization [3,36]. Acetylene glycol surfactants such as 2,4,7,9-tetramethyl-5-decyne-4,7-diol are reported to be effective surfactants that do not increase the viscosity of the surface of bubbles as much as surfactants such as alkylphenol ethoxylates [37].

A variety of additives can be used to break foam bubbles. Most depend on creating surface tension differential driven flow on the surface of bubbles. If the surface tension of a spot on the surface can be lowered, liquid from that area will flow away to try to cover neighboring higher surface tension areas. But the wall thickness of bubbles is thin, and as material flows away, the wall gets still thinner and weaker so that this spot on the surface of the bubble breaks. For example, poly(dimethylsiloxane) fluids, are effective in breaking a variety of foams, since their surface tension is low compared to almost any foam surface. Of course, as in other uses of silicone fluids, a little may be fine, but a little extra can cause problems. Other low surface tension additives, such as poly(octyl acrylates), act as defoamers. Small particle size hydrophobic SiO_2 can also act as a defomer and/or a carrier for active defoaming agents [37]. Also, a small amount of immiscible hydrocarbon will often reduce foaming of an aqueous coating.

Several companies sell lines of proprietary antifoam products and offer test kits with small samples of their products. The formulator evaluates the antifoam products in a coating with foaming problems to find one that overcomes, or at least minimizes, the problem. While it is possible to predict which additive will break a foam in a relatively simple system, such predictions are difficult for latex paints because of the variety of components that could potentially be at the foam interface. The combination of surfactants, wetting agents, water-soluble polymers, and antifoam can be critical. A variety of test methods has been used to compare foaminess of coatings [36-38].

23.9. DIRT

A common defect is dirt at the surface of a film. It has been called the most common defect of all [39]. A wide variety of solid particles can land on the wet surface of a freshly applied film. Sanding dust, floor dirt, fibers from wiping cloths or clothing worn by operators, and oven dirt are a few examples of such particles. Prevention of dirt problems requires clean raw materials, clean paint, and a clean paint shop, preferably isolated from the rest of the factory. The air supply to spray booths and spray guns must be clean. Sanding should be minimized and sanding dust cleaned up before painting. Ovens should be carefully and frequently cleaned. Lint-free protective clothing and wiping cloths reduce lint contamination. In some applications, such precautions cannot or probably will not be taken; then fast drying becomes a desirable characteristic of coatings.

GENERAL REFERENCE

P. E. Pierce and C. K. Schoff, *Coating Film Defects*, Federation of Societies for Coatings Technology, Blue Bell, PA, 1988.

REFERENCES

1. S. LeSota, Ed., *Coatings Encyclopedic Dictionary*, Federation of Societies for Coatings Technology, Blue Bell, PA, 1995.
2. M. J. Owen, "Surface and Interfacial Properties," in *Physical Properties of Polymers Handbook*, American Institute of Physics, Woodbury, NY, 1996.
3. G. P. Bierwagen, *Prog. Org. Coat.*, **3**, 101 (1975).
4. G. P. Bierwagen, *Prog. Org. Coat.*, **19**, 59 (1991).
5. R. E. Smith, *Ind. Eng. Chem. Prod. Res. Devel.*, **22**, 67 (1983); J. Schwartz, *J. Coat. Technol.*, **64** (812), 65 (1992).
6. L. E. Scriven and C. V. Sternling, *Nature*, **187** (4733), 186 (1960).
7. S. E. Orchard, *Appl. Sci. Res.*, **A11**, 451 (1962).
8. T. C. Patton, *Paint Flow and Pigment Dispersion*, 2nd. ed., Wiley-Interscience, New York, 1979, p. 554.
9. W. S. Overdiep in D. B. Spalding, Ed., *Physicochemical Hydrodynamics*, V.G. Levich Festschrift, Vol. II, Advance Publications Ltd., London, 1978, p. 683.
10. W. S. Overdiep, *Prog. Org. Coat.*, **14**, 159 (1986).
11. S. Kojima and T. Moriga, *Polym. Eng. Sci.*, **35**, 1098 (1995).
12. S. K. Wilson, *Surface Coat. Intl.*, **80**, 162 (1997).
13. O. Cohu and A. Magnin, *Prog. Org. Coat.*, **28**, 89 (1996).
14. P.-A. P. Ngo, G. D. Cheever, R. A. Ottaviani, and T. Malinski, *J. Coat. Techol.*, **65** (821), 29 (1993).
15. F. J. Hahn, *J. Paint Technol.*, **43** (562), 58 (1971).
16. P. E. Pierce and C. K. Schoff, *Coating Film Defects*, Federation of Societies for Coatings Technology, Blue Bell, PA, 1988.
17. K. Tachi, C. Okuda, and K. Yamada, *J. Coat. Technol.*, **62** (791), 19 (1990).
18. T. C. Patton, *Paint Flow and Pigment Dispersion*, 2nd. ed., Wiley-Interscience, New York, 1979, p. 572.
19. W. S. Overdiep, *Prog. Org. Coat.*, **14**, 1 (1986).
20. T. C. Patton, *Paint Flow and Pigment Dispersion*, 2nd. ed., Wiley-Interscience, New York, 1979, p. 578.
21. S. H. Wu, *J. Appl. Polym. Sci.*, **22**, 2769 (1978).
22. D. R. Bauer and L. H. Briggs, *J. Coat. Technol.*, **56** (716), 87 (1984).
23. L. W. Hill and Z. W. Wicks, Jr., *Prog. Org. Coat.*, **10**, 55 (1982).
24. W. H. Ellis, *J. Coat. Technol.*, **55** (696), 63 (1983).
25. R. M. Christenson, T. R. Sullivan, S. K. Das, R. Dowbenko, J. W. Du, and R. L. Pelegrinelli, U.S. Patent 4055607 (1977); J. M. Maklouf and S. Porter, U.S. Patents 4147688 and 4180619 (1979), M. S. Andrews and A. J. Backhouse, U.S. Patent 4180619 (1979); A. J. Backhouse, U.S. Patent 4268547 (1981); H. J. Wright, D. P. Leonard, and R. A. Etzell, U.S. Patent 4290932 (1981).
26. S. Ishikura, K. Ishii, and R. Midzuguchi, *Prog. Org. Coat.*, **15**, 373 (1988).
27. L. J. Boggs, M. Rivers, and S. G. Bike, *J. Coat. Technol.*, **68** (855), 63 (1996).
28. L. B. Brandenburger and L. W. Hill, *J. Coat. Technol.*, **51** (659), 57 (1979).
29. F. Fink, W. Heilen, R. Berger, and J. Adams, *J. Coat. Technol.*, **62** (791), 47 (1990).
30. R. Berndimaier, J. W. Du, D. R. Haff, J. C. Kaye, E. D. Kelley, R. LaGala, J. McGrath, J. M. McKeon, U. Schuster, M. Sileo, L. Waedle, S. Westerveld, and M. E. Wild, *J. Coat. Technol.*, **62** (790), 37 (1990).

31. M. Schnall, *J. Coat. Technol.*, **63** (792), 95 (1991).
32. L. R. Waelde, J. H. Willner, J. W. Du, and E. J. Vyskocil, *J. Coat. Technol.*, **66** (836), 107 (1994).
33. M. Schnall, *J. Coat. Technol.*, **61** (773), 33 (1989).
34. Z. W. Wicks, Jr. and G. F. Chen, *J. Coat. Technol.*, **50** (638), 39 (1978).
35. B. C. Watson and Z. W. Wicks, Jr., *J. Coat. Technol.*, **55** (698), 59 (1983).
36. J. Schwartz and S. V. Bogar, *J. Coat. Technol.*, **67** (840), 21 (1995).
37. W. Heilin, O. Klocker, and J. Adams, *J. Coat. Technol.*, **66** (829), 47 (1994).
38. J. Kozakiewicz, J. Zhu, and G. P. Bierwagen, *J. Coat. Technol.*, **65** (824), 47 (1993).
39. C. K. Schoff, private communication, 1997.

CHAPTER 24

Solvent-Borne and High Solids Coatings

From this chapter through Chapter 28, we discuss the principal classes of coatings: solvent-borne, water-borne, electrodeposition, powder, and radiation cure coatings. The intent is to discuss the principles involved in these classes and to compare the various types of resins applicable to each group. In Chapters 29 through 32, end use applications for these various types of coatings are discussed.

Historically, almost all coatings were solvent-borne; the other four classes of coatings have been developed in large measure to reduce solvent usage. The original motivation was to reduce fire hazards and odor, and to permit cleanup with water. Since the 1960s, a major driving force has been to reduce VOC emissions. There have been significant reductions in solvent use by shifting to other classes of coatings. Some feel that in the long run, solvent-borne coatings will disappear, but there are advantages to solvent-borne coatings over the alternatives. Capital cost for application is generally lower, especially as compared to water-borne coatings in which stainless steel equipment is needed. Electrostatic spray installations for solvent-borne coatings are less expensive than for water-borne coatings. Solvent evaporation is not dependent on humidity. There are fewer problems of air entrapment and popping. Significant reductions in VOC emissions have been made by shifting to high solids coatings and by refining application methods to minimize solvent requirements. Current research and development is aimed at still higher solids coatings and, ultimately, at *solventless* (solvent-free) liquid coatings.

For some end uses, especially when cost is particularly important, a single coating is adequate; however, in many other end uses, performance requirements can only be met by applying at least two coats. Almost always when more than one coat is to be applied, it is preferable to have a *primer* specifically designed to be the first coat and a different coating for the *top coat*. The primer is designed to adhere strongly to the substrate and to provide a surface to which the top coat adheres well. It is not necessary for the primer to meet such other requirements as exterior durability that are critical for top coat performance. Commonly, primers are less expensive than top coats.

24.1. PRIMERS

The first consideration in formulating a primer is to achieve adequate adhesion to the substrate. Adhesion is discussed in Chapter 6; the conclusions of those discussions are briefly summarized here. The substrate should be clean and preferably have a uniformly rough surface. The surface tension of the primer must be lower than that of the substrate. The viscosity of the continuous phase of the primer should be as low as possible to promote penetration of the vehicle into pores and crevices in the surface of the substrate. Penetration is also promoted by use of slow evaporating solvents, use of slow cross-linking systems, and, whenever feasible, use of baking primers. The primer binder should have polar groups scattered along the backbone of the resin that can interact with the substrate surface. Interaction between the binder and the substrate should be such that the coating will not be displaced by water when water molecules permeate through the coating to the interface. (See Section 7.2.3.) For primers over metal substrates, as well as over alkaline surfaces such as masonry, saponification resistance of the primer is an important criterion in binder selection.

The binders in many primers for metal substrates are BPA epoxy resins and their derivatives. (See Sections 11.1–4 and 15.8.) While a full explanation is not available, experience has shown that in general, BPA epoxy resin based coatings exhibit superior adhesion to clean metal. Use of a small amount of epoxy phosphate esters (see Section 11.5) can further enhance adhesion, especially wet adhesion. For baked coatings, epoxy-phenolic coatings are particularly appropriate. For air dry coatings, epoxy-amine coatings are commonly selected. Both types have excellent wet adhesion and saponification resistance, critical for long-term corrosion protection. In general, baked epoxy-phenolics are strongest in these properties, but air-dry epoxy-amines are also very good. Primers based on alkyd (see Chapter 15) and epoxy ester (see Section 15.8) resins are widely used. Material and application costs are less than with epoxy coatings, but wet adhesion is usually not as good, except possibly to metal contaminated with oily residues. Saponification resistance of alkyd resins is limited. Saponification resistance of epoxy esters is intermediate between alkyds and epoxy-amines, at intermediate cost.

The lowest cost vehicles used in metal primers are styrenated alkyds. (See Section 15.3.) They are generally used in air dry primers. The higher T_g resulting from the large fraction of aromatic rings leads to primers that give dry-to-touch films, or even dry-to-handle films, rapidly. Styrenation reduces the fraction of ester groups by dilution and, hence, may increase saponification resistance. However, styrenated alkyds cross-link more slowly than their nonstyrenated counterparts. As a result, the films develop solvent resistance slowly and have a time window within which they should not be top coated. If little or no cross-linking has occurred, or if cross-linking has proceeded to the extent that the whole film is sufficiently cross-linked to be solvent resistant, the primer can be top coated without problems. However, at intermediate times, the films have poor and nonuniform solvent resistance, and if the primer is top coated within this time interval, *lifting* is likely to occur. Lifting results in puckered areas in the top coat surface due to nonuniform degrees of swelling by the top coat solvent, which leads to uneven film shrinkage when the film finally dries.

The surface of galvanized steel is a layer of zinc hydroxide, zinc oxide, and zinc carbonate; all are somewhat water-soluble, strong bases. Especially if the metal is not phosphate treated shortly before coating, the binder in the primer must be resistant to saponification. Alkyds have been used in primers on galvanized steel, sometimes with satisfactory initial results. However, there are commonly sporadic, serious adhesion failures. Fresh galvanized steel surfaces or surfaces that have been carefully protected from dampness are relatively easy to adhere to. However, if the galvanized metal has been stored where it came in contact with water, then the surface becomes coated with zinc hydroxide and its reaction products. For air dry coatings, acrylic latex vehicles are more appropriate than alkyds.

Primers are also required for surfaces other than metal substrates. Adhesion to polyolefin plastics can be difficult to achieve; it can be enhanced by use of a primer based on a chlorinated polymer. (See Sections 6.5 and 30.3.) Primers for wood substrates have somewhat different functions. Adhesion to new wood surfaces is almost never a problem, since the substrate is porous, permitting penetration into the wood surface. However, the porosity is uneven. If one applies a solvent-borne high gloss coating directly on a wood surface, more vehicle penetrates into the more porous areas, resulting in an increase in PVC of the coating above these areas of the surface. As a result, the gloss of the coating varies from relatively high gloss over the least porous areas to relatively low gloss over the most porous areas. If one applies a solvent-borne low-gloss coating on wood, over the porous areas the penetration of vehicle can lead to an increase of PVC so that the local PVC is greater than CPVC. Latex paints minimize these problems, since the latex particles are large compared to the pores in the wood surface and do not penetrate far into the wood; hence, the PVC remains about the same over both high- and low-porosity sections of the surface.

Concrete and other masonry surfaces are alkaline and often require special surface treatments, most commonly washing with hydrochloric or phosphoric acid. The acid wash not only neutralizes the surface alkalinity, but it also etches the surface. Saponification resistant primers such as epoxy-amine or latex primers, provide longest life. Concrete blocks have very porous surfaces, and penetration of solvent-borne paints into the surface requires a relatively large volume of coating for coverage. Substantially better coverage can be achieved by using latex paints, as discussed in Section 31.1.

24.1.1. Pigmentation of Primers

Selection of pigments and their amount is critical in primer formulation. Pigmentation level affects adhesion of top coats to the primer film. Formulations with a high ratio of PVC to CPVC (see Chapter 21) are low-gloss coatings; the roughness and increased surface area of low-gloss films give improved intercoat adhesion. In some cases, it is desirable to formulate primers with PVC > CPVC. The resultant primer film is somewhat porous, permitting penetration of top coat vehicle into the pores, assuring excellent intercoat adhesion. The PVC should be only slightly higher than CPVC or the loss of vehicle from the top coat may be enough to raise the PVC of the top coat film sufficiently to reduce its gloss. Primers with PVC > CPVC can be sanded more easily and are less likely to clog sand paper than primers with PVC < CPVC. Since inert pigments are generally the least expensive components of the dry film, high PVC minimizes cost. See Section 21.2 for a discussion of factors controlling CPVC; in general terms, CPVC is higher with pigments that have a broad particle size distribution.

Since primers almost always are low gloss coatings, inexperienced formulators may not think it is necessary to have good pigment dispersions for primers. This is not so! Since CPVC is affected by the degree of dispersion and particularly by the extent of flocculation, and since primers are usually formulated to be either slightly above or slightly below

CPVC, pigment dispersion can be critical. If the pigment dispersion is not properly stabilized, CPVC decreases; instead of having a primer with a PVC slightly less than CPVC, its PVC could be greater than CPVC. Or, in the case of a primer designed to have a PVC slightly greater than CPVC, the PVC may become much greater than CPVC, leading to loss of gloss of the top coat.

If only one color top coat is to be applied over the primer, it is usually desirable for the primer to have a similar color, since this minimizes the effect of the primer color on the final top coat color. In many cases, several different color top coats are applied over the same primer; then, it is usually desirable to use a light gray primer. Gray primers have better hiding than white primers, or equal hiding at lower cost, since not as much TiO_2 is needed to hide the substrate when a pigment that strongly absorbs light, such as lamp black, is present. Light gray primers have relatively little affect on top coat color. Primers pigmented with red iron oxide provide good hiding at low cost but are harder to hide.

As discussed in Section 7.3.3, in addition to adhesion, water and oxygen permeabilities are important factors in corrosion protection of metals by barrier coatings. The pigments can strongly affect oxygen and water permeability. Up to PVC to CPVC of about 0.9, the higher the PVC of the film, the lower the gas and vapor permeability. To occupy the largest possible volume of the dry film with impermeable pigment, a combination of pigments giving a high CPVC should be used with PVC close to CPVC. Platelet-shaped pigment particles tend to provide better barriers to oxygen and water permeability. Mica and micaceous iron oxide are widely used in primers for metals because of their platelet form. One must also use resin-pigment combinations in which the resin is strongly adsorbed on the pigment surface. If the pigment has a polar surface with a weakly adsorbed resin, water permeating through the film may displace the resin from the pigment surface, leading to an increase in water permeability.

Pigments should be completely insoluble in water. If, for example, zinc oxide were used as a pigment in a primer formulation, some of the zinc oxide would dissolve in water permeating through the film, resulting in establishment of an osmotic cell that can cause blistering. As discussed in Section 7.4.2, passivating pigments can be useful for protecting steel against corrosion when the coatings on the steel substrate have been ruptured. However, passivating pigments must be somewhat soluble in water in order to passivate the steel. Furthermore, dissolving the pigments requires that the binder is able to swell with water to a degree. These characteristics mean that blistering is more likely with a primer that contains a passivating pigment. In most baking primers for OEM products, it is preferable not to use passivating pigments, but to rely on barrier properties to provide corrosion protection. However, applications such as coatings for bridges, storage tanks, ships, and offshore drilling platforms are examples of end uses for which the coating cannot be baked and, in many cases film rupture must be anticipated. As discussed in Sections 32.1.2 and 32.1.3, containing passivating pigments or zinc-rich primers are commonly used for such applications.

24.1.2. High Solids Primers

There is regulatory pressure to reduce the VOC emissions from primers. In conventional low solids primers, the effect of pigmentation on the viscosity of the coating is relatively small if the pigment is not flocculated. However, as solids are increased, the volume of dispersed phase (including both the volume of the pigment and of the adsorbed layer on the surfaces of the pigment particles) increases and becomes an important factor

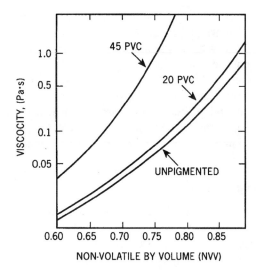

Figure 24.1. Effect of pigmentation on viscosity as a function of volume solids for an unpigmented coating and for two pigmented coatings based on the same binder with pigment loadings sufficient to give 20% PVC and 45% PVC in the dry films. See Ref. [1] for the assumptions made in the calculations. (From Ref. [1], with permission.)

controlling the viscosity of the coating, limiting the solids at which a coating can be applied. See Section 20.1.3 for discussion of factors affecting viscosity of high solids pigmented coatings.

In coatings with a nonvolatile volume (NVV) content below about 70, the effect of pigmentation levels less than that required for a PVC of 20 in the dry film is small. But pigmentation levels such as encountered in low gloss coatings with PVC of 45 or higher substantially increase the viscosity of the wet coating, requiring reduction of solids for application. Figure 24.1, based on model calculations, shows plots of viscosity as a function of volume solids for three sets of calculations: one for an unpigmented coating, one for a 20 PVC (i.e., a gloss) coating, and one for a 45 PVC (i.e., low gloss) coating. The assumptions made in carrying out the calculations are provided in Ref. [1].

With the present status of our knowledge, somewhere in the neighborhood of 60 NVV is probably an upper limit for a primer with a PVC near to or above CPVC. As a result, some high solids primers are formulated with much lower than optimum PVC. A challenge to increasing solids content above 60 NVV at optimum PVC is development of a reasonable cost means of stabilizing the pigment dispersion in a primer with thinner adsorbed layers. However, even assuming stabilization with an adsorbed layer thickness of 5 nm, the upper limit of volume solids is probably not over 70, or at most 75 NVV. As a result of such limitations, the major thrust of development of reduced VOC primers is on water-borne primers.

24.2. TOP COATS

Top coats include coatings for use both over primers and directly on a substrate. In the former case, the primer provides adhesion to the substrate and a major part of the corrosion protection for coatings on metal. The top coat must adhere well to the primer and provide the desired appearance and other properties. A single coating layer must combine both functions. In general, it is preferable to use a primer/top coat system; however, one coat application can be functional for many applications and is less expensive than multiple coat systems. Single coats are used on products that need little corrosion protection and for which the need for maintaining adhesion in the presence of water is not critical. When appearance and exterior durability requirements are minimal, a primer with excellent corrosion protection properties may be used without a top coat, for example, inside ballast tanks of ships and for the interior of structural components of aircraft.

24.2.1. Binders for Top Coats

To an important degree, the properties of a top coat is controlled by the class of resin used as the principal binder in the top coat. The chemistry of these various binders is discussed in Chapters 8 through 16; in this chapter, the advantages and disadvantages of some classes of resins used in top coats, especially for OEM product coatings on metal, are compared.

24.2.1.1. Alkyds

In the 1930s and 1940s, alkyds (see Chapter 15) were major binders for coatings. While increasingly being replaced by other binders, alkyds are still used on a large scale. We use them as a base to which to compare other binder systems. A major advantage of alkyds in many applications is lower cost. A second major advantage is that application of alkyd coatings is, in general, the most foolproof. Solvent-borne alkyd coatings are least subject to film defects of all the classes of coatings. This advantage results from the low surface tension of most alkyd coatings. Therefore, there are seldom problems with crawling, cratering, and other defects that result from surface tension driven flows or surface tension differential driven flows. (See Section 23.4.) With alkyds it is relatively easy to make pigment dispersions that do not flocculate. A further advantage is their ability to cross-link by autoxidation. This makes air dry or low temperature baking possible and avoids the need for cross-linking agents with potential toxic hazards.

The major limitations of alkyds, which vary, in degree depending on the alkyd and other coating components, are comparatively poor color retention on baking and limited exterior durability. It is also difficult to achieve very high solids in alkyd solvent-borne coatings. (See Section 15.2). A further problem can be generation of smoke in baking ovens that causes visual air pollution problems.

Oxidizing alkyd binders are used for low performance, low cost coatings such as for steel shelving, machinery, coat hangers, and exterior coatings for drums. Individual applications may be relatively small, but there is a host of products for which the performance of these coatings is adequate. Oxidizing alkyds are also used for a significant fraction of architectural gloss enamels. (See Section 31.3.1.) With metal salt driers, oxidizing alkyd coating films air dry in a few hours, force dry at 60 to 80°C in an hour, or bake at 120 to 130°C in half an hour. Baking time is shortest with an oil length of about 60 with highly unsaturated drying oil derived alkyds. Tung oil modified alkyds give the fastest cross-linking, followed in turn by mixed tung-linseed oil alkyds, and then linseed oil modified alkyds. Cured films from these alkyds are quite yellow and turn yellow brown on overbaking. Generally, they are only used for dark color coatings. Since the cured films still contain substantial amounts of unsaturation and metal driers, the films embrittle with age. Better color and color retention can be obtained by using alkyds made with fatty acids or oils such as soy or tall oil fatty acids that are less unsaturated. These alkyds are also less expensive, but they cure more slowly than the more highly unsaturated alkyds.

Medium oil alkyds have a substantial number of hydroxyl groups on each molecule; these functional groups can be cross-linked with a variety of cross-linkers. Since melamine-formalehyde (MF) cross-linkers are generally lowest cost, they are the most widely used. To achieve compatibility, butylated, isobutylated, or octylated MF resins are used, not methylated ones. Generally, the more reactive, Class II MF resins with high NH content are used. (See Section 9.2.) The solvent system should contain some butyl alcohol to promote package stability and reduce viscosity as compared to straight hydrocarbon solvents. Usually, no added catalyst is needed, since the residual unreacted carboxylic acid groups on the alkyd are sufficient catalyst for the Class II MF resin. If needed, relatively weak acid catalysts such as an alkylphosphoric acid can be used. Color and speed of cure are affected by the type of fatty acid used in the alkyd. Highly unsaturated alkyds are seldom used in top coats because of their dark color, but soy, tall oil, and dehydrated castor alkyds are used. Some of the cross-linking results from the unsaturation and the balance from the reactions of the MF resin. The color, color retention, and resistance to embrittlement are better than obtained using the oxidizing alkyd without the MF resin, since the MF cross-linking minimizes or eliminates the need for metal driers to catalyze the oxidation reactions.

The best color, color retention, and exterior durability are obtained using nonoxidizing, saturated fatty acid based alkyds. For example, solid color exterior top coats (monocoats) based on coconut oil alkyds and MF resins were used by many automobile manufacturers at one time. To minimize degradation due to hydrolysis under acidic conditions, alkyds made with isophthalic acid (or hexahydrophthalic anhydride) are used rather than those from phthalic anhydride. The exterior durability is good, but not as good as the best acrylic coatings. The coatings are said to exhibit a greater appearance of depth than is obtained with acrylic-MF coatings. The cost of these coatings is similar to polyester-melamine coatings, but film defects during application are less of a problem.

The hydroxyl groups on alkyds can also be cross-linked at room temperature or under force dry conditions with polyisocyanates. IPDI isocyanurate prepolymers are often used for this purpose, since their color stability and exterior durability are superior to TDI derivatives. (See Section 10.3.2.) In many cases, use of a polyisocyanate with an oxidizing alkyd permits sufficiently rapid development of tack free coatings so that metal driers are not needed; this further improves exterior durability.

24.2.1.2. Polyesters

Polyester resins (see Chapter 13) are one of the major classes of resins replacing alkyd resins in MF cross-linked baking enamels. Polyesters are also widely used in urethane coatings. Their cost is generally somewhat higher than oxidizing alkyds, but in some cases, less than nonoxidizing alkyds. Color, color retention, exterior durabiliity, and resistance to embrittlement are better than obtained with most alkyds, but exterior durability and resistance to saponification are generally not as good as with acrylics. Adhesion and impact resistance of polyester-based coatings without primers over clean, treated steel and aluminum substrates are comparable to alkyds and generally superior to acrylics. Polyester coatings generally have higher surface tensions than alkyd coatings and, hence, are more subject to crawling and surface tension differential driven flow defects, such as cratering. Most polyesters are hydroxy-terminated and are cross-linked with MF resins or with isocyanates. The MF resins are less expensive; methylated or mixed methylated/butylated MF resins are usually used. For lower temperature cure or air dry coatings, aliphatic polyisocyanates are used.

A major advantage of polyesters over alkyds and most acrylic resins is the relative ease of preparing polyester resins suitable for very high solids and even solventless coatings. Even low molecular weight hydroxy-terminated polyesters that have at least two hydroxyl groups on virtually every molecule can be synthesized. (See Sections 13.2 and 24.2.2.) Polyesters are also adaptable for use in powder coatings. (See Chapter 27.)

24.2.1.3. Acrylics

In general terms, the major advantages of acrylic binders are their low color, excellent color retention, resistance to embrittlement, and exterior durability at relatively modest costs. They are quite photochemically stable and very hydrolytically stable. In general, their surface tensions are intermediate between alkyds and polyesters, and as a result, the susceptibility of acrylic based coatings to film defects is intermediate. Generally, their adhesion to metal surfaces is inferior to both alkyds and polyesters coatings; therefore, they are usually used over a primer.

Thermoplastic solution acrylic resins (see Section 12.1) were widely used in OEM automotive coatings, but have been replaced with thermosetting acrylics (TSAs) (see Chapter 12) to reduce VOC emissions. Most commonly, TSAs are hydroxy-functional, with perhaps a minor amount of carboxylic acid functionality. They are cross-linked with MF resins or with polyfunctional isocyanates. Either Class I or Class II MF resins can be used; the choice usually depends on curing temperature requirements. Aliphatic isocyanate cross-linkers are more expensive than MF resins and present greater toxic hazards, but cure at lower temperatures and, with HALS stabilizers, usually provide somewhat greater exterior durability and frequently have better environmental etch resistance. Excellent environmental etch resistance can also be achieved by use of epoxy-functional acrylics cross-linked with dicarboxylic acids.

As discussed in Section 12.2.1, it is difficult to make very high solids acrylic resins because of the difficulty of assuring that substantially all of the low molecular weight molecules have at least two hydroxyl groups. While progress is being made toward increasing solids, it is unlikely that acrylic coatings will ever be made with as high solids as polyesters.

24.2.1.4. Epoxies and Epoxy Esters

While the major uses for epoxy resins (see Chapter 11) are in primers, significant volumes of epoxy-based top coats are used. Since epoxy-based coatings generally exhibit excellent adhesion to metal, especially in the presence of water vapor, as well as resistance to saponification, they are commonly used as single coats. For BPA and novolac based epoxy resins, a major limitation is their poor exterior durability. Applications such as beer and soft drink can linings, do not require exterior durability. Epoxy-functional acrylic resins cross-linked with polybasic acids give films with excellent exterior durability. (See Section 11.3.2.)

Epoxy esters (see Section 15.8) offer properties intermediate between alkyds and epoxies. An example of their use in top coats is in coating bottle caps and crowns. They exhibit the requisite combination of hardness, formability, adhesion, and resistance to water.

24.2.1.5. Urethanes

Polyisocyanate cross-linkers have been mentioned in connection with alkyds, polyesters, and acrylics. They are widely used as cross-linkers because of low temperature curing and to achieve abrasion resistance combined with resistance to swelling of cured films with solvent. They are mentioned further here because urethanes can also be backbone links in resins that have reactive groups other than isocyanates. Hydroxy-terminated urethane resins (see Section 10.7) can be cross-linked with polyisocyanates or MF resins. When MF resins are used, the coatings have the advantage of being one package coatings without the concern for toxic hazards of free isocyanate. There is no difficulty making low molecular weight resins where substantially all of the molecules have a minimum of two hydroxyl groups. Compared to polyesters, cost is higher but the hydrolytic stability and, therefore, exterior durability can be superior. However, the intermolecular hydrogen bonding of urethane groups leads to higher viscosity at equal concentration solutions of equal molecular weight compared to polyester resins. Low molecular weight, hydroxy-functional urethanes can be blended with other generic types of hydroxyl-functional resins to modify coating properties.

Urethanes, especially those with primary NH_2 groups, react with MF resins to give greater hydrolytic stability under acidic conditions than hydroxy-functional resins cross-linked with MF. These resins are frequently called carbamate-functional resins. Clear coats using this approach are finding application in automotive coatings.

24.2.1.6. Silicone and Fluorinated Resins

The highest oxidation resistance both to thermal degradation and photoxidation is obtained with silicones (see Section 16.5.2) and fluorinated resins (see Section 16.1.3). The cost of both is high, especially that of fluorinated resins. Silicone alkyds give greater exterior durability than alkyds for air dry coatings. In a compromise to obtain excellent outdoor durability at an intermediate cost, silicone-modified polyesters and acrylics (see Section 16.5.3) are relatively widely used. Fluorinated copolymer resins cross-linked with MF resin or aliphatic polyisocyanates have outstanding exterior durability.

24.2.2. Formulating Solvent-Borne Coatings for Low VOC

A significant fraction of industrial and special purpose coatings and some architectural coatings are still solvent-borne, but there is increasing pressure to reduce VOC. The purpose of this section is to consider the various types of top coats with emphasis on the potential for reduction of solvent usage while still achieving the required level of performance.

While some thermoplastic resins are still being used, their very high VOC contents mean that their usage will continue to fall. The majority of solvent-borne coatings currently applied and to be expected in the future are thermosetting coatings. Conventional thermosetting coatings have volume solids (NVV) on the order of 25–35%. Technical efforts on high solids coatings since around 1970 have accomplished varying degrees of increases in solids of various types of coatings for various applications. People ask for a definition of high solids coatings; there is no single definition. For metallic automotive top coats (or base coats) high solids means about 45 NVV. For a highly pigmented primer, high solids might be 50 NVV. For clear or high gloss pigmented coatings 75, or even higher, is possible. The situation is further complicated by the difficulty of exactly measuring, or even of defining VOC of a coating. (See Section 17.8.1.) For example, in some cases, solvents with functional groups may partially react with the cross-linker and, hence, not be evolved; there can also be volatile byproducts of cross-linking that should be included in VOC, and low molecular weight components may volatilize before cross-linking; the extent to which such emissions occur may vary with baking conditions.

A limitation on solids content is the increasing difficulty of achieving desired mechanical properties as molecular weight and average functionality per molecule $\bar{f}_n$ are decreased and molecular weight distribution is narrowed. As discussed in Section 12.2.1, in a conventional thermosetting acrylic coating, one uses a resin with an $\bar{M}_w/\bar{M}_n$ on the order of 35,000/15,000 and, depending on the end use, an $\bar{f}_n$ of 10 to 20. The cross-linker might have an $\bar{M}_w/\bar{M}_n$ of the order of 2000/800 and an $\bar{f}_n$ of 3-7. An acrylic resin that permits 45 NVV might have $\bar{M}_w/\bar{M}_n$ on the order of 8000/3000 with $\bar{f}_n$ of 3 to 6. As $\bar{f}_n$ is reduced it becomes increasingly critical and difficult to control formulations and cure conditions to keep all film properties within the desired ranges.

As one aims for still lower molecular weights and functionalities for still higher solids coatings, achieving high performance properties becomes even more difficult. For NVV of 70, the $\bar{M}_w/\bar{M}_n$ must be on the order of 2000/800 or less, with an average of little over two hydroxyl groups per molecule. It is critical that essentially all molecules have at least two functional groups per molecule. Any molecules with only one functional group cannot cross-link and will leave dangling ends in the network; any molecules with no functional groups will be plasticizers and, if low enough in molecular weight, may be partially volatilized in a baking oven. It is difficult to synthesize acrylic resins for very high solids coatings. (See Section 12.2.1.) On the other hand, it is relatively easy to make polyurethane or polyester resins with $\bar{M}_w/\bar{M}_n$ of 2000/800, or even lower and with an average of a little over two hydroxyl groups per molecule where essentially all of the molecules have at least two hydroxyl groups. (See Sections 10.7 and 13.2.)

In conventional coatings, the *cure window* is relatively large; that is, it makes little difference if the baking temperature is off by $\pm10°C$, if the baking time is off by $\pm20\%$, or if the user carelessly adds 10% too much or too little catalyst. In high solids coatings, the cure window is narrower [2]. If there are a large number of hydroxyl groups on each resin molecule and 10% are not reacted, the change in properties may be small. If, however, there are only a little over two hydroxyl groups per average molecule and 10% are left unreacted, a significant fraction of the molecules will be tied into the network in only one place, with detrimental effects on film properties. The problem can be minimized by using cross-linkers with higher $\bar{f}_n$, at least then, the problem only arises from the polyester (or other resin), not from both. Class I MF resins, because of their greater average functionality compared to Class II MF resins, generally offer broader cure windows, as discussed in Section 9.3. The extent of self-condensation reactions of MF resins is particularly dependent on time, temperature, and catalyst concentration.

In using high solids coatings, the applicator should be very careful in controlling the time and temperature in baking ovens and in following the coating supplier's recommendations. The formulator must be more careful in checking the film properties when the temperature is about 10°C above and below the standard temperature. In making recommendations of cure cycles to a customer, one should use pieces of the customer's metal for establishing the baking schedule. The critical temperature for baking is that of the coating itself, not that of the air in the oven. Coatings applied to a heavy piece of metal heat up more slowly in an oven than coatings on light gauge sheet metal. On sheet metal, the coating over a place where the sheet metal has been welded to a supporting member heats up more slowly than on the rest of the surface. In conventional coatings, variations resulting from such differences are generally small; high solids coatings are more likely to be subject to variation in properties due to differences in coating temperature and time.

To increase solids, polydispersity of resins can be decreased; however, this may narrow the breadth of the T_g transition range (see Section 4.2), which may adversely affect the mechanical properties of films [3]. An approach to overcoming this problem is to blend resins with differences in composition, but still similar enough to each other to be compatible; for example, polyester/acrylic blends using low molecular weight hydroxy-terminated polyesters to reduce the viscosity and thereby increase the solids [4]. Resins and cross-linkers for high solids coatings generally have $\bar{M}_n$ well below 5000. In contrast to high molecular weight polymers, the entropy of mixing of different low molecular weight resins is large enough to be a significant factor favoring compatibility. (See Section 17.2.) The effect of molecular weight is illustrated by 50 : 50 blends of several acrylic and methacrylic homopolymers that are compatible, when their $\bar{M}_n$ values are less than 5000 but incompatible when $\bar{M}_n$ values are above 10,000 [5]. The broader compatibility of low $\bar{M}_n$ resins permits formulation of high solids coatings based on mixtures of different generic type resins. Not all blends of low $\bar{M}_n$ resins are compatible, and there is the possibility that phase separation could occur as molecular weight increases in the early stages of cross-linking. Unpigmented films should be made and checked for transparency; haziness suggests phase separation on a scale large enough to detract from appearance or properties.

Isocyanurates, unsymmetrical trimer, biuret, and allophanate polyisocyanates (see Section 10.3.2) with lower molecular weights and, hence, lower viscosities have been made available. However, the lower viscosity of these polyisocyanates is only one factor affecting the VOC of coatings [6]. In some cases, the lower viscosity cross-linker also has a lower equivalent weight; this means that the ratio of low viscosity cross-linker to much higher viscosity polyol is decreased. The lowest VOC could be obtained using a somewhat higher equivalent weight polyisocyanate, even though it has a somewhat higher viscosity. Aldimines and hindered diamines (see Section 10.4) permit preparation of very high solids clear top coats.

Research is now focused on what are called "ultra high solids" or solvent-free coatings. These are based on the use of low molecular weight oligomers or reactive diluents; the distinction between the terms is disappearing. VOC of two package polyurethane coatings can be reduced by replacing some of the hydroxy-terminated acrylate or polyester with a reactive diluent [6]. Examples of such coreactants are aliphatic diols and low molecular weight ester or urethane diols.

Polyesters afford the major opportunity for very low VOC coatings, since low molecular weight oligomers can be made such that all molecules have at least two reactive groups, usually hydroxyl groups. For example, a 1,4-butanediol ester of mixed glutaric, adipic, and azeleic acids with $\bar{M}_n$ of 310 have a viscosity at 25°C of 270 mPa·s [7]. Solvent-free coatings can be formulated with polyisocyanate cross-linkers. With MF resins, film hardness of baked films is too low; however, the coatings can be thinned with water (see Section 17.4), permitting formulation with a variety of other components in solvent-free coatings [7]. For example, a novel reactive diluent, the reaction product of glycidyl neodecanoate and *p*-hydroxybenzoic acid (PHEA), shows promise of providing higher T_g films with better physical properties when used with polyester diols and Class I MF resins [8].

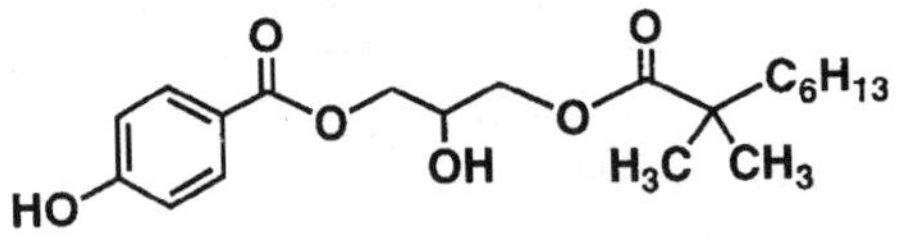

PHEA

When catalyzed by sulfonic acids, the hydroxyl group reacts with the MF resin, and MF resin also reacts by a Mannich reaction to give benzoxazine cross-links. Formation of the rigid, cyclic cross-links substantially hardens the films. In the formula below, Mel represents the melamine ring of an MF resin.

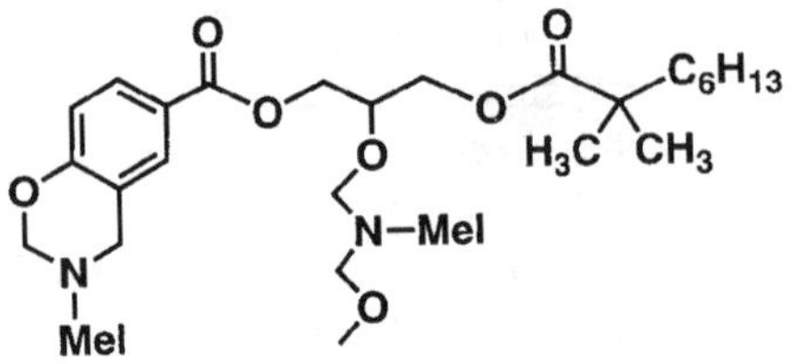

Combinations of Class I MF resins and HDI triisocyanurate have been used as cross-linkers for oligopolyesters with PHEA to permit formulation of solventless coatings with good properties [9].

Another example of a "zero" VOC coating that uses water as a diluent is an epoxy-hydroxyl system [10]. Caprolactone polyols are used with 3,4-epoxycyclohexylmethyl-3′,4′-epoxycyclohexane carboxylate at a 2 : 1 epoxy to hydroxyl ratio, and a triflic acid derivative as catalyst.

VOC reduction is being aided by application methods that permit use of higher viscosity coatings. Hot spray (see Section 22.2.4), high speed electrostatic disks (see Section 22.2.3), and supercritical fluid spray (see Section 22.2.5) are examples. Improvements in transfer efficiency are making a significant contribution to reducing VOC emissions (see Sections 22.2.1 and 22.2.3).

As solids increase, it becomes more difficult to avoid pigment flocculation [11]. As discussed in Section 20.1.3, the primary factor controlling stabilization of pigment dispersions is the thickness of the adsorbed layer on the surface of the pigment particles. The very low molecular weight resins used in very high solids coatings are incapable of providing an adequate adsorbed layer. Furthermore, as the number of functional groups per molecule decreases, the solvent can compete more effectively for adsorption sites on the pigment surface promoting flocculation. Hyperdispersants (see Section 20.1.3) have been developed for use in high solids coatings.

Another limiting factor for some high solids coatings is surface tension effects. Generally speaking, as molecular weight gets lower, the equivalent weight must be lowered even further, since the number of reactions required to achieve high molecular weight will be greater. In most coatings, the functional groups are highly polar, such as hydroxyl and carboxylic acid groups. Increased levels of such groups give higher surface tensions. Furthermore, achieving high solids at a given viscosity with such resins generally requires using hydrogen-bond acceptor solvents rather than hydrocarbon solvents. Again, this results in higher surface tensions than that of most conventional coatings, and, therefore, increased probability of film defects during application. When coating metal, it is more important to have clean surfaces. When coating plastics, one must be careful to remove mold release. A broader range of plastic materials have to be surface treated to avoid crawling and achieve adhesion. (See Sections 6.5 and 30.3.) Since the surface tension of a freshly applied high solids coating is generally higher than that of a conventional coating, a larger fraction of contaminating particles floating in the air have lower surface tensions than that of the wet high solids coating. Thus, cratering is more probable. (See Section 23.4.)

Sagging is more of a problem with high solids than with conventional coatings [12]. As discussed in Section 23.3, this results from the slower rate of evaporation of solvents from high solids coatings than from conventional coatings. While the reasons for this difference have not been completely elucidated, the consequences pose significant problems. Sagging of spray applied high solids coatings cannot be easily controlled by adjusting the evaporation rate of the solvent in the coating or by changing the distance between the spray gun and the substrate. Sagging can be minimized by using hot spray (see Section 22.2.4) or supercritical fluid spray (see Section 22.2.5). For many applications, thixotropic flow properties must be built into a high solids coating for spray application. Fine particle size SiO_2, bentonite clay pigments, zinc stearate, and polyamide gel thixotropes are additives that can be useful. The problem of sagging in metallic coatings used on automobiles is particularly serious; microgel particles have been developed that are effective. (See Sections 23.3 and 29.1.2.)

Even when sagging is not encountered during application, it can sometimes occur with high solids coatings during baking. The temperature dependence of viscosity of high solids coatings is greater than for conventional coatings, as illustrated in Figure 24.2. Before the cross-linking reactions have proceeded far enough to increase the viscosity, the tempera-

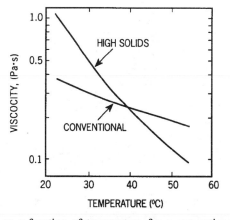

Figure 24.2. Viscosity as a function of temperature for a conventional and a high solids resin solution. The high solids solution is of a 1500 molecular weight polyester at 90% solids in ethylene glycol monoethylether acetate. The conventional solution is of a 20,000 molecular weight polyester at 25% solids in methyl ethyl ketone and ethylene glycol monoethyl ether acetate. (Adapted from Ref. [1], with permission.)

ture has already decreased the viscosity sufficiently to lead to sagging as temperature is increased during baking [1,12]. The phenomenon is called *oven sagging*.

GENERAL REFERENCE

L. W. Hill and Z. W. Wicks, Jr., *Prog. Org. Coat.*, **10**, 55 (1982).

REFERENCES

1. L. W. Hill and Z. W. Wicks, Jr., *Prog. Org. Coat.*, **10**, 55 (1982).
2. D. R. Bauer and R. A. Dickie, *J. Coat. Technol.*, **54** (685), 101 (1979).
3. S. L. Kangas and F. N. Jones, *J. Coat. Technol.*, **59** (744), 99 (1987).
4. L. W. Hill and K. Kozlowski, *J. Coat. Technol.*, **59** (751), 63 (1987).
5. J. M. G. Cowie, R. Ferguson, M. D. Fernandez, M. J. Fernandez, and I. J. McEwen, *Macromolecules*, **25**, 3170 (1992).
6. S. A. Jorissen, R. W. Rumer, and D. A. Wicks, *Proc. Water-Borne Higher-Solids Powder Coat. Symp.*, New Orleans, 1992, p. 182.
7. F. N. Jones, *J. Coat. Technol.*, **68** (852), 25 (1996).
8. V. Swarup, A. I. Yezrielev, J. L. Smith, and R. W. Ryan, *Proc. Waterborne High-Solids Powder Coat. Symp.*, New Orleans, 1997, p. 310.
9. F. N. Jones, *Proc. Waterborne High-Solids Powder Coat. Symp.*, New Orleans, 1997, p. 1.
10. R. F. Eaton and K. T. Lamb, *J. Coat. Technol.*, **68** (860), 49 (1996).
11. S. Hochberg, *Proc. Water-Borne Higher-Solids Coat. Symp.*, New Orleans, 1982, p. 143.
12. D. R. Bauer and L. M. Briggs, *J. Coat. Technol.*, **56** (716), 87 (1984).

CHAPTER 25

Water-Borne Coatings

Before about 1950, almost all coatings were solvent borne. Introduction of latex paints for architectural uses was the first major step away from solvent-borne coatings. Since 1970, there has been a further trend away from solvent-borne coatings to reduce VOC emissions, and use of water-borne coatings in the United States now exceeds the volume of solvent-borne coatings. Their use can be expected to increase further as restrictions on VOC become more stringent. Most water-borne coatings contain some organic solvent. The solvents play a variety of important roles in resin manufacture, coating production and application, and film formation. Research efforts are underway to reduce and, in some cases, eliminate the need for solvents.

The two largest classes of water-borne coatings are *water-reducible* coatings (see Section 25.1) and *latex coatings* (see Section 25.2); modest amounts of *emulsion coatings* (see Section 25.3) are also used. The terminology is not uniform [1]. We use the term water-reducible to refer to resins made in solvent that are reduced with water to form a dispersion of resin in water; but some authors use the term aqueous dispersion resins to identify these resins. Others call such resins *water-soluble resins*, but they are not really soluble in water. Some authors call latex coatings, emulsion coatings; we recommend against this practice to avoid confusion of latexes (dispersions of solid polymer particles in water) with coatings that really are emulsions (dispersions of liquids in liquids). Polyurethane latexes are called *aqueous dispersion resins*. There is growing use of blends of different types, for example, combinations of latex and water-reducible resins. Water is also used as a solvent in solvent-borne coatings; we call this use *water-thinning*, see Sections 13.4 and 24.2.2.

Some of the differences in properties of water from those of organic solvents are advantageous. Water presents no toxic hazard, is odor free, and is not flammable. Nonflammability reduces risks and insurance costs and permits use of less make-up air in baking ovens, reducing energy consumption in some cases. There are no emission or disposal problems directly attributable to use of water. With some formulations, cleanup of personnel and equipment is easier with water-borne coatings; however, in other cases, cleanup is more difficult. The cost of water is low; but it does not necessarily follow that the cost of water-borne coatings is low.

On the other hand, there are some important disadvantages to the use of water. At 25°C, the relative evaporation rate of water is low; however, the vapor pressure increases

467

relatively rapidly with increasing temperature. Whereas a wide range of solvents with different evaporation characteristics enables the formulator to fine-tune evaporation rates from solvent-borne coatings, there is only one kind of water. The heat capacity and heat of vaporization of water are high, resulting in high energy requirements for evaporation. With a given amount of energy available for evaporation, water evaporates more slowly than solvents with similar vapor pressures. Evaporation of water is affected by relative humidity (RH); as discussed in Section 25.1, variations in RH when coatings are applied can lead to problems.

The surface tension of water is higher than that of any organic solvent. In water-reducible coatings, pigments can be dispersed in solvent solutions of the resins before addition of water, and the surface tension can be reduced by use of solvents such as butyl alcohol or a glycol butyl ether. The surfactants used in latex paints reduce surface tension, improving pigment wetting and enabling the coating to wet many kinds of surfaces. The presence of surfactants tends to give films with poorer water resistance; as discussed in Section 8.1.1, various approaches are being used to reduce or even eliminate the need for free surfactants.

Water increases corrosion of storage tanks, coating lines, ovens, and so forth. This requires that corrosion resistant equipment be used for applying water-borne coatings, increasing capital cost. For example, mild steel lines may have to be replaced with stainless steel. Since it is electrically conductive, water requires the use of special adaptations of electrostatic spray equipment, which increases cost. (See Section 22.2.3.)

The majority of water-reducible resins is used in OEM product coatings with limited applications in special purpose coatings. One class of water-reducible coatings, electrodeposition coatings, has such different formulation and application procedures from other coatings that we have chosen to discuss it separately in Chapter 26. Latex systems are used in a majority of architectural coatings; while the general principles of latex coatings are covered in Section 25.2, detailed discussion is postponed to Chapter 31. Small, but increasing, fractions of both OEM product coatings and special purpose coatings are latex based.

25.1. WATER-REDUCIBLE COATINGS

As discussed in Section 12.3 on water-reducible acrylics, the resins in water-reducible coatings are not soluble in water. Most are made as high solids solutions in water-miscible solvent(s) of resins with either carboxylic acid groups that are at least partially neutralized with low molecular weight amines or amine groups that are partly neutralized with low molecular weight acids. Pigment is dispersed in the partly neutralized resin solution, cross-linker is added along with additives such as catalysts, and the coating is diluted with water to application viscosity. The resin is not soluble in the aqueous solution of the solvent that results. Instead, aggregates form with salt groups oriented at the water-particle interfaces and with low polarity parts of the resin molecules in the interior of the particles. The solvent partitions between the water phase and the particles. The particles are swollen by solvent, as well as by water that associates with the salt groups and dissolves in the solvent. The cross-linker is dissolved in the resin-solvent particles, and the pigment is also inside the aggregate particles.

When solutions of these neutralized resins are diluted with water, the change of viscosity with concentration is abnormal. See Figures 12.1–12.3 for typical relationships between resin concentration and log viscosity when water-reducible acrylic resin solutions are diluted with water. As water is added, there is an initial rapid drop in viscosity. As more water is added, the viscosity plateaus and, frequently, with further addition of water, the viscosity increases. Still further addition of water leads to a rapid drop in viscosity. The solids of the coatings at application viscosity are usually lower than solids of solvent-borne coatings. While the viscosity of the organic solvent solution of resin is Newtonian, the viscosity of the water-diluted system in the region of the peak or plateau is highly shear thinning. When diluted to application viscosity, flow is usually only slightly shear thinning. As explained in Section 12.3, this behavior is consistent with the formation of swollen aggregates. As also explained in Section 12.3, the pH of these systems is abnormal. In the case of carboxylic acid-functional resins neutralized with a low molecular weight amine, the pH is basic even though less than stoichiometric amounts of amine are frequently used. The viscosity, as application solids are approached on dilution, decreases rapidly as water is added and is very dependent on the ratio of amine to carboxylic acid. If a small excess of water is inadvertently added, the viscosity may be too low. However, this problem can usually be remedied by adding a small further increment of amine, which increases the viscosity.

Selection of the amine for neutralization is an important formulation consideration. (See Section 12.3.) Amines are expensive, and they add to VOC emissions. Volatility and base strength are important. If there is insufficient amine, there will be macrophase separation of the coating. Therefore, it is desirable to select amines that provide for the necessary stability of the aggregate dispersion at low concentration. The principal factor controlling the ratio of amine to carboxylic acid needed appears to be water solubility of the amine. For example, the amount of tripropylamine required is substantially larger (on an equivalent weight basis as well as a weight basis) than the amount of triethylamine required. Still more efficient are aminoalcohols; N,N-dimethylaminoethanol (DMAE) is probably the most widely used amine. Somewhat less efficient, but still more effective than trialkylamines, are morpholine derivatives. Since only about 20% of the volatiles are organic solvents, the VOC emitted per unit volume of coating, excluding water, is fairly low. In a water-reducible gloss acrylic coating, VOC is equivalent to the amount of solvent emitted by a solution acrylic with approximately 60 NVV. Air quality goals in the future will require still lower VOC emissions; research is underway.

Most types of resins can be chemically modified to make them water reducible. The most widely used resins are acrylics with both carboxylic acid and hydroxy functionality. (See Section 12.3.) Water-reducible polyesters are used, but their hydrolytic stability is limited and low molecular weight cyclic oligomers can volatilize in baking ovens. (See Section 13.4.) Since polyesters can be synthesized in the absence of solvent, it is possible to make solvent-free dispersions [2]. The resin is cast as a solid, pulverized, and stored as a solid until needed, delaying possible hydrolysis. When a coating is to be made, the powdered resin is stirred into a hot aqueous solution of dimethylethanolamine to make a dispersion. Water-reducible alkyds (see Section 15.8) are used to a degree, but their use is limited because of the difficulty in achieving adequate saponification resistance for package stability. Water-reducible epoxy esters (see Section 15.8) and uralkyds (see Section 15.7) have better hydrolytic stability. Water-reducible urethanes have excellent saponification resistance and a minimum of cyclic oligomers. (See Section 10.8.) Solvent-free and low solvent water-reducible urethanes have also been reported [3]. Urethane resins can give films with excellent properties, but are generally more expensive than acrylic resins.

Coatings made with water-reducible resins have advantages and limitations. An important advantage follows from the high molecular weight that can be used, as high as that of resins used in conventional solution thermosetting coatings. This is true because the viscosity at application dilution is almost independent of molecular weight. For example, one can use water-reducible acrylics with $\bar{M}_w/\bar{M}_n$ on the order of 35,000/15,000 and an average of around ten hydroxyl groups and 5 carboxylic acid groups per molecule along with a Class I melamine-formaldehyde (MF) cross-linker and achieve properties essentially equal to those of a conventional solution acrylic enamel. The window of cure is comparable to that of the conventional thermoset acrylic. The problems related to molecular weight and functionality of high solids coatings are not encountered.

A disadvantage is that the application solids are low. Typically, such coatings are applied by spraying, roll coating, or curtain coating at around 20 to 30 NVV. The low solids means more wet film thickness has to be applied in order to achieve the same dry film thickness. On the other hand, in automotive metallic coatings, the low solids is an advantage because it permits better orientation of the aluminum pigment in the film.

The viscosity of water-reducible resin coatings depends strongly on the ratio of water to solvent. Depending on humidity, there can be changes in the ratio of the two components remaining in the coating as volatiles evaporate. If the humidity is above a critical level, water evaporates more slowly than even slow evaporating solvents, such as 2-butoxyethanol (see Section 17.3.6). In some cases, the viscosity of the coating during flash off after application can decrease, rather than increase, as evaporation continues leading to delayed sagging [4]. Variations in humidity during application and drying can lead to problems. If the RH is over 70%, the rate of evaporation of water is very slow, and at 100% there is no net evaporation of water. In the mid-RH range, the effect can be reduced by a relatively modest increase in temperature, since RH decreases with temperature. However, if the humidity is very high, the only recourses have been to cool the air to condense out some of the water and then rewarm it, an expensive expedient, or to wait until the RH is lower. For factory applied coatings, it is generally desirable to formulate for best application at a relatively high RH, say 60%, since it is less expensive to increase RH than to decrease it.

Another problem when applying thick baked coatings can be popping. (See Section 23.7.) Popping during baking of water-reducible coatings is more difficult to control than with solvent-borne coatings [5]. The probability of popping increases as film thickness increases. Since the solids are low and the rate of evaporation of water during spraying and flash off is low, the wet film thickness required to apply the same solids is higher than with solvent-borne coatings. The heat of vaporization of water is high as compared to solvents. This leads to a slower rate of heating of the coating and, hence, slower evaporation. Also, until the amine evaporates, there are polar salt groups that tend to retain water. When the coated article enters an oven, water evaporates fastest from the surface, so the viscosity of the surface layer increases. Subsequently, when the water remaining in the lower layers of the film volatilizes, some of the bubbles of water vapor either cannot break through the surface layer, leaving a blister, or break through at a stage when the viscosity of the surface layer is so high that a crater cannot flow out. Popping is affected by the base strength and volatility of the amine. Popping can be minimized by spraying the coating in more, thinner coats (often called passes) so that there is a greater chance for evaporation as the film thickness is built up; by having longer flash off times before entering the baking oven; and by zoning the oven, such that the first part of the oven is at lower temperature, permitting water to diffuse out of the film before the viscosity at the surface increases unduly. In some installations, infrared ovens can be used to drive off most of the water before entering the baking oven. All these alternatives have relatively high capital or operating costs.

The greatest difficulties occur when spray applying thick coats. One cannot apply completely uniform films by spray. To assure that sufficient coating is applied on all areas, there will be some areas with substantially more than the average film thickness. Also, there is a greater chance of air entrapment with spray applied coatings. If the air bubbles do not break, the air in the bubbles expands when the coating is baked ultimately leading to popping. Air entrapment is particularly severe with airless spray, since at the high pressure more air dissolves in the coating and then comes out as pressure decreases on leaving the gun. (See Section 22.2.6.) It has been shown that the problem of air entrapment can be reduced by using carbon dioxide rather than air for spraying [6]. The carbon dioxide largely evaporates during the time of flight between the spray gun and the substrate. Popping can be minimized by using as low a T_g resin as appropriate for required film properties. Use of some slow evaporating solvent, such as 1-propoxy-2-propanol or the monobutyl ether of diethylene glycol, in the formulation assists in reducing the probability of popping.

Thin coatings applied by roll coating or curtain coating seldom have a popping problem. The films are more uniform than spray applied films, and there is no air entrapment problem. With reasonable care about flash off times, curtain coating of relatively thick films can be accomplished without major problems. Water-reducible coatings are used on a large scale for such applications as can coatings and panel coatings.

One might think that the high surface tension of water would cause serious problems with crawling and cratering. This is usually not the case, presumably, because orientation of nonpolar segments of solvent to the surface is rapid, so the coating surface tension is low.

With air dry water-reducible alkyd coatings, equivalency of properties with solvent alkyd coatings has not been achieved. Water-reducible alkyds can be, and are, made, as are modified alkyds such as silicone-modified alkyds. Their hydrolytic instability leads to limited storage life. With careful inventory control, they can be used for some industrial applications, but they are generally unsatisfactory for trade sales paints for which a shelf life of two or more years is needed. Also, until the neutralizing amine required for water dilution evaporates from the film, the films are water sensitive. The time required to lose amine is related to both the volatility and the base strength of the amine. Ammonia is the most volatile amine and is widely used. However, the last of the amine is lost at a stage when volatile loss is controlled by diffusion rate through free volume holes, not by volatility. The time for amine loss may be reduced by using a combination of ammonia and a less volatile, but relatively weak basic amine, such as N-methylmorpholine. Even after the amine has evaporated, the films contain carboxylic acid residues and are sensitive to water and, particularly, to bases.

It was widely assumed that water-borne urethane coatings that have free isocyanate groups would not be feasible due to the reactivity of isocyanates with water. However, two package (2K) urethane coatings with adequate pot life for commercial use have now been reported [7]. A water-dispersible, aliphatic, "hydrophilically-modified" polyisocyanate is used in one package, and the second package contains a hydroxy-terminated, water-reducible polyurethane with carboxylic acid groups from dimethylolpropionic acid. (See Section 10.8.) A 2 : 1 ratio of $N=C=O$ to OH is used to offset possible reaction with water. Films cross-linked within a week at 25°C when the RH was 55% or lower. At high humidity, solvent resistance did not develop at 25°C, but warming to 31°C permitted cross-linking even at 80% RH. Acrylic polyols can also be used [8]. Pot life problems can be severe, since water can diffuse into the dispersed particles of the isocyanate component, leading to cross-linking, which can preclude good coalesence. Such problems can be minimized by use of appropriate mixing and application equipment [8]. Alternatively, low viscosity polyisocyanates, such as blends of HDI dimer and trimer, can be used with aqueous polyurethane dispersions [9]. See Section 10.8 for further discussion of water-borne polyurethanes.

Since, as discussed in Section 24.1.3, reducing VOC levels in solvent-borne primer formulations is particularly difficult, water-borne primers are particularly important. For example, maleated epoxy esters (see Section 15.8) are used for spray or dip applied primers for steel. Water-borne coatings are especially suited for dip coating applications (see Section 22.3) because they eliminate fire hazards of solvent-borne dip coatings. All electrodeposition coatings are water-reducible coatings. (See Chapter 26.) Interior linings for beverage cans are water-reducible coatings, based on grafting acrylates, styrene, and acrylic acid onto BPA epoxy, as described in Section 11.4.

25.2. LATEX-BASED COATINGS

Latexes have been used for many years in architectural coatings and are the major type of vehicle for these coatings. For household applications, such as flat wall paint, the advantages of latex paints over any solvent-borne paint are so large that solvent-borne paints are seldom marketed. Important advantages of interior latex paints include rapid drying, low solvent odor, absence of odor of oxidation byproducts of drying oils and alkyds, easy cleanup, reduced fire hazard, and better long term retention of mechanical properties. For exterior paints, a major advantage is that exterior durability of high performance latex paints is superior to drying oil or alkyd paints. On wood siding, blistering is reduced, since the latex films are more permeable to water vapor. On the other hand, adhesion of latex paints to chalky surfaces is inferior to solvent-borne paints. (See Section 31.1.)

A further advantage of acrylic, styrene/acrylic, and styrene/butadiene latexes over alkyds is their excellent resistance to saponification. Latex paints generally show better adhesion to galvanized metal surfaces than do alkyd paints. They also generally show better performance than alkyds over cement and concrete surfaces, since they are alkaline surfaces. Also, for reasons discussed in Section 31.2, latex paints give better coverage over porous cement surfaces.

Latex coatings form films by coalescence of the polymer particles. (See Section 2.3.3.) Coalescence can occur only if the film formation temperature is higher than the T_g of the polymer particles. While initial coalescence proceeds rapidly at temperatures just a few degrees above T_g, completion of coalescence is relatively slow unless the temperature is significantly higher than T_g. For most architectural paints, slow final coalescence is not a real problem, so the T_g need be only a little below the film formation temperature. In baked industrial coatings, film formation should be complete by the time the coated article comes out of the baking oven; therefore, baking temperatures have to be significantly above the T_g.

There are limitations to latex paints, particularly on how low the temperature can be while still allowing proper coalescence for film formation. To have a final film with a high enough T_g to resist blocking, it is common to use *coalescing solvents* in the formulation.

The coalescing solvent dissolves in the polymer particles, reducing the T_g, permitting film formation at a lower temperature. After film formation, the coalescing solvent slowly diffuses out of the film and evaporates. Even with the use of coalescing solvents, however, there are limitations on the temperatures required for good film formation. (See Section 31.1.)

Latex paints with less organic solvent or, in some cases, no VOC are being developed and marketed. Blends of latexes of different T_g and particle sizes [10,11] and latexes with particles having gradient T_g made by sequential polymerization (see Sections 2.3.3 and 8.2) [12] have been used. Thermosetting latexes permit use of a low T_g polymer, facilitating film formation at relatively low temperatures; subsequent cross-linking gives the needed block resistance and other properties. (See Section 8.1.4.) Some thermosetting latexes are used in two-package coatings for industrial use; pot life has to be long enough to permit coalesence before a significant degree of cross-linking occurs. For one package latexes, polymers with allylic substitutions that permit autoxidative cross-linking and polymers with trialkoxysilyl groups are being developed. Also hybrid alkyd/acrylic latexes have been prepared by dissolving an oxidizing alkyd resin in the monomers used in the emulsion polymerization.

Increasing volumes of polyurethane latexes are being used; they are usually called *aqueous polyurethane dispersions*. (See Section 10.8.) An important factor promoting their usage is that films of higher T_g polymers can be formed at ambient temperatures than with acrylic and other conventional latexes, hence, reducing or eliminating the need for coalescing solvents. The water acts as a plasticizer for the polyurethane, reducing the T_g of the polymer permitting film formation [13]. In most cases, thermosetting dispersions are used.

Another limitation in the formulation of latex coatings is the difficulty of formulating high gloss latex paints. The main problem of formulating gloss coatings results from the random distribution of pigment and latex particles as the volatiles evaporate, so there is not the same chance of obtaining a pigment free, or low pigment content, upper surface of the film as with solvent-borne coatings. This problem can be minimized, but not eliminated, by using a small particle size latex. Gloss is also limited by the presence of surfactants in the dry film, which can lead to haze and blooming. Latexes with very low surfactant content or with surfactants that can polymerize are being developed to minimize or eliminate this problem. (See Section 8.1.1.) See Sections 18.10.1 and 31.3 for further discussion of gloss latex paints.

Latex coatings tend to be excessively shear thinning. When the viscosity at high shear rate is set appropriately for application, the viscosity at the low shear rates encountered in leveling tends to be high. This is one of several reasons that leveling of latex-based coatings tends to be poorer than that of solvent-borne coatings. Flat coatings with only fair leveling have acceptable appearance, but as gloss is increased, the unevenness of the films becomes undesirable. Substantial progress has been made by use of *associative thickeners*, as discussed in Section 31.3.

Use of latexes has been more limited in industrial coatings. There have been many reasons for this. The problems of evaporation of water on a conveyor line and in ovens mentioned in the last section, are part of the problem. There can also be a problem of popping with latex coatings. It is desirable to use a latex polymer with as high a T_g as possible to minimize the chances of coalescence of the latex particles at the surface of the film before the water has completely evaporated. Probably, the major limitation for industrial applications has been flow properties of latex-based coatings. Leveling requirements for many industrial coatings are more rigorous than for architectural coatings. In general, latex coatings have exhibited relatively high degrees of shear thinning and, in many cases, thixotropy. In some cases, these flow characteristics result from flocculation of latex particles. Flocculation of latex increases low shear viscosity to a major degree; furthermore, it further exacerbates the gloss problem. Use of associative thickeners minimizes this problem. Since most of the published work with associative thickeners has been done with latex paints for architectural end uses, discussion of them is deferred to Section 31.3. Combinations of latexes with water-reducible resins are increasingly used. Frequently, they have better flow properties than latex coatings.

An advantage of latexes is that their high molecular weights provide excellent mechanical properties without need for cross-linking. The viscosity of latexes is independent of molecular weight, so they can be applied at relatively high solids even though their molecular weight is high. In highly pigmented coatings, the fraction of internal phase volume becomes so large that the solids level has to be reduced, but the reduction is done primarily with water so that VOC emissions are minimal.

As pressure to reduce VOC emission increases still further, it is to be anticipated that there will be substantial increases in the use of latexes, particularly thermosetting latexes, in industrial applications of both OEM product coatings and special purpose coatings.

25.3. EMULSION COATINGS

While most water-borne coatings are based on water-reducible or latex binders, emulsion coatings have found some applications and may have potential for wider use.

Two package coatings in which one package is a BPA (or novolac) epoxy resin solution and the second is an amine-terminated cross-linking agent (see Section 11.2.6) containing a nonionic surfactant have had significant commercial applications [14-16]. The amine cross-linker package is diluted with water, and the epoxy resin solution package is added with vigorous stirring. The pot life is limited to a few hours, since the epoxy resin can react not only with the amine groups, but also, slowly, with water. While pot life of solvent-borne epoxy-amine coatings is limited by viscosity increase, emulsion systems show little, if any, change in viscosity with age, since the viscosity is controlled by internal phase concentration not by molecular weight, and because reaction with water does not lead to cross-linking. Instead pot life is limited by reduced gloss of applied coatings or by inferior film properties. Such emulsion epoxy paints are used where hard, easily cleaned wall coatings are needed, for example, in hospitals and food processing plants. The residual surfactant reduces the corrosion protective properties for application to metal surfaces. Emulsion coatings based on aliphatic epoxy resins and aliphatic carboxylic acid-functional resins are reported to give superior pot life and improved properties [17]. An approach to eliminating surfactants is to use nitroethane as one of the solvents [18,19]. As discussed in Section 11.2.6, nitroethane forms a salt with an amine group of an amine-terminated polyamide, which acts then as a surfactant. When the film dries, the nitroethane evaporates, leaving a less water sensitive film. Such coatings are used in high performance applications such as aircraft primers. (See Section 32.4.)

Acrylic copolymers such as amine salts of methacrylic acid/methyl methacrylate/ethyl acrylate/styrene (40 : 20 : 20 : 20) can be used as emulsifying agents for epoxy/phenolic coatings [20]. During baking, the carboxylic acid groups of the acrylic react with epoxy groups, incorporating the surfactant into the polymer structure, avoiding the ill effects of nonreactive surfactants on film properties.

Another example is nitrocellulose lacquers emulsified into water for use as top coats for wood furniture (see Section 30.1) [21]. The emulsions have significantly lower VOC than solvent-borne lacquers, but longer times are required to achieve print resistance.

Alkyd emulsion paints have been formulated [22]. Higher gloss can be obtained than with latex paints and there is not the problem of temperature required for film formation versus blocking resistance experienced with latex paints. However, durability is poorer due to yellowing and embrittlement of alkyds. Nevertheless, for some applications their properties are adequate, and especially in Europe, alkyd emulsion paints are used commercially. Long oil alkyds with anionic surfactants provide the most stable emulsions. Drying generally is slower than with solvent-borne alkyds. A contributing factor may be the partitioning of driers between the water and alkyd phases. Alkyds are emulsified into some latex paints to improve adhesion to chalky surfaces. (See Section 31.1.)

GENERAL REFERENCES

D. R. Karsa and W. D. Davies, Eds., *Waterborne Coatings and Additives*, The Royal Society of Chemistry, Cambridge, 1995.

J. E. Glass, Ed., *Technology for Waterborne Coatings*, American Chemical Society, Washington, DC, 1997.

REFERENCES

1. J. C. Padget, *J. Coat. Technol.*, **66** (839), 89 (1994).
2. R. Engelhardt, *Proc. Water-Borne Higher-Solids Powder Coat. Symp.*, New Orleans, 1996, p. 408.
3. J. W. Rosthauser and K. Nachtkamp, "Water-Borne Polyurethanes," in K. C. Frisch and D. Klempner, Eds., *Advances in Urethane Science and Technology*, Vol. 10, Technomic Publishers, Westport, CT, 1987, p. 121.
4. L. B. Brandenburger and L. W. Hill, *J. Coat. Technol.*, **51** (659), 57 (1979).
5. B. C. Watson and Z. W. Wicks, Jr., *J. Coat. Technol.*, **55** (732), 61 (1983).
6. M. S. Gebhard and L. E. Scriven, *J. Coat. Technol.*, **66** (830), 27 (1994).
7. P. B. Jacobs and P. C. Yu, *J. Coat. Technol.*, **65** (622), 45 (1993).
8. M. Dvorchak and Hunter, *Proc. Waterborne High-Solids Powder Coat. Symp.*, New Orleans, 1997, p. 515.
9. J. M. O'Connor, A. T. Chen, R. S. Blackwell, and M. J. Morgan, *Proc. Waterborne High-Solids Powder Coat. Symp.*, New Orleans, 1997, p. 458.
10. M. A. Winnik and J. Feng, *J. Coat. Technol.*, **68** (852), 39 (1996).
11. S. A. Eckersley and B. J. Helmer, *J. Coat. Technol.*, **69** (864), 97 (1997).
12. K. L. Hoy, *J. Coat. Technol.*, **51** (651), 27 (1979).
13. R. Satguru, J. McMahon, J. C. Padget, and R. G. Coogan, *J. Coat. Technol.*, **66** (830), 47 (1994).
14. R. Albers, *Proc. Water-Borne Higher-Solids Coat. Symp.*, New Orleans, 1983, p. 130; U. S. Patent 4352898 (1982).
15. A. Wegman, *Proc. Waterborne High-Solids Powder Coat. Symp.*, New Orleans, 1997, p. 92.
16. E. Galgoci, K. Dangayach, A. B. Pangelinan, R. J. Knabe, D. R. Denley, and G. A. York, *Proc. Waterborne High-Solids Powder Coat. Symp.*, New Orleans, 1997, p. 106.
17. D. R. Eslinger, *J. Coat. Technol.*, **67** (850), 45 (1995).

18. A. Wegmann, *J. Coat. Technol.*, **65** (827), 27 (1993).
19. J. A. Lopez, U. S. Patent 4816502 (1989).
20. S. Kojima, T. Moriga, and Y. Watanabe, *J. Coat. Technol.*, **65** (818), 25 (1993).
21. C. M. Winchester, *J. Coat. Technol.*, **63** (803), 47 (1991).
22. A. Hofland, in J. E. Glass, Ed., *Technology for Waterborne Coatings*, American Chemical Society, Washington, 1997, p. 183.

CHAPTER 26

Electrodeposition Coatings ———

For many end uses, electrodeposition is an efficient method of applying high performance coatings. The largest volume uses are for primers, but there are also uses for single and two coat systems. The general principle is relatively simple; development and long term production use are very complex [1,2]. There are two types of electrodeposition coating systems: anionic and cationic. A variety of terminology is used: E-coat, electrocoat, electropaint, ED, and ELPO are fairly common synonyms for electrodeposition coatings. Some authors call anionic and cationic coatings, anodic and cathodic coatings, respectively. Cationic E-coat is used worldwide for priming auto bodies, and its adoption in the 1970s and 1980s led to a major improvement in the corrosion resistance of cars.

In anionic coatings, negatively charged particles of coating in an aqueous dispersion are electrophoretically attracted to a substrate, which is the anode of an electrochemical cell. The coating particles are precipitated by hydrogen ions generated there by electrolysis of water. For cationic coating, the object is made the cathode and positively charged particles of coating are attracted to the cathode and precipitated on its surface by hydroxide ions. In both types, thermosetting binders are used, and coatings are almost always baked. The coatings must be designed so that all coating components are attracted to the electrode at the same rate; otherwise, the composition will change with time. A vehicle must be used in which pigments can be dispersed and cross-linkers dissolved and that forms a stable, electrically charged dispersion of aggregate particles when diluted with water. The pigment must be preferentially wet by the resin so that it does not migrate out of the resin aggregates. It is possible to electrodeposit polymer films from solutions of the salt of a polymer in water, but one cannot use a dissolved polymer as a binder for pigmented coatings; the resin, pigment, cross-linker, and other components would not be deposited in the same ratio over time. One can electrodeposit polymer films from a latex. For some purposes, unpigmented electrodeposited latex films can be useful, but they cannot be used for pigmented coatings, since pigment particles and latex particles would be attracted at different rates to the substrate.

The coating is diluted to 10 to 20% solids with water. This low solids is used for three reasons. For coating products like automobiles and appliances, the electrodeposition tanks are very large—up to 500,000 L. The tank is kept full at all times, therefore, part of the capital investment is the cost of a tank full of coating–the investment is lower with 10 to 20% solids than it would be with 50% solids. When the coated object is brought out of the tank, it carries a layer of the bath liquid with it that must be rinsed off; losses are less and rinsing is easier if the solids are low. Furthermore, ultrafiltration is easier with lower solids.

A critical requirement for electrodeposition coatings is indefinite stability after dilution. As the coating is applied, coating solids are removed from the bath, and they have to be continually replaced to maintain the same composition in the tank. Ideally, the tank would never be emptied. Some material from the original loading will be in the tank for long times and, therefore, must be very stable to hydrolysis and mechanical agitation. The cross-linker must be stable in the diluted coating at a pH over 7 for anionic coatings and under 7 for cationic coatings. Stability to oxidation is critical, since air is continually mixed into the bath by the agitation. If an oxidizing type vehicle is used, an antioxidant that volatilizes in the baking oven is an essential additive.

26.1. ANIONIC ELECTRODEPOSITION COATINGS

Resins used in anionic systems are substituted with carboxylic acid groups that have an acid number in the range of 50 to 80 mg KOH/g resin. The pigments and other components are dispersed in the resin, and the carboxylic acid groups are partially neutralized with an amine such as 2-(*N*,*N*-dimethyamino)ethanol. To load the tank, the coating is diluted to about 10% solids with water. The degree of substitution with salt groups is designed to be such that the resin is not soluble in water, but it rather forms aggregates on dilution. The aggregates are stabilized as a dispersion in water with salt groups on the outer surface of the particles. Even with less than the theoretical amount of amine to neutralize the carboxylic acid groups, the pH is above 7 due to entrapment of unneutralized COOH groups in the center of aggregate particles, see Section 12.3.

In early work on electrodeposition primers for automobiles, maleated linseed oil (see Section 14.3.5) was used as a vehicle. The anhydride moiety was bonded to the linseed oil molecules by carbon-carbon bonds that can not be hydrolyzed, and hence, the coating showed reasonable stability. However, adhesion to steel was relatively poor, and maleated linseed oil was soon replaced with maleated epoxy esters (see Section 15.8), which had even better hydrolytic stability and provided superior adhesion to steel. The cross-linking obtained through the drying oil fatty acid esters in the epoxy ester was supplemented by using some melamine-formaldehyde (MF) resin as a cross-linker. Mixed methyl ethyl ether Class I MF resins (see Section 9.2) are most appropriate for electrodeposition. The mixed ether MF resin has sufficient solubility in water to permit easy incorporation, but is more soluble in the resin aggregates than in water, so it deposits in constant ratio as the bath is used over long times. Carboxylic acid-substituted resins made by reacting moderate molecular weight polybutadiene with maleic anhydride have also been used as vehicles for anionic primers. Since the backbone linkages are all carbon-carbon bonds, there is no problem of hydrolysis in the bath.

Maleated epoxy esters are not appropriate for many top coat applications because their color stability and chalking resistance are poor. The most widely used resins for anionic electrodeposition top coats are acrylic copolymers made using acrylic (or methacrylic) acid and 2-hydroxyethyl methacrylate as comonomers. The carboxylic acid groups permit formation of stable aqueous dispersions with electronegatively charged particles; the hydroxyl groups, as well as the carboxylic acid groups, serve as sites for cross-linking with MF resin.

The primary reaction occurring at the anode in anionic E-coats is electrolysis of water to yield hydrogen ions. The hydrogen ions neutralize carboxylate ions on the resin at the surface of the anode. Neutralization removes the charge that stabilizes the aggregates against coalescence. When the salt groups are neutralized, the surface is less polar; there is less swelling of the surface with water, so stabilization is eliminated and the particles coalesce on the metal surface.

$$2H_2O \rightarrow 4H^+ + O_2 + 4e^-$$

$$RCOO^- + H^+ \rightarrow RCOOH$$

Not all the salt groups have to be neutralized for precipitation to occur. When the film is formed, some salt groups and, hence, ammonium ions are trapped within the film. Side reactions can also occur at the anode. Iron can dissolve to form ferrous ions that are oxidized to ferric ions. The ferric ions can form insoluble salts with the carboxylic acids on the resin, leading to reddish brown discoloration. Discoloration is not a problem for primers, but is a limitation for making light colored or white electrodeposition top coats. Zinc phosphate conversion coatings generally reduce discoloration to acceptable levels. Anionic top coats can be used on aluminum, since aluminum salts do not affect the color of a coating. In fact, the protective oxide layer on aluminum is enhanced when it is the anode, so anionic E-coat is preferable to cationic for coating aluminum.

An important side reaction in electrocoating of phosphate conversion treated steel is partial dissolution of the iron-zinc phosphate layer by hydrogen ions generated at the anode surface:

$$Zn_3(PO_4)_2 + 2H^+ \rightarrow 3Zn^{2+} + 2(HPO_4)^{2-}$$

This partial removal of conversion coating has two potentially serious effects. Damage to the phosphate coating can lead to poorer adhesion to the steel surface and less corrosion protection. Also, the soluble ion concentration in the bath is increased, which leads to higher conductivity of the water phase. As discussed later, maintaining constant, relatively low conductivity of the water phase is critical.

Also, electrolysis of water at the anode generates oxygen. The oxygen can be generated at the metal surface after it has been coated, leading to film rupture as the bubbles of oxygen escape through the film.

At the cathode, hydrogen gas and hydroxide ions are generated. The hydroxide ions neutralize the ammonium counterions:

$$4H_2O + 4e^- \rightarrow 4OH^- + 2H_2$$

$$R_3NH^+ + OH^- \rightarrow R_3N + H_2O$$

The amines formed are water soluble, and only a fraction of the amine is removed when the coated article is taken out of the bath. This leads to accumulation of amine in the bath. Methods of control of amine concentration are discussed in Section 26.4. An advantage of electrodeposition as compared to other water-borne systems is that only a small amount of counterion is deposited on the substrate with the rest of the coating. Some anionic electrodeposition tanks have been run using potassium hydroxide as the neutralizing base.

26.2. CATIONIC ELECTRODEPOSITION COATINGS

Cationic coatings have positively charged aggregates that are attracted electrophoretically to the cathode. The resins have amine groups neutralized by a low molecular weight, water soluble acid such as formic, acetic or lactic acid. Coatings stable at a pH a little below 7 are preferred; otherwise stainless steel or other expensive corrosion resistant piping and handling facilities should be used. Some suppliers recommend corrosion resistant equipment in all cases. Commercial cationic electrodeposition tanks are operated within a narrow pH range, 5.8 to 6.2. The resins used in the application of cationic automotive primer E-coats primers are proprietary; they are based on BPA epoxy resins reacted with polyamines to yield a resin with amine and hydroxyl groups. The resin is reacted with a polyisocyanate, which is partially blocked with an alcohol (e.g. 2-ethylhexyl alcohol). Many E-coat primers contained basic lead silicate as a catalyst and perhaps, interacting with the phosphate coating. Lead-free E-coats have been developed to replace the older formulations. Salts are formed with the amine groups with a low molecular weight carboxylic acid. The cross-linking agent is stable in the slightly acidic water system, whereas MF resins are not. During baking, the blocked isocyanate reacts with a hydroxyl group to form a urethane cross-link.

Cationic coatings are used for automotive primers because their corrosion protection is significantly better than that of an anionic electrodeposition primer. Amine-substituted resin binders provide greater corrosion protection for steel, perhaps owing to strong interaction between the amine groups and the substrate surface that increases wet adhesion. As discussed in Section 7.3.2, wet adhesion is the most critical factor in corrosion protection. There is not the problem of acid dissolving the phosphate conversion coating that exists in the case of anionic deposition. Zinc phosphate conversion coatings (hopeite) are somewhat alkali soluble so Zn-Fe phosphate conversion coatings that generate phosphophyllite crystal layers must be used. (See Section 6.4.1.) On zinc coated steels, Zn-Mn-Ni phosphate conversion treatments are used [3].

E-coats based on BPA epoxies and TDI or MDI are used only as primers. If used as top coats, they would exhibit poor color retention and exterior durability. For top coats, use of blocked aliphatic diisocyanates with acrylic resins gives better color retention and exterior durability. Use of 2-(N,N-dimethylamino)ethyl methacrylate and hydroxyethyl methacrylate as comonomers provides the needed amine groups for salt formation and hydroxyl groups for cross-linking. The curing temperature required for alcohol blocked aliphatic isocyanates to react with hydroxyl groups is high. One can use oxime blocked isocyanates that cross-link at lower temperatures, but then the bath stability may be limited. Alternatively, one can make an acrylic resin using glycidyl methacrylate as a comonomer; the pendant epoxy groups can be reacted with amines to yield secondary amines or mixed secondary and tertiary amines. Alcohol blocked aliphatic isocyanates can be used to cross-link the secondary amine groups; high temperature bakes are required, but chemical stability in the tank is excellent.

26.3. EFFECT OF VARIABLES ON ELECTRODEPOSITION

After the current is turned on, deposition of coating is not instantaneous. Some time elapses before enough hydrogen ions (anionic E-coat) or hydroxide ions (cationic E-coat) are formed to neutralize enough charges on aggregate particles to cause precipitation. After this initial time interval, the rate of deposition is affected by the rate of electrophoresis of the aggregates. This rate is importantly affected by the impressed voltage; the higher the voltage, the faster the film deposition. Coatings are designed to coat in 2 to 3 minutes with 225–400 V; the high voltage is not needed to electrolyze the water, but rather to increase the driving force for electrophoretic attraction of particles to the electrode and for good coating of recessed areas.

The first areas covered are the edges of the metal, since current density is highest there. As film thickness increases, electrical resistance increases, reducing the rate of deposition at those sections coated first. There is a limiting film thickness beyond which deposition of further coating stops or, at least, becomes very slow. After edges are coated, outer flat surfaces of the object are coated, followed by recessed and enclosed areas. The further back in a recess, the later the area is coated. Particularly for corrosion protective primers, it is desirable to have the entire surface of the steel coated, so it becomes important to coat the furthest recessed areas in the 2 to 3 minute dwell time in the tank.

The rate of deposition is also affected by the equivalent weight of the coating. The higher the equivalent weight, the greater the amount of coating precipitated by each hydrogen (or hydroxide) ion and, therefore, the faster the buildup of film thickness. On the other hand, it is critical to have a low enough equivalent weight so that there are sufficient polar salt groups to maintain stability of the dispersion of aggregates in the coating bath. The rate is also affected by the amount of soluble low molecular weight ions present in the bath. These ions are attracted to the electrodes in competition with the aggregate particles. Since they are small, they move more rapidly in the electrophoretic field. Concentration of soluble salts must be low and maintained close to constant.

A coating that deposits in recessed areas quickly is said to have a high *throw power* (called throwing power by some authors). Standard tests for throw power determine how far up into a pipe or an open-ended box the coating is applied at a standard voltage in a standard time. Throw power increases with higher impressed voltage and longer dwell times. However, if the voltage is increased to too high a level, there will be film rupture of the coating applied to the outer surfaces. At a sufficiently high voltage, the current breaks through the film leading to local generation of gas under the film (oxygen in anionic electrodeposition and hydrogen in cationic electrodeposition) and bubbles of gas can blow out through the film, leaving film defects. It has been shown that at higher voltages, electric discharges occur through the film during electrodeposition causing visible sparks [4]. These sparks may also be responsible for film ruptures. Sparking is reported to occur at lower voltages (240 V) when the substrate is galvanized steel than when the substrate is steel, about 300 V or higher. As conductivity of the deposited film increases, film rupture tends to occur at lower voltages. Direct current electricity obtained by rectifying AC electricity has a relatively wide variation in voltage; the current is said to ripple. The effect of rippling can be to break through at lower (average) voltage [5]. Model equations have been developed to permit predicting throw power and film build on complex shaped surfaces like an automobile [6].

Throw power is affected by the conductivity of the bath; the higher the conductivity, the greater the throw power. However, there is a limitation; as the conductivity due to the presence of soluble salts increases, the rate of electrophoresis of coating aggregates decreases. Increased numbers of salt groups on the resin increase conductivity and, hence, throw power, but equivalent weight drops, which decreases the rate of deposition. Entrapment of conductive material in the film increases with a corresponding increase in the likelihood of film rupture. A compromise on conductivity must be reached.

Conductivities used are in the range of 1000–1800 microsiemens (μS); the older unit, mho, is still used; 1 μS = 1 μmho.

Film rupture, and hence throw power, is affected by variations in coatings composition. If the viscosity of the aggregates precipitated on the surface is high, full coalescence is not achieved and a porous film results, giving high film conductivity and poor throw power. On the other hand, if the viscosity of the precipitated aggregates is very low, the films will be soft; then, if any electrolysis takes place below the surface, bubbles will break through easily and film rupture will be severe. A compromise must be made between the extremes. The glass transition temperature of the resin is a controlling factor, and correspondingly, the temperature at which the electrodeposition is carried out is important. Temperature of the bath must be controlled within a fairly narrow range, typically 32–35°C. Higher T_g resins can be used only if solvent is included in the formula. Many E-coats now use a small amount of solvent, but eliminating it is a desirable environmental goal. Solvents can affect the electrical conductivity of the deposited film so care must be taken in their selection. Excess solvent leads to film rupture at lower voltage, which reduces throw power. Also, the partition coefficient must be such that all the solvent is dissolved in the aggregates and none in the water, or the solvent concentration in the bath will build up over time.

Pigment concentration also affects coalescence. If the PVC is near or above CPVC and the amount of solvent, if any, is small, the deposited film will not coalesce. Even if low enough to permit coalescence, the amount of pigment has a significant effect on leveling of the film. Since the film as applied has a low solvent concentration, its viscosity is very dependent on pigment concentration. Unless PVC is quite low, in most coatings, less than half of CPVC, leveling will be poor because of the high viscosity. Because of the low level of pigmentation as compared to conventional primers, the gloss of electrodeposited primers is higher, especially if the pigment content is reduced to low enough levels to permit good leveling. The PVC also affects flow away from edges during baking; higher PVC reduces this problem, so a compromise between leveling and edge flow must be made.

26.4. APPLICATION OF ELECTRODEPOSITION COATINGS

A schematic diagram of an E-coat system is shown in Figure 26.1. The object to be coated is hung from a conveyer and carried into a dip tank. Both to return coating carried out of the tank by the object and to avoid local accumulation of excess coating on the object, the object is rinsed with ultrafiltrate as it comes out of the tank. At the rinsing stage, the coating film has not been cross-linked, but is of sufficiently high viscosity to remain intact; only the bath liquid is washed from the surface of the coating. Concentration of coating in the bath must be kept constant by adding "make-up" coating continuously to replace coating solids removed by deposition.

The heat exchanger maintains temperature within a narrow range. The bath liquid is continuously recirculated through an ultrafiltration unit. The ultrafiltration membrane permits removal of excess water and water-soluble materials, while not removing the aggregates containing the resin, pigment, and cross-linker. Ultrafiltration permits maintaining the concentration of soluble salts essentially constant so that there is constant conductivity.

Figure 26.1 also shows an electrolyte tank. This represents a system for controlling the concentration of solubilizing agent (acid for cationic coatings and amine for anionic ones).

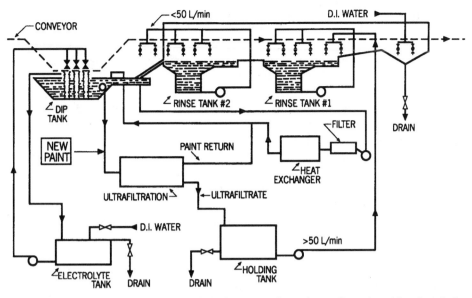

Figure 26.1. Schematic drawing of a cationic automotive primer electrodeposition installation. (Adapted from Ref. [2], with permission.)

When coating is precipitated on the substrate surface, a corresponding amount of solubilizing agent is released. The concentration of solubilizing agent must be maintained at a constant level. A small amount of solubilizing agent is carried out of the bath with the coated product and some is removed by ultrafiltration, but these losses of solubilizing agent are less than the amount being released by electrocoating. There are two ways of maintaining the balance. A sufficiently low level of solubilizing agent content is used in the make-up coating, and the remainder of the required level of solubilizing agent is the excess left in the bath. The other, more effective, approach is to have the counter electrode in a microporous polypropylene box. The pore size of the membrane must be such that the aggregate particles cannot pass through the membrane, but water and carboxylate (or ammonium) ions can easily go through. In some cases, it is necessary to use ion-selective membranes. The clear liquid in the box is recirculated to the electrolyte tank, where the concentration is monitored and corrected automatically. Maintenance of the proper level of solubilizing agent is critical. It is common to hear that the pH of the bath must be kept constant. While this is true, the pH of these coatings is insensitive to the ratio of the weak acid and weak base. (See Section 12.3.) Conductivity is a more important control criterion. The application lines can be highly automated with feedback control of rate of addition of make-up coating, solubilizing agent, and water.

26.5. ADVANTAGES AND DISADVANTAGES OF ELECTRODEPOSITION

Electrodeposition is used for applying coatings to a variety of products. All, or almost all new automobiles have electrodeposited primer coats. Many appliances are primed by electrodeposition. Aluminum extrusions, drapery fixtures, metal toy trucks, and steel furniture are a few of many examples of single coat electrodeposition applications.

Electrodeposition can be a highly automated system with low labor requirements, especially as compared to spray application. A startling example of manpower savings has been reported in the use of cationic electrodeposition to apply an epoxy coating to air conditioners [7]. The former coating system of a flow coated primer and a spray applied acrylic top coat required 50 people, including those who did the required touchup and repair. The E-coat operation required only one operator. Further savings result from elimination of coating losses from overspray. Coating utilization in excess of 95% are reported. The economic advantage of the combination of these factors is large in assembly line operations. However, the capital cost of the automated line is high, limiting applicability of highly automated lines to large production operations. Simpler installations are used for applications like coating metal toys. Development of electrodeposition in the 1960s and 1970s was a long and expensive process, and even today it is expensive to start a new system and get it running smoothly.

Solvent content of E-coats is relatively low, so VOC emissions are low and fire hazard is reduced. Another environmental advantage as compared to spray applied coatings is that there is no overspray sludge disposal. (Unless a poor job of formulating, coating production, bath maintenance, or control makes it necessary to dispose of 500,000 L of bad coating). Since the solids of the deposited film are high, only 3–5 minute flash off time is required before entering the oven, an advantage relative to spray applied coatings.

Another advantage, assuming adequate throw power, is that complete coverage of surfaces is obtained. There can be differences in film thickness; for example, the recessed areas will generally have thinner deposits than the exposed face areas, but all of the surface will have some coating. Recessed and enclosed areas that cannot be coated by spray application can be coated by electrodeposition. Objects with many edges, such as drapery fixtures, can be coated better by electrodeposition than any other way. The edges are uniformly coated, in contrast to excess coating at the bottom edges of objects coated by dipping.

Sagging can be experienced when the coated part is heated in an oven. However, due to the relatively high viscosity of the coating immediately on application, severe sagging is less likely than during spray or conventional dip application methods. Also, one is less likely to encounter the relatively large differences in film thickness between the top and bottom of the dipped object commonly experienced with conventional dip coating.

Uniform film thickness can lead to a problem, especially with relatively highly pigmented primers: the applied coating follows the surface contours of the metal closely, so a rough metal gives a rough primer surface. Some authors refer to this phenomenon of replication of the substrate surface profile as *telegraphing*. Unfortunately, the term telegraphing is also used to describe other film defects. (See Section 23.4.) If there is minimal replication of metal scratches in the surface of the primer coat, some authors say the primer shows good *metal filling*. E-coat films are relatively thin, varying from 15–30 μm, depending on coating composition and application variables; hence, it generally is not feasible to sand the surface smooth because bare metal would be exposed before the surface would be level. In some cases, defects are repaired by sanding and patching with conventional primer.

Substantial variations in smoothness of the primer surface can result from changes in E-coat primer composition [8]. Leveling can be improved by reducing the pigment content of the primer. With lower PVC, the viscosity of the coating after application and before cross-linking is lower. During baking, some leveling occurs before cross-linking increases the viscosity such that no further flow can occur. The lower viscosity can also permit undesirable flow; in some cases, edge coverage is reduced by flow away from the edges of the coated product. The edges heat up first, reducing the surface tension of the coating in that area, leading to flow over the adjoining higher surface tension area. Also, when the pigment content is decreased, gloss of the films is increased.

There can be a problem achieving adequate adhesion of top coats to the surface of electrodeposited primers. Adhesion is affected by cross-link density of the E-coat; overbaking can lead to poorer adhesion. The gloss of electrodeposited primers is relatively high, and the smooth surface makes achieving intercoat adhesion more difficult. When the PVC is further reduced to promote leveling, the surface is still smoother and intercoat adhesion is exacerbated. An approach to improving intercoat adhesion is to apply a sealer over an electrodeposited primer. A sealer is usually a top coat that has been reduced to very low solids with relatively strong, slow evaporating solvents. A top coat is applied over the wet sealer surface. This procedure permits more time for solvent to penetrate into the surface of the primer and swell it so that the sealer can penetrate into the surface of the primer. Sometimes a primer-surfacer is applied over the E-coat primer. If the solvents are carefully selected, penetration into the electrodeposited film promotes intercoat adhesion. The primer-surfacer can have a PVC > CPVC to enhance adhesion of top coat, and it can be sanded to smooth the surface without reducing adhesion. The primer-surfacer also improves adhesion of sealants and adhesives.

The substrate must be conductive; most applications have been for metal primers or for one-coat metal coatings. Two coat applications require that the first coat be a conductive film and are only used in a few instances in production.

Another limitation that electrodeposition shares with any other dipping system is the difficulty of formulation changeovers. If it is decided to change color of an automotive primer, what do you do with 500,000 L of coating? At least once when this problem was faced, the coating supplier developed the new color primer compatible with the old color primer so that the new coating could be introduced into the existing tank. Of course, for a period of time there was a slow change of primer color from the old color to the new color. This was not serious for a primer, but would not be acceptable for a top coat. Use of electrodeposition for top coats is limited to applications for which long runs of the same color are made so that a line can be dedicated to a particular color. In small installations, such as for coating toys, one coating bath can be pumped out of the electrodeposition tank into a holding tank, where it must be kept continuously agitated, and another color is pumped into the tank. But these are relatively small tanks; the procedure would not be economically feasible with large objects.

REFERENCES

1. F. Beck, *Prog. Org. Coat.*, **4**, 1 (1976).
2. M. Wismer, P. E. Pierce, J. F. Bosso, R. M. Christenson, R. D. Jerabek, and R. R. Zwack, *J. Coat. Technol.*, **54** (688), 35 (1982).
3. C. K. Schoff, *J. Coat. Technol.*, **62** (789), 115 (1990).
4. R. E. Smith and D. W. Boyd, *J. Coat. Technol.*, **60** (756), 77 (1988).
5. J. J. Vincent, *J. Coat. Technol.*, **62** (785), 51 (1990).
6. D. W. Boyd and R. R. Zwack, *Prog. Org. Coat.*, **27**, 25 (1996).
7. T. J. Miranda, *J. Coat. Technol.*, **60** (760), 47 (1988).
8. J. A. Gilbert, *J. Coat. Technol.*, **62** (782), 29 (1990).

CHAPTER 27

Powder Coatings

Powder coating is undergoing rapid growth. Worldwide production of powder coatings in 1998 was estimated to be about 527,000 metric tons [1]. Powder coatings are used most extensively in Europe and Japan; North American production in 1998 was estimated to be 120,300 metric tons. The general principle is to formulate a coating from solid components, melt mix them, disperse pigments (and other insoluble components) in a matrix of the major binder components, and pulverize the formulation. The powder is applied to the substrate, usually metal, and fused to a continuous film by baking. Thermosetting and thermoplastic powders are available; the major portion of the market (>90%) is for thermosetting types.

27.1. BINDERS FOR THERMOSETTING POWDER COATINGS

Binders for thermosetting powder coatings consist of a mixture of a primary resin and a cross-linker, often called a *hardener*. The major types of binders are grouped somewhat arbitrarily into several classes, as shown in Table 27.1. The terminology has grown historically and has become confusing. Epoxy coatings include only coatings based on BPA and novolac epoxy resins with amine, anhydride, and phenolic hardeners. Hybrid coatings also contain BPA epoxy resins, but are cross-linked with carboxy-functional polyester resins. Polyester coatings contain polyesters with various cross-linkers other than

Table 27.1. Classes of Thermosetting Powder Coatings

Common Name	Primary Resin	Cross-Linker
Epoxy	BPA (or novolac) epoxy	Polyamines, anhydrides or phenolics
Hybrid	COOH-Functional polyester	BPA epoxy
Polyester	COOH-Functional polyester	Triglycidylisocyanurate or Hydroxyalkylamide
	OH-Functional polyester	Blocked-isocyanate or amino resin
Acrylic	Epoxy-functional acrylic	Dibasic acid
	OH-Functional acrylic	Blocked-isocyanate or amino resin
UV Cure	Acrylate-functional resin	Free radical
	Epoxy-functional resin	Cationic

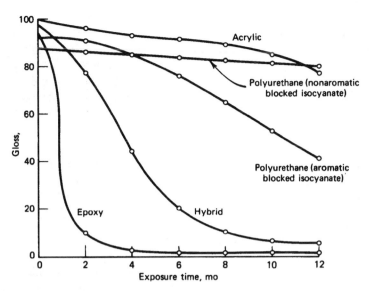

Figure 27.1. Florida outdoor exposure data on different types of powder coatings. (From Ref. [3], with permission.)

BPA and novolac epoxies; the term is used only for coatings that exhibit good to excellent exterior durability. Acrylic coatings contain acrylic resins with various cross-linkers. In addition, various blends, sometimes called *alloys*, of these classes are used. Blends are gaining increasing importance, for example, in primers for auto bodies [2]. Among the factors in choosing a class of powder coating for an application are protective properties, exterior durability, and cost. Differences in exterior durability are indicated in Figure 27.1.

In powder coatings, it is necessary to control the balance of binder T_g, $\bar{M}_n$, $\bar{f}_n$, and reactivity [4]. It must be possible to process the material without significant cross-linking, and the resultant powder must not sinter (start coalescing) or start to cross-link during storage, but it must fuse during baking, level to form a desirable film, and cross-link. In general, the primary resins are amorphous polymers with T_g high enough to avoid sintering of the powder and with $\bar{M}_n$ of a few thousand. Recommended minimum binder T_g values are 40°C in Europe and 45–50°C in the United States, reflecting the higher temperatures during shipment and storage in parts of the United States. A typical U.S. powder coating might have a T_g of 50°C; it could be melt processed at about 80°C and could be handled and stored at temperatures up to about 40°C. When heated in a baking oven, its viscosity would briefly drop to about 10 Pa·s allowing coalescence, flow, and leveling, and with continued heating for 15 minutes at a temperature in the range of 130–200°C, it would cross-link. The T_g values referred to are those of the combination of the primary resin plus the cross-linker. The required T_g for the primary resin alone varies, depending on the cross-linker used with it. UV cure powder coatings can be cured at temperatures as low as 100°C.

27.1.1. Epoxy Binders

Epoxy powder coatings are the oldest, and still one of the largest classes of thermosetting powder coatings. Decorative coatings have been based on BPA epoxy resins with *n* values of 3–5, usually made by the advancement process. (See Section 11.1.1.) The trend, however, is to lower molecular weights with some epoxy resins having *n* as low as 2.5, to provide better flow in thin film applications. For protective coatings, *n* values range up to 8. The most commonly used cross-linkers are dicyandiamide (dicy) (see Section 11.2.1) or a modified dicyandiamide. The curing reactions are complex [5]. Modified dicys are more soluble in epoxy resins and tend to form uniform films more readily. 2-Methylimidazole is a widely used catalyst.

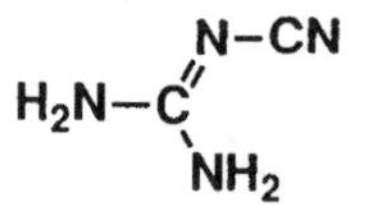

| Dicyandiamide | Modified dicyandiamide |

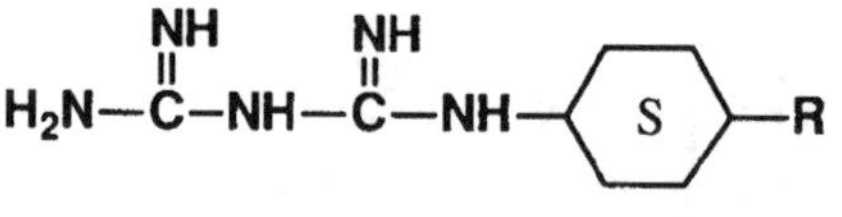

Epoxy powder coatings have good mechanical properties, adhesion, and corrosion protection; however, their exterior durability is poor. Applications for decorative types include institutional furniture, shelving, and tools. Applications of protective epoxy coatings include pipes, rebars, electrical equipment, primers, and automotive underbody parts. Where enhanced chemical and corrosion resistance is needed, phenolic resins are used as cross-linkers (see Section 11.3.1), with 2-methylimidazole as catalyst. Novolac epoxy resins (see Section 11.1.2) or blends of novolac and BPA epoxies give higher cross-link densities than BPA epoxies alone. All these coatings discolor and chalk on exterior exposure. Polycarboxylic acid anhydrides (see Section 11.3.3), such as trimellitic anhydride, are sometimes used with BPA epoxy resins in applications for which greater resistance to yellowing and to acids and solvents is needed. These latter coatings are generally being replaced with hybrid coatings that have somewhat better exterior durability and less question of toxic hazard.

27.1.2. Hybrid Binders

BPA epoxy resins are cross-linked with carboxylic acid-terminated polyester resins with $\bar{M}_n$ of a few thousand. Such formulations are called *hybrid powder coatings*, implying they are intermediate between epoxy and polyester coatings. Hybrid coatings have better color retention and UV resistance than epoxy powder coatings, but still do not have good exterior durability. Examples of end uses include water heaters, fire extinguishers, radiators, and transformer covers.

A variety of polyesters has been described. Most are derived from neopentyl glycol and terephthalic acid with smaller amounts of other monomers to adjust the T_g to the desired level and give branching to increase the $\bar{f}_n$ above two. (See Section 13.5.) An example is a polyester from neopentyl glycol (NPG) (364 parts by weight, 3.5 mol), terephthalic acid (TPA) (423 parts, 2.55 mol), adipic acid (AA) (41 parts, 0.24 mol), and trimellitic anhydride (TMA) (141 parts, 0.74 mol) [6]. The acid number of the resin is 80 mg KOH/g of resin. The relatively high trimellitic anhydride content increases $\bar{f}_n$, compensating for the low (about 1.9) $\bar{f}_n$ of the BPA epoxy.

The primary cross-linking reaction is ring opening of the oxirane groups by carboxylic acids. (See Section 11.3.2.) Esterification and transesterification reactions involving hydroxyl groups of the epoxy resin and homopolymerization reactions of oxirane groups may also play a role. A catalyst such as an ammonium or phosphonium salt, for example, tetrabutylammonium bromide or choline chloride, permits baking temperatures in the range of 160–200°C. Often, the polyester resins are supplied with the catalyst blended in.

Flow properties of powder coatings containing carboxylic acid-terminated polyesters tend to be poorer than those made with hydroxy-terminated polyesters. A proprietary, modified BPA epoxy that exhibits greater flow with comparable sintering resistance compared to conventional BPA epoxies has been reported [7].

27.1.3. Polyester Binders

Further improvement in exterior durability can be attained by replacing BPA epoxies with other cross-linkers. Depending on the cross-linker, carboxylic acid-functional or hydroxy-functional polyester resins are used. Triglycidylisocyanurate (TGIC) (see Section 11.1.2) has been widely used as a cross-linker for carboxylic acid-terminated polyesters with basic catalysts. TGIC based powder coatings have good exterior durability and mechanical properties. Examples of end uses are outdoor furniture, farm equipment, fence poles, and air conditioning units. While TGIC is expensive, the amounts required are relatively small because of its low equivalent weight. Typical binders contain 4 to 10 wt% TGIC and 90 to 96 wt % of carboxylic acid-terminated polyester. The polyesters used are generally less branched than those used in hybrid coatings because of the higher functionality of TGIC as compared to BPA epoxies. One polyester, for example, is made from NPG (530 parts by weight, 5 mol), TPA (711 parts, 4.3 mol), isophthalic acid (IPA) (88 parts, 0.47 mol), pelargonic acid (58 parts, 0.37 mol), and TMA (43 parts, 0.22 mol); the acid number is 35 [6]. Such resins are prepared in a two-stage process to minimize the problems caused by the high melting point and low solubility of TPA. (See Section 13.3.) Alternatively, dimethyl terephthalate is used in place of TPA by transesterification. (See Section 13.5.) High equivalent weight (low acid number) is desirable, since this reduces the required amount of TGIC, but cross-link density decreases with the higher equivalent weight; so there is an optimum for each application.

There has been increasing concern about toxic hazards of TGIC. Partly as a result, tetra(2-hydroxyalkyl)bisamides (see Section 16.6.1) are being used as cross-linking agents for carboxylic acid-functional polyesters in exterior durable coatings [8]. These coatings also have good mechanical properties and flow. Water is evolved from the cross-linking reaction, which may limit the film thickness that can be applied without popping.

Other polyester coatings use blocked aliphatic isocyanates (see Section 10.5) as cross-linkers for hydroxy-functional polyesters. The coatings have exterior durabilities equal to, or somewhat better than, TGIC cross-linked polyesters, and the excellent mechanical properties and abrasion resistance typical of polyurethane coatings. Blocked isocyanate-polyester powder coatings generally show better flow than most powder coatings, perhaps because the unreacted cross-linkers or blocking agents released by unblocking are good plasticizers. Examples of their end uses are automobile wheels, lighting fixtures, garden tractors, fence fittings, and playground equipment.

Derivatives of isophorone diisocyanate (IPDI), bis(4-isocyanatocyclohexyl)methane (H_{12}MDI) isocyanurate, and tetramethylxylidene diisocyanate (TMXDI) low molecular weight prepolymers (see Section 10.3.2) are examples of blocked isocyanates that are solids. Blocked isocyanates from sterically crowded isocyanates such as TMXDI have the potential advantage of unblocking at a somewhat lower temperature [9,10]. Probably, the most widely used blocking agent is ε-caprolactam. A disadvantage of caprolactam is build up of volatilized caprolactam in ovens. Also there is pressure to reduce curing temperatures. Oxime blocking agents are being used, since they react at lower temperatures [10]. The highly crowded blocked isocyanates from diisobutyl and diisopropyl oximes give the lowest cure temperatures [10]. Oxime-blocked isocyanate coatings tend to yellow during cure, especially if overbaked. Also there is concern about toxic hazards with oximes. 3,5-Dimethyl pyrazole and 1,2,4-triazole (and mixtures of the two) give blocked isocyanates that combine lower cure temperatures and freedom from yellowing [11]. Isocyanate dimers (uretdiones, see Section 10.5) can also be used [12]. Oligomeric uretdiones made from IPDI and diols cleave thermally to generate isocyanates that cross-link hydroxy-functional binders without release of a volatile blocking agent.

The polyesters are prepared with excess hydroxyl monomer with a mixture of diol and triol. A representative polyester is made from NPG (436 parts by weight, 4.2 mol), TPA (552 parts, 3.33 mol), IPA (117 parts, 0.7 mol), sebacic acid (25 parts, 0.12 mol), and trimethylolpropane (TMP) (28 parts, 0.2 mol); the hydroxyl number is 35 mg KOH/g resin [13]. With the low ratio of TMP, this polyester is only slightly branched with an $\bar{f}_n$ only a little over 2. It would probably be used with a blocked isocyanate with an $\bar{f}_n$ of 3 or more. Polyesters have to be designed for specific blocked isocyanates, since flow properties depend on the viscosity of the combined melt.

An amino resin, tetramethoxymethylglycoluril (see Section 9.4.3), is also used as a cross-linker for hydroxy-functional resins [14]. Since methyl alcohol is generated as a byproduct, film thickness may be limited by popping or retention of bubbles in the cured film. Release of the methyl alcohol from the film can be aided by a variety of approaches that slow cross-linking, such as use of methyltoluenesulfonimide as a catalyst with a solid amine such as tetramethylpiperidinol as an inhibitor. Toluene sulfonamide-modified melamine-formaldehyde resins can also serve as cross-linkers for hydroxy-functional polyesters [15]. The reaction with these resins releases volatile byproducts also; it is reported that buffering by a cure rate regulator, 2-methylimidazole, minimizes popping [16].

27.1.4. Acrylic Binders

A variety of acrylic resins can be used in powder coatings: Hydroxy-functional acrylics can be cross-linked with blocked isocyanates or glycolurils [13], and carboxylic acid-functional acrylics can be cross-linked with epoxy resins, hydroxyalkylamides [17], or carbodiimides [18]. But the greatest interest has been in epoxy-functional acrylics made with glycidyl methacrylate (GMA) (see Section 11.1.2) as a comonomer and cross-linked with dicarboxylic acids, such as dodecanedioic acid [$HOOC(CH_2)_{10}COOH$] [19], or carboxylic acid-functional resins [20]. An epoxy-functional acrylic for automotive primer-surfacer is said to require an $\bar{M}_n$ below 2500, a calculated T_g above 80°C, and a monomer composition such that melt viscosity is less than 40 Pa·s at 150°C [21]. Such a resin can be made with 15 to 35% GMA, 5 to 15% butyl methacrylate (BMA), with the balance being methyl methacrylate (MMA) and styrene. Such an acrylic evaluated for automotive clear coats had $\bar{M}_n$ 3,000, $\bar{M}_w/\bar{M}_n$ 1.8, and a T_g of 60°C [20]. Automotive primers made with acrylic powders have been introduced [22]. Acrylic powder coatings generally have superior detergent resistance and are used for applications such as washing machines. Acrylics tend to be incompatible with other powder coatings, requiring caution when changing coating types to avoid contamination, which can result in cratering. Acrylic coatings tend to have poorer impact resistance than polyester coatings [23].

27.1.5. UV Cure Powder Coatings

Powder coatings have been developed that are cured by UV. (See Chapter 28 for discussion of UV curing.) This process permits rapid cure at lower temperatures [24,25]. Since the powder is stable in the dark, premature reaction during powder production is minimized. Both free radical and cationic cure coatings have been made. Free radical cure coatings use acrylated epoxy resins (see Section 16.4) and/or acrylated polyesters or unsaturated maleic polyesters (see Section 16.3) as binders. Cationic UV cure coatings use BPA epoxy resins as binders. Photoinitiators must be incorporated in the formulation. After application, the powders are fused by passing under infrared lamps and then are cured by passing under UV lamps. Film formation with infrared lamps can be carried out at film temperatures below 120°C and films cured while still hot with UV in 1 second or less. This permits use on heat sensitive substrates such as wood and some plastics. Good leveling is possible because their viscosity does not begin to increase until cure is initiated by UV. As with any other UV cure system, pigments can interfere with curing, since the pigments may absorb UV, limiting the film thickness that can be cured. While some pigmented coatings can be cured, the main interest has been for clear coatings.

27.2. BINDERS FOR THERMOPLASTIC POWDER COATINGS

The first powder coatings were thermoplastic powder coatings, but they now account for less than 10% of the U.S. market. Thermoplastic coatings have several disadvantages compared to thermosetting coatings. They are difficult to pulverize to small particle sizes; thus, they can only be applied in relatively thick films. Due to the high molecular weights of the binders required, they are viscous and give poor flow and leveling, even at high baking temperatures.

Vinyl chloride copolymers (PVC) and, to a more limited extent, polyamides (nylons), fluoropolymers, and thermoplastic polyesters are used as binders. High vinyl chloride content copolymers (see Section 16.1.1) are formulated with stabilizers and a limited amount of plasticizer, often a phthalate ester, so that the T_g is above ambient temperature. The partial crystallinity of PVC may help stabilize the powder against sintering. Vinyl powders are generally applied as quite thick films, 0.2 mm and higher, by fluidized bed application, discussed in Section 27.5.2. Dishwasher racks, handrails, and metal furniture are examples of end uses.

Nylon-11 and nylon-12 based powder coatings exhibit exceptional abrasion and detergent resistance. They are used as antifriction coatings and as coatings for hospital beds, clothes washer drums, and other applications that must withstand frequent cleaning or sterilization and have good toughness and wear resistance. Fluoropolymers, such as poly(vinylidene fluoride) and ethylene/chlorotrifluoroethylene copolymers, are used for coatings requiring exceptional exterior durability, such as aluminum roofing and window frames, and also for resistance to corrosive environments, such as equipment for chemical plants.

Thermoplastic polyester coatings are sometimes made using scrap or recycled poly(-ethylene terephthalate). Polyolefin-based powder coatings are used, for example, in carpet backing, but the volume used in metal applications has been limited by generally inferior adhesion. Ethylene/acrylic acid (EAA) and ethylene/methacrylic acid copolymer resins that give coatings with substantially better adhesion [26,27] are available. (See Section 27.5.2.)

27.3. FORMULATION OF POWDER COATINGS

The challenge facing formulators is satisfying a combination of conflicting needs: (1) minimization of premature cross-linking during production; (2) stability against sintering during storage; (3) coalescence, degassing, and leveling at the lowest possible baking temperature; and (4) cross-linking at the lowest possible temperature in the least possible time. Furthermore, flow and leveling must be balanced to achieve acceptable appearance and protective properties over the range of expected film thicknesses. Coatings that flow readily before cross-linking can form smooth films, but they may flow away from edges and corners because of the surface tension differential driven flows that result from the faster heating of the edges. (See Section 23.3.)

If the T_g of a coating is high enough, sintering can be avoided. However, coalescence and leveling at the lowest possible temperature are promoted by low T_g. Short baking times at low temperatures are possible if the resins are highly reactive and if the baking temperature is well above the T_g of the final cross-linked film. However, such compositions may cross-link prematurely during extrusion, and the rapid viscosity increase as the particles fuse in the oven limits the ability of the coating to coalesce and level. Compromises are needed. With current technology, a crude rule of thumb is that the lowest feasible baking temperature is 50°C above the melt extrusion temperature and 70–80°C above the T_g of the uncured powder. Thus, minimum baking temperatures are about 125–135°C for a powder with a T_g of 55°C. Use of supercritical CO_2 as a solvent during extrusion permits processing at lower temperatures and, hence, use of more reactive formulations. (See Section 27.4.1.)

Several studies have addressed the changes in viscosity during film formation [14,17,28,29,30]. Nakamichi used a rolling ball viscometer to measure viscosities of powder coatings on a panel during heating [28]. Results for three types of coatings are shown in Figure 27.2. In each case, viscosity is high immediately after fusion of the powder, but drops off sharply with increasing temperature. Viscosity levels off as cross-linking reactions begin to increase the molecular weight and then increases rapidly as the coating approaches gelation. Flow is governed by the lowest viscosity attained and by the length of time that the coating stays near that viscosity, called the *flow window*. In Figure 27.2, coating 2 will flow more than coating 1 even though the lowest viscosity is about the same because the reaction of coating 2 is slower and the flow window is longer.

Some authors discuss the temperature dependence of viscosity in terms of Arrhenius type relationships and discuss the activation energies for viscous flow. As indicated in Section 3.4.1, the dependency of viscosity on temperature does not actually follow Arrhenius relationships, but rather is dependent on free volume availability. The most important factor controlling free volume availability is $(T - T_g)$.

Dynamic mechanical analysis (DMA) (see Sections 4.1 and 4.4) is useful for characterizing cured films [15,26]. A DMA study showed that the T_g values of cured films of a series of decorative powder coatings, including a hybrid coating, a TGIC-polyester coating, and a blocked isocyanate-polyester coating, were all in the range of 89–92°C [15]. The average molecular weights between cross-links, $\bar{M}_c$ (see Section 4.2), for the cured films were in the narrow range of 2500 to 3000. It is noteworthy that years of trial

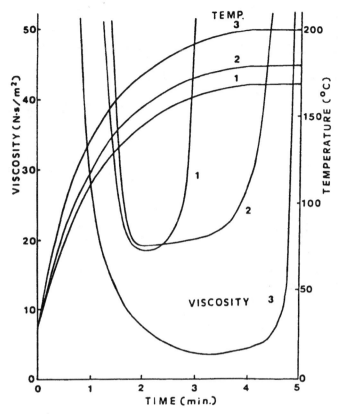

Figure 27.2. Nonisothermal viscosity behavior of powder coatings during film formation, as functions of time and panel temperature. (1) Acrylic-dibasic acid type; (2) Polyester-blocked isocyanate type; (3) Epoxy-dicy type. (From Ref. [28], with permission.)

and error formulation in different laboratories with such disparate binders led to such similar T_g and $\bar{M}_c$ values. On the other hand, a protective epoxy powder coating with a modified dicy cross-linker gave cured films with T_g of 117°C and $\bar{M}_c$ of 2200. These studies show that powder coatings are formulated with higher T_g and lower cross-link densities than liquid coatings for the same end uses. Similar, but not identical, mechanical properties are attainable. DMA studies also showed that pigmentation with TiO_2 had only a weak reinforcing effect and had essentially no effect on T_g, suggesting that pigment-binder interaction is weaker in powder coatings than in liquid coatings [15]. DMA can be a powerful tool for helping suggest starting points for development work on new binders for powder coatings for similar applications.

Two different factors control the T_g of the resins: chemical composition and molecular weight. It has been reported advantageous to use higher molecular weight, more flexible resins since these can have adequate package stability, and also flow more easily during baking than a lower molecular weight resin of similar T_g that has more rigid chains [3]. DSC is a very useful tool for T_g determination and cure response; its use for studying powder coatings is reviewed in Ref. [31].

An area that needs more study is the important question of the driving forces for coalescence and leveling. Coalescence of latex particles has been extensively studied (see

Section 2.3.3), but not coalescence of powder coatings. The main driving force for coalescence may be the reduction of surface area driven by surface tension; in keeping with this hypothesis, it has been suggested that high surface tension promotes coalescence [29].

Low melt viscosity promotes leveling, but the mechanism involved is not well understood. It has been proposed that the driving force for leveling is surface tension as reflected by the Orchard equation (see Section 23.2) [29]. The data obtained fit the Orchard equation reasonably well for relatively thin and/or fluid films, but thicker and/or somewhat more viscous films leveled better than predicted. The possible effect of surface tension driven flow on leveling, which Overdiep has shown to be important in leveling of liquid paints (see Section 23.2), has apparently not been considered [32]. It has been reported that surface tension differential flow can cause cratering of powder coatings and that addition of small amounts of additives such as poly(octyl acrylate) derivatives may overcome this problem [29]. In the same paper, it was reported that epoxidized soy fatty acids and hydroabietyl alcohol reduce orange peel, but do not affect cratering.

Many powder coatings contain on the order of 0.1–1% benzoin (melting point 133–134°C) as an additive. It is said to improve the appearance of films and to act as an antipinholing agent and a degassing aid. One study showed benzoin plasticizes the melt and increases the flow window of polyester-glycoluril formulations, indicating it can improve leveling; high levels (1.4–2.4%) were used [14]. Little information has been published about the mechanism of action of benzoin. Perhaps it melts during the fusing step, lowering T_g, thus promoting coalescence and degassing, and improving leveling. Other solid plasticizers such as aluminum stearate and a solid acetyleneic diol surfactant dispersed on a solid silica support are said to have effects similar to those of benzoin.

Benzoin

Since there is no volatile solvent, the volume of pigment in powders approaches the PVC of the final film. At PVC near CPVC, viscosity of the fused powder would be far too high for acceptable leveling. Thus, a common method for making low gloss liquid coatings by using a PVC that approaches CPVC (see Section 21.1), can only be used to a limited degree in powder coatings. It has been shown that as pigmentation increases above a PVC of about 20, the problems of leveling increase due to the increase in melt viscosity [29]. Low gloss and semigloss powder coatings have been prepared using approaches other than high PVC [33]. Some reduction in gloss can be achieved by incorporating polyethylene micronized wax with as much inert pigment as flow permits. In hybrid polyester powders, addition of organo-metallic catalysts along with wax is said to reduce gloss somewhat further. Low gloss hybrid polyester coatings are made by using a large excess over the stoichiometric ratio of epoxy resin with a high acid number polyester and curing at high temperatures. Another approach is to blend two different primary resins or two different cross-linkers with substantially different reactivities or with poor compatibility. By selection of catalysts such as cyclohexylsulfamic acid (cyclamic acid) and stannous methane sulfonate, it has been found that smooth matte finishes can be obtained with tetramethoxymethyl glycoluril (TMMGU) cross-linked polyester coatings [14]. However, as a general rule, it is more difficult to achieve low gloss powder coatings than with liquid coatings, and it is more difficult to control intermediate gloss levels reproducibly.

Wrinkle powder coatings have also been formulated with TMMGU cross-linkers using amine blocked catalysts such as 2-dimethylamino-2-methylpropanol blocked *p*-toluene-sulfonic acid [14].

27.4. MANUFACTURING OF POWDER COATINGS

Manufacture of powder coatings poses production and quality control issues quite different from those of liquid coating manufacture.

27.4.1. Production

Most powder coatings are manufactured by the same process: premixing, melt extrusion (see Figure 27.3), and pulverization (see Figure 27.4). All major ingredients must be solids; some liquid additives are used, but they must first be melted into one of the solid components to make a *masterbatch* that is then granulated. The granulated ingredients, resins, cross-linkers, pigments, and additives are premixed in a batch process. A variety of premixers is used; it is essential that they provide a uniform, intimate mixture of the ingredients. The premix is fed through a hopper to an extruder in a continuous process. The barrel of the extruder is maintained at a temperature moderately above the T_g of the binder. In passing through the extruder, the primary resin and other low melting or low T_g materials are fused, and the other components dispersed in the melt. The extruder operates at a high rate of shear so that it can be very effective in separating pigment aggregates. The melt can be extruded through a die—either a slot to produce a flat sheet or a series of round orifices to produce *spaghetti*. More commonly, to reduce heat exposure, the melt is extruded through a die with larger bore and the *sausage* is fed between chilled rollers to flatten it into a sheet and cool it. The extruded material is deposited on a conveyor for further cooling. At the end of the conveyor, the material has hardened enough to be coarsely granulated. At this stage, it is a brittle, uncross-linked solid.

Extruders have developed into sophisticated and rather expensive pieces of equipment. Two types are commonly used: single screw and twin screw; in both types, a powerful motor turns screws to drive the material through a barrel. The screws and barrel are configured to mix the material thoroughly and apply a high rate of shear. A popular single screw extruder uses a reciprocating action in addition to radial turning of the screw to effect mixing and dispersion. Twin screw extruders use a combination of screw segments and kneading segments. Both types of extruders are capable of excellent dispersion of most pigments. They operate with relatively high viscosity formulations at high shear rates and, hence, efficiently separate pigment aggregates. (See Section 20.4.5.) However, there is a trade-off between separating pigment aggregates and production rate. Production capacity can be increased by pushing more material through the extruder per unit time. Residence time in the extruder is sometimes reduced to 10 seconds or less, but at some point, pigment dispersion, especially with some organic pigments, begins to suffer. Poor color development and color variability may result. (See Section 27.4.2.)

The granules are then pulverized. A variety of pulverizers is used. Some, such as *pin disk mills* and *hammer mills*, work on the principle of striking airborne granules with metal dowels or hammers mounted on a rapidly spinning disk. The newer *opposed jet mills* work by causing high velocity collisions of granules with one another. Opposed jet mills perform well for small (<12 μm) diameter powders needed for thin film application.

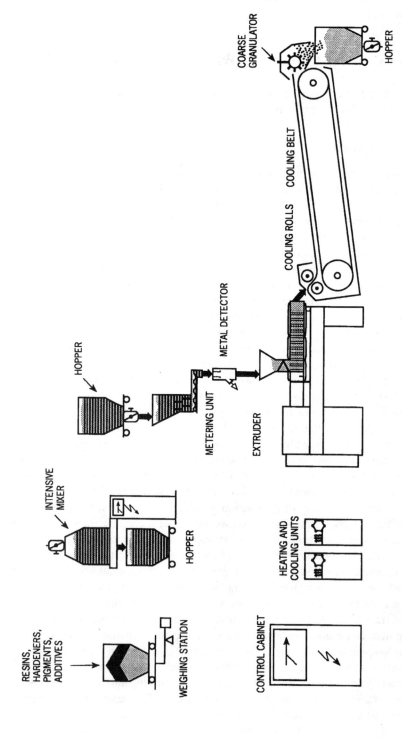

Figure 27.3. Schematic diagram of a line for premixing, melt extrusion, and granulation. (Adapted from Ref. [6], p. 242, with permission.)

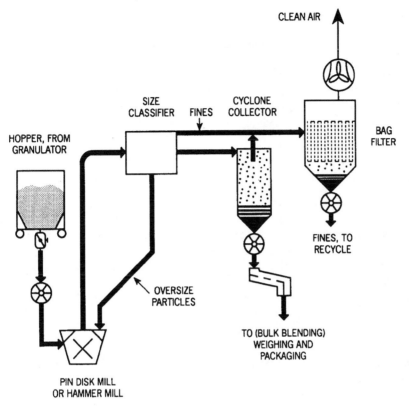

Figure 27.4. Schematic diagram of a line for pulverization and classification. (Adapted from Ref. [6], p. 252, with permission.)

Thermosetting extrudates are brittle and relatively easy to pulverize, but thermoplastics are generally quite tough and can be difficult to pulverize. In some cases, it is necessary either to cool the mill with liquid nitrogen or to grind dry ice along with the granules so that the temperature is kept well below the T_g of the binder to offset the heating effect of the milling process. Even so, thermoplastic powders are generally available only in large particle sizes.

Some mills partially classify (i.e., fractionate) the powder, automatically returning oversize particles for further pulverization. Further size classification, by sieving and/or by an air classifier, is usually needed. The coarse particle fraction is sent back into the mill for further reduction in size. The fines are collected in a bag filter and recycled in the next batch of the same coating to be processed through the extruder. Finally, the classified powder is bulk blended for uniformity, packaged, and shipped to the customer.

A relatively new process uses supercritical CO_2 as a solvent during extrusion [34]. (See Section 22.2.5 for discussion of the use of supercritical CO_2 in other applications.) The powder coating materials and CO_2 are premixed under pressure and then fed to an extruder still under pressure. The extruder can be operated at lower temperatures than conventional processing permits, reducing the risk of premature cross-linking in the extruder. When the compound comes out of the extruder, the CO_2 flashes off, reducing the temperature and pulverizing a major part of the product. The product is classified; only the larger particle size fraction needs to be pulverized in the conventional manner. It is said that the process affords excellent pigment dispersion.

A suspension method of preparing powders has been recommended for making narrow size distribution GMA epoxy-functional acrylic/dicarboxylic acid powders for automotive clear coat powders [35]. An aqueous dispersion of a solution of the resins and cross-linkers is prepared, and the solvent is distilled off. The resulting suspension is centrifuged, washed with water, and dried. The particle size distribution is said to be much narrower than that of powders made by the conventional pulverizing process.

27.4.2. Quality Control

Close quality control of all components is required. In solvent-based coatings, the effect of small differences in molecular weight or molecular weight distribution can be readily adjusted for by small variations in the solids of the coating. In powder coatings, there is no solvent in the formula with which to make such adjustments. The only way of maintaining consistent quality of powder coatings is to assure that the raw materials have no significant variation in molecular weight and molecular weight distribution, as well as monomer composition.

In processing thermosetting compositions, care must be exercised so that no more than a minimal amount of cross-linking occurs at the elevated temperatures in the extruder. The rate of travel through the extruder should be as rapid as possible, consistent with achieving the necessary mixing and dispersion of pigment aggregates. Reprocessing should be minimized. In an extreme case, reprocessed material could gel in the extruder. More commonly, reprocessing can lead to increases in molecular weight, due to some cross-linking reactions giving powders that show incomplete coalescence or poor leveling after application. Only a limited amount of fines from the micropulverizer should be put in any one batch of coating. If reprocessing is required, it is best to use a limited amount of the batch to be reprocessed in each of several new batches rather than reprocessing the batch alone. These problems are minimized using supercritical CO_2.

This reprocessing limitation particularly affects color matching. Color matching is more difficult with powder coatings than with liquid coatings. One cannot blend batches of powder to carry out color matching; it must be done using the appropriate ratio of pigments in the extruder mix. In laboratory processing, it is feasible to make an extruder mix with an estimate of the ratio of colorants needed to match a color. The color of the coating is then checked against a standard. Color can be adjusted to match the standard by mixing into the initial batch the estimated additional quantities of pigments and running the coating through the extruder and pulverizer again. In the laboratory, a third hit might be possible. The number of hits required is kept to a minimum by using computer color matching programs. (See Section 18.9.) In production, however, it is almost essential to have the mix right the first time. Potential problems can be minimized by extruding a small fraction of the production batch and then checking its color. If the color match is satisfactory, processing is continued. If the color needs adjusting, the initial fraction is granulated and returned to the hopper along with the required additional amounts of pigments. Some batch to batch variation of color is inevitable in pigment manufacture. For closely color matched powder coatings, the pigment manufacturer is requested to supply selected batches with narrow tolerance limits. In the case of large production runs, several extruder batches can be blended before pulverizing in order to average out batch to batch differences. With sufficient care, color reproducibility can be satisfactory for all but the most demanding end uses.

Other important quality control issues are particle size and particle size distribution. These important variables become increasingly critical as particle size is reduced for thin film (<50 μm) applications. Particle size distributions can be measured by passing the powder through a stack of graduated sieves and weighing the fraction retained on each sieve or by a variety of instrumental methods, such as laser diffraction particle sizing [36]. The important effects of particle size and its distribution on the handling and application characteristics of the powder are discussed in Section 27.5.1.

27.5. APPLICATION METHODS

Almost all thin-film powder coatings are applied by *electrostatic spray*. Other application methods, important for protective and thick-film powder coatings, include *fluidized bed*, *electrostatic fluidized bed*, and *flame spray*. See the general references listed at the end of the chapter for more extensive discussions.

27.5.1. Electrostatic Spray Application

Electrostatic spray (see Section 22.2.3) is the major process for applying powder coatings. An electrostatic spray gun for powder consists of a tube to carry airborne powder to an orifice at which an electrode is located. The electrode is connected to a high voltage (40–100 kV), low amperage power supply. Electrons emitted by the electrode react with molecules in the air, generating a cloud of ions, called a *corona*, around the orifice. The corona probably consists predominantly of HO^- and, perhaps, $O_2{}^-$ ions. Powder particles come out of the orifice, pass through the corona, and pick up anions. The object to be coated is electrically grounded. The difference in potential attracts the powder particles to the surface of the part. They are attracted most strongly to areas are not already covered, forming a layer of powder even on irregularly shaped objects. During application, size segregation of the particles has been observed, with the finer particles more prevalent near the substrate, presumably due to a combination of the higher charge on small particles and their fitting between larger particles [37]. The particles cling to the surface strongly enough for the object to be conveyed to a baking oven, where the powder particles fuse to a continuous film, flow, and cross-link. Heat can be furnished by either conventional convection ovens or by infrared (IR) lamps; use of induction heating has also been proposed. One comparison of alternative heat sources concluded that IR is least expensive and can be adjusted to provide more rapid cure [38].

The powder particles that do not adhere to the substrate (overspray) are recovered as a dry powder. This powder is usually recycled by blending with virgin powder. A schematic diagram of electrostatic spray application of powder coatings, including recovery of overspray, is given in Figure 27.5. Production spray guns are usually mounted on automated reciprocators; such equipment functions smoothly with minimal worker attention as long as a single color and type of powder is being applied.

A limitation of powder coatings is the difficulty of changing colors. When spraying liquid coatings, one can flush the gun with solvent and shift the feed lines to the guns from those feeding one color to those feeding another. In this way, successive objects on a conveyor line can easily be painted different colors. However, if changeover of powder

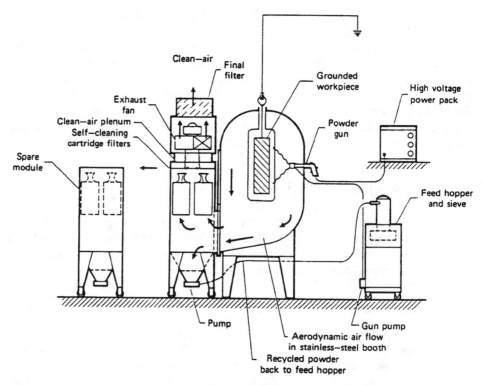

Figure 27.5. Production equipment for electrostatic spray application of powder coating showing the collection of overspray powder. (From Ref. [3], p. 18, with permission.)

coatings was attempted in this way, there would be sufficient dust in the air in the spray booth for color contamination both on the product and in the overspray collecting units. In production, the operation must be closed down, the booth cleaned, and the overspray collecting units changed to collect the next color. While spray booths and overspray units have been designed to minimize cleanup time, it is still only economically feasible to make a reasonably long run of one color and then shift over to another color. Many important applications of powder coatings, such as fire extinguisher cases, are single color end uses, or are applications, such as metal furniture, for which runs of single colors are long enough to dedicate a spray booth to each of the limited number of colors involved.

Water-washed spray booths have been suggested to avoid the cost of installing special spray booths for shorter runs [39]. The overspray collected in the water is dispersed by a low foaming nonionic surfactant and flocculated with a melamine resin. (See Section 22.2.7.).

In electrostatic spray, the charged powder particles wrap around the grounded object to a degree and coat exposed surfaces not in direct line with the spray gun. Nevertheless, in production, it is often desirable to use spray guns on both sides of the object making it possible to uniformly coat intricately shaped objects such as automobile wheels and tube-and-wire metal furniture. The process is strongly affected by the Faraday cage effect. (See Section 22.2.3.) As a result, it is difficult to get full coverage of areas such as interior corners of steel cabinets; the interior of pipes can only be coated with a spray gun inside the pipe.

Film thickness increases with increasing voltage and decreasing distance between the spray gun and the product being coated. Larger particle size powders tend to give increased film thickness. One can apply thicker coatings by heating the object to be coated before applying the powder. However, film thickness is limited by the fact that once a certain film thickness has been reached, the powder coating acts as an insulator and does not attract further particles. The insulating properties of the powder coating mean that defective coated parts generally cannot be recoated, and must be stripped to be recoated. Coating over other coatings and over plastics frequently requires application of a conductive primer coating.

There is considerable room for improvement of the process and need for a better understanding of how it works. For example, there is no satisfactory physical explanation for why the powder clings to the object as well as it does [40]. One opportunity for improvement is to increase efficiency of charging by the corona of the electrostatic gun. It is estimated that only about 0.5% of the anions in the corona become attached to powder particles [41]. The rest are attracted to the nearest grounded object, where they at best do no good and at worst may reduce deposition efficiency and increase the Faraday cage effect [42]. Research on ways to improve this situation is underway; some experts think that if charging efficiency could be raised to 10%, deposition efficiency might become high enough to reduce overspray sufficiently that it would no longer be necessary to collect it.

Particle size and distribution have a critical effect on powder coating. The range of particle sizes must be limited; as a rule of thumb, the predominant particle diameter should be somewhat less than the intended film thickness and the largest particles should be no more than twice the film thickness. Figure 27.6 shows a typical distribution, presumably for a powder coating to be applied at a thickness of around 50 μm.

Very fine particles do not flow properly in hoppers and feed lines. In general only 6–8 wt% of the particles are smaller than 10 μm diameter. Small particles have a higher ratio of surface area to volume and, hence, acquire a higher charge to mass ratio as they pass through the corona of an electrostatic gun. After charging the particles are affected by three forces: the electrostatic field, air movements, and gravity. Theoretical calculations predict that gravity should be the predominant effect on very large particles and that air flow should predominate for the very small ones [44]. These theoretical predictions were

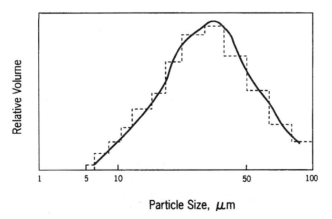

Particle Size, μm

Figure 27.6. Particle size distribution of a typical white epoxy powder coating. (From Ref. [43], with permission.)

confirmed by particle size analysis of overspray; with a typical powder, the overspray was rich in particle sizes below 20 and above 60 μm, indicating that the intermediate sizes were transferred most effectively. Small particles probably penetrate best into Faraday cages. Large particles flow better through the application system and retain their charge longer, and therefore cling to the object better between application and baking. The dusts of very small (<1 μm) particles may present a toxic hazard if they are inhaled. On balance, it appears that particle sizes concentrated in the 20–60 μm range are best for powder coatings that are to be applied at film thicknesses of 30–60 μm. Particle size of solid cross-linkers also affects appearance and film properties. It was found that particle size of dodecanedioic acid for epoxy-functional acrylic resin should be 5 μm or less [19].

Automotive clear coats with maximum smoothness are reported to result with fine particle size powders with a narrower size distribution powders [35,45]. However, application of fine particle size (averaging 10 μm) powders presents problems. Modification of application equipment has been reported to solve many of the problems [45]. Constant feeding of powder was accomplished by controlling temperature and humidity in the powder feed system and incorporating an agitator in the system. Spitting of aggregates of particles was minimized by redesigning the gun nozzle. Transfer efficiency was improved by increasing charging efficiency, minimizing free ions, and adding an external electrode of intermediate charge to increase the electric field intensity near the part being coated.

Another means of applying electrostatic powder employs *triboelectric charging* of the powder particles. Instead of a high voltage source generating a corona at the gun orifice, the particles are charged by the friction generated from streaming through a poly(tetra-fluoroethylene) spray tube in the gun. The mechanism is analogous to the buildup of a static charge on a comb when combing your hair. Since there is not the large differential in charge between the gun and the grounded work being sprayed, no significant magnetic field lines are established and Faraday cage buildup is minimal, facilitating coating hollows in irregularly shaped objects. Smoother coatings are obtained. On the other hand, throughput is slower and stray air currents can more easily deflect the particles between the gun and the object being coated. Triboelectric charging is widely used in Europe and is gaining popularity in North America. Guns are available that combine triboelectric and corona charging.

27.5.2. Other Application Methods

Fluidized beds are the oldest method of applying powder coatings. The equipment consists of a dip tank, the bottom of which is a porous plate. Air is forced through the porous plate and acts to suspend the powder in the dip tank in air. The flow behavior of air suspensions of powders resembles that of fluids, hence the term fluidized bed. The object to be coated is hung from a conveyor and heated in an oven to a temperature well above the T_g of the powder. The conveyor then carries the part into the fluidized bed tank. Powder particles fuse onto the surface of the object. As the thickness of fused particles builds up, the coating becomes a thermal insulating layer so that the temperature at the surface of the coating becomes lower, finally reaching the stage at which further particles do not stick to the surface. The last particles that attach to the coated surface are not completely fused, so the conveyor must then carry the object into another oven, where the fusion is completed. Film thickness depends on the temperature to which the part is preheated and on the T_g of the powder. Thin films cannot be applied in this fashion. Most commonly, the method is used for applying thermoplastic coating materials.

Electrostatic fluidized beds are similar, but electrodes are added to generate ions in the air before the object passes through the powder. The object to be coated is grounded. The process is illustrated schematically in Figure 27.7 [46]. Powder is attracted to the object by electrostatic force, as in electrostatic spray. The object can be heated when thick films are desired, but heating is not necessary. This method is used to apply thermoplastic and some thermosetting powders, for example, electrical insulation coatings. There is no overspray, and powder losses are minimal; color change-over is also easier. Thinner films can be applied compared to conventional fluidized bed coating. However, very thin films are hard to apply, there is a strong Faraday cage effect, and it is difficult to coat large objects uniformly.

Powder coil coating is also beginning to be practiced. In some cases, the powder is applied to the coil strip by electrostatic spray [25]. In another process, an electrostatically charged strip is passed through a cloud of powder then to the oven for fusing [47]. The method makes possible coating of perforated or pre-embossed metal and has the advantage

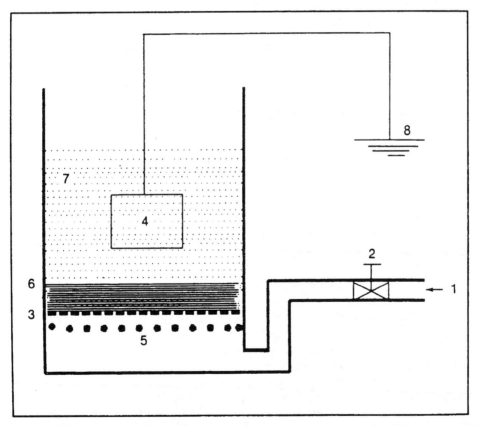

Figure 27.7. An electrostatic fluidized bed coating apparatus. 1. Air inlet. 2. air regulator. 3. porous membrane. 4. object to be coated. 5. electrodes. 6. fluidized powder. 7. cloud of charged powder. 8. ground. (From Ref. [46], with permission.)

of no VOC emissions. Capital cost is said to be lower; but line speeds are somewhat slower than with conventional coil coating.

Flame spray is another technology for applying thermoplastic powder coatings [26,27,48]. In a flame spray gun, the powder is propelled through a flame, remaining there just long enough to melt. The molten powder particles are then directed at the object to be coated. The flame heats and melts the polymer and heats the substrate above the melting temperature of the polymer so that the coating can flow into irregularities in the surface to provide an anchor for adhesion. The combination of flame temperature (on the order of 800°C), residence time in the flame (small fractions of a second), T_g of the coating, particle size distribution, and substrate temperature must be carefully balanced. Particle size distribution must be fairly narrow; very small particles pyrolyze at 800°C before larger particles can melt.

In contrast to other methods of application, flame spray permits application in the field, not just in a factory. Examples of commercial or experimental applications include drum linings, metal light poles, bridge rails, concrete slabs, and grain railcars, among others. Since the application is not electrostatic, nonconductive substrates such as concrete, wood, and plastics can be coated. The coating need not be baked, so substrates that will withstand just the temperature from the impinging spray can be coated by this method. Since the coatings are thermoplastic and are not applied electrostatically, it is possible to repair damaged areas of coating, which is generally not possible with other powder coating methods.

Disadvantages of flame spray application include limitations on the service temperatures of objects coated with thermoplastic coatings (in the case of EAA-based coatings (see Section 27.1) 75°C, and the need for careful control of application variables [26]. Overheating the polymer can lead to thermal degradation and, hence, to poor coating performance, with no visual indication of degradation until the polymer begins to pyrolyze. Adhesion to steel is also affected by application variables [27]. The carboxylic acid groups promote adhesion, but coatings are subject to cathodic delamination (disbonding) because of the acid groups.

27.6. ADVANTAGES AND LIMITATIONS

In this section, advantages and limitations of powder coatings are compared to solvent-borne or water-borne coatings. The primary advantages may be summarized as: VOC emissions are low; flammability and toxicity hazards are substantially reduced; thick films of 100–500 μm can be applied in a single operation; and energy consumption can be reduced. At first blush, the low energy requirements of powder coatings are surprising, since their baking temperatures are generally higher than those of most solvent-borne baked coatings. However, with little or no volatiles being emitted into the oven, the air in the oven can be recirculated with almost no make-up air. In contrast, with solvent-borne coatings, the solvent concentration in the air in the oven must always be kept well below the lower explosive limit, resulting in the need to heat a substantial volume of make-up air. Also air flow through the spray booth can be lower with powder coatings because it is no longer necessary to keep the solvent concentration in the air below a safe concentration for people to be in the spray booth. In winter, the cost of heating air flowing at a high rate through a spray booth can be high.

An important further factor reducing overall cost of powder coatings is that the overspray powder can be collected in filter bags from the spray booth and reused. This not only increases paint utilization, but it also eliminates the cost and difficulty of disposing of the sludge obtained from the water wash spray booths used with liquid paints.

Important limitations of powder coatings are:

1. *Explosive hazards.* While the absence of solvents eliminates the flammability problem, suspensions of powder in air can explode. Consequently, manufacturing and application facilities must be designed to avoid powder explosions. With good engineering and good housekeeping, the processes can be operated safely. Triboelectric charge systems are less likely to initiate explosions than corona charge spray guns, which can spark if inadvertently brought near a grounded conductor.

2. *Inability to coat large or heat sensitive substrates.* With electrostatic spray and fluidized bed methods, only baked applications can be considered, and since the baking temperature must be fairly high, only substrates that can withstand the baking cycle can be used. This limits applications of conventional thermosetting powder coatings almost entirely to metal substrates. UV cure powder coatings can be used on some heat sensitive substrates, and as noted in Section 27.5.2, flame spray can be used for some other applications.

3. *Appearance limitations.* Broadly speaking, powder coatings can have good appearance, but some of the appearance effects attainable with liquid coatings are difficult or impossible to match with powders. Since there is no solvent loss, there is not the degree of shrinking during film formation required for metallic coatings to show the color flop typical of automotive coatings. One can make powder coatings that contain aluminum flake pigments; they sparkle, but display little color flop. As indicated in Section 27.3, a compromise between leveling and edge coverage is inevitable. The usual result is that the formulator must settle for a film with some orange peel to get acceptable edge coverage. The thinner the film, the more difficult the problem becomes. As also discussed in Section 27.3, there are problems making and controlling gloss of low and medium gloss powder coatings. Also, as discussed in Section 27.4.2, close color matching is more difficult than with conventional types of coatings.

Achieving satisfactory appearance is a major obstacle to development of powder clear coats for autos and trucks [49]. Uniform film thickness must be applied to the entire car body because leveling of powder coatings is quite sensitive to variations in film thickness. This requires excellent control of powder flow rates; which is an engineering challenge. The entire operation, including reclamation of overspray, must be scrupulously clean to avoid dirt particles, which are especially noticeable in clear coatings. At least one automotive plant is applying powder clear coat commercially in 1998.

4. *Materials limitations.* Since all major components must be solids, a smaller range of raw materials is available to the formulator. Furthermore, it is not possible to make thermosetting powder coatings for which the T_g of the final film is low. This limits the range of mechanical properties that can be formulated into a powder coating.

5. *Limits of production flexibility.* Economics of production and application suffer badly whenever frequent color changes are needed. Cleanup between color changes is time consuming. Powder coatings are best suited to reasonably long production runs of the same type and color of powder. Water-washed spray booths permit shorter runs, but the advantage of recycling overspray is lost.

Some of the limitations of powder coatings can be overcome by making aqueous dispersions (slurries) of powders. This eliminates the powder explosion problem, broadens the range of applicable application methods, and reduces storage stability problems. The T_g of the powder no longer needs to be high to avoid sintering, and therefore, more flexible coatings can be formulated and lower baking temperatures can be utilized. Settling problems can be minimized and appearance improved by using smaller particle size, less than 10 μm, in the slurries. However, new challenges are added: Aqueous dispersions must be stabilized without undue sacrifice of properties due to the presence of surfactant, and spray rheology must be controlled. As of 1998, serious efforts are being made in Europe to use aqueous powder dispersions as automotive clear coats.

GENERAL REFERENCES

D. A. Bate, *The Science of Powder Coatings*, two Vols., Selective Industrial Training Associates, Ltd., London, 1990.

J. H. Jilek, *Powder Coatings*, Federation of Societies for Coatings Technology, Blue Bell, PA, 1991.

T. A. Misev, *Powder Coatings Chemistry and Technology*, Wiley, New York, 1991.

T. A. Misev, and R. van der Linde, *Prog. Org. Coat*, **34**, 160 (1998).

REFERENCES

1. R. Higgins, *Powder Coatings*, Campden Publishing Ltd., U.K., 1998, p. 77.
2. P. R. Gribble, *Proc. Intl. Waterborne, High-Solids, & Powder Coat. Symp.*, New Orleans, 1998, p. 218.
3. D. S. Richert, "Powder Coatings," in *Kirk-Othmer Encyclopedia of Chemical Science and Technology*, 3rd. ed., Vol. 19, Wiley, New York, 1982, p. 11.
4. L. Kapilow and R. Samuel, *J. Coat. Technol.*, **59** (750), 39 (1987).
5. M. Fedtke, F. Domaratius, K. Walter, and A. Pfitzmann, *Polymer Bulletin*, **31**, 429 (1993).
6. T. A. Misev, *Powder Coatings Chemistry and Technology*, Wiley, New York, 1991, p. 48.
7. G. C. Fischer and L. M. McKinney, *J. Coat. Technol.*, **60** (762), 39 (1988).
8. K. Kronberger, D. A. Hammerton, K. A. Wood, and M. Stodeman, *J. Oil Colour Chem. Assoc.*, **74**, 405 (1991).
9. S. P. Pappas and E. H. Urruti, *Proc. Water-borne Higher-Solids Coat. Symp.*, New Orleans, 1986, p. 146.
10. J. S. Witzeman, *Prog. Org. Coat.*, **27**, 269 (1996).
11. Th. Engbert, E Konig, and E. Jurgens, *Farbe Lack*, **102**, 51 (1996); M. Sohoni and P. Figlioti, *Proc. Waterborne High-Solids Powder Coat. Symp.*, New Orleans, 1998, p. 267.
12. R. M. Guida, *Modern Paint Coat.*, July, 34 (1996); F. Schmidt, A. Wennig, and J.-V. Weiss, *Prog. Org. Coat.*, **34**, 227 (1998).
13. T. A. Misev, *Powder Coatings Chemistry and Technology*, Wiley, New York, 1991, p. 146.
14. W. Jacobs, D. Foster, S. Sansur, and R. G. Lees, *Prog. Org. Coat.*, **29**, 127 (1996).
15. H. P. Higginbottom, G. R. Bowers, J. S. Grande, and L. W. Hill, *Prog. Org. Coat.*, **20**, 301 (1992).
16. H. P. Higginbottom, G. R. Bowers, and L. W. Hill, *Polym. Mater. Sci. Eng.*, **67**, 175 (1992).
17. S. G. Yeates, T. Annable, B. J. Denton, G. Ellis, R. M. D. Nasir, D. Perito, and I. Parker, *J. Coat. Technol.*, **68** (861), 107 (1996).
18. J. W. Taylor, M. J. Collins, and D. R. Bassett, *J. Coat. Technol.*, **67** (846), 43 (1995).
19. J. C. Kenny, T. Ueno, and K. Tsutsui, *J. Coat. Technol.*, **68** (855), 35 (1996).
20. T. Agawa and E. D. Dumain, *Proc. Waterborne High-Solids Powder Coat. Symp.*, New Orleans, 1997, p. 342.

21. C. D. Green, *Paint Coat. Ind.*, September 1995, p. 45.
22. E. R. Messerly and K. Lis, *Powder Coating*, **6** (3), 45 (1995).
23. R. van der Linde and B. J. R. Scholtens, *Proc. 6th Ann. Intl. Conf. Cross-Linked Polym.*, Noordwijk, Netherlands, June 1992, p. 131.
24. S. F. Thames and J. W. Rawlins, *Powder Coating*, **7** (6), 19 (1996).
25. T. A. Misev and R. van der Linde, *Prog. Org. Coat.*, **34**, 160 (1998); K. Blatter, B. MacFadden, M. Strid, and F. Niggemann, *Coatings World*, **3** (2), 26 (1998).
26. T. Glass and J. Depoy, *Soc. Mfg. Eng. Finishing Conf.*, FC91-384, Cincinnati, OH, September 1991.
27. T. Sugama, R. Kawase, C. C. Berndt, and H. Herman, *Prog. Org. Coat.*, **25**, 205 (1995).
28. T. Nakamichi and M. Mashita, *Powder Coatings*, **6** (2), 2 (1984).
29. P. G. de Lange, *J. Coat. Technol.*, **56** (717), 23 (1984).
30. M. J. Hannon, D. Rhum, and K. D. Wissbrun, *J. Coat. Technol.*, **48** (621), 42 (1976).
31. L. Gherlone, T. Rossini, and V. Stula, *Proc. Intl. Conf. Org. Coat.*, 1997, p. 127.
32. W. S. Overdiep, *Prog. Org. Coat.*, **14**, 159 (1986).
33. D. S. Richart, *Proc. Waterborne, High-Solids Powder Coat, Symp.*, New Orleans, 1995, p. 1.
34. P. M. Koop, *Powder Coating*, **6** (2), 58 (1995).
35. H. Satoh, Y. Harada, and S. Libke, *Prog. Org. Coat.*, **34**, 193 (1998).
36. G. H. Barth and S. T. Sun, *Anal. Chem.*, **61**, 143 (1989).
37. Z. Huang, L. E. Scriven, and H. T. Davis, *Proc. Waterborne High-Solids Powder Coat. Symp.*, New Orleans, 1997, p. 328.
38. R. J. Dick, K. J. Heater, V. D. McGinniss, W. F. McDonald, and R. E. Russell, *J. Coat. Technol.*, **66** (831), 23 (1994).
39. S. F. Kia, D. N. Rai, J. C. Simmer, and C. Wilson, *J. Coat. Technol.*, **69** (875), 23 (1997).
40. E. F. Meyer, III, *Polym. Mater. Sci. Eng.*, **67**, 220 (1992).
41. J. F. Hughes, *Electrostatic Spraying Of Powder Coatings*, Research Studies Press, Ltd., Letchworth, England, 1984.
42. J. F. Hughes, *Ytbehandlingdagar*, Stockholm, May 1992, p. A-1.
43. Anonymous, *Tech. Bull. SC:586–82*, Shell Chemical Co., 1989.
44. H. Bauch, *Polym. Mater. Sci. Eng.*, **67**, 344 (1992).
45. K. Yanagida, M. Kumata, and M. Yamamoto, *J. Coat. Technol.*, **68** (859), 47 (1996).
46. T. A. Misev, *Powder Coatings Chemistry and Technology*, Wiley, New York, 1991, p. 346.
47. F. D. Graziano, *Proc. Intl. Conf. Org. Coat.*, Athens, 1997, p. 139.
48. T. A. Misev, *Powder Coatings Chemistry and Technology*, Wiley, New York, 1991, pp. 347–349.
49. J. Maty, *Paint & Coat. Ind.*, January 1988, p. 24.

CHAPTER 28

Radiation Cure Coatings

Radiation cure coatings cross-link by reactions initiated by radiation, rather than heat. Such coatings have the potential advantage of being indefinitely stable when stored in the absence of radiation, while after application, cross-linking occurs rapidly at ambient temperature on exposure to radiation. Rapid cure at ambient temperature is particularly significant for heat sensitive substrates, including paper, some plastics, and wood. See Section 27.1.5 for discussion of UV cure powder coating.

Two classes of radiation cure coatings are: (1) UV cure coatings in which the initial step is excitation of a photoinitiator (or photosensitizer) by absorption of photons of UV-visible electromagnetic radiation and (2) EB (electron beam) cure coatings in which the initial step is ionization and excitation of the coating resins by high energy electrons. Cross-linking is initiated by reactive intermediates that are generated from the photoexcited photoinitiator in UV curing and from excited and ionized resins in EB curing. Infrared and microwave radiation are also used to cure coatings, but these systems are not included here, since the radiation is converted to heat, which initiates thermal curing.

While the physical volume of radiation cure materials used in the United States is relatively small, perhaps 50,000 metric tons per year, their economic importance is disproportionate to their volume. They are essential to the production of computer chips, optical fibers, and printed circuit boards, and they are used in a variety of other economically important applications.

28.1. UV CURING

Two classes of polymerization reactions are used in UV curing: free radical and cation initiated chain-growth polymerization. While there have been attempts to use photoreactions in which the radiation leads to generation of a reactive functional group, this approach is not useful for coatings. In such reactions, each photon absorbed can effect one cross-linking reaction. In chain reactions, absorption of a single photon can lead to formation of many cross-links.

A key requirement for UV curing is a UV source that produces high intensity UV radiation at low cost without generating excessive infrared radiation. The major sources in commercial use are medium pressure mercury vapor lamps. Such *electrode lamps* are tubes up to 2 meters long; power outputs of 80 watts cm^{-1} are in wide use, and lamps with outputs up to 325 watts cm^{-1} are available. The radiation has continuous wavelength distribution with major peaks at 254, 313, 366, and 405 nm, among others. Visible radiation and a minor, but not insignificant, amount of infrared radiation that causes some heating are also emitted. The wavelength distribution of a medium pressure mercury vapor electrode lamp is shown in Figure 28.1.

Radiation is emitted in all directions around tubular lamps, and its intensity drops off with the square of the distance from the source. To increase efficiency of use of radiation, lamps are mounted in an elliptical reflector with a focal length such that the maximum intensity of radiation is focused at the distance between the lamp and the surface being coated. A limitation of UV curing is that the distance between the lamp and the coating on various parts of the object being coated must be fairly uniform. Hence, UV curing is most easily applicable to coating flat sheets or webs that can be moved under the UV lamps or cylindrical objects that can be rotated under or in front of the lamps. Since thermal energy is also produced, the lamp housing must be water and air cooled. UV radiation is hazardous and can lead to severe burns. It is essential to avoid exposure of eyes to the radiation. The lamps are housed in enclosures; when the enclosure is opened, the current to the lamps is automatically turned off. Depending on the radiation source, a greater or lesser amount of ozone is generated. Since ozone is toxic, the UV unit must be ventilated. Reference [1] reviews these and other safety considerations in radiation curing.

In some cases, there are advantages of using other wavelength distributions, especially those with increased fractions of radiation in the very near UV-visible region. Changes in distribution can be made by doping the lamp with traces of other elements besides mercury or by having fluorescent coatings on the lamp tube that absorb short UV and emit longer UV radiation.

Electrodeless lamps, powered by microwaves, also enjoy substantial commercial use. Electrodeless lamps are more suitable for doping, since the lifetime of lamp electrodes is

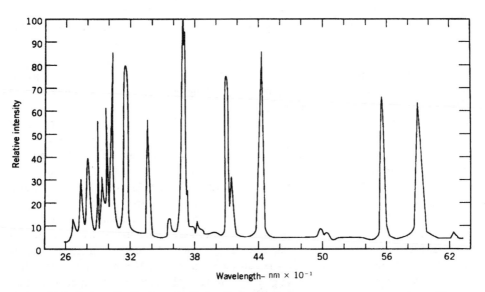

Figure 28.1. Energy distribution of radiation from a medium pressure mercury vapor lamp. (Source: U.S. Patent 3650699 as cited in Ref. [2]), with permission.)

generally reduced by dopants. Electrodeless lamps have the further advantage of essentially instantaneous start-up and restart. On the other hand, electrodeless lamps are more expensive. For further discussion of UV sources, see Ref. [2].

Excimer lamps emit very narrow bands of UV [3,4]. They use silent electrical discharge through two quartz tubes and an enclosed gas volume. Electronically activated molecules are produced in the gas phase and decompose within nanoseconds to produce photons of high selectivity. The narrowness of distribution is comparable to lasers, but the radiation is incoherent in contrast to laser sources and can therefore be used for large area applications. Different lamps are available or in commercial development that have principal emissions at 172, 222, 308, and 351 nm. The high intensity of very narrow wavelength distribution can provide rapid cures. The 172 nm excimer lamps are reported to activate acrylates directly [4]. Xenon chloride excimer lamps that produce energy at 308 nm have been shown to effect UV cure in shorter times than mercury vapor lamps at equal photoinitiator concentration or at equal speed with lower initiator concentration [5]. The 308 nm excimer lamps are reported to be particularly effective for cationic curing [6].

For initiation via any lamp source, there must be absorption of radiation by the photoinitiator (or, as discussed later, by some substance that leads to the generation of an initiator). The fraction of radiation absorbed, I_A/I_0, is related to molar absorptivity ε, concentration of photoinitiator C, and optical path length of radiation in the film X. Assuming there is no other absorber present and neglecting surface reflection, the fraction of radiation of a given wavelength absorbed is expressed by Eq. 28.1. The molar absorptivity ε can be concentration dependent, so the range of concentration over which the equation is valid may be limited.

$$\frac{I_A}{I_0} = 1 - 10^{\varepsilon C X} \tag{28.1}$$

Molar absorptivity varies with wavelength, so the fraction of radiation absorbed also varies with wavelength. Also, the intensity of radiation from the source varies as a function of wavelength. The total number of photons absorbed per unit time depends on the combination of these factors. When a photoinitiator absorbs a photon, it is raised to an excited state. Some reaction of this excited state leads to generation of an initiating species, but there are also other possible fates of the excited state photoinitiator. For example, it may emit energy of a longer wavelength; that is, it may fluoresce or phosphoresce. It may be quenched by some component of the coating or by oxygen. It could undergo other reactions besides those that lead to initator generation. The efficiency of generation of initiating species is an important factor in selection of a photoinitiator.

The rate of polymerization reactions is related to the concentration of initiating radicals or ions. It would, therefore, seem that the higher the initiator concentration, the faster the curing. As one increases from very low concentrations to somewhat higher concentrations, the rate of cure increases. However, there is an optimum concentration. The rate of cure in the lower part of the film decreases above this concentration. If the concentration is high enough, such a large fraction of the radiation is absorbed in the upper few micrometers of the film that little radiation reaches the lower layers. Since the half-life of free radicals is short, they must be generated within a few nanometers of the depth in the film where they are to initiate polymerization. Although the half-life of the acid species in cationic cure coatings is substantially longer, migration may be limited by diffusional constraints as polymerization and cross-linking proceed.

Optimum photoinitiator concentration is dependent on film thickness: The greater the film thickness, the lower the optimum concentration. Lower concentration is favorable from the standpoint of cost, since the photoinitiator is generally an expensive component. However, comparing the cure speeds of films of different thicknesses, each containing the optimum concentration of photoinitiator for its thickness, the time required to cure a thick film is longer than to cure a thin film. This follows since less radiation is absorbed in any volume element with increasing film thickness. The problem is further exacerbated when surface cure is oxygen inhibited, as is the case with free radical polymerization. (Oxygen inhibition is discussed in Section 28.2.3.) In general, one should determine the concentration of photoinitiator just sufficient to give the required extent of cure at the surface; this concentration gives the maximum rate of cure for that film thickness of that system at the lowest possible cost.

The problem of achieving both surface and through cure can be ameliorated by using a photoinitiator or a mixture of photoinitiators that has two absorption maximums with distinctly different molar absorptivities near different emission bands of the UV source. The emission band that is highly absorbed by the photoinitiator(s) is absorbed more strongly near the surface, and less UV is available for absorption in the lower layers. This band is most important for counteracting oxygen inhibition (see Section 28.2.3), but does not contribute substantially to through cure. The weaker absorption of a second emission band is more uniform throughout the film to provide through cure.

Another factor that affects absorption of UV by a photoinitiator is the presence of competitive absorbers or materials that scatter UV radiation. In designing vehicles for UV cure systems, it is desirable to minimize their absorption of UV in the range needed for the excitation of the photoinitiator. Pigments can absorb and/or scatter UV radiation; the effects of pigmentation are discussed in Section 28.4.

While film thickness is one variable that affects path length, it is not the only one. If the same coating is applied over a black substrate and over a highly reflective metal substrate, the rate of cure over the metal substrate is close to twice as fast as over the black substrate. UV radiation that passes through the film to the black substrate is absorbed, but that reaching a smooth metal substrate is reflected and passes through the film twice; the path length is twice the film thickness, so there is almost twice the opportunity for absorption.

28.2. FREE RADICAL INITIATED UV CURE

In these coatings, free radicals are photogenerated and initiate polymerization by adding to vinyl double bonds, primarily acrylates. Two classes of photoinitiators are used: those that undergo unimolecular bond cleavage and those that undergo bimolecular hydrogen abstraction from some other molecule.

28.2.1. Unimolecular Photoinitiators

A range of unimolecular photoinitiators has been studied [7]. The first ones used on a large scale commercially were ethers of benzoin. Benzoin ethers undergo cleavage to form benzoyl and benzyl ether radicals:

Both these radicals can initiate polymerization of acrylate monomers [8]. There is doubt whether the benzyl ether radical can initiate the polymerization of styrene [7]. The package stability of UV cure coatings containing such benzoin ethers tends to be limited. Apparently, this is due to the ease of abstraction of the hydrogen on the benzyl ether carbon. Any organic material contains hydroperoxides that slowly decompose. The resulting radicals can abstract the benzylic hydrogen, leading to initiation of polymerization and, hence, poor package stability. Package stability is improved if the benzylic carbon is fully substituted. Accordingly, the ketal, 2,2-dimethoxy-2-phenylacetophenone, is an effective photoinitiator with good package stability. Photocleavage produces benzoyl and dimethoxybenzyl radicals. The latter is a sluggish initiator; however, it can undergo further cleavage to the highly reactive methyl radical; the extent of this cleavage increases with increasing temperature.

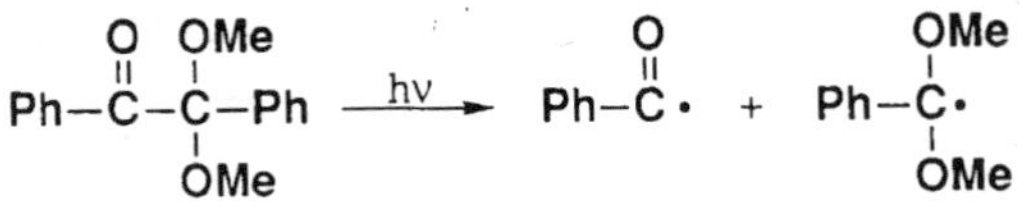

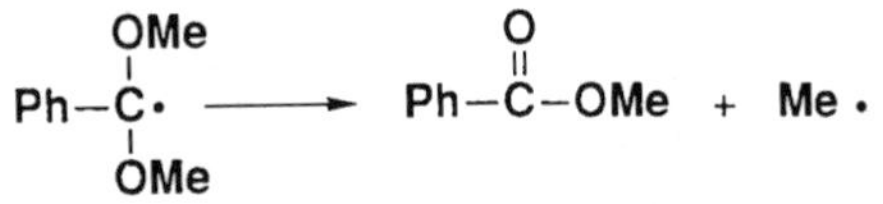

$$\underset{\underset{OMe}{|}}{\overset{\overset{O\ \ OMe}{||\ \ |}}{Ph-C-C-Ph}} \xrightarrow{h\nu} Ph-\overset{O}{\overset{||}{C}}\cdot \ + \ \underset{\overset{|}{OMe}}{\overset{\overset{OMe}{|}}{Ph-C}}\cdot$$

$$\underset{\overset{|}{OMe}}{\overset{\overset{OMe}{|}}{Ph-C}}\cdot \ \longrightarrow \ Ph-\overset{O}{\overset{||}{C}}-OMe \ + \ Me\cdot$$

2,2-Dialkyl-2-hydroxyacetophenones are also commercially important photoinitiators with good package stability. These photoinitiators tend to give less yellowing than phenyl substituted acetophenones, including benzoin ethers as well as the above ketal, probably because benzylic radicals are not generated by photocleavage.

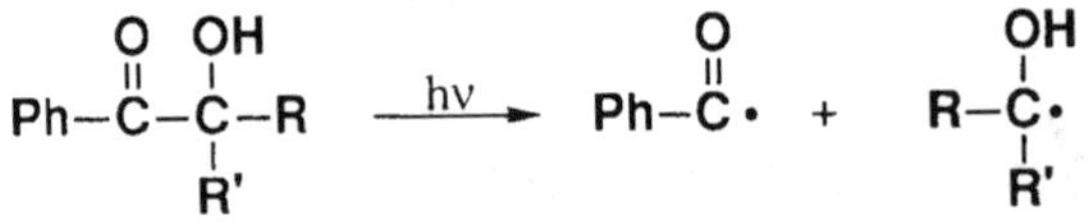

$$\underset{\overset{|}{R'}}{\overset{\overset{O\ \ OH}{||\ \ |}}{Ph-C-C-R}} \xrightarrow{h\nu} Ph-\overset{O}{\overset{||}{C}}\cdot \ + \ \underset{\overset{|}{R'}}{\overset{\overset{OH}{|}}{R-C}}\cdot$$

All these photoinitiators have an acetophenone chromophore, the absorptivity of which is enhanced by electron donating substituents on the benzoyl ring. This is exemplified by the morpholino-substituted photoinitiator 2-dimethylamino-2-benzyl-1-(4-morpholinophenyl)-butan-1-one, which is recommended for pigmented coatings [9]. Yellowing can be a problem in clear coatings.

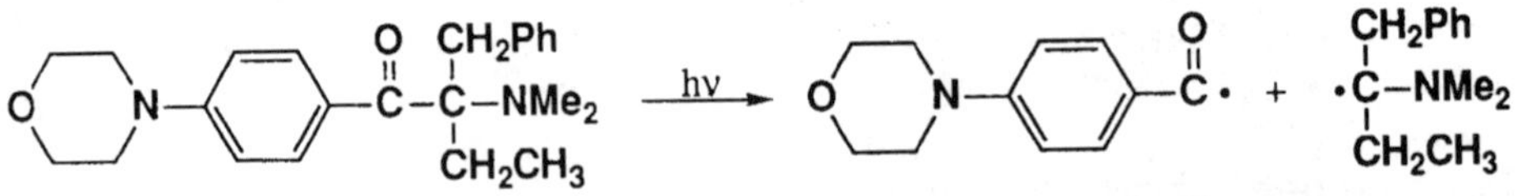

Acylphosphine oxides are another class of unimolecular photoinitiators. Several acylphosphine oxides are available, for example, diphenyl-2,4,5-trimethylbenzoylphosphine oxide. Irradiation generates the corresponding benzoyl and phosphinyl radicals. Acylphosphine oxides tend to be nonyellowing and give good package stability.

Bis(2,6-dimethoxybenzoyl)-2,4,4-trimethylpentylphosphine oxide (BAPO) on irradiation is reported to generate four radicals that can initiate polymerization [10]. The strong absorption of near UV radiation and short wavelength light makes possible curing of relatively thick films of white pigmented coatings. While BAPO is yellow, photocleavage results in reduced absorptivity (bleaching), so BAPO can be used in white coatings. The absorption spectrum can be broadened by using a blend with 2-hydroxy-2-methyl-1-phenylpropan-1-one (HMMP), providing combined surface and through cure at lower cost.

28.2.2. Bimolecular Initiators

Photoexcited benzophenone and related diarylketones, such as xanthone and thioxanthones, do not cleave to give free radicals, but can abstract hydrogens from a hydrogen donor to yield free radicals that initiate polymerization. Thioxanthones, such as 2-isopropylthioxanthone, are used when their high absorption in the very near UV is desirable to permit absorption in the presence of pigments or dyes that absorb UV strongly in the shorter UV range.

Benzophenone   2-Isopropylthioxanthone

Widely used hydrogen donors are tertiary amines with hydrogens on α carbon atoms, such as 2-(dimethylamino)ethanol (DMAE) and, in printing inks, methyl p-(dimethylamino)benzoate. It has been shown that the accompanying ketyl free radical does not initiate polymerization.

An advantage of bimolecular initiators with amine coinitiators is reduced oxygen inhibition, as discussed in the next section. A disadvantage is that the excited states of these initiators are more readily quenched by oxygen, as well as by vinyl monomers with lower triplet energies.

28.2.3. Oxygen Inhibition

Oxygen inhibits free radical polymerizations. In coatings, this inhibition is particularly troublesome, since coating films have such a high ratio of surface area, where oxygen exposure is high, to total volume. Oxygen reacts with the terminal free radical on a propagating molecule to form a peroxy free radical. The peroxy free radical does not readily add to another monomer molecule; thus, the growth of the chain is terminated. The terminal radical from methyl methacrylate has been shown to have a rate constant for reaction with oxygen that is 10^6 times that for reaction with another monomer molecule. Furthermore, the excited states of certain photoinitiators are quenched by oxygen, thereby reducing the efficiency of generation of free radicals.

Several approaches are available to minimize this problem. Curing can be done in an inert atmosphere, but this is relatively expensive. One can incorporate paraffin wax in the coating. As the coating is applied and cured, a layer of wax comes to the surface of the coating, shielding the surface from oxygen. While effective, the residual wax detracts from the appearance of the film and makes recoating difficult. High intensity UV sources minimize, but do not eliminate, the problem with fast curing systems. In effect, free radicals can be generated so rapidly that their high concentration can deplete the oxygen at the film surface, permitting other radicals to carry on the polymerization before more oxygen diffuses to the surface.

Free radicals on carbon atoms alpha to amines react especially rapidly with oxygen. In the case of DMAE, there are eight potentially abstractable hydrogens on each molecule, and each DMAE molecule can potentially react with eight oxygen molecules. Thus, benzophenone-amine initiating systems substantially reduce oxygen inhibition. Oxygen inhibition can also be reduced using unimolecular photoinitiators by adding small amounts of an amine. This process is useful when low intensity lamps or short exposure times are employed.

28.2.4. Vehicles for Free Radical Initiated UV Cure

In the first UV cure coatings, styrene solutions of unsaturated polyesters (see Section 16.3) were used along with benzoin ether photoinitiators. However, styrene is sufficiently volatile that a significant amount evaporates between application and curing. Furthermore, the rate of polymerization is relatively slow as compared to acrylate systems. However, the cost is low, and they continue to find some application.

Most current coatings use acrylated reactants. (See Section 16.4.) Acrylate, rather than methacrylate, esters are used, since acrylates cure more rapidly at room temperature; they are also less oxygen inhibited. Furthermore, polymerization of acrylates tends to terminate by combination, whereas methacrylate polymerization terminates largely by disproportionation. The extent of cross-linking, as well as higher molecular weights are favored, when termination of growing radicals occurs by combination.

In general, the vehicle consists of two types of acrylate esters: multifunctional acrylate-terminated oligomers and acrylate monomers. The monomers range from mono- to hexafunctional, most commonly mixtures of mono-, di-, and trifunctional acrylates. The monomers are also called *reactive diluents*. Multifunctional oligomer contributes to high rate of cross-linking, owing to its polyfunctionality and in large measure, controls the properties of the final coating due to the effect of the backbone structure on such properties as abrasion resistance, flexibility, and adhesion. Their viscosity is too high alone and monomers are required to reduce viscosity for application. Multifunctional acrylate monomers also give fast cross-linking owing to polyfunctionality, but have lower viscosities than an oligomer. Viscosity is lowered further by monofunctional acrylates, but the rate of cross-linking is reduced. The extent conversion of the double bonds is also affected by the choice of components; in general terms, extent conversion increases with an increase in monofunctional monomer, presumably because the small monomer molecules can diffuse through the film during the reaction. Extent conversion is also affected by free volume availability; in general, conversion is increased by use of components that give low T_g films. As polymerization proceeds, T_g of the binder increases. If, as is commonly the case, T_g approaches the temperature at which curing is being carried out, the rate of reactions slows. Reactions cease at temperatures only a little above $T_g = T_{cure}$. Since there is heat from the radiation source and a reaction exotherm, T_{cure} is somewhat above ambient temperature. See Ref. [3] for a review of cross-linking rates and extent conversions of various systems.

Two principal types of oligomers are derived from epoxy-functional and isocyanate-functional starting materials. Low molecular weight BPA epoxy resins can be reacted with acrylic acid to yield predominantly difunctional acrylate derivatives. (See Section 16.4.) Epoxidized soybean or linseed oil acrylates have been used where the higher functionality and lower T_g are desirable. Urethane oligomers are by reacting isocyanate-terminated resins with 2-hydroxyethyl acrylate. (See Section 16.4.) Any hydroxy-terminated resin can be reacted with a diisocyanate to make the isocyanate-terminated resin. Polyesters have been the most widely used base resins but polyurethanes, acrylics, and other hydroxy-functional resins are also used. If color retention is important, aliphatic diisocyanates are used; if not, aromatic diisocyanates are used because of their lower cost.

Many multifunctional acrylate monomers have been used; examples are trimethylol-propane triacrylate, pentaerythritol triacrylate, 1,6-hexanediol diacrylate, and tripropyleneglycol diacrylate. Care must be used in handling them because many are skin irritants and some are sensitizers [1]. A range of monofunctional acrylates have been used. Those with lowest molecular weight tend to reduce viscosity most effectively, but may be too volatile. Ethylhexyl acrylate has sufficiently low volatility. Ethoxyethoxyethyl acrylate, isobornyl acrylate, 2-carboxyethyl acrylate, and others are also used. Small amounts of acrylic acid as a comonomer promote adhesion. 2-Hydroxyethyl acrylate has low volatility, high reactivity, and imparts low viscosities, but the toxic hazard is too great in many applications. N-Vinylpyrrolidone (NVP) is an example of a nonacrylate monomer that copolymerizes with acrylates at speeds comparable to acrylate polymerization; NVP is particularly useful because the amide structure promotes adhesion to metal and reduces oxygen inhibition. But it introduces a possible toxic hazard. It has been found that monomers with carbamate, oxazolidine, or carbonate ester groups in addition to the acrylate functionality give faster curing and more complete conversion than simple acrylates [9].

Acrylated melamine-formaldehyde resins (see Section 16.4) have been reported that permit formulation of coatings that can cure by two different routes: UV curing through the acrylamide double bonds and thermal curing through residual alkoxymethylol groups [11]. If a coating containing an multifunctional acrylated MF resin is first UV cured and then thermally cured, it is reported that films with increased hardness, improved stain resistance, and greater durability are obtained.

The combination of unsaturated polyesters with vinyl ether resins has been reported to UV cure at rates comparable with acrylates [12]. Maleated resins [13] as well as maleimides [3] can also be used for copolymerization with vinyl ethers. Maleimide-vinyl ether compositions have also been reported to cure at rates comparable to acrylates without the use of photoinitiators. This is possible because maleimides absorb in the 300–310 nm region. Initiating free radicals are reported to arise from H-abstraction by the photoexcited maleimide. These compositions are particularly efficient with high intensity lamps that emit primarily at 308 nm, where the maleimides have peak absorbance [3].

Other UV cure copolymer compositions are amine-enes [14] and thiol-enes, which have been advanced by using highly reactive norbornene derivatives as ene components [15]. However, applications have been limited, in the former case by relatively slow cure rates and, in the latter, by the odor of monothiols.

28.3. CATIONIC UV CURE

In cationic UV cure coatings, strong acids are photogenerated to initiate chain-growth polymerization. When the idea of using cationic polymerization for cross-linking coatings is mentioned, some polymer chemists assume the idea is not feasible. They think of cationic polymerization being carried out at very low temperature in the rigid absence of water to achieve high molecular weight. However, in coatings, the objective is not to make high molecular weight linear polymers, but to cross-link polyfunctional reactants. The reactions that terminate cationic polymerizations, still give cross-links. Therefore, cationic polymerization reactions can be carried out in coating films even in the presence of some water.

28.3.1. Cationic Photoinitiators

Cationic photoinitiators are onium salts of very strong acids [3,16,17]. Iodonium and sulfonium salts of hexafluoroantimonic and hexafluorophosphoric acids are examples. Irradiation of diaryliodonium and triarylsulfonium salts yields strong protic acids of the corresponding counter anions, as well as radical cations; both initiate cationic polymerization. Biscumyliodonium tetrakis(pentafluorophenyl)borate is reported to give substantially faster cure rates than other onium salts [18].

Onium salts can also be utilized as photoinitiators for free radical polymerization, as well as for concurrent cationic/free radical polymerization, since free radical species are also formed in their photolysis. These reactions are shown for a triphenylsulfonium salt (anion omitted) following the primary unimolecular bond cleavage.

$$Ph_3S^+ \longrightarrow [Ph_2S \cdots Ph]^+$$

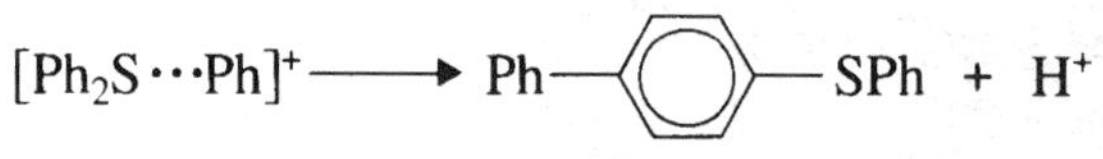

$$[Ph_2S \cdots Ph]^+ \longrightarrow Ph \!-\!\!\!\bigcirc\!\!\!-\! SPh \; + \; H^+$$

+ ortho and meta isomers

$$[Ph_2S \cdots Ph]^+ \longrightarrow Ph_2S^{\cdot+} \; + \; Ph\cdot$$

Diaryliodonium and triarylsulfonium salts absorb radiation only weakly above 350 nm; however, their spectral response can be extended into the UV-visible, as well as into the midvisible, range by use of photosensitizers [17]. 308 nm Excimer lamps are reported to be particularly effective using triarylsulfonium salts as photoinitiators [6]. Diphenyliodonium salts have limited solubility and are toxic. Substituted derivatives such as bisdodecylphenyliodonium hexafluoroantimonate have both higher solubility and lower toxicity [18].

28.3.2. Vehicles for Cationic UV Cure

Homopolymerization of oxirane groups (see Section 11.3.5) is the major type of cationic polymerization used commercially. The counterions must be very weak nucleophiles; in other words anions of very strong acids. Hexafluoroantimonates, hexafluorophosphates, and the recently introduced borate derivatives are particularly effective. In contrast to free radicals, cations do not react with each other. Consequently, in the absence of nucleophilic anions, cation initiated cross-linking can continue after exposure to the radiation source until the reactive cations become immobilized. Reactions with water and alcohols terminate polymer growth. Nevertheless, cross-linking occurs even if only one pair of epoxy groups reacts before reaction with water; furthermore, the termination reaction is accompanied by regeneration of a proton, corresponding to chain transfer. The absence of oxygen inhibition further distinguishes cationic from free radical polymerization.

Bisphenol A epoxy resins react slowly at ambient temperatures. Higher cure temperatures promote the reaction; optimum cure temperatures are about 70–80°C. Rapid curing of coatings based on BPA epoxy resins can be effected by using a combination of UV and infrared sources. In some cases, a UV curable epoxy coating is applied to one side of a coil strip and a thermally cured epoxy coating to the other. The initial partial UV cure of the UV curable coating is sufficient to permit that coated side to go over a roller without sticking during application of the thermal cure coating. Subsequent thermal exposure also advances the cure of the UV curable epoxy. More flexible films are obtained using epoxidized polybutadiene [19].

Cycloaliphatic epoxides such as 3,4-epoxycyclohexylmethyl-3′,4′-epoxycyclohexane carbonate (1) show high reactivity. This results from the added ring strain due to the fused ring system that promotes ring opening of the oxonium ion during the propagation step shown previously. The low viscosity makes such epoxides useful as reactive diluents.

1

Epoxy substituted silicone polymers have been recommended for use in UV cure release coatings [18].

Vinyl ethers and styrenes rapidly polymerize cationically. Cationic photopolymerization of vinyl ethers is more rapid than polymerization of epoxy-functional reactants [3,16]. Less photoinitiator is required; moreover multifunctional vinyl ether monomers are said to show a very low order of toxicity. Vinyl ether monomers are interesting in their own right and are also useful as a highly reactive component in epoxy resin coatings. An example of a divinylether monomer, reported to have a very high rate of polymerization, is the divinylether derived from the reaction product of chloroethyl vinyl ether with bisphenol A [16].

28.4. EFFECTS OF PIGMENTATION

Since many pigments absorb and/or scatter UV radiation, they generally inhibit UV curing to some degree [20]. Scattering leads to reflection back out of the film, and absorption also reduces UV availability to the photoinitiator. The effect becomes more serious as film thickness increases, since the pigment plus the photoinitiator reduce the amount of UV that can reach the lower layers of the film. With a strongly absorbing pigment like carbon black, the films that can be cured, even with very large amounts of photoinitiator, are limited to 1–2 μm in thickness. Hence, black printing inks can be UV cured, but not black coatings.

The most widely used pigment in coatings is rutile TiO_2. However, rutile TiO_2 absorbs some violet light and essentially all but the very nearest UV even at thin film thickness. Rutile white coatings with good hiding can be cured using a blend of 25% BAPO and 75% HMMP (see Section 28.2.1) [10]. To permit flow, low levels of pigmentation are used (PVC of 6) requiring film thicknesses of 50–100 μm for hiding. Anatase TiO_2 does not absorb near UV to a great extent; its major absorption is only at wavelengths less than about 360 nm. Therefore, thicker films can be cured than with rutile but hiding by anatase is not as efficient.

Equations have been developed that permit calculation of the fraction of UV of each wavelength that will be absorbed in the presence of pigments that absorb and scatter UV [21]. The equations permit calculation of the total fraction of radiation absorbed as a function of wavelength and, perhaps more importantly, permit calculation of absorption in the bottom layer of films of different film thicknesses. Calculation of absorption by all wavengths is not feasible. The database needed and the calculations would be very extensive, so it appears more practical to make model calculations using the equations to illustrate the effect of variables that can be useful in guiding a formulator in considering selection of pigments, photoinitiators, and their concentrations.

Figure 28.2 shows the results of model calculations of the effect of different pigments on UV absorption by photoinitiator in the bottom 0.1 μm of films as a function of film thickness. It must be emphasized that these calculations were based on a single wavelength value and with several simplifying assumptions. (See Ref. [21] for details.) For comparison, the calculated absorption by 1% benzoin ether (BE) in a pigment-free coating is shown as line 8. This concentration permits curing of unpigmented UV cure coatings. The corresponding curve for 1.2% of 2-chlorothioxanthone (CTX) is shown as line 7.

Figure 28.3 illustrates the effect of concentration of photoinitiator on absorption of UV by photoinitiator in the bottom 0.1 μm of 15 μm films. The calculations were based on assumptions intended to illustrate the effect of 20 PVC of rutile TiO_2 in the film. The photosensitizer absorption was based on CTX. With the assumptions used, the appropriate concentration of CTX for this system would be about 0.33%.

Curing is favored by using UV sources with higher contents of very near UV radiation and by using photoinitiators, such as CTX, with higher absorption coefficients in the very near UV wavelengths. As previously discussed, there is an optimum photoinitiator concentration that decreases as film thickness increases. Pigmented coatings are more sensitive to this effect than unpigmented coatings. The effect of reflection by the substrate can be critical in determining the film thickness that can be cured.

A further problem of pigmentation in curing arises from the large differential of absorption of UV at the surface and at the bottom of the films, which can result in wrinkling. (See Section 23.6.) If the surface layer of a film cures while the bottom layer is still fluid, then when the bottom layer cures it shrinks, causing the top layer to wrinkle. This effect is particularly likely to be seen with free radical cure done in an inert atmosphere or with cationic cure. In both cases, there is no inhibition of the surface cure by oxygen.

Pigmentation also affects flow properties. In almost all cases, UV cure coatings are applied without solvent, which means that the volume of pigment in the wet coating is almost as high as the PVC of the final film. Without solvent, viscosity tends to be higher

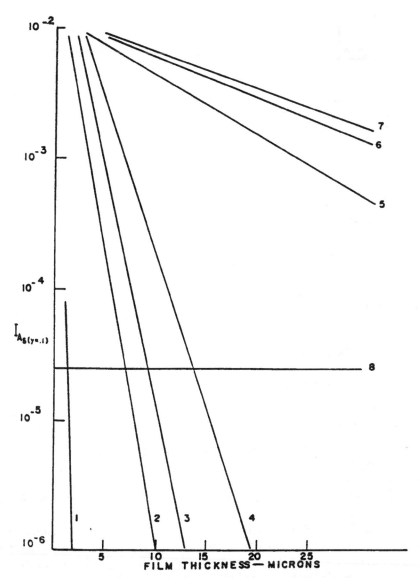

Figure 28.2. Fraction of radiation entering films on 85% reflectance substrates that is calculated to be absorbed by photoinitiator in the bottom 0.1 μm of a film as a function of film thickness. Coatings 1 through 7 contain 1.2% CTX. Coating 8 contains 1% BE. Pigments in coatings 1 through 6 are: 1) carbon black, 2) molybdate orange, 3) lemon chrome yellow, 4) anatase TiO_2, 5) madder lake, 6) ultramarine blue. See Ref. [21] for assumptions. (From Ref. [21], with permission.)

than desirable for good flow in the short time between application and cure. Even at 20 PVC, there is a significant increase in viscosity due to the presence of the pigment. By using sufficient monomer replacing oligomer, the viscosity can be reduced to application viscosity, but this reduces the curing speed. As one tries to make higher PVC coatings the problem becomes more difficult. As PVC approaches CPVC, viscosity of a wet coating approaches infinity. The problem is compounded by the difficulty of stabilizing the

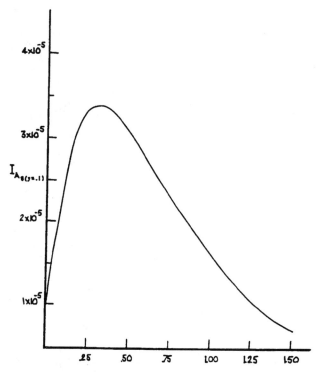

Figure 28.3. Absorption by CTX of 375 nm radiation in the bottom 0.1 μm of 15 μm films with 20 PVC TiO_2 over an 85% reflectance substrate as a function of CTX concentration. (From Ref. [21], with permission.)

pigment dispersion against flocculation with the relatively low molecular weight vehicle. (See Section 20.1.3.) Flow problems are just as serious for cationic systems as for free radical systems.

For many wood finish applications, transparent low gloss coatings are desired, but they are difficult to achieve with radiation cure coatings. Low gloss is attained in transparent lacquers used for finishing furniture and paneling by adding low concentrations of small particle size SiO_2. (See Section 30.1.) When the lacquer is applied, the solvent evaporates, setting up convection currents in the drying lacquer film. The SiO_2 particles are carried up to the surface of the film by these convection currents. As the viscosity of the surface increases, the SiO_2 particles are trapped at the surface. The resultant high PVC at the surface gives low gloss. In UV cure coatings, there is no solvent to evaporate and, therefore, no mechanism to concentrate the pigment at the surface. If the level of pigmentation is increased to the extent necessary to provide low gloss, the viscosity is too high for application.

Some progress has been made in reducing gloss by use of a dual cure method. A UV cure coating containing as much small particle size SiO_2 as viscosity permits and designed so that its cure is strongly oxygen inhibited is applied and cured in air. The lower layer of the coating cures; the top layer does not. When the lower layer cures, it shrinks, exerting an unbalanced stress on the pigment particles in the upper parts of the film, forcing them toward the surface. An alternative explanation is that reactive monomer or oligomer undergoes net migration to the polymerizing regions, thereby concentrating pigment in the nonpolymerizing region. Then, the coating is cured again, this time under an inert atmosphere. Now, the surface layer cures resulting in further shrinkage and, further increasing the PVC near the surface. In this way, medium gloss transparent coatings can be made. Several production lines have operated using this dual cure method.

28.5. ELECTRON BEAM CURE COATINGS

High energy electron beams (EB) can be used to polymerize acrylate coatings. In EB cure compositions, the high energy electrons cause direct excitation of the coatings resins (P), as well as their ionization into radical cations ($P^+ \cdot$) and secondary electrons (e^-).

$$P + EB \longrightarrow P^* + P^+_\bullet + e^-$$

$$P^+_\bullet + e^- \longrightarrow P^*$$

$$P^* \longrightarrow I\cdot$$

The major fate of the radical cations is recombination with the relatively low energy secondary electrons, which can yield additional excited state resins. The excited state resins (P*) primarily undergo homolytic bond cleavage to free radicals that initiate polymerization of the acrylated resins.

The high energy electrons are generated by charging a tungsten filament at a high negative potential, 150–300 keV. The electrons are directed by magnets in a curtain through a metal "window" to the coatings to be cured. (The term *electron curtain* is also used.) The types of vehicles used are the same as those used in free radical initiated UV curing. Since the polymerization reactions are oxygen inhibited and there is little differential in absorption between the top and the bottom of the film, it is essential to do EB curing in inert atmospheres.

Cationic polymerization of vinyl ethers [3,22], as well as highly reactive epoxides [23], can be induced by EB radiation in the presence of diaryliodonium and triarylsulfonium salts. The readily reducible onium salts have been implicated in the generation of cationic initiators by oxidation of free radical species, as well as by capture of secondary electrons. Scavenging of secondary electrons by onium salts tends to lengthen the lifetime of cations and to improve the prospects for cationic polymerization [22].

The principal advantages of EB curing over UV curing are that no photoinitiator is needed and that pigments do not interfere with the curing. These advantages are often offset by the higher capital cost of the electron beam or curtain generating equipment, the need to use inert atmospheres, and the need for shielding to protect workers from the electron beam. The advantage of being able to cure pigmented systems is real, but is of limited importance in coatings, since the flow problem is not alleviated by EB curing.

28.6. ADVANTAGES, DISADVANTAGES, AND APPLICATIONS

A recent survey of 130 people involved in radiation curing assessed the advantages, disadvantages, and prospects for the technology [23]. 89 percent of the respondents reported that reduced solvent emission is an important motivation for use of radiation cure. Generally, the formulations contain no solvent and emissions are negligible.

Very short cure times are possible at ambient temperature with package stable coatings. In the case of clear acrylate compositions, curing times required are fractions of a second. Since the curing can be done at ambient temperatures, the coatings are applicable to heat sensitive substrates such as paper, some plastics, and wood. Coatings can be formulated without solvent and, therefore, lead to minimal VOC emission.

The energy requirement is minimal. The major losses in energy encountered in thermal curing systems are not experienced. The cure is done at or near to ambient temperature, so no energy is required to heat the object being coated. (The IR radiation from the UV lamps does cause a modest temperature increase.) The heat loss from air flow through a baking oven required to keep the solvent concentration below the lower explosive limit is eliminated. The extent of this advantage is, of course, dependent on the cost of energy. The use of radiation curing in Europe and Japan has been greater than in the United States, at least in part because the cost of energy is higher in Europe and Japan than in the United States.

Capital cost for UV curing is low. This results primarily from the small size of the curing units as compared with ovens for baking. A UV cure coating only needs about 0.5 second exposure for curing. A line being run at a speed of 60 m min^{-1} need be exposed to UV over a distance of only 0.5 meters. A UV curing unit with 4 lamps can do this in a total length of about 2 meters. In contrast, a thermal cure oven has to be a hundred or more meters long to provide time for curing at this speed. In addition, there has to be space for flash off of at least some of the solvent from conventional coatings before the coated object enters the oven and space for it to cool back down to handling temperature. The savings in building space can thus be large.

Half of the end users in the survey report "improved physical properties/product performance" as a major motivation for using radiation cure [23].

There are, however, limitations and disadvantages to radiation curing. It is most applicable to flat sheets or webs for which the distance to the UV source or the window of the EB unit can be approximately constant. Cylindrical, or nearly cylindrical objects, can be rotated in front of the radiation source and fibers can be pulled through a circle of sources, but irregularly shaped objects are not easy to expose uniformly.

In the case of UV curing, pigmentation can limit the thickness of films that can be cured. In the extreme case of carbon black pigmented films, the limit is about 2 μm. Electron beam cure coatings do not have this limitation because there is no gradient of initiation from top to bottom of the film. Owing to the effects of pigmentation on flow, the level of pigmentation is limited in both UV cure and EB cure coatings. Furthermore, low gloss coatings cannot be applied. These limitations can be overcome by using solvent containing coatings, but this eliminates many of the advantages.

Shrinkage during curing can lead to adhesion problems on metal and some plastic substrates. (See Section 6.2.) When polymerization results from chain addition of free radicals to double bonds, there is a substantial decrease in volume, since the length of the carbon-carbon bonds formed is shorter than the intermolecular distance of monomer units. The degree of shrinkage is directly related to the number of double bonds reacted. In acrylic UV or EB cure coatings, the potential shrinkage is 5 to 10%. Since cure occurs in less than a second, there is not time for much of the volume shrinkage to take place before the free volume in the coating is so limited that movement becomes restricted. The stress from the restricted shrinkage applies a force to offset the adhesion forces holding the coating to a substrate so that less external force is required to remove the coating. Shrinkage can be minimized by using higher molecular weight oligomers or higher ratios of oligomer to low molecular weight reactive monomers, but these expedients increase viscosity and tend to increase cure time. Curing of epoxy systems by cationic photo-initiators leads to less shrinkage. The decrease in volume resulting from the formation of polymers from monomers is partially offset by an increase in volume from the ring opening. Typical epoxy UV cure compositions shrink less than 3%. Adhesion of cationic cure coatings can be further enhanced by surface treatment of the substrate [24].

Porous substrates such as paper and wood give good adhesion, since the mechanical effect of penetrating into the surface of the substrate holds the coating on the substrate. However, there can be a curling problem with thin substrates as a result of the tension from shrinkage on one side of the substrate.

The photoinitiator in free radical initiated coatings is only partly consumed in curing, so substantial amounts are left in films. The photoinitiator can accelerate photodegradation of films exposed outdoors. Conventional UV stabilizers and antioxidants tend to reduce UV cure rates, which limits their usage. These constraints limit the use of UV cure coatings to interior applications. While there has been work done using precursors, including photoinitiators that convert to UV stabilizers on exposure, the results do not appear to be satisfactory. Cationic coatings have the advantage that they do not generate free radicals when films are exposed outdoors. Electron beam cure coatings have the distinct advantage that there is no photoinitiator in the system. However, even with EB curing, the coatings used must have excess unsaturation to cure on rapid schedules and residual unsaturation can adversely affect exterior durability. Also, there are trapped free radicals that can react with oxygen after the cure and generate peroxides.

Radiation cure coatings tend to be more expensive than conventional coatings so that commercial applications are primarily those where the advantages of UV cure permit unique advantages.

UV cure printing inks, particularly lithographic inks, are widely used. The viscosity of lithographic inks is relatively high, on the order of 5–10 Pa·s, so there is not a flow problem. The films are thin, 2 μm or less, so interference of pigment with curing can be offset by use of highly absorbing photoinitiators and/or high photoinitiator concentrations. Since the films are not continuous, adhesion or curling difficulties resulting from shrinkage are not a major problem. The advantage of fast curing is large, it allows the webs to be wound up quickly without offsetting on the back of the substrate. In printing of tin plate and aluminum for cans, there is the advantage that the four colors can be printed in line by simply having a UV lamp after each printing station; operating costs are lower than with heat cure inks. Ink jet printing is another rapidly growing market for UV inks [23].

Clear coatings for paper, plastic film, and foil are probably the largest volume market for radiation cured coatings [23]. The unpigmented coatings cure very fast at low temperatures.

An early commercial use of UV curing was in Europe for top coating wood furniture. A large part of European furniture is manufactured in flat sheets and assembled after coating. (See Section 30.1.) Furthermore, the finish on much European furniture is glossy. UV curing is ideal for such an application. The fast cure at low temperature is appropriate for wood. Early coatings were based on styrene-polyester resins, but acrylic coatings were later adopted because they cure more rapidly and emission of styrene is avoided. There has been relatively little use of UV cure top coats on wood furniture in the United States because a major fraction of U.S. furniture is assembled and then coated. Furthermore, most

U.S. furniture has a low gloss top coat that, as discussed previously, is difficult with solvent-free UV cure coatings. Wood kitchen cabinets and flooring are appropriate applications for UV cure coatings.

An important furniture application is filling of particle board. Particle board has a rough surface, and before use in furniture tops, it must be coated with a filler and sanded smooth to give an even surface for the application of a base coat and printing. (See Section 30.1.) This was originally done with highly pigmented solvent-borne coatings applied at about 30% volume solids. However, the low solids requires two or three coats to give sufficient film thickness to permit sanding to a smooth surface without cutting down into the substrate. With UV cure, the volume loss is only a few percent, so an adequate layer can be applied in one coat. This together with the fast cure means that UV cure fillers are less expensive to apply. The pigments used are clays and silica that absorb or scatter only minor amounts of UV. Aluminum trihydroxide absorbs still less UV and may be a particularly effective inert pigment [25]. The high level of pigmentation leads to poor leveling, but it does not matter, since the surface is sanded. Furthermore, the pigmentation results in less resin being at the surface so air inhibition is less of a problem. Originally, the fillers were styrene/polyester systems, but the higher speed of acrylics has led to their adoption despite their higher cost.

Another large scale application whose success results from the fast curing at low temperature possible with UV cure coatings is top coating of vinyl flooring. It is especially applicable to flooring with patterns made by expanded foams, since the foam pattern is heat sensitive. Acrylated urethane oligomer coatings have the high abrasion resistance needed in a floor covering; acrylated polyester coatings have superior stain resistance.

Abrasion resistant clear coatings for plastics have been developed. The poor abrasion resistance of clear plastics is a distinct disadvantage in applications, such as replacing glass in uses ranging from glazing to eyeglasses. Silicone-based coatings have been used for many years to improve the abrasion resistance. (See Section 4.3.) These coatings require long thermal cure schedules and have high VOC levels. Radiation cure coatings based on trialkoxysilyl-functional acrylate monomers and colloidal silica cure in seconds and have low VOC emission [26]. They have excellent adhesion to polycarbonates and other plastics and have abrasion resistance equal or superior to solvent-based coatings [26,27].

An application for cationic UV cure epoxy coatings is for coil coating on tin plate for the exterior side of can ends. The superior adhesion of epoxy coatings makes them preferable for this application, as compared to acrylate coatings. In some cases, a UV cure coating is applied on one side of the strip and UV cured. Then, a thermally cured epoxy coating is applied on the other side. When the strip is put through the oven to cure the thermally cured coating, the cross-linking of the UV cure coating is also advanced.

A large use of UV curing is for coating optical fibers for wave guides used in telecommunication cables. The process is carried out at high speeds in towers up to 25 meters tall. The glass fiber is pulled from a heat softened blank of glass and then coated through a die coater, that is, a reservoir with holes on each end through which the glass is pulled. The UV lamps are aligned parallel to the fiber as it comes out of the coater, with elliptical reflectors both behind the UV source and on the opposite side of the fiber from the source. The fiber is at the focal point of the pair of reflectors. In this way, the coating is given very high dosages of UV, permitting cure speeds of the order of 20 m s^{-1} without an excessive bank of UV lamps. Acrylated urethane oligomers have been used since abrasion resistance is a critical requirement. Two UV cure coatings are applied, a soft first coat and a harder outer coat.

UV curing is also the basis for important photoimaging processes used in the printing and electronic industries [28,29]. In these applications, UV curable coatings on selected substrates are exposed through an image bearing transparency, such as a film negative. Cross-linking occurs selectively under the transparent regions. The unexposed regions are subsequently removed, commonly by washing out with aqueous or organic solvent. The resulting image transfer into the coated substrate is a critical step in making printing plates as well as printed circuit boards and integrated microcircuits (e.g., computer chips).

GENERAL REFERENCES

S. P. Pappas, Ed., *Radiation Curing: Science and Technology*, Plenum, New York, 1992.

P. K. T. Oldring, Ed., *Chemistry & Technology of Formulating UV Cure Coatings, Inks, and Paints*, (Vols. 1–5), SITA Technology, London, 1991–1994.

J. P. Fouassier and J. F. Rabek, *Radiation Curing in Polymer Science and Technology*, (Vols. 1–5), Elsevier Applied Science, London and New York, 1993.

A. B. Scranton, C. N. Bowman, and R. W. Peiffer, Eds., *Photopolymerization Fundamentals and Applications*, American Chemical Society, Washington, DC, 1997.

REFERENCES

1. R. Golden, *J. Coat. Technol.*, **69** (871), 83 (1997).
2. A. J. Bean in S. P. Pappas, Ed., *Radiation Curing: Science and Technology*, Plenum, New York, 1992, p. 308.
3. S. Jonsson, P.-E. Sundell, J. Hultgren, D. Sheng, and C. E. Hoyle, *Prog. Org. Coat.*, **27**, 107 (1996).
4. A. Roth, *RadTech Report*, September/October, 1996, p. 21.
5. K. Sharp, J. Mattson, and S. Jonsson, *J. Coat. Technol.*, **69** (865), 77 (1997).
6. J. K. Braddock, *RadTech '96 Proc.*, Vol. 1, p. 478.
7. H. J. Hageman, *Prog. Org. Coat.*, **13**, 123 (1985).
8. L. H. Carlblom and S. P. Pappas, *J. Polym. Sci., A: Polym. Chem.*, **15**, 1381 (1977).
9. C. Decker and K. Moussa, *J. Coat. Technol.*, **65** (819), 49 (1993).
10. W. Rutsch, K. Dietliker, D. Leppard, M. Kohler, L. Misev, U. Kolczak, and G. Rist, *Prog. Org. Coat.*, **27**, 227 (1996).
11. J. J. Gummeson, *J. Coat. Technol.*, **62** (785), 43 (1990).
12. C. B. Friedlander and D. A. Diehl, U.S. Patent 5536760, (1996).
13. G. K. Noren, *Polym. Mater. Sci. Eng.*, **74**, 321 (1996).
14. G. K. Noren, *J. Coat. Technol.*, **65** (818), 59 (1993).
15. A. F. Jacobine, D. M. Glaser, P. J. Grabek, D. Mancini, M. Masterson, S. T. Nakos, M. A. Rakas, and J. G. Woods, *J. Appl. Polym. Sci.*, **45**, 471 (1992).
16. J. V. Crivello, *J. Coat. Technol.*, **63** (793), 35 (1991).
17. S. P. Pappas, in S. P. Pappas, Ed., *Radiation Curing: Science and Technology*, Plenum, New York, 1992, p. 1.
18. C. Priou, A. Soldat, and J. Cavezzon, *J. Coat. Technol.*, **67** (851), 71 (1995).
19. F. Cazaux, X. Coqueret, B. Lignot, C. Loucheux, and P. Rousseau, *J. Coat. Technol.*, **66** (838), 27 (1994).
20. S. P. Pappas and Z. W. Wicks, Jr., in S. P. Pappas, Ed., *UV Curing: Science and Technology*, Vol. I, Technology Marketing Corp., Norwalk, CT, 1978, pp. 78–95.
21. Z. W. Wicks, Jr. and W. Kühhirt, *J. Paint Technol.*, **47** (610), 49 (1975).

22. S. C. Lapin, in S. P. Pappas, Ed., *Radiation Curing: Science and Technology*, Plenum, New York, 1992, p. 241.

23. K. Lawson, *RadTech Report*, **12** (1), 25 (1998).

24. F. Molenaar, P. Buijsen, and C. N. Smit, *Prog. Org. Coat.*, **22**, 393 (1993).

25. A. A. Parker and E. S. Martin, *J. Coat. Technol.*, **66** (829), 39 (1994).

26. J. D. Blizzard, J. S. Tonge, and L. J. Cottington, *Proc. Water-Borne Higher-Solids Powder Coat. Symp.*, New Orleans, 1992, p. 171.

27. L. N. Lewis and D. Katsamberis, *J. Appl. Polym. Sci.*, **42**, 1551 (1991).

28. B. M. Monroe in S. P. Pappas, Ed., *Radiation Curing: Science and Technology*, Plenum, New York, 1992, p. 399.

29. D. Funhoff, H. Binder, S. Nguyen-Kim, and R. Schwalm, *Prog. Org. Coat.*, **20**, 289 (1992).

CHAPTER 29

Product Coatings for Metal Substrates

In 1996 the volume of U.S. shipments of product coatings for OEM (original equipment manufacture) application was 1.41×10^9 L with a value of $5.4 billion [1]. This was approximately 31% of the volume and 36% of the value of all U.S. shipments of coatings. In this and the following chapter on coatings for nonmetallic substrates, we discuss some of the major OEM product coating end uses to illustrate the factors involved in selecting coatings for particular applications. A large fraction of industrial coatings is used on metals. Multitudes of products are coated; space only permits discussion of a few. We have selected four of the larger end uses: automotive, appliance, container, and coil coatings. The coatings used are generally proprietary and are continually evolving. Thus, these discussions focus on the principles involved, not on specific formulations.

29.1. OEM AUTOMOTIVE COATINGS

OEM automotive coatings are those used for painting cars and trucks on assembly lines. The coatings for repair and refinishing of cars are discussed in Section 32.3. This discussion is further limited to coatings applied to the exterior bodies of automobiles. Many other coatings are used on car interiors, wheels, trunk linings, air filters, and so on. Coatings for plastics are discussed in Section 30.3.

There are two major reasons for coating an automobile: appearance and corrosion protection. Initial appearance can be a critical factor in the sale of a car; the coating is the first thing a potential buyer sees. The purchaser can be attracted by good appearance of the coating; furthermore, if the coating is poor, the purchaser may assume that the manufacturer is careless not only in applying the coating, but also in other aspects of manufacturing the automobile. Maintaining excellent appearance of the coating over many years of service is also critical. If the coating does not continue to look good, not only may the owner of the car be unhappy, but also, others may decide not to buy that make of car because they have seen that the coating does not stand up well. High-gloss coatings are used, and gloss retention is a critical requirement. Thirty or more years ago, many people waxed their cars frequently; since performance of the coating has been

527

substantially improved waxing is no longer necessary. Corrosion protection is another critical performance requirement. The use of salt on icy highways provides an environment that can lead to rapid corrosion of steel. In the 1970s and 1980s, coatings were developed that provide protection against corrosion for the lifetime of the car.

Automobile bodies are generally fabricated from steel; see Ref. [2] for a discussion of various types of steel and their treatments. The single most important factor affecting corrosion is the steel [3,4]. There are large variations in cold-rolled steel that can lead to greater or lesser probabilities of corrosion failures. The principal differences result from the annealing process. When steel is cold rolled, oil is used as a lubricant; the oily, rolled steel is then annealed. There are two processes of annealing. In one, coils of the steel are put in annealing ovens and heated to about 500°C to relieve stresses set up during the rolling process. In the other, called strip annealing, the steel is run as a single layer through an annealing oven. The oils can get baked into the surface of the steel, especially in the coil annealing process. This is likely to happen if the coils are wrapped tightly so that oils cannot volatilize away from the steel surface. The oils partly decompose, leaving organic carbon compounds embedded in the surface that are not removed by subsequent cleaning steps; these organic compounds make uniform conversion coating more difficult. The effect can be irregular; only one side of the coil or one end of the coil may be affected. To minimize corrosion, coated steels such as electrogalvanized, nickel-zinc alloy electro-coated, and aluminum-zinc alloy coated steels (see Ref. [2]) are often used for part or most of the car body.

Plastic and rubber components are used for parts of the exterior. In some cases, plastic and rubber parts are completely coated before assembly; in others, they are primed and are attached to the car body before application of the top coat. Some fillers and sealants are applied to the original steel; others are applied after E-coating, but before top coating; and others are applied after top coating. Furthermore, there has been increasing use of decals and trim items that are adhesively bonded to the top coat. All these complexities impose additional requirements on the coating process and the properties of the coatings. Automobile producers want to use the lowest possible baking temperatures to avoid deformation or deterioration of polymeric components, but there is a trade off between film performance and reduced baking temperatures.

The steel body of the car is fabricated, the doors, hood, and trunk lid are attached, and the assembly is cleaned to remove dirt and oil. (See Section 6.4.1.) As discussed in Section 6.4.1, the surface is next conversion coated by dipping into an aqueous solution containing primary zinc phosphate, $Zn_2(H_2PO_4)_2$, as the major component. This treatment results in the deposition of a layer of mixed crystals of $Zn_3(PO_4)_2 \cdot 4H_2O$ and $Zn_2Fe(PO_4)_2 \cdot 4H_2O$ [5]. The weight of the phosphate layer is about 2 g m^{-2} and the thickness is about 5 μm. Zinc-manganese-nickel conversion coatings are recommended for treating zinc-coated metal [6].

After treatment, the metal must be thoroughly rinsed to remove any residual soluble salts that can accelerate corrosion, as explained in Section 7.3.2 and loosely adhering crystals. The last rinse water contains chromic acid, which reacts with the surface to deposit a small amount of chromate to act as a passivating agent against corrosion until the surface is coated. The treated surface must be kept clean until primer is applied. If, for example, someone should touch the car body with bare hands, oil and salts could be deposited on the surface, potentially leading to blisters and inferior corrosion protection.

29.1.1. E-Coat Primers and Primer-Surfacers

As discussed in Section 7.3.2, adhesion is the primary key to corrosion protection, particularly adhesion that resists displacement by water. The primer surface must be smooth and permit good adhesion of top coats. Almost all automobiles are primed with cationic electrodeposition (E-coat) primers, as discussed in Section 26.2. The E-coat primers provide excellent adhesion and resist displacement by water. The strong driving force of the electrophoretic application aids in penetration into the phosphate crystal mesh on the steel surface. The amine groups on the binder resin provide strong interaction with the phosphate coating and the steel surface. The binders are designed to be nonsaponifiable; even if the coating is scraped through, undercutting will be very slow, and the coating will continue to protect the metal next to the gouge against corrosion. The E-coat covers all metal surfaces, including recessed areas, such as the interior of rocker panels below the doors, that cannot be coated by spray application. The process provides a relatively uniform film thickness on all metal surfaces of about 25 μm of dry film.

While E-coat primers provide excellent corrosion protection, there are two problems involved in their use: (1) The film thickness can be so uniform that surface irregularities in the metal and conversion coating are copied in the surface of the primer film, and (2) it can be difficult to achieve the required adhesion of top coat to the primer surface. E-coat primers have been formulated that provide greater smoothness (the term *metal filling* is often applied) by reducing the pigment content of the primer. The viscosity of the uncured primer film is lower, and when the primed car body is baked, more leveling can occur than if the PVC were higher. While this aids in obtaining surface smoothness, adhesion of top coat to the smooth, glossy E-coat surface is more difficult.

Surface roughness and adhesion problems can be minimized by spray applying a primer-surfacer (also called a tie coat, a guide coat, a primer, or a full-body anti-chip) over the E-coat before top coating. The binder in the primer-surfacer can be a polyester or an epoxy ester that provides better adhesion than an acrylic top coat to the epoxy based E-coat primer. In some cases, a two package (2K) polyurethane primer-surfacer is used. The primer-surfacer has a lower cross-link density than the primer providing greater opportunity for the solvent in the top coat to penetrate into this surface than into the surface of the E-coat primer. The primer-surfacer is formulated with a higher PVC so that surface roughness provides a greater opportunity for top coat adhesion. The PVC of primer-surfacers is commonly greater than CPVC (see Chapter 21), providing for penetration of a little of the vehicle from the top coat into the primer-surfacer film. Also, such a highly pigmented primer-surfacer can be relatively easily sanded so that local areas of excessive roughness can be sanded smooth and film thickness is sufficient to minimize the danger of sanding through to bare metal. Powder primer-surfacers are being used over E-coat on some cars to reduce VOC emissions. They also provide improved chip resistance. Adhesion to the E-coat surface is presumably obtained by the high baking temperature, 150°C, used with the powder primer-surfacer. Another way to improve adhesion of top coats is to apply a so-called *sealer* to the E-coat surface. A sealer is a low solids top coat with a high content of relatively strong solvents. It is applied just before the top coat; the solvent softens the surface of the primer slightly, permitting penetration of top coat binder into the surface, providing increased adhesion.

A current trend is to use color key primers; the colors are picked for use under a group of top coats with related colors. Use of several different color primers is a significant expense, but can improve appearance when top coats have thin areas with less than 100% hiding. A further advantage is that colored primers can be sprayed under the hood, in the trunk, and on the inside of doors. Then a relatively thin layer of the more expensive top coat gives satisfactory appearance. The primer must have good exterior durability; otherwise, where there are thin spots in the top coat, UV radiation will degrade the surface of the primer, causing loss of adhesion of top coat to primer and, ultimately, *delamination*. UV absorbers and HALS stabilizers (see Section 5.2.1) can be added to the primer-surfacer to improve its UV resistance. A UV absorber in the top coat that strongly absorbs UV in the wavelength range of 290–380 nm also helps protect the primer from degradation. Partial migration of UVA and HALS among the coats is to be expected.

A *chip-resistant primer*, called anti-chip, is frequently applied over the E-coat primer on the lower parts of the car body, some manufacturers apply anti-chip to the whole outer body of their cars. It is designed to be especially resistant to impact by stones thrown up from a road against the car body. The coatings are organosols or polyurethanes using blocked isocyanates as cross-linkers. Reference [7] gives a discussion of the mechanical properties involved in stone chip resistance and approaches to testing.

29.1.2. Top Coats

The primary purpose of top coats is appearance. Top coats are high gloss and must maintain their appearance for long periods. While the primary factors in maintaining appearance are resistance of the resin binder to photoxidation and hydrolysis, durability of the pigment in the film is also a critical factor. (See Chapter 5.) There are many other requirements including: resistance to car wash brushes, bird droppings, acid rain, sudden thunder showers on a car that has been sitting in the hot sun, the impact of pieces of gravel striking the car, gasoline spillage, and so on. Until the early 1980s all top coats were *monocoats*, a single coating composition applied in several coats. Monocoats have been largely supplanted by *base coat-clear coat systems*; a base coat containing the color pigments covered by a transparent coating. Base coat-clear coat systems provide better gloss and gloss retention than monocoats.

In both monocoat and base coat-clear coat systems, a majority of top coats are metallic or other polychromatic colors. As discussed in Sections 18.4 and 19.2.5, metallic coatings give an attractive appearance due to the change in color with the angle of viewing. When viewed at an angle near the perpendicular, the color is light, and when viewed at larger angles, the color is darker. The extent of this phenomenon is called the degree of *color flop*, or just *flop*. High color flop depends on three factors: minimal light scattering by the coating matrix between the aluminum flakes; a smooth, high gloss surface; and orientation of the aluminum flakes parallel to the surface.

Minimal light scattering by the coating matrix requires that pigment selection and dispersion provide transparent films when prepared in the absence of the aluminum flake. Also, all resins and additives in the coating must be compatible so that there is no haziness in the unpigmented film. Surface roughness can be affected by many factors. It has been shown that the single most important factor affecting surface roughness is the roughness of the initial metal substrate phosphate layer [8]. Other important factors are leveling after spraying and, in the case of base coat-clear coat systems, the thickness of the clear coat.

Orientation of aluminum flakes parallel to the surface has been achieved in everyday production for decades, but it has never been entirely clear why orientation is as good as it is [9,10]. Most workers have concluded that an important factor is shrinkage of the film after application; shrinkage is accompanied by increasing viscosity of the film most rapidly near the surface as the solvent evaporates, leading to a viscosity gradient. This viscosity gradient causes the upper edge of an aluminum platelet to be immobilized before the lower edge, and the platelet swivels towards parallel orientation as the film shrinks. In general, the lower the solids of the coating as it arrives at the substrate, the better the alignment will be. Some workers think that spray droplets spread out when they strike the surface, and the resulting flow forces tend to align the flake particles parallel to the surface. It has also been suggested that the spray droplets must penetrate the surface of the wet film and crash on the substrate [10]. These and other factors may influence orientation. Changes in atomization, air flow, solvent evaporation rates, and gun to surface distance during spraying can give substantial variations in flake alignment [11]. Transfer efficiency is greatly improved by electrostatic spraying; however, it is more difficult to obtain good surface smoothness and metal orientation using electrostatic spray. It has been suggested that there may be some alignment of the metal flakes parallel to the lines of force in the electrostatic field that are perpendicular to the substrate. Use of automatic high-speed rotary bell spray guns further complicates the problem of obtaining uniform appearance.

Most monocoats are formulated with thermosetting acrylic and melamine-formaldehyde (MF) resins. To achieve high gloss, pigmentation levels are low. In solid colors, PVCs of 8–9% are used; in the case of metallic colors, PVCs are much lower—on the order of 2–4%. Owing to the low pigment content, relatively thick films (on the order of 50 µm) are required for hiding. Since the coatings are spray applied, some areas have significantly over 50 µm of dry coating, for example, around the "A post" between the windshield and the front door openings.

The highest solids at which it has been possible to formulate high performance solvent-borne monocoats is about 45 NVV. This has also been near the upper limit of solids that provides outstanding exterior durability. If the molecular weight of the acrylic resin is further reduced in order to increase the solids above 45 NVV, the average number of functional groups per molecule and molecular weight get so low that the fraction of molecules with single functional groups becomes too large to permit good properties with thermosetting acrylic resins synthesized by free radical polymerization. (See Section 12.2.1.)

Clear coats can be sprayed at somewhat higher solids than monocoats. As discussed in Section 23.3, as solids increase, the problems of controlling sagging increase. Automotive coatings are particularly vulnerable to sagging, since the film thickness applied is large and variable. As solids have been increased, it has become necessary to add thixotropic agents to increase the viscosity at low shear rates to minimize sagging [12,13]. Conventional thixotropic agents make a coating hazy due to light scattering thereby reducing gloss. Therefore, viscosity modifiers were designed with refractive indexes similar to those of the acrylic binder. Those most widely used are acrylic microgels, which are highly swollen gel particles that are lightly cross-linked so that they can swell, but not dissolve, in the liquid coating. (See Section 23.3.) The swollen particles flocculate, giving a thixotropic coating; the viscosity decreases with the high shear going through the spray gun, stays low long enough for leveling, but then increases, reducing sagging. The rheology of microgel dispersions is discussed in Refs. [12] and [13].

Application of a clear coat over a pigmented base coat gives higher gloss. Historically, clear coats were not used in automotive finishes because of the cost of an extra coating step and because the exterior durability of available clear coatings was inadequate. However, with development of better binders and light stabilizers, especially combinations of HALS and UV absorbers (see Section 5.3.3), clear coats now have long-term exterior durability. Total combined film thickness is only a little greater than of a single coat system. The base coat contains roughly twice the PVC of a monocoat. Thus, a base coat with a dry film about 20–30 μm thick, depending on the color, has about the same hiding as a 50 μm monocoat. The clear coat thickness should be about 40–50 μm, but considerable variation is experienced in production. Thin (less than 30–35 μm) areas appear satisfactory, but have inferior durability. Properly formulated and applied base coat clear coat systems have high gloss and long-term gloss retention. The advantage is particularly marked for metallic coatings. The same overall VOC emission can be achieved using a lower solids base coat and higher solids clear coat. The lower solids in the base coat permits better orientation of the aluminum in the coating and, hence, superior color flop. The higher level of pigmentation leads to a somewhat rougher surface of the base coat as compared to a single coat, but this is covered by the clear coating. Water-borne base coats generally give superior metal orientation, since they are applied at 20–25 NVV.

While it would be possible to bake the car body after application of the base coat, then apply the clear coat, and bake the car again, it is more economical to apply the clear over a wet base coat and bake the overall system just once. Part of the solvent must evaporate from the base coat film before clear coat is sprayed over it; otherwise, the force of the spray of the clear coat would distort the base coat. On the other hand, if too much solvent has evaporated, intercoat adhesion may be adversely affected. The solvent must be permitted to flash off until distortion will not occur, and then the top coat is applied. The flash-off time required is about two minutes, but varies with spray booth ventilation conditions as well as coating formulations. In some systems, to permit application of the clear coat in two minutes, it is necessary to incorporate a small amount of wax, zinc stearate, or cellulose acetobutyrate in the base coat formulation [14]. The additive apparently orients at the surface and minimizes distortion of the layer. Care must be exercised, since excess additive could interfere with intercoat adhesion.

In most cases, base coat binders are thermosetting acrylic resins with MF cross-linkers. It has been reported that hydroxy-functional urethane modified polyesters are also useful in base coats [14]. There is a trend towards use of water-borne base coats with high solids solvent-borne clear top coats [15]. The lower film thickness of the base coat and the flash-off time required before applying the clear top coat reduce the popping problem (see Sections 23.7 and 25.1) that tends to occur with water-borne monocoats [16]. Control of sagging during application requires that the water-borne base coat be shear thinning, which also reduces surface distortion during subsequent application of the clear coat [17]. The low solids (15-20 NVV) of water-borne base coats facilitates aluminum orientation. Opaque colors can be applied at 25–35 NVV. Water-borne base coats are formulated for application at relative humidities (RH) of about 60%, since it is less expensive to increase RH than to decrease it. Published papers on water-borne base coats mention MF cross-linkers with water-reducible acrylics, water-reducible polyester-polyurethanes, and acrylic latexes [15–18]. Use of acrylic latexes appears to be gaining favor as experience is gained in designing the complex combinations of rheological characteristics needed for application and flake orientation [9,18]. Water-borne base coats require a dehydration bake in which most of the water and some of the solvent is driven off before application of top coat.

Many types of clear coats are being used and others evaluated. There are many requirements in addition to high gloss: VOC, resistance to environmental etching, mar resistance, and cost are major considerations in selecting a coating. Most clear coats are acrylic resins with various cross-linkers. The acrylic resin is the major factor controlling VOC content. As discussed in Section 12.2.1, volume solids obtainable is limited by the necessity that essentially all of the resin molecules must have at least two functional groups. Considerable research has been aimed at blending low molecular weight polyester and polyurethane polyol resins with acrylic resins to increase solids without reducing the molecular weight of acrylic resins below the level that provides adequate film properties.

Various cross-linking reactions are being worked on. Hydroxy-functional acrylic polymers can be cross-linked with MF resins (see Section 9.3), polyisocyanates (see Section 10.4), or combinations of the two. Epoxy-functional acrylic resins can be cross-linked with dicarboxylic acids. (See Section 11.3.2.) Trialkoxysilyl-functional acrylic resins can be cross-linked by reaction with atmospheric water [19]. Use of auxiliary cross-linkers such as blocked polyisocyanates and/or MF resins further enhances the properties obtained with silyl-functional acrylics [20]. 3,5-Dimethylpyrazole or 1,2,4-triazole blocked polyisocyanates are preferred, since they permit curing at somewhat lower temperature and do not have the yellowing problem of methyl ethyl ketone blocked isocyanates. (See Section 10.5.)

Soon after introduction of base coat-clear coats, a serious problem was encountered. After a few days or weeks in some locations, shallow, unsightly pits developed in the clear coat. Since acid hydrolysis of cross-links in the top coat was the suspected cause, coatings were said to have poor *acid etch resistance*. Since other factors are now known to contribute to the problem (see Section 5.5), the terminology now used is *environmental etch resistance* [21]. Various laboratory tests have been devised, but none is completely satisfactory. Outdoor tests of environmental etch resistance, performed near Jacksonville, FL give useful, but still not very reproducible, results.

To have good environmental etch resistance, a coating must have good hydrolytic stability under acidic conditions. The activated ether groups in hydroxy-functional acrylic-MF cross-linked films are subject to hydrolysis. (See Section 9.3.) Environmental etch resistance of MF cross-linked coatings can be improved by increasing T_g, which minimizes penetration of water into the film. Since urethanes are more resistant to hydrolysis under acidic conditions, isocyanate cross-linked acrylics generally have superior environmental etch resistance. Most urethane clear coats have been two package (2K) coatings (see Section 10.4); however, weighing against the use of 2K coatings are the need for dual spray systems, concern about toxicity, and cost. Another approach is to use hydroxy-functional urethanes. For example, a low molecular weight hydroxy-functional urethane derived from the reaction of a triisocyanate and a diol such as neopentyl glycol has been reported to give good clear coat properties when cross-linked with MF resin [22]. Moisture cure trialkoxysilylacrylic resins provide excellent etch resistance [19,20]. Clear coats based on a combination of MF resin and trialkoxysilyl cross-linking tend to have better environmental etch resistance than those based on MF cross-linking alone. Epoxy-functional acrylics cross-linked with dicarboxylic acids or anhydrides also provide excellent environmental etch resistance. Since reduced solubility of water in films increases environmental etch resistance, highly fluorinated resins (see Section 16.1.3) improve environmental etch resistance, but are expensive. Another approach is to use blocked isocyanates; most blocked isocyanates require too high a cure temperature, but tris(alkoxycarbonylamino)triazine (see Section 10.5) reacts at conventional baking temperatures and provides environmental etch resistance [23]. Higher solids is possible by using MF resin to cross-link carbamate-functional resins (see Section 9.3.4) [24]. For example, a coating made with the reaction product of hydroxypropyl carbamate with an isocyanate prepolymer derived from IPDI, together with an MF resin and dodecylbenzenesulfonic acid catalyst, has 85% weight solids. The coating is said to combine excellent environmental etch resistance with excellent mar resistance.

Mar resistance is discussed in Section 4.3.2; Reference [25] provides a review of mechanical properties required for mar resistance. An important cause of marring is the action of automatic car washes. Mar resistance requires a tough film (resistant to breaking) that also is hard; this combination requires compromises. Films with low coefficients of friction improve mar resistance. MF cross-linked coatings generally provide excellent mar resistance, but as previously noted generally have inadequate environmental etch resistance. Urethane cross-linked coatings have good environmental etch resistance, but poor mar resistance. Also as noted previously, cross-linking with a combination of urethane and MF and cross-linking silyl-functional acrylics with auxiliary cross-linkers can provide both mar and environmental etch resistance. Mar resistance can be further improved by incorporating some fluoroalkyl acrylate in the acrylic resins to decrease coefficient of friction. Carboxylic acid cross-linked epoxy-functional acrylics suffer from inferior mar resistance. Water-borne clear coats are also being used. For example, a hydroxy-functional styrene/acrylic latex with an $H_{12}MDI$ blocked isocyanate cross-linker has been reported [26].

Lowest emissions could be achieved if powder coatings (see Chapter 27) could be used for exterior automotive coatings. Powder automobile top coats have been in a development stage since the late 1960s. In Japan in the early 1980s, powder coatings were used as nonmetallic monocoats. Their major limitations are the need for a separate spray facility for each color (or very expensive cleanup between color changes) and the very poor color flop of metallic powder coatings. These limitations do not apply to clear coatings. Major development effort is being applied to powder clear coats, and one automotive plant started using one in 1998. Fine particle size powder is a requirement to achieve the smooth films needed for clear coats. For example, a powder coating that had an average particle size of the acrylic resin of 10 μm with a narrow distribution of particle size of an epoxy-functional acrylic, formulated with dodecanedioic acid with a particle size of about 3 μm, gave good surface smoothness [27]. Special application equipment is required for application over nonconductive surfaces and to handle the small particle size [28].

The increasing use of plastics in the exterior of automobiles leads to difficulties in designing coatings. (See Section 30.3 for discussion of coating plastics.) Some plastic moldings distort at temperatures normally required for baking top coats. Some plastics are flexible and some have rigidities similar to steel. Coatings designed for flexible plastics are generally too soft for car exteriors, and coatings designed for steel or rigid plastics are too rigid and are subject to cracking when used over flexible substrates. In many cases, plastic components are coated and then installed on the car body. There would be substantial advantages to a "universal coating" that could be applied to flexible and rigid plastics and steel in one operation. Reference [29] discusses the design criteria for a universal coating.

29.1.3. Factory Repair Procedures

During assembly of a car, it is common for the coating to be damaged or to have blemishes caused by dirt, resulting in need for repair. Reducing the frequency and cost of repairs is a major goal in auto production. Repairs are made at several stages in the process: after application of the primer-surfacer to the car body, after the base coat and clear coat have been applied, after assembly, and after shipment to the dealer. Once the glass, upholstery, tires, and the like have been installed, the car can no longer be baked at the temperatures for which regular production coatings are designed. Either the whole car can be baked at 80°C or the area repaired can be heated somewhat above 80°C with infrared lamps.

When cars were coated with acrylic lacquer, repair was relatively simple, since the thermoplastic systems stay soluble in the solvent in the repair lacquer. However, thermosetting enamels are more difficult to repair. Achieving adhesion to the surface of the cross-linked coating is more difficult, and when properly done, the whole panel in which the damage occurred has to be refinished. The top coat is removed, any bare metal is primed, and special repair base coat and clear coats are applied to the whole panel. Since the coating cannot be baked at a high a temperature, additional strong acid catalyst must be added to coatings with MF resins to allow curing at lower temperature. The excess catalyst remains in the film and can lead to more rapid hydrolysis. Durability of such repairs is good, but not as good as the original coating. Urethane 2K repair coatings are being used increasingly, since they cure at relatively low temperatures without loss in long term durability. It is unlikely that corrosion resistance of repaired areas is equal to unrepaired areas because the E-coat is often sanded through during the repair process.

29.2. APPLIANCE COATINGS

Major markets for OEM appliance coatings are for washing machines, dryers, refrigerators, air conditioners, and ranges. In some cases, single coats are used, but a major part of the market is for primer-top coat systems. For applications for which corrosion protection is required, such as washing machines and air conditioners, cationic E-coat primers (see Section 26.2) are used on the highest quality products. Nonelectrodeposition primers are frequently applied by flow coating. (See Section 22.4.) To minimize VOC, water-reducible epoxy ester based primers are appropriate.

Cationic E-coat epoxy coatings are used for some applications as single coats. Although epoxy coatings chalk badly on exterior exposure, the drum of a dryer or the interior of an air conditioner does not get exposed outdoors, but they do need a relatively high degree of corrosion protection. The uniform coverage of edges attainable with E-coat permits use of 12 μm E-coats to replace 50 μm solution epoxy coatings on air conditioners, with substantial reduction in manpower required for application, while maintaining the necessary performance [30]. White E-coats have been applied to appliances. Anionic acrylic E-coats can be used on aluminum, but for steel, as discussed in Section 26.1, discoloration results from iron salt formation. Cationic E-coat single coats based on acrylic resins (see Section 26.2) avoid this problem on steel.

Thermosetting acrylic coatings are generally used over primers. For single coat compositions, polyesters are more commonly used because they tend to exhibit better flexibility and adhesion than acrylics to treated steel or aluminum. The most commonly used cross-linkers are amino resins. For washers and dishwashers benzoguanamine-formaldehyde resins (see Section 9.4.1) are used because they impart greater resistance to alkaline detergents. For other applications, conventional MF resins are used because of lower cost. In end uses for which performance requirements are not severe, such as hot water heaters, lower cost semioxidizing alkyd-MF based coatings may be most appropriate.

Use of powder coatings as top coats for appliance applications has been growing rapidly. (See Chapter 27.) The long runs of single colors make powder coatings a natural choice for appliances. A limitation is the greater difficulty of achieving good leveling with powder coatings as compared to liquid coatings. While some orange peel can be desirable to conceal metal irregularities, powder coatings tend to have too much orange peel. However, in some cases, particularly in Europe, consumers are accustomed to the orange peel typical of porcelain enamels, and the finish from powder coatings is readily accepted. Low VOC emissions, low fire risk (with proper precautions in handling powders), low energy requirements, and reuse of overspray powder are strong economic and environmental reasons for using powders. Hydroxy-functional polyester or acrylic resins are used as binders with a blocked isocyanate or tetramethoxymethylglycoluril (see Section 9.4.3) as cross-linker. Alternatively, carboxylic acid-functional resins are used together with triglycidyl isocyanurate (see Section 11.1.2) or tetra(hydroxyethyl)adipamide (see Section 16.6.1) as cross-linker.

Another approach to reducing VOC emissions at the appliance manufacturer's factory is to use coil coated metal. (See Section 29.4.) The solvent emissions occur at the coil coating factory, where they can be burned to provide part of the fuel for curing the coatings. A potential problem with coil coated metal is that the edges are bare. The appliance must be designed so that cut edges are turned under and protected by a sealant. In some cases, it is possible to weld through the coating in areas that are not visible.

29.3. CONTAINER COATINGS

Container coatings were historically called *metal decorating coatings*, since a major portion of the business was in coating flat sheets, then lithographically printing, followed by a finishing varnish (clear top coat) to protect the ink. There were many uses for such sheets—metal boxes, trays, wastepaper baskets, bottle caps and crowns, and, most importantly, cans. Plastics have replaced coated metal in most applications other than bottle caps, crowns, and cans. The field is now usually referred to as container or can coatings. In the United States alone, about 135 billion cans are produced annually, of which about 100 billion are beverage cans.

Most cans are food or beverage containers, and one of the key requirements is that there be no possibility of introducing toxic compounds into the foods or beverages. In the United States, all can linings must be acceptable to the Food and Drug Administration (FDA) or, in the case of meat products, the Department of Agriculture. Contrary to what many people seem to believe, the FDA does not approve coatings; it lists acceptable ingredients. Rules are published in the Code of Federal Regulations (CFR), Title 21, Part 175. The most important is 21CFR175.300, which deals with resinous and polymeric coatings. In most cases, if all of the components of a new coating have already been used in can coatings, a new coating will be acceptable. In some cases, it is necessary to prove that no material is extracted into any food or beverage that will be packed in the can. However, new raw materials must pass extensive tests. Toxicity considerations predominantly affect interior can coatings, but in some cases, there are also restrictions for possible contamination from exterior can coatings. This is particularly so when metal sheets coated on both sides are stacked, in which case the exterior coating on one sheet is in direct contact with the interior coating on the next sheet in the stack. In this configuration, migration of low molecular weight components between coatings is possible.

Effect of coatings on the flavor of the food or beverage packed in the container is critical. While flavor requirements are particularly important for interior coatings, care must also be exercised with coatings for exterior application. Flavor changes can result from either extraction of some contaminant from the coating; absorption of flavor agents from the packed food into the coating, reducing the flavor of the food; or failure to isolate the food or beverage from the metal of the can. Flavor can be affected by minute amounts of substances. To assure that all residual solvent and other volatile flavor detractors are driven out of the coatings, high baking temperatures are used. The only way to evaluate effects of coatings on flavor is by making test packs of the food or beverage in the container and tasting the food or beverage. As a result, major suppliers of can coatings maintain flavor panels of people trained to taste and, particularly, to use consistent words to describe flavors.

For beer cans, a major aspect of the flavor problem is to prevent contact between the beer and the can because metals catalyze flavor changes in beer. For this reason, the final interior coating is spray applied after formation of the can to avoid potential problems from breaks in the can lining resulting from stresses during can forming operations. Linings are spray applied to the interior of soft drink cans not only to protect flavor, but also to protect the can; the acid present in most soft drinks could eat through the metal without a coating barrier. An interesting sidelight on history is that in the early days of packing pineapple products and grapefruit juice in cans, coatings for lining these cans that would resist their high acidity were not available. They were packed in cans with heavy tin plate linings to protect the steel bodies. Tin affects the flavor of the pineapple products and grapefruit juice and acts as a bleaching agent, allowing light color fruits and juices to retain their light color. Even though organic coatings that could be used with these products are now available, most are still packed in heavily tinned cans. Apparently, consumers are used to and prefer canned pineapple and grapefruit products that have a tinny flavor.

There are two major classes of cans: three-piece and two-piece cans. In a three-piece can, one piece is the body and the other two are the ends. Metal sheets are coated or continuous strips of metal are coil coated. Blanks for the can bodies are stamped out of the coated metal, formed, and sealed into the cylindrical body. The body is formed, usually into a cylinder, and sealed by soldering, by welding, or with an organic adhesive. Soldering was widely used, but is now restricted to nonfood cans because of concern about lead toxicity. The side seam is sprayed with a fast drying coating called a *side striper* to cover exposed metal resulting from soldering or welding. Solvent-borne side stripers are predominantly used; some water-borne stripers are used, and for containers for aggressive products, powder coatings are used. The coatings are applied while the metal is still hot, resulting in curing of the coating. The ends are made separately; they are stamped out of coated sheets or coil coated metal, formed, and equipped with a formed-in-place rubber gasket. One end is put on by the can maker; the other end is put on after the can has been filled.

Coatings on metal sheets are generally applied by direct roll coating. (See Section 22.4.) In some cases, transfer rollers with sections cut out are used so that the coating is only applied in selected areas. For example, the edges that will be soldered or welded at the sides of the body of the can are commonly not coated. On coming out of the coater, the sheets are fed onto wickets attached to a conveyor. This permits the sheets to go through the baking oven in an almost vertical position, reducing the necessary length of the oven. Baking schedules vary with the application. Interior food can coatings may be cured in 10 minutes at 200–210°C while exterior white and varnish coatings may receive only 10 minutes at 160–175°C. Inks are applied by offset lithography; in which the ink is transferred from a lithographic plate to a rubber blanket and is offset to the sheet.

There are two processes for making two-piece cans. Drawn and wall-ironed (DWI or D&I) cans are formed by drawing a cup from a flat blank and then ironing the walls to thinner thickness and greater depth. In draw-redraw (DRD) cans, a coated blank is formed into a shallow cup and then drawn one or two more times to achieve the desired height and configuration and shape of the bottom of the can. The other piece of a two-piece can is an end similar to the ends of the three-piece can.

Examples of DRD cans are shallow cans for tuna fish and taller cans for vegetables and pet foods. Flat sheets are coated and then the can is drawn and formed. In some cases, such as shoe polish and auto wax cans, printed sheets are formed into two-piece cans. The design must be distortion printed; that is, the print must be designed so that it looks correct *after* the distortion that results from forming. In general terms, it costs less to coat and print flat sheets before forming than to coat and print a formed can. The ability of the coating to withstand this degree of formation depends on the depth of draw. It is not just a matter of how deep the can is, but also of how wide. Forming a large diameter can involves less distortion than a narrow can; a deep can involves more distortion than a shallow can. Bottle caps and crowns are made from coated and distortion printed sheets and then are punched out and formed.

Beverage cans, and some food cans are DWI cans; uncoated metal is drawn and formed before coatings and inks are applied to the exterior, and a lining is sprayed into the interior of the can. The exterior coating is applied by rotating the can against the transfer roll of a small coater, and inks are transferred from a litho plate to a soft rubber roller that in turn transfers the inks to the can surface. Baking is done in short time cycles at high temperature. Some lines produce over 2000 cans per minute. It is estimated that the coating reaches a peak temperature of about 205°C for only about 1 second. Only partial cross-linking occurs in the short time. The cross-linking is completed when the can is baked again after the interior lining is applied. Interior coatings are applied by spray. A small spray gun is automatically inserted into a spinning can; the gun sprays, and then is pulled out of the can. To assure removal of all solvent, the final part of the curing is done with air directed into the can bodies. Typically, the cure schedule is about 2 minutes at 200°C. Final forming of the top of the can to fit the end is done after the coating and printing are completed.

There are three major types of metal used in cans: tin plate, steel, and aluminum. The choice of metal depends on the end use. Most food cans are three-piece cans made from tin-plated steel. A large fraction of pet food cans are made from treated steel—so-called black plate—since it is less expensive and the highly reflective tin coating is not needed. Beverage cans are two-piece aluminum cans. Pressure from the carbonation of either soft drinks or beer keeps the thin walled aluminum cans sufficiently rigid. In the case of food cans, which do not have significant interior pressure, aluminum does not compete with steel because thicker walls of aluminum would be required to achieve adequate rigidity. There has been some discussion of using aluminum for food packing by putting a small piece of dry ice into the pack before sealing. Vaporization of the CO_2 provides the pressure necessary for rigidity. However, people associate slightly bulging food cans with spoilage, so consumer acceptance has been slow.

29.3.1. Interior Can Linings

The composition of interior can coatings depends on the food or beverage to be packed in the can. In most cases, if the food is to be cooked, it is cooked in the can. A common cooking cycle is 60 minutes at 121°C. Most beer is pasteurized in the can at somewhat lower temperature. In both cases, the interior and exterior coatings must maintain their adhesion and integrity through the cooking or pasteurizing process.

Most vegetables and fruits are packed in cans with an interior coating called an *R enamel*. Historically, R enamel was a phenolic varnish. Now, it is more common for R enamel to be a phenolic resin modified with some drying oil derivatives during its synthesis or an epoxy-phenolic coating—a resole phenolic (see Section 11.6.1), with a BPA epoxy and phosphoric acid catalyst. In order to achieve adequate flexibility, high molecular weight epoxy resins [1007 or 1009 type (see Section 11.1.1)] were formerly used. New resole phenolics have been developed that permit use of lower molecular weight 1001 or 1004 BPA resins thus decreasing VOC emissions [31]. For packing vegetables, that give off hydrogen sulfide during cooking, such as corn, fine particle size ZnO pigment is dispersed in the coating, and it is called a *C enamel*. The ZnO reacts with H_2S to form white ZnS. This prevents or conceals formation of unsightly black tin sulfide by reaction of tin oxides with H_2S.

Water-borne coatings have also been developed to reduce VOC emissions. Epoxy-phenolic emulsions with nonionic surfactants have been used, but tend to have marginal properties. Water-borne coatings based on amine salts of a carboxylic acid-functional epoxy phenolic are said to have properties equal to solvent-borne coatings [32].

Fish cans are generally lined with a resole phenolic resin. (See Section 11.6.1.) To obtain sufficient formability to permit drawing of two-piece fish cans, cross-link density is reduced by using a mixture of *p*-cresol and phenol in making the resole phenolic resin. Poly(vinyl butyral) is commonly incorporated in the formulation to promote adhesion and act as a plasticizer. The extent of cross-linking is adjusted for flexibility together with resistance to swelling and softening by oils from the fish (and, in some cases, additional oil in which the fish is packed), during processing and storage. Thin film thickness is used; this helps permit forming without film rupture. Aluminum flake pigment is sometimes incorporated in the coating to minimize permeability. In linings for meats such as ham, one of the key requirements is that the coating permit the meat to slide out of the can easily after the top is removed. This requires incorporation of a release agent, such as petroleum wax, in the can lining.

An increasing fraction of food cans are DRD or DWI two-piece cans. The draw-redraw process requires that coatings have a greater degree of ductility than the coatings used for interiors of three-piece food cans. Most commonly, vinyl organosol (see Section 16.1.2) coatings lightly cross-linked with phenolic resins or MF resins have been used since the vinyl chloride copolymer binder shows ductility below T_g [33]. (See Section 4.2.) In some European countries, there is concern about toxic emissions from factories recycling vinyl chloride copolymer coated cans [34]. This has led to work on other more distensible coatings. For example, epoxy resins have been designed that will take considerably deeper draws than the conventional BPA epoxy coatings [35].

Large volumes of coatings are used for lining beverage cans. Historically, most spray applied interior coatings were solvent-borne vinyl chloride copolymers or high molecular weight epoxy/phenolic coatings. The volume solids of these coatings at application viscosity were very low, 12–15% NVV. The high cost of solvent and the high level of VOC emissions forced a change. A water-borne coating based on a graft copolymer is now widely used in lining both beer and soft drink cans [36,37]. Styrene/ethyl acrylate/acrylic acid side chains are grafted onto a BPA epoxy resin. The resin is "solubilized" with dimethylaminoethanol (DMAE) in glycol ether solvent and reduced with water. (See

Section 11.4.) To lower the cost of some soft drink can linings, it has been possible to use a special latex binder blended with small amounts of the graft copolymer and epoxy phosphate as adhesion promoters.

Concern has been expressed about use of BPA epoxy resins in interior coatings, since BPA is an estrogen mimic. Procedures for more accurate analysis of BPA extractables have been published [38]. Studies of extractables from beverage can linings have shown no detectable BPA using analysis methods sensitive to 5 ppb. Average migration from food can linings was 37 ppb, a level that is estimated to lead to a maximum potential dietary exposure of approximately 2.2 ppb, which is less than the levels considered to be of concern [39].

29.3.2. Exterior Can Coatings

Most food cans are not coated on the outside; rather, paper labels are used. The appearance of coated and printed cans is more attractive than paper labels. Some food and all beverage cans are coated and printed. Printed metal cans are not affected by water such as from ice chests or condensation when cans are brought out of a refrigerator into humid air.

The general procedure used on the exterior of sheets for the bodies of three-piece cans is to apply a base coat, often called an *enamel*, print on the base coat up to four colors by offset lithography, and finally top coat with a finishing varnish. The most common color for base coats is white, but a variety of colors is used for specific products. Color stability and lack of color change on baking are critical requirements. To minimize VOC emissions, water-borne acrylic coatings cross-linked with MF resins are widely used vehicles. The lithographic printing inks are either baking or UV cure inks. The vehicles for the baking inks are long oil alkyd resins with some MF resin cross-linker, while the vehicles for the UV cure inks are blends of acrylated epoxidized soy (or linseed) oil with acrylated epoxy resin and acrylate reactive diluents (see Section 28.2.4).

The finishing varnishes are also acrylic-MF binder compositions. To decrease friction in forming machines and conveyors, the finishing varnish commonly contains a small amount of petroleum wax or fluorinated surfactant to reduce surface tension so that the cans will have a low coefficient of friction. Two inks are applied, the sheets are baked, the other two inks are applied, the finishing varnish is applied, and the sheets are baked again. This is called *wet-ink varnishing*, since the varnish is coated on the printed surface before the last two inks are cured. With UV cure inks, each ink is partially cured by passing under a UV lamp before the next ink is applied. Commonly, finishing varnishes are not used over UV cure inks.

For 2-piece cans, white water-borne acrylic base coats are used. The film thickness varies from 8 to 15 μm; even at 15 μm, hiding is less than 100%, but the coating is definitely white. An increasing proportion of two-piece cans is simply printed with inks 2–3 μm in thickness with no base coat. Appearance of these cans is inferior, but they cost less than cans that are coated with a base coat before printing. A transparent yellow-brown base coat is used on some cans to give a bronze or gold color to the relatively shiny metal surface.

UV cure coatings are also used on some beverage cans. (See Chapter 28.) Exterior finishing varnishes are acrylated resin coatings cross-linked by photogeneration of free radicals. Coatings for the exterior side of can ends are epoxy resins cross-linked by photogeneration of acid. The UV cure end coating is applied to the side of the sheets that will be on the outside of the cans and passed under the UV lamps, giving a partially cured film; the other side, which will be on the inside of the can is coated with an FDA listed epoxy-phenolic thermal cure coating. When the epoxy-phenolic is cured in an oven, the cure of the UV cure epoxy coating is advanced to completion by the photogenerated acid still present in the films. This thermal complement to the UV cure also enhances adhesion of the coating.

29.4. COIL COATING

Steel and aluminum are manufactured in long strips that are rolled into coils. In many cases, the coils are cut, formed, and fabricated before they are coated, as in production of automobiles, for example. In other cases, it is possible to coat metal coils and later fabricate the final product from the precoated metal. When precoated metal can be used, there can be substantial advantages (see Section 29.4.1), and coil coating has grown into a major industrial coating process [40].

Coil coating started in 1935 as a process for coating venetian blind slats. The strips of metal were about 5 cm wide, and the line was run at a rate of about 10 m min^{-1}. Modern coil lines can coat metal up to 1.8 meters wide at rates as high as 250 m min^{-1}. Most lines run at a rate of around 100–200 m min^{-1}. A schematic drawing of a coil line is shown in Figure 29.1.

The metal is shipped from the steel or aluminum mill in coils weighing up to 25,000 kg that are 0.6–1.8 meters wide and 600–1800 meters long. In some lines, as shown in Figure 29.1, the first step is precleaning; brushes and sanding remove any physical contaminants. The strip then goes to the entry accumulators. The rollers of the accumulators move apart to accumulate a significant length of coil so that when one coil is about to run out, the next coil can be stitched (welded), on while the accumulator rolls move together, supplying strip to the line without interruption of the process. After the stitching is done, the new coil is fed to the coating line. As the process continues, the accumulator rolls gradually separate to store strip for the next change of coils.

Next, the strip is carried through the metal treatment area. Detergent washing and rinsing are followed by application of various conversion coatings including phosphates, complex oxides and chromates, water rinse, and finally chromic acid or other rust inhibiting rinse. Because of concerns about toxicity of chromates, proprietary chromate-free conversion coatings are being adopted. All the cleaning and conversion coating procedures must be designed to work at high speeds. Since the metal is moving at 100–200 m min^{-1} or more, the total time for cleaning and treatment is on the order of a minute or less. Next, the strip is carried through a dry-off oven and finally to coaters and a baking oven. In Figure 29.1, a laminator is also shown, although laminating of film in line with coating is not common. It is more usual to have two coating stations, each followed by a baking oven. Often, a primer and a top coat are applied on one side of the strip and another coat on the back side. It is fairly common to coat, cure, and then print one or more colors.

Most coatings are applied by reverse roll coating, but sometimes direct roll coating is used for thinner films. (See Section 22.4.) Reverse roll coating is used for thicker films for which close control of thickness is critical, and for which the flow advantages that result from the wiping action, in contrast to the film splitting action in direct roll coating, is important. To achieve even film thickness across a strip 2 meters wide, the rolls have to be crowned; that is, designed so that their diameter is greater in the middle than on the edges, since the pressure involved tends to bow the rolls to a small degree. To avoid damage to the

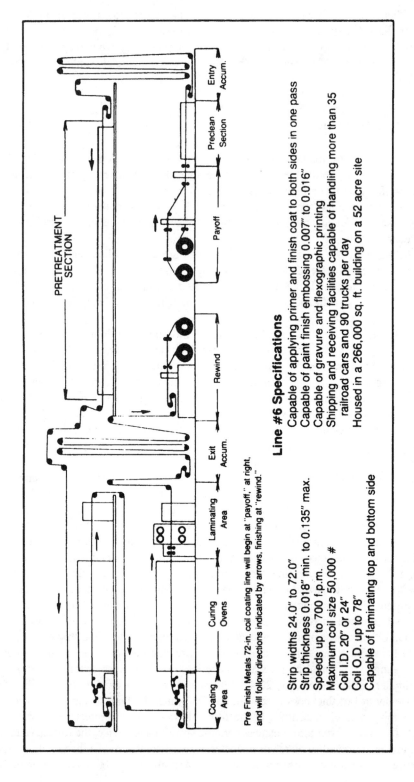

PRETREATMENT SECTION

Entry Accum.

Preclean Section

Payoff

Rewind

Exit Accum.

Laminating Area

Curing Ovens

Coating Area

Pre Finish Metals 72-in. coil coating line will begin at "payoff," at right, and will follow directions indicated by arrows, finishing at "rewind."

Strip widths 24.0" to 72.0"
Strip thickness 0.018" min. to 0.135" max.
Speeds up to 700 f.p.m.
Maximum coil size 50,000 #
Coil I.D. 20" or 24"
Coil O.D. up to 78"
Capable of laminating top and bottom side

Line #6 Specifications

Capable of applying primer and finish coat to both sides in one pass
Capable of paint finish embossing 0.007" to 0.016"
Capable of gravure and flexographic printing
Shipping and receiving facilities capable of handling more than 35 railroad cars and 90 trucks per day
Housed in a 266,000 sq. ft. building on a 52 acre site

Figure 29.1. Schematic drawing of a coil coating line. (From Ref. [40], with permission.)

rollers, the line is programmed so that very shortly before a stitched section joining two coils is going to pass through the coating nip between the rollers; the rollers are automatically separated slightly and then almost immediately returned to normal operating pressure.

At the high line speeds, the dwell time in the ovens is less than a minute, even with long ovens. In some cases, the dwell time is as low as 10 seconds, but more commonly is in the range of 15–40 seconds. After a short initial period, hot air is directed over the surface of the coatings at high velocity. The air temperature can be as high as 400°C. The temperature reached by the coating on the metal is the critical temperature for curing the coating. This temperature cannot be directly measured, but is closely related to the metal temperature, which can be measured. The temperature considered most important is the *peak metal temperature* (PMT); this can be as high as 270°C. After the coating is baked, the strip passes through the exit accumulator to the rewind. The exit accumulator stores coated strip during removal of a coated coil. In some lines, the strip passes over chilling rolls, or through water quenching, to reduce the temperature before the strip is rewound into a coil. The pressure in the center of the rewound roll is very high; consequently, the T_g of the coating on the metal must be very high and/or it must be adequately cross-linked to avoid blocking.

The exhaust air from the hoods over the coaters and particularly from the oven contains solvents. On most lines, the exhaust air streams are used as part of the air used to burn the gas to heat the ovens. In this way, part of the residual heat from the oven exhaust is recycled and the solvent is burned. Burning the solvent essentially eliminates VOC emissions, and the fuel value of the solvent is recovered. As a result, there has been less incentive to change to water-borne or high solids coatings in coil coating applications than in other applications. There are still pressures to reduce solvent content because the fuel value of solvents is low compared to the cost of the solvents and because some lines are not equipped to burn solvent.

Coatings on aluminum are frequently single coats, but on steel, primer-top coat systems are most widely used. Binders for primers were traditionally based on BPA epoxy resins; epoxy esters and epoxy/MF resins are examples. However, polyurethane, polyester, and water-borne latex primers are being used increasingly.

Many types of coating binders are used for top coats. Oxidizing alkyds with MF resin are the lowest cost and are sometimes used on the reverse side of the coated strip as a *backer*; polyesters are supplanting alkyds in this application. Backer coatings may be pigmented or unpigmented, and they contain a small amount of incompatible wax. The purpose of the backer is to avoid metal marking of the top surface coating by rubbing against a bare metal reverse side of the coil. Alkyd-MF coatings are also used as top surface coatings for which corrosion resistance requirements and/or exterior durability requirements are modest. Polyester-MF binders are widely used, especially as single coats; exterior durability and corrosion protection obtained is generally superior to that obtained with alkyd coatings. Polyester-blocked isocyanate coatings have been used to a degree for applications for which abrasion resistance and flexibility are particularly important. Close temperature control in the ovens is important with urethane coatings, since urethanes may discolor and decompose relatively rapidly at the baking temperatures involved in coil coating. Thermosetting acrylic-MF coatings are used, usually over primers.

For greater exterior durability, one can use silicone-modified polyesters and silicone-modified acrylic resins. (See Section 16.5.3.) For example, one might use 30% silicone-modified polyester resin with a small amount of MF resin as a supplemental cross-linker as the binder for color top coats for high performance residential or industrial siding. In the same quality line, the white might well be just a polyester-MF coating. While after many years of outdoor exposure, the white might start to chalk slightly, this would not adversely affect the appearance. However, even a small amount of chalking of a color coating makes an easily seen change of color due to the change in surface reflection. Such changes are particularly serious in exterior siding, since the exposure varies depending on the location on the building. The resulting nonuniform chalking of the color coating is very evident. For the highest exterior durability, fluorinated resin coatings are used. (See Section 16.1.3.) In some cases, such coatings show only slight indication of change after exposure outdoors for more than 25 years.

Organosol and plastisol coatings are used for some coil coating. The relatively low viscosity organosols are applied at about 25 μm, and the higher viscosity plastisols are used for film thicknesses of 50 μm or more. As discussed in Section 16.1.2, the vehicle in these coatings is a dispersion of vinyl chloride copolymer in plasticizer and solvent. Such coatings provide reasonable exterior durability with excellent fabrication properties. They have the advantage that they need not be cross-linked and, hence, can be run with as little as 15 second dwell time in the oven. Solution vinyl resins are used for coil coating metal for can ends for beverage containers.

Increasing use is being made of latex vehicles for coil coating. They have the advantage of high molecular weight, so mechanical properties can be achieved without need for much, if any, cross-linking. High-gloss coatings cannot be made and there can be greater problems of leveling than with solvent-borne coatings. Leveling problems are minimized by using reverse roll coating and by using associative thickeners to control viscosity. Associative thickeners minimize latex particle flocculation relative to conventional water-soluble polymeric thickeners such as hydroxyethyl cellulose. (See Sections 23.2 and 31.3.)

Powder coatings are also being applied to coil stock. One approach, which is in limited commercial use, is to electrostatically spray the strip with automatic spray guns. A disadvantage of this process is that line speeds are slow. Another process being introduced is to run the coil strip through a "cloud" of electrically charged powder particles; the strip then passes into an induction heating oven for fusing and curing [41]. Since there is no contact of rollers with the coil, embossed or perforated metal can be coated. It is projected that line speeds can be higher than for conventional coil coating.

29.4.1. Advantages and Limitations of Coil Coating

Important advantages of coil coating have promoted the growth of the business to a major component of industrial coatings. For long production runs, the cost is low compared to coating preformed metal. The rate of application of coating is faster, and labor cost is lower. Coating utilization is essentially 100%. Oven designs are such that energy usage in curing is much more efficient. In general, floor space requirement is much less so capital cost for buildings is lower. Since solvent is restricted to the immediate area of a roll coater, fire hazards and toxicity hazards are reduced as compared to spray application. Since in most cases, solvent is incinerated, VOC emissions are generally very low. Film thickness of the applied coatings is more uniform than is generally obtained in coating preformed products. Since the coatings are applied on uniform thickness metal, curing of all parts of the coating tends to be more uniform than in curing coatings on fabricated products. For many applications, the performance of coil coatings is superior. This difference is particularly evident when comparing high quality precoated exterior siding to house paint. A large part of this difference results from the difference between performance of baked coatings as compared to air dry coatings. The manufacturer using precoated metal gains some substantial advantages. The VOC emissions and fire hazards associated with coatings application are eliminated, and insurance costs drop. There is no waste disposal problem with sludge from spray booths. There may be a substantial saving in floor space.

There are limitations to coil coating. Capital cost of a line is very high; therefore, lost production time in shutdowns is expensive. It is economical only for fairly long runs of the same color and quality of coating. The cost of changing color is high because the coater must be shut down for cleanup. However, many modern lines have multiple coating heads, which substantially reduce downtime due to color changes. In a line of coated metal with several colors, the inventory cost can be high, since inventory of several colors must be kept. If a stylist changes colors, the obsolete inventory cost can be high or, said another way, the flexibility of changing colors is more limited than when the assembled product is coated.

A challenge to a coil coating supplier is color matching. Generally, very close color matches are needed. Hiding is less than complete, and the color of the metal or primer affects the color of the coating. Color can also be affected by the high temperature baking schedule. It is not possible to duplicate a curing schedule of 30 seconds with high velocity 400°C air in the laboratory. A color matcher must learn to compare color differences that can be expected in the laboratory with what will happen on a particular coil coating line and then do the color matching taking the difference into consideration.

The coated metal must be able to withstand fabrication into the final product without film rupture. This may require acceptance of somewhat softer, more flexible films than could be specified if the product were coated after fabrication. When the coated metal is die cut to make the eventual product, bare edges of metal are exposed. Welding of coated metal can be a problem. Coatings that do not interfere with the welding process are available, however, the appearance of the coated surface is destroyed at and near the welded area. In some applications, bare edges are not a problem and the product is not welded. In other cases, it is critical for the designer of the finished product to ensure that the cut edges do not show and are not in locations where they will be subject to corrosion and that welding, if any, will be done in areas that will not show.

Examples of large applications for coil coated metal are: siding for residential use, original siding for mobile homes, venetian blinds, rain gutters and downspouts, fluorescent light reflectors, appliance cabinets, can ends, and can bodies for fruits and vegetables.

GENERAL REFERENCES

B. N. McBane, *Automotive Coatings*, Federation of Societies for Coatings Technology, Blue Bell, PA, 1987.

J. E. Gaske, *Coil Coating*, Federation of Societies for Coatings Technology, Blue Bell, PA, 1987.

G. Fettis, Ed., *Automotive Paints and Coatings*, VCH Verlagsgesellschaft, Weinheim, Germany and New York, 1995.

REFERENCES

1. M. S. Reich, *Chem. Eng. News*, **75**, 36 (1997).
2. B. M. Perfetti, *Metal Surface Characteristics Affecting Coatings*, Federation of Societies for Coatings Technology, Blue Bell, PA, 1994.
3. J. J. Wojtkowiak and H. S. Bender, *J. Coat. Technol.*, **50** (642), 86 (1978).

4. S. Maeda, *J. Coat. Technol.*, **55** (707), 43 (1983).
5. B. N. McBane, *Automotive Coatings*, Federation of Societies for Coatings Technology, Blue Bell, PA, 1987.
6. C. K. Schoff, *J. Coat. Technol., 62* (789), 115 (1990).
7. E. Ladstadter and W. Gessner, *Proc. Intl. Conf. Org. Coat. Sci. Technol.*, Athens, 1986, p. 203.
8. C. D. Cheever and P.-A. P. Ngo, *J. Coat. Technol.*, **61** (770), 65 (1989).
9. D. C. van Beelen, R. Buter, C. W. Metzger, and J. W. Th. Lichtenbelt, *Proc. Intl. Conf. Org. Coat. Sci. Technol.*, Athens, 1989, p. 39.
10. K. Tachi, C. Okuda, and S. Suzuki, *J. Coat. Technol.*, **62** (782), 43 (1990).
11. G. T. Weaks, *Proc. ESD/ASM Adv. Coat. Technol. Conf. 1991*, 201 (1991).
12. D. R. Bauer, L. M. Briggs, and R. A. Dickie, *Ind. Eng. Chem. Prod. Res. Dev.*, **21**, 686 (1985).
13. L. G. Boggs, M. Rivers, and S. G. Bike, *J. Coat. Technol.*, **68** (855), 63 (1996).
14. M. Broder, P. I. Kordemenos, and D. M. Thomson, *J. Coat. Technol.*, **60** (766), 27 (1988).
15. C. B. Fox, *Proc. ESD/ASM Adv. Coat. Technol. Conf. 1991*, 161 (1991).
16. A. J. Backhouse, *J. Coat. Technol.*, **54** (693), 83 (1982).
17. Z. Vachlas, *J. Oil Colour Chem. Asoc.*, **72**, 139 (1989).
18. I. Wagstaff, *Proc. ESD/ASM Adv. Coat. Technol. Conf. 1991*, 43 (1991).
19. M. J. Chen, A Chaves, F. D. Osterholtz, E. R. Pohl, and W. B. Herdle, *Surface Coat. Intl.*, **79**, 539 (1996).
20. J. D. Nordstrom, *Proc. Waterborne Higher Solids, Powder Coat. Symp.*, New Orleans, 1995, p. 492.
21. B. V. Gregorovich and I. Hazan, *Prog. Org. Coat.*, **24**, 131 (1994).
22. J. L. Gardon, *J. Coat. Technol.*, **65** (819), 24 (1993).
23. A. Essenfield and K. J. Wu, *Proc. Waterborne High-Solids Powder Coat Symp.*, New Orleans, 1997, p. 246.
24. J. W. Rehfuss and D. L. St. Aubin, U.S. Patent 5356669 (1994); J. W. Rehfuss and W. H. Ohrbom, U.S. Patent 5373069 (1994).
25. J. L. Courter, *J. Coat. Technol.*, **69** (866), 57 (1997).
26. R. A. Ryntz, *Proc. Waterborne High-Solids Powder Coat. Symp.*, New Orleans, 1997, p. 259.
27. J. C. Kenny, T. Ueno, and K. Tsutsui, *J. Coat. Technol.*, **68** (855), 35 (1996).
28. K. Yanagida, M. Kumata, and M. Yamamoto, *J. Coat. Technold.*, **68** (859), 47 (1996).
29. R. A. Ryntz, *SAE Technical Paper Series 880597*, Society of Automotive Engineers, Warrendale, PA, 1988.
30. T. J. Miranda, *J. Coat. Technol.*, **60** (760), 47 (1988).
31. C. Nootens, P. C. Bouuaert, S. A. Stachowiak, *Surface Coat. Intl.*, **80**, 50 (1997).
32. P. Oberressl, T. Burkhart, D. Chambers, and A. Slocki, *Proc. Waterborne High-Solids Powder Coat. Symp.*, New Orleans, 1997, p. 296.
33. P. J. Palackdharry, *Polym. Mater. Sci. Eng.*, **65**, 277 (1991).
34. M. Hickling, *Polym. Mater. Sci. Eng.*, **65**, 285 (1991).
35. R. A. Dubois and P. S. Sheih, *J. Coat. Technol.*, **64** (808), 51 (1992).
36. J. T. K. Woo, V. Ting, J. Evans, R. Marcinko, G. Carlson, and C. Ortiz, *J. Coat. Technol.*, **54** (689), 41 (1982).
37. J. T. K. Woo and R. R. Eley, *Proc. Water-Borne Higher-Solids Powder Coat Symp.*, New Orleans, 1986, p. 432.
38. R. J. Wingender, P. Niketas, and C. K. Switala, *J. Coat. Technol.*, **70** (877), 75 (1998).
39. S. R. Howe, L. Borodinsky, and R. S. Lyon, *J. Coat. Technol.*, **70** (877), 69 (1998).
40. J. E. Gaske, *Coil Coatings*, Federation of Societies for Coatings Technology, Blue Bell, PA, 1987.
41. F. D. Graziano, *Proc. Intl. Conf. Org. Coat.*, Athens, 1997, p. 139.

CHAPTER 30

Product Coatings for Nonmetallic Substrates

Many products made from wood and plastics are coated in factories; examples are wood furniture and paneling, hardboard paneling and siding, and plastic auto body parts and computer cases. Do-it-yourselfers paint wood products at home, but the products they use are generally different and are discussed in Chapter 31.

30.1. COATINGS FOR WOOD FURNITURE

There are many styles of furniture, and furniture coatings are affected by styling. Furniture styling and manufacture are quite different in the United States than in most other countries in the world; most of our discussion deals with the U.S. market. A large fraction of wood furniture made in the United States is styled to resemble antiques, especially French, English, and Spanish antique furniture styles. Styling is generally initiated in the high end of the market, the expensive wood furniture that involves a great deal of hand labor and artistry in its manufacture. A representative finishing process is described in some detail since a major part of furniture is finished to resemble high style furniture, but uses manufacturing and finishing techniques that permit lower costs.

Fine quality furniture is made from a combination of woods. The tops and sides are made from five-ply plywood, the legs and rails are made from solid wood, and carved wood decorative pieces are commonly attached to the furniture. The center ply of the plywood can be wood, but other cores such as chipboard are used. Chipboard is made by pressing wood chips and an adhesive binder into sheets and curing the sheets. Chipboard is a desirable core material because it is more dimensionally stable than wood. A ply of wood veneer is laid up with glue on each side of the core. Another layer of veneer is laid up on each side, cross-grain to the first layer. The top, or face, veneer, the side seen on the furniture, is usually from a hardwood selected for the beauty of its grain pattern.

Grain patterns are affected by the kind of tree from which the wood is obtained and by the way the veneer is cut from a log. The top of a table is generally too large for the face veneer to be made from only one piece; it is laid up from several pieces. They are carefully selected and put together to give a particularly beautiful pattern. Adhesive is needed not only between the layers in the plywood, but also to connect the edges of the individual pieces of veneer in the top face. Each tabletop is unique because the grain patterns in the separate pieces of veneer are different. In many cases, elaborate patterns are laid up. Many kinds of wood are used: pecan, walnut, and oak are among the most commonly used; rosewood is an example of a rarer, more expensive wood popular because of esthetic grain patterns. All these hardwoods have open pore structures and prominent grain patterns. In some styles, mahogany, which has straighter grains and shallower pores, has been popular. Woods that have little pore structure, like maple and birch, have a more limited market than those with bold patterns.

In U.S. furniture manufacturing, the various components are cut for a run of the same style and sets of furniture. The furniture is assembled before it is finished.

The various components are of different colors; if the final color of the furniture is to be lighter than any part of any of the component pieces, the wood is *bleached* with a solution of 30% hydrogen peroxide; the bleaching solution is activated with sodium or potassium hydroxide.

In earlier times, a dilute solution of hide glue was sprayed on the surface. Now, low solids poly(vinyl alcohol) solutions are used. After the size has dried, the wood fibrils that stick up from the surface are stiffened so that the surface can be smoothed by sanding. *Glue sizing* minimizes concentration of stain at the fibrils, which would give an unattractive appearance. With some woods, sanding without glue sizing gives satisfactory results.

Next, the wood is colored with a solution of acid dyestuffs in solvent or dispersions of transparent pigments. The *stain* is selected and applied to give the overall base color to the furniture the stylist has selected. Spraying stain requires skill to obtain even coloration.

A *wash coat* is a low solids, less than 12 wt% (NVW), low viscosity lacquer. A thin layer is applied to minimize displacement of the stain, stiffen wood fibers for sanding, and prepare the surface for the next step, filling.

The purpose of *filling* is to color the pores of the wood to emphasize the grain pattern and to fill the pores to near the same level as the rest of the wood. The pores must be filled with the colored filler without leaving filler on the surfaces between the pores. Usually, the color of the filler is a dark brown and the stains are lighter yellowish or reddish browns, but for special effects, one can use white filler with black stain or other color combinations. The vehicle is linseed oil and/or a linseed long oil alkyd with a hard resin such as limed rosin and driers with aliphatic hydrocarbon solvents. The color pigments used for dark colors are natural earth pigments with a high loading of inert pigments at a PVC over 50%. The whole piece of furniture is sprayed with a liberal coat of filler, the filler is "padded" into the pores by rubbing vigorously with a pad of cloth, and then excess filler is wiped off. If the wood has been properly wash coated and the filler is wiped evenly when the degree of solvent flash off is right, the pores of the wood are filled and no filler is left between the pores. Adequate drying of the filler is critical because residual solvents can result in shrinkage, graying, and loss of adhesion of succeeding coats.

A *sanding sealer* is then applied. The purpose of the sanding sealer is to prepare the surface for application of top coat. The sealer must be easy to sand smooth without "gumming" the sandpaper. A formulation for a typical sanding sealer includes nitrocellulose; a hard resin, such as a maleated rosin esterified with a polyol; and a plasticizer such as blown soy bean oil (see Section 14.3.1). Sanding sealers contain 3–7% of zinc stearate, based on the lacquer solids to aid sanding. After diluting with lacquer thinner to spray viscosity (about 20% volume solids), the sealer is applied and, after drying, is sanded smooth.

Wood is not naturally as uniform in color as the overall staining makes it. By spraying different shades of stains to selected areas, the color can be varied and the grain highlighted. Shading is a skilled art; in recent years *shading stains* have been replaced in large measure by *padding stains*. Padding stains are made with similar dyes, but have some binder and somewhat slower evaporating solvents; they are applied by hand with a rag moistened with the stain, and less application skill is required.

The furniture is then sprayed with *top coat*. Nitrocellulose is used as the primary binder in high-style furniture lacquers for three major reasons. First, the appearance of furniture finished with nitrocellulose lacquers is outstanding; the lacquers provide an appearance of depth, fullness, and clarity of the grain pattern that has not been matched by any other coating. Second, lacquers dry quickly, so they can be rubbed a short time after application and then packed and loaded into a truck for shipment without *printing*. The term printing in this context means surface imperfections resulting from wrapping material denting the surface of a coating that is too soft. Third, lacquers are thermoplastic and permanently soluble, so they are easily repaired in case of damage during shipment. The mechanical properties of the final film are better with higher molecular weight nitrocellulose, but solids are reduced. A hard resin such as an esterified maleated rosin is blended with the nitrocellulose. A combination of a short oil coconut alkyd (see Section 15.5) and a plasticizer provide the required flexibility. The balance of nitrocellulose, plasticizer, and hard resin is critical. If the coating is too soft, it will be difficult to rub; if it is too hard, the lacquer will be subject to cracking as the wood expands and contracts with changes in moisture content and exposure to rapid temperature changes. If the lacquer is applied over a pale colored wood, UV absorber (see Section 5.2.1) is added to reduce yellowing of the wood. Citric acid is also commonly added to chelate iron salts that produce reddish colors with phenolic compounds naturally present in wood.

After the initial coat of top coat is applied the next step is *distressing*. Distressing uses a variety of techniques to give an antique appearance to furniture. Stylists are not trying to make fake antiques; rather, the furniture is styled to have an antique appearance. In the old days, quill pens with black india ink were used, resulting in drops of ink falling on tabletops. Now little drops of black pigment glazing stain are applied. Sometimes, a colonial forefather was a little tipsy, when he sat down to write and he put the feather end of the quill pen into the inkwell and then set it down on the tabletop. Sometimes, colonials walked on their tables with hob nailed boots or banged the tabletop with their pewter beer mugs. It is startling to walk by the finishing line in a fine furniture factory and see someone swinging a chain at the tops of those carefully prepared pieces of furniture. Another common thing to see is a little black pigmented stain rubbed into inside corners. Over the years in antique furniture, dirt accumulated in corners, so the new furniture is made permanently dirty in these areas. Proper distressing requires artistic skill.

Finally, another coat of top coat is applied and the lacquer is dried, usually in a force dry oven at 40–60°C. Then the lacquer surface is *rubbed*, first by sanding with fine sandpaper and a lubricant and then rubbing with cloth and rubbing compound. The result is a soft appearing, low gloss finish.

While solubility of nitrocellulose lacquer finishes is an advantage in that it permits easy repair, it can also be a disadvantage. If solvents, such as nail polish thinners are spilled on a lacquer surface, the finish will be marred. A way of solving this problem and still obtaining the beauty of a nitrocellulose finish is to add a polyisocyanate cross-linker to the lacquer just before application. The isocyanate cross-links with the hydroxyl groups on the nitrocellulose. All solvents must be urethane grade, that is, contain no more than traces of alcohols and water, and the nitrocellulose must be wet with plasticizer rather than with isopropyl alcohol.

Wood furniture produced by this process is beautiful, but expensive. A much larger part of the market is for furniture designed to look as closely as possible to this high-style furniture, but which can be manufactured and finished at lower costs. There are many approaches with varying degrees of cost savings; an intermediate cost approach is described next.

Instead of using plywood tops and sides made with expensive hardwood veneers that must be painstakingly finished, one can use printed tops and sides with solid wood legs and frames. *Particle board*, that is, pressed wood chips and particles with a resin binder, is first filled with a UV cure filler (see Section 28.3). Note that this is a very different filler than referred to above for filling the pores in wood. Particle board filler is applied over the entire surface, and after curing is sanded smooth. A UV cure filler consists of an unsaturated polyester-styrene (or acrylated) vehicle pigmented with inert pigments that absorb little UV radiation. The filler is applied to the particle board with a roll coater that has a brush roller. Shrinkage occurs during UV curing, but the shrinkage is much less than results from solvent evaporation from a filler with a high volatile content. As a result, only one coat of UV cure filler is needed. The inert pigment minimizes the problem of oxygen inhibition.

The filled, sanded panel is next coated with a *base coat*, a highly pigmented nitrocellulose lacquer whose color becomes the overall underlying color of the piece of furniture; corresponding to the color of the stain in high-style furniture. The stain used on the solid wood parts approximately matches the color of the base coat. The base coat is then ready for offset gravure printing. The gravure printing cylinders are made from photographs taken of carefully selected hardwood top veneer plywood that was stained, filled, shaded, and distressed by highly skilled finishers. One cylinder prints the darkest tones of the original finished wood, a second prints the medium depth tones, and the third prints the lightest tones. The printing ink colors for the three cylinders are selected to approximate the colors of the three depths of color on the original. The inks are pigment dispersions in plasticizer, so they adhere well to the base coat and to subsequent lacquer coats. A light coat of lacquer is applied to protect the print.

The frame and legs of the furniture are assembled and finished up to the stage of applying top coat, followed by assembly of the printed tops and sides onto the furniture. Then, the whole piece of furniture is coated with semigloss nitrocellulose lacquer top coat. Semigloss lacquer, not gloss lacquer, is used to give a finish with a *hand-rubbed effect* without extensive hand rubbing. The semigloss is obtained by pigmenting with a low level of fine particle size SiO_2, such that the PVC of the final film averages 2–4%. When the solvent evaporates after application, convection currents resulting from solvent loss carry the pigment to the surface of the film, where it is "trapped" in the viscous surface layer. As a result, the PVC of the top layer of the film is high enough that low gloss can be obtained, while the overall PVC is low enough that only a minor degree of light scattering occurs.

A disadvantage of furniture finishing systems discussed so far is that they require use of low solids coatings with very high VOC emissions. The low solids requires multiple coats, or at least more passes with a spray gun, so application cost is high. Hot spray (see Section 22.2.4) is frequently used. A temperature of 65°C permits increasing solids from 20–24 NVW to 28–34 NVW. This gives a significant reduction in VOC emissions, but still far short of probable future requirements.

There have been efforts for many years to replace nitrocellulose lacquers to achieve higher solids. Alkyd-urea-formaldehyde (UF) top coats, frequently called *conversion varnishes* or *catalyzed finishes*, have been used for many years. Tall oil alkyds with a butylated UF resin are typical. Just before use, one adds about 5% of *p*-toluenesulfonic acid catalyst, based on UF resin solids. A flatting pigment dispersion and a small amount of poly(dimethylsiloxane) to minimize orange peel (see Section 23.2) are usually added. The top coated furniture is force dried at 65–70°C for 20–25 minutes. The NVW of this formulation is 38.5%, about twice that of lacquers. Hot spray can further increase the solids, but pot life must be carefully watched, and as the solids increase, low gloss becomes more difficult to achieve. The appearance of depth is not as good as obtained with a nitrocellulose lacquer, but is still presentable. The coating is solvent resistant and more heat and gouge resistant than lacquer coatings. Repair is more difficult, but less frequently needed. High solids coatings based on isocyanate chemistry are being developed. Aldimine/isocyanate two package (2K) 90% solids gloss clear coats (see Section 10.4) have been formulated that have a pot life of about 3 hours and a dry time of about $\frac{1}{2}$ hour [1]. Conversion varnishes have been most widely used on furniture for commercial use and for kitchen cabinets.

Considerable effort has been expended on water-borne finishes for wood furniture. Application of a water-borne coating directly on wood leads to excessive grain raising, which may limit their use to applications where there is already a solvent-borne sealer on the wood. Substantial reductions of solvent usage are possible by emulsifying nitrocellulose lacquers into water [2]. Lacquers with VOC of 300–420 g L^{-1}, excluding water, can be made compared with VOC on the order of 750 g L^{-1} for conventional NC lacquers. Solids of the NC solution internal phase can be maximized by using only true solvents, esters and ketones. Short oil alkyds using ester solvents (other than the small amount of xylene needed for azeotroping in producing the alkyd) are selected as plasticizers. Water wet NC is used, rather than conventional isopropyl alcohol wet NC. A surfactant is incorporated in the lacquer as an emulsifying agent; sodium alkylphosphate surfactants are recommended. The advantage of easy strippability for repair is retained. However, long dry time and slow development of print resistance have limited their use. Combined nitrocellulose-acrylic latex-based lacquers have been recommended. Furniture sealer and top coat formulations with VOC in the 240–400 g L^{-1} range have been described [3]. Film physical and appearance properties are reported to be superior to acrylic latex counterparts.

Water-reducible acrylic resin based coatings (see Section 12.3) have been used. At a sufficiently high molecular weight, they can be used as thermoplastic coatings. Thermosetting resins can be cross-linked with polyaziridine cross-linkers. (See Section 16.6.3.) There is concern about the toxic hazard of aziridine cross-linkers in spray application. Methylated urea-formaldehyde resins can be used as cross-linkers under force dry conditions. Thermosetting latex top coats are also used. (See Section 8.1.4.)

Two package water-borne polyurethane top coats (see Section 10.8 for discussion of various types) are increasingly being used, since they provide low VOC emissions with a minimum of HAP solvents. The films have excellent abrasion resistance. The cost is relatively high, and special application equipment is required [4].

Solvent-free, water-borne 2 K epoxy-amine furniture coatings were compared to solvent-borne coatings in a study sponsored by the EPA [5]. Many of the film properties were comparable, but appearance would limit the types of application for which this water-borne coating would be satisfactory.

In Europe and to a more limited degree, in the United States, UV cure acrylic top coats are used. (See Section 28.3.) They have the advantage of little or no VOC emission. They are applied before assembly by roll coating on tops already finished up to the top coat. The coatings are gloss or high semigloss, are solvent resistant, and have excellent mechanical properties. Use in the United States is limited, since most furniture there uses low semigloss coatings and is finished after assembly.

Limits of 275 g L^{-1} of VOC for coatings for wood furniture took effect in California in 1994. As VOC regulations become more stringent, it seems probable that a larger fraction of furniture tops will be high pressure laminates with wood grain prints. Laminates have been used for many years on commercial and institutional furniture; now their use in household furniture is increasing.

30.2. PANEL AND SIDING FINISHING

Large quantities of coatings are applied to plywood and hardboard sheets to make precoated interior paneling and exterior siding. A small part of the market is for finishing expensive, high quality hardwood plywood used for paneling executive offices. This paneling is finished in essentially the same way used for the highest end of the wood furniture market. A variety of other products constitutes the bulk of the market.

One class of wall paneling is made using three-ply plywood for which the top veneer is a relatively featureless, low cost luan. The panel is dried to remove surface water, grooved to make a plank effect, and then sanded. A dark colored lacquer is applied to the grooves by pinpoint automatic spray guns and the panel is dried. A lacquer sealer colored with dyes is applied. The panel is sanded again and then printed with two or three prints to give the appearance of walnut, rosewood, or some other attractive wood with prominent grain patterns. Finally, a low gloss top coat is applied. The coatings have generally been nitrocellulose lacquers. While this business was first developed in the United States, most production has shifted overseas, initially to South Korea and more recently to the Philippines. The predominant source of the wood is the Philippines, so finished product rather than raw material can be shipped; also air pollution regulations in South Korea and the Philippines are less stringent.

A related, but quite different, application is for "door skins." In the manufacture of interior doors, a common approach is to assemble the door and then laminate a top veneer (the *skin*) to the particle board surface. The skins are veneer, laid up and adhered to kraft paper. The veneer is commonly birch or luan, sometimes printed with a hardwood grain pattern. Commonly, the coating on door skins is gloss or high semigloss. Whereas low gloss is preferred on wall paneling to avoid glare from reflection of lights, higher gloss is preferred on doors because it is easier to clean. Since the surface area of doors is relatively small, glare is not a serious problem. This combination of factors has led to the widespread use of UV cure coatings for door skins.

Part of the interior paneling market is precoated hardboard. Hardboard is made by mixing wood fibers and shreds then curing in hydraulic presses. The lignin in the wood acts as a binder to hold the wood fibers together; sometimes the lignin is supplemented with phenolic resins. Depending on the pressure applied, the density of hardboard can be varied. Furthermore, if the surface of the steel platen against which the upper face of the hardboard is formed has a pattern, the negative of that pattern is embossed into the surface of the hardboard. A variety of hardboards is made. Using smooth surface hardboard, one can apply a base coat, prints, and a semigloss top coat to make 4 by 8 foot panels that look like any kind of wood. Smooth surface hardboard is usually coated by curtain coating (see Section 22.5), since this gives level films.

Tongue and groove paneling can be simulated by routing out grooves and painting them a dark brown color. The paneling is finished by roll coating a base coat that does not flow appreciably into the grooves, followed by printing and top coating. A pattern of ticks (short line dents scattered over the surface, as in hardwoods) can be embossed. When the board is finished, it not only has the grain of the wood, it also has the little dips of a tick pattern as well as the grooves to give the effect of planked paneling.

One can simulate wood with holes, like pecky cypress. The hardboard is embossed with the hole pattern copied from real pecky cypress paneling. The *hole coat*, generally a dark brown, is applied using roller brushes to assure that the coating completely covers surfaces inside the holes. Then a base coat is applied by *precision coating*. A precision coater is a roll coater in which the application roll is a gravure printing roll uniformly covered with cells so that the upper surface of the panel is coated, but no coating is applied in the holes. Generally, the color is a relatively light brown to provide a contrasting background color. Then, grain pattern prints are applied, followed by a semigloss top coat. One is not restricted to replicating pecky cyprus or wormy chestnut; brick, stone, travertine marble, etc. can be simulated. Other interior hardboard paneling is designed primarily for bathrooms. The coatings are high gloss pigmented coatings. Sometimes, joints are embossed in to make the panels look like tiles. Sometimes, prints such as marble are used with a clear gloss top coat.

Since, in contrast to plywood, hardboard can withstand high temperatures, baking coatings, with all their advantages in performance properties, are used. Since hardboard does not undergo grain raising like wood, water-reducible acrylic-MF finishes are used. Although average film thickness of a base coat is relatively high, popping is not a major problem, since the coating thicknesses applied by roll coating and curtain coating are much more uniform than when coatings are applied by spray. (See Section 23.7.)

High-density hardboard is also used for exterior siding. The largest volume of siding is factory primed with a primer designed to have at least 6 months exterior durability before exterior house paint is applied. (See Section 31.1.) Acrylic-MF binders are used in the primers, generally water-borne to reduce VOC. Challenges in formulating these primers are to be sure that the coating cross-links sufficiently that the coated boards coming out of the oven can be stacked without blocking and also that the paint to be applied to them in the field will have good adhesion. The combination requires careful control of cross-link density. The coatings are low gloss because high loading with inert pigment reduces cost and enhances adhesion of paint to the surface. The largest volume is used in siding for tract houses. The primed siding is erected on to the house, but painting is postponed until the house is sold so that the buyer can pick the paint color. An extensive study of exterior durability of different hardboards with different combinations of coatings has been published [6]. The best results were obtained when preprimed board was field coated with an alkyd primer and acrylic latex top coats. Petrolatum and other oily substances are present in some hardboards. If the paint applied on the siding is porous, over time, some substances can migrate to the upper surface of the paint, leading to discoloration or sticky areas that will pick up dirt to give a blotchy appearance. The problem can be minimized by using only paints with PVC < CPVC.

Fully prefinished exterior siding is manufactured on a smaller scale than preprimed siding. The primed board is top coated in the factory with a baking acrylic enamel or, for greater exterior durability, a silicone-modified acrylic resin based coating. (See Section

16.5.3.) The durability is superior to field applied paint, but flexibility in color selection for tract homes is sacrificed. Fully prefinished siding is more commonly used on commercial buildings.

30.3. COATING OF PLASTICS

Plastics have become a major substrate for coatings. Design of the coatings is complicated by the variety of polymers used in plastics and the range of approaches that are possible for decorating the surfaces. Reference [7] has several useful chapters discussing various approaches; Reference [8] reviews coatings for plastics. For molded plastics, there are two alternatives: (1) *in-mold coating*, in which the inside of a mold is coated and then the plastic material put into the mold-when the part is taken out of the mold, the outer surface is the coating; or (2) *post-mold coating*, in which the molded part is coated.

30.3.1. In-Mold Coating

Polar thermosetting plastics can be in-mold coated; a variety of coating materials is used for in-mold coating, depending on the plastic. To assure good adhesion to the plastic, it is desirable to use compositions related to the composition of the plastic, for example, the use of *gel coats* with glass-reinforced plastic parts. Gel coats are pigmented styrene-polyester resins (see Section 16.3) that are cured along with the resin in the molding compound so that the coating is chemically bound to the main body of the plastic. The molding resin is usually a styrene solution of a phthalic anhydride, maleic anhydride, propylene glycol polyester. For some products, similar polyesters are used in the gel coats. More expensive isophthalic neopentyl glycol resins are used in gel coats when superior hydrolytic and exterior durability are needed. In-mold coating with gel coats is used in making glass reinforced styrene-polyester molded products such as boats and shower stalls.

With polyurethane RIM (reactive injection molding) parts, in-mold coatings with free hydroxyl groups permit reaction with isocyanate groups in the molding compound. Rigid urethane steering wheels are made by coating the inside of the mold with the appropriate color enamel. Rigid urethane foam parts to replace wood carvings for furniture are in-mold coated. The inside of the mold is sprayed with a lacquer, which is color matched to the base coat of the furniture. (See Section 30.1.) Many in-mold coatings were solvent-borne coatings, frequently low solids lacquers. These high VOC coatings are being replaced with higher solids coatings, water-borne coatings, and powder coatings.

For some plastic automobile parts that are to be assembled onto the steel body of the car and painted with the same top coat as the rest of the body, the interior of the mold can be coated with a primer to provide adhesion to the plastic and intercoat adhesion with the top coat. Sometimes, such a primer is made with conductive pigments, such as acetylene carbon black, so that electrostatic spray will be effective over the plastic parts.

A further alternative approach is to use coated films as a laminate for in-mold application of coatings [9]. The coating is reverse roll coated on a heat resistant, smooth, high gloss polyester film. A size coat is applied to the surface of the coating, and then a film of plastic of the same polymer (or a compatible polymer) as the plastic part to be coated is laminated on. The resulting coated film is vacuum formed into a mold, followed by a sheet of the plastic. The part is removed from the mold and the polyester film is stripped off, leaving the coating on the surface of the plastic substrate. Multiple layers of coating can be applied. For example, for finishing plastic parts for use with base coat-clear coat systems, the polyester film would be coated first with the clear coat, followed by the base coat and then laminated to the plastic film and molded.

30.3.2. Post-Mold Coating

Coatings are also applied to plastic articles after they are fabricated. Many of the coatings used for metals can be adapted for coating plastics, and many of the selection criteria are the same as for metals. However, there are important differences. Adhesion, discussed in Section 6.5, is usually the central problem in coating plastics. The coatings must be more flexible than those for metal coatings. This follows from the easier deformability of many plastics and all elastomers; the coatings must be at least as easily deformed as the substrate. As a rough guideline, elongation-at-break (see Section 4.1) of the coating should be greater than that of the substrate.

To obtain good adhesion, the surface of the plastic must be clean. Machining oils, sanding dust, finger prints, etc. must be removed. Many plastic parts have residual mold release on the surface. If a mold release must be used in making the plastic article, it is desirable to use a water-soluble soap such as zinc stearate as a mold release, since it is the relatively easy to remove. Wax mold releases are more difficult to remove, and silicone or fluorocarbon mold releases should not be used for plastic parts that are to be painted. Surface contaminants and mold release are removed by spray washing in three stages. First, a detergent wash is used followed by a water rinse, and finally by a deionized water rinse. At the end of the washer, droplets are removed by compressed air jets, and the part goes into a drying oven before coating.

After the surface is clean, the surface tension varies widely, depending on the type of plastic. In general, thermosetting plastics and thermoplastics with polar structural groups, such as nylons, have relatively high surface tensions, although lower than metals. Often, they can be coated without further surface treatment. However, the surface tension of some less polar plastics, particularly polyolefins, is lower than that of most coatings, and the polarity of the surfaces is very low. Primers that have very low surface tensions and that can also penetrate the surface of the plastic are required. Alternatively, the surface of the plastic can be treated to increase its surface tension and provide polar groups that can hydrogen bond with components of the coating to promote adhesion. (See Section 6.5.)

In some cases, it is possible to promote adhesion to untreated polyolefins by spraying with a thin *tie coat* of chlorinated polyolefin [10]. The chlorinated polymer presumably dissolves partly in the surface of the polyolefin, resulting in a surface to which a broader range of coatings can adhere. Comparisons of results with a variety of treatments and tie coats have been published [11]. Sometimes, the tie coats are pigmented so that the sprayer can judge whether all areas have been tie coated. Conductive tie coats permit use of electrostatic spray in application of subsequent coats. To reduce VOC, higher solids and water-borne tie coats have been developed [9,12].

Instead of applying a special primer, the surfaces can be oxidized to yield polar groups; there are several approaches to oxidation treatment [13]. Oxidizing agents such as chromic acid/sulfuric acid baths are effective and have been used for many years. However, disposal of chromate containing wastes is now tightly controlled. To avoid chromium, sodium hypochlorite with a detergent has been recommended for treating the surface of polyolefin polymers to improve adhesion [14]. The surface can be oxidized by directing an oxidizing flame from propane or butane onto its surface. Care is required to ensure that all surfaces are adequately treated. Flame treatment is widely used in Europe for surface treating plastic parts for automobiles. Another approach to oxidation of the surface is by *corona discharge*. The plastic parts are passed through a cloud of ionized air generated, as described in Section 6.5, by electrodes with many wire ends at high voltage. The ions oxidize the surface of the polyolefin. For irregular shaped parts, the corona discharge can be carried out in a vacuum chamber; this method is frequently called *plasma discharge*. Still another approach is to spray a solution of benzophenone on the surface of a polyolefin and then expose the part to UV radiation. This leads to generation of free radicals that initiate autoxidation of the surface. This process is particularly effective with rubber-modified polyolefins.

From the standpoint of adhesion, it is preferable for the plastic part to have a somewhat roughened surface. Plastics that are highly filled with pigments have rougher surfaces to which it is easier to adhere. Alternatively, the surface of the mold can be designed to impart some roughness to the molded pieces. Adhesion can be promoted by baking at temperatures above the T_g of the plastic. However, in many cases heating, above T_g leads to heat distortion of the molded part. This is one of the advantages of in-mold coating: adhesion is promoted by the heat involved in molding the plastic, avoiding the problem of heat distortion when baking post-coated plastic moldings.

Adhesion can also be promoted by using solvents in the coating that are soluble in the plastic. This reduces the T_g of the plastic at the surface, permitting penetration of binder from the primer into the surface of the substrate. However, very volatile strong solvents must be avoided on high T_g thermoplastics such as polystyrene and poly(methyl methacrylate); such solvents cause *crazing*, that is, the development of a network of microcracks in the surface of the plastic. A possible explanation for crazing is that the solvent penetrates into the plastic; then, if it is very volatile, it evaporates rapidly from the surface while there is still solvent left just below the surface. When the solvent diffuses out from below the surface, there is a reduction in volume, leading to stresses sufficient to crack the surface layer of the plastic. Another potential problem with using penetrating solvents is the possibility of solvent popping caused by release of solvents from the plastic after cross-linking of the coating is well advanced. In a study of factors that affect popping on sheet molding compound parts (molded fiberglass reinforced polyester plastics), it was found that the problem can be minimized by using primers with the lowest possible solvent permeability [15]. Permeability can be reduced by using highly cross-linked or partially crystalline binders in the primer. While solvents can improve adhesion, care must be exercised in the selection of solvents for plastics such as TPO (thermoplastic polypropylene blended with rubber) so that the crystallinity of the plastic is not disturbed, since that would affect the cohesive strength of the plastic [16]. The crystallinity and cohesive strength of TPO is also affected by the baking temperature used for curing the coatings [17].

When painted TPO plastic automobile bumpers rub together or scrape against a rigid object, delamination has been a serious problem. In many cases, the friction induced damage results from cohesive failure near the surface of the plastic. New tests have been developed to investigate the problem [18]. (See Section 6.6.) Coating composition, application conditions, and type of TPO plastic are among the variables that affect performance. TPO in which the rubber in the compound is chemically bonded to the polyolefin was less susceptible to damage as compared to TPO in which the rubber is physically mixed with the polyolefin [18].

Electrostatic spraying of plastics is difficult, since the plastics are usually insulators and, therefore, cannot be adequately grounded to give the charge differential needed to attract the charged spray droplets to the surface. Surface resistivity of plastics and time for charges on their surfaces to dissipate can be measured and correlated with the feasibility of electrostatic spraying [19]. The charge dissipation time is affected not only by the composition of the plastic, but also by the humidity in the spray booth. Conductive tie coats or primers can be applied to the plastic; the effect of a range of conductive primers on the charge dissipation time has been studied [16]. It has also been shown that the effect of surface charges on the plastic substrate in the electrostatic spraying can be reduced by placing a continuous grounded metal backing in contact with the plastic substrate [20]. When the object being sprayed has both plastic and metal components there can be nonuniform deposition of coatings even when a conductive primer has been applied to the plastic, since the metal can distort the electrostatic field near the plastic-metal interfaces.

The lower curing temperatures possible with two package polyurethane coatings make them particularly attractive for many applications to minimize the possibility of heat distortion of the plastic product [21]. The higher abrasion resistance and flexibility available with urethane coatings are frequently also advantageous. While acrylic resins can be cross-linked with polyisocyanates, higher solids are possible with hydroxy-functional polyesters. For example, a polyester resin prepared from adipic acid, isophthalic acid, neopentyl glycol, and trimethylolpropane in a mole ratio of $1:1:2.53:0.19$ with $\bar{M}_n$ of 730 and $\bar{f}_n$ of 2.09 hydroxyl groups per molecule was cross-linked with various trifunctional aliphatic isocyanates at a $1.1:1$ NCO:OH ratio [21]. The coating provided a good balance of properties on a variety of plastics, including good impact resistance at $-29°C$.

A further approach to coating plastics subject to heat distortion is radiation curing. (See Chapter 28.) The temperatures involved in radiation curing are generally not substantially above ambient and can be further reduced by filtering out IR radiation. For example, UV cure gloss top coats are applied to plastic flooring. The low temperature curing is particularly important in coating flooring, which has an embossed foam decorative layer as the upper surface; elevated temperatures could collapse the foam.

In many applications, the function of the coating is decorative, such as for color, but there are many examples of functional coatings for plastics. The UV cure top coat applied to plastic flooring mentioned earlier has the functions of increasing wear-life, stain resistance, and gloss retention. Coatings are applied to polyethylene tanks to reduce permeability. Magnetic coatings are applied to tapes and sheets to make recording tapes.

Transparent plastics can serve as replacements for glass for applications ranging from window glazing to eyeglasses, but they are less abrasion resistant than glass. (See Section 4.3.) Surface coatings to improve abrasion resistance have been developed based on alkoxysilanes and colloidal silica [22]. The coatings provide excellent abrasion resistance to plastics such as polycarbonate, but have relatively high VOC levels and long curing times. Radiation cure coatings have been developed with low VOC that cure in a few seconds. (See Section 28.3.)

GENERAL REFERENCE

R. A. Ryntz, *Adhesion to Plastics: Molding and Paintability*, Global Press, Moorhead, MN, 1998.

REFERENCES

1. M. J. Dvorchak, S. D. Hicks, T. D. Wayt, and D. A. Wicks, *Coat. World*, November–December, 28 (1997).

2. C. M. Winchester, *J. Coat. Technol.*, **63** (803), 47 (1991).

3. H. F. Haag, *J. Coat. Technol.*, **64** (814), 19 (1992).

4. M. J. Dvorchak, *J. Coat. Technol.*, **69** (866), 47 (1997).

5. E. W. Huang, R. Guan, and R. C. McCrillis, *Coat. World*, November–December 21, 1997.

6. W. Bailey, S. Bussjaeger, N. F. Dispensa, G. Early, M. Froese, R. Haines, A. Moser, L. J. Murphy, and M. A. Trigg, *J. Coat. Technol.*, **62** (789), 133 (1990).

7. D. Satas, Ed. *Plastic Finishing and Decoration*, Van Nostrand Reinhold, New York, 1986.

8. R. A. Ryntz, *Painting of Plastics*, Federation of Societies for Coatings Technology, Blue Bell, PA, 1994.

9. C. H. Fridley, *Proc. Soc. Mfg. Eng, Finishing '91 Conf.*, FC91-374, 1991.

10. R. J. Clemens, G. N. Batts, J. E. Lawniczak, K. P. Middleton, and C. Suss, *Prog. Org. Coat.*, **24**, 43 (1994).

11. R. A. Ryntz, *J. Coat. Technol.*, **63** (799), 63 (1991).

12. J. E. Lawniczak, P. J. Greene, R. Evans, and C. Suss, *J. Coat. Technol.*, **65** (827), 21 (1993).

13. J. M. Lane and D. J. Hourston, *Prog. Org. Coat.*, **21**, 269 (1993).

14. H. F. Haag, U.S. Patent 5053256 (1991).

15. R. A. Ryntz, W. R. Jones, and A. Czarenecki, *J. Coat. Technol.*, **64** (807), 29 (1992).

16. R. A. Ryntz, Q. Xie, and A. C. Ramamurthy, *J. Coat. Technol.*, **67** (843), 45 (1995).

17. R. A. Ryntz, *Prog. Org. Coat.*, **27**, 241 (1996).

18. A. C. Ramamurthy, J. A. Charest, M. D. Lilly, D. J. Mihora, and J. W. Freese, *Wear*, **203–204**, 350 (1997); D. J. Mihora and A. C. Ramamurthy, *Wear*, **203–204**, 362 (1997).

19. D. P. Garner and A. A. Elmoursi, *J. Coat. Technol.*, **63** (803), 33 (1991).

20. A. A. Elmoursi and D. P. Garner, *J. Coat. Technol.*, **64** (805), 39 (1992).

21. S.H. Shoemaker, *J. Coat. Technol.*, **62** (787), 49 (1990).

22. J. D. Blizzard, J. S. Tonge, and L. J. Cottington, *Proc. Water-Borne Higher-Solids Powder Coat. Symp.*, New Orleans, 1992, p. 171.

CHAPTER 31

Architectural Coatings

In 1996, U.S. shipments of architectural coatings were 2.42×10^9 L with a value of $6.2 billion [1]. This was about 52% of the volume and 42% of the value of all U.S. coatings shipments. Architectural coatings are also called *trade sales paints*. There are two overlapping markets for these paints: contractors and *do-it-yourselfers*, who paint their own homes and furniture. The contractor is particularly concerned with cost of application, which is greater than the cost of the paint. For example, contractors want paints that cover in one coat. The do-it-yourselfer is more likely to be concerned about cost of the paint, ease of cleanup, odor, and the range of colors available. In the United States, paint is sold through several distribution systems: large merchandisers, hardware and lumber stores, and paint stores owned or franchised by paint companies. Some paint is sold directly to large contractors by paint manufacturers. There are three, sometimes overlapping, classes of paint companies: large companies that do extensive national advertising to promote trade name paints; small companies that sell only locally, primarily to local contractors, but commonly with one or more factory stores for sales to individuals; and companies that sell private label paints to large merchandisers or hardware chains. The large national companies usually make the same lines of paint for sale all over the country. Local and regional paint companies design their paints to be most suitable for the climate and style trends in their market area.

Large companies usually make three lines of products of major items: good, better, and best. The good grade is designed for the individual whose principal criterion for selecting paint is price per unit volume. Such paint is usually adequate for the application with fair coverage. The best paint is usually designed to give the longest life and to have as good coverage as can be designed into that class of paint. The better paint is a compromise between the two.

While some colored paints are manufactured, the majority are manufactured and distributed as white paints. The manufacturing company supplies an extensive line of color chips with tinting colors and formulations to tint white paints to match the color chips. This makes it possible for the paint store to have thousands of colors available with only a limited inventory.

A multitude of products is made for the architectural market. We restrict discussion to the three largest classes of paints: exterior house paint, interior flat wall paint, and gloss enamels. Space limitations do not allow inclusion of other smaller, but still significant, product lines such as stains, varnishes (see Sections 14.3.2 and 15.7), floor paints, and many specialty products.

31.1. EXTERIOR HOUSE PAINT

Reference [2] provides an extensive discussion about finishing wood with many types of coatings. This section is limited to exterior house paint, that is, paint applied to exterior wall surfaces. Paint for trim around windows, doors, and on shutters is generally gloss enamel, discussed in Section 31.3. The majority of exterior house paint sold in the United States is latex paint. The exterior durability of latex paints is superior to that of air dry, solvent-borne house paints. Latexes are made by emulsion polymerization (see Chapter 8) and undergo film formation by coalescence (see Section 2.3.3). Solvent-borne house paints are used if outside painting must be done when the temperature is below 1–3°C (the minimum application temperature is given on the label), since latex paints do not coalesce properly at such low temperatures. Latex paints do not adhere well to a chalky surface, that is, a surface with a layer of poorly adherent pigment and eroded old paint, requiring careful washing to remove chalk.

Latex paints perform better than oil or alkyd paints on wood siding. When oil and alkyd paints were widely used on wood house siding, paint failures by blistering were common. Water can get into siding from the back side. Since films from oil and alkyd paints have low water vapor permeability, when water reaches the back side of a film and starts volatilizing in heat from the sun, blisters are blown. Latex paints are not cross-linked and have T_g values below summer temperatures, and therefore, have relatively high moisture vapor permeabilities. When water gets to the back of a latex paint film, the water vapor can pass through the film; blistering is unusual.

Latex paints are more resistant than oil or alkyd paints to grain cracking. Wood expands and contracts as its moisture content increases and decreases. A coating applied with an acrylic latex binder maintains its extensibility after many years of exterior exposure and can expand and contract with the wood. Highly unsaturated oil and alkyd coating films continue to cross-link with exterior aging, becoming less extensible and more likely to crack as the wood expands and contracts. Oil and alkyd paints also commonly fail by chalking after exposure. High quality latex paints are less likely to fail by chalking or, at least, have longer lifetimes before chalking occurs to a serious extent.

Leveling of latex paints is generally not as good as solvent-borne paints. This is not viewed as too serious a problem in exterior house paints, but it is with gloss paints, as discussed in Section 31.3. This disadvantage is offset in exterior house paints by the accompanying superior sag resistance of latex paints compared to solvent-borne paints. Since they continue to be thermoplastic, latex paints are more prone to retain dirt on their surface than are oil or alkyd paints. Dirt pickup is undesirable anywhere, but is a particular problem in places where lignite and soft coal are used for heating and cooking, white latex paint turns unsightly, blotchy grey after only a few months exposure. As discussed in Section 2.3.3, film formation of latex paints occurs by coalescence of the polymer particles. Coalescence requires that the temperature during film formation be above the T_g of the particles. If the paint is to be applied at low temperature, the T_g must be correspondingly lower. However, this means that when the temperature is high during the summer, the film is soft and susceptible to dirt pickup, even when the dirt content in the air is not excessive. Dirt pickup decreases as T_g is increased from 0° to 10°C and styrene/acrylic latex paints show less dirt pickup than straight acrylic latex paints [3]. Lower gloss paints generally have lower dirt pickup than higher gloss ones.

The condition of the surface of wood can be a factor in the performance of paint. Contamination with dirt and oil can interfere with adhesion. When wood is exposed outdoors, the surface degrades. Painting degraded surfaces can result in what looks like adhesive failure, but is actually cohesive failure of the wood. It is reported that even a three to four week exposure of freshly cut wood to weather can adversely affect adhesion [2]. It is recommended that joints and cracks be caulked and that the wood be treated first with a paintable water-repellent preservative and then primed. On many kinds of wood, a latex paint can be used for both the primer and top coat. However, woods like redwood, cedar, and some pines contain water-soluble materials that extract into a latex paint leading to reddish brown stains. The extractives are naturally occurring phenolic compounds. Special *stain-blocking* latex primers are made for use over redwood and cedar. Formerly, these primers were formulated with a somewhat soluble lead pigment, which formed insoluble salts with the phenolic compounds. To avoid lead compounds, other approaches for insolubilizing the phenolics have been found. One approach is to include in the formulation a cationic ion exchange latex. Many houses are sided with preprimed hardboard siding; see Section 30.2 for discussion of painting this siding.

As mentioned earlier, another potential problem in applying latex house paints can be poor adhesion to chalky surfaces. The surface of chalky paint is covered with a layer of loosely held pigment and paint particles. When latex paint is applied over such a surface, the continuous phase of the paint penetrates among the particles down to the substrate, but latex particles are large compared to the interstices among the chalk particles and do not penetrate significantly. When the water evaporates and coalescence occurs, the paint film is resting on top of the chalk particles with nothing binding the chalk particles together or to the substrate. As a result, adhesion is poor. Latex chalk sealers made with very small particle size latexes are available.

Alternatively, one can use an oil or alkyd based primer. With oil or alkyd paints, the continuous phase penetrates among the chalk particles, the binder from the paint surrounds the chalk particles, and penetrates through them down to the substrate surface minimizing the adhesion problem. Adhesion of latex paints can be improved by careful cleaning of the surface to be painted to remove all chalk. However, it is difficult to do this over an entire house. Another approach is to replace about 15% of the latex polymer binder with long oil alkyd resin or synthetic drying oil. The alkyd or oil is emulsified into the latex paint. After application, when the water evaporates, the emulsion breaks and some of the alkyd or oil can penetrate between the chalk particles, providing improved adhesion. For the paint to have good storage life, the alkyd or oil should be hydrolytically stable. The drying oil or alkyd reduces the exterior durability of the paint as compared to pure latex paints. As the use of oil-based paints and "cement paints" has decreased, the problem of adhesion over chalky surfaces is becoming less common.

Table 31.1 gives a formulation recommended by a latex manufacturer for a "high quality" exterior white house paint [4]. This formulation is not given to recommend it or these raw materials over others; rather, it provides a framework for discussing the reasons for the many components of a latex paint and for the indicated order of addition during making of the paint. Note that the formulation is given in terms of both volume and weight. The total of the volume is approximately 100. This is done since quantities manufactured are generally multiples of 100 units of volume, and also, one can compare percentage volumes of ingredients. On the other hand, many of the ingredients added on a weight basis.

Table 31.1. Exterior White House Paint

Materials	Weight	Volume
Natrosol 250 MHR (Aqualon) (2.5%)	120.0	14.40
Ethylene glycol	25.0	2.68
Propylene glycol	35.0	4.04
Tamol 1124 (Rohm & Haas) (50%)	4.6	0.47
Triton CF-10 (Union Carbide)	1.0	0.11
Colloid 643 (Rhodia)	2.0	0.26
Ti-Pure R-902 (duPont)	150.0	4.50
Minex 4 (Indusmin, Inc.)	50.0	2.30
Icecap K (Unimin Specialty Minerals, Inc.)	15.0	0.68
Celite 281(Johns Manville)	45.0	2.34

Disperse for 20 minutes with a high speed impeller; then let down at slower speed with:

Ropaque OP-96 (36.5%)	120.0	13.96
Rhoplex Multilobe 200 (53.5%) (Rohm & Haas)	336.8	37.96
Texanol (Eastman Chemical)	11.2	1.41
Colloid 643 (Rhodia)	2.0	0.26
NH$_4$OH (28%)	0.6	0.08
Natrosol 250 MHR (2.5%) (Aqualon)	49.0	5.88
Water	72.3	8.67
	1039.5	100.00

Properties		
PVC, %	47.0	
Volume solids, %	36.4	
Weight solids, %	46.7	
pH	8.8–9.0	
Stormer viscosity, equilibrated, KU	88	
ICI Viscosity, equilibrated, Pa · s	0.095	
VOC (g L^{-1} excluding water)	196	

Natrosol 250 MHR is hydroxyethylcellulose (HEC), which is in the formulation for two reasons: to increase the viscosity of the external phase of the paint during production and application and to control the viscosity of the final paint. Viscosity of the paint as a function of shear rate particularly affects ease of brushing, film thickness applied, leveling, sagging, and settling. The viscosity of the external phase controls its rate of penetration into a porous substrate, such as wood. If penetration is rapid, the viscosity of the paint above the porous surface increases rapidly, resulting in poorer leveling. The HEC is added at this stage to provide the viscosity necessary for pigment dispersion.

The ethylene and propylene glycols are in the formulation for two reasons. First, they act as antifreeze to stabilize the paint against coagulation during freezing and thawing. The expansion of water as it freezes exerts substantial pressure on the latex particles and can push them together with enough force to overcome the repulsion by the stabilizing layer, resulting in coagulation. The glycols lower the freezing point; even if the temperature gets low enough to freeze the solution, it freezes to slush so that less pressure is exerted on the latex particles. The second reason for the glycols is to control the rate of drying of the paint to permit *wet lapping* without disruption of the edge of the film. As water evaporates from a latex paint film, viscosity increases rapidly as a result of the increase in volume fraction of internal phase. When paint is applied with a brush or roller, the edge of the wet paint film is painted over (lapped) so that no substrate is left unpainted or with only a thin layer of paint. With oil paint, there is no problem with wet lapping. With a latex paint, by the time lapping occurs, the viscosity of the applied paint may have increased enough so that the film is semisolid, but since only limited coalescence has occurred, the film is weak. The pressure of a brush or a roller can break up the film, resulting in irregular chunks of paint film along the lapped edge. This can be minimized by slowing the rate of evaporation of the continuous phase by incorporating slow evaporating glycols. The combination of glycols controls the rate of evaporation.

Tamol 1124 and Triton CF-10 are surfactants. Their function is to stabilize the pigment dispersion while not interfering with the stability of the latex dispersion or the stability of tinting color dispersions that may be added to the paint. Tamol 960 is an anionic dispersing agent. Triton CF-10, a nonionic surfactant, helps stabilize the pigment dispersion and also is effective in reducing the surface tension of the latex paint so that it will wet out low surface tension substrates. It has been proposed that dynamic surface tension may be more important in wetting the substrate than equilbrium surface tension [5]. The combination of dispersing agents used in a formulation is generally empirically determined by trial and error. Careful records should be kept of the results of using different combinations with particular latexes, pigments, and surfactants, especially records of failures. This information serves to guide formulators in selecting dispersion agent combinations for future formulations.

Colloid 643 is an antifoam agent. (See Section 23.8.) The minimum antifoam necessary for controlling foam should be used; excess antifoam can result in crawling when paint is applied. Selection of antifoams is empirical; the manufacturers of proprietary antifoams offer test kits with samples of various antifoams. The formulator tries the samples to find which is most effective with the particular formulation.

Ti-Pure R-902 is a rutile TiO_2 pigment. While not stated in the formulation constants given with the formulation, the PVC of TiO_2 in the formulation is approximately 10%. This level of TiO_2 is insufficient to provide good hiding; the Ropaque OP-62LO later in the formulation provides the additional light scattering for adequate hiding. In this formulation, dry powder TiO_2 is called for. In most cases, large manufacturers of latex paints use TiO_2 "slurries," that is, dispersions of TiO_2 in water supplied by the pigment manufacturer for more economical material handling. (See Section 19.1.)

Minex 4 is a sodium potassium aluminosilicate inert pigment, and Icecap K is an aluminum silicate inert pigment. Celite 281 is diatomaceous earth. (See Section 19.3.) These inert pigments have refractive indexes sufficiently similar to that of the binder in the paint that they contribute little directly to hiding. Use of some fine particle size inert pigment, called *spacer* pigment, is generally believed to increase the hiding efficiency of the TiO_2. This subject is discussed further in Section 31.2.

A variety of types of inert pigments can be used. Particle size, cost per unit volume, and color are major selection criteria. Calcium carbonates are inexpensive and are sometimes used as inert pigments in exterior latex paints, but there is a potential problem. Latex paint films have quite high water permeability; water and carbon dioxide, which are in equilibrium with carbonic acid, can permeate into the film. Calcium carbonate dissolves in carbonic acid to give a calcium bicarbonate solution that can diffuse out of the film. On the surface, the water evaporates, leaving a deposit of calcium bicarbonate that then reverts to insoluble calcium carbonate. The white deposit, called *frosting*, on the surface of the film is undesirable, especially with colored paints. Calcium carbonate pigmented paints are subject to degradation by acid rain; not only can frosting result, but also film properties deteriorate and the films become more subject to mildew due to porosity developed in the films by loss of calcium carbonate [6].

The gloss of this, and most, latex exterior house paints is quite low. Gloss is controlled by the PVC to CPVC ratio. (See Chapter 21.) Exterior durability, as well as many other properties, are also affected by PVC. It is desirable to maximize CPVC, since the PVC of the paint at the same PVC to CPVC ratio increases with CPVC. With a given latex, a major factor controlling CPVC of latex paints is the particle size distribution of the pigments. Multiple inert pigments maximize the breadth of particle size distribution, which increases CPVC and permits higher pigment loading at the same PVC to CPVC, decreasing cost. (See Sections 21.2 and 21.4.)

After the pigment is dispersed, the high-speed impeller is slowed and the latexes, Ropaque OP-62 LO and Rhoplex Multilobe 200, are added. Latexes can coagulate at the shear rates involved when the impeller is running at high speed. Ropaque latex minimizes the TiO_2 requirement; it is a high T_g latex that contains small bubbles of water within the particles (see Section 19.1.1) [7]. When the paint film dries, the water diffuses out of the particles, leaving microvoids that scatter light, reducing the TiO_2 requirement. In calculating PVC, one includes the volume of these particles, since they do not coalesce with the latex binder particles during film formation.

Rhoplex Multilobe 200 is an all acrylic latex with a T_g of 9°C that is designed for exterior durability [4,8]. The latex particles have a lobed structure. The lobed structure permits formulation of paints with a relatively low viscosity at low shear rates and a relatively high viscosity at high shear rates, combining reasonably good leveling with the thicker films required for good hiding. The latex is also said to give better adhesion to chalky surfaces than conventional spherical latexes [4]. It is reasonable to guess that the latex is a copolymer of methyl methacrylate (MMA) with ethyl or butyl acrylate. Such copolymers have excellent exterior durability. Latexes with part of the MMA replaced by styrene have lower costs, lower dirt pickup, and reasonably satisfactory exterior durability. Vinyl acetate/acrylate copolymers are less expensive. While their hydrolytic stability is inadequate for exterior use in climates with high humidities and rainfall, they are used outside in arid climates. Vinyl versatate copolymers (see Section 8.3) have better exterior durability than vinyl acetate copolymers and are widely used in Europe as vehicles for exterior paint [9,10].

Colloids 643 is added again in an amount necessary to control foaming. Texanol is a coalescing agent; it is a mixture containing the monoisobutyrate ester of 2,2,4-trimethyl-pentane-1,3-diol. The high degree of steric hindrance of the ester group of Texanol provides hydrolytic stability. The ammonium hydroxide is added to adjust the pH to 8.8–9. The high pH assures the stability of the anionic dispersing agents. While latex paints are packed in lined tin cans, it is possible to have breaks in the lining, and the high pH minimizes the possibility of corrosion of the cans. In many cases, the latex polymer is made with some acrylic acid as a comonomer; the viscosity of paints made with such a latex is very pH dependent. (See Section 8.2.)

The last two items in the formulation are water and a 2.5% solution of HEC. The amounts of the two determine the viscosity and solids of the final paint. There is variation from batch to batch of the viscosity of any paint formulation. The ratio of water and HEC solution can be varied so that the viscosity and solids of the paint both come out at the standard level.

While not listed in the formula, addition of 1.5% Kathon LX (5-chloro-2-methyl-4-isothiazolin-3-one and 2-methyl-4-isothiazolin-3-one), a bactericide, is recommended, as well as the addition of a fungicide such as Skane M-8. Bactericides are needed to control bacterial growth in the can of paint. There are three adverse effects of bacterial growth: A putrid odor can develop; the metabolic processes release gases that can build up enough pressure to blow off the can lid; and since bacteria can digest cellulose derivatives like HEC, the viscosity of the paint may drop. Fungicides are needed to minimize mildew growth on the paint film after it is applied. In many exterior paints, zinc oxide is used as one of the white pigments. It contributes somewhat to hiding, but its scattering efficiency is poor because the refractive index difference between it and the binder is relatively small. Zinc oxide is used primarily as a fungicide. Zinc oxide can cause large viscosity increases during storage of some latex paints [11]. The effect of ZnO is system dependent; Reference [11] discusses the variables and possible formulation approaches to minimize the instability problems.

Many other compounds are used as fungicides and bactericides. Examples of biocides are substituted 1-aza-3,7-dioxabicyclo(3.3.0)octanes, substituted 3,5,7-triaza-1-azonia adamantane chloride, 1,2-benzthiazolone, and methylchloro/methylisothiazolone. Benzothiophene-2-cyclohexylcarboxamide-S,S-dioxide is said to be a broader spectrum biocide [12]. References [13] and [14] provide a review of other biocides. Testing is difficult because fungal and bacterial growth are dependent on ambient conditions and because many fungicides and bactericides are effective against only a limited number of organisms. ASTM D-2574-96 is a standard test method for resistance of latex paints in the container to attack by microorganisms. Diluted paints are more likely to support bacterial growth than undiluted paints; a test method based on inoculating diluted latex paint with several bacteria and fungi has been reported to be reliable [15]. A rapid impedimetric test procedure has been reported [16].

Housekeeping in a latex paint factory should be as good as it is in a food processing factory. The best control of bacterial growth is to avoid contamination of the paint with bacteria. Furthermore, bactericides kill bacteria, but they do not deactivate enzymes that have been produced by bacteria. If the factory has places where bacteria can grow, enzymes can get into paint. The enzymes split the HEC molecules and the viscosity drops, even though there is adequate bactericide. Some thickeners are not subject to attack by bacteria.

The PVC of the paint is 47%; it is a relatively low gloss paint as are most exterior latex paints. Commonly, gloss of latex paints is lower than oil paints. The lower PVC reduces cost while still providing excellent exterior durability. The volume solids of the paint is 36.4%, substantially lower than exterior oil or alkyd-based paints. As a result, coverage per unit volume of latex paints is lower than oil-based paints, but the cost per unit volume of latex paints is lower. The grams of VOC per liter of paint, excluding water, is 196, lower than that of most alkyd- and oil-based house paints.

Note that in spite of the widely recognized shortcomings of the Stormer viscometer discussed in Section 3.3.5, in this case (and in most paint labs in the United States), it is the viscometer used. The "equilibrated viscosity" signifies the viscosity after the thixotropic structure has time to build up again. The ICI viscosity refers to the viscosity run on an ICI viscometer that measures viscosity at a high shear rate similar to that experienced in application by brush or roller. It is important to have the proper high shear viscosity because this is a major factor controlling the film thickness of the paint that is applied.

31.2. INTERIOR FLAT WALL PAINT

The largest volume of trade sales paint is interior flat wall paint; almost all is latex paint. Since exterior durability is not needed, lower cost vinyl acetate copolymer latexes (see Section 8.3) are the principal binders. Since the T_g of vinyl acetate homopolymer is about 32°C, some comonomer such as butyl acrylate is used to reduce T_g. For interior applications for which greater water resistance is needed, acrylic latexes or vinyl versatate copolymer latexes are used.

The major advantages of latex paints as compared to the older oil-based flat wall paints are:

Fast drying and less sagging. If desired, two coats can be applied to the walls of a room during a day, the furniture moved back, and the room used that night. The rapid increase in low-shear viscosity in the early stages of film formation reduces vulnerability to sagging, even when relatively thick films are applied.

Low odor. The odor of mineral spirits and byproducts from oxidation of drying oils in solvent-borne paints is unpleasant for days after walls are painted. Although the odors are less offensive, some odor from the slow evaporation of coalescing solvents from latex paints persists for up to a week. A reason for developing zero VOC paints is to eliminate this odor problem.

Ease of cleanup. Spills, dripped spots, brushes, and rollers are easily cleaned with soapy water in the case of latex paints; solvent is required with oil paints. However, cleaning up must be done promptly because once latex has coalesced, cleaning it up is more difficult than with oil paints.

Low VOC emissions. Latex paints were widely adopted before there was concern about VOC emissions. Progress is being made in reducing the already low VOC, as discussed in Section 31.3. There is also the advantage with water-borne paints of substantially reduced fire hazard. Not only is solvent-borne paint flammable, but also rags wet with oil-based paints in a confined space can undergo *spontaneous combustion*, that is, ignite as a result of heat generated by autoxidation.

Less yellowing and embrittlement. White and light color vinyl and vinyl/acrylic latex paints retain their color better than alkyd paints, which gradually turn yellow with age. Alkyds made with oils that contain very little linolenic acid, such as sunflower or safflower oils (see Section 15.1.1.), yellow more slowly than alkyds based on soybean oil or tall oil fatty acids, but they still yellow with time. Alkyd films also eventually become brittle as autoxidative cross-linking continues far beyond the point of optimum film properties.

Flat wall paint is usually stocked as white paint, sometimes called a tint base, and tinting colors are added to make a color chosen by the customer from a large array of color cards. This requires that equal white tinting strength be maintained through quality control or the colors obtained will differ. Furthermore, any new formulation must have the same white tinting strength as the formulation being replaced or else the color cards and formulations in dealer stores will have to be replaced. In each quality line, two, or sometimes three, white base paints are included. The base white paint is used alone as a white or tinted to make pastel colors. A deep tone base containing little TiO_2 is used for tinting to deep colors that could not be made if the regular base paint were used. Frequently, a third base is in the line that is used for intermediate depths of shade.

Intermediate depths of shade could be matched using the base white paint, but the cost would be excessive because more tinting color would be needed to match the colors; hiding would be greater than needed.

Users are sometimes confused by the change in color of latex paint as it dries; the color of a dry film is darker than the color of the wet paint. In the wet paint, the interfaces between the water ($n = 1.33$), polymer particles ($n =$ approximately 1.5), TiO_2 ($n = 2.73$), and inert pigments ($n =$ approximately 1.6) scatter light to a greater degree than when the paint is dry. The dry paint has fewer interfaces as a result of coalescence of the latex particles, and also there are smaller refractive index differences since the pigment particles are in a polymer matrix instead of water. Since light scattering decreases as water evaporates and the latex particles coalesce, the color gets darker; hiding also decreases as the paint dries.

When painting ceilings, one is particularly anxious to get hiding in one coat, since painting over one's head and moving the ladder is more of an effort than when painting walls. The problem is particularly challenging because ceiling paints are commonly plain white to reflect light diffusely. Since there are no color pigments in a white paint to absorb light, the hiding by white paints is poorer than any color paint made from it. The problem is compounded by the decrease in hiding when a latex paint film dries. The user thinks he or she has applied enough paint to hide marks on the ceiling, but comes back an hour or so later and finds that marks show through the dry paint. Special ceiling paints minimize this problem by formulating with PVC above CPVC. Dry paint films with PVC above CPVC have voids of air with $n = 1$ that add additional light scattering by the new interfaces between air and polymer as well as air and pigment. Formulations can be adjusted so that wet hiding and dry hiding are approximately equal. The films do not have as high mechanical strength as films of paint with PVC < CPVC, and resistance to staining is poorer, but neither property is important for ceiling paints.

The most expensive major component of any white flat paint on a volume basis is the TiO_2. Several approaches are used to minimize the TiO_2 content required at a given level of hiding. The efficiency of hiding by TiO_2 is affected by the choice of inert pigments used with it. While there is some controversy, most workers accept that inert pigments with particle sizes less than that of the TiO_2, called spacer inerts, increase the efficiency of the TiO_2 [17]. A mathematical model has been developed that can be used to improve the spacing of TiO_2 by predicting the optimum inert pigment size distribution and concentration for a given formulation [18].

Another approach to greater hiding at lower cost to use a high T_g latex, such as polystyrene, as a pigment [19]. When the latex binder coalesces, the high T_g latex does not coalesce, but the particles remain separate, as with any other pigment. Including the dry volume of the high T_g latex particles as part of the pigment volume, paints can be formulated with a PVC greater than CPVC resulting in air bubbles that increase hiding, but without making the surface of the film porous. The paints provide equal hiding at lower TiO_2 content while retaining good enamel holdout and stain resistance. No convincing explanation of why an intact surface film forms in such paints has been published, but the method is said to be used on a large commercial scale.

Yet another approach to minimizing the TiO_2 requirement is to use as pigments special high T_g latexes, such as Ropaque (Rohm and Haas Co.) [7] as is shown in the formula for an exterior house paint given in Table 31.1 and explained in the text after the table.

A large fraction of latex flat wall paints is applied by roller. During roller application, latex paints *spatter*, some to a major degree. Paints with high extensional viscosity are likely to spatter severely (see Section 22.1.2) [20]. Extensional viscosity increases when high molecular weight water-soluble polymers with very flexible backbones are used as thickeners in a latex paint [21]. Spattering can be minimized by using low molecular weight water-soluble thickeners with rigid segments in the polymer backbone, such as low molecular weight HEC. Since the selling price of HEC is the same for various molecular weights, and more low molecular weight polymer is needed to reach the same shear viscosity, paint cost increases when low molecular weight HEC is used. It is common to compromise by using an intermediate molecular weight or a mixture of high and low molecular weight grades of HEC.

While in many applications, the increase in external phase viscosity resulting from use of water-soluble polymeric thickeners is desirable, an application for which it is undesirable is on concrete block walls. Concrete block surfaces contain holes that are large compared to the pigment and latex particles, as well as pores that are small compared to them. When a solvent-borne paint is applied to concrete block, the coverage is low, since so much paint penetrates into the holes. When regular latex paint is applied, the coverage is better. The lower viscosity of the external (continuous) phase relative to that of solvent-borne paint leads to more penetration of continuous phase into the small pores of the block so that the viscosity of the remaining paint increases rapidly and there is less penetration into the holes. The coverage with latex paint can be further improved by omitting water-soluble thickener from the formulation. This further reduces the viscosity of the external phase, which can then penetrate more rapidly into the small pores and, hence, give even more rapid buildup of viscosity of the remaining paint. The effect is enhanced by the requirement that the volume fraction of internal phase in the modified paint must be significantly higher than when water-soluble polymer is present in order to have equal starting viscosity. With increasing volume fraction internal phase, the viscosity increases more rapidly as the continuous phase is drained off in the small pores, so penetration into the holes is reduced. A further advantage of latex paints, especially acrylic, styrene/acrylic, or styrene/butadiene latexes, over oil or alkyd paints is that they are not subject to saponification by the basicity of the concrete block.

When latex paint films form, surfactants sometimes bloom out of the film, collecting on the surface. Usually, this is not noticeable; however, if water condenses on the paint surface, surfactant on the surface can dissolve. Then, when the water evaporates, surfactant concentrates in the last remaining water droplets, leaving brown spots of surfactant on the surface. The problem can be minimized by avoiding dark-colored surfactants. However, surfactant can leave whitish spots on dark color paints. It has been shown that nonionic surfactants are more compatible with latex polymers and less likely to surface than anionic surfactants [22].

While latex paints have always had low VOC, stricter regulations are requiring even further reduction. Blends of high and low T_g latexes and large and small particle size latexes have been proposed as ways of reducing or eliminating VOC. (See Section 31.3.)

31.3. GLOSS ENAMELS

The term *enamel* connotes a hard, glossy surface analogous to porcelain enamel. In the United States, gloss enamels are used both indoors and outdoors for trim around windows and doors, for shutters, for wood furniture, and on kitchen and bathroom walls. In some other countries, they are also used for walls. For gloss enamels, the merits and drawbacks of alkyd paints and latex paints are about evenly balanced, and latexes have partly, but not completely, displaced alkyds. Mounting pressure to reduce VOC may eventually tilt the balance toward latexes, unless there is a technical breakthrough with alkyds.

31.3.1. Alkyd Gloss Enamels

As described in Chapter 15, alkyd enamels cross-link by autoxidation to form tough films that adhere well to a variety of surfaces and are block resistant and water resistant. An advantage of gloss alkyd enamels is that they have higher gloss than latex enamels. As discussed in Section 18.10.1, during film formation from solution vehicle paints a polymer layer with very low pigment content forms at the upper surface of the coating; this does not happen with latex paints.

The principal advantage of gloss alkyd paints is that they afford good hiding with one coat when used over a surface without sharp color contrasts. One-coat hiding is especially important to painting contractors for whom the cost of application is higher than the cost of paint. Several factors account for the difference in hiding between alkyd and latex gloss paints. The NVV of a typical alkyd gloss paint can be 66% or even higher, while the NVV of a latex gloss paint is limited to about 33%. To obtain the same dry film thickness, one has to apply twice as thick a wet film of the latex paint.

Another factor is that affects hiding is leveling. Assume that a uniform dry film thickness of, say, 50 μm of a paint provides just satisfactory hiding. If the paint levels poorly, there will be streaks of thinner film, say, 35 μm, and thicker film, say, 65 μm, and the hiding of the uneven film will be poor. Hiding is likely to be inferior to that of a uniform 35 μm dry film of the same paint. Since the 35 and 65 μm areas are immediately next to each other, the contrast emphasizes the poor hiding. Furthermore, not only is the hiding adversely affected, but also the contrast in colors resulting from the poor hiding in the valleys compared to the ridges emphasizes the poor leveling. Alkyd paints for brush application generally level better than latex paints because the solvent is slow evaporating mineral spirits. The difference in leveling is amplified when the paint is applied under warm, dry conditions. Yet another factor is that the volume fraction of internal phase in latex paints is higher than in alkyd paints since both the latex and pigment particles are dispersed phases. Therefore, even if the volatile material evaporates at equal rates, the viscosity of a latex paint increases faster.

As discussed in Section 23.2, Overdiep demonstrated that leveling of solvent-borne, brush-applied paints is promoted by surface tension differential driven flows in the wet paint film [23]. When solvent evaporation starts, the fraction of solvent lost from the valleys of the brush marks is greater than from the thicker ridges. Since the surface tension of the solvent is lower than that of the alkyd, the surface tension of the more concentrated alkyd solution in the valleys becomes higher than that in the ridges. The resulting differential in surface tension causes the paint to flow from the ridges into the valleys to minimize overall surface tension, thus promoting leveling. This driving force may be absent in leveling of latex enamels. (See Section 31.3.2.)

When selecting paints for repainting, it should be remembered that latex paints can be used over alkyds or latexes as long as the surface is properly prepared, but use of alkyds over latexes is risky. The solvent in alkyds may penetrate uncross-linked latex films and cause lifting.

The drawbacks of alkyd gloss enamels are largely the same as those of alkyd flat wall paints. (See Section 31.2.) They include slow drying, odor, yellowing and embrittlement with age, and the need to use solvent for cleanup. In the future, perhaps the most important consideration will be VOC. There is disagreement as to what level of VOC can be achieved with alkyd paints while retaining reasonable application and film properties. Many in the coatings field doubt that VOC can be reduced below 250 g L^{-1} without a technical breakthrough. The problems of low-VOC alkyds include application characteristics, through dry, color change, and durability.

High solids alkyds are discussed in Section 15.2. Solids can be increased some by solvent selection, especially the use of hydrogen-bond acceptor solvents to reduce intermolecular hydrogen bonding. While solids can be increased by reducing molecular weight and by using narrower molecular weight distributions, both of these approaches lead to inferior film properties and durability if carried very far.

A promising approach to increasing solids is the use of reactive diluents. (See Section 15.2). Such additives are low molecular weight materials designed to reduce viscosity almost as efficiently as solvent, but also to coreact with the oxidizing alkyd during film formation. This permits reduction in VOC while maintaining film properties. An example is dicyclopentenyloxyethyl methacrylate. This reactive diluent has both an acrylate double bond and an activated allylic position. In the presence of driers, it can coreact with an alkyd such as a long oil linseed alkyd. Using this reactive diluent with specially designed alkyds is reported to permit formulation of gloss alkyd paints with a VOC of 155 g L^{-1}. The properties of the films are said to approach those from an alkyd paint with a VOC of 350 g L^{-1} [24]. Other types of reactive diluents are condensation products of mixtures of drying oil acid amides and acrylamide with hexaalkoxymethylmelamine [25] and oligomeric polyallyl ethers [26]. While use of reactive diluents is attractive, their cost and performance characteristics require further improvement. Use of polyallyl ethers has been questioned because of the potential for formation of toxic, volatile acrolein during cross-linking.

Two approaches to water-borne alkyds for trade sales paint have been extensively investigated. Efforts to develop gloss enamels based on water-reducible alkyd resins (see Section 15.3) have not been very successful. A key obstacle has been the difficulty of making alkyds with sufficient hydrolytic stability for the required two-year shelf life. Water-reducible alkyds generally contain significant levels of solvent and may not reduce VOC much below 250 g/L^{-1}. Emulsions of alkyds (see Section 25.3) in water can also be used to prepare gloss enamels [27]. They are used in Europe, but have not been widely adopted in the United States. The VOC content can be lower, but the surfactants required for emulsification adversely affect film properties, and the shortcomings of odor and poor resistance to yellowing and embrittlement remain.

31.3.2. Latex Gloss Enamels

Development of latex gloss enamels has proven to be one of the most challenging problems of coatings technology. Key problems and potential solutions are discussed in this section.

One problem is that of obtaining high gloss. As noted in Section 31.3.1, the clear polymer layer that forms on the surface of alkyd enamels does not readily form as latexes coalesce. The ratio of pigment to binder at the surface of a latex paint film can be reduced somewhat by using a finer particle size latex, but there is still a difference from alkyd paint films. Gloss is affected by flocculation of either pigment or latex. The choice of dispersing surfactants and thickeners and their order of addition can affect gloss. In one study, it was shown that when a mixture of polyacrylic acid (PAA) and hydroxyethylcellulose (HEC) was used, gloss was higher if the HEC was added first than when PAA was added first [28]. Nonionic surfactants resulted in higher gloss than anionic surfactants. Other factors tending to reduce the gloss of latex paint films are the haze that results from incompatibility of surfactants (and perhaps other components) and the blooming that results when surfactants migrate to the surface of the film. These problems can be minimized by making latexes with as low monomeric surfactant content as possible and by selecting surfactants for pigment dispersion that are as compatible as possible.

Considerable work has been done using blends of compatible water-soluble resins with latexes to achieve high gloss. In floor wax applications, morpholine salts of styrene/acrylic acid copolymers have been used. When the film forms, the solution resin concentrates in the surface layer. When the film dries, the morpholine evaporates, leaving the free carboxylic acid groups. The film has sufficient water resistance for normal use, but it can be removed by mopping with ammonia water. This level of resistance is an advantage for floor wax, but is not adequate for paint films. There have been some proprietary resins marketed that give better gloss with smaller loss of resistance properties.

In exterior and, to a lesser degree, interior applications, the advantage of high initial gloss alkyd enamels is more than offset by the better gloss retention and resistance to cracking exhibited by latex gloss paints. Depending on location, alkyd enamels exposed outdoors lose so much gloss in a year or two that the coatings become flat. While a latex enamel starts out with a lower gloss, it retains most of its initial gloss for several years. In interior applications, the low odor, superior color retention, and greater resistance to cracking are important advantages of latex gloss paints.

The principal limitation of gloss latex paints is not their lower gloss; the major problem is achieving adequate hiding in one coat. The factors involved are discussed in Section 31.3.1. Not much can be done about the relatively low NVV of latex paints. The formulating challenge is to make latex enamels that can be easily applied in films thick enough for one-coat hiding. For maximum hiding, a latex enamel should be formulated to have a higher viscosity at high shear rate than an alkyd enamel. In practice, as is discussed later, the high shear rate viscosity of latex paints has traditionally been formulated to be lower than that of alkyd paints, further compounding the problem. Wet film thickness can be controlled to a degree by the extent to which the painter brushes out the paint. The painter tries to judge how much he or she should brush out by how well the wet paint is covering. In the case of alkyd paints, there is relatively little difference between the wet and dry hiding power of the paint. However, the wet hiding power of latex paints is greater than their dry hiding power. This increases the difficulty of judging how far to brush out a latex paint.

The need for good leveling is also described in Section 31.3.1. The driving forces for leveling of latex paints do not seem to have been considered in the literature. It is not obvious how surface tension differentials could develop during film formation of latex enamels. The surface tension of the water phase of latex paints is controlled primarily by the surfactants in the paint, which suggests that there probably is little change in surface tension as water evaporates. If this hypothesis is right, the driving force for leveling of latex paints is the relatively small force of surface tension driven leveling rather than the larger force of surface tension differential driven leveling thought to operate with alkyd enamels. Another factor that may be important is dynamic surface tension. It has been shown that some surfactants reach equilibrium surface tension more rapidly than others [5]. The need for research is evident.

However, the major factor affecting the leveling and, therefore, the hiding of gloss latex paints is probably their rheological properties. As they have traditionally been formulated with water-soluble thickener polymers like HEC, latex paints have exhibited a much greater degree of shear thinning than alkyd paints. This has led to latex paints having too low a viscosity at high shear rate, so the applied film thickness tends to be too thin, and too high a viscosity at low shear rate to permit adequate leveling. The problem is especially severe because the rate of recovery of viscosity after exposure to high shear rates is generally rapid with latex paints. The use of Stormer viscometers (see Section 3.3.5) has been at least partly responsible for the prolonged time before the problem was well defined. This so-called viscometer measures something related to viscosity in a midrange of shear rates, but gives no information about viscosity in either of the critical regions—at high and low shear rates.

The reasons for this greater dependency of viscosity on shear rate in latex coatings have not been fully elucidated: at least two factors may be involved, flocculation of latex particles and/or pigment particles in the presence of HEC, and possible entanglement of chains of swollen high molecular weight HEC [29]. Progress in minimizing the problem has been made by using *associative thickeners*; of which many kinds are available. They all are moderately low molecular weight, hydrophilic polymers with two or more long chain nonpolar hydrocarbon groups spaced along the backbone. Examples include hydrophobically-modified ethoxylated polyurethanes (HEUR), styrene-maleic anhydride terpolymers (SMAT), and hydrophobically-modified alkali-swellable emulsions (HASE). Use of such thickeners permits formulation of latex paints that exhibit less shear thinning so that viscosity at high shear rates can be higher; as a result, thicker wet films can be applied [29,30]. Formulation with associative thickeners also reduces viscosity at low shear rates, so leveling is also improved. (The thicker wet film in itself helps promote leveling too, since the rate of leveling depends on wet film thickness, as discussed in Section 23.2.)

Reynolds has reviewed possible mechanisms of thickening by associative thickeners and factors involved in their use in formulating latex paints [29]. He emphasizes that greater care and skill is required in formulating with associative thickeners than with conventional water-soluble thickeners. The results obtained can be very sensitive to the combination of the particular latex and thickener and to the amounts and types of surfactant present in the formulation [29,31]. It has been shown that formulations with associative thickeners not only level better, but also give somewhat higher gloss [32]. Higher gloss, as well as better rheological properties, have been reported by use of a combination of associative thickeners [33]. It is said the combination reduces flocculation of the TiO_2. Another factor in the improved gloss may be that associative thickeners are effective with small particle size latexes, which give the highest gloss [29]. It has also been shown that paints thickened with associative thickeners spatter less when applied by roller than those thickened with HEC and that most of the thickeners resist viscosity loss by bacterial action [29]. Sag control of latex paints is more difficult with associative thickeners but still easier than with alkyd paints; leveling and gloss are still not equal to most alkyd paints [32].

Another shortcoming of latex paints, particularly evident in gloss formulations, is the time required to develop final film properties. Part of the problem is that users are, in a sense, fooled by the drying properties of latex paints. They dry to touch more rapidly than alkyd paints and can be handled sooner. However, they require a longer time to reach their ultimate properties. For example, even though latex paints dry more rapidly than alkyd paints, longer time is necessary to develop block resistance required to prevent windows and doors from sticking or to permit putting heavy objects on a newly painted shelf. The initial film formation of latex particles is rapid, but full coalescence is limited by the availability of free volume. Since $(T - T_g)$ must be small, free volume is small. The situation is helped by using coalescing solvents. However, loss of these solvents is controlled by diffusion rate, which is also limited by $(T - T_g)$. It has been reported that latex particles that have inner layers of relatively high T_g with a gradient down to a relatively low T_g on the outer shell of the particles can provide film formation at low temperatures, but yet relatively quickly achieve block resistance [34]. It has been recommended that a high T_g latex be used with larger amounts of carefully selected coalescing agents [35]. The dimethylether of dipropylene glycol is reported to be an effective coalescing agent with a relatively rapid rate of evaporation from films [36]. There is need for further progress, but as this problem is unlikely to ever be completely solved, there is need for better educating consumers on the limitations of latex paints.

Progress has been made in developing gloss latex paints with adequate hiding by one coat, but further efforts lie ahead. The problem is made more difficult by the lack of adequate laboratory tests to measure gloss (see Section 18.10.2) or absolute hiding (see Section 18.3). The lack of adequate test procedures is particularly troublesome because many of the companies supplying raw materials to the paint industry, many of the people establishing specifications and regulations, and, sadly, even people working as paint formulators are not aware that the laboratory tests are inadequate.

Another problem of some gloss latex paints is poor adhesion to old gloss paint surfaces when the new dry paint film gets wet. After wetting with water, some latex paint films can be peeled off the old paint surface in sheets. Such a film is said to exhibit poor *wet adhesion*. There is always a problem of achieving adequate adhesion when repainting old gloss paint surfaces, even with alkyd paints, but the problem with latex paints is more severe. It is essential to wash any grease off the surface and to roughen the surface by sanding, but even with such surface preparation, many latex paints do not show good wet adhesion. Wet adhesion improves with age, but it can remain seriously deficient for several weeks or even months. Latexes that minimize the problem of wet adhesion have been developed by several manufacturers. Incorporation of small amounts of hydrogen bonding, polarizable comonomers such as methacrylamidoethyleneurea (see Section 8.1) improves wet adhesion and wet scrub resistance.

Although VOC of latex paints is lower than that of alkyd paints, latex enamels are usually formulated with higher VOCs than other latex paints. Extra coalescing solvent is used to permit use of latexes with the relatively high T_g needed for block, scrub, and stain resistance. Their VOC is often only slightly less than the maximum (currently 250 g L^{-1}) allowed by the strictest regulations. There is mounting pressure to further reduce VOC emissions. One approach is to select the most efficient coalescing solvents to minimize the amount required. For example, the acetate of propylene glycol n-butylether has been recommended as an efficient coalescent [36]. Another approach is to use blends of high and low T_g latexes [37]. The binder must be transparent, which requires that the difference in refractive index be small and that the particle size of the high T_g latex be small. Still another approach is to use thermosetting latexes (see Section 8.1.4); this permits use of low T_g latexes that form films without need for coalescing solvents, but that reach adequate hardness through cross-linking. For example, a proprietary latex that cross-links by autoxidation is reported to permit formulation of zero VOC gloss latex paints [38].

As previously mentioned, solvents such as propylene glycol are included in the formulation to improve freeze-thaw stability and to permit wet lapping. It has been found that associative thickeners of hydrophobically modified polyacrylic acid ammonium salts increase open time (that is, they permit wet lapping) and improve freeze-thaw resistance. However, such large amounts are needed that the alkali resistance of the films becomes inadequate. By combining the use of these associative thickeners and a cross-linking latex, satisfactory gloss enamels having low VOC have been reported [38].

GENERAL REFERENCE

W. C. Feist, *Finishing Exterior Wood*, Federation of Societies for Coatings Technology, Blue Bell, PA, 1996.

REFERENCES

1. M. S. Reich, *Chem. Eng. News*, **75**, 36 (1997).
2. W. C. Feist, *Finishing Exterior Wood*, Federation of Societies for Coatings Technology, Blue Bell, PA, 1996.
3. A. Smith and O. Wagner, *J. Coat. Technol.*, **68** (862), 37 (1996).
4. Anonymous, *Technical Bulletin Multilobe 200*, Rohm & Haas, Philadelphia, 1997.
5. J. Schwarz, *J. Coat. Technol.*, **64** (812), 65 (1992).
6. J. W. Hook, III, P. J. Jacox, and J. W. Spence, *Prog. Org. Coat.*, **24**, 175 (1994).
7. D. M. Fasano, *J. Coat. Technol.*, **59** (752), 109 (1987).
8. C.-S. Chou, A. Kowalski, J. M. Rokowski, and E. J. Schaller, *J. Coat. Technol.*, **59** (755), 93 (1987).
9. R. A. Prior, W. R. Hinson, O. W. Smith, and D. R. Bassett, *Prog. Org. Coat.*, **29**, 209 (1996).
10. F. Decocq, D. Heymans, M. Slinckx, S. Spanhove, and C. Nootens, *Proc. Waterborne High-Solids Powder Coat. Symp.*, New Orleans, 1997, p. 168.
11. I. V. Mattei, R. Martorano, and E. A. Johnson, *J. Coat. Technol.*, **63** (803), 39 (1991).
12. P. Wachtler and F. Kunisch, *Polym. Paint Colour. J.*, October, 2 (1997).
13. J. W. Gillatt, in D. R. Karsa and W. D. Davies, Eds., *Waterborne Coatings and Additives*, The Royal Society of Chemistry, Cambridge, 1995, pp. 202–215.
14. W. B. Woods, *Paint Coat. Ind.*, **3** (5), 25 (1987).
15. P. A. Jaquess and M. C. McLaurin, *J. Coat. Technol.*, **65** (823), 77 (1993).
16. P. K. Cooke, U. R. Gandhi, E. S. Lashen, and E. L. Leasure, *J. Coat. Technol.*, **63** (796), 33 (1991).
17. J. H. Braun, *J. Coat. Technol.*, **60** (758), 67 (1988).
18. J. Temperley, M. J. Westwood, M. R. Hornby, and L. A. Simpson, *J. Coat. Technol.*, **64** (809), 33 (1992).
19. A. Ramig, Jr. and F. L. Floyd, *J. Coat. Technol.*, **51** (658), 63, 75 (1979).
20. D. B. Massouda, *J. Coat. Technol.*, **57** (722), 27 (1985).
21. J. E. Glass, *J. Coat. Technol.*, **50** (640), 53, 61 (641), 56 (1978).
22. K. W. Evanson and M. W. Urban, *J. Appl. Polym. Sci.*, **42**, 2309 (1991).
23. W. S. Overdiep, *Prog. Org. Coat.*, **14**, 159 (1986).
24. D. B. Larson and W. D. Emmons, *J. Coat. Technol.*, **55** (702), 49 (1983).
25. Anonymous, *Technical Bulletin, Resimene AM-300 and AM-325*, Monsanto Chemical Co. (now Solutia, Inc.), St. Louis, 1986.
26. I. Badou and S. Dirlikov, *Polym. Mater. Sci. Eng.*, **70**, 334 (1994).
27. A. Hofland, "Making Paint from Alkyd Emulsions," in J. E. Glass, Ed., *Technology for Water-Borne Paints*, American Chemical Society, Washington, DC, 1997, p. 183.

28. M. Hulden, E. Sjoblom, and A. Saarnak, *J. Coat. Technol.*, **66** (836), 99 (1994).
29. P. A. Reynolds, *Prog. Org. Coat.*, **20**, 393 (1992).
30. R. H. Fernando, W. F. McDonald, and J. E. Glass, *J. Oil Colour Chem. Assoc.*, **69**, 263 (1986).
31. M. Chen, W. H. Wetzel, Z. Ma, and J. E. Glass, *J. Coat. Technol.*, **69** (867), 73 (1997).
32. J. E. Hall, P. Hodgson, L. Krivanek, and P. Malizia, *J. Coat. Technol.*, **58** (738), 65 (1986).
33. D. J. Lundberg and J. E. Glass, *J. Coat. Technol.*, **64** (807), 53 (1992).
34. K. L. Hoy, *J. Coat. Technol.*, **51** (651), 27 (1979).
35. A. Mercurio, K. Kronberger, and J. Friel, *J. Oil Colour Chem. Assoc.*, **65**, 227 (1982).
36. C. Geel, *J. Oil Colour Chem. Assoc.*, **76**, 76 (1993).
37. M. A. Winnik and J. Feng, *J. Coat. Technol.*, **68** (852), 39 (1996).
38. G. Monaghan, "Formulating Techniques for a Low VOC High Gloss Interior Latex Paint," FSCT Symposium, Louisville, KY, May 1996.

CHAPTER 32

Special Purpose Coatings

The term *special purpose coatings* designates industrial coatings that are applied outside a factory. In 1996, they accounted for 17% (7.9×10^8 L) of the volume and 22% ($3.2 billion) of the value of U.S. coatings shipments [1]. While the volume is the smallest of the three classes of coatings, it is substantial and the value per unit volume is the highest. Separate profit figures are not available, but it is probable that the operating profit percent is also the highest. Many different end uses are involved; our discussion centers on four of the larger markets: maintenance, marine, automobile refinish, and aircraft coatings.

32.1. MAINTENANCE PAINTS

The term *maintenance paints* is generally taken to mean paints for field application, including highway bridges, refineries, factories, power plants, and tank farms; not usually included are paints for office buildings or retail stores, which are classified as architectural coatings. For many maintenance paints, a major requirement is corrosion protection. (See Chapter 7.) Sometimes, the term *heavy duty maintenance paints* is used, implying that the paints must perform more effectively in aggressive environments than trade sales paints. Although selling prices are important, the major economic demands of customers emphasize proven performance, the time interval to be expected between repainting, and service, rather than the cost of the paint. The frequency of repainting in a factory is especially critical, since production may have to be shut down for repainting.

Since no laboratory tests adequately predict field performance of maintenance coatings, potential customers want to inspect actual field use examples of a coating system being recommended to them. The Federal Department of Transportation, state highway departments and the larger oil, chemical, and construction companies employ specialized engineering groups responsible for working both with coating suppliers to select appropriate coatings for their company's needs and with applicators to specify the application parameters. These groups also serve as inspectors to see that the coatings are properly applied. Records are kept of surface preparation, application conditions, and coating composition, coating supplier; the performance of coatings on the various installations is monitored. Composition of the coatings is a major variable, but surface preparation and application procedures are also critical to performance. The *Steel Structures Painting*

Council is a valuable source of information about effects of such variables such as that provided in its *Steel Structures Painting Manual* [2]. The problems of predicting service life of coatings are discussed in the introductory paragraphs of Chapter 4, and the status of corrosion testing and performance prediction in Section 7.5. As discussed in those sections, reliability theory is a potentially powerful tool [3].

The most commonly used method for surface preparation has been sandblasting. However, as discussed in Section 6.4, many modifications and alternatives to sandblasting are being studied and used because of potential hazard and contamination problems with sandblasting. Dry abrasive blasting can be very effective for removing surface rust; it also roughens the surface, increasing the contact area with the coating, which in turn promotes adhesion. Ultra high pressure hydroblasting at pressures above 25,000 psi is more effective for removing oil and other surface contaminants. Only a relatively limited area should be cleaned at a time, and primer applied as soon as possible. One of the authors was asked to explain why there had been a massive failure of coatings on a Taiwan oil tank farm located only a few hundred meters from the sea. The tanks had been carefully sandblasted, and a high performance coating system had been used. However, there was a delay of some days between sandblasting and coating application. In the meantime, spray from the sea resulted in salt deposits on the surface; the primer was sprayed over salt-contaminated steel. The salt crystals dissolved in water that permeated though the coating, leading in a few weeks to blistering and massive failure.

The coating system must be carefully selected for the particular installation. Most systems include at least two types of coatings: a primer and a top coat. Frequently, more than one layer of primer and/or top coat is applied, and in some cases, a combination of primer, an intermediate coat, and top coats is used. Primers provide the primary corrosion control, but intermediate and top coats also have significant effects on corrosion protection by reducing oxygen and water permeability of the combined films. Top coats also serve to protect the primer and provide other properties such as gloss, exterior durability, and abrasion resistance. Three classes of primers are used: barrier primers, zinc-rich primers, and primers containing passivating pigments.

32.1.1. Barrier Coating Systems

As discussed in Section 7.3, a requirement for a barrier system primer is that it have excellent wet adhesion. The primer should have a low viscosity continuous phase with slow evaporating solvents to permit as rapid and complete penetration as possible into microscopic cracks and crevices in the metal surface. Amine-substituted binders are particularly resistant to water displacement. Phosphate esters such as epoxy phosphates have also been shown to enhance wet adhesion. (See Section 11.5.) The T_g of the fully reacted binder should generally be only a little above the ambient temperature at which the curing is to be carried out. If the T_g of the fully reacted cross-linked film is too high, the rate of the cross-linking reaction becomes mobility controlled, and the reaction may stop prior to complete conversion. (See Section 2.3.2.)

Two package (2K) primers formulated with BPA and/or novolac epoxy resin in one package and a polyfunctional amine in the other (see Section 11.2) are widely used, since they provide good adhesion combined with excellent saponification resistance. Pot life is increased by using relatively high molecular weight epoxy resins and amine cross-linkers at relatively low solids, since this minimizes the concentrations of reactive groups. Especially when the solvents are slow evaporating, the vehicle can penetrate well into surface irregularities. Pigmentation should be high, but somewhat below critical pigment volume concentration (CPVC), to reduce permeability to oxygen and water and give a low gloss surface that enhances adhesion of top coat to the primer. Intercoat adhesion is further enhanced by applying top coat before cure of the primer is complete. Only a thin coat of primer is needed, but it is generally desirable to apply multiple coats of primer to assure that the entire surface of the metal has been coated. Sometimes, two different epoxy formulations are used: one as primer and the other for intermediate coats. To reduce VOC emissions, higher solids coatings are being adopted. As solids increase the challenge of formulating coatings with adequate pot life becomes more difficult. The lower molecular weight reactants in high solids coatings increase possibilities for toxic hazards. Water-borne epoxy-amine primers (see Section 11.2.6) are used in some applications when performance requirements are less stringent.

The poor exterior durability of BPA and novolac epoxy-amine coatings limits their use as top coats to interior applications, for which their generally excellent chemical resistance makes them useful. An exception is that most epoxy-amine coatings do not resist acetic acid (or similar organic acids) very well. In contrast to inorganic acids, acetic acid dissolves in the films; this effect is promoted by the presence of amines, especially if cross-link density is inadequate. Novolac epoxy resins have a higher average functionality and generally provide greater organic acid resistance than BPA epoxies.

Generally, top coats have a different composition than that of the primer. Chlorinated polymers such as vinyl chloride copolymer resins and chlorinated rubbers are used as top coat vehicles, since they have low moisture vapor and oxygen permeabilities. (See Section 16.1.) Chlorinated resins require stabilization against photodegradation. (See Section 5.4.) Since they are not cross-linked, they remain solvent sensitive and are, therefore, not appropriate for applications such as petroleum refineries and chemical plants. Use of chlorinated resins is limited by the high VOC required due to their high molecular weight. Polypropylene has low water permeability and is being used for top coats on pipelines [4]. A system of an epoxy primer, an intermediate coat of a polar-modified polypropylene copolymer, and a polypropylene top coat is used. Since polypropylene is subject to photoxidation, UV stabilization is required.

Two package urethane coatings (see Section 10.4) are increasingly used as top coats due to lower VOC and high solvent resistance of the cured films; urethanes are particularly useful when abrasion resistance is an important requirement. Moisture cure urethanes (see Section 10.6) have the advantage of being single package coatings. Two package water-borne urethane coatings (see Section 10.8) have the advantage of lower VOC. Alkyds are used, although their saponification resistance and exterior durability are inferior. Coatings based on alkyds are generally lower cost and have intermediate VOC emissions. As a result of their low surface tension, alkyd-based coatings are less likely to develop film defects during application. Latex-based top coats are being used more and more frequently; their use can be expected to increase as limitations on VOC emissions become more stringent.

Corrosion protection generally increases as water vapor and oxygen permeability of the coatings are reduced. (See Section 7.3.3.) High levels of pigmentation reduce permeability and cost, but also reduce gloss. Sometimes, low gloss intermediate coats are applied over the primer, followed by a high gloss top coat. Platelet pigments, such as mica and micaceous iron oxide, that can orient parallel to the surface of the coating as the solvent in the coating evaporates, are particularly effective in reducing permeability. Pigmentation of the final top coat with leafing aluminum flake pigment is desirable because an almost continuous layer of aluminum forms at the surface.

Since zinc rapidly develops strongly basic corrosion products, alkyds are not suitable vehicles for galvanized steel. An extensive study showed that epoxy-polyamide primer with a urethane top coat gave the best performance on galvanized steel when exposed for $4\frac{1}{2}$ years at Cape Kennedy, Florida [5].

32.1.2. Systems with Zinc-Rich Primers

As discussed in Section 7.4.3, zinc-rich primers provide excellent performance either when the surface cannot be completely cleaned of rust or when, as is the case with latex paints, complete penetration into surface irregularities cannot be achieved. The zinc acts as a sacrificial metal, protecting the steel from corrosion. To be effective, the level of zinc pigmentation must be such that PVC > CPVC. The high level of pigmentation provides for electrical contact between the zinc particles, and the porosity permits water to enter the film, establishing a conductive circuit with the steel surface. Even after a significant fraction of the zinc is consumed, the primer continues to provide protection, perhaps because the pores fill with $Zn(OH)_2$ and $ZnCO_3$, whose alkalinity may provide passivation [6]. In most applications, top coats are required to protect the zinc against corrosion, reduce probability of mechanical damage, and provide a desired appearance. See Ref. [7] for a review of the types of zinc-rich primers, their application and performance.

There are three classes of zinc-rich primers: inorganic, organic, and water-borne. In inorganic zinc-rich primers, the binder is a prepolymer derived by reacting tetraethyl orthosilicate with a limited amount of water. The solvent is ethyl or isopropyl alcohol, usually with another alcohol that has a lower evaporation rate to permit better flow. After application, the prepolymer reacts with water from the air to give a polysilanol network, which is partially converted to zinc salts by the zinc oxide on the surface of the pigment. (See Section 16.5.5.) Cross-linking is affected by the relative humidity when the coating is applied. At low humidities, especially if the temperature is high, film properties such as abrasion resistance may be adversely affected [8]. If it is necessary to paint under hot and dry conditions, the coating should be mist sprayed with water immediately after application.

Inorganic primers generally provide better protection than organic primers. It has been reported that in a seacoast environment, a six-year service life is estimated for inorganic primers as compared to three years for the organic type [9]. The binder in organic zinc-rich primers is usually an epoxy resin. Such organic primers command a share of the market by their advantages of greater tolerance to incomplete removal of oils from the substrate, easier spray application, and better compatibility with some top coats. For example, they are said to perform well on bridges in Michigan, an inland environment, but one in which salt is used to de-ice bridge decks [10]. In response to the need to reduce VOC, water-borne zinc-rich primers have been developed. The binder is a combination of potassium, sodium, and/or lithium silicates with a dispersion of colloidal silica [11]. Excellent performance on oil and gas production facilities in marine environments has been reported [12].

A challenge in using zinc-rich primers is proper selection and application of the top coat. Activity of the primer depends on maintaining its porosity by having PVC > CPVC. If top coat vehicle penetrates through pores in the primer, PVC of the primer is reduced to a level approximately equal to CPVC, reducing the effectiveness of the primer. Penetration of continuous phase of a top coat into pores is primarily controlled by viscosity of the external phase. Penetration can be minimized by applying a *mist coat* (*i.e.* a very thin coating) of top coat first. The solvent evaporates rapidly from a thin coating, so the viscosity of the continuous phase increases rapidly, minimizing penetration and sealing the pores. Proper application requires considerable skill, since excess mist coat in an area leads to increased penetration into the pores, but lack of coverage of an area means that the surface of the primer is not sealed in that area, so the vehicle of the next coat of top coat will penetrate into the pores. Sealing the pores minimizes pinholing or bubbling during application of thick layers of top coat. It is desirable for the color of the mist coat to contrast with that of the zinc-rich primer in order to aid the sprayer in applying complete, but not excessive, coverage. After the mist coat is applied, application of further coats requires no special spraying skill. Since alkaline zinc oxide, hydroxide, and carbonate are present on the zinc surface, top coats in contact with the primer coat must be stable against saponification. Two package urethane, vinyl, or chlorinated rubber coatings are used. Sometimes, an intermediate epoxy coat is applied to the primer, followed by a urethane top coat.

Since the continuous phase of latex paints contains no binder to penetrate into the pores, it would appear that latex paints should be ideal for at least the first coat over a zinc-rich primer. Latex polymers also meet the need for saponification resistance. However, latex paint films have high moisture vapor and oxygen permeability. Possibly, high permeability is desirable for the first coat in order to permit the water to diffuse out of the primer pores relatively rapidly. However, the balance of the top coats should have as low a permeability as possible. Permeability can be reduced by use of platelet pigments. Use of vinylidene chloride/acrylate copolymer latexes (see Section 8.2) as the binder for latex paints appears to be a promising approach [13]. Moisture vapor permeability through such films is decreased by the vinylidene chloride. However, stabilization of vinylidene chloride copolymer films against photodegradation over long time periods can be a major problem. Use of latex paints for corrosion protection is discussed further in Section 32.1.3.

32.1.3. Systems with Passivating Pigment Containing Primers

When extensive film damage must be anticipated and when the substrate cannot be completely cleaned (especially when oily rust will remain), or when the coating can not penetrate into the surface irregularities, passivating pigment primers are the primers of choice. (See Section 7.4.2.) In contrast to the situation with baked OEM product coatings, for which passivating pigments are seldom appropriate, there are many cases in field applications for which they are preferred despite the fact that blistering is more likely to occur.

A range of primer vehicles is used. Alkyds have the advantage that they are relatively low cost and wet oily surfaces, but they are deficient in saponification resistance. Epoxy-amine primers are used because of greater saponification resistance and good wet adhesion. Epoxy ester binders are intermediate in cost and performance.

Zinc yellow (see Section 7.4.2) was, for many years, the passivating pigment of choice. However, zinc yellow is carcinogenic, and care must be taken to avoid inhaling or ingesting spray dust, sanding dust, or welding fumes; in some countries, its use has been prohibited. Zinc yellow is being replaced by other passivating pigments; possible replacement pigments are discussed in Section 7.4.2.

For application over oily, rusty steel, red lead-in-oil primers have been used for many years. The surface to be painted should be cleaned as well as possible, but the primers can be used over sections from which oily rust has not been removed. The drying oil vehicle has a low surface tension and can wet the oily surface of the rust. Furthermore, the drying oil can dissolve the thin layers of oil from the rust surface. It is important that the viscosity of the external phase is low, the solvents used have slow evaporation rates, and the rate of cross-linking is slow so that the vehicle can surround the rust particles and penetrate through the rust aggregates to the substrate surface. There are no U.S. regulations prohibiting the use of red lead in such applications, but there is a strong trend to avoid lead pigments. Not only may lead pigments be banned by future regulations, but also elaborate, expensive precautions must be taken when the time comes to remove lead-pigmented paint.

In recent years, the use of latex paint systems for maintenance applications has increased. When water is applied to a freshly abrasive blasted steel surface, there is almost instantaneous rusting, called *flash rusting*. Flash rusting occurs with latex paints. To avoid it, the formulation should include an amine such as 2-amino-2-methyl-propan-1-ol (AMP). Use of mercaptan-substituted compounds as additives to prevent flash rusting has also been recommended [14].

Since latex particles are large compared to the size of many of the crevices in the surface of steel, and since the viscosity of the coalesced latex polymer is extremely high, complete penetration into the crevices cannot be expected. Therefore, it is essential to use a passivating pigment. The passivating pigment must be chosen such that the concentration of polyvalent ions is low enough so that package stability of the latex is not adversely affected, but still high enough that it can serve its passivating function. Strontium chromate is less soluble than zinc yellow and is preferred. Zinc phosphate has been used. Among the newer passivating pigments, zinc-calcium molybdate and calcium borosilicates have been recommended.

Acrylic, styrene/acrylic, and vinylidene chloride/acrylic latex polymers are completely resistant to saponification. Special proprietary latexes are sold that provide enhanced adhesion to metal in the presence of water. It has been suggested that amine-substituted latexes are especially appropriate for wet adhesion. Use of 2-(dimethylamino)ethyl acrylate or methacrylate as a comonomer is a way to incorporate amine groups on the latex polymer. Methacrylamidoethylethyleneurea has also found use as a wet adhesion promoting monomer. Another approach to enhancing wet adhesion is to use an alkyd or a modified drying oil to replace part of the latex polymer. The alkyd is emulsified into the coating. After application, as the water evaporates, the emulsion breaks and some of the alkyd can penetrate into the fissures in the steel surface. Epoxy esters are more hydrolytically stable than alkyds and would be expected to provide better corrosion protection.

Owing to the high moisture vapor and oxygen permeability of most latex paint films, it is desirable to use some platelet pigment in the formulation. Mica is used in both primers and top coats. Final top coats using leafing aluminum (see Section 19.2.5) are particularly appropriate. To avoid problems of reactivity of aluminum with water, a special grade of leafing aluminum is mixed into a latex base paint just before application.

The need for reducing VOC emissions has been particularly critical in California. Since commercial paint suppliers did not have latex paints available for maintenance coatings for bridges, the California State Office of Transportation Laboratory undertook the formulation and application of latex paints to several highway bridges in order to evaluate their potential utility [15]. Latex primers were used in some cases and inorganic zinc-rich primers in others; in all cases, latex top coats were used. It is reported that some systems were still standing up well after over five years field exposure. Based on the experience obtained, it was recommended that application should only be done when the temperature is above 10°C and when the relative humidity is less than 75%. Use of latex systems can be expected to expand as requirements for reduced VOC emissions become more stringent.

32.2. MARINE COATINGS

The marine coatings market includes coatings for pleasure craft, yachts, as well as for naval and commercial ships. Coatings for pleasure craft are commonly sold on a retail basis to the individual owner. A range of products is available, one for wood is spar varnish. The original spar varnish (see Section 14.3.2) was a phenolic-tung oil varnish; the tung oil provides high cross-linking functionality, and the phenolic resin imparts hardness, increased moisture resistance, and exterior durability. While some phenolic-tung oil is still used by traditionalists, the bulk of this market has shifted to uralkyds (see Section 15.7), which provide greater abrasion and water resistance.

The larger part of the market is for commercial and naval shipping and some seaside installations that are exposed to marine environments. A variety of substrate surfaces are coated, and many require special coatings. Reference [16] provides a useful overview of the market and products.

Coatings for exterior surfaces above the waterline are generally analogous to those supplied for heavy duty maintenance as discussed in the previous section. Ultra high pressure hydroblasting is used to clean the surface; salt is a common contaminant in shipyards and hydroblasting removes all salt from the surface. A primer must be applied soon after blasting to avoid contamination. Inorganic zinc-rich or epoxy primers are used. In a study of the performance of a range of coatings over zinc-rich primers on panels exposed for 10 years in Jacksonville, FL, inorganic zinc-rich primer with an intermediate epoxy coating and an aliphatic polyurethane top coat gave the best overall performance [17]. Inorganic zinc-rich primer with an epoxy second coat and a vinyl top coat received the next highest rating for overall exterior performance. Where the requirements are somewhat less demanding, a two component epoxy primer with a silicone alkyd top coat may perform adequately.

The exterior durability requirements of coatings for ship superstructures are severe, since in addition to the direct UV, further UV is reflected off the water surface, and the humidity is also generally high. Alkyd coatings have been the standard; however, urethane coatings are being used increasingly because of their greater durability. Especially for interior paints, fire retardancy is important; chlorinated alkyd coatings have been recommended [18]. Latex paints are used to a minor extent; although initial gloss is lower with latex than with alkyd coatings, gloss retention and resistance to cracking are superior. The low solvent content of latex paints reduces fire hazard of paint in storage on a ship. Deck paints have the further requirement of being skidproof. 2K aliphatic urethane coatings satisfy the exterior durability and abrasion resistance requirements. Skid resistance is obtained by mixing coarse sand into the paint just before application.

For many ship bottom areas, 2K coal tar-epoxy coatings have been found to be particularly effective; they have also been used in ballast areas. The coatings must be handled with care in view of the potential carcinogenicity of components of coal tar. Coal tar is being replaced by petroleum derived hydrocarbon resins; in addition to lower toxicity hazard, the lighter color of hydrocarbon resins facilitates inspection in ballast areas. One package contains hydrocarbon resin, amine-terminated polyamide resin, and pigment; the other contains the epoxy resin. Multiple coats are applied to give a total film thickness on the order of 400 μm. The coatings have high chemical resistance and high dielectric strength. This latter property is especially important for ships that use zinc or magnesium sacrificial metal anodes to provide cathodic protection against corrosion. (See Section 7.2.2.) Over properly prepared surfaces, the expected lifetime in seawater immersion is on the order of seven years [16]. Pigmentation with leafing aluminum enhances performance. To protect against environmental disasters, most oil tankers entering American coastal waters are required to have a double-hull construction. This has increased the surface area of ballast spaces, which is as much as six times the outer hull area. The difficulty and high cost of repair work in double-hull construction dictates that the coating systems offer very long term protection against corrosion.

Fluorinated urethane coatings are used in tanks such as fuel tanks, septic tanks, and sometimes bilge tanks [19]. Fuel tanks must be protected against corrosion, since they are frequently filled with seawater ballast after the fuel is consumed. The lower surface free energy surface of the coatings simplifies cleaning.

The most challenging class of marine coatings is *antifouling* coatings. Spores and larvae of a variety of plants and animals, ranging from algae to barnacles, can settle on the underwater hull of a ship. The growth of plants and animals on ship bottoms increases roughness of the hulls, which in turn increases turbulence of water flow over the surfaces and drag; hence, speed decreases and fuel consumption increases. Removal of such growths requires putting the ship into dry dock. The economic penalty of fouling is very large.

The major approach has been to apply coatings that contain biocides that leach out of the coating. The biocide must be a very general toxic agent. The leaching rate must be above a critical level required to kill all the organisms settling on the surface over a long period of time. Leaching rates decay exponentially in accordance with a first order release of biocide. The first biocide used on a large scale was cuprous oxide. Typical binders are solution vinyl resins containing rosin salts. The binder is water sensitive, so loss of cuprous oxide continues after the surface layer is depleted of cuprous oxide. To have sufficient cuprous oxide at the end of the life of the coating, the rate of loss of cuprous oxide in the early stages of use has to be excessive. The antifouling lifetimes of such coatings in actual use are on the order of 7 to 24 months.

The economic need was for a coating with a lifetime of 48 months or longer. Organotin toxicants like tributyltin oxide initially looked promising, but were found in practice to be little, if any, better than cuprous oxide. A breakthrough came with the development of organotin substituted polymers that release tin compounds on hydrolysis [20,21]. Copolymers of organotin esters of methacrylic acid with conventional acrylates gave excellent antifouling paints for aluminum pleasure craft when used in white paints pigmented with zinc oxide. Unfortunately, it was found that such coatings disappeared from the bottoms of commercial seagoing vessels, sometimes in a single voyage, because they were too soluble in seawater. Copolymers were developed having lower solubility, but still sufficient solubility to be washed off the surface of the film after release of the tin biocide. The rate of hydrolysis is controlled by the design of the copolymer and also by the incorporation of a slightly water-soluble pigment. Leaching of the pigment from near the surface of the film increases the contact area between polymer and water, increasing hydrolysis rate. Part of the pigment can be cuprous oxide to provide further biocidal activity. Since the rate of loss of biocide is controlled by the rate of hydrolysis at the surface, the rate of leaching is approximately linear with time, rather than exponential. Furthermore, slow dissolution of polymer keeps the surface of the film smooth, which favorably influences speed and fuel consumption. These coatings are called *self-polishing coatings*. Depending on conditions, service lifetimes are up to 5 years.

When ships are in port, leaching of toxicants continues. In some harbors, concentrations of toxicant can build up sufficiently to affect marine life. There is concern about potential heavy metal entrance into the food chain, affecting fish that might be caught for human consumption. Therefore, regulations controlling biocide release are getting more stringent. In the mid-1980s use of tributyltin compounds was banned on vessels less than 25 meters in length, affecting most pleasure craft. The International Maritime Organization is drawing up legislation to restrict its use on all commercial vessels [22].

Usage has increased again of copper based antifouling paints, self-polishing copper acrylate coatings have been made available [22]. Copper release by hydrolysis of the polymer is not sufficient to provide antifouling on its own; therefore, copper oxide and a biocide are used with it. An example of an organic toxicant is 4,5-dichloro-2-*n*-octyl-4-isothiazolin-3-one. It is reported to be an effective toxicant, but it degrades rapidly in seawater so that there is no bioaccumulation [23]. Reference [20] provides further discussion of antifouling paints, regulations, and approaches to alternative toxicants.

Research in recent years has been aimed at attempts to find means of controlling fouling other than by use of toxicants. A promising approach is the development of coatings to which adhesion of fouling organisms is so limited that ship movement through the water or underwater brushes or hoses are sufficient to remove any fouling without dry docking. Studies have shown that silicone elastomers with added methyl phenyl silicone fluids give surfaces that exhibit considerable fouling resistance over a period in excess of 10 years [20,24]. However, abrasion and tear resistance are somewhat limited. Research is directed at improving abrasion resistance and tear strength of silicone elastomer coatings. Use of volatile methylsiloxanes, which are exempt under VOC regulations, as a solvent permits use of higher molecular weight elastomeric silicones in the binder, thereby improving the physical properties [25]. A silicone elastomer system has been commercialized for use in the newly emerging fast ferry market, for which the high speeds (over 30 knots per hour) displace fouling organisms that manage to attach to the coating [26]. The coating is applied over an epoxy primer for corrosion protection. Service life is over three years, and the excellent leveling properties give a smooth surface, which reduces drag. The U.S. Navy has been evaluating this technology for its fast ships. Other approaches to release coatings are possible, and it seems probable that antifouling coatings of the future will be based on principles other than the use of toxic agents.

32.3. AUTOMOBILE REFINISH PAINTS

The value of coatings for cars and trucks outside of assembly plants is almost as large as that for OEM automotive coatings, discussed in Section 29.1. Some refinish paints are used for overall finishing of cars and new trucks in special colors, but the largest segment of the market is for repair. When a car has been in an accident and a fender is straightened or a new door is installed, these parts must be coated so that the color matches the color of the original coating. There are major technical problems meeting the demanding application and performance requirements, and also major marketing and distribution problems. There are hundreds of kinds of cars made each year, each in as many as a dozen different colors. While most cars on the road are five years old or less, many are well over ten years old. If there is an accident involving a 15-year old Jaguar, the owner expects to be able to bring it to a repair shop and have that section of the car painted to match the rest of the car. Furthermore, the *body shop* expects to be able to call the paint distributor and have the necessary liter of paint delivered by the next day. According to EPA figures, refinish shops in the United States emit almost twice as much VOC than the paint shops of all auto and truck assembly plants in the United States. The refinish industry is under increasing pressure to reduce emissions, complicating the technical challenge.

In some cases, coating manufacturers make and stock small containers to match the colors. These are ready to be shipped to dealers and repair shops before the first new car comes off the assembly line. However, these "factory packaged" colors are limited due, to inventory costs and the variations in color shade at the OEM manufacturing site. In most cases, the manufacturers provide formulations and tinting color bases to distributors (or larger repair shops) to permit matching any original color. Establishing the formulations requires great color matching skill, especially for metallic colors.

Both air dry and force dry (65–80°C) coatings are used. The heat in force dry coatings makes a large difference in enhancing the quality of the repair. In Europe and Japan, most shops are equipped with force dry ovens. In the United States, use of force dry ovens is increasing, but there is still a significant fraction of repair shops that use air dry coatings.

Repair of damage after an accident is usually accomplished with replacement parts, as labor costs to bump out dents have become too high. Replacement parts are usually received from the manufacturer coated with electrodeposition primer. In some cases, the damaged panel is repaired, then surface preparation is critical. The old coating surface must be cleaned to remove dirt, tar, and wax by scrubbing with detergent, rinsing thoroughly, and drying; in some cases solvent cleaners are required. Scuff sanding may be required to remove chalky pigment and degraded polymer. If the old paint is cracked, it must be sanded down to bare metal. Any breaks through to metal require that the edge of the area be feathered out, that is, sanded with a bevel so that there is a smooth change in film thickness. When bare metal is exposed, it must be washed free of grease with solvent.

A primer-surfacer is applied after checking whether the coating on the car can withstand the primer-surfacer solvent. Current OEM coatings give no problem, but if the car has been refinished in the relatively recent past, the coating might be only partially cross-linked and lift when solvent is applied to it. In the past in the United States and still in South America and Asia, the most common primer-surfacers are made with nitrocellulose-alkyd binders. To achieve fast dry, a medium oil length, rosin-modified, tall oil alkyd is used [27]. The rosin modification increases the T_g of the alkyd. The primers are highly pigmented; for example, with a PVC of 38% and sometimes even higher. The use of nitrocellulose with rosin-modified alkyd and high levels of pigmentation permit sanding the primer-surfacer within 30 minutes after application. The primer is sanded smooth with fine grit paper, and the spray dust is removed. Application solids of such primers is low. Stricter VOC controls and OEM warranty repairs are forcing the use of epoxy and urethane primers, which also give superior performance. In areas with the most restrictive VOC regulations, water-borne undercoats are being used. The initial products have been formulated with acrylic latexes or polyurethane dispersions. Drying of these coatings is dependent on humidity, and the time required for drying in high humidity are excessive. They are being replaced by thermosetting 2K water-borne epoxy (see Section 11.2.6) or urethane (see Section 10.8) primers.

Two broad classes of refinish top coats are used: lacquer (thermoplastic) and enamel (thermosetting). A major application advantage of lacquer is fast dry; they have short *out-of-dust* times (i.e. the time required for the film to become sufficiently dry that dust particles do not adhere to the surface). There tends to be a great deal of dust in the air in repair shops, and fast dry is a major advantage, reducing contamination of the freshly applied coating. A disadvantage of acrylic lacquers for refinish is that when dried at room temperature, their gloss is not high enough. To match the gloss of the OEM applied coating, refinish lacquers must be polished with rubbing compound, adding to the cost.

In acrylic refinish lacquers, thermoplastic acrylic polymers with high methyl methacrylate content (see Section 2.2.1) are blended with cellulose acetobutyrate (CAB) and a plasticizer, such as butyl benzyl phthalate. Gloss retention is good, but inferior to current OEM coatings. Application solids are low (10–12 NVV) and VOC emissions are very high. Lacquers are now used in North America only by shops that do not have spray booths, for antique car restoration, and by the do-it-yourself market.

The other broad class of refinish top coats is enamels; there are several types. The oldest and lowest cost are alkyd enamels; the alkyd is a medium oil, isophthalic soy or tall oil alkyd. (See Section 15.1.) These enamels have several advantages besides low cost. Fewer film defects such as crawling or cratering occur during application, due to their low surface tension. Their gloss is high enough to match OEM monocoats without polishing. VOC emissions are substantially less than with lacquer. On the other hand, out-of-dust time approaches half an hour, and if recoating is necessary, it must be done within four hours after application or after 24 hours. At intermediate times, when the film is partially cross-linked, a second coat of enamel can lead to lifting. Gloss retention of alkyd coatings is poorer than that of OEM coatings or refinish lacquer.

The properties of alkyd enamels have been improved in several ways. Triisocyanates, such as isophorone isocyanurate (see Section 10.3.2), can be added just before spraying. They serve as auxiliary cross-linkers, reacting with free hydroxyl groups of the alkyds. This supplemental cross-linking reduces out-of-dust time and recoating problems. Another approach is to use methacrylated alkyds. (See Section 15.4.) These alkyds give shorter out-of-dust times, but methacrylated alkyds cross-link less rapidly, so the films remain sensitive to gasoline for a protracted time. Another type resin is made by reacting an acrylic resin made with glycidyl methacrylate as a comonomer with drying oil fatty acids. Films from such a vehicle show out-of-dust times and durability approaching acrylic lacquers and superior to alkyds. Since the relatively high T_g acrylic backbone provides the necessary initial dry, driers are not needed to accelerate the oxidative cross-linking. The absence of driers improves exterior durability.

With the increasing use of base coat-clear coat finishes (see Section 29.1.2), corresponding coatings for use in refinishing are required. The base coat must be able to hide with two coats. Relatively low solids are required to permit alignment of aluminum or other flake pigments. Solvent evaporation must be relatively fast to allow clear coat to be applied within an hour without disturbing the flake orientation. Acrylic lacquer was initially used for refinish base coats, but is being replaced with a binder based on a polyester with cellulose acetobutyrate to accelerate drying and a polyethylene-polyvinyl acetate wax suspension as a rheology control to assist in fixing the flake. Such base coats have a relatively high VOC and are used with very high solids clear coats (up to 85% weight solids) so that the combined VOC is within regulations. Water-borne base coats based on either acrylic latexes or polyurethane dispersions are being made available. These water-borne coatings have a volume solids of 15%, which permits flake orientation comparable to OEM coatings. Treated aluminum flake must be used to avoid hydrogen evolution. (See Section 19.2.5.) Rheology control additives are also required. Relative humidity conditions affect the drying speed; the coatings can be force dried or air dried with high volume dry-air flow to remove most of the water before clear coating.

Two package urethane enamels have become the most important for clear coats. Thermosetting hydroxy-functional acrylic resins similar to those used in OEM coatings can be used together with a polyfunctional isocyanate cross-linker. Polyfunctional aldimines and hindered amines are being used to further increase solids. (See Sections 10.4 and 24.2.2.) Monocarboxylic acid-modified polyesters cross-linked with HDI or IPDI isocyanurate have also been recommended [28]. Durability of these coatings is in the same range as OEM coatings, and they give high gloss without polishing. Their out-of-dust times are intermediate between lacquers and alkyd enamels. The out-of-dust times can be reduced by changing the amount of catalyst (often dibutyltin dilaurate), but pot life is also reduced. Refinish shops cannot generally afford dual-mixing guns, so the pot life has to be at least long enough to permit complete spraying of a car—over an hour. See Section 24.2.2 for discussion of approaches to formulating coatings with reduced VOC and increased pot life.

VOC emissions have been reduced by changing from conventional air spray guns to high volume, low pressure air guns (HVLP). HVLP guns permit better transfer efficiency, and, hence, lower paint usage and lower VOC emissions. (See Section 22.2.1.)

While paint suppliers emphasize the need to wear masks and to mix and apply the paint in a well-ventilated spray booth, some refinishers do not follow recommended safety precautions. Especially if cross-linkers with some volatile diisocyanate components are supplied, sprayers can develop respiratory problems. Some shops will not use 2K urethane coatings. Refinish coating suppliers are pursuing nonisocyanate cross-linking systems. It is important to bear in mind that any reactant that can cross-link hydroxyl, carboxyl, or amine groups on a synthetic polymer can also cross-link proteins. Any such cross-linker will be toxic. A new system must be designed so that there is no volatile reactive component and so that the molecular weight is high enough to minimize permeation through the skin and membranes. Silylated acrylic resins that are said to combine good appearance, mar resistance, acid resistance, and exterior durability have been developed [29]. The silyl groups are introduced using 3-trimethoxysilylpropyl methacrylate as a comonomer. After application, the resins moisture cure through the silyl group. (See Section 16.5.4.)

32.4. AIRCRAFT COATINGS

The U.S. market for aircraft and aerospace coatings is estimated to be $60 million [30]. The majority of the business is for exterior primer and top coats for aircraft—roughly 50–50 for commercial and military aircraft. Excellent adhesion and corrosion protection are required for a range of substrates, primarily aluminum and aluminum alloys, and increasingly, composite plastics. The high strength aluminum alloys used for aircraft structures are more vulnerable to corrosion than ordinary aluminum. The coatings must resist swelling by phosphate ester hydraulic fluids, lubricating oils, and fuel. Swelling resistance is usually obtained by relatively high cross-link density. However, the coatings must also have good flexibility at low temperatures and excellent abrasion resistance while the aircraft is flying through dust, rain, or sleet at high speed. Flexibility and abrasion resistance tend to decrease with increasing cross-link density.

Epoxy-amine (see Section 11.2) primers are generally used. For interior areas, such as in hold or cargo areas, internal structural members, and wheel wells, the primers are not top coated. The interior coatings are expected to last for the life of the aircraft and are not stripped and recoated as exterior coatings are. They are generally formulated with polyfunctional novolac epoxies (see Section 11.1.2) with amido-amines or amine adduct curing agents to give relatively high cross-link density and maximum resistance to lubricating and hydraulic fluids. Passivating pigments are used to control corrosion in case of breaks through the film. Strontium chromate is preferred over zinc yellow (see Section 7.4.2), since its solubility in water is somewhat lower and, therefore, the rate of leaching of the pigment from the film by water is lower. Strontium chromate also has greater heat resistance than zinc yellow. Chrome free primers are being evaluated.

Primers for exterior surfaces are designed with lower cross-link density by using only, or primarily, BPA epoxy resins. Amine-terminated polyamides (see Section 11.2.2) are usually used as the cross-linking agents. The long chain fatty acid parts of the molecule provide good wetting to the metal surface and flexibility to the final film. Cross-linking is catalyzed with 2,4,6-tris(dimethylamino)phenol. The lower cross-link density provides greater flexibility and permits easier stripping of the coatings for repainting. Strontium chromate is used as a pigment, and again, chrome free primers are being evaluated.

Top coats are almost always 2K urethane coatings. Most commonly, hydroxy-terminated polyesters are used with isocyanurate trimers from hexamethylene diisocyanate or isophorone diisocyanate; commonly, a high ratio of isocyanate to hydroxyl is used so that the coatings are partially moisture cured. (See Sections 10.4 and 10.6.) Cross-linking is catalyzed by organotin compounds.

Aircraft coatings have had relatively high solvent contents, but there is pressure to reduce VOC emissions. Water-borne epoxy-amine primers are increasingly used [30]. (See Sections 11.2.6 and 25.3.) Some reduction of solvent in top coats has been achieved by using lower molecular weight resins and polyisocyanates, but progress has been limited by the requirements for rapid dry at ambient temperatures, which is harder to achieve as molecular weight is reduced. Aldimine based coatings are being evaluated. (See Section 10.4.)

The exteriors of airplanes are frequently repainted. The biggest problem is not application of the new paint, but removal of the old paint. The paint strippers that have been used for many years contain substantial amounts of methylene chloride, which is toxic and environmentally undesirable. Studies are underway to eliminate the use of strippers by using mechanical paint methods for removal. The removal method must not weaken the metal or composite plastic substrate. For example, sandblasting will remove coatings, but it also erodes the metal surface. Alternative blasting media like plastic particles, walnut shell granules, dry ice pellets, and crystalline starch are being investigated. (See Section 6.4.1.)

REFERENCES

1. M. S. Reich, *Chem. Eng, News*, **75**, 36 (1997).
2. J. D. Kearne, Ed., *Steel Structures Painting Manual*, Vol. I, *Good Painting Practices*, 3rd. ed., 1993; Vol. II, *Systems and Specifications*, 7th ed., Steel Structures Painting Council, Pittsburgh, PA, 1995; C. H. Hare, *Protective Coatings*, Steel Structures Painting Council, Pittsburgh, PA, 1995.
3. J. W. Martin, S. C. Saunders, F. L. Floyd, and J. P. Wineburg, *Methodologies for Predicting Service Lives of Coating Systems*, Federation of Societies for Coatings Technology, Blue Bell, PA, 1996.

4. G. P. Guidetti, G. L. Rigosi, and R. Marzola, *Prog. Org. Coat.*, **27**, 79 (1996).

5. R. W. Drisko, *J. Prot. Coat. Linings*, **12** (9), 27 (1995).

6. S. Feliu, Jr., M. Morcillo, J. M. Bastidas, and S. Feliu, *J. Coat. Technol.*, **65** (826), 43 (1993).

7. H. H. Kline, *J. Prot. Coat. Linings*, **13** (11), 73 (1996).

8. G. Eccleston, *J. Prot. Coat. Linings*, **15** (1), 36 (1998).

9. J. Paert, *J. Prot. Coat. Linings*, **9** (2), 50 (1992).

10. E. Phifer, *J. Prot. Coat. Linings*, **9** (2), 48 (1992).

11. E. Montes, *J. Coat. Technol.*, **65** (821), 79 (1993).

12. A. Szokolik, *J. Prot. Coat. Linings*, **12** (5), 56 (1995).

13. H. R. Friedli and C. M. Keillor, *J. Coat. Technol.*, **59** (748), 65 (1987).

14. G. Reinhard, P. Simon, and U. Remmelt, *Prog. Org. Coat.*, **20**, 383 (1992).

15. R. Warness, *Low-Solvent Primer and Finish Coats for Use on Steel Structures*, Technical Report, FHWA/CA/TL-85/02 (1985).

16. H. R. Bleile and S. Rodgers, *Marine Coatings*, Federation of Societies for Coatings Technology, Blue Bell, PA, 1989.

17. C. M. Sghibartz, *FATIPEC Congress Book*, Vol. IV, (1982) p. 145.

18. L. V. Wake, J. R. Brown, and Z. Mathys, *J. Coat. Technol.*, **67** (844), 29 (1995).

19. R. F. Brady, Jr., *Polym. Mater. Sci. Eng.*, **78**, 249 (1998).

20. E. B. Kjaer, *Prog. Org. Coat.*, **20**, 339 (1992).

21. C. D. Anderson, *Proc. Ship Repair Conversion Conf.*, 1993, p. 93.

22. J. E. Hunter, *Protective Coat. Eur.*, November, 16 (1997).

23. L. Raber, *Chem. Eng. News.*, **74**, (29), 9 (1996).

24. W. A. Finzel, and H. L. Vincent, *Silicones in Coatings*, Federation of Societies for Coatings Technology, Blue Bell, PA, 1996.

25. G. G. Bausch and J. S. Tonge, *Proc. Water-Borne High-Solids Powder Coat. Symp.*, New Orleans, 1996, p. 340.

26. J. Millett and C. D. Anderson, *Proc. Fast Ferries '97 Conf.*, Sydney, 1997, p. 493.

27. B. N. McBane, *Automotive Coatings*, Federation of Societies for Coatings Technology, Blue Bell, PA, 1987.

28. T. F. Wilt, J. M. Carney, S. J. Thomas, J. A. Claar, and W. J. Birkmeyer, U.S. Pat. 5468802 (1995).

29. M. J. Chen, A. Chaves, F. D. Osterholz, E. R. Pohl, and W. B. Herdle, *Surface Coat. Intl.*, **79**, 539 (1996).

30. A. K. Chattopadhyay and M. R. Zentner, *Aerospace and Aircraft Coatings*, Federation of Societies for Coatings Technology, Blue Bell, PA 1990.

CHAPTER 33

Perspectives on Coatings Design

Formulation of coatings is a challenging assignment. Sometimes, scientists look down their noses at the lowly formulator without realizing that the task of formulating a new coating can be more technically challenging than much so-called pure research. Any coating must meet a multitude of requirements. There are innumerable possible raw materials, combinations, and proportions. Test methods are generally subject to large ranges of error and frequently do not give results that predict use performance well. The formulator is faced with variable substrates and application methods. There are commonly severe cost constraints. Frequently, the volume of any one coating is too limited to justify expenditure of large amounts of time; furthermore, time available to solve the problem is usually limited. In fact, as one looks at the complexities of the field, one sometimes wonders how a useful coating ever gets formulated.

Historically, the problems of formulation were somewhat eased by following a procedure of making small modifications of coatings known to be satisfactory. Over time, excellent coatings were formulated in this way. This approach of continuous improvement, with a focus on the customer's needs and close contact between the formulator and the customer's engineering group, continues to be important. However, it has become necessary to make major changes in formulations in less time than formerly used to make even small changes. This need results from a variety of factors, but particularly from the introduction of VOC emission controls and the increasing number of raw materials identified as having potentially serious toxic hazards.

Most of this chapter is written from the perspective of the United States, but our experience teaching courses in many countries around the world indicates that the differences in the challenges to coatings design in different countries are small compared with the common denominators.

The challenge of formulation is intensified by the need for increasing productivity and creativity. We all want more of the good things of life, but the only way this desire is made possible for the population as a whole is to increase the aggregate productivity. Some politicians and some uninformed segments of the population tend to associate productivity problems with production line workers. Clearly, there is need for increased productivity of labor, but overall productivity depends on the efficiency with which management, sales personnel, accountants, clerical personnel, and, yes, laboratory personnel, including formulators, work. In the inimitable words of Pogo, we "have met the enemy and they is us."

At least as important, and perhaps more important, is the need for increased creativity in formulation. Some of the problems facing the coatings formulator seem "impossible." Some of them are, in fact, impossible, but many so-called impossible problems aren't really impossible, but require creative visualization of ways around the supposedly insuperable hurdles. Enhanced creativity should be a critical component of most aspects of people's lives, and the need for creativity is especially important in technical people, including formulators.

The coatings industry has been known as an industry with a relatively low profit margin. It is commonly said, "we cannot invest more money in research and development because the profit margins are so low." Actually, the inverse is probably truer: Profit margins are low because a large fraction of the technical effort is spent trying to copy competitors' products and/or applying the same old ideas that every other formulator is trying, rather than on innovative research and development.

There is no magical single route to increasing creativity and productivity. This chapter provides some ideas based on the accumulated experience of three technical people who have worked in different aspects of the coatings field. Other people will have other suggestions; the particular approach is not important, but the end result is critical.

There are several important aspects involved in working on a formulating problem. While, to a degree, these are sequential steps, continuous review and reevaluation of these aspects is essential as the work progresses.

Define the problem. The first stage in working on a formulating problem is to define it. This seems so obvious that one would think that it need only be mentioned in passing. However, experience shows that inadequate definition of the problem is a large factor in unsuccessful technical projects. Appropriately defining a formulating problem is in itself a difficult and relatively time consuming effort, but effort invested at this stage often saves time overall. It is particularly critical to define a coating problem in terms of performance requirements. The statement "Formulate a coating to meet Specification Number XXXX with a 30% reduction in VOC emissions" does not define the performance requirements in most cases. Very often, the specification is based on quality control tests that a satisfactory previous coating happened to pass, but quality control tests are not designed to predict performance of a new product. Consequently, an effective formulator must know or find out what the actual end use requirements are.

One should also question the statement of a problem that he or she receives. On one occasion, a laboratory was asked to formulate a "harder" coating; this was done and the customer accepted the new formulation, but after some months of use, the customer complained loudly that the new coating had a poor wear lifetime. The customer really wanted improved abrasion resistance and assumed that greater hardness would provide it. The customer was mistaken, but the fault was not really his, the fault was that of the coating formulator who didn't ask, "Why does the customer think he wants a harder coating?" There is great need, if at all possible, for direct contact and interaction between the customer and the formulator who works on a problem.

It is not adequate to define a problem as: "Match competitor A's coating No. YYY at a lower cost." First of all, this definition propagates a self-fulfilling prophecy—that profits in the paint industry will continue to be low. Second, by the time the work has been finished, the competitor may have produced a better product than YYY, and thus, a match for YYY is no longer adequate to get the business. Third, the difficulties of analysis, especially for additives, are such that one can seldom analyze a competitor's formulation precisely enough to duplicate it. Matching the laboratory test results of the competitor's product does not assure that the product performance will be equal. The only satisfactory definition of the technical aspects of a project is a listing of detailed performance requirements. In any listing of performance needs, there is a range of degrees of importance of the various needs. Some are essential requirements, others are important, still others would be "nice if we could do it," and still others are in the category of "as long as we are pipe dreaming, why not put this goal down too." The laboratory worker should clearly understand how each of the performance needs fit into such a scale.

Applicable regulations must be known—not just current regulations, but best estimates of what they will be over the lifetime of the potential project. No one can accurately predict future regulations, but a choice as to goals from this point of view must be made. After all, assuming that current regulations are going to continue unchanged is also a prediction of the future. Toxicity problems should be assessed, taking into consideration possible future developments. Starting a new study of a corrosion protective primer based on only zinc yellow pigment, with a target date for significant sales five years or so, ahead is risky, since it is known that zinc yellow is a human carcinogen.

Test methods must be agreed on. Few laboratory tests are adequate predictors of use performance. Since coatings are complex compositions and end use requirements are variable, there is a real danger in relying on any one test. Some widely used tests, such as salt spray tests for corrosion protection, have been repeatedly shown not to correlate with end use results. (See Section 7.5.) It is absurd to base decisions for a major research project on whether or not the product will or will not pass a salt spray test. The problem of evaluation should be faced in advance. A variety of tests relating to mechanical, spectroscopic, and thermal properties should be also carried out using coatings with known field performance as standards for comparison. Dickie has published a methodology for integrating the results of laboratory performance tests, field history, environmental factors, design parameters, and the fundamentals of degradation to predict service performance [1]. The problems involved with predicting service life of coatings have generally been underestimated. See Ref. [2] for a review of the problems of prediction and the proposed use of reliability theory for developing better methodologies.

Cost requirements are an integral part of the definition of a project. The formulator must know what the real permissible upper cost is, not just what some salesperson hopes the cost can be kept down to. Similarly, the timing requirement must be known. Some projects have to be finished by a certain date. Some do not have a specific deadline, although obviously, the sooner they are finished, the better. Unrealistic cost and time goals can lead to wrong decisions in project planning. The technical person has the obligation to let any possible time slippage or cost increase be known to others as soon as he or she is aware of it.

The potential value of a project should be compared with the estimated total cost of the project. Some companies use elaborate discounted cash flow methods of analyzing the potential value of a project. Our experience has been that such analyses have killed more good projects than bad ones. It should be recognized that estimates can be used to "prove" that any project will or will not be economically sound. On the other hand, there have been frequent occasions when a little thought would have shown there was no sensible possibility of a return approaching the potential cost of the project.

Unfortunately, misunderstandings about the definitions of projects are all too common. A way of minimizing such misunderstandings is to have the person who will be doing the work write out his or her understanding of the project. It is far better to take the time required to reach consensus on a project definition at the outset than to risk disagreements midway or even near the project's end.

Background search. Too often, laboratory workers jump on their horses and ride off in all directions. First, one should assess the available knowledge. Review the pertinent scientific literature, review (with appropriate concern for bias) suppliers' technical data bulletins; review any pertinent background with fellow workers, in the company's files, and on the internet. Discussions with the customers' technical and engineering personnel can be useful. Increasingly, companies are accumulating computer data banks that compare actual field performance with composition variables; such data can provide useful ideas. It is particularly critical to be sure that a problem is not impossible. Some problems *are* impossible, and there is no point in working on them. No amount of wishful thinking will ever permit the development of a flat, jet black paint or a white paint with gloss equal to that of a high gloss jet black paint. (See Section 18.10.1.) No one can match the color of a gloss coating at all angles of illumination and viewing with that of a low gloss coating. (See Section 18.9.1.) No one will ever make a kinetically controlled one package coating with 6 months package stability at 30°C that will cure in 30 minutes at 80°C. (See Section 2.3.2.) Stop such projects before they start.

Identify Approaches. Commonly, a technical person starts working on a project based on the most obvious approach to the problem. This approach is probably the same approach with which his or her counterpart in a competitor's laboratory will start. The greatest opportunity to apply creativity comes early in a project. Devise all of the approaches you can to solving a problem; solicit ideas from fellow workers. Don't be dissuaded by veterans who say, "We tried that 20 years ago, and it didn't work." Assess the merits of their comments, but in 20 years, many other things have changed; what didn't work then might work now. Set the problem aside for a few days and try again to come up with really different ways by which the problem might be solved. Open up your mind; get out of ruts. Having accumulated a variety of possible approaches on which to work, try to assess their merits and pick the one or two most promising. Sometimes, the one picked will be the first approach identified, but commonly, it is not.

As the basic understanding of the factors controlling the performance of coatings increases, the opportunity for basing experimental approaches on sound scientific understanding increases. It is sad to see the failures to apply known understanding. To take an obvious example, it is well established that in room temperature cure coatings, cross-linking may become limited by the availability of free volume. If the T_g of the fully reacted system is significantly above room temperature, the cross-linking reaction will slow and probably stop before completion. Why bother to prove this again? Why not start out with combinations of raw material that can react fully? Yet, it is common for paint formulators to be given a sample of a new raw material—say a cross-linker—and substitute it in a formulation they are using, only to find out that it doesn't work. In many cases, just looking at the formulation will tell you not to bother even running the experiment, but rather to change the resin being cross-linked to fit with the characteristics of the new cross-linker.

Understanding the principles that control exterior durability permits a better estimate of the durability of a new coating than any laboratory tests available. Using this knowledge permits one to concentrate efforts on compositions that have a reasonable chance of being appropriate. A person who understands recent work on corrosion protection can make a better prediction of corrosion protection by some new formulation by looking at the formulation than can be predicted by running salt fog chamber tests. Understanding the scientific principles gives a head start on formulation of a coating with improved performance.

It is appropriate to repeat a theme that has recurred through the text: volume relationships, rather than weight relationships, are almost always the critical values. A good motto for a coatings formulator is: "Think volume."

Understanding relationships between properties and compositions is approaching the stage at which one will be able to design the binder for a formulation from first principles. We know what factors control T_g; we know what factors control cross-link density; we have some good leads as to the factors that control the breadth of the glass transition region; and we have a fairly good understanding of the relationships of these three characteristics with the performance of films. The day is not far off when it will be possible to design a resin and cross-linker for an end use without having to rely as much as is necessary today on trial and error. But, this will only be true for the formulator who studies the advances in basic understanding and then tries to apply the principles.

Experimental approaches. Except in the simplest projects, two major factors face the experimental worker: Most test methods used in the coatings industry are subject to considerable variation in results, and there are many possible variables and different possible responses in many different end uses. There are almost always conflicting requirements, so compromises are needed. Statistical experimental design and statistical analysis of data are particularly applicable to coatings formulation. A great contribution to increasing productivity can be made by increasing the use of statistics. Proper experimental design permits learning more information, with a higher degree of confidence in the results, with fewer experiments.

People are still being taught in school that "thou shalt *never* change more than one variable at a time." This commandment should have just one word changed so that it reads "thou shalt *always* change more than one variable at a time." There are two major problems with changing only one variable at a time. First, there are so many potential variables and levels of those variables that one could still be changing single variables years later. Second, changing one variable at a time does not permit identification of interactions among variables. For example, the "best" catalyst at one temperature is not necessarily the "best" catalyst at another temperature. The "best" pigment for a coating that must have excellent exterior durability made with one class of binder is not necessarily the "best" pigment to use with a different class of binder.

The strength of experimental design is that it permits changing multiple variables simultaneously in ways that permit separation of the effects of the different variables and of interactions between the variables. Many texts are available to provide background in experimental design. Reference [3], although relatively old, is a useful introduction because it uses examples from the coatings industry to illustrate the advantages and limitations of various types of experimental designs. Reference [4] provides more extensive coverage. Courses in experimental design are available; every coatings formulator should take such a course. Any plan for a project of significant size should include statistical experimental design.

Data analysis is also of critical importance. How many replicates are needed to obtain a test result with a 90 or a 95% confidence limit? Chemists, particularly, are used to standard deviations and do not seem to realize that 10 ± 2 means that there is about a 1 in 3 chance that the "real value" is greater than 12 or less than 8. Commonly, people use test results subject to wide variations to decide between two formulations, the differences between which are small compared to the errors in the test results. The worrisome thing is not the poorer materials that are accepted for further development, but that good ideas are discarded because of erroneous test results. It is desirable to analyze data obtained from statistical experimental designs, since this gives the opportunity of allocating the differences in results between the different variables, the different interactions, and "error." Error is the unexplained remaining difference. If it is large, there can be two possible explanations: The test methods may have error ranges larger than the differences being investigated, or there may be one or more important uncontrolled variables. In either case, before proceeding further, one should do something about the test results, either by increasing the number of replications and/or identifying the other variable(s).

Work. One should think first, but one must also work, and work efficiently. One can think through the day before and what has to be done the next day so as to mesh as many different tasks together as possible. In general, it is desirable to identify the most difficult goal of the project and concentrate initial efforts on this aspect of the problem. If the most difficult problem cannot be solved, there is no need for solutions to the relatively easy parts of the task. Time management is one of our most critical needs. It is fashionable to attend seminars on time management, and indeed, they may help, but primarily, each individual needs to think through how he or she can spend his or her time most effectively. Plans and the problem definition should be frequently reviewed; the situation may have changed, and you may end up solving a problem that isn't there any more. In complex problems involving several people from different disciplines, planning procedures that permit monitoring progress in all aspects of the project should be used.

Reports. Most technical people detest writing reports, but in a discipline like coatings, there is need for continuous accumulation of data. Writing a report forces one to review the work so far and plan the next steps. The most valuable parts of reports can be those on the experiments that did not work. The hundreds and thousands of unsuccessful experiments represent a wealth of information that can be used to solve production problems with a current product, to minimize future work, or to meet the requirements of a coating for a different application. It is particularly critical to get the results of actual field uses and the performance obtained into the database.

The report on a new formulation should spell out the reasons for inclusion of each of the components in the formulation. When others look at the formulation years later, they may not have any idea why a certain component is there. This situation will complicate reformulation necessitated by a new development.

The coatings field is frustrating because there are so many variables to deal with, but this is also what makes it fun and challenging. The primary factor controlling success is enthusiasm to tackle and solve complex problems.

REFERENCES

1. R. A. Dickie, *J. Coat Technol.*, **64** (809), 61 (1992).
2. J. W. Martin, S. C. Saunders, F. L. Floyd, and J. P. Wineburg, *Methodologies for Predicting the Service Lives of Coatings Systems*, Federation of Societies for Coatings Technology, Blue Bell, PA, 1996.
3. H. Grinsfelder, *Resin Review*, **19** (4), 20 (1969); **20** (1), 20; (2), 25; and (3), 16 (1970).
4. H. E. Hill and J. W. Prane, *Applied Techniques in Statistics for Selected Industries, Coatings, Paints, and Pigments*, Wiley, New York, 1984.

APPENDIX

Sources

In this section, we list selected works about coatings science and technology. Many excellent books that address specific topics are not listed here, but are listed as general references in the appropriate chapters.

BOOKS ON COATINGS

Annual Book of ASTM Standards, Paints, Related Coatings, and Aromatics, Vols. 6.01, 6.02, and 6.03, ASTM (formerly American Society for Testing and Materials), Philadelphia, revised annually. Provides standard procedures for carrying out tests. It is critical to remember that these are, in general, quality control tests for which precision has been checked, not performance predicting tests.

J. V. Koleske, Ed., *Paint and Coating Testing Manual*, (*Gardener-Sward Handbook*, 14th ed.), ASTM, Philadelphia, 1995. Very useful review of paint testing with summaries of test methods and chapters on many aspects of coatings by recognized specialists.

J. J. Brezinski, Ed., *Manual on Determination of Volatile Compounds in Paints, Inks, and Related Coating Productsd, ASTM Manual Series: MNL.4*, 2nd ed., ASTM, Philadelphia, 1993. Periodic new editions have been promised.

The SSPC (Steel Structures Painting Council), Pittsburgh, PA publishes bulletins and books on maintenance coating. For example: *Steel Structures Painting Manual*, Vol. I: *Good Painting Practice*, 1993 and Vol. II: *Systems and Specifications*, 1995.

Federation of Societies for Coatings Technology, Monograph Series on Coatings, Blue Bell, PA. New series began in 1986; monographs on various aspects of coatings continue to appear at a rate of about four per year. Many are excellent.

S. LeSota, Ed., *Coatings Encyclopedic Dictionary*, Federation of Societies for Coatings Technology, Blue Bell, PA, 1995.

D. Burghardt, *Coatings: The Dictionary, Das Worterbuch, Le Dictionnaire, El Diccinario*, Vincentz, Hannover, 1990. Terms are cross-listed in four languages.

T. C. Patton, *Paint Flow and Pigment Dispersion*, 2nd. ed., Wiley-Interscience, New York, 1979. Engineering flavor, somewhat outdated but still very useful. A classic.

S. Paul, *Surface Coatings Science & Technology*, 2nd. ed., Wiley, New York, 1996.

C. H. Hare, *Protective Coatings*, Steel Structures Painting Council, Pittsburgh, PA, 1996. A collection of articles from *J. Prot. Coat. Linings*.

D. Stoye and W. Freitag, Eds., *Resins for Coatings*, Hanser, Munich, New York, 1996.

L. J. Calbo, Ed., *Handbook of Coating Additives, Marcel Dekker, New York, 1987.*

G. E. Weismantel, *Paint Handbook*, McGraw-Hill, New York, 1981. Written mainly from the end-users point of view for application of architectural paints. Much practical information on application.

A. R. Marrion, Ed., *The Chemistry and Physics of Coatings*, Royal Society of Chemistry, London, 1994.

E. B. Gutoff and E. D. Cohen, *Coating and Drying Defects: Troubleshooting Operating Problems*, Wiley-Interscience, New York, 1995.

M. Ash and I. Ash, *Handbook of Paint and Coating Raw Materials*, Gower, Brookfield, VT, 1996. Available in CD-ROM or printed form.

M. W. Urban, *Laboratory Handbook of Organic Coatings, Global Press, Moorhead, MN, 1997.*

See also, Section on Conferences, Conventions and Symposia for proceedings.

POLYMER SCIENCE: GENERAL (NOT SPECIFIC TO COATINGS)

J. Brandrup, E. H. Immergut, and E. A. Grulke, *Polymer Handbook*, 4th ed., Johmn Wiley & Sons, New York, 1998. Reference work for polymer and polymerization data.

J. E. Mark, Ed., *Physical Properties of Polymers Handbook*, American Institute of Physics, Woodbury, NY, 1996. Contains many valuable articles.

H.-G. Elias, *An Introduction to Polymer Science*, VCH, Weinheim, New York, 1997.

H. R. Allcock and F. W. Lampe, *Contemporary Polymer Chemistry*, 2nd ed., Prentice Hall, Engelwood Cliffs, NJ, 1990. A widely used textbook.

F. W. Billmeyer, Jr., *Textbook of Polymer Science*, 3rd. ed., Wiley-Interscience, New York, 1984. Older, but still a valuable textbook.

G. Odian, *Principles of Polymerization*, 3rd. ed., Wiley-Interscience, New York, 1991. Good text written from a polymer synthesis mechanism point of view.

L. H. Sperling, *Introduction to Physical Polymer Science*, 2nd. ed., Wiley-Interscience, New York, 1992. Readable introductory text on polymer chemistry from a physical chemistry point of view. Complements Odian.

P. A. Lovell and M. S. El-Aasser, Eds., *Emulsion Polymerization and Emulsion Polymers*, Wiley, New York, 1997.

D. H. Everett, *Basic Principles of Colloid Science*, Royal Society of Chemistry, Letchworth, England, 1988.

L. E. Nielsen, *Polymer Rheology*, Marcel Dekker, New York, 1977. Most readable book on subject.

F. Garbassi, M. Morra, and E. Occhiello, *Polymer Surfaces from Physics to Technology*, John Wiley & Sons, New York, 1993; revised and updated 1998.

P. A. Lewis, Ed., *Pigment Handbook*, 2nd. ed., 4 volumes, Wiley-Interscience, New York, 1988–1994.

Symposium Series and *Advances in Chemistry Series*, American Chemical Society. Edited books derived from ACS symposia. Many cover polymer topics and some are directly relevant to coatings. For example: R. W. Tess and Poehlein, Eds., *Applied Polymer Science*, ACS Symposium Series No. 285, 1985; T. Provder, M. A. Winnick, M. W. Urban, Eds., *Film Formation in Waterborne Coatings*, ACS Symposium Series No 648, 1996.; J. E. Glass, Ed., *Technology for Water-borne Coatings*, ACS Symposium Series No. 663, 1997. The *Advances in Chemistry Series* was discontinued in 1997.

ENCYCLOPEDIAS

Encyclopedia of Polymer Science and Engineering, 2nd. ed., Wiley-Interscience, New York, 1985–1990.

Kirk-Othmer Encyclopedia of Chemical Technology, 4th ed., Wiley-Interscience, New York, 1991–1998.

J. C. Salamone, Ed., *Polymeric Materials Encyclopedia*, CRC Press, Boca Raton, FL, 1996.

ABSTRACTS

World Surface Coatings Abstracts, Paint Research Association, Pergamon Press, Helmsford, NY. Invaluable and well organized. Covers most literature and many patents relating to coatings. About 10,000 abstracts per year. Available on CD-ROM.

Chemical Abstracts, American Chemical Society, Columbus, OH. Extensive coverage and indexing. Section 42 covers coatings, but abstracts of interest in the coatings field are also in several other sections. Less convenient than WSCA, but more comprehensive. One can subscribe to individual sections and to on line computer search services.

CPI Digest, 2117 Cherokee Parkway, Louisville, KY 40204. Brief abstracts of recent articles and patents, many related to coatings. Useful.

JOURNALS: REFEREED SCIENTIFIC JOURNALS

Articles in these journals have been peer reviewed. While reviewing is inconsistent, the peer-review process improves the quality of published articles by encouraging authors to submit good articles and screening out poor ones.

Journal of Coatings Technology, Federation of Societies for Coatings Technology, Blue Bell, PA. Formerly named *Journal of Paint Technology* and before that *The Official Digest*. Leading North American journal in the field. Since pagination starts anew with each issue, citations must include issue number as well as page and year.

Progress in Organic Coatings, Elsevier Science Publishers, Lausanne, Switzerland. Started as a review journal, now also publishes research papers. Annually publishes papers from Athens Conference.

Farbe & Lack, Vincentz, Hannover, Germany. In German, with summaries in English. Often has excellent papers.

Surface Coatings International. (formerly, *Journal of the Oil and Colour Chemists Association*), Wembley, UK. Interesting variety of papers.

Color—Research and Application, John Wiley & Sons, New York.

Most general scientific journals covering polymers include papers relevant to coatings science.

JOURNALS: TRADE

Many articles in trade journals may have been carefully written and edited and are informative, but they have not been peer reviewed. Subscription prices are relatively modest, and many trade journals are free to students or to people involved in the coatings industry.

American Paint Journal, Douglas Publications, Richmond, VA. Weekly. Large circulation among paint industry people, newsy, good coverage of industry happenings.

Paint & Coatings Industry, Business News Publishing, Northbrook Park, IL. Carries a variety of technical and trade articles and news.

Journal of Protective Coatings & Linings, Technology Publishing Co., Pittsburgh, PA (affiliated with the Steel Structures Painting Council). Excellent source of information about industrial maintenance coatings. Carries a good series of articles on coating fundamentals.

Protective Coatings Europe, Technology Publishing Co., Pittsburgh, PA. European counterpart of *JPCL*.

Modern Paint and Coatings (formerly *Paint and Varnish Production*), PTN Publishing Co., Melville, NY.

Industrial Paint and Powder, Hitchcock Publishing, Wheaton, IL.

Coatings World, Rodman Publishing Corp., Ramsey, NJ. Worldwide coverage of developments and conferences.

Coatings Magazine, Kay Publishing Co., Oakville, Ontario, Canada. Canadian counterpart of *Modern Paint and Coatings*.

European Coatings Journal, Vincentz Verlag, Hannover, Germany. In English and French; some useful articles.

Polymer Paint and Colour Journal, DMG Business Media, Redhill, Surrey, England.

MARKET AND PRODUCT INFORMATION

Chemical Economics Handbook, SRI International, Palo Alto, CA. Contains useful information on coatings. Expensive.

Skeist Reports, Skeist, Inc., Whippany, NJ. Detailed profiles of the coatings market. Expensive.

Kline Guide to the Paint Industry, Charles H. Kline & Co., Inc., Fairfield, NJ. A new edition of this market research report appears every few years.

Paint Red Book, Modern Paint and Coatings, P. O. Box 470, Fort Atkinson, WI 53538. Annual. Lists suppliers of raw materials and equipment manufacturers.

Powder Coating Institute, 1800 Diagonal Road, Suite 370, Alexandria, VA 22314. Market information and technical bulletins on powder coatings.

U.S. Department of Agriculture, Forest Products Laboratory, Madison, WI, publishes bulletins on painting wood and wood products.

U.S. Department of Commerce, Bureau of Census, *Current Industrial Reports—Paint, Varnish, and Lacquers*, M-88-F, issued monthly. Provides statistics on production of various classes of coatings.

Technical Bulletins: Resin, solvent, pigment, and additive suppliers issue technical bulletins on their products. Many of them provide excellent background information and comparisons as well as suggested uses and starting formulations. One must expect some bias in such publications; there is a wide range of the extent of bias. In the case of resin suppliers, remember that in most cases, their preliminary evaluations are of clear coatings or, at most, of white gloss coatings. Increasingly, this information is accessible on the World Wide Web.

SAFETY AND TOXIC HAZARDS

Material Safety Data Sheets. In the United States, all producers of chemicals and chemical formulations are required to provide Material Safety Data Sheets (MSDS) to their customers. MSDSs contain reliable safety information on each substance and formulation in a standard format.

National Institute of Occupational Safety and Health. *Recommendations for the Control of Occupational Safety and Health Hazards*; *Manufacture of Paint and Allied Coating Products*, U.S. Dept. of Health and Human Services, 1984, DHHS (NIOSH), Pub. No. 84–115.

R. L. Lewis, Sr., Ed., *Sax's Dangerous Properties of Industrial Materials*, 9th ed., 3 Vols. Van Nostrand Reinhold, New York, 1996. A comprehensive encyclopedia of toxic and other hazards of industrial chemicals.

B. Waldo and R. deC. Hinds, Esq., *Chemical Hazard Communication Guidebook*, Executive Enterprises, New York, 1988. Outlines U.S. Safety and Health Administration, Environmental Protection Agency, and Department of Transportation requirements.

CONFERENCES, CONVENTIONS, AND SYMPOSIA

The American Chemical Society, 1155 Sixteenth St., SW, Washington, DC 20036 holds large meetings each spring and fall. Papers directly concerning coatings are often in symposia of the Division of Polymeric Materials, Science and Engineering and sometimes in other divisions' symposia. Some divisions publish the papers presented; *Polymeric Materials: Science and Engineering Proceedings* and *Polymer Preprints* are published for each meeting. While the papers are not refereed, they are timely and some are valuable.

The Federation of Societies for Coatings Technology, 492 Norristown Rd., Blue Bell, PA 19422 holds a national meeting in late October or early November each year. Technical papers are presented, although the number is declining; many are subsequently printed in the *Journal of Coatings Technology*. There is also a trade exhibit called the "Paint Show" with extensive displays by raw material and equipment manufacturers. Technical bulletins are available. Each of the 26 individual societies hold monthly meetings at which one or two papers are presented, most commonly by raw material suppliers. Several societies also present annual symposia.

Gordon Conferences on The Chemistry and Physics of Coatings and Films are held on odd-numbered years in late summer. By invitation, but you can ask for an invitation. Programs are published in *Science* and elsewhere, also at http://www.grc.uri.edu. Roughly 18 presentations on research in progress. No proceedings or summaries are published. Ample time for discussion of presentations and for informal discussions with the hundred-odd participants.

Waterborne, High-Solids, and Powder Coatings Symposium organized by the Polymer Science Department at The University of Southern Mississippi, Hattiesburg, MS. Held annually just before Mardi Gras in New Orleans. Popular conference. Proceedings are published in large, useful volumes.

International Conference on Organic Coatings Science and Technology. Organized by Prof. A.V. Patsis, of SUNY–New Paltz held each summer near Athens, Greece. Proceedings are published, and later, the papers are published in *Progress in Organic Coatings*.

FATIPEC Congresses are held on even numbered years in various European cities. Many papers are presented. Proceedings published.

European Coatings Show is held on odd numbered years in Nuremberg, Germany. Cosponsored by Vincentz Verlag and Paint Research Association.

Paint Research Association, 8 Waldegrave Rd., Teddington, Middlesex, TWII 8LD England sponsors a variety of specialized conferences in North America, Europe, and Asia. Proceedings are sometimes published.

Oil and Colour Chemists Association Convention. Held on odd numbered years in England. Smaller version of FATIPEC Congress.

Society of Manufacturing Engineers, PO Box 930, Dearborn, MI. Organizes symposia oriented toward industrial end-users. Symposium topics include powder coatings, radiation curing, and, more generally, "finishing." Proceedings of symposia are often published.

RadTech Conferences on radiation curing are held annually in North America, Europe, and Asia by RadTech, 60 Revere Drive, Northbrook, IL, 60062. Proceedings are published.

ELECTRONIC INFORMATION SOURCES

Coatings scientists and technologists have access to enormous amounts of information on the Internet, the World Wide Web, and in proprietary databases. Because methods of using electronic information resources change quickly and lists of resources rapidly go out of date, we must trust our readers to learn to use these resources as they become available. Two references that may be useful:

S. M. Bachrach, Ed., *The Internet: A Guide for Chemists*, American Chemical Society, Washington, DC, 1996.
D. D. Ridley, *Online Searching: A Scientist's Perspective: A Guide for the Chemical and Life Sciences*, John Wiley & Sons, New York, 1996.

Web sites we have found useful are http://patent.womplex.ibm.com (displays abstracts and claims of patents back to 1971 for free; search capabilities) and http://www.epa.gov (for regulatory and environmental information).

Index

AA, *see* Acrylic acid; Adipic acid
Abrasion resistance, 76–77, 85–86, 578
  of coatings for plastics, 77, 82, 524, 556–558
  effect of CPVC, 363
  of polyurethanes, 76–77, 180, 202, 408, 489, 524, 557, 578
  testing, 76–77, 85–86, 556
Abrasive particle blasting, 113, 577, 588
Absorption of light, 340–341, 344–347
Absorption of UV radiation, 89–93, 510–513, 516–520, 523
Acetaldehyde, 159
Acetic acid, 54, 159, 194, 213, 220, 240, 480, 578
Acetoacetoxyethyl methacrylate, 301
Acetoacetylated resins, 119, 154, 163, 301–302
Acetone, 54, 120, 202, 271, 275, 307, 316, 329
Acetylene black, 380, 554
Acetylene glycol surfactants, 398, 494
Acid-base interactions in adsorption, 116, 389
Acid etching, *see* Environmental etching
Acid number, 238, 247, 254, 275, 280–282, 478
Acid precursors, 172
Acid rain, 89, 100, 102, 139, 564.*See also* Environmental etching
Acrolein, 282
Acrylamide, 162, 178, 236–237, 293, 570
Acrylamide copolymers, 178, 236–237, 379
Acrylated oligomers, 53, 292–293, 486, 491, 514–515, 523, 540, 550
Acrylic acid (AA), 15, 564
  comonomer in latexes, 145, 148, 153, 156–157, 303
  comonomer in TSAs, 233
  comonomer in water-reducible acrylics, 237–243, 478
  reaction with epoxy groups, 293, 515
Acrylic esters, T_g of homopolymers, 15
Acrylic lacquers, *see* Lacquers, acrylic
Acrylic latexes, 154–158, 472–473, 532, 534, 487, 495, 498, 500, 512, 516
  hybrid alkyd/acrylic latexes, 154, 275
  thermosetting, 153–154, 158, 299, 303, 534, 551, 573
Acrylic microgels, 158, 441, 465, 531
Acrylic powder coatings, *see* Powder coatings, thermosetting, acrylic
Acrylic resins, 15–20, 231–244, 460
  amide-functional, 178, 236–237
  amine-substituted, 480
  blends with polyesters, *see* Polyesters, acrylic blends

carbamate-functional, 78, 100, 176, 237, 461, 533
carboxylic acid-functional, 54, 224, 236, 265, 478, 490
drying oil esters from, 586
for electrodeposition, 478, 480, 535
epoxy-functional, 199, 214, 224, 237, 460, 486, 490, 498, 533–534
glass transition temperatures, 15–16, 152–158, 231–235
group transfer polymerization, 236, 391, 396
high solids, 233–236, 460, 462
hydroxy-functional, 193, 196, 200, 202, 232–236, 297–298, 460, 469, 486, 533
isocyanate-functional, 193, 237
latexes, *see* Acrylic latexes
organocopper substituted, 584
organotin substituted, 583
silicone-modified, 91, 101, 297–298, 461, 543
thermoplastic (TPA), 231.*See also* Lacquers, acrylic
thermosetting (TSA), 232–237
  functionality of, 233–235, 460
trialkoxysilyl-functional, 237, 299, 533–534, 587
water-reducible, 172–173, 203, 237–244, 469–470, 478, 480, 532, 540, 551
  amine selection for, *see Amines for water-reducible coatings*
  pH on dilution, 240, 469
  silicone-modified, 262
  viscosity changes on dilution, 238–243, 468–469
Activation energy, 28–31
Activation energy for viscous flow, 49
Activity coefficient in solvent mixtures, 281–282
Acylphosphine oxides, 512
N-Acylurea cross-links, 303
Addition polymerization, *see* Chain-growth polymerization
Additives, 4, 81–88, 390, 396, 442, 444, 446, 450
Additive color mixing, 355
Adhesion, 109–123, 131–133
  to chalky surfaces, 152, 417, 420, 561, 564
  cooperative, 117
  covalent bonding effects, 118–119
  effect of amine groups on, 117, 120, 132, 480, 529, 577
  effect on corrosion, *see* Corrosion, effect of wet adhesion
  effect of dodecylbenzenesulfonic acid, 113, 154